Regulation of
Chloroplast Biogenesis

NATO ASI Series

Advanced Science Institutes Series

A series presenting the results of activities sponsored by the NATO Science Committee, which aims at the dissemination of advanced scientific and technological knowledge, with a view to strengthening links between scientific communities.

The series is published by an international board of publishers in conjunction with the NATO Scientific Affairs Division

A	Life Sciences	Plenum Publishing Corporation
B	Physics	New York and London
C	Mathematical and Physical Sciences	Kluwer Academic Publishers
D	Behavioral and Social Sciences	Dordrecht, Boston, and London
E	Applied Sciences	
F	Computer and Systems Sciences	Springer-Verlag
G	Ecological Sciences	Berlin, Heidelberg, New York, London,
H	Cell Biology	Paris, Tokyo, Hong Kong, and Barcelona
I	Global Environmental Change	

Recent Volumes in this Series

Series A: Life Sciences

Regulation of Chloroplast Biogenesis

Edited by

Joan H. Argyroudi-Akoyunoglou

Institute of Biology
National Centre for Scientific Research "Demokritos"
Athens, Greece

Plenum Press
New York and London
Published in cooperation with NATO Scientific Affairs Division

Proceedings of a NATO Advanced Research Workshop
on Regulation of Chloroplast Biogenesis,
held July 28–August 3, 1991,
in Crete, Greece

Library of Congress Cataloging-in-Publication Data

Regulation of chloroplast biogenesis / edited by Joan H. Argyroudi
  -Akoyunoglou.
       p.   cm. -- (NATO ASI series. Series A, Life sciences ; v.
  226)
     "Proceedings of a NATO Advanced Research Workshop on Regulation of
  Chloroplast Biogenesis, held July 28-August 3, 1991, in Crete,
  Greece"--T.p. verso.
     "Published in cooperation with NATO Scientific Affairs Division."
     Includes bibliographical references and index.
     ISBN 0-306-44184-5 (hard)
     1. Chloroplasts--Formation--Regulation--Congresses.   I. Argyroudi
  -Akoyunoglou, J. H.   II. NATO Advanced Research Workshop on
  Regulation of Chloroplast Biogenesis (1991 : Crete, Greece)
  III. North Atlantic Treaty Organization. Scientific Affairs
  Division.   IV. Series.
  QK725.R517 1992
  581.87'33--dc20   92-2952
                                                                  CIP

ISBN 0-306-44184-5

© 1992 Plenum Press, New York
A Division of Plenum Publishing Corporation
233 Spring Street, New York, N.Y. 10013

Printed in the United States of America

This book is dedicated to the memory of
George Akoyunoglou

George Akoyunoglou 1927-1986

It is most appropriate to open this conference on "Regulation of Chloroplast Biogenesis" with words of remembrance and honour of our late colleague and my personal friend George Akoyunoglou.

Without going into details, a few stepstones of the scientific career of George Akoyunoglou shall be mentioned. He obtained his Ph.D. with a study on the carboxydismutase reactions under the guidance of Melvin Calvin. He was called back to his home country to become Head of the Photobiology Laboratory and one of the pioneers of the Nuclear Research Center "Demokritos" in Athens. He acted as Head of the Biology Department for many years.

He has spent longer research visits in the laboratories of C. Sironval in Liege, Bill Siegelman in Brookhaven, Itzhak Ohad in Jerusalem and ours in Marburg. George's scientific work is marked by the fortunate and successful collaboration with his wife Joan. He represented the unique combination of a Biologist and a Biochemist.

During the 25 years of his scientific work George in collaboration with his wife Joan has set landmarks for the understanding of chloroplast biogenesis. During the transition of etioplasts to chloroplasts they investigated the chlorophyll synthesis, the formation of PS II and PS I and contributed significantly to the understanding of the development of the photosynthetic mechanism. They introduced the intermittent illumination of developing plants for suppressing the formation of the light-harvesting Chl a/b complex, a technique used worldwide to study the relation between LHC and RC. They elaborated the technique of gel electrophoreses of pigment protein complexes and disclosed the sequential binding of chlorophylls to the various integral polypeptides of the thylakoid membrane. Always taking into consideration the biology of the plants they studied, George and Joan Akoyunoglou's work emphasizes the close correlation between the developmental stage of the plant from which the chloroplasts derived and the post-translational regulation of the chloroplast biogenesis.

But even to those who had no intimate knowledge of his scientific work, George Akoyunoglou is known as the outstanding organizer of several most memorable meetings. I mention the "International Symposium on

Chloroplast Development" at Spetsai (1978), the "5th International Photosynthesis Congress" at Halkidike (1980), the "Symposium on the Autonomy and Biogenesis of Mitochondria and Chloroplasts" at Delphi (1982) and the "International Meeting on the Regulation of Chloroplast Differentiation" at Rhodos (1985). Those of us having had the privilege to participate in these meetings experienced George as the perfect organizer. He was everywhere where help was needed, but he never placed himself into the foreground. Perfect organization, the warm and friendly atmosphere and the combination of a program of highest scientific standards with relaxing social events and inspiring cultural excursions became a trade mark of those meetings. During all these meetings George could count on the help of his family and we appreciated their cheerful help during the meetings. George, his family and his coworker Mr. Pastellacos appeared as an admirable entity during the past meetings.

As products of the meetings George has edited with various coeditors a number of valuable books. Among them the volumes of the Proceedings of the 5th International Photosynthesis Congress still the most comprehensive work in photosynthesis literature.

George Akoyunoglou has served in many national and international committees. He took considerable part in the promotion of biology and biochemistry of his own country. He was a founding member of the Hellenic Biochemical and Biophysical Society. Last but not least he inspired the foundation of the "International Society of Chloroplast Development" for which he was elected as first chairman.

My first personal encounter with George Akoyunoglou was, when he was my chairman at the 3rd International Photosynthese Congress in Rehovot (1974). Thereafter we met at various meetings and discussed our mutual research interest. Impressed by his scientific accomplishments and by his personality, I could persuade him to accept a one year guest professorship at our university. During this time we became friends. My co-workers and students could not only experience him as an inspiring, stimulating scientist and skillful experimentor but also as a cheerful and charming person who loved to exchange and laugh with them. But he could also be a serious and witty counterpart in long discussions about science, philosophy and Greek mythology.

I had the privilege to help George in organizing the last meeting in Rhodes. When most participants were gone, we sat in the lobby of the Capsis Hotel and did the last accounting. After we finished, George said to me: Horst, that meeting was fun, let's do it again, and he already planned ahead for this meeting.

We have to be most thankful to George's wife Joan that she fulfilled the plan of George and took all the trouble of organizing this International Meeting on the "Regulation of Chloroplast Biogenesis".

We dedicate this meeting to the memory of George Akoyunoglou and I am sure it will be held in his sense to serve the scientific progress and friendship in good spirits.

Horst Senger
President, International Society
for Chloroplast Development

PREFACE

From July 28 to August 3, 1991, an International Meeting on the
REGULATION OF CHLOROPLAST BIOGENESIS was held at the Capsis Beach Hotel
in Aghia Pelaghia, on the island of Crete, Greece. The Meeting (Advanced
Research Workshop-Lecture Course) was co-sponsored by NATO, FEBS and IUB,
and was held under the auspices of the International Society for Chloroplast Development, the Greek Ministry of Industry, Research and Technology, and the National Center for Scientific Research "Demokritos".

The Meeting focused on recent advances in the field of chloroplast
biogenesis and the regulatory mechanisms underlined, and brought together
over 120 experts and students of the field from 22 countries. The subject
of chloroplast biogenesis has experienced great progress in recent years
mainly thanks to the application of Molecular Biology techniques and
methodology. New findings that emerge gradually unravel the regulatory
mechanisms involved in the assembly, stabilization and growth of the
photosynthetic units in thylakoids, the signal transduction chain leading
from photoreception to gene expression, the transport of nuclear-coded
proteins into stroma-soluble supramolecular enzyme complexes as well as
thylakoid-bound supramolecular complexes, involved in light-energy
transduction.

It was the aim of this meeting to bring together experts and students
coming from diverse disciplines (ranging from Botany and Plant Physiology
to Molecular Biology, Biophysics and Biotechnology), to discuss the
recent advances in the field so that thorough exchange of ideas and
working hypotheses would be achieved.

The current volume contains near to 90 papers given as lectures or
poster presentations in the areas of:

I.   Regulation of gene expression

II.   Import of nuclear-coded proteins into chloroplasts and mecha-
      nisms of chloroplast protein translocation

III. Biosynthesis and origin of chloroplast pigments

IV.   Regulation of pigment-protein complex formation and assembly
      into photosynthetic units

V.   Organization and interactions of photosynthetic units

VI.   Adaptation mechanisms in plants and algae-Shade adaptation-State
      transition

VII. Genetic manipulation as a means to monitor assembly and function

I want to express my thanks to all who contributed to the Meeting and to this volume. My sincere thanks are also due to members of the International Organizing Committee, and especially to Prof. Dr. Horst Senger, for their help, to my collaborators in "Demokritos", and especially to Mr. Leonidas Pastelakos and Ms Mitsy Akoyunoglou for their efficient and never expiring assistance.

August 1991   Joan Argyroudi-Akoyunoglou
Athens, Greece

CONTENTS

REGULATION OF GENE EXPRESSION

PHOTOSYNTHETIC UNIT ASSEMBLY AND ITS DYNAMIC REGULATION

ORGANIZATION AND INTERACTIONS OF PHOTOSYNTHETIC UNITS

REGULATION OF GENE EXPRESSION

CHAIRMAN'S INTRODUCTION – STRUCTURE AND ORGANIZATION OF

PLASTIDAL GENES AND THE FEATURES OF THEIR EXPRESSION

Rudolf Hagemann

Institute of Genetics, Martin-Luther-University
Domplatz 1, D-O-4020 Halle (Saale) Germany

INTRODUCTION

The introductory remarks of the Chairman of a session have – according to my opinion – in general a twofold task. One task, of course, is the introduction for the topics, to be dealt with by the following speakers of the symposium. The other task is to try to characterize the general topic of the symposium and to point also to results, problems and trends, which will not be discussed in the following lectures, but which are of important relevance for the topic of the meeting in general and especially for the topic of the symposium "Gene structure and regulation of gene expression".

In the printed version of my introduction I shall concentrate on the second aspect, just mentioned, and shall deal with four topics.

1. Classification of the plastid genes: 'Genetic system genes' and 'Photosynthesis genes'.

A decisive breakthrough in the field of molecular biology and genetics of plastids was the complete sequencing of the plastid genomes of three species in the years 1986 and 1989: liverwort Marchantia polymorpha (Ohyama et al. 1986, 1988, Umesono et al. 1988, Fukuzawa et al. 1988, Kohchi et al. 1988), tobacco Nicotiana tabacum (Shinozaki et al. 1986, 1988, Shinozaki and Sugiura 1986, Sugiura 1987, 1989 a,b) and rice Oryza sativa (Hiratsuka et al. 1989, Sugiura 1989 a,b).

These sequence data are nowaday the most valuable reference system for all studies on sequence, organization, gene structure and coding capacity of the plastid DNA.

Shimada and Sugiura (1991) have recently made a detailed comparison of the fine structural features of the three plastid genomes of tobacco, rice and liverwort, and came to interesting general results. The analysis of homology, of GC content and codon usage of the plastid genes allowed their classification into at least two groups, the 'Photosynthesis genes' and the 'Genetic system genes'.

Regulation of Chloroplast Biogenesis
Edited by J.H. Argyroudi-Akoyunoglou, Plenum Press, New York, 1992

The group of the 'Photosynthesis genes' consists of the plastid genes for the photosystem I and photosystem II thylakoid proteins, for the subunits of the cytochrome b/f and the H^+-ATPase complex and the gene for the large subunit of rubisco.

To the group of the 'Genetic system genes' belong the plastid genes for the plastidal rRNAs (23S, 16S, 5S, 4,5S) and tRNAs, and moreover the plastid protein genes for the ribosomal proteins (for the 30S and the 50S subunits), the RNA polymerase subunits, and the initiation factor IF; obviously the plastid genes for the plastidal NADH dehydrogenase subunits belong to this group (although their characteristics are given separately in Table 1).
The comparison of the protein coding genes of these two groups reveals interesting differences (Table 1):
The 'Genetic system genes' are characterized by a lower GC content of the genes (compared with the 'Photosynthesis genes'). The values of the NADH dehydrogenase genes is even lower than that of the other 'Genetic system genes'.

Nuclear genes of higher plants very often have G or C in their third codon position, monocots more than 80%. In contrast, plastid genes in general have a high usage of A or U in the third codon position. Moreover, again the 'Genetic system genes' have a still higher frequency of A and U in the third codon position (and vice versa a lower G+C content) than the 'Photosynthesis genes'; this is especially striking for liverwort (Table 1).

These criteria allow to assign still unidentified Open Reading Frames (ORFs) to one of these two groups; this may give hints at the function of the gene products of the ORFs.

The fact that within the plastid genome there exist two distinct groups of genes is interesting per se; moreover it leads to questions about the evolutionary basis of these differences.

Very interesting in this context are the findings in two unusual taxa: the parastic flowering plant Epifagus virginiana (studied by J. Palmer's group) and the colourless euglenoid flagellate Astasia longa (studied by W. Hachtel's group). The plastid DNAs of both taxa have undergone large deletions which caused the loss of great parts of the plastid DNA. Epifagus virginiana has a plastid genome of only 71 kilobases (dePamphilis and Palmer 1990) and Astasia longa of only 73 kilobases (Siemeister and Hachtel 1989) – compared with 156 kilobases of tobacco.

Remarkable is the finding that in both – rather unrelated – taxa those genes, which have been lost, are (almost) all 'Photosynthesis genes' for the proteins of photosystem I and II, of the cytochrome b/f and the ATPase complexes. Still present in these white plants are the genes for the ribosomal RNAs and a number of 'Genetic system genes'. This finding gives indications that at least some of the 'Genetic system genes' may fulfil house-keeping functions for the life of the plant cell as a whole, even when it is white,

Table 1

G+C content of genes

	Genetic system genes	NADH dehydrogenase genes	Photosynthesis genes
Rice	G+C 38,3 %	G+C 36,5 %	G+C 40,0 %
Tobacco	G+C 38,0 %	G+C 36,3 %	G+C 40,1 %
Liverwort	G+C 29,5 %	G+C 27,0 %	G+C 33,3 %

Codon usage in the third position

	Genetic system genes	NADH dehydrogenase genes	Photosynthesis genes
Rice	A+U 4599 G+C 2099 = 31,3 %	A+U 2207 G+C 876 = 28,4 %	A+U 4517 G+C 2149 = 32,2 %
Tobacco	A+U 4430 G+C 1714 = 27,9 %	A+U 2265 G+C 839 = 27,0 %	A+U 4642 G+C 2078 = 30,9 %
Liverwort	A+U 5617 G+C 666 = 10,6 %	A+U 2706 G+C 347 = 11,4 %	A+U 5668 G+C 949 = 14,2 %

3

and/or may be responsible for the interaction of plastids
and nucleus.
There are some differences between both genera regarding the
genes which are still present and those which have been lost.

2. Arguments for complete sequencing of the plastid DNA of
   several higher and lower plants

So far the plastid DNA has been completely sequenced
for only three species: liverwort, tobacco and rice. But
other species will soon follow. The plastid DNA of maize
will soon be totally sequenced, and the same seems to be
true for Euglena. I hold the view that the plastid genomes
of other lower and higher plants should be completely sequenced, because it is highly probable that with the increase of complete sequence data from taxonomically divergent taxa we may gain a lot of new insights. M. Sugiura's
research group is now sequencing the plastid DNA of the
gymnosperm Pinus thunbergii (black pine), one of the most
popular and economically important trees in Japan and northern Asia. So far five regions have been sequenced, which
represent about 40% of the plastid DNA altogether; 35 genes
were mapped through their homology with the tobacco plastid
genes. The black pine genome is highly rearranged in comparison with the genomes of monocots and dicots; it has no inverted repeat. One unique feature is that black pine lacks
the ribosomal protein gene rps16, which is present in tobacco and rice, but has the ORF510 instead of rps16. In this
respect, black pine resembles liverwort which also lacks
rps16 but contains ORF513 in the corresponding region. Both
ORFS are highly homologous to each other (M. Sugiura, personal comm.).

I tend to predict that the sequencing of the plastid
genomes of taxonomically divergent taxa may considerably
increase our knowledge about the structure and the coding
capacity of the plastid DNA and special features of the
transcription and translation in plastids and the processing
following both steps of gene expression.

One of such unexpected features is outlined in the following paragraph.

3. The first two examples of RNA editing in plastids

In mitochondria of higher plants and in mitochondria/
kinetoplasts of protists the phenomenon of RNA editing was
described. The main characteristic of RNA editing is the
secondary change of the nucleotide composition and sequence
of a pre-mRNA, hereby changing the primary nucleotide sequence coding for a non-functional polypeptide into an edited secondary sequence coding for a functional polypeptide.
The essential experimental procedure is the comparison of
the sequence of the genomic DNA with that of the cDNA from
a mature mRNA; the changes caused by RNA editing thus become evident.

The most frequent change in the process of RNA editing
in mitochondria of higher plants is the transition of C into
U (much more seldom are the changes C → G or G → C or G

$\rightarrow$ A or U $\rightarrow$ A). In the mitochondria/kinetoplasts of protists (Trypanosoma, Leishmania, Crithidia) even more complicated editing events have been found: In addition to C$\rightarrow$ U changes pre-mRNAs are edited by single or multiple insertions of U residues into the RNAs (or deletions of U). Until August 1991 processes of RNA editing had been unknown for plastids. But now two examples of RNA editing in plastids have been elucidated in Hans Kössel's lab.

The sequence of the maize plastid gene for the ribosomal protein 2 of the large subunit of plastidal ribosomes (rpl 2) was determined by Kavousi et al. (1990). Its start triplet in plastid DNA is ACG/TGC, which is transcribed into ACG in the pre-mRNA. But in the mature mRNA for this ribosomal protein the initiation codon is AUG. This means that in the start codon C is edited into U (Hoch et al. 1991). Interestingly this is the same type of change as was frequently found in editing mitochondrial mRNAs.

One of my PhD students, Jörg Kudla, who has been working for several months in Hans Kössel's lab, found during the work in this lab another case of RNA editing in plastids: The photosynthesis gene psbL of Nicotiana tabacum has the start triplet ACG/TGC (Shinozaki et al. 1986). In most species, in which the psbL gene was sequenced, the start triplet is ATG/TAC: Antirrhinum majus, Euglena gracilis, Hordeum vulgare, Marchantia polymorpha, Oryza sativa, Pisum sativum, Triticum aestivum, Zea mays; the initiation codon of their mRNA is AUG. Only tobacco and spinach have the start triplet ACG/TGC. So far it was assumed that the codon ACG may in some plastid genes act as the initiation codon as in Sendai viruses (Wolfe and Sharp 1988). However, it could be proved that in Nicotiana tabacum the triplet ACG is edited to AUG (Kudla et al. 1991 submitted).

Thus RNA editing is taking place in plastids. It is well possible that the frequency of RNA editing in mitochondria is higher than in plastids. But it is likely that this interesting phenomenon will be found in the future in several more than just these two plastid genes. The genes rpl 2 of rice and psbL of spinach seem to be very good candidates. Several more general conclusions can be drawn from these data: RNA editing in plastids is taking place in monocots as well as in dicots, and it occurs both in 'Genetic system genes' and in 'Photosynthesis genes'. Obviously one cannot automatically predict the amino acid sequence and length of a plastidal polypeptide from the DNA sequence determined for specific plastid genes or ORFs, because RNA editing may cause several types of changes as described in mitochondria and kinetoplasts (change of codons, insertion of additional amino acids, creation of initiation or termination codons or altering termination codons).
Finally, it should be emphasized that hints at special features of the processes of expression of plastidal genes, e.g. RNA editing, can often only be obtained by comparing a larger number of plastid DNA sequences. This gives - as already mentioned in paragraph 2 - one more argument in favor of sequencing the plastid DNAs of a considerable number of lower and higher plants.

4. Gene transfer into plastids with a particle gun

Many different techniques are now available for gene transfer into the nuclei of eukaryotic plant cells. However, the experimental transfer of genes into plastids (and mitochondria) of plants was for many years very difficult, de facto impossible. A stimulating change took place when Klein et al. (1987, 1988) were able to demonstrate that small DNA-coated tungsten (or gold) particles can be accelerated to high velocities, and that these microprojectiles penetrate the membranes of living cells and deliver the DNA into the cells. Following this strategy many research groups are nowadays using microprojectiles for the delivery of specific DNA molecules into the nuclei of several plant species in order to obtain transient expression or stable integration and expression of transferred genes.

Of great importance and value for plastid genetics is the finding that this method can be successfully applied and used for direct gene transfer into plastids. Already in 1988 J. Boynton's and N. Gillham's group has in cooperation with the group of Klein and Sanford proved that microprojectiles coated with plastid DNA lead to stable genetic transformation of chloroplasts of Chlamydomonas reinhardtii. In the meantime several research groups have contributed many interesting results to the further analysis of the transfer of plastid genes and the processes of transient expression or stable integration of genes into the plastid genome. Several contributions to this symposium deal with that topic. I wish to emphasize how important the use of the particle gun is for the transfer of plastid genes. For gene transfer into plant nuclei the use of the particle gun is one method among several ways for gene transfer. But for a gene transfer into plastids the use of the particle gun is at present the only successful method. Therefore it is of great importance for further progress of recombination genetics of plastids.

Literature

Boynton, J. E., Gillham, N. W., Harris, E. H., Hosler, J. P., Johnson, A. W., Jones, A. R., Randolph-Anderson, B. L., Robertson, D., Klein, T. M., Shark, K. B., and Sanford, J. C., 1988, Chloroplast transformation in Chlamydomonas with high velocity microprojectiles, Science 240: 1534 - 1538.
dePamphilis, C. W. and Palmer, J. D., 1990, Loss of photosynthetic and chlororespiratory genes from the plastid genome of a parasitic flowering plant, Nature 348: 337 - 339.
Fukuzawa, H., Kohchi, T., Sano, T., Shirai, H., Umesono, K., Inokuchi, H., Ozeki, H., and Ohyama, K., 1988, Structure and organization of Marchantia polymorpha chloroplast genome - III. Gene organization of the large single copy region from rbcL to trn I (CAU), Journ. Mol. Biol. 203:333 - 351.

Hiratsuka, J., Shimada, H., Whittier, R., Ishibashi, T., Sakamoto, M., Mori, M., Kondo, C., Honji, Y., Sun, R., Meng, B. Y., Li, Y. Q., Kanno, A., Nishizawa, Y., Hirai, A., Shinozaki, K., and Sugiura, M., 1989, The complete sequence of the rice (Oryza sativa) chloroplast genome: Intermolecular recombination between distinct tRNA genes accounts for a major plastid DNA inversion during the evolution of the cereals, Molec. General Genetics 217:185 - 194.

Hoch, B., Maier, R. M., Appel, K., Igloi, G. L., and Kössel, H., 1991, Editing of a chloroplast mRNA: creation of an AUG initiation codon by a C to U conversion, Nature in press.

Kavousi, M., Giese, K., Larrinua, I. M., McLaughlin, W. E., and Subramanian, A. R., 1990, Nucleotide sequence and map position of the duplicated gene for maize (Zea mays) chloroplast ribosomal protein L2, Nucl. Acids Res. 18:4244.

Klein, T. M., Wolf, E. D., Wu, R., Sanford, J. C., 1987, High velocity microprojectiles for delivering nucleic acids into living cells, Nature 327:70 - 73.

Klein, Th. M., Harper, E. C., Svab, Z., Sanford, J. C., Fromm, M. E., Maliga, P., 1988, Stable genetic transformation of intact Nicotiana cells by the particle bombardment process, Proc. Natl. Acad. Sci. USA 85:8502 - 8505.

Kohchi, T., Shirai, H., Fukuzawa, H., Sano, T., Komano, T., Umesono, K., Inokuchi, H., Ozeki, H., and Ohyama, K., 1988, Structure and organization of Marchantia polymorpha chloroplast genome. IV. Inverted repeat and small single copy regions, Journ. Mol. Biol. 203:353 - 372.

Kudla, J., Igloi, G. L., Metzlaff, M., Hagemann, R. and Kössel, H., 1991, RNA editing in chloroplasts: Formation of a translatable mRNA of the Tobacco psbL gene by C to U editing within the initiation codon (submitted).

Ohyama, K., Fukuzawa, H., Kohchi, T., Shirai, H., Sano, T., Sano, S., Umesono, K., Shiki, Y., Takeuchi, M., Chang, Z., Aota, S., Inokuchi, H., Ozeki, H., 1986, Chloroplast gene organization deduced from complete sequence of liverwort Marchantia polymorpha chloroplast DNA, Nature 322:572 - 574.

Ohyama, K., Fukuzawa, H., Kohchi, T., Sano, T., Sano, S., Shirai, H., Umesono, K., Shiki, Y., Takeuchi, M., Chang, Z., Aota, S.-J., Inokuchi, H., and Ozeki, H., 1988, Structure and organization of Marchantia polymorpha chloroplast genome. I. Cloning and gene identification, Journ. Mol. Biol. 203:281 - 298.

Siemeister, G. and Hachtel, W., 1985, A circular 73 kb DNA from the colourless flagellate Astasia longa that resembles the chloroplast DNA of Euglena: restriction and gene map, Current Genetics 15:435 - 441.

Shimada, H. and Sugiura, M., 1991, Fine structural features of the chloroplast genome: comparison of the sequenced chloroplast genomes, Nucl. Acids Res. 19:983 - 995.

Shinozaki, K., Hayashida, N. and Sugiura, M., 1988, Nicotiana chloroplast genes for components of the photosynthetic apparatus, Photosynthesis Research 18:7 - 31.

Shinozaki, K., Ohme, M., Tanaka, M., Wakasugi, T., Hayashida, N., Matsubayashi, T., Zaita, N., Chunwongse, J., Obokata, J., Yamaguchi-Shinozaki, K., Ohto, C., Torazawa, K., Meng, B. Y., Sugita, M., Deno, H., Kamogashira, T., Yamada, K., Kusuda, J., Takaiwa, F., Kato, A., Tohdo, N., Shimada, H. and Sugiura, M., 1986, The complete nucleotide sequence of the tobacco chloroplast genome: its gene organization and expression, EMBO Journ. 5: 2043 - 2049.

Sugiura, M., 1987, Structure and function of the tobacco chloroplast genome, Bot. Mag. Tokyo 100:407 - 436.

Sugiura, M., 1989 a, Organization and expression of the tobacco and rice chloroplast genomes, in: "Highlights Modern Biochem.", Kotyk, A. et al. (Eds.), Vol. 1, pp. 695 - 704.

Sugiura, M., 1989 b, The chloroplast chromosomes in land plants, Ann. Rev. Cell Biol. 5:51 - 70.

Umesono, K., Inokuchi, H., Shiki, Y., Takeuchi, M., Chang, Z., Fukuzawa, H., Kohchi, T., Shirai, H., Ohyama, K. and Ozeki, H., 1988, Structure and organization of Marchantia polymorpha chloroplast genome. II. Gene organization of the large single copy region from rps 12 to atp B, Journ. Mol. Biol. 203:299 - 331.

Wolfe, K. H. and Sharp, P. M., 1988, Identification of functional open reading frames in chloroplast genomes, Gene 66:215 - 222.

CHLOROPLAST GENE EXPRESSION AND REVERSE GENETICS IN

CHLAMYDOMONAS REINHARDTII

J.D. Rochaix, M. Goldschmidt-Clermont, Y. Choquet,
Y. Takahashi, M. Kuchka, J. Girard-Bascou[+] and
P. Bennoun[+]

Departments of Molecular Biology and Plant Biology,
University of Geneva, Geneva, Switzerland
[+]Institut de Biologie Physico-Chimique, Paris, France

I. INTRODUCTION

It is well established that the biosynthesis of the photosynthetic apparatus in higher plants and eukaryotic algae is achieved through the concerted action of the chloroplast and nucleo-cytosolic genetic systems. Although numerous studies have appeared on this subject the molecular mechanisms underlying this interaction are still poorly understood. The green unicellular alga Chlamydomonas reinhardtii is particularly well suited for studies of this sort. First, photosynthetic function is dispensable provided a reduced carbon source, e.g. acetate, is added to the growth medium. Thus mutants deficient in photosynthesis can be generated and maintained with ease. Since the photosynthetic apparatus of C. reinhardtii is very similar to that of higher plants, this alga provides an excellent model system for studying the biosynthesis and function of the various components of the photosynthetic machinery. Second, standard genetic analysis is feasible in C. reinhardti. Third, reliable transformation procedures have been established recently both for the chloroplast and nuclear compartments of C. reinhardtii (1,2,3,4).

Both nuclear and chloroplast photosynthetic mutants have been isolated from C. reinhardtii. These mutations can easily be distinguished through their particular inheritance patterns during crosses. Nuclear mutations can be grouped into two major classes: The first affects nuclear genes encoding the various components of the photosynthetic system (structural genes). The second class of mutations defines nuclear genes whose products are required for the proper expression of chloroplast genes. These nuclear encoded factors may act at various levels such as chloroplast RNA maturation and splicing, chloroplast RNA stability and translation. Here we review some of the salient features of several of these photosynthetic mutants and we also discuss recent results obtained through chloroplast reverse genetics.

Regulation of Chloroplast Biogenesis
Edited by J.H. Argyroudi-Akoyunoglou, Plenum Press, New York, 1992

II. MUTANTS AFFECTED IN CHLOROPLAST GENE EXPRESSION

a. Chloroplast RNA maturation

The psaA gene of C. reinhardtii which codes for one of
the photosystem I (PSI) reaction center subunits consists of
three exons that are widely separated on the chloroplast
genome (5, fig. 1). Each of these exons is flanked by group
II intron consensus sequences and is transcribed
independently (6). Exon ligation is presumably mediated
through two trans-splicing reactions. To understand how the
psaA RNA is assembled to its mature form we have isolated a
large number of mutants deficient in photosystem I activity.
Surprisingly, about one third of these mutants were affected
in psaA mRNA maturation. Based on the psaA RNA precursors
which accumulate in the mutants, 3 classes could be
distinguished. Mutants unable to splice exons 1 and 2 and
exons 2 and 3 (class B); mutants unable to splice exons 2 and
3, but able to splice exons 1 and 2 (class A) and finally
mutants unable to splice exons 1 and 2, but able to splice
exons 2 and 3 (class C). All these mutations fall into 14
nuclear complementation groups: 5 for class A, 2 for class B
and 7 for class C (7). Since several of these complementation
groups contain a single allele it is likely that many other
nuclear genes are required for the maturation of the psaA
RNA. These mutations appear to act specifically on psaA.
Expression of other chloroplast genes containing group I
introns is unaffected. However, no other chloroplast gene of
C. reinhardtii with group II introns has been identified
besides psaA. While so far classes A and B contain
exclusively nuclear mutants, class C includes both nuclear
and chloroplast mutants.

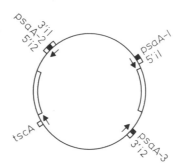

Fig. 1   Split structure of psaA of C. reinhardtii
The two large boxes represent the two segments of
the chloroplast inverted repeat. The three exons psaA-1, -2,
-3 and their flanking intron sequences (5'i1, 3'i1, 5'i2
3'i2) and tscA are shown on the circular chloroplast genome.
Arrows indicate the direction of transcription.

One might have expected that chloroplast class C mutations would map close to either exons 1 or 2 in the 5'i1 or 3'i1 regions. However these mutations map in a region, tscA, that is far apart from either exon, close to one of the chloroplast inverted repeats as first shown by Roitgrund and Mets (8, fig. 1). One of our mutants, H13, has a 2.8 kb deletion in this region. To map tscA more precisely we used particle gun mediated chloroplast transformation (9). Cells of H13 were bombarded with various fragments covering part of the deletion. Fragments disrupted at various sites with the Ω cassette (10) were also used for transformation. These experiments allowed us to map tscA within a region of 700 bp. The sequence of this region revealed the presence of short open reading frames. Disruption of any of these ORFs still allowed for phototrophic growth of the transformants. These results suggested that the tscA product is not a protein, but most likely a RNA. Indeed, Northern hybridizations with a tscA specific probe revealed the presence of a 400 base RNA in wild-type that is absent in the H13 mutant.

How does the tscA RNA function ? Fig. 2 shows a model in which this RNA completes the group II catalytic core of intron 1: The 5' region of the tscA RNA base pairs with the 5' part of intron 1 thereby forming domain I while the

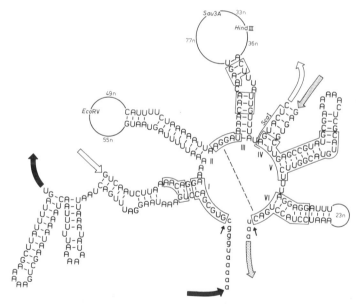

Fig. 2.   Tripartite structure of intron 1 of psaA
          Large arrows indicate the 5'-3' orientation of the RNAs. Intron and exon sequences are shown in uppercase and lowercase letters, respectively. The two exon-intron junctions are marked by small arrows. In the model tscA RNA (open arrows) base pairs with the 5' region of intron 1 (black arrows) thereby forming domain I. The tscA RNA completes the catalytic core by forming domains II and III and it base pairs with the 3'part of intron 1 (stippled arrows) within domain IV.

3'region of the tscA RNA base pairs with the 3' part of
intron 1 to form the stem of domain IV. The intron of psaA
with its typical 6 domain wheel structure is therefore made
of 3 discrete parts (fig. 2). The three Ω insertions which
map within the peripheral loops of domains II and III still
produce transformants of H13 with restored PSI activity
although the maturation of psaA mRNA is slower. These
insertions do therefore not appear to inactivate the
catalytic core of the intron. The only insertion which blocks
splicing occurs at the ScaI site within the pairing region of
domain IV (fig. 2). In contrast to introns 1, intron 2 of
psaA does not require a tscA equivalent since the 5' and 3'
portions of this intron can complete the intact group II core
structure (9). This may explain why we have never observed
faulty assembly of the three exons of psaA.

It is interesting to consider the origin of the unusual
structure of the first intron of psaA. One possibility is
that this discontinuous intron represents an ancient
structure which has been conserved in the chloroplast genome
of C. reinhardtii. It should be noted that in higher plants
another gene, rps12, also has split introns (11,12). Another
possibility is that this intron structure arose later through
some transposition event. In this view the tscA product may
be considered as a precursor of a snRNP particle, in
agreement with the notion that group II introns are the
likely predecessors of the introns of higher eukaryotes (13).

b. Chloroplast RNA stability

A nuclear mutant of C. reinhardtii, nac2-26 has been
described that has no detectable amount of psbD RNA which
specifies the reaction center polypeptide D2 of photosystem
II (14). The mutation acts specifically as other chloroplast
messages accumulate to wild-type levels. RNA pulse labeling
experiments show that the transcription of psbD occurs
normally. It is therefore likely that the mutation leads to
an accelerated degradation of the psbD RNA. Precursor
transcripts that cover both psbD and exon 2 of psaA (which
accumulate in class B psaA mutants, cf. above) are also
destabilized by this mutation indicating that the free 3'end
of psbD is not required for the accelerated breakdown of the
RNA. To test for possible target sites of message
degradation, a chimeric gene consisting of the atp B coding
sequence fused to the 3'untranslated region of psbD was
introduced into the chloroplast genome by transformation.
This chimeric RNA was stable in the nac 2-26 nuclear
background indicating that the 3'untranslated psbD region is
not sufficient to promote RNA destabilization. Similar
mutants have been described for psbB and psbC that are
affected in RNA stability (15,16; C. Monod, M. Goldschmidt-
Clermong, J.D. Rochaix, unpublished results). It therefore
appears that stability of several chloroplast messages is
determined by specific nuclear encoded factors.

c. Chloroplast translation

Both chloroplast and nuclear mutants affected in the
translation of the chloroplast psbC RNA have been isolated.

Two nuclear mutants, F34 and F64, belonging to distinct complementation groups and one chloroplast mutant Fud34 have been examined (17). Furthermore, a chloroplast suppressor that partially suppresses the nuclear mutation of F34 has been found. Analysis of the 500 nucleotide 5'untranslated region of psbC has revealed the presence of a large stem-loop structure near its middle with a free energy ΔG of - 21 kcal.This stem has been altered in the chloroplast mutant Fud34 in such a way that the stem is tighter (ΔG = - 30 kcal). In the chloroplast suppressor a new single base change has occurred in the same stem which does not appreciably alter the stability of the stem (ΔG = - 20 kcal) relative to wild-type. These findings suggest the existence of a nuclear factor (defined by the F34 mutation) that may interact directly or indirectly with the stem region. The observation that the F34 mutation is recessive suggest that this factor may act as an activator of translation, possibly by opening the stem-loop structure and allowing movement of the ribosome through this region.

Taken together these results reveal the existence of a large number of nuclear encoded factors that are required for the expression of individual chloroplast genes. Although the number of these factors is surprisingly high, this may be a means to tightly integrate the organelle within the cell by making it highly dependent on nuclear gene activity. It is possible that some of these factors may have a dual role as has been demonstrated for some fungal mitochondrial tRNA synthetases which also function as splicing factors (18,19). To understand how these factors work, it will be necessary to clone their genes, a task which may be achieved through transformation of the mutants with genomic wild-type libraries or through gene tagging.

III. CHLOROPLAST REVERSE GENETICS

The establishment of a reliable chloroplast transformation system in C. reinhardtii (1) has opened new approaches for probing the function and expression of the chloroplast genome. Disruption or alteration of chloroplast genes requires the availability of a foreign selective marker that is expressed in the chloroplast compartment. An expression cassette conferring spectinomycin resistance in the chloroplast has recently been designed by fusing the coding sequence of the bacterial aadA gene (aminoglycoside adenyl transferase) to the promoter and 5'untranslated region of atpA and the 3'flanking region of rbcL (20).

In a first series of experiments this aadA cassette was inserted into the psaC gene gene of C. reinhardtii (21). psaC encodes the apoprotein of the Fe-S clusters FA and FB that act as electron acceptors of PSI. Wild-type cells were bombarded with a plasmid containing the disrupted psaC gene and transformants were selected for drug resistance. After restreaking these cells several times on spectinomycin containing plates individual colonies were tested by fluorescence induction kinetics. Several clones that showed a characteristic PSI phenotype were isolated. Analysis of the DNA of these transformants revealed that all psaC genes had been disrupted. Northern hybridizations with the RNA from the transformants showed that the psaC transcript was no longer present. Instead transcripts corresponding to aadA were detectable. Analysis of the thylakoid proteins of the transformants showed that the PSI subunits accumulate at greatly reduced levels. It appears therefore that the absence of the psaC product leads to a drastic destabilization of the PSI complex. Hence the psaC product plays a key role both in electron transfer and in the stabilization of PSI. Using the same strategy it is now in principle possible to perform disruption and site directed mutagenesis on any of the chloroplast photosynthetic genes.

About 120 genes have been identified on the chloroplast genomes of higher plants (22). Close to 90 of these genes code for chloroplast proteins which include 21 ribosomal proteins, 4 RNA polymerase subunits, one initiation factor and close to 30 proteins involved in photosynthesis. The function of the remaining 30 ORFs has not yet been elucidated. Amongst these ORFs of unknown function some appear to be unique to each plant examined. One example is ORF 472 of C. reinhardtii which does not have any counterpart in the chloroplast genomes of either tobacco, rice and liverwort and in the current DNA data banks (21). To obtain more insights into the role of this ORF, it was disrupted with the aadA cassette and introduced into wild type cells by transformation selecting for spectinomycin resistance (20). Transformants were subcultured repeatedly and the DNA was analyzed by Southern hybridization. In contrast to the psaC disruption, it was not possible in this case to replace all wild-type copies of ORF 472 with the disrupted ones. Instead a heteroplasmic chloroplast DNA population was obtained suggesting that ORF 472 encodes a factor that is indispensable for cell growth even in the presence of a reduced carbon source.

IV. CONCLUSIONS AND PERSPECTIVES

Analysis of several photosynthetic mutants in C. reinhardtii has revealed the existence of a large number of nuclear encoded factors that are required for the expression of chloroplast genes and that appear to act mostly at the post-transcriptional level. A striking example is provided by psaA where at least 14 nuclear encoded factors are required for the synthesis of mature message. Perhaps as many as several hundred genes of this sort may exist in the nucleus. Understanding their exact function is an important task for the future.

Chloroplast transformation provides a new powerful tool for studying chloroplast gene expression. The use of chimeric genes may allow us to rapidly identify the target sites of some of these factors on chloroplast transcripts. Site directed mutagenesis may further identify important regulatory chloroplast sequences and provide new insights into the structure-function relationship of proteins involved in photosynthesis. Chloroplast research has now reached a stage were close interactions between molecular biology, genetics, biochemistry, biophysics and physiology are likely to occur and to yield new insights into chloroplast function and biogenesis.

Acknowledgements : We thank O. Jenni for drawings and photography. This work was supported by grant 31-26345.89 from the Swiss National Research Fund.

References

1. J.E. Boynton, N.W. Gillham, E.H. Harris, J.P. Hosler, A.M. Johnson, A.R. Jones, B.L. Randolph-Anderson, D. Robertson, T.M. Klein, K.B. Shark, and J.C. Sanford, Science 240: 1534-1538 (1988).

2. R. Debuchy, S. Purton, and J.D. Rochaix, EMBO J. 8: 2803-2809 (1989).

3. K.L. Kindle, R.A. Schnell, E. Fernandez and P.A. Lefebvre, J. Cell Biol. 109: 2589-2601 (1989).

4. S.P. Mayfield and K.L. Kindle, Proc. Natl. Acad. Sci. USA 87: 2087-2091 (1990).

5. U. Kück, Y. Choquet, M. Schneider, M. Dron and P.Bennoun, EMBO J. 6: 2185-2195 (1987).

6. Y. Choquet, M. Goldschmidt-Clermont, J. Girard-Bascou U. Kück, P. Bennoun and J.D. Rochaix, Cell 52: 903-913 (1988).

7. M. Goldschmidt-Clermont, J. Girard-Bascou, Y. Choquet and J.D. Rochaix, Mol. Gen. Genet. 223: 417-425 (1990).

8. C. Roitgrund and L.J. Mets, Curr. Genet. 17: 147-153 (1990).

9. M. Goldschmidt-Clermont, Y. Choquet, J. Girard-Bascou, F. Michel, M. Schirmer-Rahire and J.D. Rochaix, Cell 65: 135-143 (1991).

10. P. Prentki and H. Krisch, Gene 29: 303-313 (1984).

11. H. Fukuzawa, T. Kohchi, H. Shirai, K. Ohyama, K. Umesono, H. Inokuchi and H. Oseki, FEBS Lett. 198, 11-15 (1986).

12. N. Zaita, K. Torazawa, K. Shinozaki and M. Sugiura, FEBS Lett. 210: 153-156 (1987).

13. T.R. Cech, Cell 44:207-210 (1986).

14. M. Kuchka, M. Goldschmidt-Clermont, J. van Dillewijn and J.D. Rochaix, Cell 58: 869-876 (1989).

15. K.H. Jensen, D.L. Herrin, F.G. Plumley and G.W. Schmidt, J. Cell Biol. 103: 1315-1325 (1986).

16. L.E. Sieburth, S. Berry-Lowe and G.W. Schmidt, The Plant Cell 3: 175-189 (1991).

17. J.D. Rochaix, M.R. Kuchka, M. Schirmer-Rahire, J. Girard-Bascou and P. Bennoun, EMBO J. 8: 1013-1021 (1989).

18. A.A. Akins, and A.M. Lambowitz, Cell, 50: 331-345 (1987).

19. J.H. Herbert, M. Labouesse, G. Dujardin and P.P. Slonimski, EMBO J. 7: 473-483 (1988).

20. M. Goldschmidt-Clermont, Nucl. Acids Res. (1991), in press.

21. Y. Takahashi, M. Goldschmidt-Clermont, S.Y. Soen, L.G. Franzen and J.D. Rochaix, EMBO J. 10: 2033-2040 (1991).

22. H. Shimada and M. Sugiura, Nucl. Acids Res. 19: 983-995 (1991).

23. J.M. Vallet, M. Rahire and J.D. Rochaix, EMBO J. 3: 415-421 (1984).

TRANSCRIPTIONAL AND POST-TRANSCRIPTIONAL DETERMINANTS
OF CHLOROPLAST BIOGENESIS

Jörg Nickelsen, Kai Tiller, Claudia Fiebig,
Andrea Eisermann and Gerhard Link

Univ. of Bochum, Plant Cell Physiol. & Mol. Biol.
D-4630 Bochum, FRG

INTRODUCTION

Chloroplast biogenesis requires coordinated activities of both the organelle and the nucleo-cytoplasmic compartment. This is evident in the synthesis of the photosynthetic apparatus and other multi-component complexes, including those of the organellar gene expression system. The flow of genetic information from DNA to final gene products involves a complex network of basic and regulatory cis/trans-interactions between DNA (RNA) sequence elements and their cognate protein factors, both inside and outside the organelle. We have investigated functional determinants that play a role in the transcription and RNA processing of genes for chloroplast constituents.

In mustard (Sinapis alba) and related crucifers, most plastid RNAs accumulate during seedling development in a light-independent way. Transcripts of the psbA gene for the D1 protein, however, reach much higher levels in light-grown than in dark-grown seedlings. The precursor transcript of the trnK gene in front of psbA accumulates transiently, with peak levels at the time when psbA transcripts begin to show a light/dark difference (Link, 1988). What are the molecular mechanisms that underlie these various patterns of plastid gene expression?

PLASTID TRANSCRIPTION

Most sequenced chloroplast promoters seem to contain 'prokaryotic' -35 and -10 sequence elements. The psbA promoter, however, has an additional sequence motif resembling the 'eukaryotic' TATA-box, i.e. the typical promoter element of nuclear genes transcribed by RNA polymerase II (Link, 1988). In vitro transcription assays using chloroplast extracts and mutant promoters (Eisermann et al., 1990) have provided evidence that, in addition to the -35/-10 regions, the TATA-box-like element is indeed required for full functional activity. Moreover, a comparison of transcription in chloroplast and etioplast extracts has indicated a regulatory 'switch' involving differential usage of promoter elements in these two plastid types. Whereas both the -35 and TATA-box-like elements play a role in chloroplast transcription, only the latter appears to be used in etioplasts (Eisermann et al., 1990). To investigate this promoter element

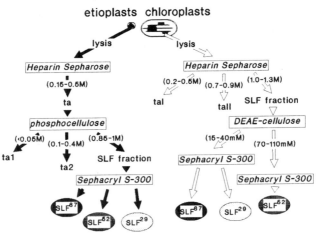

Fig. 1. Purification of sigma-like factors (SLF) from mustard chloroplasts and etioplasts. M: molarity of ammonium sulphate; ta: transcriptionally active fractions.

switch in more detail, we have focussed on the proteins involved in DNA-binding and transcription.

The differential usage of promoter elements could be the result of (1) different RNA polymerases in chloroplasts and etioplasts, (2) stage-specific transcription factor(s) supplementing the same 'core' enzyme, or (3) a combination of these two possibilites. Purified plastid RNA polymerase from mustard has a polypeptide pattern similar to that of other plant species and the enzymes prepared from either chloroplasts or etioplasts lack significant stage-specific differences in their polypeptide composition (not shown). On the other hand, we have obtained evidence for at least three chloroplast transcription factors resembling bacterial sigma factors. These 'sigma-like' factors of 67 kDa (SLF67), 52 kDa (SLF52) and 29 kDa (SLF29) are loosely associated with chloroplast core RNA polymerase and can be separated on Heparin Sepharose (Tiller et al., 1991). We investigated whether the same or different SLFs could be purified from etioplast extracts. Heparin Sepharose did not separate SLF activity from the etioplast polymerase, but this could be achieved by phosphocellulose (Fig. 1). Gel filtration and SDS-gel electrophoresis showed etioplast SLF polypeptides of 67, 52 and 29 kDa, i.e. the same molecular sizes as for chloroplasts.

SLF affinity for three plastid promoters was tested by gel shift assays, using E. coli core enzyme to reconstitute a full binding complex (Fig. 2). Each factor, regardless of whether it was obtained from chloroplasts or etioplasts, conferred efficient interaction with the psbA promoter. Differences were noted in the affinity for the trnQ and rps16 promoters. Among chloroplast factors it was high for SLF29, while it was low for all etioplast factors. The preference of the latter for the psbA promoter reflects that of etioplast holoenzyme (Eisermann et al., 1990).

Although it cannot be excluded that the chloroplast factors differ from those of etioplasts in their primary sequence, another possibility would be that SLF specificity is related to modifications such as phosphorylation. To test this, plastid RNA polymerases from the Heparin Sepharose stage (Fig. 1) were incubated with gamma-^{32}P-ATP and bovine heart protein kinase. SDS-gel electrophoresis showed labelled bands at positions of putative subunits, including those of SLFs. Most signals were stronger

Library Request Form

(Book) or Journal	8/31
Circle One	Today's Date

Title: Regulation of chloroplast biogenesis

Article Title:

Vol. No.		Year: 1992

Author: Joan H. Argyroudi-Akoyunoglou
(Include first initial)

Call. No. QK 725. R517 1992 (in Ag library)

Your NAME: Jennifer Robison

EMAIL and EXT. jrobison@   x4379

If you need copies you must supply the account info.

ACCT. NO.

ACCT. NAME

ILL Instructions:

From Library Home Page www.lib.udel.edu, click on Forms, click ILL Request and fill in form. BE SURE TO CHECK THE E-MAIL BOX and fill in your e-mail address. If you fill out one of these forms, your article will be e-mailed directly to you within days.

USE BACK OF FORM IF NEEDED

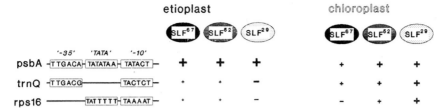

Fig. 2. Affinity of chloroplast and etioplast SLF polypeptides for the psbA, trnQ and rps16 promoters. Results of gel shift experiments with labelled psbA promoter fragment and excess of unlabelled competitor fragments. Binding reactions contained E. coli core RNA polymerase and SLF.

with preparations from chloroplasts than from etioplasts. When core enzyme from either plastid form was added to SLF fractions and the mixture then treated with kinase, no significant free SLF activity could be detected upon re-chromatography on Heparin Sepharose. On the other hand, treatment of etioplast holoenzyme with calf intestine alkaline phosphatase led to release of free SLF on Heparin Sepharose (data not shown). Hence, the 'tight' (etioplast-type) core polymerase-SLF complex and the 'loose' (chloroplast-type) complex seem to differ in their phosphorylation status and can be 'interconverted' by kinase/phosphatase treatment. This structural change might have important functional consequences: The etioplast holoenzyme is capable of interacting efficiently with the psbA promoter. However, the tight association of SLF could interfere with subsequent steps in RNA chain initiation and/or elongation. This 'transcription arrest' model is consistent with our finding that the etioplast enzyme has high DNA binding but low transcription activity (Eisermann et al., 1990). This model does not, however, explain the presence of the multiple SLF polypeptides and further work will be required to define their exact role in plastid transcription as well as their intracellular coding site and site of synthesis. Regulation of plastid gene expression takes place within the context of the entire cell. Modification of nuclear-encoded factors would be an efficient mechanism for trans-compartmental signal transduction.

PLASTID RNA MATURATION

Despite the evidence for transcriptional control of plastid gene expression (Klein and Mullet, 1990; Haley and Bogorad, 1990; Eisermann et al., 1990), it has become clear that major control points reside at the post-transcriptional level (Klein and Mullet, 1990; Gruissem, 1990). This is illustrated for the mustard trnK and psbA genes in Fig. 3, which de-

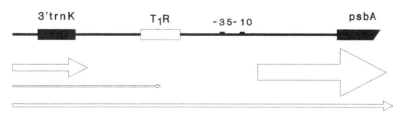

Fig. 3. Scheme of the trnK-psbA intergenic region on mustard chloroplast DNA. T1R: target site for 54-kDa RNA-binding protein. Transcripts are indicated by horizontal arrows.

picts the various transcripts within the intergenic region. The main trnK RNA 3' end is located near the 3' exon. A second 3' end maps at a region referred to as T1R, which is the target site for a 54 kDa RNA-binding protein. Finally, a fraction of transcripts extends into the psbA region, suggesting trnK-psbA cotranscription (Nickelsen and Link, 1988, 1991).

The purified 54 kDa RNA-binding protein contains associated endonucleolytic activity (not shown) and hence appears to act on read-through transcripts, suggesting that it might be involved in the control of expression of the downstream psbA gene. To test for stage-dependent differences in RNA-protein complex formation at T1R, we carried out RNA gel shift experiments using chloroplast and etioplast extracts (Nickelsen and Link, 1989). As shown in Fig. 4A, RNA-protein complexes of the same mobility were detected if the RNA probe spanning this region was incubated with either chloroplast (lane 2) or etioplast (lane 3) proteins. UV-crosslinking (Fig. 4B) revealed the labelled 54 kDa polypeptide with extracts from either plastid type (lanes 5 and 6). The second signal, representing a 32 kDa protein that binds RNA in a non-sequence-specific manner (Nickelsen and Link, 1991), is also visible in comparable intensity with either extract. These data, together with our finding that both plastid types contain comparable amounts of transcripts spanning the intergenic region (not shown), do not support the idea that psbA mRNA accumulation might be controlled by RNA-protein interactions at the T1R site.

Instead, the expression of the trnK-psbA intergenic region appears to resemble that of trnK, transcripts of which show light-independent transient peak levels during seedling development. Although we do not know if the read-through molecules represent the entire psbA coding region, this raises the possibility of functional mRNAs with a 'trnK-like' mode of expression. Such putative psbA transcripts would differ from those initiated at the psbA promoter by a longer 5' leader region. The leader region of plastid mRNAs contains signals necessary for translation and, in addition, target sites for translational modulators (Rochaix et al., 1989). Hence, psbA transcripts with a different 5' region might be an element in the known translational control of D1 formation (Boschetti et al., 1990).

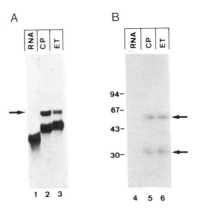

Fig. 4. RNA-protein interactions in extracts from mustard chloroplasts and etioplasts. A RNA gel shift assays with labelled RNA probe from trnK 3' region. RNA incubated alone lane 1), with 30 µg protein extract from chloroplasts (lane 2), or from etioplasts (lane 3). Arrow, RNA-protein complex. B Labelled RNA-binding proteins (54 and 32 kDa) after UV-crosslinking and SDS-PAGE.

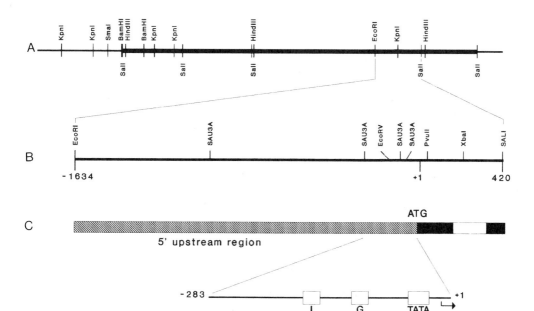

Fig. 5. A Lambda clone carrying a rbcS gene from Brassica napus. Heavy
bar: 16-kb insert. B 2.1-kb EcoRI/SalI region of the sequenced rbcS gene.
C Same region as in B with exon I and II (black), intron I (white), and
the 5' region (hatched). On a larger scale: transcription start site
(arrow), TATA-box, I- and G-box-like elements.

NUCLEAR DETERMINANTS OF CHLOROPLAST BIOGENESIS

Ribulose-1,5-bisphosphate carboxylase/oxygenase (RUBISCO) has become
a paradigm for the coordinated action of nuclear and chloroplast determinants in plastid development. Both transcriptional and post-transcriptional mechanisms are important for the regulated expression of the chloroplast rbcL gene for the large subunit as well as the nuclear rbcS gene
family for the small subunit (Boschetti et al., 1990). Present knowledge
on the determinants of rbcS transcription is based on transgenic plant and
transient in vivo assays in combination with DNA-protein binding studies
in vitro. A number of cis-regulatory elements and their cognate trans-
factors have been identified within the 5' upstream region of rbcS genes
from various plant species (Gilmartin et al., 1990).

We have examined 5' regions of rbcS genes from Brassica napus. These
genes fall into at least three classes (I-III) with regard to their mode
of expression during seedling development (Fiebig et al., 1990a, b). We
have analyzed the 5' region of a class III gene, which shows high transcript levels in the light and low levels in the dark. We concentrated our
interest on a region containing I-box-like elements (Giuliano et al.,
1988) (Fig. 5). In DNA-protein binding studies using cell-free extracts it
was found that this region interacts more efficiently with trans-factor(s)
from light-grown than with those from from dark-grown seedlings. By using
deletion mapping, competition gel shift assays, and DNase I footprinting,
we could assign the binding signal to a 20-bp sequence 110-130 bp upstream
of the transcription start site. This sequence shows symmetry and contains
two half-elements, 'CAC'and 'GTGG', which together resemble the G-box
motif (Giuliano et al., 1988). The I-box-like element also present in this
region did not significantly interact with extract proteins (not shown).

```
Bras.:    -160  TACACCTTTCTTTAATCCTGTGGCAGTTAACGACGATATCATGAAA -110
ATS1B:    -152  AACACCTTTCCTTAATCATGTGGTAGTGAACGA GTTATCATGAAT -108
ATS2B:    -175  AACACCTTTCCTTAATCCTGTGGCAATTAACGACGTTATCATGAAT -130
ATS3B:    -176  AACACCTTTCCTTAATCCTGTGGCAGTAAACGACGTTATCATGAAT -131
ATS1A:    -216  AACACCTCTCCTTAGGCCTTTGTGGCC                    -190
```

Fig. 6. Alignment of the Brassica napus rbcS 5' region (Bras.) with those
of the rbcS genes ATS1B - 3B and ATS1A from Arabidopsis thaliana (Krebbers
et al., 1988). Bold letters: half-elements of the split G-box.

Comparison of the 5' sequence of the Brassica rbcS gene with those
of the four rbcS genes from Arabidopsis thaliana (Krebbers et al., 1988)
revealed almost complete sequence identity within the region responsible
for binding (Fig. 6). All four Arabidopsis genes contain the 'split'
G-box-like motif of the Brassica gene. However, the 5' region of the
Arabidopsis ATS1A gene, as well as that of a second Brassica gene which
has recently been sequenced (not shown), contain in addition a contiguous
G-box motif. These data raise the possibility that sets of related cis-
elements are used in the regulation of rbcS gene expression. Progress in
this area can be expected to help define developmental 'cascades' of
nuclear protein-DNA interactions that might be important for chloroplast
biogenesis.

REFERENCES

Boschetti, A., Breidenbach, E., and Blättler, R. ,1990, Plant Sci. 68:
    131-149.
Eisermann, A., Tiller, K., and Link, G., 1990, EMBO J. 9: 3981-3987.
Fiebig, C., Neuhaus, H., Teichert, J., Röcher, W., Degenhardt, J., and
    Link, G., 1990a, Planta 181: 191-198.
Fiebig, C., Kretzschmar, F., Sprenger, I., and Link, G., 1990b, Bot. Acta
    103: 258-265.
Gilmartin, P. M., Sarokin, L., Memelink, J., and Chua, N.-H., 1990, Plant
    Cell 2: 369-378.
Giuliano, G., Pichersky, E., Malik, V. S., Timko, M. P., Scolnik, P. A.,
    and Cashmore, A. R., 1988, Proc. Natl. Acad. Sci. USA 85: 7089-7093.
Gruissem, W., 1990, Cell 56: 161-170.
Haley, J., and Bogorad, L., 1990, Plant Cell 2: 323-333.
Klein, R. R., and Mullet, J. E., 1990, J. Biol. Chem. 265: 1895-1902.
Krebbers, E., Seurinck, J., Herdies, L., Cashmore, A. R., and Timko, M.
    P., 1988, Plant Mol. Biol. 11: 745-759.
Link, G., 1988, Plant, Cell Environ. 11: 329-338.
Nickelsen, J., and Link, G., 1988, Nucleic Acids Res. 17: 9637-9648.
Nickelsen, J., and Link, G., 1991, Mol. Gen. Genet., in press.
Rochaix, J.-D., Kuchka, M., Mayfield, S., Schirmer-Rahire, M., Girard-
    Bascou, J., and Bennoun, P., 1989, EMBO J. 8: 1013-1021.
Tiller, K., Eisermann, A., and Link, G., 1991, Eur. J. Biochem. 198:
    93-99.

REGULATION OF EXPRESSION

OF THE PEA PLASTOCYANIN GENE

J.C Gray, K.-H. Pwee, R.E. Slatter and P. Dupree

Botany School, University of Cambridge
Downing Street, Cambridge CB2 3EA, U.K.

INTRODUCTION

Plastocyanin is a 10kDa copper protein located in the lumen of the chloroplast thylakoid membrane system where it transfers electrons from cytochrome f to the primary donor P700 of photosystem I. Plastocyanin is confined to chloroplast-containing tissues, and is not found in roots (Katoh et al., 1961; de Boer et al., 1988; Last and Gray, 1990). Similarly, plastocyanin mRNA is found only in photosynthetic tissues or tissues destined to become photosynthetic (Nielsen and Gausing, 1987; Barkardottir et al., 1988). In most plants the accumulation of plastocyanin is light-induced and appears to be under the control of phytochrome. Haslett and Cammack (1974) showed that plastocyanin accumulation in bean leaves was stimulated by red light and this stimulation was reversed by subsequent illumination with far-red light. A 30-fold increase in the amount of the plastocyanin polypeptide was observed during a 72 hour greening period in pea, wheat and barley (Takabe et al., 1986), although previously the amount of plastocyanin active in electron transfer in barley seedlings has been found to remain constant during greening (Plesnicar and Bendall, 1972), perhaps suggesting a post-translational regulation of the amount of the plastocyanin Increased amounts of plastocyanin mRNA on illumination of etiolated leaves have been detected in barley, spinach and pea (Nielsen and Gausing, 1987; Bichler and Herrmann, 1990; Last and Gray, 1990). The aim of our research is to understand the mechanisms responsible for tissue-specific and light-regulated expression of the pea plastocyanin gene.

Plastocyanin is encoded by a single-copy nuclear gene in spinach, *Arabidopsis thaliana* and pea (Rother et al., 1986; Vorst et al., 1988; Last and Gray, 1989). The genes encode a precursor form of plastocyanin with a long N-terminal extension of 58-71 amino acid residues (see Fig. 1). This presequence has been shown to be essential for targeting plastocyanin to the lumen of the thylakoid membrane system from its site of synthesis in the cytosol (Grossman et al., 1982; Smeekens et al., 1986). The presequence consists of two separable domains, an N-terminal chloroplast import domain and a C-terminal thylakoid transfer domain (Smeekens et al., 1986; Hageman et al., 1990). The chloroplast import domain is required for binding and translocation across the chloroplast envelope into the stroma (Smeekens et al., 1986; Hageman et al., 1990), where it is removed by the stromal processing peptidase (Robinson and Ellis, 1984). The remaining presequence of the intermediate form is required for binding and translocation across the thylakoid membrane into the lumen, where it is removed by a thylakoidal processing peptidase (Hageman et al., 1986; Kirwin et al., 1988) generating the mature plastocyanin polypeptide. Comparison of the presequences from six species reveals similarities in the N-terminal 20 amino acid residues and in the C-terminal 30 amino acid residues, but with little homology in the region between these sequences (Fig. 1). The conserved region at the N-terminus has been shown to be required for binding to the chloroplast envelope (Hageman et al., 1990), whereas the conserved C-terminal region contains the stromal processing site and the thylakoid transfer domain.

```
ps    MATVTSTT-VAIPSFSGLKTNAAT--KVS--AMAKIPTSTSQSPRLCVRASL--KDIGVALVATAASAVLASNALA V
      :::::::   :::::::.:::      :..          : . .     ::....:::   ::.:....::..:: :::::.:
le    MATVTS-AAVAIPSFTGLKAGASSSSRVSTGASAKVAAAPV--ARLTVKASL--KDVGAVVAATAVSAMLASNAMA L
      :::::: :::::::::: ::::  ::..:. :     :::.:        :....::::  ::::.:::::::. ...:: ::::
sp    MATVTSSAAVAIPSFAGLKA--SSTTRAATV---KVAVAT---PRMSIKASL--KDVGVVVAATAAAGILA-NAMA A
      ::::  ::::::::.::: ::::    :      :      :.. :     ::.:.::::  :.::..:.::::::::.:: ::::
so    MATVASSAAVAVPSFTGLKASGSIKPTTA-----KIIPTTTAVPRLSVKASL--KNVGAAVVATAAAGLLAGNAMA V
      ::. . :  : .:::::::  : :: :      .       : :.:..: ::  :.:...:::::   ::::::: .:::::::
at    MAAITSAT-VTIPSFTGLKLAVSSKPKTL-STISRSSSATRAPPKLALKSSL--KDFGVIAVATAASIVLAGNAMA M
      :::. :   :..::: .       . ..    ::::             ..... ::  :  . ..  :. ..:.: :::
hv    MAAL-SSAAVSVPSFAA-----ATPMRS-----SRSS-------RMVVRASLGKKAASAAVAMAAGAMLLGGSAMA Q
                                                                                     ▲
```

Fig. 1. Comparison of presequences of plastocyanin precursors. Sequences were obtained from pea (ps), Last and Gray (1989); tomato (le), Detlefsen et al. (1989); *Silene pratensis* (sp), Smeekens et al. (1985); spinach (so), Rother et al. (1986); *Arabidopsis thaliana* (at), Vorst et al. (1988); barley (hv), Nielsen and Gausing (1987). Sequences are compared with the sequence above; identical residues are marked with a colon (:) and conservative changes with a stop (.). Gaps introduced into the sequence to aid alignment are marked with a dash (-). The cleavage site for the thylakoidal processing peptidase is marked with an arrowhead.

EXPRESSION OF THE PEA PLASTOCYANIN GENE IN TRANSGENIC TOBACCO

To examine the expression of the pea plastocyanin gene in a foreign plant, a 3.5kbp *Eco*RI fragment containing the coding region with approximately 1kbp upstream sequence and 2kbp of downstream sequence (see Fig. 2), was tranferred to tobacco by *Agrobacterium*-mediated leaf disc transformation (Last and Gray, 1990). Primary transformed plants containing an intact pea plastocyanin gene were shown to contain pea plastocyanin by non-denaturing polyacrylamide gel electrophoresis of leaf extracts. Electrophoresis under non-denaturing conditions resolves the pea plastocyanin from the endogenous tobacco plastocyanin in sufficient amounts to be detected by staining with Coomassie blue (Last and Gray, 1990). Quantitative estimates of the amounts of pea plastocyanin in several transgenic tobacco plants, obtained by scanning densitometry of the stained bands, indicated that the mass ratios of pea plastocyanin to tobacco plastocyanin were in the range 0.23-1.84. Quantitative analysis of ten kanamycin-resistant plants obtained by self-pollination of plant A3, which contained a single pea plastocyanin gene per nucleus, indicated that the mean value of the ratio of pea plastocyanin to tobacco plastocyanin for hemizygous plants was 0.54±0.05 whereas for homozygous plants it was 0.85±0.07. Doubling the plastocyanin gene-copy number in the homozygous plants therefore gave an almost two-fold increase in the amount of pea plastocyanin, but had no effect on the amount of tobacco plastocyanin. The amount of pea plastocyanin in leaves of these transgenic plants was in the range 2-10mg per g fresh weight, although there was considerable variation depending on the age and growth conditions of the plants.

Pea plastocyanin was localised to the thylakoid lumen of chloroplasts from transgenic tobacco plants and had exactly the same electrophoretic mobility in non-denaturing polyacrylamide gels as authentic pea plastocyanin suggesting that it had been correctly processed and copper had been inserted. This was confirmed by analysis of pea plastocyanin purified from leaves of transgenic tobacco plants by ion-exchange chromatography and gel filtration (Last and Gray, 1990).The purified protein was blue in the oxidised state and showed an identical absorption spectrum to the protein purified from pea plants. The *N*-terminal sequence of the isolated plastocyanin was found to be Val-Glu-Val-Leu-Leu-Gly-Ala-Ser-Asp-Gly, identical to the published *N*-terminal sequence of pea plastocyanin (Boulter et al., 1979). This indicated that the *N*-terminal presequence of the precursor had been correctly removed in transgenic tobacco plants, and the isolation of a blue protein indicated that copper had been incorporated into the protein structure. The pea plastocyanin was shown to be correctly folded by ^{1}H nmr spectroscopy; the spectra of oxidised pea plastocyanin from transgenic tobacco and from pea were identical (Last and Gray, 1990).

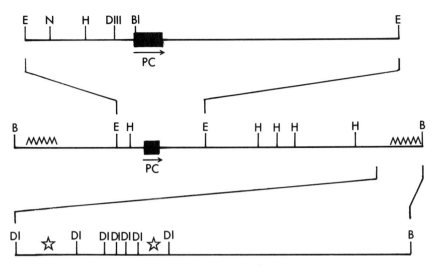

Fig. 2. Restriction maps of pea nuclear DNA in the vicinity of the plastocyanin gene. The middle map shows the 15.3kbp *Bam*HI fragment isolated from a pea genomic library. The position of the gene is shown with a filled box, and the direction of transcription with an arrow. Repetitive sequences are marked ΛΛΛ. The upper map is an expansion of the 3.5kbp *Eco*RI fragment showing the restriction sites used for construction of promoter deletions. The lower map is an expansion of the region containing the SAR, showing the *Dra*I sites. Fragments associated with the scaffold in a binding assay *in vitro* are marked with a star. E, *Eco*RI; H, *Hin*dIII; B, *Bam*HI; N, *Nde*I; DI, *Dra*I; DIII, *Dra*III, Bl, *Bal*I.

Pea plastocyanin purified from transgenic tobacco plants has been found to be active in electron transfer *in vitro* by measuring its rate of reduction in the presence of digitonin-treated pea chloroplasts as a source of the cytochrome *bf* complex (D.S. Bendall, personal communication). Second-order rate constants of approximately 10^8 M^{-1}s^{-1} for both authentic pea plastocyanin and pea plastocyanin from transgenic plants indicate that the proteins were essentially identical. The high-level expression of the pea plastocyanin gene in transgenic tobacco and the easy purification of pea plastocyanin suggests the use of tobacco as an expression system for the production of altered plastocyanins created by mutagenesis *in vitro*.

REGULATION OF EXPRESSION OF THE PEA PLASTOCYANIN GENE IN TRANSGENIC TOBACCO PLANTS

The pea plastocyanin gene, contained within the 3.5kbp *Eco*RI fragment, was expressed in the same light-regulated and organ-specific manner in transgenic tobacco plants as in pea plants, as shown by northern hybridisation and western blotting (Last and Gray, 1990). To identify the *cis*-acting sequences required for the regulated expression of the gene, regions of the 5' upstream sequence were fused to the β-glucuronidase (GUS) reporter gene and the constructs transferred into tobacco by *Agrobacterium*-mediated transformation of leaf discs. All constructs were translational fusions between the *Bal*I site (+4bp from the translation start) of the pea plastocyanin gene and the *Sma*I site just upstream of the GUS gene in the binary vector pBI101.2 (Jefferson, 1987). Different amounts of the 5' upstream region were included by digestion with *Eco*RI (at -992bp), *Nde*I (at -784bp), *Hin*dIII (at -444bp) and *Dra*III (at -176bp)(see Fig. 2). Assays of GUS activity in leaf extracts of rooted, kanamycin-resistant plants revealed significant differences in gene expression between the different constructs (Fig. 3). The full-length -992 promoter directed a level of GUS expression which

25

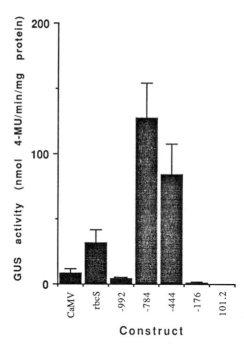

Fig. 3. GUS activities in leaf extracts of tobacco plants transformed with the promoter-GUS fusions. Plants were grown in tissue-culture. GUS activity was measured fluorimetrically using 4-methylumbelliferyl glucuronide as substrate. Activities are expressed as means with the standard error denoted with a bar. The plastocyanin promoter constructs are denoted -992, -784, -444 and -176. Control plants transformed with a promoterless GUS construct are denoted 101.2.

was lower, but not significantly different, to that directed by the cauliflower mosaic virus (CaMV) 35S promoter and a tobacco *rbc*S promoter. However, deletion from -992 to -784 resulted in a 40-fold increase in expression, indicating removal or disruption of a negative *cis*-element or silencer element. A further deletion to -444 decreased GUS activity slightly, but this was not significantly different from the -784 deletion. Deletion to -176 resulted in only low levels of GUS activity, indicating the removal of positive *cis*-element(s) located between -444 and -176. The region of the pea plastocyanin gene from -784 to +4 is the most active photosynthesis-gene promoter described to date.

The tissue-specificity of GUS expression was examined by histochemical staining of thin hand-sections of various organs with the chromogenic substrate 5-bromo 4-chloro 3-indolyl glucuronide (X-gluc). In the high-expressing promoter deletions (-784 and -444), GUS activity showed a close association with green chlorophyll-containing cells in leaves, stems, sepals, petals, anthers, stigmas and ovaries. Activity was located in the epidermal cells of leaves, which were shown to contain chloroplasts by light- and electron-microscopy (Dupree et al., 1991). GUS activity was not expressed in roots or non-chlorophyllous cells of the vascular tissue. In the low-expressing constructs (-992 and -176), GUS activity was detectable histochemically only in the palisade and mesophyll cells of the leaves, but fluorimetric assays of extracts of leaves, sepals, anthers and stigmas indicated the presence of GUS activity in the same relative proportions to that in the organs of high-expressing plants. No activity was detected in roots. This suggests that the *cis*-elements responsible for tissue-specific expression are located in the region from -176 to +4.

The effects of light on the expression of GUS were examined on seedlings from the self-pollinated primary transgenic plants. Seeds (150-200 per sample) were germinated and grown for 5 days in the dark and then for a further 2 days either in white light or the dark,

Table 1. GUS activities in dark-grown and light-exposed seedlings of transgenic tobacco. Seedlings (150-200) of self-pollinated primary transgenic plants were germinated and grown in the dark for 5 days and then subjected to a futher 2 days of light (L) or dark (D). Extracts of complete seedlings were assayed for GUS activity with 4-methylumbelliferyl glucuronide as substrate. Results are expressed as mean ± standard error (number of individual primary transgenic plants). The ratios (L/D) of activities of progeny of individual primary transgenic plants are also expressed as mean ± standard error. The range of L/D ratios is also given.

| Construct | GUS activity, pmol 4-MU min^{-1}mg protein^{-1} | | L/D | |
	Light (L)	Dark (D)	Mean	Range
PC -992	1480 ± 240 (6)	1960 ± 240 (6)	0.77 ± 0.09	0.6- 1.2
PC -784	80570 ±24210 (6)	44250 ±16460 (6)	2.08 ± 0.34	1.1- 3.5
PC -444	81330 ±25220 (6)	50400 ±15560 (6)	1.67 ± 0.14	1.1- 2.1
PC -176	240 ± 50 (3)	350 ± 200 (3)	1.30 ± 0.53	0.5- 2.3
CaMV 35S	13160 ± 2970 (4)	8740 ± 1450 (4)	1.45 ± 0.13	1.2- 1.8
rbcS	9900 ± 2290 (6)	2070 ± 260 (6)	5.08 ± 1.05	1.0- 7.9
FNR	510 ± 143 (9)	90 ± 20 (9)	6.34 ± 1.40	1.9-12.9

fluorimetric assay of seedling extracts for GUS activity. None of the constructs containing 5' upstream regions of the pea plastocyanin gene directed highly light-regulated expression of GUS, although the highly expressing -784 and -444 constructs gave a slight light-stimulation. However, these highly expressing constructs directed high levels of GUS expression in dark-grown seedlings. The light response was indicated by the ratio (L/D) of GUS activity in the light-exposed seedlings to that in the 7-day old dark-grown seedlings (Table 1). The L/D ratios of the promoter deletions were not significantly different from the ratio with seedlings containing a CaMV-GUS construct. In contrast seedlings with the rbcS-GUS construct and with a pea ferredoxin-NADP+ reductase (FNR) promoter fusion to GUS showed a strong light induction (Table 1). This suggests that the 5' upstream region of the pea plastocyanin gene does not contain all the cis-elements necessary for light regulation in young seedlings. Additional elements within the coding region or the 3' downstream region may be required for light regulation; these elements may act by decreasing the relatively high levels of expression in the dark directed by the 5' upstream region.

CHROMOSOMAL ORGANISATION OF THE PEA PLASTOCYANIN GENE

A complete understanding of the expression of the pea plastocyanin gene will require a knowledge of the arrangement of the gene in the three-dimensional organisation of the chromosome. It is now recognised that chromosomes are arranged with long (5-200kbp) loops of DNA attached to a central proteinaceous scaffold. The torsional stress and replication of each loop may be independently regulated, thus providing an additional level of factors affecting gene expression (Slatter and Gray, 1991). We have attempted to define the extent of the chromosomal loop containing the plastocyanin gene in pea. Scaffold-associated regions of DNA (SARs) have been isolated following restriction enzyme digestion of pea leaf chromatin treated with 6mM lithium diiodosalicylate (LIS) to remove histones. LIS has been widely used to remove histones, without affecting the attachment of the DNA to the chromosome scaffold (Mirkovitch et al., 1984). Restriction enzyme digestion released the loop DNA and the scaffold-associated material was collected by centrifugation at 2400g for 2 minutes. The isolated DNA from the pellet (scaffold-associated) and supernatant (loop) fractions were labelled with ^{32}P-dATP and hybridised separately to Southern blots of restriction fragments of cloned pea genomic DNA in the vicinity of the plastocyanin gene. Restriction fragments containing an SAR hybridise to the pellet DNA rather than to the supernatant DNA. This method has indicated the presence of an SAR in a region 8-9kbp downstream of the pea plastocyanin gene, closely linked to a copy of a repetitive sequence present 300 times in the pea haploid genome (see Fig. 2).

To confirm that this region of the pea genome could associate with the scaffold *in vitro*, ^{32}P-end-labelled restriction fragments of the 2.6kbp *Hind*III-*Bam*HI fragment in pUC18 were added to LIS-extracted chromatin during restriction enzyme digestion. Scaffold-associated material was collected by centrifugation as described above, the extracted DNA was separated by agarose gel electrophoresis and labelled fragments detected by autoradiography. Two *Dra*I fragments of 230bp and 110bp were preferentially bound to the pea scaffolds, even in the presence of 200µg ml^{-1} competitor pea DNA. Nucleotide sequencing showed that these fragments were in a region of 540bp containing 77% A/T-rich DNA. A/T-rich DNA is a feature of SARs characterised from other organisms (Slatter and Gray, 1991). The pea SAR contains homology to to a number of sequences, such as the *Drosophila* topoisomerase II consensus (Sander and Hsieh, 1985), A-box and T-box regions (Gasser and Laemmli, 1986) and nucleation sites for strand separation (Kohwi-Shigematsu and Kohwi, 1990), which are present in SARs from other organisms.

The identification of the SAR located upstream of the pea plastocyanin gene will define the size of the chromosomal loop. However, at present, the presence of another repeat sequence approximately 5kbp upstream of the plastocyanin gene is dominating hybridisation signals from this region. This repeat sequence is present 15000 times in the genome and constitutes 0.2% of the genomic DNA. It is largely present on 2.0kbp and 0.7kbp *Bam*HI fragments.

ACKNOWLEDGEMENTS

This work was supported by grants from the Science and Engineering Research Council and the Gatsby Charitable Foundation. K-H.P. was supported by a Government of Singapore Scholarship and P.D. was supported by an SERC Research Studentship and The Leathersellers Company Graduate Scholarship at Fitzwilliam College, Cambridge.

REFERENCES

Barkardottir, R.B., Jensen, B.F., Kreiberg, J.D., Nielsen, P.S. and Gausing, K. (1987) Expression of selected nuclear genes during leaf development in barley. *Dev. Genet.* 8, 495-511.
Bichler, J. and Herrmann, R.G. (1990) Analysis of the promoters of the single-copy genes for plastocyanin and subunit δ of the chloroplast ATP synthase from spinach. *Eur. J. Biochem.* 190, 415-426.
Boulter, D., Peacock, D., Guise, A., Gleaves, J.T. and Estabrook, G. (1979) Relationships between the partial amino acid sequences of plastocyanin from members of ten families of flowering plants. *Phytochemistry* 8, 603-608.
de Boer, D., Cremers, F., Teertstra, R., Smits, L., Hille, J., Smeekens, S. and Weisbeek, P.(1988) *In vivo* import of plastocyanin and a fusion protein into developmentally different plastids of transgenic plants. *EMBO J.* 7, 2631-2635.
Detlefsen, D.J., Pichersky, E. and Pecoraro, V.L. (1989) Preplastocyanin from *Lycopersicon esculentum. Nucl. Acids Res.* 17, 6414.
Dupree, P., Pwee, K-H. and Gray, J.C. (1991) Expression of photosynthesis gene-promoter fusions in leaf epidermal cells of transgenic tobacco plants. *Plant J.* 1, 115-120.
Gasser, SM. and Laemmli, U.K. (1986) Cohabitation of scaffold binding regions with upstrean/enhancer elements of three developmentally regulated genes of *D. melanogaster. Cell* 46, 521-530.
Grossman, A.R., Bartlett, S.G., Schmidt, G.W., Mullet, J.E. and Chua, N.-H. (1982) Optimal conditions for post-translational uptake of proteins by isolated chloroplasts. *J. Biol. Chem.* 257, 1558-1563.

Hageman, J., Robinson, C., Smeekens, S. and Weisbeek, P. (1986) A thylakoid processing protease is required for complete maturation of the lumen protein plastocyanin. *Nature* 324, 567-569.

Hageman, J., Baecke, C., Ebskamp, M., Pilon, R., Smeekens, S. and Weisbeek, P. (1990) Protein import into and sorting inside the chloroplast are independent processes. *Plant Cell* 2, 479-494.

Haslett, B.G. and Cammack, R. (1974) The development of plastocyanin in greening bean leaves. *Biochem. J.* 144, 567-572.

Jefferson, R.A. (1987) Assaying chimeric genes in plants: the GUS gene fusion system. *Plant Mol. Biol. Rep.* 5, 387-405.

Katoh, S., Suga, I., Shiratori, I. and Takamiya, A. (1961) Distribution of plastocyanin in plants, with special reference to its localization in chloroplasts. *Arch. Biochem. Biophys.* 94, 136-141.

Kirwin, P.M., Elderfield, P.D. and Robinson, C. (1987) Transport of proteins into chloroplasts. Partial purification of a thylakoidal peptidase involved in plastocyanin biogenesis. *J. Biol. Chem.* 262, 16386-16390.

Kohwi-Shigematsu, T. and Kohwi, Y. (1990) Torsional stress stabilises extended base unpairing in suppressor sites flanking immunoglobulin heavy chain enhancer. *Biochemistry* 29, 9551-9560.

Last, D.I. and Gray, J.C. (1989) Plastocyanin is encoded by a single-copy gene in the pea haploid genome. *Plant Mol. Biol.* 12, 655-666.

Last, D.I. and Gray, J.C. (1990) Synthesis and accumulation of pea plastocyanin in transgenic tobacco plants. *Plant Mol. Biol.* 14, 229-238.

Mirkovitch, J., Mirault, M.E. and Laemmli, U.K. (1984) Organisation of the higher-order chromatin loop: specific DNA attachment sites on nuclear scaffold. *Cell* 39, 223-232.

Nielsen, P.S. and Gausing, K. (1987) The precursor of barley plastocyanin. Sequence of cDNA clones and gene expression in different tissues. *FEBS Lett.* 225, 159-162.

Plesnicar, M. and Bendall, D.S. (1973) The photochemical activities and electron carriers of developing barley leaves. *Biochem. J.* 136, 803-812.

Robinson, C. and Ellis, R.J. (1984) Transport of proteins into chloroplasts. Partial purification of a chloroplast protease involved in the processing of imported precursor polypeptides. *Eur. J. Biochem.* 142, 337-342.

Rother, C., Jansen, T., Tyagi, A., Tittgen, J. and Herrmann, R.G. (1986) Plastocyanin is encoded by an uninterrupted nuclear gene in spinach. *Curr. Genet.* 11, 171-176.

Sander, M. and Hsieh, T-S. (1985) Drosophila topoisomerase II double-stranded DNA cleavage: analysis of DNA sequence homology at the cleavage site. *Nucl. Acids Res.* 13, 1057-1072.

Slatter, R.E. and Gray, J.C. (1991) Chromatin structure of plant genes. *Oxford Surv. Plant Mol. Biol. Cell Biol.*, in press.

Smeekens, S., Bauerle, C., Hageman, J., Keegstra, K. and Weisbeek, P. (1986) The role of the transit peptide in the routing of precursors toward different chloroplast compartments. *Cell* 46, 365-375.

Smeekens, S., De Groot, M., Van Binsbergen, J. and Weisbeek, P. (1985) Sequence of the precursor of the chloroplast thylakoid lumen protein, plastocyanin. *Nature* 317, 456-458.

Takabe, T, Takabe, T. and Akazawa, A.T. (1986) Biosynthesis of P700-chlorophyll *a* protein, plastocyanin, and cytochrome b_6-f complex. *Plant Physiol.* 81, 60-66.

Vorst, O., Oosterhoff-Teerstra, R., Vankan, P., Smeekens, S. and Weisbeek, P. (1988) Plastocyanin of *Arabidopsis thaliana*: isolation and characterisation of the gene and chloroplast import of the precursor protein. *Gene* 65, 59-69.

REGULATION OF GENES ENCODING PLASTID PROTEINS DURING CHLOROPLAST BIOGENESIS

IN *EUGLENA*

D. E. Buetow[1,2], L.S.H. Yi[1] and G. Erdös[1]

Departments of Physiology and Biophysics[1] and of Plant
Biology[2], University of Illinois 61801, USA

INTRODUCTION

Multiple and complex molecular mechanisms activate and coordinate the
expression of genes that encode chloroplast proteins. These genes themselves are encoded both in the organelle's own genome and in the nuclear
genome. At least some of the nuclear genes are regulated at the transcriptional level.[1] In other cases, and especially during chloroplast biogenesis, posttranscriptional regulation of nuclear genes appears to occur.[2] For
some chloroplast-encoded genes, a specific transcriptional control results
from variations in the strengths of their promoter elements.[3] Even so, the
expression of these latter genes also appears to be strongly influenced by
postranscriptional regulatory mechanisms occurring at the levels of mRNA
stability and translation.[3,4]

The alga *Euglena gracilis* is a particularly useful model system for
studies on chloroplast biogenesis because, unlike higher plants, it contains "conditional" chloroplasts. In the dark, *Euglena* can grow heterotrophically, but its chloroplasts dedifferentiate and remain as rudimentary
proplastids. When dark-grown *Euglena* are placed in white light, chloroplast
biogenesis occurs during which the proplastids differentiate ("green") into
mature chloroplasts over about a 3-day period. Considerable evidence indicates that, during light-induced chloroplast biogenesis in *Euglena*, gene
expression is primarily regulated posttranscriptionally.[5-10] Here we further
investigate the regulation of expression of eight chloroplast-encoded genes
and one nuclear-encoded gene in *Euglena* during light-induced chloroplast
biogenesis. All nine genes encode plastid proteins.

MATERIALS AND METHODS

E. gracilis strain Z was grown and maintained continuously in the dark
on a defined medium containing 0.1 M ethanol as a carbon source.[11] Cultures
at densities of 1-2 x 10^6 cells/ml were placed in continuous white light
($20\mu Em^{-2}s^{-1}$) to induce chloroplast biogenesis. Dark-grown cells and cells
exposed to the light for various times up to 72 h were collected.

To measure levels of the *atp*B, *atp*E and *atp*H transcripts, internal
fragments of the maize *atp*B and *atp*E (from L. Bogorad) and spinach *atp*H
(from R. Herrmann) genes were used. For *psa*A, *psb*A, *psb*B, *psb*C and *rbc*L,
the *Euglena* chloroplast genes[12] were used.   Genes were inserted into

plasmids, amplified in *E. coli*, recovered and radiolabelled with ^{32}P-CTP by random primer extension.[13]   Total cell RNA was isolated and hybridized to the ^{32}P-labelled gene probes by dot hybridization.   Dots were cut out and counted in a scintillation counter.

Protein products of chloroplast genes were isolated (e.g., Freyssinet et al.[14]; Yi et al.[15]) and used to generate monospecific antibodies in New Zealand rabbits.   The antibodies were used to quantitate chloroplast proteins by a highly-sensitive dot-immunoassay[16] which detects as little as 10 fmoles of specific protein which can be quantitated directly in a whole cell preparation.   The *psb*A gene product ("D1-protein") has a short half-life[17] and its steady-state concentration in light-grown *Euglena* is too low for use in generating antibodies.   Therefore, its level was quantitated by the ^{14}C-atrazine binding method.[18]

Pulse-chase studies on protein synthesis and turnover were done with ^{3}H-phenylalanine.[19] Proteins were precipitated with specific antibodies.

RESULTS

Following growth in the dark with ethanol as carbon source, *Euglena* cultures show little or no cell division when placed in the light for up to 72 h (Fig. 1).   During this time, chloroplast biogenesis occurs (proplastids differentiate into mature chloroplasts).   When chloroplast biogenesis is nearly completed (about 72 h), some cell division may resume.

Transcripts from all genes examined were detected at low levels in dark-grown cells (e.g., Figs. 2,4,5).   Transcripts from *rbc*L, *psa*A, *psb*C, *atp*E transiently increased in amount per cell during light-induced chloroplast biogenesis (Fig. 2, Table 1), but only to a small extent (1.5X to 6.0X depending upon the transcript).   Each transcript peaked in amount at 12 to 36 h and then decreased by 72 h to a level near the dark-grown cell level.   In contrast, *psb*B transcripts increased slowly throughout chloroplast biogenesis (Fig. 4, Table 1) whereas the *psb*A and *atp*H transcripts increased to 12-18 h and then remained relatively constant (Fig. 5, Table 1).   Again, the increases in the latter three transcripts were small (Table 1).

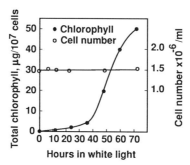

Fig. 1.   Chloroplast biogenesis (greening) measured by the increase in total chlorophyll per cell in dark-grown *Euglena* exposed to light.   Zero h is the beginning of the light exposure.

Table 1. Comparison of the Concentration of Plastid-encoded mRNAs Versus the Number of
Plastid Genomes per *Euglena* Cell during Light-induced Chloroplast Biogenesis

Time[a]	Chloroplast Genomes/Cell[b]		Relative Amount/Cell[c] in Transcripts Encoded by							
	Number	Relative Amount[c]	psaA	psbA	psbB	psbC	atpB	atpE	atpH	rbcL
3-4 h	1400	1.0	1.0	1.0	1.0	1.0	1.0	1.0	1.0	1.0
6	--	--	1.2	1.4	1.0	1.4	1.4	2.8	1.4	1.0
8	1500	1.1	--	--	--	--	--	--	--	--
12	2100	1.5	1.5	3.2	1.0	1.7	3.0	4.2	2.1	1.1
18	--	--	--	3.5	--	2.1	2.0	4.3	2.7	--
24	--	--	1.0	1.9	1.5	2.6	1.4	3.2	2.4	1.6
36	--	--	--	1.9	--	6.0	0.8	2.9	1.5	--
48	2900	2.1	0.9	2.3	2.2	4.2	0.4	1.7	1.9	1.0
72	2500	1.8	0.9	2.3	4.8	3.7	0.2	1.7	2.0	0.7

[a] During chloroplast biogenesis (greening) in *E. gracilis*; dark-grown cells exposed to continuous white light at zero h.
[b] From Chelm et al20.
[c] Amount per cell relative to that of a cell greened for 3-4 h; in each column, the maximal relative amount observed during greening is double-underlined.

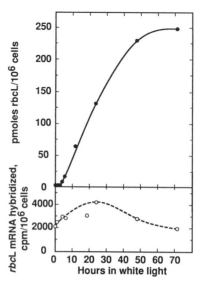

Fig. 2. Accumulation of the *rbc*L mRNA (lower panel) and protein product
(upper panel) during light-induced chloroplast biogenesis in
Euglena. Zero h is the beginning of the light exposure.

Though the exact number of chloroplast genomes per *Euglena* cell
depends on the previous nutritional history of the cells,[12] this number does
increase up to about 48 h of light-induced chloroplast biogenesis and then
levels out or somewhat decreases.[20] Levels of seven of the above transcripts correlate with their gene levels only during the first one-third to
one-half of the chloroplast biogenesis period and then decline as their gene
levels remain high (Table 1). A second pattern is shown by *psb*B whose transcript level increases and more closely mimics its gene level throughout the
72 h.

Protein products of the chloroplast-encoded *rbc*L, *psa*A, *psb*B, *psb*C and
*atp*H and the nuclear-encoded *rbc*S are detectable in dark-grown cells, and
the level of each increases during light-induced chloroplast biogenesis
(Figs. 2-5, Table 2), but to different extents, i.e., from about 5-fold
(*atp*H) to about 800-fold (*rbc*L). Patterns of accumulation differ
considerably among the different proteins (e.g., Figs. 2,4,5). Even the two
subunits of ribulose-bisphosphate carboxylase accumulate differentially
(Fig. 3). Accumulation of the small subunit (rbcS) was detected by 1 h of
light exposure and continued up to 12 h. Then, the rbcS remained constant
to at least 48 h and then accumulated again. In contrast, the large subunit
(rbcL) remained at the dark-cell level for 2-3 h and only then began to
accumulate which it continued to do throughout the remaining chloroplast-
biogenesis period (Fig. 3). The lag in the rbcL accumulation correlates
with its rate of synthesis in dark-grown *Euglena* exposed to light and which
remains low up to 12 h and then dramatically increases.[9] The differential
accumulation of rbcL and rbcS leads to a shifting ratio of L:S during
chloroplast biogenesis (Fig. 3). However, as the organelle approaches the
mature chloroplast stage, the L:S ratio nears 1.0.

Table 2. Increase in Amount of Plasmid Proteins During Light-induced Chloroplast Biogenesis

| Gene[a] | Protein Product | Amount (pmoles/10^6cells) Present[b] | |
		0h	72 h (fold-increase)
psaA	P700 Apoprotein, Subunit 1a	7.7	80.0 (10.4X)
psbA	D1 Reaction Center Polypeptide	0.8[c]	10.7 (13.4X)
psbB	CP47 Chlorophyll a Apoprotein	4.5	34.5 (7.7X)
psbC	CP43 Chlorophyll a Apoprotein	3.8	82.1 (21.6X)
atpH	CF_O Membrane-portion of the ATPase Complex, Subunit III	78.8	388.9 (4.9X)
rbcL	Ribulose-bisphosphate Carboxylase, Large Subunit	0.3	248.4 (828X)
rbcS	Ribulose-bisphosphate Carboxylase, Small Subunit	0.3	223.6 (745X)

[a]All genes are plastid-encoded except for rbcS which is nuclear-encoded.
[b]Content per 10^6 Euglena cells at the beginning (0h) and after 72 h of light-induced chloroplast biogenesis.
[c]Zero h amount not available; value is amount at 3 h of light.

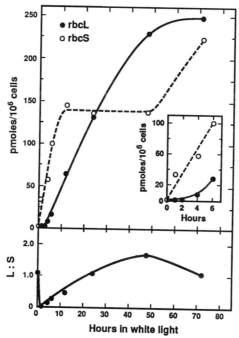

Fig. 3. Accumulation of the rbcL and rbcS protein products (upper panel) during light-induced chloroplast biogenesis in Euglena. The inset gives an enlarged view of the first 6 h of the accumulation. The lower panel shows the molar ratio of rbcL to rbcS during chloroplast biogenesis. Zero h is the beginning of the light exposure.

Table 3. Accumulation of Specific and Total Plastid
          Proteins and of Total Chlorophyll and Carotenoids
          during Light-induced Chloroplast Biogenesis

| | Hours in the Light | | |
	Oh	12h	60h
Protein coded by			
psbA[a]	7.7[b]	10.9	56.0
psbA	0.8[c]	2.0	10.9
psbB[a]	4.5	10.7	24.4
psbC[a]	3.8	7.9	48.5
atpH	78.8	166.7	332.0
rbcL	0.3	65.8	245.0
rbcS	0.3	144.6	178.0
Other Substances			
Total Proteins	327[d]	349	342
Total Chlorophylls	0[e]	370	2,250
Total Carotinoids	150[e]	110	590

[a]Product is a chlorophyll-binding protein.

[b]Cellular content in pmoles/10^6 cells.

[c]Zero h amount not available; value in amount at 3 h
   of light.

[d]Cellular content in μg/10^6 cells.

[e]Cellular content in pg/10^6.

     Levels of proteins in most cases correlated with levels of their mRNAs
only during the early portion of the chloroplast-biogenesis period (e.g.,
Figs. 2,5). After this, the proteins continued to accumulate as the levels
of the respective mRNAs either declined or became stable. Only the psbB
product accumulated as its mRNA accumulated throughout the 72-h chloroplast-
biogenesis period (Fig. 4).

     The total protein content of Euglena changed little during light-
induced chloroplast biogenesis while plastid-protein contents increased
(Table 3).   Thus, as expected, there is a differential synthesis and
accumulation of plastid proteins during chloroplast biogenesis. The length
of the lag phase in accumulation of a plastid protein may be shorter, the
same as, or longer than that of its mRNA (Table 4).   However, plastid
proteins are detected in dark-grown cells when no chlorophyll is present
(Table 3). Also, these proteins, including the chlorophyll-binding proteins
coded by psaA, psbB and psbC, begin to accumulate before chlorophyll is
detected in the cells (Table 4). Therefore, though at least some of the

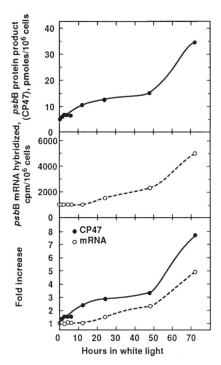

Fig. 4. Accumulation of the *psb*B mRNA (middle panel) and its "CP47" protein product (upper panel) during light-induced chloroplast biogenesis in *Euglena*. The lower panel shows the fold-increase in the *psb*B mRNA and protein product during chloroplast bioge-nesis. Zero h is the beginning of the light exposure.

plastid proteins ultimately may be stabilized by binding chlorophyll[21], the absence of chlorophyll precludes the formation of protein-chlorophyll complexes during early chloroplast biogenesis (Table 4). Carotenoids are always detectable in the cells (Tables 3,4). Perhaps these help stabilize, at least to some extent, the chlorophyll-binding proteins when chlorophyll is lacking, as indicated by the studies of Plumley and Schmidt.[22] In any case, the dark cell contains substantial amounts of the relevant mRNAs (e.g., Figs. 2,4,5), but the level of the corresponding proteins is quite low at least compared to cells exposed to the light (Table 3). This suggests that these proteins are subject to a relatively high level of turnover in the dark.

Pulse-chase studies were done on the in vivo protein synthesis and turnover in cells grown continuously in the dark or in the light or in cells first grown in the dark and then placed in the light to induce chloroplast biogenesis. Cells were pulsed for 40 min. with ³H-phenylalanine and then chased with unlabelled phenylalanine. Newly-synthesized plastid proteins encoded by *psb*B showed a rapid turnover in the dark, but were stable during chloroplast biogenesis (Fig. 6). Similar results were shown with the products of *psa*A, *psb*C and *atp*H. Pulse-chase studies with the rbcL and rbcS also showed them to be more stable in light-grown cells and in those undergoing chloroplast biogenesis than in dark-grown cells (Figs. 7,8). Importantly, the pulse-chase studies show that cellular levels of plastid proteins are regulated at the posttranslational level of protein turnover with the proteins being stabilized as they accumulate during chloroplast biogenesis.

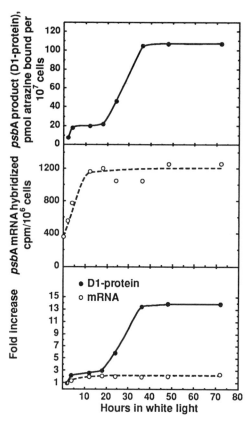

Fig. 5. Accumulation of the *psb*A mRNA (middle panel) and its "D-1" protein
product (upper panel) during light-induced chloroplast biogenesis
in *Euglena*. The lower panel shows the fold-increase in the *psb*A
mRNA and protein product during chloroplast bioge-nesis Zero h is
the beginning of the light exposure.

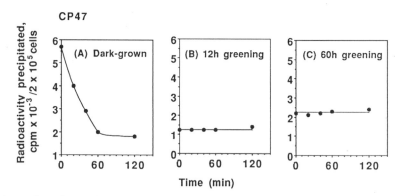

Fig. 6. Kinetics of the in vivo turnover of the "CP47" protein product
coded by *psb*B in *Euglena*: (A) grown in continuous darkness, (B)
after 12 h of light-induced chloroplast biogenesis ("greening"),
and (C) after 60 h of light-induced chloroplast biogenesis.

rbcL

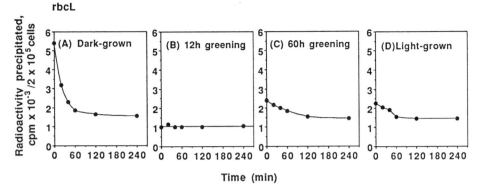

Fig. 7. Kinetics of the in vivo turnover of the large subunit of ribulose bisphosphate carboxylase (rbcL) in *Euglena*, (A) grown in continuous darkness, (B) after 12 h of light-induced chloroplast biogenesis ("greening"), and (C) after 60 h of chloroplast biogenesis, and (D) grown in continuous light.

rbcS

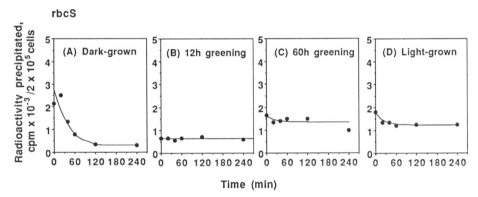

Fig. 8· Kinetics of the in vivo turnover of the small subunit of ribulose bisphosphate carboxylase (rbcS) in *Euglena*: (A) grown in continuous darkness, (B) after 12 h of light-induced chloro-plast biogenesis ("greening"), (C) after 60 h of chloroplast biogenesis, and (D) grown in continuous light.

CONCLUSIONS

During light-induced chloroplast biogenesis in *Euglena*: (a) levels of transcripts encoding most plastid proteins correlate with their gene levels only during the early portion of chloroplast biogenesis. An exception is the *psb*B product whose transcript level mimics its gene level throughout chloroplast biogenesis; (b) patterns of accumulation differ considerably among the different plastid proteins; (c) levels of plastid proteins correlate with levels of their mRNAs only during the early portion of chloroplast biogenesis; (d) plastid proteins, including chlorophyll-binding proteins begin to accumulate before chlorophyll is detectable in the cells; and (e) the cellular levels of plastid proteins are regulated at the posttranslational level of protein turnover with the proteins being stabilized as they accumulate. The latter regulation is important because the turnover rate (degradative rate) not only influences the final steady-state

Table 4. Length of Lag Phase in Accumulation of
Plastid Proteins and Carotenoids During
Light-induced Chloroplast Biogenesis

Gene or Compound	Lag (h)[a] in Accumulation of	
	mRNA	Protein Product
psaA[b]	2	<1
psbA	<1	<3
psbB[b]	12	<1
psbC[b]	5	5
atpH	<3	5
rbcL	<1	3
rbcS	--	<1

	Lag (h) in Accumulation
Total Chlorophyll	6-7
Total Carotenoids	12

[a]Hours of exposure to light.
[b]Product is a chlorophyll-binding protein.

of a protein but also how fast changes in the regulation of gene expression
are reflected in changes in the cellular level of the protein.

ACKNOWLEDGEMENT

Supported in part by grants from the National Science Foundation and
the McKnight Foundation.

REFERENCES

1. E. M. Tobin and J. Silverthorne, Light regulation of gene expression
    in higher plants, Ann. Rev. Plant Physiol. 36:569 (1985).
2. S. D. Swartzbach, Photocontrol of organelle biogenesis in Euglena,
    Photochem. Photobiol. 51:231 (1990).
3. W. Gruissem, Chloroplast gene expression: how plants turn their
    plastids on, Cell 56:161 (1989).
4. J. E. Mullet, Chloroplast development and gene expression, Ann. Rev.
    Plant Physiol. Plant Mol. Biol. 39:475 (1988).
5. M. E. Miller, J. E. Jurgenson, E. M. Reardon, and C. A. Price,
    Plastid translation in organello and in vitro during light-induced
    development in Euglena, J. Biol. Chem. 258:14478 (1983).

6. S. A. McCarthy and S. D. Schwartzbach, Absence of photoregulation of abundant mRNA levels in Euglena, Plant Sci. Lett. 35:61 (1984).

7. R. Schantz, Mapping of the chloroplast genes coding for the chlorophyll a-binding proteins in Euglena gracilis, Plant Sci. 40:43 (1985).

8. C. Bouet, R. Schantz, B. Pineau, G. Dubertret, and G. Ledoight, Regulation of protein synthesis during plastid development in Euglena gracilis, in "Regulation of Chloroplast Differentiation", G. Akoyunoglou and H. Senger, eds., A. R. Liss, New York, p. 587 (1986).

9. A. Rikin, A. Monroy, and S. D. Schwartzbach, Evidence for translational regulation of Euglena protein synthesis by light, Prog. Photosynth. Res. 4:581 (1987).

10. C. Weiss, G. Houlné, M.-L. Schantz, and R. Schantz, Photoregulation of the synthesis of chloroplast membrane proteins in Euglena gracilis, J. Plant Physiol. 133:521 (1988).

11. C. W. Gilbert and D. E. Buetow, Two-dimensional gel analysis of polypeptide appearance in forming thylakoid membranes, Biochem. Biophys. Res. Comm. 107:649 (1982).

12. R. B. Hallick and D. E. Buetow, Chloroplast DNA, in "The Biology of Euglena", Vol. IV, D. E. Buetow, ed., Academic Press, San Diego, p. 351 (1989).

13. A. P. Feinberg and B. Vogelstein, A technique for radiolabeling DNA restriction endonuclease fragments to high specific activity, Anal. Biochem. 132:6 (1983).

14. G. Freyssinet, R. L. Eichholz, and D. E. Buetow, Kinetics of accumulation of ribulose-1,5-bisphosphate carboxylase during greening in Euglena gracilis: Photoregulation, Plant Physiol. 75:850 (1984).

15. L.S.H. Yi, C. W. Gilbert, and D. E. Buetow, Temporal appearance of chlorophyll-protein complexes and the N,N^1-dicyclohexylcarbodiimide-binding coupling factor -subunit III in forming thylakoid membranes of Euglena gracilis, J. Plant Physiol. 118:7 (1985).

16. R. Jahn, W. Schiebler, and P. Greengard, A quantitative dot-immunobinding assay for proteins using nitrocellulose membrane filters, Proc. Natl. Acad. Sci. USA 81:1684 (1984).

17. M. Edelman and A. Reisfeld, Characterization, translation and control of the 32000 dalton chloroplast membrane protein in Spirodela, in "Chloroplast Development", G. Akoyunoglou and J. H. Argyroudi-Akoyunoglou, eds., Elsevier/North-Holland, Amsterdam, p. 641 (1978).

18. D. R. Paterson and C. J. Arntzen, Detection of altered inhibition of photosystem II reactions in herbicide-resistant plants, in "Methods in Chloroplast Molecular Biology", M. Edelman, R. B. Hallick, and N.-H. Chua, eds., Elsevier, Amsterdam, p. 609 (1982).

19. E. S. Kempner and J. H. Miller, The molecular biology of Euglena gracilis. III. General carbon metabolism, Biochemistry. 4:2735 (1965).

20. B. K. Chelm, P. Hoben, and R. B. Hallick, Cellular content of chloroplast DNA and chloroplast ribosomal RNA genes in Euglena gracilis during chloroplast development, Biochemistry 16:782 (1977).

21. J. Bennett, G. I. Jenkins, and M. R. Hartley, Differential regulation of the accumulation of the light-harvesting chlorophyll a/b complex and ribulose bisphosphate carboxylase/oxygenase in greening pea leaves, J. Cell Biochem. 25:1 (1984).

22. F. G. Plumley and G. W. Schmidt, Reconstitution of chlorophyll a/b light harvesting complexes: Xanthophyll-dependent assembly and energy transfer, Proc. Nat. Acad. Sci. USA 84:146 (1987).

CELLULAR DIFFERENTIATION PROCESSES DURING CHLOROPLAST

DEVELOPMENT IN *EUGLENA GRACILIS*

Steffen Reinbothe

Institute for Plant Biochemistry, Weinberg 3,
D-0-4050 Halle/Saale, FRG, Present address:
Institute for Plant Sciences, ETH Zentrum,
Universitätsstrasse 2, CH-8092 Zürich,
Switzerland

INTRODUCTION

The unicellular phytoflagellate *Euglena gracilis* is a well-suited organism for the study of regulatory circuits that underly differentiation processes in response to changing environmental conditions, e.g. light or carbon sources (Parthier, 1981; Kempner, 1982; Reinbothe and Parthier, 1990; Schwartzbach, 1990; for reviews). Light brings about chloroplast formation from proplastids in a gene expression program termed chloroplast development (Schiff and Schwartzbach, 1982). After light induction, the prothylakoids and the prolamellar body of the proplastids desintegrate, thylakoids emerge and are organized into lamellae. The pyrenoid region appears (Osafune and Schiff, 1980; Pellegrini, 1980). The whole morphological differentiation process depends on qualitative and quantitative changes of the plastid and nuclear genomes (summarized in: Reinbothe and Parthier, 1990). However, little is known how the expression of plastid and nuclear genes is coordinated. Since light-induced chloroplast development is repressed by certain organic carbon sources, e.g. ethanol (App and Jagendorf, 1963; Monroy and Schwartzbach, 1984; Monroy et al, 1986; Monroy et al., 1987), photo- and metabolite-regulated steps in gene expression must be taken into account.

In this review I will summarize recent information concerning multiple changes in gene expression of the nucleo-cytoplasmic, mitochondrial and plastid compartment during light-induced chloroplast development in *Euglena gracilis*. I will demonstrate that glucose affects almost all processes normally occuring during chloroplast development. Special emphasis will be placed on nuclear and plastid encoded chloroplast proteins and their mRNAs. The functional role and putative regulatory significance in chloroplast gene expression of cytoplasmically synthesized and post-translationally imported plastid aminoacyl-tRNA synthetases (AaRS) and their cognate plastid tRNAs will be discussed.

LIGHT-INDUCED CHANGES IN THE CELLULAR PROTEIN COMPOSITION DURING CHLOROPLAST DEVELOPMENT

When *Euglena gracilis*, Z. (no. 1224-5/25 of the algae collection of the university of Göttingen) grown in the dark (Cramer and Myers, 1952) in the presence of glucose (chemoorganotrophic growth) is exposed to light, two effects can be observed. If carbon dioxide instead of glucose is contained in the medium, chloroplast development is initiated as seen by the appearence of chlorophyll (autotrophic greening leading to photoautotrophic cells). If glucose remains the basic carbon source (photoorganotrophic cultivation), chlorophyll synthesis is significantly delayed suggesting that chloroplast formation is inhibited.

Regulation of Chloroplast Biogenesis
Edited by J.H. Argyroudi-Akoyunoglou, Plenum Press, New York, 1992

We have analyzed by two-dimensional gel electrophoresis the polypeptide composition of cells under autotrophic and organotrophic conditions of growth (Reinbothe et al., 1991a; Reinbothe et al., 1991b). Comparison of the Coomassie-stained protein pattern from chemoorganotrophic and photoautotrophic cells with that of isolated mitochondria from the plastid-free *Euglena* mutant $W_{10}BSmL$ (Osafune and Schiff, 1983) and of isolated chloroplasts from photoautotrophic wild-type cells, respectively, demonstrated that mitochondrial proteins predominate in dark-grown cells, whereas light-grown cells are mainly composed of chloroplast proteins. Besides the large number of light-regulated proteins whose amount changed, a few constitutive polypeptides could be found (Reinbothe et al.,1991a). These results make evident that a drastic change in the polypeptide composition must occur when dark-grown cells are illuminated in the presence of carbon dioxide. Due to the different routes of carbon utilization by either conversion of the organic carbon into cellular metabolites or fixation of carbon dioxide via photosynthesis, different cell compartments predominantly function in the two cell types as evidenced by the prevalence of mitochondrial and chloroplast proteins, respectively.

GLUCOSE EFFECTS ON MITOCHONDRIAL, NUCLEO-CYTOPLASMIC AND PLASTID GENE EXPRESSION

The effect of glucose on protein synthesis was investigated following pulse-labeling of polypeptides by L-[³⁵S]methionine and two-dimensional gel electrophoresis (Reinbothe et al., 1990a, Reinbothe et al., 1991a). The autoradiograms shown in Figure 1 reveal a differnt pattern of radioactive polypeptides of chemoorganotrophic (Fig.1A) and photoautotrophic cells (Fig.1C). Inclusion of glucose in the greening process prevented the reprogramming of synthesis of almost all proteins. For mitochondrial and cytoplasmic proteins whose synthesis depended on the organic carbon source or was repressed by light and thus was drastically lowered in photoautotrophic cells (Reinbothe et al., 1991a; Reinbothe et al., 1991b), a nearly unchanged rate of synthesis was observed under photoorganotrophic conditions (Reinbothe et al., 1991a). On the other hand, the synthesis of chloroplast proteins normally occuring under photoautotrophic conditions was strongly repressed under photoorganotrophic conditions (Fig.1C versus B). These results imply that the preexsisting set of enzymes of dark-grown cells is likewise used to convert the organic carbon source in the illuminated cells. Since chloroplast formation is not required under these circumstances, the synthesis of chloroplast proteins is almost completely inhibited.

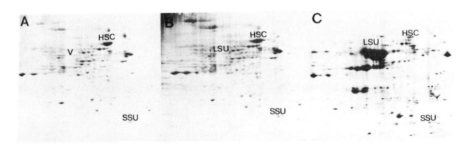

Fig.1. *In vivo* labeled polypeptides of 60h old chemoorganotrophic cells (A) and of cells illuminated for 72h under either organotrophic (B) or autotrophic conditions (C). Autoradiograms of two dimensional gels separating equal cpm of L-[³⁵S]methionine-labeled proteins are shown. LSU and SSU designate the large and small subunits of ribulose-1,5-bisphosphate carboxylase, respectively. HSC marks heat-shock cognate protein of Mr 70 000. Positions of ¹⁴C-labeled heat-shock marker proteins from *Lycopersicon peruvianum* are included (taken from Reinbothe et al., 1991a; with permission of Gauthier-Villars, Paris).

GLUCOSE INHIBITION OF RIBULOSE-1,5-BISPHOSPHATE CARBOXYLASE/OXYGENASE AND 5-ENOLPYRUVYLSHIKIMATE-3-PHOSPHATE SYNTHASE FORMATION

As found for higher plants (Miziorko and Lorimer, 1983), ribulose-1,5-bisphosphate carboxylase/oxygenase (RuBPCase, E.C. 4.1.1.39) in *Euglena gracilis* is composed of eight large and eight small subunits (Pineau, 1982) that are of plastid and cytoplasmic origin, respectively. The holoenzyme formation (Tab.1) results from the assembly of light-induced large subunits and constitutive small subunits (cf.Fig.1). The lowered amount of accumulated small subunits under photoorganotrophic conditions in comparison to photoautotrophic conditions thus seems to be primarily the result of their assembly with large subunits whose level is significantly reduced (Reinbothe et al., 1991a). The constitutive, light-independent synthesis of RuBPCase small subunits apparently reflects the constant rbcS mRNA content (Reinbothe et al., 1990a) and, thus, is a peculiarity for *Euglena gracilis*, since rbcS expression in higher plants is at least in part controlled transcriptionally (Kuhlemeier et al., 1987). Regarding the expression of the rbcL gene, an additional glucose effect on the steady state transcript level became evident from Northern blot hybridization experiments with a cloned corresponding spinach gene (Reinbothe et al., 1991a). During the first ten hours of illumination, a period of rapid initial rise in the amount of the rbcL transcripts in autotrophically greening cells, only half of the message level was reached under organotrophic conditions of greening (Tab.1).

As another plastid enzyme, we have investigated 5-*enol*pyruvyl-shikimate-3-phosphate synthase (EPSP-S). This nuclear-encoded chloroplast protein is involved in the biosynthesis of aromatic amino acids (Haslam, 1974) and acts as the target molecule of N-(phosphonomethyl)glycine, better known as glyphosate, in higher plants (Amrhein et al., 1987) and in *Euglena gracilis* (Reinbothe et al., 1991c). The enzyme activity increases with an appreciable lag-phase and becomes detectable in autotrophically greening *Euglena* cells after 9 to 24 hours (Tab.1). Thereafter, the activity is raised continously up to 72 hours and levels off after 96 hours of illumination (not shown). Similarly to RuBPCase formation, glucose addition to the greening cells results in a slower rise of enzyme activity which is caused by a lowered message content for EPSP-S (Tab.1).

Table 1. Ribulose-1,5-bisphosphate carboxylase/oxygenase (RuBPCase) and 5-*enol*pyruvylshikimate-3-phosphate synthase (EPSP-S) expression under autotrophic [light (L) + CO_2] and organotrophic [light (L) + glucose (Glc)] conditions of greening. The relative enzyme and mRNA contents refer to the maximal levels found in autotrophic cells after 72 h.

	RuBPCase Activity[a]		rbcL mRNA[b]		EPSP-S Activity[c]		mRNA[d]	
	L+CO_2	L+Glc	L+CO_2	L+Glc	L+CO_2	L+Glc	L+CO_2	L+Glc
0	5	5	7	7	n.d.[e]	n.d.	10	10
3	12	8	50	35	n.d.	n.d.	80	14
9	12	8	80	38	2	1	95	32
24	35	18	85	45	9	5	97	38
48	63	35	90	57	50	25	99	38
72	100	55	100	65	100	45	100	40

[a]The enzyme activity was determined in 108 000 x *g* supernatans of ultrasonicated cells (data taken from Reinbothe et al., 1991a, with permission of Gauthier-Villars, Paris).
[b]The rbcL steady state transcript content was estimated from Northern blot hybridizations with a cloned spinach gene probe (data taken from Reinbothe et al., 1991a, with permission of Gauthier-Villars, Paris).
[c]The enzyme activity was determined after non-denaturing polyacrylamide gel electrophoresis of proteins from 15 000 x *g* supernatants of ultrasonicated cells as described by Reinbothe et al., (1991c).
[d]The transcript content was determined by Northern blot hybridization with a synthetic oligonucleotide probe complementary to a nucleotide sequence motif of the *Saccharomyces cerevisiae* EPSP-S gene encoding the amino acid stretch N_{81} Thr Val Val Val Glu Gly C_{86} (Kishore and Shah, 1988).
[e]n.d. = not detectable.

Chloroplast AaRS (E.C. 6.1.1.x) charge plastid-encoded tRNAs with their cognate amino acids. We have focussed our interest on one particular AaRS, namely the leucine-specific enzyme (LeuRS). Like all other plastid AaRS, LeuRS is encoded in the nucleus and synthesized *de novo* on cytoplasmic ribosomes (Parthier, 1973; Nover, 1976; Weil and Parthier, 1982). Its accumulation in photoautotrophically greening cells (Fig.2C) is preceded by a transient rise of its low abundance (Reinbothe et al., 1990c) translatable mRNA (Fig.2A). Confirming earlier results from our laboratory obtained by determination of the enzyme activity after cell-free translation (Krauspe et al., 1987), here we present the immunological technique we have previously used (Reinbothe et al., 1990a) to quantitate the message content (inset of Fig.2A). This approach also allowed us to analyze the post-translational uptake of the *in vitro* formed precursor enzyme into isolated plastids in a homologous *Euglena* system (Reinbothe et al., 1990a). As indicated in Figure 2B, the disappearance of precursor LeuRS of 112 kDa from the incubation medium, i.e. outside the plastids (P in Fig.2B, b versus a) coincides with the appearance of mature, protease-resistant chloroplast LeuRS of 105 kDa inside the chloroplasts (L in Fig.2B, d versus c). Obviously, the wheat germ system itself is able to process the precursor enzyme (Fig.2B, a, and Fig.2A), but the resulting mature enzyme is not sequestered into the chloroplasts (Fig.2B, b). The 7 kDa-chloroplast transit peptide is thus necessary and sufficient for targeting the precursor LeuRS into *Euglena* chloroplasts *in vitro* and very likely *in situ*.

With respect to the enzyme formation, it seems to be sensitive towards glucose at the transcription level, since a direct correlation was found between the considerably reduced maximal mRNA content (Fig.2A) and the accumulated enzyme level (Fig.2C).

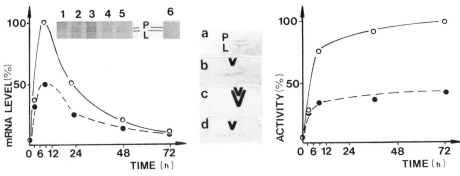

Fig.2. Leucyl-tRNA synthetase expression under organotrophic (●) or autotrophic conditions (O) of greening: A. Amount of *in vitro* translatable LeuRS mRNA per µg of total RNA as determined from the activity of enzyme or from the amount of immuno-detected proteins (inset) formed by cell-free synthesis of RNAs from dark-grown cells (lane 1) or cells illuminated for 3h (lane 2), 10h (lane 3 and 5) or 72h (lane 4) under autotrophic (lanes 1 - 4) or organotrophic conditions (lane 5). P and L designate precursor LeuRS of 112 kDa and mature LeuRS of 105 kDa, respectively, that neither were detectable if preimmune serum was used instead of the antiserum (lane 6). B. *In vitro* transport of precursor LeuRS into *Euglena* chloroplasts. Parts from autoradiograms of two-dimensional gels separating immuno-detected proteins before (a,c) or after a 60 min incubation (b,d) with chloroplasts outside (a,b) or inside (c,d) the plastids are reproduced. C. LeuRS activity on a cell basis as determined in 108 000 x *g* supernatants of ultrasonicated cells (taken from Reinbothe et al. , 1990a; and Reinbothe et al., 1991a; with permission of Gauthier-Villars, Paris).

TRANSLATIONAL REGULATION OF PLASTID GENE EXPRESSION BY THE AVAILABILITY
OF CHARGED GLUTAMYL-tRNA^{Glu} MOLECULES - A HYPOTHETICAL MODEL

Plastid protein synthesis depends on charged tRNA molecules which
are formed by acylation of plastid tRNAS with the cognate amino acids by
means of AaRS. There is one exceptional aminoacyl-tRNA, i.e. glutamyl-
tRNA$_{GAA}$, which is used not only for protein synthesis, but also for
chlorophyll biosynthesis (Gomez-Silva et al., 1985; Kannangara et al.,
1988). Its charging depends on GluRS which is a light-induced enzyme
(Krauspe and Parthier, 1974; Mayer and Beale, 1990) like LeuRS (cf.
Fig.2). During the first hours of illumination both the tRNA^{Glu} and the
GluRS (Krauspe et al., 1987) are not yet fully induced and, for this
reason, probably limiting. Due to the contemporary need of glutamyl-
tRNA^{Glu} in chlorophyll and protein synthesis, this particular tRNA is very
likely depleted. The limited availability of charged glutamyl-tRNA
molecules could explain the differential rates of synthesis of proteins
if polysome-bound messages contain different frequencies of glutamate
codons (Reinbothe and Parthier, 1990). Indeed, from the sequence analysis
of known *Euglena* protein genes coded for by plastid DNA (Nickoloff et
al., 1989; Reinbothe and Parthier, 1990; Christopher and Hallick, 1989;
Yepiz-Plascencia et al., 1990) a striking difference in the glutamate
codon frequency and a pronounced bias towards the GAA codon could be
inferred. Three groups of genes referred to as low, intermediate and high
frequency were discernible (Reinbothe and Parthier, 1990). All known
genes for ribosomal proteins except rps7 and rpl5 as well as all known
psa genes as well as the psbF gene share a low frequency of glutamate
codons being in the range of 1.1 to 3.6 per cent. The protein genes psbA,
psbE and atpH contain an intermediate percentage of glutamate codons (4.1
% to 5.1 %), whereas the genes for tufA, rbcL, rps7, rpl5, rpoB and rpoC
contain approximately three times more glutamate codons (6.0 - 7.3 %)
than the members of the low frequency group. In the rbcL, psbA, rpl5 and
rpoB genes, an additional accumulation of the glutamate codons in direct
repeats of two or three codons is observed (Reinbothe and Parthier, 1990;
Yepiz-Plascencia et al., 1990). Assuming that equal proportions of psaA
and tufA transcripts, for instance, are bound to polysomes (Fig.3), the
rate of their translation should depend on the frequency of glutamate
codons under the conditions of glutamyl-tRNA^{Glu} limitation. Translation of
mRNAs with a low number of glutamate codons such as the psaA mRNA would
be less affected than that of transcripts with a high frequency of
glutamate codons such as the tufA mRNA (Fig.3). Whereas the translating
ribosome would stall at the GAA codons of the tufA gene, it could proceed
in reading the psaA message. If one compares the number of read codons at
two different times (designated t$_1$ and t$_2$ in Fig.3), obvious differences
in the length of the translated polypeptides become evident. Finally,
different patterns of appearance of the encoded products can be
predicted. Synthesis of membrane proteins coded for by the psa and psb
genes would precede accumulation of RuBPCase large subunits (rbcL gene
product) or elongation factor Tu (tufA gene product), for instance. In
the first hours of chloroplast development, the photosynthetic reaction
centers I and II and the proton translocating ATPase (the atpH gene
encodes one of its subunits) required for electron transport and thus
energy conversion would be established successively and before proteins
involved in carbon fixation and metabolism would appear.

Indeed, this predicted pattern of appearence of polypeptides has
been demonstrated experimentally for the psbA, psbB, psbC and rbcL genes
(summarized in: Reinbothe and Parthier, 1990). In spite of the
concomitant polysomal association of the corresponding transcripts,
synthesis of plastid polypeptides started with formation of thylakoid
membrane proteins required for chlorophyll binding and was followed by
accumulation of stroma proteins which are involved in carbon
assimilation. Coincident with the formation of thylakoid membrane
proteins, most plastid encoded ribosomal proteins (the rps and rpl gene
products) would appear. Their accumulation is thought to prepare new
ribosomes for the assembly with newly transcribed messages in a process
termed light-induced polysome formation (Heizmann et al., 1972). But the
final assembly of ribosomal proteins and mRNAs seems to depend on the
rps7 and rpl5 gene products, whose late appearance apparently triggers
the whole process. In later phases of chloroplast development when the
amount of glutamyl-tRNA^{Glu} is no longer limiting, plastid polypeptides of
the gene expression apparatus like elongation factor Tu or the ß and ß'
subunits of the soluble RNA polymerase are formed. Their appearance might

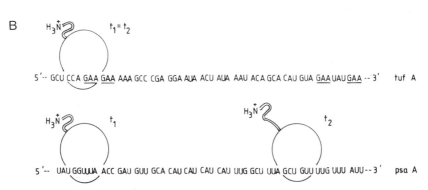

Fig.3. Glutamate codon frequency in selected plastid mRNAs encoded by the respective genes (A) and its implications on translational elongation of ribosome-bound tufA and psaA mRNA (B). Only parts of mRNA sequences derived from the known sequences of the *Euglena* psaA gene (Cushman et al., 1988a), psbE gene (Cushman et al., 1988b) and tufA gene (Montandon and Stutz, 1983) are shown in which the glutamate codon frequency is representative for the different groups of plastid genes. The GAA codons are underlined. The position of the translating ribosome on the tufA or psaA mRNA is schematically shown for two different points of time t_1 and t_2.

be an indication for a general shift-up in the capacity of plastid protein synthesis.

SUMMARY

Euglena gracilis is unique in its ability to utilize for cell metabolism a vast variety of organic carbon sources and anorganic carbon in the form of carbon dioxide as well. Adaptation of cells from one organic carbon source to another requires only minor adjustments to a preexisting enzyme set. In contrast, the switch from chemoorganotrophic to photoautotrophic carbon metabolism depends on complex changes in gene expression of the nucleo-cytoplasmic, mitochondrial and plastid compartment (Fig.1.). During light-induced chloroplast development, chloroplast proteins like RuBPCase, EPSP-S (Tab.1) and AaRS (Fig.2) are formed at the expense of certain mitochondrial and cytoplasmic proteins whose synthesis declines due to the lacking organic carbon or is repressed by light (Fig.1). In cells that are illuminated in the presence of glucose, no such drastic alteration in the cellular polypeptide composition is observed (Fig.1). Suggesting that apparently the same set of enzymes can be used for carbon metabolism as before under chemoorganotrophic conditions, chloroplasts are not needed under these circumstances and thus a repression of synthesis of nuclear- and plastid-encoded chloroplast proteins is observed (Figs.1 and 2; Tab.1). In the particular case of AaRS, their lowered level under photoorganotrophic conditions (Fig.2) might have direct consequences on plastid protein synthesis which depends on aminoacyl-tRNAs formed by these enzymes. The depletion of one particular tRNA, i.e. glutamyl-tRNA[Glu] that is competitively used for chlorophyll and protein synthesis, has regulatory implications on the translation of individual plastid proteins from polysome-bound messages (Fig.3). In accordance to this idea, the synthesis of thylakoid membrane proteins seems to be less affected than translation of proteins involved in photosynthetic carbon fixation.

REFERENCES

Amrhein,N., Holländer-Czytko,H., Johänning,D., Schulz,A., Smart,C.C. and
    Steinrücken,H., 1987, Overproduction of 5-*enol*pyruvylshikimate-3-
    phosphate synthase in glyphosate-tolerant plant cell cultures, in:
    " Plant tissue and cell culture," C.F. Green, D.A. Somers, W.D.
    Hackett and D.D. Biesboer, eds., Alan R. Liss, New York.
App, A.A. and Jagendorf, A.T., 1963, Repression of chloroplast
    development in *Euglena gracilis* by substrates, J. Protozool.,
    10:340-343.
Christopher, D.A. and Hallick, R.B., 1989, *Euglena gracilis* chloroplast
    ribosomal protein operon: a new chloroplast gene for ribosomal
    protein L5 and description of a novel organelle intron category
    designated group III, Nucl. Acids Res., 17:7591-7608.
Cramer, M. and Myers, J., 1952, Growth and photosynthetic characteristics
    of *Euglena gracilis*, Arch. Mikrobiol., 17:384-402.
Cushman, J.C., Hallick, R.B. and Price, C.A., 1988a, The two genes for
    the P_{700} chlorophyll a apoproteins on the *Euglena gracilis*
    chloroplast genome contain multiple introns, Curr. Genet., 13:159-
    172.
Cushman, J.C., Christopher, D.A., Little, M.C., Hallick, R.B. and Price,
    C.A., 1988b, Organization of the psbE, psbF, orf 38 and orf 42 gene
    loci on the *Euglena gracilis* chloroplast genome, Curr. Genet.,
    13:173-180.
Gomez-Silva, B., Timko, M P. and Schiff, J.A., 1985, Chlorophyll
    biosynthesis from glutamate or 5-aminolevulinate in intact *Euglena*
    chloroplasts, Planta, 165:12-22.
Haslam,E., 1974, "The shikimate pathway", Butterworths, London.
Heizmann, P., Trabuchet, G., Verdier, G., Freyssinet, G. and Nigon, V.,
    1972, Influence de l'eclairement sur l'evolution des polysomes dans
    cultures d'*Euglena gracilis* etiolees, Biochem. Biophys. Acta
    277:149-160.
Kannangara, C.G., Gough, S.P., Bruyant,P., Hoober, K., Kahn, A. and von
    Wettstein,D., 1988, tRNA[Glu] as cofactor in delta-aminolevulinate
    biosynthesis: steps that regulate chlorophyll synthesis, Trends
    Biol. Sci., 13:139-143
Karabin, G.D., Farley, M. and Hallick, R.B., 1984, Chloroplast gene for
    Mr 32000 polypeptide of photosystem II in *Euglena gracilis* is
    interrupted by four introns with conserved boundary
    sequences, Nucl. Acids Res., 12:5801-5812.
Kempner, E.S., 1982, Stimulation and inhibition of the metabolism and
    growth of *Euglena gracilis*, in: "The biology of *Euglena*, vol. III,"
    D.E. Buetow, ed., Academic Press, New York.
Kishore, G.M. and Shah, D.M., 1988, Amino acid biosynthesis inhibitors as
    herbicides, Ann. Rev. Biochem., 57:627-663.
Krauspe, R. and Parthier, B., 1974, Chloroplast and cytoplasmic
    aminoacyl-tRNA synthetases of *Euglena gracilis*. Chromatographic
    separation and some properties. Biochem. Physiol. Pflanzen,
    165:18-36.
Krauspe, R., Lerbs, S., Parthier, B. and Wollgiehn, R., 1987, Light-
    induction of translatable mRNAs for chloroplastic leucyl- and
    valyl-tRNA synthetases of *Euglena gracilis*, J. Plant Physiol.,
    130:327-342.
Kuhlemeier, C., Green, P.J. and Chua, N.-H., 1987, Regulation of gene
    expression in higher plants, Ann. Rev. Plant Physiol., 38:221-257.
Mayer, S.M. and Beale, S.I., 1990, Light regulation of delta-
    aminolevulinic acid biosynthetic enzymes and tRNA in *Euglena*
    gracilis, Plant Physiol., 94:1365-1375.
Miziorko, H.M. and Lorimer, G.H., 1983, Ribulose-1,5-bisphosphate-
    carboxylase-oxygenase, Ann. Rev. Biochem., 52:507-535.
Monroy, A.F. and Schwartzbach, S.D., 1984, Catabolite repression of
    chloroplast development in *Euglena*, Proc. Natl. Acad. Sci. USA,
    81:2786-2790.
Monroy, A.F., Gomez-Silva, B., Schwartzbach, S.D. and Schiff, J.A., 1986,
    Photocontrol of chloroplast and mitochondrial polypeptide levels in
    Euglena, Plant Physiol., 80:618-622.
Monroy, A.F., McCarthy, S.A. and Schwartzbach, S.D., 1987, Evidence for
    translational regulation of chloroplast and mitochondrial
    biogenesis in *Euglena*, Plant Sci., 51:61-77.
Montandon, P.-E. and Stutz, E., 1983, Nucleotide sequence of a *Euglena*
    gracilis chloroplast genome region coding for the elongation factor
    Tu; evidence for a spliced mRNA, Nucl. Acids Res., 11:5877-5892.

Nickoloff, J.A., Christopher, D.A., Drager, R.G. and Hallick, R.B., 1989, Nucleotide sequence of the *Euglena gracilis* chloroplast genes for isoleucine, phenylalanine and cysteine transfer RNAs and ribosomal protein S14, <u>Nucl. Acids Res.</u>, 17:4882.

Nover, L., 1976, Density labeling of chloroplast-specific leucyl-tRNA synthetase in greening cells of *Euglena gracilis*, <u>Plant Sci. Lett.</u>, 7:403-407.

Osafune, T. and Schiff, J.A., 1980, Stigma and flagellar swelling in relation to light and carotenoids in *Euglena gracilis* var. *bacillaris*, <u>J. Ultrastruct. Res.</u>, 73:336-349.

Osafune, T. and Schiff, J.A., 1983, $W_{10}BSmL$, a mutant of *Euglena gracilis* var. *bacillaris* lacking plastids, <u>Exp. Cell Res.</u>, 148:530-536.

Parthier, B., 1973, Cytoplasmic site of synthesis of chloroplast aminoacyl-tRNA synthetases in *Euglena gracilis*, <u>FEBS Lett.</u>, 38:70-74.

Parthier, B., 1981, Photocontrol of chloroplast development in *Euglena*: Regulatory aspects, <u>in</u>:" Biochemistry and physiology of protozoa, vol IV", M. Levandowsky and S.H. Hutner, eds., Academic Press, New York.

Pellegrini, M., 1980, Three-dimensional reconstruction of organelles in *Euglena gracilis* z. I. Qualitative and quantitative changes of chloroplasts and mitochondrial reticulum in synchronous photoautotrophic culture, <u>J. Cell Sci.</u>, 43:137-166.

Pineau, B., 1982, Biosynthesis of ribulose-1,5-bisphosphate carboxylase in greening cells of *Euglena gracilis*. The accumulation of ribulose-1,5-bisphosphate carboxylase and its subunits. <u>Planta</u> 156:117-128.

Reinbothe, S. and Parthier,B., 1990, Translational regulation of plastid gene expression in *Euglena gracilis*. <u>FEBS Lett.</u>, 265:7-11.

Reinbothe, S., Krauspe, R. and Parthier, B., 1990a, In-vitro-transport of chloroplast proteins in a homologous *Euglena* system with particular reference to plastid leucyl-tRNA synthetase, <u>Planta</u> 181:176-183.

Reinbothe, S., Krauspe, R. and Parthier, B., 1990b, Partial purification and analysis of mRNAs for chloroplast and cytoplasmic aminoacyl-tRNA synthetases from *Euglena gracilis*, <u>J. Plant Physiol.</u>, 137:81-87.

Reinbothe, S., Reinbothe, C. and Parthier, B., 1991a, Glucose repression of chloroplast development in *Euglena gracilis*, <u>Plant Physiol. Biochem.</u>, in press.

Reinbothe, S., Reinbothe, C., Krauspe, R. and Parthier, B., 1991b, Changing gene expression during dark-induced chloroplast dedifferentiation in *Euglena gracilis*, <u>Plant Physiol. Biochem.</u>, in press.

Reinbothe, S., Nelles, A. and Parthier, B., 1991c, N-(Phosphonomethyl)glycine (glyphosate) tolerance in *Euglena gracilis* acquired by either overproduced or resistant 5-*enol*pyruvyl-shikimate-3-phosphate synthase, <u>Eur. J. Biochem.</u>, in press.

Schiff, J.A. and Schwartzbach, S.D., 1982, Photocontrol of chloroplast development in *Euglena*, <u>in</u>: "The biology of *Euglena*, vol. III", D.E. Buetow, ed., Academic Press, New York.

Schwartzbach, S.D., 1990, Photocontrol of organelle biogenesis in *Euglena*, <u>Photochem. Photobiol.</u>, 51:231-254.

Weil, J.-H. and Parthier, B., 1982, Transfer RNA and aminoacyl-tRNA synthetases in plants, <u>in</u>:" Encyclopedia of plant physiology, N.S. Vol. 14A: Nucleic acids and proteins in plants I", D. Boulter and B. Parthier, eds., Springer, Berlin Heidelberg New York.

Yepiz-Plascencia, G.M., Radebaugh, C.A. and Hallick, R.B., 1990, The *Euglena gracilis* chloroplast rpo B gene. Novel gene organization and transcription of the RNA polymerase subunit operon, <u>Nucl. Acids Res.</u>, 18:1869-1878.

PLASTID GENE EXPRESSION DURING CHLOROPLAST DEVELOPMENT

BY TWO ALTERNATIVE PATHWAYS IN BARLEY SEEDLINGS[*]

Karin Krupinska, Anke Schmidt and Jon Falk

Institut für Allgemeine Botanik
der Universität Hamburg,
Ohnhorststrasse 18, D-2000 Hamburg 52, FRG

INTRODUCTION

Photosynthetically competent chloroplasts in angiosperms may develop by two alternative pathways: either by etioplast-to-chloroplast transition during light-dependent greening of etiolated seedlings or by differentiation of proplastids during normal leaf development in a daily light-dark regime (Boffey et al., 1980; Leech, 1984).

The present study aims at contributing to a comparison of the regulation of plastid gene expression during chloroplast biogenesis via the two pathways. For this purpose transcription of the plastid genome was analyzed on one hand in etioplasts and chloroplasts obtained by light-dependent transformation of etioplasts and on the other hand in proplastids and chloroplasts differing in age and developed in a daily light-dark regime.

Transcripts synthesized during each developmental stage were analyzed by hybridization of radioactive run-on transcripts synthesized by lysed plastids to immobilized plastid DNA fragments.

While light-induced transformation of etioplasts to chloroplasts apparently occurs without qualitative changes in plastid DNA transcription (Krupinska and Apel, 1989), biogenesis and maturation of chloroplasts in a daily light-dark regime are accompanied by specific changes in the transcript pattern.

DNA protein complexes capable of RNA synthesis (transcriptional active chromosomes, TAC) were isolated from young and mature chloroplasts to re-examine the results obtained with run-on transcripts and for further analysis of the molecular mechanisms governing transcriptional regulation during chloroplast maturation. The results obtained by hybridization of

---

[*] this paper is dedicated to Prof. Dr. Horst Senger on the occasion of his 60th birthday

Regulation of Chloroplast Biogenesis
Edited by J.H. Argyroudi-Akoyunoglou, Plenum Press, New York, 1992

TAC derived transcripts to specific plastid DNA fragments
indicate that TAC preparations from barley chloroplasts are
capable of synthesizing the same transcript species as lysed
plastids in run-on transcription assays and also retain in
principal the development specific differences in transcription
of the plastid genome observed by analysis of run-on transcripts.

MATERIAL AND METHODS

Plant material: Barley (Hordeum vulgare L., var. Carina)
seedlings were grown for 5 or 7 days in moist vermiculite in
controlled environment chambers at 21-23 C. Seedlings were
either kept under a photoperiod of 16 hours light (8000 lux)
and 8 hours darkness for 5 or 7 days or in complete darkness
for 5 days. For analysis of etioplast derived chloroplasts
etiolated seedlings were illuminated for the last 16 hours
before harvest. For the preparation of proplastids 0.5cm basal
sections of 5d old seedlings grown in a daily light-dark regime
were used. Etioplasts and chloroplasts of different age were
prepared from 2cm apical segments after cutting off the leaf
tips.

Isolation of plastids: For run-on transcription assays
plastids were prepared as described (Krupinska and Apel, 1989).
For the preparation of transcriptionally active chromosomes
(TAC) chloroplasts were obtained from 250-500g of leaf segments
by the method of Poulsen (1983).

Transcription in lysed plastids: Based on the method of
Mullet and Klein (1987) transcription in lysed plastids was
carried out as described previously (Krupinska and Apel, 1989).
In addition heparin was included in run-on assays at a final
concentration of 0.5mg/ml (Klein and Mullet, 1990).

In vitro transcription with TAC: Transcriptionally active
chromosomes (TAC) were isolated in the presence of 0.2M ammonium sulphate following the protocol of Rushlow et al. (1980)
from basal leaf segments enriched in proplastids and apical
segments of barley primary foliage leaves of seedlings grown
for 5 or 7d under a light-dark regime. In vitro transcription
assays were performed as described (Rushlow et al., 1980).

Preparation and hybridization of DNA filters: Southern
blots of cloned DNA fragments representing in sum the complete
plastome were obtained as described in previous papers (Krupinska and Apel, 1989, Krupinska, 1991). For DNA dot blots recombinant plasmid DNA with the specific plastid DNA fragments
inserted were dotted on a nylon filter in a dilution series of
640(1x), 160(4x) and 40(16x)ng. The hybridization protocol follows the zetaprobe membrane manufactorer´s (Bio-Rad) instructions as described (Krupinska and Apel, 1989).

RESULTS AND DISCUSSION

To test whether chloroplast biogenesis by one or the other pathway is regulated at the transcriptional level run-on transcription assays were performed on one hand with etioplasts and etioplast derived chloroplasts and on the other hand with proplastids and chloroplasts obtained from barley seedlings grown for 5 or 7 days in a daily light-dark regime. The relative transcription rates of all different genes in the barley plastid genome were analyzed by hybridization of the radiolabelled run-on transcripts to immobilized plastid DNA fragments in sum representing the complete plastome.

The autoradiograms obtained after successive hybridization of the same Southern blot with run-on transcripts derived from the various plastid preparations are shown in Fig.1. The autoradiographic patterns are quite similar in case of etioplasts and etioplast derived chloroplasts of 5d old seedlings as well as in the case of chloroplasts derived from seedlings grown for 5 days in a natural daily light-dark regime. However,

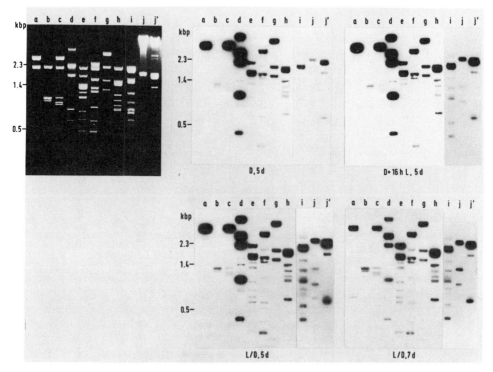

Fig. 1. Analysis of barley plastid run-on transcripts by hybridization to barley plastid DNA fragments in sum representing the complete plastome (for location of the fragments see Krupinska, 1991). The ethidium bromide stained gel is shown on the left top. Autoradiograms were obtained after successive hybridization of the same filter with run-on transcripts derived from etioplasts (D) and etioplast derived chloroplasts (D+16hL) of 5 d old seedlings and from chloroplasts of different age (5d, 7d) prepared from seedlings grown in a light-dark regime (L/D).

transcripts derived from chloroplasts of 7 days old barley
seedlings grown in a light-dark regime yield marked differences
in the hybridization pattern compared to the autoradiograms
belonging to the other stages. A decrease is obvious in the
hybridization intensity of bands belonging to regions a,c and d
and representing parts of the rrn operon (Krupinska, 1991) with
increasing age of the chloroplasts (Fig.1). In contrast, the
relative transcription rate of a 1.1 kbp EcoRI fragment which
is specific for the psbD gene in region j (Krupinska, 1991)
seems to increase during chloroplast maturation.

Based on these results we conclude that transcriptional
regulation plays a more important role during biogenesis and
maturation of chloroplasts in a natural daily light-dark regime
than during light-dependent transformation of etioplasts. To
further analyze these age dependent specific changes in
plastome transcription during development of chloroplasts in a
light-dark regime plastid run-on transcripts synthesized by
proplastids and chloroplasts of different age were hybridized
to DNA dot-blots carrying various gene specific probes. Again
the relative hybridization intensities among various DNA fragments differ between run-on transcripts derived from pro-

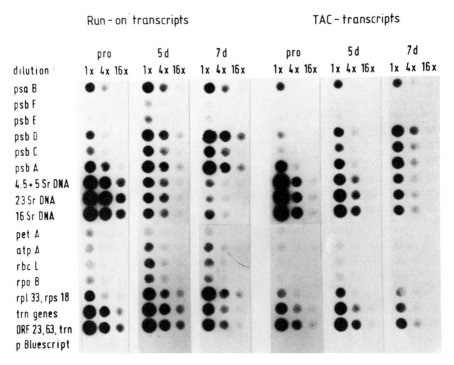

Fig. 2. Autoradiograms obtained after hybridization of radiolabelled transcripts to DNA dot-blots. The transcripts
were obtained either by run-on transcription in lysed
proplastids (pro) and chloroplasts of different age
(5d, 7d) or by in vitro transcription with TAC preparations of the same plastid stages (pro, 5d, 7d).
Plasmid DNAs with individual gene specific inserts
were dotted in a dilution series of 640(1x)ng to
40(16x)ng.

plastids and chloroplasts of 5d and 7d old barley seedlings grown in a daily light-dark regime (Fig. 2). The observed differences in the hybridization patterns suggest that in proplastids rrn transcription is predominant and that during maturation of chloroplasts rrn transcription preferentially decreases while inversely the relative transcription of psbD and psbA increases. Changes in rrn transcription most probably coincide with changes in the translational activity of the plastids. A preferential decrease in rrn transcription during chloroplast maturation might be the consequence of free ribosomes (Laboure et al., 1988) or inversely might restrict the protein synthesis capacity of the chloroplasts. A differential maintenance of psbD and psbA transcription rates during ageing of illuminated leaves might relate to the light-enhanced turnover of the corresponding D2 and D1 proteins, which constitute the reaction center of photosystem II and are supposed to prevent photoinhibition by their high turnover (Schuster et al., 1988).

Such development specific differences in the transcription pattern of the plastome could either be due to changes in the specificity and activity of the RNA-polymerase or could be caused by modifications of the DNA itself. Since the observed variations in plastid gene transcription during chloroplast maturation are not caused by gross changes in the methylation of plastid DNA (Krupinska, 1991), we investigated the RNA polymerase activity in young and mature barley chloroplasts in more detail. For this purpose transcriptionally active chromosomes (TAC) were prepared from both young and mature chloroplast populations and were tested for their transcriptional activity. Radiolabelled transcripts derived from TACs of young and mature chloroplasts were hybridized to the same DNA dot-blot as the run-on transcripts synthesized by chloroplasts of the same developmental stages. The autoradiograms are shown in Fig. 2. Obviously, barley TAC preparations are capable to synthesize RNAs of all kinds as lysed plastids do. Though less pronounced the developmental stage specific variations in the transcription of the plastome which have been observed by hybridization of run-on transcripts to immobilized plastid DNA fragments (Fig. 2) are in principal retained by TAC preparations (Fig. 2). Similarly as the incorporation of radioactive nucleotides into elongating transcripts in lysed plastids the overall transcriptional activity of TAC declines during maturation of barley chloroplasts (data not shown).

Development specific variations in the transcriptional activity of TAC and in the composition of the transcript population have also been observed during blue light dependent chloroplast differentiation in cultured plant cells (Richter and Ottersbach, 1990). Reiss and Link (1985) observed dramatic changes in the overall transcriptional activity of TAC during greening of mustard seedlings. However they did not describe differences in the quality of TAC derived transcripts between etioplasts and chloroplasts of mustard. While other studies with Euglena report a preferential transcription of the rrn operon by TAC (Rushlow et al., 1980), taken together these and our investigations with higher plant plastids demonstrate that TAC is capable to synthesize both rRNA and mRNAs (Reiss and Link, 1985; Richter and Ottersbach, 1990) besides tRNAs (Fig.2).

The results of this comparison between the transcript patterns obtained from barley TAC preparations and by run-on transcription in lysed plastids (Fig. 2a,b) which has been proven to be a useful system to study the regulation of plastid gene transcription during developmental changes (Gruissem, 1989) indicate that barley TAC preparations provide an excellent material to study the molecular basis of transcriptional regulation of rRNA synthesis and of the expression of D1 and D2 proteins during barley chloroplast maturation.

REFERENCES

Boffey, S.A., Sellden, G., and Leech, R.M., 1980, Influence of cell age on chlorophyll formation in light-grown and etiolated wheat seedlings, Plant Physiol., 65:680-684.

Gruissem, W., 1989, Chloroplast RNA: Transcription and processing, in:"The Biochemistry of Plants, Vol.15," A. Marcus, ed., Academic Press Inc., pp 151-191.

Krupinska, K., and Apel, K., 1989, Light-induced transformation of etioplasts to chloroplasts of barley without transcriptional control of plastid gene expression, Mol. Gen. Genet., 219: 467-473.

Krupinska, K., 1991, Transcriptional control of plastid gene expression during development of barley primary foliage leaves under a daily light-dark regime, Planta, submitted.

Laboure, A.-M., Lescure, A.-M., and Briat, J.-F., 1988, Evidence for a translation-mediated attenuation of a spinach chloroplast rDNA operon, Biochimie, 70:1343-1352.

Mullet, J.E., and Klein, R.R., 1987, Transcription and RNA stability are important determinants of higher plant chloroplast RNA levels, EMBO J., 6: 1571-1579.

Klein, R.R., and Mullet, J.E., 1990, Light-induced transcription of chloroplast genes. psbA transcription is differentially enhanced in illuminated barley, J. Biol. Chem., 265:1895-1902.

Leech, R.M., 1984, Chloroplast development in angiosperms: current knowledge and future prospects, in: "Chloroplast biogenesis", Baker, N.R., and Barber, J., eds., Elsevier Science Publishers B.V., pp 1-21.

Poulsen, C., 1983, The barley chloroplast genome: Physical structure and transcriptional activity in vivo, Carlsberg Res. Commun., 48: 57-80.

Reiss, T., and Link, G., 1985, Characterization of transcriptionally active DNA-protein complexes from chloroplasts and etioplasts of mustard (Sinapis alba L.), Eur. J. Biochem., 148:207-212.

Richter, G., and Ottersbach, N., 1990, Blue light-dependent chloroplast differentiation in cultured plant cells: evidence for transcriptional control of plastid genes, Bot. Acta 103:168-173.

Rushlow, K.E., Orozco, E.M., Lipper, C., and Hallick, R.B., 1980, Selective in vitro transcription of Euglena chloroplast ribosomal RNA genes by a transcriptionally active chromosome, J. Biol. Chem., 255:3786-3792.

Schuster, G., Timberg, R., and Ohad, I., 1988, Turnover of the thylakoid photosystem II proteins during photoinhibition of Chlamydomonas reinhardtii, Eur. J. Biochem., 177:403-410.

CIS-REGULATORY ELEMENTS RESPONSIBLE FOR THE TISSUE-SPECIFIC

EXPRESSION OF THE WHEAT CAB-1 GENE

M. Széll, M. Szekeres, É. Ádám, E. Fejes, F. Nagy

Institute of Plant Physiology, Biological
Research Center, Hungarian Academy of Sciences
Szeged, Hungary

INTRODUCTION

Previously the characterisation of plant cells with different functions were based on morphological and physiological studies. Using the methods of molecular biology, i.e. reporter gene systems and plant transformation technology plant cells can be further defined by the genes they express (Edwards and Coruzzi, 1990). Photosyntetic genes such as those encoding the chlorophyll a/b binding protein provide an ideal experimental system to study tissue and cell specific gene expression in plants.

The light-harvesting chlorophyll a/b pigment-protein complex of PSII is the predominant integral protein component of the thylakoid membrane in green algae and higher plants. The protein is encoded in the nucleus, synthesised on cytoplasmic ribosomes and transported into the chloroplast (Karlin-Neumann et al., 1985). In wheat the protein is encoded by a 7-member gene family among which the Cab-1 gene is expressed at the highest level. Recent studies demonstrated that the gene is expressed in a highly regulated manner:
- the expression of the gene is induced by light, the induction being mediated by at least two photoreceptors: by the red absorbing phytochrome and by an as yet unidentified blue absorbing photoreceptor;
- the expression of the gene is further modulated by an endogenous rhythm which results in the 24-hour periodicity of the expression;
- the gene is expressed only in certain cell types, like in leaf mesophyll cells and in the guard cells of the stomata.

Earlier studies showed that these characteristics of the expression are maintained in transgenic tobacco plants (Nagy et al., 1988). In order to define cis-acting elements responsible for the regulation of the Cab-1 gene expression we carried out experiments with a series of 5' deletion mutants and internal deletions of the Cab-1 promoter. We found that in the case of the 5' deletion mutants (from -1816 to -127) the -244 mutant still exhibits maximum level expression. Further removal of the upstream sequences (-230) resulted in a tenfold decrease, while additional deletions from -211 to -190 lowered the level of the expression below detection level in pooled samples. The -211 deletion mutant still showed circadian clock-regulated gene expression.

The analysis of tobacco plants transformed with chimeric genes (CAT and GUS constructs) showed that
- the removal of upstream sequences from -1816 to -357 does not change the level and the pattern of the expression,
- the internal deletion between -357 and -127 completely eliminates the expression (Fejes et al., 1990). From the charac-terisation of the light-induced, phytochrome-regulated expression of the wheat Cab-1 promoter (Nagy et al., 1987) and from the results summarized above we can conclude that the promoter region from -244 to -90 contains all the cis-acting elements responsible for the light-induced, circadian clock-responsive expression of the Cab-1 gene.

RESULTS

In order to identify cis-regulatory elements responsible for the tissue-specific expression of the Cab-1 gene we constructed chimeric genes containing various regions of the Cab-1 promoter fused to the bacterial GUS gene and NOS 3' end. Using standard methods (Fraley et al., 1985) the constructs were introduced into tobacco plants. The transgenic tobacco plants were grown up and seeds were collected. After a 15 min white light induction GUS histochemical staining of 7 day old etiolated seedlings was performed according to Jefferson (1987).
First we analysed two constructs:
     1. (-1816 - +31)Cab/GUS/NOS   2. (-357 - +31)Cab/GUS/NOS
We found no difference between the expression of the two chimeric genes, either in level or in pattern: the expression was high in the cotyledon and in the first foliage leaf, and there was no expression in the root of the seedling. We also prepared and stained mature leaf cross sections from transgenic tobacco plants transformed with the (-357 - +31)Cab/GUS/NOS chimeric gene. We observed high-level staining in the cells of the palisade parenchyma and a lower level of staining in the cells of the spongy parenchyma. There was no staining in the epidermis cells, except in the chloroplast-containing guard cells of the stomata. By this experiment we could show that I/ the removal of the upstream sequences from -1816 to -357 does not effect the level and the tissue-specificity of the expression and íí/ the promoter region of the Cab-1 gene from -357 to +31 contains all the cis-regulatory elements necessary for tissue-specific expression.

In order to further analyse the promoter region we constructed chimeric genes containing various regions of the Cab-1 promoter fused to the truncated 35S promoter (from -90 to +8), the bacterial GUS gene and NOS 3' end. As it was reported previously by Benfey et al. (1989), the 35S promoter region from -90 to +8 confers strong root-specific expression in transgenic tobacco plants.
We analysed the following constructs:

    1.   35S/GUS/NOS
    2. (-357 - -127)Cab/35S/GUS/NOS
    3. (-357 - -90)Cab/35S/GUS/NOS
    4. (-357 - -30)Cab/35S/GUS/NOS
As expected, the analysis of the 35S/GUS/NOS construct showed strong staining in the root apex and no staining in the cotyledon. In the case of the promoter region from -357 to -127 we detected relatively low-level expression in the cotyledon and high-level expression in the root apex. We observed a clearly different pattern of expression in the seedlings carrying the constructs of the promoter elements from -357 to -90 and from-357 to -30: strong staining in the cotyledon, no staining in the root apex. From the above results one can conclude that the promoter region from -357 to -90 is

able to maintain the tissue-specific expression of the Cab-1 gene and the sequence element from -127 to -90 may be involved in the silencing of the root-specific expression of the truncated 35S promoter.

To further characterise the promoter region from -357 to -127 we constructed the following chimeric genes and analysed their expression in transgenic plants:
1. $(-357 - -232)_4$Cab/35S/GUS/NOS
2. $(-211 - -147)_4$Cab/35S/GUS/NOS
3. $(-244 - -179)_4$Cab/35S/GUS/NOS
4. $(-179 - -127)_4$Cab/35S/GUS/NOS

From the four promoter regions tested, the sequence element from -357 to -232 brought about the strongest staining in the cotyledon. In the case of the regions from -244 to -179 and from -179 to -127 we could observe a lower level of staining. Staining of the cotyledon was lowest with the region from -211 to -147. All of the constructs showed high-level expression in the root tip. These results suggest that all of these four sequence elements are able to maintain expression in the leaves and do not affect the root-specific expression of the truncated 35S promoter to any measurable extent.

In order to identify nuclear proteins that interact with specific elements of the Cab-1 promoter we carried out footprint experiments. Using nuclear extracts prepared from tobacco and wheat leaf tissue we identified five binding sites on the Cab-1 promoter region from -357 to -90:
- Cab-A: -244 - -220   - Cab-B: -216 - -200
- Cab-C: -190 - -170   - Cab-D: -170 - -150
- Cab-E: -120 - -100

These regions were cloned and used in gel retardation and competition assays. As specific competitors we used elements definied as binding sites for nuclear proteins. We found that the as-2 element of the 35S promoter (Lam et al., 1989) and the conserved G-box sequence of the rbcS promoters (Giuliano et al., 1988) cannot compete effectively (100-fold excess) the binding of nuclear proteins by the above described Cab-1 promoter elements. On the other hand we clearly established that bindings by the Cab-D and Cab-E elements of the Cab-1 promoter can be competed off with the boxII sequence (Green et al., 1988) of the pea rbcS promoter (50-fold excess). None of the above mentioned specific competitors could compete with the Cab-C region.

To define the function of the above identified binding sites we constructed chimeric genes:

1. $(-244 - -220)_7$Cab/35S/GUS/NOS
2. $(-218 - -198)_4$Cab/35S/GUS/NOS
3. $(-170 - -150)_4$Cab/35S/GUS/NOS
4. $(-190 - -170)_4$Cab/35S/GUS/NOS
5. $(-123 - -100)_4$Cab/35S/GUS/NOS

We carried out histochemical localisation of the GUS gene product in the seedlings of transgenic tobacco plants containing the above described constructs. Untill now we have finished the analysis of chimeric genes No. 1., 2. and 3. respectively. All of the analysed constructs were expressed in the cotyledon: we observed high-level staining in the case of the -218 - -198 and the -170 - -150 regions and a lower level of staining in the case of the -244 - -220 region. Staining was uniformly high-level in the root tip of the seedlings containing these chimeric genes. These data indicate that these cis-acting elements of the Cab-1 promoter probably contribute to the expression in the cotyledon and none of these three regions has a detectable effect on the expression of the 35S promoter in the root tip.

CONCLUSIONS

We have shown that the region of the Cab-1 promoter from -357 to -90 contains not only the cis-regulatory elements for light-induced and circadian clock responsive gene expression but also the promoter sequences required for the tissue-specific expression of the gene. In agreement with earlier studies on several Cab promoters (Castresana et al., 1988, Gidoni et al., 1988, Ha et al., 1989) we found that the wheat Cab-1 promoter contains multiple cis-regulatory elements. The contribution of these elements to the regulated expression of the Cab-1 gene is not yet fully understood.

Based on results reported above we suggest that the promoter region from -127 to -90 is able to modify the expression of the GUS reporter gene driven by the truncated 35S promoter in the root tip. This fact may indicate the presence of a putative root-specific silencer in this region. Similar results have been published by Simpson et al. (1986). As described above, this binding site of the Cab-1 promoter could be competed by the boxII sequence of the pea rbc-3A promoter. Interestingly, when the rbcS-3A box II tetramer was fused to the 35S promoter region (from -90 to +8), the construct did not modify the strong root-specific expression of the truncated 35S promoter (Lam et al., 1990). The question whether this feature of the Cab-1 promoter region discussed is a unique phenomenon is still unanswered.

Further characterisation of the five factor binding sites of the Cab-1 promoter, using them as specific probes to isolate trans-acting factors is in progress. We hope that the charac-terisation of the putative tran-scriptional factor involved in the regulation of expression may lead us to understand the molecular mechanism of the complex regulation of the wheat Cab-1 gene.

REFERENCES

Benfey, P.N., Ren, L., Chua, N-H., 1989, The CaMV 35S enhancer contains at least two domains which can confer different developmental and tissue-specific expression pattern, EMBO J., 8:2195

Castresana, C., Garcia-Luque, I., Alonso, E., Malik, V.S., Cashmore, A.R., 1988, Both positive and negative regulatory elements mediate expression of a photoregulated CAB gene from Nicotiana plumbaginifolia, EMBO J., 7:1929

Edwards, J.W., Coruzzi, G.M. 1990, Cell-specific gene expression in plants, Annu. Rev. Genet., 24:275

Fejes, E., Páy, A., Kanevsky, I., Széll, M., Ádám, É., Kay, S.A., Nagy, F., 1990, A 268 bp upstream sequence mediates the circadian clock-regulated transcription of the wheat Cab-1 gene in transgenic plants, Plant Mol. Biol., 15:921

Fraley, R., Rogers, S., Horsch, R., Eicholz, D., Flick, F., Hoffman, N., Sanders, P., 1985, The SEV system : a new disarmed vector system for plant transformation, Biotechnology, 3:629

Gidoni, D., Brosio, P., Bond-Nutter, D., Bedbrook, J., Dunsmuir, P., 1989, Novel cis-acting elements in Petunia Cab gene promoters, Mol. Gen. Genet., 215:337

Giuliano, G., Pichersky, E., Malik, V.S., Timko M.P., Scolnik, P.A., Cashmore, A. R., 1988, An evolutionarily conserved protein binding seqence upstream of a plant light-regulated gene, Proc. Natl. Acad. Sci. USA, 85:7089

Green, P. J., Yong, M. H., Couzzo, M., Kano-Murakami, Y., Silverstein, P., Chua, N-H., 1988, Binding site requirements for pea nuclear protein factor GT-1 correlate with sequences required for ligh-dependent transcriptional activation of the rbcs-3A gene, EMBO J., 7:4035

Hu, S. B., An, G., 1989, Identification of upstream regulatory elements involved in the developmental expression of the Arabidopsis cab1 gene, Proc. Natl. Acad. Sci. USA, 85:8017

Karlin-Neumann, G. A., Kohorn, B. D., Thornbern, P., Tobin, E. M., 1985, A chlorophyll a/b protein encoded by a gene containing an intron with characteristic of a transposable element, J. Mol. Appl. Genet., 3:45

Lam, E., Chua, N-H., 1989, ASF-2: A factor that binds to the Cauliflower Mosaic Virus 35S promoter and a conserved GATA motif in Cab promoters, The Plant Cell, 1:1147

Lam, E., Chua, N-M., 1990, GT-1 binding site confers light responsive expression in transgenic tobacco, Science, 248:471

Nagy, F., Boutry, M., Hsu, M. Y., Wong, M., Chua, N-H., 1987, The 5' proximal region of the wheat Cab-1 gene contains a 286 bp enhancer-like sequence for phytochrome response, EMBO J., 6:2537

Nagy, F., Kay. S. A., Chua, N-H., 1988, A circadian clock regulates transcription of the wheat Cab-1 gene, Genes. Dev., 2:376

Simpson, J., Schell, J., Van Montagu, M., Herrera-Estrella, L., 1986, Light inducible and tissue-specific pea lhcp gene expression involves an upstream element combining enhancer- and silencer-like properties, Nature, 323:551

CHARACTERIZATION OF cDNAs WHICH ENCODE ENZYMES INVOLVED IN CHROMOPLAST DIFFERENTIATION AND CAROTENOID BIOSYNTHESIS IN *CAPSICUM ANNUUM*

S. Römer, A. Saint-Guily, F. Montrichard, M.L. Schantz, J.H. Weil, R. Schantz, M. Kuntz and B. Camara*

Institut de Biologie Moléculaire des Plantes du CNRS, Université Louis Pasteur, 67084 Strasbourg, France, and
*Université de Bordeaux 1, 33405 Talence, France

INTRODUCTION

During plant development the plastid compartment undergoes important structural and biochemical changes. The different types of plastids and their precursors (proplastids, etioplasts, chloroplasts, chromoplasts,...) are often interconvertible but each type exhibits its own structure and biological function (Sitte et al., 1980; Akoyunoglou, 1984). Chloroplast differentiation represents a complex process involving a coordinate interaction between the nucleus and the plastid (Mullet, 1988; Taylor, 1989). It is well known that the etioplast/chloroplast development requires the formation of pigments and proteins which are, together with other components, necessary for the assembly of an intact photosynthetic apparatus. (Wild , 1978; Humbeck et al., 1989).

In ripening fruits prochromoplasts or photosynthetic chloroplasts differentiate into non-photosynthetic chromoplasts. During this process changes in nuclear and plastid gene expression (Grierson, 1986; Kuntz et al., 1989) take place leading for example to remarkable differences in RNA and protein levels (Bathgate et al., 1985; Piechulla et al., 1987; Wrench et al., 1987; Livne and Gepstein, 1988). At the same time, chlorophylls are degraded and carotenoids are synthesized in large amounts (Camara and Monéger, 1978; Britton, 1988; Camara, 1985a). Therefore, carotenoid biosynthesis is of special importance in chloroplast/chromoplast conversion.

We have chosen *Capsicum annuum* (bell pepper) as a model system to study chromoplast differentiation (Camara et al., 1989). Several enzymes of the carotenoid biosynthetic pathway (Fig. 1) have been isolated from *C. annuum* fruits (Camara, 1985a; Dogbo and Camara, 1987; Dogbo et al., 1988). One of the key enzymes in carotenoid synthesis is the geranylgeranyl pyrophosphate synthase (GGPPS), a stromal multifunctional enzyme of dimeric structure. It catalyzes the synthesis of geranylgeranyl pyrophosphate, a common precursor of all terpenoids in the plastid (Kleinig, 1989). This enzyme has been purified to homogeneity by Dogbo and Camara (1987) from *C. annuum* chromoplasts. Polyclonal antibodies raised against GGPPS were used to screen a cDNA library from *C. annuum* fruits.

Regulation of Chloroplast Biogenesis
Edited by J.H. Argyroudi-Akoyunoglou, Plenum Press, New York, 1992

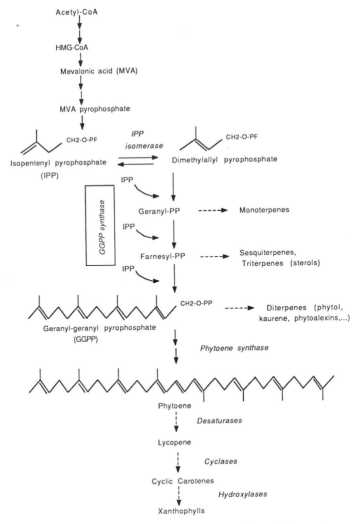

Fig. 1. Simplified scheme of the carotenoid biosynthetic pathway.

RESULTS AND DISCUSSION

Chromoplast Differentiation in Capsicum annuum Involves Induction of Specific Genes as well as Constitutive Expression of Others

It has been shown previously that specific proteins accumulate during chromoplast differentiation in *C. annuum* ripening fruits (Hadjeb et al., 1988; Newman et al., 1989). It has also been demonstrated that *C. annuum* chromoplasts incorporate isopentenyl pyrophosphate into carotenes and UDP-galactose into galactosyl diglyceride (Camara, 1985b). Other data have demonstrated that there is an increased synthesis of tocopherols during chromoplast differentiation in this plant (Camara, 1985b).

Thus, despite the loss of the photosynthetic activity, *C. annuum* chromoplats remain an anabolically active system. This is further illustrated by the following observations.

We could clone a cDNA for the plastid located O-acetyl-L-serine sulfhydrylase (EC.4.2.99.8), an enzyme which catalyzes the synthesis of cysteine. RNA gel blot experiments revealed that this gene is expressed in leaves and in fruits at all developmental stages, including the fully ripe fruit. Comparison of the deduced peptide sequence of this clone (not shown) to the corresponding sequence from *E. coli* (Sirko et al., 1990) revealed that the *C. annuum* enzyme is synthesized as a protein containing an N-terminal extension resembling a transit peptide.

We have also cloned a fruit ripening-specific cDNA. This cDNA clone encodes an acidic protein of 35 kDa whose N-terminal end most likely represents a transit peptide: this domain is positively charged and contains a number of hydroxylated amino acids (not shown). The size of the mature protein would therefore be about 29 kDa. The function of this protein is presently unknown and will be a matter of further investigations. Significantly, RNA gel blot experiments revealed that the corresponding mRNA is absent in young seedlings, in leaves and in green fruits, whilst it accumulates to high levels during chromoplast differentiation.

The GGPPS Gene Expression is Greatly Stimulated during Chromoplast Differentiation in *C. annuum*

We have isolated a 1.3 kbp cDNA clone encoding the full-length GGPPS polypeptide. The deduced peptide sequence of the cDNA clone was 369 amino acids long which corresponds to a 40 kDa protein. This protein possess an N-terminal domain resembling a transit peptide. Therefore, this result confirms that the GGPPS is located in the plastids. A significant homology has been found between this sequence and others deduced for enzymes encoded by the bacterial *crtE* genes (Armstrong et al., 1989 and 1990a; Misawa et al., 1990). In addition, limited but significant sequence similarity was also detected between the *C. annuum* GGPPS sequence and the protein encoded by the *Neurospora crassa al*-3 gene (Carattoli et al., 1991). Especially, four regions (I-IV) have been well conserved (Fig. 2).

The *N. crassa al*-3 gene has been demonstrated to encode GGPPS (Harding and Turner, 1981). However, the bacterial *crtE* gene from *R. capsulatus* has been identified as a phytoene synthase gene (Armstrong et al., 1990b). This discrepancy prompted us to further characterize the enzymatic activity of the protein encoded by the cloned *C. annuum* cDNA. Therefore, we expressed the GGPPS cDNA in *E. coli* and analyzed the prenyl transferase activities of the transformed cells. In presence of radiolabelled isopentenyl pyrophosphate the non-transformed *E. coli* cells synthesized only geranyl pyrophosphate (GPP) as the major product. In contrast, the transformed *E. coli* cells (expressing the GGPP synthase cDNA) catalyzes the synthesis of an additional prenyl pyrophosphate. This compound was identified as GGPP by normal and reversed phase chromatography (data not shown).

```
      I                    II                  III           IV
1..GGKRVRP..AVEMIHTMSLIHDDLPCMDNDDLRRGK..KTAALL..IGLLFQVVDDILDV..
2..PGKDIRS.:VISMLHTASLLVDDVE..DNSVLRRGF..KTGGLF..IGLIFQIADDYHNL..
3..GGARIRP..ALEMVHAASLILDDMPCMDDAKLRRGK..KTSTLF..LGQAFQLLDDLTDG..
4..PGKRIRP..AVELMHCASLVHDDLPAFDNADIRRGK..KTGALF..IGSAFQIADDLKDA..
5..GGKRLRP..AVECIHAYSLIHDDLPAMDDDDLRRGL..KTGALI..IGLAFQVQDDILDV..
6..GGKYNRG..CVELLQAFFLVADDIM..DSSLTRRGQ..KTAFYS..MGEFFQIQDDYLDL..
7..gGkr.Rp..avem.h..sLihDDlp.mDn.dLRRGk..KT.al...iGl.FQ..DD.ld...
```

Fig. 2. Conserved regions in the GGPPS sequences from *C. annuum* (1) and *Neurospora crassa* (2), in the CRTE sequences from *Erwinia uredovora* (3) and *Rhodobacter capsulatus* (4) and in the farnesyl pyrophosphate synthase from *E. coli* (5) and human (6). The four blocks of conserved sequences are termed I-IV. Dots separating these blocks represent stretches of variable lengths and amino acid sequences. Consensus (7): upper case letter for amino acid conserved in all 6 sequences, lower case letter for amino acid conserved in 3 out of 6 sequences (including that of *C. annuum* GGPPS)

We have compared the GGPPS sequence to another prenyltransferase, namely farnesyl pyrophosphate synthase (FPPS), which catalyzes the synthesis of a C15 isoprenoid compound (farnesyl pyrophosphate). This cytosolic enzyme has been purified from *C. annuum* (Hugueney and Camara, 1990) but its sequence has not been determined yet. However, the FPPS sequence is known from *E. coli*, yeast, rat and human (Anderson et al., 1989; Shingo et al., 1990; Teruya et al., 1990; Wilkin et al., 1990) and these sequences also possess the conserved regions depicted in Fig. 2. In addition, the hydropathy plots of the *C. annuum* GGPPS, the CRTE proteins and the FPPS show striking similarities (Fig. 3). This is particularly visible when the 4 conserved regions are represented on these plots. It will be interesting to try to understand why, despite their strong similarity, GGPPS and FPPS catalyze the synthesis of different molecules.

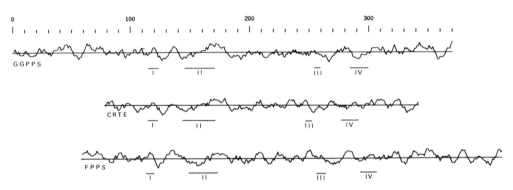

Fig. 3. Hydrophilicity profiles of the *C. annuum* GGPPS, the *Rhodobacter capsulatus* CRTE protein and the human farnesyl pyrophosphate synthase (FPPS). The four conserved regions shown in fig. 2 are depicted.

RNA gel blot blot analysis was performed using the cDNA clone coding for GGPPS as hybridization probe. A single hybridization signal corresponding to a 1.3 kb mRNA was observed in similar amounts in dark grown seedlings, in young and scenescent leaves and in very young green fruits (Fig. 4). The mRNA level decreased when fruits were growing further up to the mature stage. Nevertheless, a dramatic transient increase in mRNA level was found in the early fruit ripening stage correlating with the visible accumulation of carotenoids.

These results suggest that the GGPPS gene is expressed in all tissues were carotenoid synthesis occurs and that the expression of this gene is tightly coordinated with chromoplast differentiation in fruits. Carotenoid biosynthesis is influenced by several factors including light. In the case of the GGPPS gene from *N. crassa* a light regulation was demonstrated (Carattoli et al., 1991). A light-regulation has not been observed in preliminary experiments for the GGPPS gene from *C. annuum*, because it was already expressed at low levels in dark grown seedlings (which are yellow).

CONCLUSIONS

We have shown previously (Kuntz et al., 1989) that during the chloroplast/chromoplast transition in *C. annuum* fruits, regulation of plastid gene expression occurs at a post-transcriptional/ translational level. We have also shown that nuclear genes coding for chloroplast proteins, namely *rbcS* and *cab*, are turned off before the onset of ripening. Our present results demonstrate that this is not the case for other nuclear genes, such as the O-acetyl-L-serine sulfhydrylase gene which is constitutively

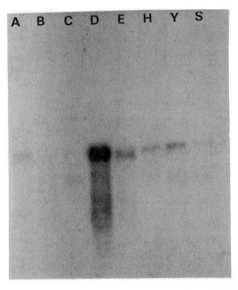

Fig. 4. RNA gel blot analysis of the GGPPS cDNA clone. Total RNAs were extracted from young green fruits (A), expanding green fruits (B), mature green fruits (C), fruits at an early ripening stage (D), fully mature fruits, dark-grown hypocotyls and cotyledons (H), young leaves (Y) and senescing leaves (S).

expressed. The expression of other genes coding for proteins involved in chromoplast differentiation, such as the gene encoding GGPPS, are strongly induced during the early stages of chromoplast differentiation, most likely by a transcriptional regulation mechanism.

Thus, a complex pattern of gene expression occurs during chromoplast differentiation. One of our future goals is to study the regulatory mechanisms controlling this differentiation process.

ACKNOWLEDGEMENTS

The authors wish to thank A. Klein for excellent technical assistance. S.R. is a recipient of a long-term EMBO fellowship (ALTF 191-1990).

REFERENCES

Akoyunoglou, G., 1984, Thylakoid biogenesis in higher plants. Assembly and reorganization, in: "Advances in Photosynthesis Research", Vol. IV, C. Sybesma, ed, M. Nijhoff and W. Junk Publ., The Hague.

Anderson, M. S., Yarger, J. G., Burck, C. L., and Poulter, C. D., 1989, Farnesyl diphosphate synthetase: molecular cloning, sequence and expression of an essential gene from *Saccharomyces cerevisiae*, J. Biol. Chem., 264:19176.

Armstrong, G. A., Alberti, M., Leach, F., and Hearst, J. E., 1989, Nucleotide sequence organization and nature of the protein products of the carotenoid biosynthesis gene cluster of *Rhodobacter capsulatus*, Mol. Gen. Genet., 216:254.

Armstrong, G. A., Alberti, M., and Hearst, J. E., 1990a, Conserved enzymes mediate the early reactions of carotenoid biosynthesis in nonphotosynthetic and photosynthetic prokaryotes, Proc. Natl. Acad. Sci. USA, 87:9975.

Armstrong, G. A., Schmidt, A., Sandmann, G., and Hearst, J. E., 1990b, Genetic and biochemical characterization of carotenoid biosynthesis mutants of *Rhodobacter capsulatus*, J. Biol. Chem., 265:8329.

Bathgate, B., Purton, M. E., Grierson, D., and Goodenough, P. W., 1985, Plastid changes during the conversion of chloroplasts to chromoplasts in ripening tomatoes, Planta, 165:197.

Britton, G., 1988, Biosynthesis of carotenoids, in: "Plant Pigments", T.W. Goodwin, ed, Academic Press, London.

Camara, B., 1985a, Carotenogenic enzymes from *Capsicum annuum* chloroplasts, Pure Appl. Chem., 57:675.

Camara, B., 1985b, Carotene synthesis in *Capsicum* chromoplasts, Methods Enzymol., 110:244.

Camara, B., Bousquet, J., Cheniclet, C., Carde, J. P., Kuntz, M., Evrard, J. L., and Weil, J. H., 1989, Enzymology of isoprenoid biosynthesis and expression of plastid and nuclear genes during chromoplast differentiation in pepper fruits (*Capsicum annuum*), in: "Physiology Biochemistry and Genetics of Nongreen Plastids", C.D. Boyer, J.C. Shannon and R.C. Hardison, eds, Amer. Soc. Plant Physiol.

Camara, B., and Monéger, R., 1978, Free and esterified carotenoids in green and red fruits of *Capsicum annuum*, Phytochemistry, 17:91.

Carattoli, A., Romano, N., Ballario, P., Morelli, G., and Macino, G., 1991, The *Neurospora crassa* carotenoid biosynthetic gene (Albino 3) reveals highly conserved regions among prenyltransferases, J. Biol. Chem., 266:5854.

Dogbo, O., and Camara, B., 1987, Purification of isopentenyl pyrophosphate isomerase and geranylgeranyl pyrophosphate synthase from *Capsicum* chromoplasts by affinity chromatography, Biochem. Biophys. Acta, 920:140.

Dogbo, O., Laferriere, A., d'Harlingue, A., and Camara, B., 1988, Carotenoid biosynthesis: isolation and characterization of a bifunctional enzyme catalyzing the synthesis of phytoene, Proc. Natl. Acad. Sci. USA, 85:7054.

Grierson, D., 1986, Molecular biology of fruit ripening, Oxford Surveys of Plant Molecular and Cell Biology, 3:363.

Hadjeb, N., Gounaris, I., and Price, C. A., 1988, Chromoplast-specific proteins in *Capsicum annuum*, Plant Physiol., 88:42.

Harding, R. W., and Turner, R. V., 1981, Photoregulation of the carotenoid biosynthetic pathway in albino and white collar mutant of *Neurospora crassa*, Plant Physiol., 68:745.

Hugueney, P., and Camara, B., 1990, Purification and characterization of farnesyl pyrophosphate synthase from *Capsicum annuum*, FEBS Lett., 273:235.

Humbeck, K., Römer, S., and Senger, H., 1989, Evidence for an essential role of carotenoids in the assembly of an active photosystem II, Planta, 179:242.

Kleinig, H., 1989, The role of plastids in isoprenoid biosynthesis, Annu. Rev. Plant Physiol., 40:39.

Kuntz, M., Evrard, J. L., d'Harlingue, A.,Weil, J. H., and Camara, B., 1989, Expression of plastid and nuclear genes during chromoplast differentiation in bell pepper (*Capsicum annuum*) and sunflower (*Helianthus annuus*), Mol. Gen. Genet., 216:156.

Livne, A., and Gepstein, S., 1988, Abundance of the major chloroplast polypeptide during development and ripening of tomato fruits, <u>Plant Physiol.</u>, 87:239.

Misawa, N., Nakagawa, M., Kobayashi, K., Yamano, S., Izawa, Y., Nakamura, K., and Harashima, K., 1990, Elucidation of the *Erwinia uredovora* Carotenoid Biosynthetic Pathway by Functional Analysis of Gene Products Expressed in *Escherichia coli*, <u>J. Bacteriol.</u>, 172:6704.

Mullet, J. E., 1988, Chloroplast development and gene expression, <u>Annu. Rev. Plant Physiol.</u>, 39:475.

Newman, L. A., Hadjeb, N., and Price, C. A., 1989, Synthesis of two chromoplast-specific proteins during fruit development in *Capsicum annuum*, <u>Plant Physiol</u>

Piechulla, B., Glick, R. E., Bahl, H., Melis, A., and Gruissem, W., 1987, Changes in photosynthetic capacity and photosynthetic protein pattern during tomato fruit ripening, <u>Plant Physiol.</u>, 84:911.

Shingo, F., Hara, H., Nishimura, Y., Horiuchi, K., and Nishimo, T., 1990, Isolation and characterization of an *Escherichia coli* mutant having temperature-sensitive farnesyl diphosphate synthase, <u>J. Biochem.</u>, 108:995.

Sirko, A., Hryniewicz, M., Hulanicka, D., and Boeck, A., 1990, Sulfate and thiosulfate transport in *Escherichia coli* K-12: nucleotide sequence and expression of the cysTWAM gene cluster, <u>J. Bacteriol.</u>, 172:3351.

Sitte, P., Falk, H., and Liedvogel, B., 1980, Chromoplasts, <u>in</u>: "Pigments in Plants", F.C. Czygan, ed, Fischer, Stuttgart.

Taylor, W. C., 1989, Regulatory interactions between nuclear and plastid genomes, <u>Annu. Rev. Plant Physiol.</u>, 40:211.

Teruya, J. H., Kutsunai, S. Y., Spear, D. J., Edwards, P. A., and Clarke, C. F., 1990, Testis-specific transcriptional initiation sites of rat farnesyl pyrophosphate synthetase mRNA, <u>Mol. Cell. Biol.</u>, 10:2315.

Wild, A., 1978, Studies on chloroplast development of a mutant of *Chlorella fusca*, <u>in</u>: "Chloroplast Development", G. Akoyonoglou and J.H. Argyroudi-Akoyunoglou, eds, Elsevier, Amsterdam.

Wilkin, D. J., Kutsunai, S. Y., and Edwards, P. A., 1990, Isolation and sequence of the human farnesyl pyrophosphate synthase cDNA: Coordinate regulation of the mRNAs for farnesyl pyrophosphate synthetase, 3-hydroxy-3-methylglutaryl coenzyme A reductase, and 3-hydroxy-3-methylglutaryl coenzyme A synthase., <u>J. Biol. Chem.</u>, 265:4607.

Wrench, P. M., Olive, M., Hiller, R. G., Brady, C., and Speirs, J., 1987, Changes in plastid proteins during ripening of tomato fruits, <u>J. Plant Physiol.</u>, 129:89.

PHOTORECEPTORS AND THEIR ACTION ON CHLOROPLAST DEVELOPMENT

PHOTOREGULATED GENE EXPRESSION

Gerhard Richter

Institut für Botanik, Universität Hannover
D-3000 Hannover 21, Germany

INTRODUCTION

The profound influence of light on growth and the development of plants has been demonstrated in many cases. As an environmental factor light is mediated mainly by photoreceptors. From these the reversible red/far red sensing phytochrome is the best characterized system in plants. Here, the pioneer work by Mohr and coworkers as well as that by Borthwick and Hendricks should be mentioned. In addition, the involvement of UV/blue photoreceptor systems of yet unknown nature has been found to be effective in plants.

PHYTOCHROME

One mechanism through which light controls the differentiation of chloroplasts seems to be modulation of the expression of nuclear genes which encode prominent chloroplast proteins. When barley seedlings and fronds of Lemna gibba are exposed to pulses of red light the complete pattern of morphological changes characteristic for chloroplast development does not take place - however, several substantial changes in gene expression occur: The expression of the genes encoding the small subunit (SSU) of rubisco (rbcS) and the main apoprotein of the light-harvesting chlorophyll a,b complex (LHC II; Cab) is positively regulated as shown by Apel (1979) respective by Silverthorne and Tobin (1984), the latter using an isolated nuclei system to address the question of whether phytochrome exerts its influence on transcription of the two nuclear genes. In case of the NADP-protochlorophyllide oxidoreductase and phytochrome itself the corresponding genes are apparently repressed by light mediated by phytochrome as demonstrated by Mösinger et al. (1985) respective Lissemore and Quail (1988).

Obviously, the phytochrome control of these nuclear genes occur on the level of transcription. Nevertheless, a similar control on the post-transcriptional level cannot be excluded. This was shown by Silverthorne and Tobin (1990) with Lemna gibba where the expression of members of the rbcS gene family differed between the organs of light-grown plants as a consequence of individual organ-specific developmental programs.

UV / BLUE LIGHT SYSTEMS

Blue light control of the transcription for nuclear genes in pea plants was demonstrated recently by Marrs and Kaufman (1989): Seedlings were grown in continous red light to saturate any phytochrome response, then treated with a single pulse of blue light at definite stage of growth, i.e. 6 d after planting. The blue light treatment resulted in an increase in the steady-state level and transcription rate of the Cab genes. These results could explain previous observations indicating blue light control of essential processes along the line of chloroplast development in higher plants like pea and transgenic petunia (Fluhr and Chua, 1986; Fluhr et al., 1986) as well as in the green alga Scenedesmus (Senger, 1982).

MOLECULAR LIGHT SWITCHES

Detailed analysis of the rbcS gene of pea and tomato has revealed short DNA regions within the 5'-upstream region that can act as light-responsive elements,i.e. that seem to function as cis-acting elements (Benfey and Chua, 1989; Ueda et al. 1989; Davis et al., 1990). The specific light response is mediated by phytochrome. The presence of similar sequence elements was demonstrated for other light-responsive genes, e.g. Cab, in other plant species (Sun and Tobin, 1990). However, there is evidence that other regions of a gene can mediate changes in transcript abundance in response to light, too.

Recent work has focused on nuclear proteins that interact with these specific sequence elements. Several factors - GT-1, GBF (CG-1), 3AF1, AT-1, GAF-1 - were identified and their binding sites characterized (for ref. see Gilmartin et al., 1990).

PLANT CELL CULTURES

Another type of light-dependency in chloroplast differentiation has been found in plant cell cultures: This process respective the expression of genes involved was found to be controlled by blue light - to a certain degree or exclusively. In several cell lines of Chenopodium rubrum the transformation of leukoplast-like precursors is not only induced but also maintained by this light quality; at no stage red light can replace blue light. In contrast to the phytochrome mediated light response here the substained exposure to blue light of the cells is mandatory (Richter et al., 1987). Recent data support the notion that blue light exerts its influence on plastidic genes by enhancing their transcription rate (Richter and Ottersbach, 1990): The activity of TAC (="transcriptionally active chromosome or complex") increased steadily with the onset of blue irradiation following closely the time course of chloroplast differentiation; analysis of the RNA sequences synthesized in vitro revealed that genes coding for stroma and membrane proteins as well as plastid rRNA genes were specifically transcribed (Fig. 1).

For nuclear genes a differential regulation by blue light was observed: Three early light-induced genes (see below) exhibit an in-vitro transcription rate which is in good accordance with the steady-state level of their transcripts. Apparently, the blue light-dependent rapid and transient transcript accumulation in vivo is caused by a temporary increase in the transcription rate of the corresponding genes. In contrast, for the rbcS and Cab genes the changes in transcription rate do not follow those in steady-state concentration, thus the observed increase of the latter cannot be explained as an enhancement in the transcription rate of both genes, but is likely a consequence of post-transcriptional events (Fig. 2, 3) as shown by Bockholt et al. (1991).

72

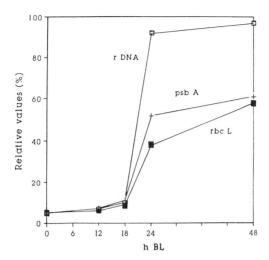

Fig. 1. Time course of in vitro transcription rate of individual plastid genes in TAC isolated from plastids of cell
cultures irradiated with blue light for the times indicated.

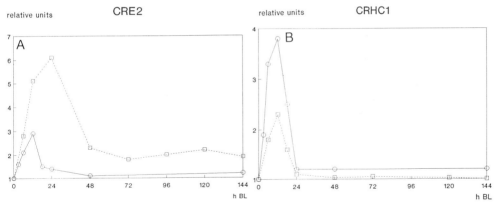

Fig. 2. Time course of blue light-induced changes in the steady-state
concentration (□--□) of nuclear RNAs and in transcription
rate of the corresponding genes (O—O) in isolated nuclei.
A: Clone CRE2, B: Clone CRHC1, both representing rapidly blue
light-induced genes (see Fig. 4). Total RNA as well as nuclei
were isolated from blue light-irradiated cells at the times
indicated. Level of RNA was determined by dot hybridization
using the specific antisense RNA of each clone as labelled
probe. Quantitation of in vitro transcripts was achieved by
dot hybridization; the antisense RNA of each clone served as
immobilized unlabelled probe.

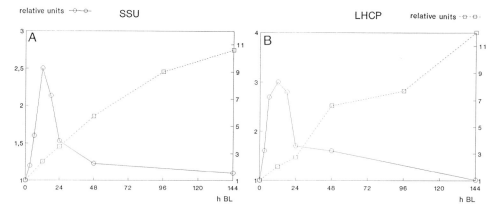

Fig. 3. As described for Fig. 2 except: Steady-state concentration
         of nuclear RNAs and transcription rate of the corresponding
         gene in isolated nuclei; A: SSU mRNA and rbcS; B: LHC II
         mRNA and Cab. Total RNA as well as nuclei were isolated from
         blue light-irradiated cells at the times indicated.

SIGNAL TRANSDUCTION

    Little is known about the signal transduction events that intervene
between photoreceptor activation and gene expression. Three strategies have
been developped recently to tackle this problem. The aim of the first one
was the identification of genes which are rapidly and transiently expressed
during the initial phase of blue light-induced chloroplast differentiation
in cultured plant cells (Chenopodium rubrum), and to evaluate the role of
their products with regard to light perception, signal transduction and
response. A cDNA library was established using polyA-RNAs from cells exposed to blue light for 6, 12 und 24 h; most of the mRNA species representative of dark-grown and fully greened cells had been eliminated prior
to the cloning procedure. By differential screening several clones
corresponding to genes rapidly induced by blue light were identified
(Kaldenhoff and Richter, 1990). For most of these a temporary accumulation
of the specific mRNA between 30 min and 72 h of blue light irradiation was
observed (Fig. 2; see also Fig. 4). With regard to the nucleotide sequence
and the respective deduced amino-acid sequence a glycine-rich protein, a
ß-tubulin-like protein and one species with a substantial overall homology
to the sequence of RLAO proteins (= regulatory "acidic ribosomal proteins"
of yeast and human cells) were among the products of the early light-
induced genes.

    Since typical DNA-binding motifs are absent in the amino acid sequen-
ces deduced from the cDNA insertions of the clones identified an involvement of the putative proteins in other mechanisms of regulation must be
envisaged. With regard to the effect of light as inducing factor the
observed changes could be interpreted as either a sequence of events in
close connection with the induced process of chloroplast differentiation
or as a general adaptation of cellular metabolism to the impact of sudden
blue light irradiation. Support for the first possibility comes from the
results of similar experiments with carrot suspension cultures: During the
induction of somatic embryogenesis a number of genes are rapidly and trans-

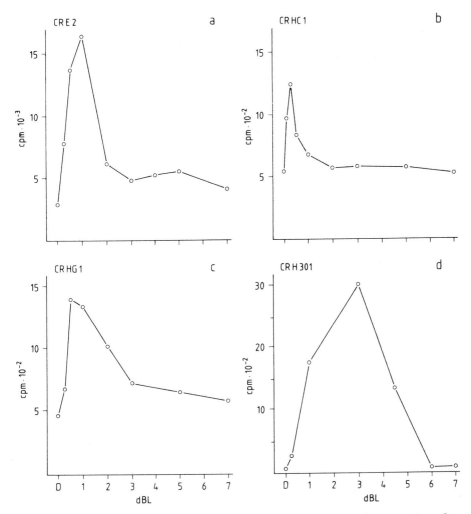

Fig. 4. Time course of blue light-induced changes in the amounts of
RNAs homologous to clones CRE2 (a; RLAO-like protein), CRHC1
(b; glycine-rich protein), CRHG1 (c; ß-tubuline-like protein)
and CRH301 (d). Total RNA or polyA-RNAs were isolated from
cultured cells after the onset of illumination at the times
indicated. Hybridization was to radiolabelled antisense RNA
transcribed from the cDNA insertions of the clones. The relative signal intensities were obtained by counting the radioactivity bound to the hybrids; clones without light induction
were used as internal standards. BL, blue light.

iently expressed; at least the products of two share homologies with
proteins whose genes are activated during the initial phase of blue light-
induced chloroplast differentiation: a glycine-rich protein and a ß-tubulin-
like species (Aleith and Richter, 1991).

The objective of recent studies was to identify genes of the plastid
which are rapidly expressed during the initiation of blue light-induced
chloroplast differentiation in suspension cultured cells of Chenopodium
rubrum and to analyze the encoded proteins. From a cDNA library constructed from total RNA of plastid preparations of dark-grown cells exposed to
blue light for 3, 6 and 12 h three clones were identified which evidently
correspond to rapidly light-induced plastidic genes. For these a tem-

porary accumulation of the specific mRNA between 12 and 48 h of blue light exposure was registered (Specht and Richter, 1991).

The two other strategies taking advantage of the light-induced developmental changes in etiolated seedlings aim also for the elucidation of unknown signal transduction events that follow excitation of phytochrome. Chory et al. (1991) identified new Arabidopsis mutants, DET and det (="de-etiolated") in which the light developmental program is constitutively active: In complete darkness the mutant seedlings behave in many respects like light-grown wild-type seedlings, except they cannot synthesize chlorophylls. Apparently, the det-gene product normally represses the light developmental program in the dark; photoactivation of phytochrome would cause the product of the det-gene to be inactivated thus allowing the light program to proceed.

A different approach has been chosen by Karlin-Neumann et al. (1991): The promoter of the phytochrome-regulated Cab-gene was fused to a selective marker gene, the tms2 gene from Agrobacterium tumefaciens which encodes the enzyme for the conversion of nontoxic auxin amides into plant auxins becoming toxic at high concentrations. Seedlings carrying the gene construct grow normally on the auxin amides in the dark, but are severely inhibited in their growth when exposed to red light. Seedlings of Arabidopsis that have mutations in genes which encode components of the phytochrome signal transduction pathway should grow normally even after red light treatment; they would be readily distinguishable from the growth-inhibited plants. With this strategy the selection for mutants that fail to respond to phytochrome should be successful.

REFERENCES

Aleith, F., and Richter, G., 1991, Planta 183:17.
Apel, K., 1979, Eur.J.Biochem. 97:183.
Benfey, P.N., and Chua, N.-H., 1989, Science 244:174.
Bockholt, S., Kaldenhoff, R., and Richter, G., 1991, Bot.Acta 104:245.
Chory, J., Nagpal, P., and Peto, C.A., 1991, Plant Cell 3:445.
Davis, M.C., Yong, M.-H., Gilmartin, P.M., Goyvaerts, E., Kuhlmeier, C., Sorokin, L.,and Chua, N.-H., 1991, Photochem. Photobiol., in press.
Fluhr, R., and Chua, N.-H., 1986, Proc.Natl.Acad.Sci. USA 83:2358.
Fluhr, R., Moses, P., Morelli, G., Coruzzi, G., and Chua, N.-H., 1986, Embo J. 5:2063.
Gilmartin, Ph. M., Sarokin, L., Memelink, J., and Chua, N.-H., 1990, Plant Cell 2:369.
Kaldenhoff, R., and Richter, G., 1990, Planta 181:220.
Karlin-Neumann, G.A., Brusslan, J.A., and Tobin, E.M., 1991, Plant Cell 3:573.
Lissemore, J.L., and Quail, P.H., 1988, Mol.Cell Biol. 8:4840.
Marrs, K.A., and Kaufman, L.S., 1989, Proc.Natl.Acad.Sci. USA 86:4492.
Mösinger, E., Batschauer, A., Schäfer, E., and Apel, K., 1985, Eur.J. Biochem. 147:137.
Richter, G., and Ottersbach, N., 1990, Bot.Acta 103:168.
Richter, G., Dudel, A., Einspanier, R., Dannhauer, I., and Hüsemann, W., 1987, Planta 172:79.
Senger, H., 1982, Photochem. Photobiol. 35:911.
Silverthorne, J., and Tobin, E.M., 1984, Proc.Natl.Acad.Sci. USA 81:1112.
Silverthorne, J., and Tobin, E.M., 1990, Plant Cell 2:1181.
Specht, U., and Richter, G., 1991, J.Photochem. Photobiol. 11: in press.
Sun, L., and Tobin, E.M., 1990, Photochem. Photobiol. 52:51.
Ueda, T., Pickersky, E., Malik, V.S., and Cashmore, A.R., 1989, Plant Cell 1:217.

REGULATION OF GENE EXPRESSION IN PLASTIDOGENESIS BY LIGHT,

NITRATE AND PLASTIDIC FACTOR

Hans Mohr

Biological Institute II
University of Freiburg
D-7800 Freiburg, Germany

SCOPE OF THIS ARTICLE

In higher plants light affects chloroplast formation in several ways. A photoreaction within the pathway of chlorophyll synthesis is well documented: formation of chlorophyll(ide) from protochlorophyll(ide). Moreover, light absorption by sensor pigments is crucial as well. These are phytochrome (operating predominantly in the red and far-red spectral range) and cryptochrome (operating in the blue and UV-A spectral range).

In the present article the following problems will be considered: First, mode of coaction among the different photosensors (photoreceptors). - Second, coaction of phytochrome and nitrate in controlling expression of nuclear genes involved in plastidic nitrate/ammonium assimilation. - Third, obligatory dependency of plastid-bound expression of nuclear genes on a 'plastidic factor'. It will become obvious that full gene expression of nuclear genes involved in chloroplast formation requires a subtle coaction of phytochrome, cryptochrome, nitrate and 'plastidic factor'.

Using different plant materials and different approaches, it has been shown that the expression of nuclear genes encoding proteins which function in the chloroplast - such as the small subunit of ribulose-1,5-bisphosphate carboxylase and the light-harvesting chlorophyll a/b-binding protein of photosystem II - depends on the integrity of the plastids (see Oelmüller 1988, for review). It was inferred from these findings that a plastid-derived factor is involved in the transcriptional control of nuclear genes coding for proteins destined for the chloroplast (Fig. 1). The result of damage done to the plastid, e.g. photooxidation of the plastid, would be to destroy the ability of the organelle to send off this signal. The nature of the postulated plastidic signal has so far remained elusive, but it is probably not a plastidic protein (Oelmüller et al., 1986).

Regulation of Chloroplast Biogenesis
Edited by J.H. Argyroudi-Akoyunoglou, Plenum Press, New York, 1992

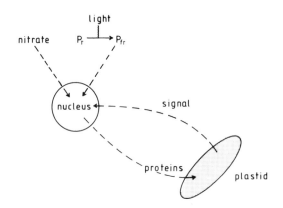

Fig. 1.Scheme to illustrate the significance of plastidic
factor for expression of nuclear genes coding for proteins destined for the   chloroplast

MODE OF COACTION BETWEEN PHYTOCHROME AND CRYPTOCHROME

In recent molecular studies (Gilmartin et al., 1990) it
was noticed that phytochrome and cryptochrome control the
light-induced expression of nuclear genes encoding plastidic
proteins. The question was raised of whether genes that respond to more than one photoreceptor do so through distinct
cis-acting elements or whether the signal-transduction pathways converge to act upon the same regulatory sequence. The
presently available data from transgenic plants do not allow
to discriminate between the alternatives. Rather, an approach
focusing on the photoreceptors was timely. Light-mediated synthesis of ferredoxin-dependent glutamate synthase (Fd-GOGAT)
in the cotyledons of Scots pine (Pinus sylvestris L.) was investigated (Elmlinger and Mohr, 1991). Fd-GOGAT is a nuclear
encoded plastidic protein which plays a crucial role in plastidic ammonium assimilation. Appearance of Fd-GOGAT is mainly
controlled by light (Fig. 2, Table 1). The conspicuous feature
is that up to 10 d after sowing the light effect on Fd-GOGAT
synthesis can fully be attributed to the operation of phytochrome whereas between 10 and 12 days after sowing synthesis
breaks down and degradation begins if no blue light (B) is
provided. It appeared that the operation of phytochrome was
replaced by the operation of cryptochrome (B/UV-A photoreceptor). However, dichromatic experiments (simultaneous treatment
of the seedlings with two light beams to vary the level of the
far-red-absorbing form of phytochrome [Pfr] in blue light)
showed that B does not affect enzyme synthesis if the Pfr level is low.

In dichromatic light experiments high-irradiance RG9-light
($\varphi_{RG9} < 0.01$) or R ($\varphi_R = 0.8$) were given in addition to B to
determine the Pfr/Ptot ratio (φ) and thus the Pfr level during the B treatment. The major results are shown in Table 1.

For any increase of the enzyme level beyond 6 d after the
onset of the experimental period light is indispensable (lines
1,2). In seedlings pretreated with B (6 d B) a similarly
strong increase of enzyme level is observed in R and B (lines
3,4). When the level of Pfr was kept very low in B - with si-

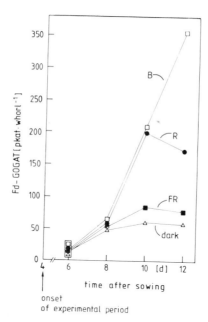

Fig. 2.Time course of Fd-GOGAT (= Ferredoxin-dependent glutamate synthase) levels in the first (cotyledonary) whorl
of Scots pine seedlings. Seedings were grown in perlite
on 15 mM KNO_3 and kept in continuous red (R, 6.8 Wm^{-2}),
blue (B, 10 Wm^{-2}) or far-red (FR, 3.5 Wm^{-2}) light.
(From Elmlinger and Mohr, 1991)

multaneously given RG9-light ($\varphi_{sim} < 0.05$) - no inductive effect of B was seen (line 6). When the level of Pfr was kept
high in B - with simultaneously given R ($\varphi_{sim} > 0.7$) - the
same high rate of enzyme synthesis was observed as in B and R
alone (line 5). These results show that B does not affect enzyme synthesis if the Pfr level is low. On the other hand, in
a seedling which has never seen B, phytochrome is not capable
to maintain enzyme synthesis beyond 10 d after sowing. It is
concluded that B/UV-A light has no direct effect on Fd-GOGAT
gene expression. Rather, its action appears to be restricted
to establishing responsiveness towards Pfr. This type of coaction between phytochrome and B/UV-A light appears to be widespread since it was observed previously in quite a number of
photoresponses in angiosperms (see Mohr, 1986, for review).
    While our data exclude the indepedent operation of B and R
at the level of *cis*-acting factors, we cannot distinguish at
present between a coaction of *cis*-elements and a coaction during signal transduction, even though the latter appears very
probable.

Table 1. Action of dichromatic irradiation on appearance of Fd-GOGAT (= Fd-dependent glutamate synthase) in the cotyledonary whorl of Scots pine seedlings (*Pinus sylvestris* L.). Seedlings were grown in perlite without nitrate and pretreated with continuous blue light (B) between 4 and 10 days after sowing (see Fig. 2). Dichromatic treatment was then applied for 2 days. Whorls were assayed 8 d after the onset of the experimental period (12 d after sowing). The values in the table are means $\pm$ SE from five to eight independent experiments. (From Elmlinger and Mohr, 1991)

Light treatment (after onset of experimental period)	Fd-GOGAT [pkat $\cdot$ whorl^{-1}]
(1) 6 d B	156 $\pm$ 6
(2) 6 d B + 2 d dark	130 $\pm$ 2
(3) 6 d B + 2 d B	251 $\pm$ 3
(4) 6 d B + 2 d R	259 $\pm$ 5
(5) 6 d B + 2 d [B + R]	254 $\pm$ 3
(6) 6 d B + 2 d [B + RG9]	155 $\pm$ 4
(7) 6 d B + 2 d RG9	147 $\pm$ 2
(8) 6 d R	152 $\pm$ 5
(9) 6 d R + 2 d R	144 $\pm$ 6
(10) 12 dark	32 $\pm$ 2

B, blue light (10 Wm^{-2}); R, red light (20 Wm^{-2}); RG9, long wavelength far-red light (20 Wm^{-2}); [B + R], φ > 0.7; [B + RG9], φ < 0.05

COACTION OF PHYTOCHROME AND NITRATE IN PLASTIDIC NITRITE REDUCTASE (NIR$_2$) GENE EXPRESSION IN MUSTARD (SINAPIS ALBA L.)

This key enzyme in the nitrate-assimilation pathway catalyzes the reduction of nitrite to ammonia inside the plastid. The enzyme is nuclear-encoded. While synthesis of NIR protein depended largely on nitrate (Fig. 3), the levels of in-vitro-translatable NIR$_2$ mRNA were found to be controlled by phytochrome only (Fig. 4). It appears that phytochrome strongly stimulates the level of NIR$_2$ mRNA while significant enzyme synthesis takes place only in the presence of relatively large amounts of nitrate. Since an increased enzyme level was strictly correlated with an increase of immunoresponsive NIR protein it is improbable that activation of a precursor plays a role. Rather, it is concluded that, in situ, nitrate controls translation. As expected, a photooxidative treatment of the plastids (Fig. 4, R-NF/KNO$_3$, $\square$) decreases the level of NIR$_2$ mRNA below the limits of detectability. Thus, the plastidic factor dominates the scene also in case of NIR$_2$, i.e. in the absence of the plastidic factor gene expression is blocked, and phytochrome is totally ineffective.

Regarding the mode of coaction between phytochrome (Pfr) and nitrate in bringing about NIR$_2$ synthesis a sequential action is obvious: coarse control of the appearance of translatable NIR$_2$ mRNA operates through phytochrome while fine tuning of gene expression, i.e. the actual appearance of the NIR$_2$ protein, is controlled by nitrate.

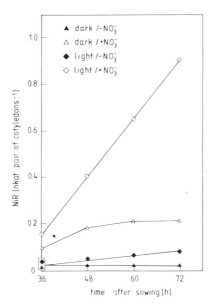

Fig. 3. Time course of nitrite reductase (NIR) isolated from the
cotyledons of mustard seedlings between 36 and 72 h
after sowing. One sees that significant enzyme synthesis
takes only place if nitrate is provided. However, *full
gene expression* of NIR requires the coaction of nitrate
and light (from Schuster and Mohr, 1990). Full gene
expression means the apearance of a final direct gene
product, a protein, active at its physiological site of
action (Lamb and Lawton, 1983)

COACTION OF LIGHT, NITRATE AND A PLASTIDIC FACTOR IN CONTROLLING NITRITE-
REDUCTASE GENE EXPRESSION IN SPINACH

The question has been whether the results obtained with
mustard can be generalized. It was decided to study control of
NIR synthesis in spinach (*Spinacia oleracea* L.) seedlings. Spinach is unrelated to mustard, and some information on spinach
NIR was already available. Back et al. (1988) reported the
first isolation of a cDNA clone coding for spinach NIR and
showed that the steady-state mRNA level of this NIR gene was
markedly increased in response to nitrate and light. In the
present study we used the NIR cDNA, pCIB 400, from spinach
(Back et al., 1988) as a probe.
    The major results we have obtained are the following: (i)
The light effect on the appearance of NIR activity occurs
through phytochrome. No specific blue-light effect is involved.
(ii) Immunotitration data indicate that light affects the
appearance of NIR by inducing the de-novo synthesis of the NIR
protein. (iii) A multiplicative relationship exists between the
action of nitrate and light on NIR apearance (Fig. 5). This
indicates that the actions of light and nitrate are indeed
independent of each other but that both factors operate on the
same causal sequence. (iv) Anion-exchange chromatography
revealed only a single form of NIR in spinach. Experiments
involving plastid photooxidation indicate that this NIR is
exclusively plastidic. (v) Northern blot analysis of NIR mRNA
showed a strong increase of the steady-state level in the

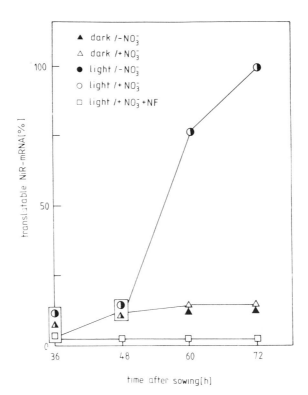

Fig. 4. Time course of the amounts of translatable NIR_2 mRNA in
the cotyledons of mustard between 36 and 72 h after
sowing. One sees that nitrate has no effect on the NIR_2
mRNA level even though nitrate is required to allow synthesis of the NIR protein. □ , in the presence of Norflurazon, i.e. when chloroplast becomes photodamaged,
the NIR_2-mRNA remains below detectability. (From Schuster and Mohr, 1990)

presence of nitrate whereas light had no effect (Table 2). NIR
mRNA was almost undetectable when the plastids were damaged by
photooxidation. It is concluded that NIR gene expression in
spinach requires positive control by a 'plastidic factor'.
Moreover, nitrate exerts a coarse control at the mRNA level
whereas fine tuning of NIR-protein synthesis is post-
transcriptional and is exerted by light, operating via
phytochrome.

Regarding the site of action of light and nitrate on NIR_2
gene expression it appeared that in mustard phytochrome controls transcription while nitrate controls translation
(Schuster and Mohr, 1990). The data from spinach rather indicate that in this species nitrate controls transcription while
the action of light (phytochrome) is strictly post-transcriptional (presumably translational).

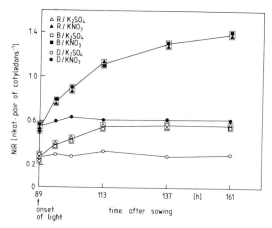

Fig. 5. Time courses of NIR activity isolated from the
cotyledons of spinach seedlings between 89 h and 161 h
after sowing. The seedlings were grown in darkness (D),
in red (R) or blue (B) light in the absence (○,△,□)
or presence (●,▲,■) of nitrate in the substrate. For
the first 89 h after sowing all seedlings were kept in
darkness. (From Seith et al., 1991)

Table 2. Quantitation of NIR-mRNA steady state levels after Northern blot analysis. RNA was extracted from the spinach
cotyledons 113 h after sowing. Seedlings were grown in
the absence or presence of nitrate (15 mM) in the substrate and kept in continuous darkness (D) or received
a 24-h red light treatment between 89 and 113 h after
sowing (R). (From Seith et al., 1991)

Treatment of seedlings	Filter-bound radioactivity [%]
113 h D/K$_2$SO$_4$	100
113 h D/KNO$_3$	360 ± 12
89 h D + 24 h R/K$_2$SO$_4$	97 ± 7
89 h D + 24 h R/KNO$_3$	360 ± 13

REFERENCES

Back, E., Burkhart, W., Moyer, M., Privalle, L., Rothstein, S., 1988, Isolation of cDNA clones codig for spinach nitrite reductase: Complete sequence and nitrate induction, <u>Mol. Gen. Genet.</u> 212:20

Elmlinger, M.W., Mohr, H., 1991, Coaction of blue/ultraviolet-A light and light absorbed by phytochrome in controlling appearance of ferredoxin-dependent glutamate synthase (= Fd-GOGAT) in the Scots pine (*Pinus sylvestris* L.) seedlings, <u>Planta</u> 183:374

Gilmartin, P.M., Sarokin, L., Memelink, J., Chua, N.-H., 1990, Molecular switches for plant genes, <u>Plant Cell</u> 2:369

Lamb, C.J., Lawton, M.A., 1983, Photocontrol of gene expression, <u>in</u>: "Encyclopedia of plant physiology", vol. 16 A, W. Shropshire and H. Mohr eds, Springer, Heidelberg, New York

Mohr, H., 1986, Coaction between pigment systems, <u>in</u>: "Photomorphogenesis in plants", R.E. Kendrick and G.H.M. Kronenberg eds, Nijhoff/Junk, Dordrecht

Oelmüller, R., 1989, Photooxidative destruction of chloroplasts and its effect on nuclear gene expression and extraplastidic enzyme levels, <u>Photochem. Photobiol.</u> 49:229

Oelmüller, R. Levitan, I., Bergfeld, R., Rajasekhar, V.K. Mohr, H., 1986, Expression of nuclear genes as affected by treatments acting on the plastids, <u>Planta</u> 168:482

Schuster, C., Mohr, H., 1990, Appearance of nitrite-reductase mRNA in mustard seedling cotyledons is regulated by phytochrome, <u>Planta</u> 181:327

Seith, B., Schuster, C., Mohr, H., 1991, Coaction of light, nitrate and a plastidic factor in controlling nitrite-reductase gene expression in spinach, <u>Planta</u> 184:74

ACTION SPECTRUM OF LHC-II PRECURSOR APOPROTEIN TRANSCRIPTION

AND COORDINATED TRANSLATION

Kostas Triantaphyllopoulos and Joan H. Argyroudi-Akoyunoglou

Institute of Biology, NRC " Demokritos ", Athens, Greece

INTRODUCTION

Light perception in higher plants involves, apart from Chlorophyll (Chl) and Protochlorophyll (PChl) pigments, active in photosynthetic light-energy transduction, one of the most extensively characterized regulatory photoreceptors, phytochrome. This binary molecule switches "on" and "off" plant gene expression in response to light (1). Phytochrome controls in a positive or negative way the expression of numerous genes, including its own (2-4). Phytochrome also acts as a synchronizer of the circadian clock regulating the accumulation of a number of mRNAs for chloroplast proteins (5). Since a heat-shock has been recently shown to act in a similar manner regarding the synchronization of the accumulation of LHC-II mRNA (6), we studied the action spectrum of LHC-II mRNA accumulation, as well as its reversal by Far-Red light, in an effort to ascertain the nature of the photoreceptor involved. In addition, and in order to study the possible correlation between transcription and translation of the LHC-II apoprotein, we also followed the appearance of the mature protein product (25 kDa) in LDS-solubilized total leaf protein extracts by immunoblot analysis.

MATERIALS AND METHODS

Nine to 10 day-old etiolated bean (Phaseolus vulgaris, red kidney var) leaves, attached to one cotyledon, were (a) irradiated with monochromatic light of equal total dose (1.17 mmoles/m^2) at various wavelengths, and then kept in the dark for 5 hours; or (b) were irradiated as in (a), kept in the dark for 5 hours, then irradiated by Far-Red light (1.17 mmoles/m^2), and kept in the dark for 19 additional hours. The leaves were then immersed in liquid Nitrogen, and their poly(A)mRNAs were isolated and translated in vitro on a wheat germ system in the presence of S-35 met (7,8). The poly(A)-mRNAs were also hybridized with nick-translated LHC-II c DNA (9). Aliquots of the translation products (containing equal TCA precipitable radioactivity) were immunoprecipitated against LHC-II antibody, and the immune complexes were analyzed by SDS-PAGE (10), and fluorography. Quantitation of fluorograms or autoradiograms was performed by scanning in a Joyce Loeble Chromoscan. -Leaves of the same age were treated as in (a), then kept in the dark for 24 hours, and then solubilized in 0.04 M Na_2CO_3-12% sucrose-2% LDS in an Omni-mixer (30 sec at 50% of the line voltage; 1 g fr w tissue/ml). The solubilized material was analyzed by Western blot analysis (11).

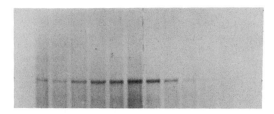

Fig. 1. Expression of the p-LHC-II apoprotein, as monitored by immunoprecipitation of in vitro translation products, obtained from poly(A)mRNA
isolated from leaves pretreated by light of equal total dose (1.17
mmoles/m^2) at the wavelengths indicated, and then kept in the dark
for 5 hours. Aliquots of translation products containing equal TCA-
precipitable radioactivity were used for immunoprecipitation.

PChl(ide) and Chl(ide) were determined in 80% acetone extracts of leaves. Monochromatic light was was provided by interference filters (Porschke
GmbH, 7-12 nm bandwidth).

RESULTS AND DISCUSSION

Figure 1 shows the immunoprecipitated 32 kDa precursor LHC-II apoprotein in in vitro translation products of poly(A)mRNA samples isolated from
leaves exposed to a light pulse of equal total dose at various wavelengths.
Etiolated leaves detached from plants 9-10 days after sowing were used,
since in such leaves the Red-light induced accumulation of LHC-II mRNA is
most pronounced (12). As shown, the LHC-II mRNA expression is most pronounced in the red-region of the spectrum.
Similar results were obtained in hybridization experiments of the isolated poly(A)mRNAs with LHC-II c DNA, as well as in fluorograms of SDS-PAGE
electropherograms of total in vitro translated products.

The action spectrum obtained from a number of such experiments is shown
in Figure 2. This action spectrum, as shown, coincides closely with the absorption spectrum of the Pr form of the phytochrome protein. Furthermore,
the action spectrum of the reversal of the LHC-II transcript accumulation by
Far-Red (applied to the plants kept in the dark for 5 hours after the initial irradiation at various wavelengths), coincides with the difference absorption spectrum of Pr minus Pfr (ΔA). This action spectrum was estimated by
subtracting the % LHC-II mRNA accumulated after the Far-Red light pulse from
that accumulated prior to the treatment by Far-Red, at the respective wavelengths. All values were normalized to the value found in the Red (650 nm).
As shown, the action spectrum of the transcription reversal by Far-Red light
is similar to the difference absorption spectrum reported by Viestra and
Quail (1983) for purified phytochrome protein (Pr-Pfr, 13).

These results suggest that light-induced LHC-II mRNA accumulation is
strongly wavelength dependent, the action spectrum of transcription resembling closely the absorption spectrum of phytochrome. Thus, phytochrome is
most likely the mediator involved in CAB gene expression.

In order to answer the obvious question as to whether the translation
of the protein is in any way correlated to the accumulation of its trans-

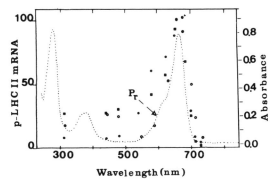

Fig. 2. Action spectrum of LHC-II mRNA transcription, as reflected by the
expression of the p-LHC-II apoprotein in poly(A)mRNA total _in vitro_
translation products (o--o); immunoprecipitates of such products
(o--o); and dot-hybridized complexes with nick-translated LHC-II
c DNA (z--z). The absorption spectrum of Pr is included (13).

cript, we further studied by Western blot analysis the appearance of the
mature 25 kDa apoprotein in total LDS-solubilized leaf protein extracts.
Figure 4 shows the results of such analyses for total LDS solubilized leaf
proteins of etiolated bean leaves (E) prior to light exposure, of such leaves after exposure to a Red light pulse (650 nm, 1.17 mmoles/m2) and then
kept in the dark for 24 additional hours (R), and of mature green leaves (G).
It is evident that the LHC-II immunodetected protein is very low in the etiolated leaf and appears after the red-light pulse. The immunodetected product
has similar molecular weight to that present in mature bean leaf extracts.
This suggests that the LHC-II apoprotein translation closely follows the
transcription of the CAB gene, and in addition, that its import into the
plastid as mature 25 kDa product takes also place. In addition, we found
that the translated protein abundance depends highly on wavelength of preexposure, and correlates closely with the accumulation of its transcript (see
Figure 5), As shown, translation is enhanced at 650 nm.

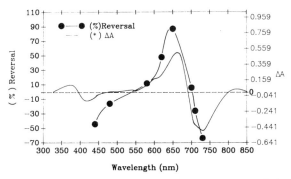

Fig. 3. Action spectrum of LHC-II transcription reversal by Far-Red light.
The difference absorbtion spectrum of purified phytochrome (Pr-Pfr)
is shown for comparison (13). The action spectrum was estimated by
subtracting the % LHC-II mRNA accumulated after Far-Red irradiation
from that accumulated prior to Far-Red exposure, in plants pretreated at various wavelengths. (see Materials and methods).

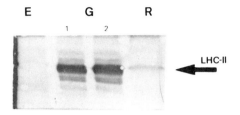

Fig. 4. Western blot analysis of LHC-II apoprotein in total LDS-extracts
of leaf proteins obtained from etiolated bean leaves prior to (E)
or after exposure to a red-light pulse (650 nm, 1.17 mmoles/m^2),
and then kept in the dark for 24 hours (R), and mature green leaves
(G_1, G_2). E,R: 100 ug leaf protein; G_1: 50 ug, G_2: 100 ug protein.

Since, as described above, the mature apoprotein appears in the leaf
extracts, and assuming that the mature protein is a thylakoid component,
the question arises whether the Western blot analysis monitors the stabilization of the LHC-II protein in thylakoids by the Chlide formed during the
light pulse, rather than its translation. This question was further strengthened by experiments showing that the action spectrum of PChlide phototransformation also follows closely the action spectrum of CAB transcription and of the LHC-II apoprotein appearance. Figure 5 shows representative
experiments of PChlide phototransformation in leaves exposed to various wavelengths. The PChlide photoconversion and Chlide formation are enhanced at
wavelengths of light in the red region of the spectrum. It is therefore,
difficult to conclude, solely on the grounds of these experiments, whether
the appearance of the apoprotein in the leaf extracts under the conditions
of these experiments, refelects translation of the protein or stabilization
by the Chlide formed.

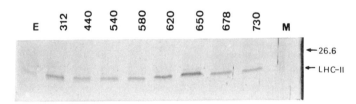

Fig. 5. Action spectrum (effect of Wavelength of light exposure) of LHC-II
apoprotein appearance in total LDS-solubilized leaf protein extracts
Etiolated leaves exposed to a pulse of light (1.17 mmoles/m^2) of
indicated wavelengths, and then kept in the dark for 24 hours were
used. All samples represent 100 ug leaf protein. E: etiolated leaf.
M: Marker proteins.

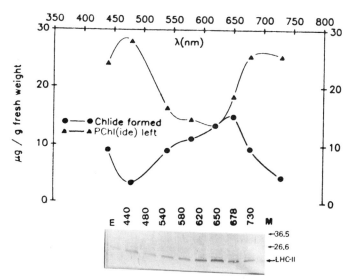

Action spectrum of PChlide phototransformation and LHC−II stabilization

Fig. 6. Action spectrum of PChlide photoconversion and Chlide formation, in etiolated bean leaves upon exposure to light (1.17 mmoles/m^2) of various wavelengths. The leaves were kept in the dark after illumination for 24 hours, and then the pigments were extracted and their concentration was determined.

Results in our laboratory (14), however, have shown that the appearance of the LHC-II 25 kDa apoprotein in LDS-solubilized total leaf protein extracts of etiolated leaves, after a Red-light pulse and subsequent incubation in the dark, oscillates in a circadian rhythm, in a way similar to that of its transcript. Since in this case, the Chlide formed <u>via</u> the Red-light induced photoconversion of PChlide is expected to be constant throughout the subsequent dark period, it is most likeley that the appearance of the LHC-II apoprotein is independent of the presence of Chl. This allows us to conclude that the abundance of the protein in Western blots of leaf extracts, under the conditions of our experiments, reflects the translation of the protein rather than its stabilization by the Chlide vailable. Translation of the LHC-II mRNA seems, therefore, to be closely coordinated to the transcription of its gene.

ACKNOWLEDGEMENT: ' The partial support by a Volkswagen grant to JHA and K. Kloppstech is gratefully acknowledged.

REFERENCES

1. E. Tobin and J. Silverthorne, Light regulation of the gene expression in higher plants, Ann. Rev. Plant Physiol. 36: 569 (1985).
2. A. H. Christensen and P.H. Quail, Structure and expression of a maize phytochrome-encoding gene, <u>Gene</u>, 85: 381 (1989)
3. K. Apel, Phytochrome-iduced appearance of mRNA activity for apoprotein of the light-harvesting chlorophyll a/b protein of barley (<u>Hordeum vulgare</u>), <u>Eur</u>. <u>J</u>. <u>Biochem</u>., 97: 183 (1979).

4. K. Apel, The Protochlorophyllide holochrome of barley. Phytochrome induced decrease of translatable mRNA coding for the NADPH-protochlorophyllide oxidoreductase, Eur. J. Biochem, 120: 89 (1981).

5. P. Tavladoraki, K. Kloppstech and J. H. Argyroudi-Akoyunoglou, Circadian rhythm in the expression of the mRNA coding for the apoprotein of the light-harvesting complex of Photosystem II. Phytochrome control and persisten Far-Red reversibility, Plant Physiol., 90: 665 (1989).

6. K. Kloppstech, B. Otto and J. Beator, Heat-induced "photo"-morphogenesis in dark-grown plants and circadian rhythmicity.-Is there a connection?, NATO ASI, this volume.

7. K. Apel and K. Kloppstech, Light-induced appearance of the mRNA for the apoprotein of the light-harvesting chlorophyll a/b protein, Eur. J. Biochem., 85: 581 (1978).

8. B. E. Roberts and B.M. Paterson, Efficient translation of tobacco mosaic virus RNA and rabbit globin 9S RNA in a cell free system from commercial wheat germ, Proc. Natl. Acad. Sci., USA, 70: 2330 (1973).

9. T. Maniatis, E.F. Fritsch and K.J. Sambrook "Molecular cloning, A laboratory manual", Cold Spring Harbor Laboratory Press, N.Y. (198).

10. S. Cullen and B.D. Schwartz, An improved method for isolation of H-2 and Ia alloantigens with immunoprecipitation induced by Protein-A-bearing Staphyloccoci, J. Immunol., 117: 136 (1976).

11. M. S. Blake, K.H. Johnston, G.J. Russell-Jones and E.C. Gotschlich, A rapid, sensitive method for detection of alkaline phosphatase-conjugated anti-antibody on Western blots, Anal Biochem., 136:175 (1984)

12. P. Tavladoraki, G. Akoyunoglou, A. Bitsch, G. Meyer and K. Kloppstech, Age and phytochrome-induced changes at the level of the translatable mRNA coding for the LHC-II apoprotein of Phaseolus vulgaris leaves, in: "Regulation of chloroplast differentiation", G. Akoyunoglou and H. Senger, eds., Alan R. Liss, N.Y., (1986).

13. R. D. Viestra and P. Quail, Purification and initial characterization of 124-kilodalton phytochrome from Avena, Biochemistry, 22: 2498 (1983).

14. T. Bei-Paraskevopoulou and J.H. Argyroudi-Akoyunoglou, Circadian Rhythm in light-harvesting protein II mRNA transcription, translation, and Protochlorophyllide regeneration, NATO ASI, this Volume.

HEAT–INDUCED "PHOTO"MORPHOGENESIS IN DARK GROWN PLANTS AND CIRCADIAN RHYTHMICITY – IS THERE A CONNECTION?

Klaus Kloppstech, Beate Otto and Jens Beator

Institut für Botanik
Universität Hannover
3000 HANNOVER 21 FRG

INTRODUCTION

It can be deduced from the multiplicity of observations in this direction that circadian rhythms will have great impact on plant development although this is not self–evident. It may be asked why it should not be sufficient to synchronize plant development with the oscillating environmental factors and with the demands of plants for an interaction with the living surrounding via diurnal regulation. The observation of fluctuations at the level of mRNAs for three light–regulated genes (Kloppstech, 1985) added at the first glance just one more parameter to measure: a molecular handle whose oscillations were somewhat more complicated and more expensive to analyze but also closer to the basal level of gene expression. Therefore, a lot of data have been collected along the traditional line of circadian measurements although for instance phase–shifting had been used for carrying out the original experiments. Essential contributions in this new field were the observation that run–off transcripts for light–inducible genes are under circadian control, indicating that it is primarily but not exclusively transcription control which is exerted by the clock (Giuliano et al., 1988).

Originally it was observed that the increase in the levels of mRNAs occurred already prior to the dark–light–transition. This indicated that phytochrome control (Pfr) of mRNA levels as shown before (Apel, 1979; Tobin, 1981) could not be a direct one. Of great impact are therefore the measurements by Tavladoraki et al., (1989) in which it was shown that one signal: continuous light or alternatively a single red light pulse could induce circadian oscillations in the CAB gene products and that light of 730 nm given immediately after the pulse could reverse the effect. Moreover, a 660 nm pulse given out of phase could induce a phase shift in the oscillations. This is a clear indication that phytochrome not only interferes with the handles of the clock but with the circadian oscillator itself. This finding may offer new access not only to the understanding of the phytochrome control including signal transfer between the receptor and the genes but also to the mechanism of the circadian clock itself. This option was seen in several laboratories. Consequently the promoter deletion constructs existing in transgenic plants which were originally designed to unravel light–regulated promoter elements have been reused to analyze for circadian control elements of promoters (Nagy et al. 1988, Fejes et al., 1990). It was found that light enhancement of mRNA levels and circadian control of transcription are located on different regions of the promoter.

More recently, it was observed that light regulated mRNAs do not oscillate in parallel but with individual maxima which are typical for the nuclear coded members of one particluar photosynthetic complex like PS I or PS II (Schneiderbauer et al., unpublished data). Another example in this direction is nitrate reductase, the first example of an oscillating nuclear coded non–plastid gene product (Deng et al, 1990).

Light control of the oscillator itself and simultaneously of transcript levels caused the problem to discriminate between the two effects of light. To circumvent this obstacle heat shock was chosen as another Zeitgeber as it was known from the early work of Bünning (Stern and Bünning, 1929) that temperature shifts can cause phase shifts in the circadian rhythm. The

Regulation of Chloroplast Biogenesis
Edited by J.H. Argyroudi-Akoyunoglou, Plenum Press, New York, 1992

effects of heat-shock but also temperature-shift treatments of dark-grown plantlets were unexpected in that considerable morphogenetic changes occured which resembled closely photomorphogenetic effects obtained after irradiation of plants at 730 nm. Furthermore, in these plantlets mRNA levels of light-regulated proteins were under circadian control (Kloppstech et al., 1991). Here we describe in more detail the effects of heat-shock on morphology and light-induced greening.

METHODS

Growth of plantlets

Pea seeds (Pisum sativum, cv Rosa Krone) were imbibed for 20 h, sown on vermiculite and kept in the dark for up to 7 days. Plantlets received 1 hr heat-shock treatments from 7 to 8 am. Applied temperatures and number of treatments were as indicated in the results section. Illumination started 24 h after the last heat-shock at 20 W/m^2 for the times indicated in the legends. Plantlets were harvested in aliquots of 6 g, frozen in liquid nitrogen and stored at -80°C.

Extraction of proteins

Plantlets were ground under liquid N$_2$, transferred into extraction buffer (1mM MgCl$_2$, 10 mM Hepes, 20 mM Tricin/KOH, pH 7.2, 2% SDS) and heated for 10 min at 65°C. Cell debris was removed by centrifugation (20.000 x g for 10 min at 20°C) and proteins precipitated 2x with 2 vol 70% acetone at 4°C. The dried pellet was solubilized at 65°C in sample buffer (6% lithium dodecylsulfate, 150 mM Tricine/ KOH pH 7.8, 150 mM DTE, 0.02% bromophenol blue, 30% glycerol) and separated on 12.5% PAGE (Neville, 1971). Proteins were transferred to nitrocellulose membranes. Western blotting was performed as described (Adamska and Kloppstech, 1991).

Extraction of poly(A)RNA and dot blotting

Extraction of poly(A)RNA, quantification, transfer to nylon membranes and dot blotting have been described (Kloppstech et al., 1991).

Extraction and separation of chlorplast pigments

The frozen plant material was ground in liquid N$_2$ and extracted 3x with acetone and transferred into ether (Evangelatos and Akoyunoglou, 1981), dried in vacuo, solubilized in chloroform, and separated on TLC (HPTLC Kieselgel 60, Merck) in petrolether, ether, chloroform, methanol, acetone (40:10:10:5:5 per volume). The pigment bands were removed with a spatula and pigments extracted from the silica gel in acetone. The pigments were characterized with regard to quantity and absorption maxima by spectrophotometry.

Electron microscopy

Samples were treated for electron microscopy as described previously (Kloppstech et al., 1991).

RESULTS

Heat-induced morphogenesis is a universal trait of plant development

The "photo"morphogenetic effect of heat-shock treatments in the dark is not restricted to pea but can be evoked in barley (Beator and Kloppstech, this volume) or even more impressively in maize (Fig. 1). The size of the heat-shocked, dark-grown plantlets is that of the green controls and the leaves are developed as in the green controls although they do not contain chlorophyll. The tremendous effect that light has on the reduction of the length of the mesokotyl is also observed after heat-shock. Taken together the observations indicate that heat can substitute for light in a considerable part of plant development but not in others as for instance the synthesis of chlorophyll. The latter finding somehow excludes the possibility that synthesis of chlorophyll in the dark, as observed in other species (Adamson et al., 1985), does occurs after the application of heat-shocks.

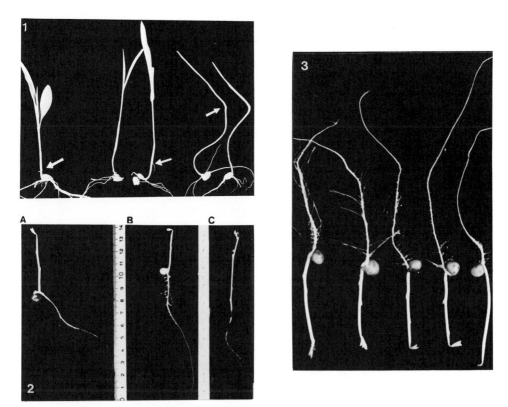

Fig. 1. Changes in morphology of etiolated maize seedlings after cyclic heat–shock treatment.
Maize was grown for 7 days in the dark (right), in the light (left), or in the dark with a
daily 1 h heat–shock at 40°C (middle). The heat–shocked plants are of the same length
as the light grown plantlets, have well developed leaves and a mesokotyl of the same
length as the green plants.

Fig. 2. The minimal heat–shock conditions for induction of morphogenesis
Pea seeds were sown on vermiculite and grown in the dark for 7 days. The seedling on
the left received a 1 hour treatment at 35°C (A). The seedlings in the middle (B) and on
the right (C) were treated by a daily heat–shock of 15 min or 30 min at 40oC,
respectively. For. an etiolated control see figure 3.

Fig. 3. The effect of the number of heat–shock treatments
Pea seedlings were grown as before. The plant on the left is a control grown at constant
conditions in the dark at 25°C for 7 days. Plants 2–5 (counted from the left) received 1,
2, 3, or 4 heat–shock treamtents, respectively. The treatments were started at day one
after sowing. Thereafter the temperature was kept constant.

Investigation of the conditions for induction of morphogenesis by heat–shock

Reduction in the length of the daily heat-shock for induction of morphogenesis was possible to 15 to 30 min per day and caused comparable effects. This is the shortest time manageable with the equipment (Fig. 2). The temperature at which the response could be evoked was around 35°C that is about 10°C above the normal growth temperature (Fig. 2). This is about the same temperature difference that induced morphogenesis in 12h:12h temperature shift experiments in which the temperature changed between 25°C and 15°C (data not shown).

Another question in this connection was the number of heat treatments necessary to induce the effect. As shown if figure 3 a single heat-shock at day one after sowing has a slight effect on morphogenesis, however, with an increasing number of heat treatments the effect on morphogenesis becomes more pronounced. There is also a developmental aspect in the induction of morphogenesis as a single heat-shock during a 5 day period of growth exerts its greatest effect at days 3 and 4 after sowing (data not shown). At this particular time of development chloroplast biogenesis starts and transcription of stress genes by heat-shock becomes inducible.

Acceleration of light–induced chloroplast development after heat–shock treatments

Higher levels of mRNAs for light-inducible proteins after heat-shock might lead to an acceleration of greening. One would expect that this should be accompanied by correspondent changes in the ultrastructure. This is the case. In figure 5 plastid ultrastructure is compared between etiolated and heat-shocked, dark grown plantlets. The difference in the structure after one hour of illumination is only a slight, but visible one. While after 6 hours of light plastids from heat-treated plantlets have developed grana stacks which resemble very closely those of the green controls. Plastids of etiolated plantlets after the same period of illumination still show typical etiochloroplasts.

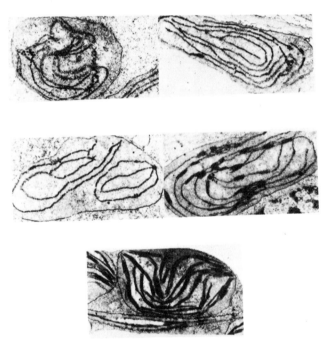

Fig. 4. Light–induced changes of chloroplast morphology after cyclic heat–shock.
Pea seedling were grown for 7 days in the dark either at constant temperature or with a daily heat-shock and thereafter transferred to white light. Samples were taken after 1 hour of illumination (upper row) or after 6 hours (middle row) and prepared for electron microscopy. Untreated controls are shown on the left samples from heat-treated plantlets on the right. Lower section: green control. In comparison to the etiolated controls membrane stacking occurs much faster in the heat-shocked plants.

prominent change occurs in a band of 8 kDa which is negatively regulated by light. This translation product is high in the 0 h samples of controls and declines with time while the correspondent band is almost absent from the heat-shocked samples (data not shown). Examination by dot blotting (Fig. 5) confirms these findings. In contrast to LHC II and ssRubisco ELIP mRNAs are non-induced after cyclic heat-shock but can be induced by 1 hr of illumination. There is almost no effect during this 1 hr illumination in the levels of LHC II and ssRubisco mRNAs.

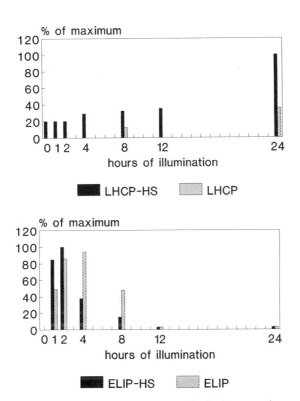

Fig. 5. Influence of cyclic heat-shock on mRNA levels during greening.
Plants were grown as indicated in figure 2 either under constant conditions or with a daily heat-shock treatment. Plant material was collected at the indicated times, poly(A)-RNA extracted and hybridized to the random primed cDNAs. The maximal values obtained with each of the probes were set to 100%.

Accumulation of light-inducible mRNAs during greening of heat-shocked plantlets

Acceleration of chloroplast formation should go together with an accelerated accumulation of light-inducible mRNAs in the heat-shocked plantlets. The translation pattern of mRNA as obtained by electrophoretic separation of translation products shows that mRNA for LHC II is already present in the 0 h sample of the heat-shocked plants and that the level for small subunit of Rubisco is higher in the heat-shocked plantlets than in the controls. The most

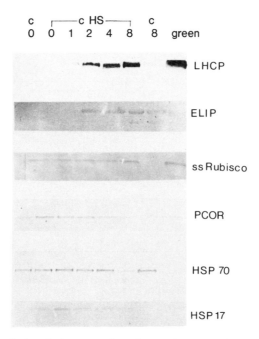

Fig. 6. Protein accumulation during greening after cyclic heat-shock.
Samples were taken during greening, proteins extracted, separated by PAGE, transferred to nitro-cellulose and quantified with the correspondei t antibodies. Samples were loaded according to equal amounts of fresh matter. C: etiolated control and CHS: cyclically heat-shocked plantlets at the indicated times of illumination. Green: 7-d-old green control.

However, upon illumination for prolonged periods mRNA for LHC II and ssRubisco are increasing much faster in the heat-shocked plantlets than in the etiolated controls.

Protein accumulation during greening

Elevated levels of mRNA in the dark and during illumination are indicative for an accelerated accumulation of the correspondent proteins. The western blot (Fig. 6) shows that this is the case. Levels for LHC II are much higher after 8 hrs of light in the cyclically heat-shocked plantlets than in the etiolated controls after the same period of illumination. Similar findings have been obtained for ELIP. The ssRubisco is already high in the dark in cyclically heat-shocked plantlets while even after 8 hrs of light in the etiolated plantlets ssRubisco is hardly detectable. Immunoreactive HSP 70 is high in both types of plantlets but decreases with light while HSP 17 seems to be specific for the heat-shocked plantlets. This indicates that the daily heat-shock leads to the expression and accumulation of the correspondent mRNA and its protein.

Pigment accumulation during greening

Together with the increasing levels of protein also an accelerated accumulation of chlorophylls a and b could be observed (Fig. 7). The lag which is typical for the etiolated plantlets is completely abolished. Furthermore a rapid decrease in the ratio of chlorophyll a/b is found indicating that chlorophyll b accumulates much faster after heat-shock.

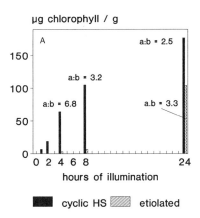

μg chlorophyll / g

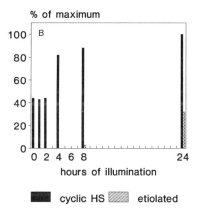

% of maximum

Fig. 7. Changes in pigment composition during greening after cyclic heat–shock
Etiolated and cyclically heat-shocked plants were illuminated for the indicated times and pigments extracted, separated and qunatified as described in the methods section. A: Chlorophylls; the values of chlorophyll a/b are included. B: Lutein.

The leaves of heat-shocked plantlets appear heavily yellow. Chromatography on silica gel revealed that the enriched pigment is lutein. Its level increases twofold during greening. In the etiolated controls lutein starts to be detectable after 8 hours but after 24 hrs of greening the amount is only one third of that of the heat-shocked plantlets.

DISCUSSION

Heat shock can serve as a substitute for light as circadian signal at least as far as starting the rhythm is concerned, whether heat-shock can also cause phase-shifts remains to be shown. As circadian "Zeitgeber" heat entrains a rhythmicity for positively light regulated genes (Kloppstech et al., 1991) as well as for negatively regulated genes (Deng et al., 1990; Beator and Kloppstech, this volume). The consequences of heat-shock on plant development seem to be of high relevance as can be concluded from the fact that the morphology of plants is influenced into the direction of light-induced development. To achieve similar morphological changes by light daily 2 minute pulses at a light intensity of 10 W/m^2 are necessary, in the latter cases the morphology of the leaves but not that of the stem is changed to the same extent as via a heat-shock. So far the mRNA levels for most of the genes tested oscillate in a circadian manner either in phase (positively light regulated proteins) or with an adversed phase (negatively light regulated genes). Although the number of the investigated probes is still small it seems likely to assume that all genes that have been found under circadian control so far will also oscillate in the heat-shocked plantlets. Therefore we would like to suggest as a working hypothesis that the primary function of light during development at least of etiolated plants is to set and phase-shift the circadian clock. There exists ample evidence from many laboratories that light will have other effects, too. Evidence for this is also presented in this and the accompanying paper (Beator and Kloppstech). While the capacity to synthesize chlorophyll oscillates in a circadian way its synthesis remains light-dependent. Similarly, at least in pea ELIP mRNA levels are under the control of light, although again the capability for light inducibility changes with the day.

Furthermore, as is known for a long time, light has also an enhancing effect on the levels of transcripts. This effect of light is retained in the heat-shocked plantlets, however, the light-stimulated increase is much faster. Again heat-shocked plantlets behave as plantlets that have received a pre-illumination signal. After a short light signal, however, circadian rhythm will be started or synchronized as has been shown by Tavladoraki et al (1989). Taken together all the available evidence it seems suggestive to propose a central role of circadian rhythmicity during plant development. This view is supported by recent publications in which it has been shown recently (Deng et al., 1990; Adamska et al., 1991) that not only the mRNA levels but also the levels of the correspondent proteins oscillate. Messengers for the light-inducible genes that were investigated are initiated into polysomes and translated into proteins. The consequence of this oscillation is that the pattern of protein synthesis in a cell changes dramatically during the day and our data and those of Deng et al. (1990) suggest that at night other proteins are synthesized at a high rate.

Finally, it should be emphasized that also the capacity for the synthesis of chlorophylls during a given period of illumination oscillates tremendously and in phase with the maximum of LHC II expression. This clearly indicates that the control of the assembly of chlorophyll protein complexes is coupled to the circadian oscillator.

The picture is even more complex as we have shown a long time ago in the barley leaf that the expression of proteins is also under developmental control (Viro and Kloppstech, 1980). This means that within this barley leaf LHC II is only expressed in cells 2 to 5 days after the emergence from the meristem and only during day time; thereafter the predominant expression of this gene is switched off. One might assume that a fine tuning according to the demands of the plant (light versus shade leaves, high light stress) is still possible. But the existence of such fine tuning remains to be proven.

REFERENCES

Adamska I, Kloppstech K (1991) Plant Mol Biol 16: 209–223.
Adamska I, Scheel B, Kloppstech K (1991) Plant Mol Biol: in press.

Adamson H, Griffiths T, Packer N, Sutherland M (1985) Physiol Plant 64: 345-352.

Apel K (1979) Eur J Biochem 97: 183-188 .

Deng MD, Moreaux T, Leydecker MT, Caboche M (1990) Planta 180: 257-261.

Evangelatos GP, Akoyunoglou GA (1981) In Akoyunoglou GA (ed), Photosynthesis V. Chloroplast Development. Balaban Intern Science Services, Philadelphia.

Fejes E, Pay A, Kanevsky I, Szell M, Adam E, Kay S, Nagy F (1990) Plant Mol Biol 15: 921-932.

Giuliano G, Hoffman NE, Ko K, Scolnik PA, Cashmore T (1988) EMBO J 7: 3635-3642.

Kloppstech K (1985) Planta 165: 502-506.

Kloppstech K, Otto B, Sierralta W (1991) Mol Gen Genet 225: 468-473.

Nagy F, Kay SA, Chua NH (1988) Genes Dev 2: 376-382.

Neville DM (1971) J Biol Chem 246: 6328-6334.

Stern K, Bünning E (1929) Ber. Dtsch. Bot. Ges. 47: 565-584.

Tavladoraki P, Kloppstech K, Argyroudi-Akoyunoglou J (1989) Plant Physiol 90: 665-672.

Tobin EM (1981) Plant Mol Biol 1: 35-51

Viro M, Kloppstech K (1980) Planta 150: 41-45.

CIRCADIAN CONTROL OF GENE EXPRESSION AND MORPHOGENESIS BY HEAT SHOCK IN BARLEY

Jens Beator and Klaus Kloppstech

Institut für Botanik
Herrenhäuser Straße 2
3000 HANNOVER 21, FRG

INTRODUCTION

Light is probably the most important environmental factor for plant productivity and development. On one hand it serves as an indispensable energy source in photosynthesis whereas on the other hand it provides an important feedback from the environment for development and gene expression. Expression of many of these genes has been shown to be controlled by phytochrome (Tobin and Silverthorne, 1986). In recent years, however, it has become increasingly clear that gene expression is governed by a circadian oscillator (Kloppstech, 1985). Oscillations of mRNA levels have been shown to be controlled primarily at the level of transcription (Giuliano et al., 1988) and are pertained in transgenic plants (Nagy et al., 1988). Furthermore it was found by Tavladoraki et al. (1989) that in etiolated beans circadian oscillations in LHC II mRNA can be evoked and phase-shifted by the phytochrome system. As the physiological clock might be of major importance in control of gene expression we wanted to study the effect of other "Zeitgeber" than light. We therefore took advantage of the fact that circadian rhythms can be induced by changes in the ambient temperature (Stern und Bünning, 1929; Frosch and Wagner, 1973).

METHODS

Plant Material: Barley seeds (Hordeum vulgare) were grown in the dark at 25°C with a daily 1-hour heat shock (HS) of 40°C between 7 a.m. and 8 a.m.. The HS was omitted in etiolated controls. Green plants were grown under a 12h light/12h dark regime. For illumination plants were placed under fluorescent light at an intensity of 15 W/m^2.

Dot Blot Analysis: Poly(A)RNA isolation and hybridisation conditions were essentially as described by Kloppstech et al. (1991). Replica filters were prepared and autoradiography signals linear in intensity with exposure time were scanned for quantification. Homologous cDNA clones of LHC II, SSU, plastocyanin and thionin are a generous gift from Dr. K. Gausing, Aarhus (Barkadottir et al., 1987). The early light inducible protein (ELIP) was probed with the cDNA for a small ELIP (Grimm et al., 1989).

HPLC analysis: Pigments from the apical 6.5cm of weighed leaf material were extracted quantitatively with acetone, transferred into petrolether, washed with 4M NaCl, dried with Na_2SO_4 and evaporated under vacuum. Pigments were redissolved in acetone and analysed by reversed phase HPLC as described by Humbeck et al. (1988). The system was calibrated by injecting samples of increasing concentration, collecting fractions corresponding to peaks of interest and determining the amount of pigment present spectrophotometrically. Regression analysis showed good linearity with regression coefficients better than 0.99 in all cases.

<u>Western blot:</u> Weighed leaves were extracted quantitatively in sample buffer (56mM Na$_2$CO$_3$, 0.1M DTT, 2% SDS, 10% glycerol), heated at 65°C for 20 min and boiled for 5min. After separating insoluble debris by centrifugation bromophenol-blue was added to a final concentration of 0.05%. Electrophoresis was performed according to Neville (1971) and transfer of proteins to nitrocellulose as described by Towbin (1978).

RESULTS

<u>Morphological changes</u>: When plants are grown in the absence of light they respond by etiolation. In monocotyledonean species this results in an increased growth rate of the coleoptile and the primary leaf compared to light-grown plants. In addition leaves remain enrolled. Treatment of dark-grown barley seedlings with daily 1-hour heat shocks provokes considerable changes in their morphological appearance. Fig. 1 is a photograph taken after 6 days of growth. Except for the lack of pigmentation the cyclically heat shocked plants are strikingly similar to normally developed green plants: The lengths of leaves and coleoptiles are comparable to those of green plants and the primary leaves are partially unrolled. For comparison dark-grown barley seedlings are also depicted.

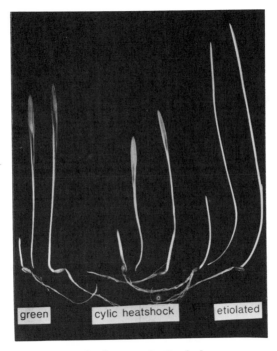

Fig. 1. Effect of cyclic HS on development in the dark.
Barley was grown as described in the methods section and representative plants of each group were taken for photography at day 6 after sowing.

Abbreviations: **LHC II**: light harvesting complex II, Chlorophyll a/b binding protein; **SSU**: small subunit of the Ribulose-1,5-Bisphosphate Carboxylase; **HS**: heatshock, **PCOR**: Protochlorophyllide – Oxidoreductase; **ELIP**: early light inducible protein, **HPLC**: high performance liquid chromatograghie

<u>Circadian oscillations of light-regulated mRNAs</u>: The effects on morphogenesis shown in Fig. 1 indicate that in etiolated barley heat can partially compensate for light. Light pulses operating via phytochrome are known to elevate mRNA levels of positively light-regulated genes, e.g. LHC II (Apel, 1979) and it has been shown by Tavladoraki et al. (1989) that in case of LHC II this is not a transient response but rather the beginning of persistent, i.e. circadian oscillations. Fig. 2 shows changes in mRNA levels of nuclear coded plastid proteins after cyclic HS treatment. After the last HS on the 5th day plants were kept in the dark at constant temperature. The circadian oscillations in the mRNA levels of LHC II, plastocyanin and SSU persist for at least 3 days. Maximal levels of LHC II are 5 to 10-fold higher than in etiolated plants (data not shown). The mRNA for the early light inducible protein (ELIP) shows a somewhat different pattern: the oscillation is rapidly dampened and after a strong signal 12 hours after the HS only a faint peak 24-28 hours after the first maximum is visible; no mRNA could be detected in etiolated controls (not shown). The first maximum is well defined in all cases and is reached either 12h (Plastocyanin, SSU, ELIP) or 16h (LHC II) after the last HS corresponding to 8 p.m. (= 20 circadian time) and 12 p.m. (= 24 circadian time), respectively. Except for ELIP and LHC II the following peaks are less well defined and appear to be biphasic. The genes described so far are positively light regulated and their products are localized in the chloroplast.

We also studied the influence of cyclic HS on the mRNA of thionin, a cytosolic/extracellular protein which is negatively light regulated (Barkadottir et al., 1987) and thought to be involved in plant defense (Bohlmann et al., 1988). The results (Fig. 3) show that the steady state level of this mRNA also undergoes circadian changes. Moreover, the phase of this oscillation is opposed to those observed for the positively light-regulated genes. Besides nitrate reductase (Deng et al., 1990) this is the only non-plastid gene product known so far to be under circadian control. The results described above were obtained with replica filters. As the oscillations of LHC II and thionin do not peak at identical times they serve as mutual controls that the changes are not due to mistakes in the quantification of mRNAs.

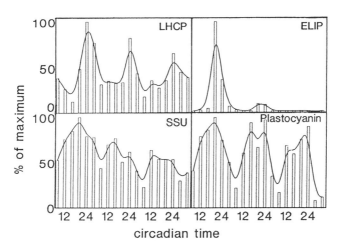

Fig. 2. Oscillations of mRNA levels of nuclear coded chloroplast proteins after cyclic HS in the dark.
After the the last HS at day 5 plants remained in the dark at constant temperature. Samples were taken every 4 hours beginning at 8 a.m. (= 8 circadian time) directly after the last HS, continuing for 3 days. Poly(A)RNA was extracted and dot-blot analysis performed. Autoradiography signals were scanned and the maximum set to 100% in each case.

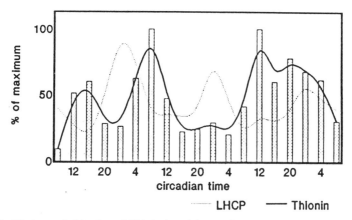

Fig. 3. Oscillations of thionin mRNA induced by cyclic heatshock treatment. Samples and filter preparation was identical to the description in the legend to Fig. 2.. The curve of LHC II is included for comparison.

Changes in the capacity of greening: In pea it was observed that cyclic HS treatment markedly accelerated light-induced accumulation of both pigments and thylakoid proteins of etiolated plants (Knack et al., 1990, Otto et al., unpublished). This effect could not be observed in identically treated barley (not shown). However, as the oscillation of mRNA levels in pea appear opposed to those observed in barley we studied greening at different times of the day. Fig. 4 shows that the capacity to accumulate chlorophylls during a 4 hour period of illumination changes dramatically during the day: Chlorophyll a varies about 2-fold during the day while the amount of Chlorophyll b differs more than 15-fold. Carotenoids do not show significant variations (not shown). In Fig. 5 we show that this observation can be extended to the LHC II complex: the maximal amount of protein accumulated coincides with maxima of chlorophylls a and b indicating the coordinate synthesis of the protein and its stabilizing pigments. As the maximum of the LHC II coincides with that of its mRNA circadian oscillations are apparently of high physiological importance.

We also studied the fate of a negatively light-regulated protein, Protochlorophyllide – Oxidoreductase (PCOR), under these conditions. Interestingly lower levels of PCOR are reached at times of maximal LHC II accumulation suggesting opposed regulation by the circadian oscillator as is the case for the Thionin mRNA.

DISCUSSION

Subjecting etiolated plants to daily heat shocks results in morphological changes similar to those induced by light, thereby mimicking to a large extent a typical "photomorphogenesis". These results are qualitatively similar to those described by Kloppstech et al. for pea (1991) where daily heat shocks also induced "photomorphogenesis"-like effects. Circadian oscillations of mRNA levels are by now a well established fact (Kloppstech, 1985; Nagy et al., 1988; Deng et al., 1990; Giuliano et al., 1988) and in addition to transcriptional control recent results point also to an involvement of differentially controlled turnover of mRNAs during dark or light phases (Fritz et al., 1991).

The induction of circadian oscillations of positively light-regulated mRNAs in the complete absence of light clearly underlines the importance of the physiological clock in gene expression. Starting or synchronizing the circadian oscillator is sufficient to activate these genes. Moreover, the results can be extended to negatively light-regulated genes showing oscillations opposed to those observed for "positively" regulated ones. This was shown at the mRNA level for thionin and at the protein level for PCOR.

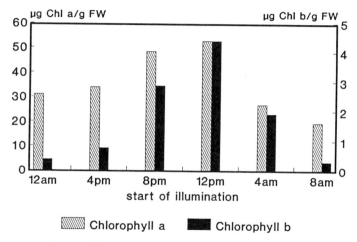

Fig. 4. **Changes in the capability to accumulate chlorophylls a and b after cyclic heat shock treatment.**
Five day old cyclically heat shocked plants were placed under fluorescent light of 15 W/m^2 for 4 h at the indicated times, harvested and frozen in liquid N_2. Pigments were extracted and analysed by reversed phase HPLC.

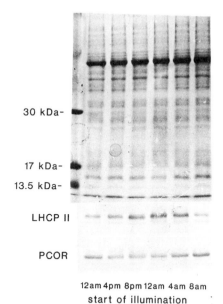

12am 4pm 8pm 12am 4am 8am
start of illumination

Fig. 5. **Changes in the capability to accumulate plastid proteins after cyclic heat shock treatment.**
Plants were identical to those described in Fig. 4, total proteins were extracted, separated by SDS–PAGE and transferred to nitrocellulose. A coomassie stained gel is shown in the upper part of the figure. Marker proteins are indicated. Filters were probed with antibodies against LHC II and PCOR.

The importance of a circadian synchronization of plants is also clearly shown in the greening experiments. This is evident especially in case of chlorophyll b, where the amount of pigment accumulated changes by a factor of 15 during the day and in parallel to the rise in the protein level (Adamska et al, 1991). These results suggest that there will be other important physiolgical responses changing on a circadian time scale as well. If this view is accepted our findings indicate that the concept of "photomorphogenesis" (Mohr, 1972) might need to be supplemented or re-defined. Our working hypothesis demands that at least part of the normal growth pattern (probably differing from species to species) is induced by light via setting or synchronizing of the circadian clock (Kloppstech et al., 1991). With the appropriate "Zeitgeber" "photo-"morphogenesis should be inducible in other plants as well. Another consequence of this hypothesis is that the effects of different illumination protocols described in the literature are in many instances a manifestation of a beginning oscillation pattern and might have to be reconsidered including aspects of circadian control.

REFERENCES

Adamska, I., Scheel, B. and Kloppstech, K. (1991) Plant Mol. Biol., in press.
Apel, K. (1979) Eur. J. Biochem. 97: 183–188.
Barkadottir, R.B., Jensen, B.F., Kreiberg, J.D., Nielsen, P.S. and Gausing, K. (1987) Develop. Genetics 8: 495–511.
Bohlmann, H., Clausen, S., Behnke, S., Giese, H., Hiller, C., Reimann-Philipp, U., Schreder, G., Barkholt, V. and Apel, K. (1988) EMBO J. 7:1559–1565.
Deng, M.-D., Moureaux, T., Leydecker, M.-T. and Caboche, M.(1990) Planta 180: 257–261.
Fritz, C.C., Herget, T., Wolter, F.P., Schell, J. and Schreier, P.H. (1991) Proc. Natl. Acad. Sci. USA 88: 4458–4462.
Frosch, S. and Wagner, E. (1973) Can. J. Bot. 51: 1529–1535.
Giuliano, G., Hoffman, N.E., Ko, K., Scolnik, P. and Cashmore, A.R. (1988) EMBO J. 7: 3635–3642.
Grimm, B., Kruse, E. and Kloppstech, K. (1989) Plant Mol. Biol. 13: 583–593.
Humbeck, K., Römer, S. and Senger, H. (1988) Botanica Acta 101: 220–228.
Kloppstech, K. (1985) Planta 165: 502–506.
Kloppstech, K., Otto, B. and Sierralta, W., (1991) Mol. Gen. Genet. 225: 468–473.

COORDINATION IN LHC-II mRNA TRANSCRIPTION, TRANSLATION AND CAPACITY FOR CHLOROPHYLL SYNTHESIS: CIRCADIAN RHYTHM IN ALL PROCESSES

Thalia A. Bei-Paraskevopoulou and Joan H. Argyroudi-Akoyunoglou

Institute of Biology, NRC Demokritos
Athens, Greece 153 10

INTRODUCTION

The stabilization and accumulation of the LHC-II (light-Harvesting) apoprotein in thylakoids during greening of chloroplast is under both transcriptional and post-translational control (Gallagher and Ellis, 1982; Akoyunoglou and Argyroudi-Akoyunoglou, 1986). Control at the transcriptional level is of dual nature, exerted via phytochrome photoreception and via synchronization of the endogenous circadian clock (Tavladoraki et al. 1989; Tavladoraki and Argyroudi-Akoyunoglou, 1989; Adam et al., 1990). On the other hand, control at the post-translational level is exerted via pigment-protein complex formation, and thus, it depends highly on Chl synthesis and the relative rate of its accumulation (Tzinas and Argyroudi-Akoyunoglou, 1987).

This raises the question whether such a circadian rhythm, as that observed in LHC-II, or LHC-II mRNA transcription, may also be observed in the translation of the mature (25 KDa) apoprotein, and whether the transcription/translation of the apoprotein is in any way coordinated with Chl synthesis. We attempted to answer these questions by irradiating etiolated bean leaves with a Red-light pulse, transferring them thereafter to the dark, and studying the appearance of the apoprotein in total LDS-extracts of the leaves by immunoblot analysis.

MATERIALS AND METHODS

Five to eleven day-old etiolated bean leaves (Phaseolus vulgaris), grown and handled as in Argyroudi-Akoyunoglou and Akoyunoglou 1970, were exposed for 2 min in Red light ($2W/m^2$) and then transferred to the dark for 4 to 78 hrs, or to 2 min Red light followed immediately by 5 min Far Red light ($3W/m^2$). The filters used were obtained from PORSCHKE GmbH, FRG, (interference filters with maximum transmittance at 650 nm (11.4 nm bandwidth) or 730 nm (12.9 nm bandwidth).

Total leaf protein extracts were obtained by homogenizing 1g fresh weight 'leaf tissue in 10 ml LDS buffer (50 mM Na_2CO_3 - 2% LDS - 12% sucrose - 0.05M DDT) in an Omni mixer at full speed for 30 sec and centrifuging thereafter at 10,000 X g for 10 min. The supernatant, containing the total LDS-solubilized leaf proteins, was then used for Western blot analysis.

Thylakoids were isolated from chloroplasts as described before (Akoyunoglou and Akoyunoglou, 1985). Solubilization of thylakoid proteins was done according to Argyroudi-Akoyunoglou et al., 1982, or to Chua, 1986.

Electrophoretic analysis of total leaf or thylakoid proteins was performed by SDS-PAGE using a mini-gel of 12.5% polyacrylamide (Laemmli, 1970). Western blot analysis was performed according to Towbin et al., 1979, and Blake et al., 1984. Polypeptides separated on mini-SDS-gels (50-200 µg/slot) were electrophoretically transferred onto nitrocellulose filter and incubated with LHC-II antibody (1:1000 to 1:300 dilution). Immune complexes were visualized using a second Ab-AP (goat-anti rabbit IgG-alkaline phosphatase-conjugated) (1:1000 to 1:500 dilution) and BCIP/NBT (5-Bromo-4-Chloro-3-Indolyl Phosphate/Nitro Blue Tetrazolium) substrates (Blake et al., 1984).

RESULTS

As earlier shown (Tavladoraki et al., 1989) the mRNA of LHC-II apoprotein in leaves kept in the dark after a Red light (2-3 min) pulse, follows a circadian rhythm with a phase of 24 hrs, after an initial peak of about 6 hrs. Western blot analysis against the LHC-II antibody, (Figure 1), shows a similar accumulation pattern for the mature 25 KDa apoprotein in "total" LDS-solubilized leaf protein extracts. It is evident that the mature apoprotein is present in a very small amount in the etiolated leaves (6 day-old bean leaves), while its synthesis increases during maturation (8-11 day-old bean leaves); its synthesis after a Red light pulse is enhanced approximately 4-5 times, but nevertheless it remains many times lower than that in a green leaf.

The Red light-induced accumulation of the LHC-II apoprotein is higher in leaves kept in the dark for 6, 24 or 48 hrs after the pulse, but lower in leaves kept in the dark for 12 or 36 hrs.

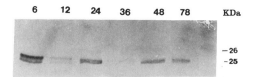

Fig. 1. Circadian rhythm in LHC-II apoprotein (25 KDa) appearance in total LDS-solubilized leaf protein extracts (200 µg per lane) obtained from 6-day old etiolated bean leaves exposed to a Red light pulse and then kept in the dark for 6 to 78 hrs. Immunoblot results are shown.

This suggests that the transcribed poly A mRNA is also translated in vivo and the rhythmical transcription of the LHC-II mRNA is followed by its translation; in addition, since the protein detected in Western blots has the molecular weight of the mature (25 KDa) apoprotein and no p-LHC-II apoprotein of 32 KDa is detected, the results also indicate that the import of the p-apoprotein into plastids is of rhythmical nature, and that the imported protein is stabilized in prothylakoids. Figure 2 shows that indeed this is the case; the Western blot analysis of prothylakoid proteins shows that the 25 KDa apoprotein is present in prothylakoid membranes, and it appears in a similar rhythmical manner. Assuming that the 25 KDa apoprotein is a Chl-stabilized thylakoid component, the results further suggest that the small amount of Chla formed after the Red light pulse (18 ug/g fr.w.) is sufficient to rescue the protein from protease digestion.

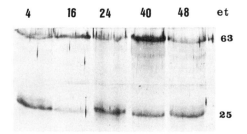

Fig. 2. Effect of a Red light pulse irradiation on the appearance of LHC-II apoprotein in prothylakoids. 6 day-old bean leaves were treated with the Red light pulse and kept in the dark for 4 to 48 hrs thereafter. 100 µg prothylakoid protein per lane. Immunoblot analysis results are shown.

The amount of the 25 KDa LHC-II apoprotein is gradually reduced as time in the dark is prolonged. This occurs in spite the fact that the LHC-II mRNA accumulation is continuously enhanced (Tavladoraki et al., 1989). This probably reflects the synthesis of the reaction center proteins during this period of time in the dark, which remove Chl from LHC-II complexes, reducing the amount of Chl available for LHC-II binding, rendering them unstable (Tzinas and Argyroudi-Akoyunoglou, 1988; Shimada et al., 1990).

Figure 3 shows that Far-Red light immediately following the Red light pulse reverses the Red light-induced accumulation of LHC-II apoprotein in "total" leaf protein extracts and in prothylakoid proteins, in a way similar to that already reported for mRNA (Tavladoraki et al., 1989). These results suggest that translation of the LHC-II mRNA is closely coupled to its transcription, and thus, it appears to be also under phytochrome control.

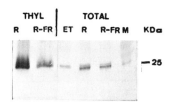

Fig. 3. Effect of Far-Red light, immediately following the Red
light pulse, on the reversal of the accumulation of
LHC-II apoprotein, in "total" leaf and prothylakoid
protein. 8 day-old bean leaves were irradiated by RL+FR
light, and then were kept in the dark for 4 hrs. 200 μg
leaf protein or 50 μg prothylakoid protein was used
per lane. Immunoblot results are shown.

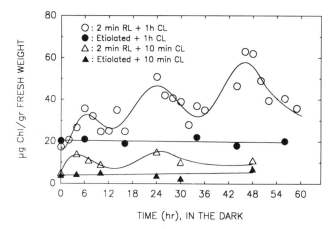

Fig. 4. Circadian rhythm in PChlide resynthesis in the dark,
following exposure of 6 day-old etiolated bean leaves
to a Red light pulse. Pchlide was monitored by the Chl
formation in continuous light for 10 min or 1 h.

For the stabilization of the LHC-II apoprotein in the
membrane, (a) Chla, formed during the RL-induced photoconversion of preexisting PChlide, (b) PChlide, regenerated in the
subsequent dark period, or finally, (c) Chla possibly formed by
dark synthesis (Walmsley and Adamson, this volume), may act as
stabilizing anchors. We therefore wondered whether the capacity
of the plants to resynthesize PChlide in the dark, following
the RL-pulse, may also be under control of the circadian clock.
Studies on this line showed that indeed the resynthesized
PChlide (as monitored by Chl formation in continuous light for
10 min or 1 h), also follows a similar circadian rhythm (Figure
4).

Since this implies that gene transcription may be mediated
by phytochrome photoreception _via_ induction of PChlide resynthesis, we further studied the effect of Levulinic acid (LA,
the PChlide synthesis inhibitor) administration on the LHC-II

apoprotein appearance, in "total" leaf or prothylakoid protein. In these experiments, the etiolated plants were immersed into 150 mM LA solution for 3 min, then dried on a filter paper and kept in the dark for 3 additional hrs. Thereafter the leaves were exposed to a RL-pulse, and kept in the dark for 4 hrs. The proteins were analyzed by immunoblotting and the PChlide present determined. It was found that, eventhough the PChlide regeneration is completely inhibited by LA, the expression of the protein is not affected. We may therefore conclude that LHC-II gene transcription is not triggered by PChlide resynthesis nor by new Chla synthesis.

We conclude that the LHC-II mRNA transcription is closely coordinated with its translation and closely follows a parallel circadian rhythm of the plants capacity to resynthesize PChlide.

ACKNOWLEDGEMENT

The partial support by a Volkswagen grant to J. H. Argyroudi-Akoyunoglou and K. Kloppstech is gratefully acknowledged.

REFERENCES

Adam, E. et al., 1990, In: "Phytochrome properties and Biological Action", NATO ASI series, in press.
Akoyunoglou, A., and Akoyunoglou, G., 1985, Reorganization of thylakoid components during chloroplast development in higher plants after transfer to darkness, Plant Physiol., 79:425.
Akoyunoglou, G., and Argyroudi-Akoyunoglou, J. H., 1986, In: "Regulation of Chloroplast differentiation" (Akoyunoglou, G. and Senger, H. eds), Alan R. Liss, N.Y.
Argyroudi-Akoyunoglou, J. H., and Akoyunoglou, G., 1970, Photoinduced changes in the chlorophyll a to chlorophyll b ratio in young bean leaves, Plant Physiol., 46:247.
Argyroudi-Akoyunoglou, J. H., Akoyunoglou, A., Kalosakas, K., and Akoyunoglou, G., 1982, Reorganization of the photosystem II unit in developing thylakoids of higher plants after transfer to darkness. Changes in chlorophyll b, light-harvesting chlorophyll protein content, and grana stacking, Plant Physiol., 70:1242.
Blake, M. S., Johnston, K. H., Russel, G. J., and Gotschlich, E. C., 1984, A rapid, sensitive method for detection of alkaline phosphatase-conjugated anti-antibody on western blots, Anal. Biochem., 136: 175.
Chua, N.-H., 1980, Electrophoretic analysis of chloroplast proteins, Methods in Enzymology, 69:434.
Gallagher, T. F., and Ellis, R. J., 1982, light stimulated transcription of genes for two chloroplast polypeptides in isolated pea leaf nuclei, EMBO J., 12:1493.
Laemmli, U. K., 1970, Cleavage of structural proteins during the assembly of the head of bacteriophage T4, Nature 227:680.
Shimada, Y., Tanaka, A., Tanaka, Y., Takabe, T., Takabe, T. and Tsuji, H., 1990, Formation of chlorophyll-protein complexes during greening 1. Distribution of newly synthesized chlorophyll, Plant Cell Physiol., 31:639.

Tavladoraki, P., and Argyroudi-Akoyunoglou, J. H., 1989, Circadian rhythm and phytochrome control of LHC-I gene transcription, <u>FEBS</u> <u>Lett.,</u> 255:305.

Tavladoraki, P., Kloppstech, K., and Argyroudi-Akoyunoglou, J H., 1989, Circadian rhythm in the expression of the mRNA coding for the apoprotein of the light-harvesting complex of photosystem II:, <u>Plant Physiol.</u>, 90:665.

Towbin, H., Staehelin, T., and Gordon, J., 1979, Electrophoretic transfer of proteins from poluacrylamide gels to nitrocellulose sheets: procedure and some applications, <u>Proc. Natl. Acad. Sci.</u>, USA, 76:4350.

Tzinas, G., and Argyroudi-Akoyunoglou, J. H., 1987, The effect of the dark interval in intermittent light on thylakoid development: photosynthetic unit formation and light harvesting protein accumulation, <u>Photosythesis Res.</u>, 14:241.

Walmsley, J., and Adamson, H., Chlorophyll synthesis and degradation in germinating barley seedlings, This volume.

INDUCTION OF THE EARLY LIGHT-INDUCIBLE PROTEIN (ELIP) IN MATURE PEA PLANTS

DURING PHOTOINHIBITION

Iwona Adamska, Itzhak Ohad and Klaus Kloppstech*

Department of Biological Chemistry, The Institute of Life
Sciences, The Hebrew University of Jerusalem, Jerusalem 91
904, ISRAEL
*Institut fuer Botanik, Universitaet Hannover, 3000 Hannover
21, GERMANY

INTRODUCTION

ELIPs are nuclear-encoded chloroplast proteins which are expressed in
the first hours of the greening process of etiolated plants. The ELIP mRNAs
appear transiently reaching a maximum level after 2-4 hours of illumination. The corresponding protein in pea accumulates between 4 and 8 hours
from the onset of the greening process. The mRNA is short-lived in the
greening plants (Meyer and Kloppstech, 1984; Grimm and Kloppstech, 1987).

In developing green plants ELIP mRNAs and proteins are transcribed and
translated under the control of the circadian oscillator (Kloppstech,
1985: Otto et al., 1988) triggered by light absorbed by the photoreceptor
phytochrome (Tavladoraki et al., 1989).

ELIPs are synthesized as precursor polypeptides, transported into the
chloroplasts and integrated into the thylakoid membranes. During this
process the 24 kDa ELIP precursor in pea is processed into a 17 kDa mature
protein.

The function of ELIP is not yet known. However, the presence of ELIPs
in pea (Meyer and Kloppstech, 1984), barley (Grimm and Kloppstech, 1987)
and bean plastid (Tavladoraki, unpublished data) indicates a widespread
occurence and possibly a similar function(s) in mono and dicotyledones
plants. A protein family related to the higher plant ELIPs and apparently
involved in carotene biosynthesis has been reported to occur also in the
green alga Dunaliella bardawil (Lear et al., 1991).

After crosslinking and fractionation of thylakoid membrane complexes,
ELIP was found in the fraction enriched in photosystem II (Adamska and
Kloppstech, 1991). The immunoprecipitation data suggested that D1 protein
of photosystem II reaction center is one of the ELIP partners in the
crosslinked products.

Photoinhibition (Prasil et al., 1991) occurs when oxygenic photo-
photosynthetic organisms or isolated photosynthetic membranes are exposed
to strong illumination. The light-induced changes correlates with reduction

Regulation of Chloroplast Biogenesis
Edited by J.H. Argyroudi-Akoyunoglou, Plenum Press, New York, 1992

in the rate of electron flow through photosystem II (PS II) followed by degradation of D1 protein. Return of the photoinhibited plants to low light intensity and replacement of the lost D1 protein by a newly synthesized molecule permit recovery of the photosynthetic activity.

In this work the influence of strong light illumination on ELIP mRNA and protein levels were studied. The results indicate a strong correlation between the process of photoinhibition and induction of ELIP synthesis.

MATERIALS AND METHODS

Pea seeds (Pisum sativum L., cv Alaska) were grown for 12-14 days on vermiculite at 25°C and a light intensity of 40 μE/m²s in a 12 h light :12 h dark regime (light from 08.00-20.00).

The high light treatment was performed on whole plants or detached mature leaves floating on water. The illumination was performed by 2 tungsten-halogen lamps (2000 W) and the plants were protected from the heat by a 3 cm double walled glass screen cooled by a water circulating system. After illumination the plant material was immediately frozen in liquid nitrogen and kept at -70°C for further analysis.

Poly(A)-rich RNA and in vitro translation in the wheat germ system were carried out as described (Adamska and Kloppstech, 1991).

In vivo labelling of whole plants was carried out as described before (Ohad et al., 1985).

For protein isolation the plant material was ground in liquid nitrogen in a mortar and the tissue powder resuspended in 2 ml double disitilled water in the presence of protease inhibitors (1 mM phenylmethylsulfonyl fluoride, 1 mM benzamidine and 5 mM ε-aminocaproic acid). The suspension was filtred through two layers of nylon nets (125 μm and 40 μm) to separate the crude material and centrifuged at 12.000 rpm for 10 min. The pellet was washed two times with water in the presence of protease inhibitors and resuspended at 10-15 mg protein/ml in sample buffer prior to electro-phoresis. Western bloting was carried out using anti-ELIP antibody kindly supplied by Prof. F. Herzfeld as described (Adamska and Kloppstech, 1991).

The degree of photoinhibition was monitored by mesurements of changes in variable fluorescence as described (Schuster et al., 1988).

RESULTS AND DISCUSSION

So far ELIP induction was reported to occur only in etiolated plants which were exposed to strong light intensity relative to the concentration of light absorbing pigments in the etioplast. Since these are actually photoinhibitory conditions the question rises whether strong light could induce ELIP synthesis also in mature green plants.

To answer this question pea plants or detached leaves were illuminated at 2500 μE/m²s for 1-4 hours and the plants were analysed for ELIP mRNA and protein content. As a control, plants were kept under low light conditions (40 μE/m²s) and harvested at the same time. The auto-radiography of in vitro translated RNA shows that the ELIP mRNA level increases with the time of photoinhibition reaching a plateau after 3 hours of high light treatment (Fig.1A). Among the known light-regulated nuclear-encoded proteins, only ELIP mRNA was induced under these conditions. The levels of the small subunit of ribulose 1,5-bisphosphate carboxylase (SSU) and the light-harvesting chlorophyll a/b protein (LHCP) remained unchanged.

Western blotting analysis showed that ELIP amount in the thylakoid membranes increases during the photoinhibition process (Fig.1B). The level of the chloroplast-encoded D1 protein and LHCP did not change significantly during the 4 hours of high light treatment. The stability of the D1 protein level under high light treatment could be explained by its rapid rate of resynthesis. Under more prolonged illumination the level of this protein decreased progresively (data not shown). The increase in the ELIP level could be demonstrated also by radioactive labelling of high light treated whole plants or detached leaves (Fig.1C).

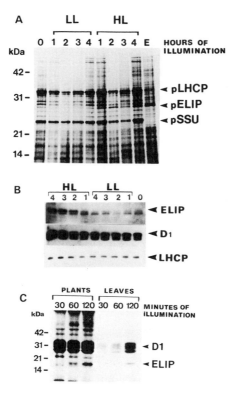

Fig. 1. Changes in ELIP mRNA and protein levels as a function of time in high light treated green mature pea plants. Pea plants or detached leaves were exposed to low light (LL, 40 µE/m²s) or high light (HL, 2500 µE/m²s) for different times. (A) Poly (A⁺)-RNA was isolated, in vitro translated and [³⁵S]-methionine labelled translation products separated by SDS-PAGE followed by autoradiography. As a control the translation products of mRNA isolated from etiolated and 4 h illuminated pea seedlings were separated (lane E) Arrows indicate the positions of precursors for three light-regulated proteins; (B) Western blot; (C) In vivo labelling of whole plants or detached leaves with [³⁵S]-methionine.

To establish the relation between light intensity and the level of ELIP mRNA induction in mature green plants, pea leaves were illuminated for 3 hours at different light intensities. The amount of translatable ELIP mRNA increased with the light intensity and reached a plateau at 1000-1500 µE/m²s (data not shown).

The increse of the ELIP amount in the thylakoids is related to that of the mRNA and is light intensity dependent. The amounts of D1 protein and LHCP remain unchanged at all light intensities used in this experiment (Fig.2). Photoinhibition and recovery of photosynthetic activity in leaves exposed to the various light intensities were estimated by measuring the variable fluorescence of isolated thylakoids (Fig.2). The degree of photoinhibition was related to the light intensity and ELIP induction.

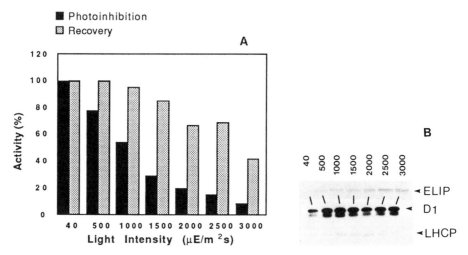

Fig. 2. Changes in ELIP level in thylakoid membranes during exposure of detached pea leaves to various light intensities. Detached pea leaves were illuminated for 3 hours at different light intensities as indicated, the membranes were isolated and their proteins resolved by SDS-PAGE. (A) Kinetics of light intensity dependent inactivation and recovery of PSII measured as a decrease of variable fluorescence after 3 hours of high light illumination and after 4 hours of recovery in low light (40 µE/m²s). Values shown as % of the control; (B) Western blot.

Recovery from photoinhibition occured in leaves subjected to all light intensity treatments and was 40 % after 4 hours of low light exposure even in leaves photoinhibited up to 90%. The question thus arrived as to the fate of ELIP mRNA and protein levels during recovery from photo-inhibition. After illumination at different light intensities for 2 hours leaves were transferred to low light conditions (40 µE/m²s) and samples were taken at various times up to 20 hours. The initial high level of ELIP mRNA decreased rapidly in the low light exposed leaves. One hour after the transfer of photoinhibited leaves to low light conditions only traces of ELIP mRNA were detected (data not shown). This corroborates the earlier findings obtained with greening plants (Meyer and Kloppstech, 1984) showing that ELIP mRNA is short-lived and undergoes very fast turnover.

As oposed to the fast decrease in the ELIP mRNA level, the protein was found to be more stable, its level being practically unchanged up to 6 of hours recovery in leaves photoinhibited at 2500-3000 µE/m²s. The life

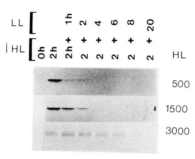

Fig.3. ELIP level in thylakoid membranes during recovery from
photoinhibition. Detached pea leaves were illuminated for 2 hours
with different light intensities indicated in the figure and then
transferred for 20 hours to low light conditions (40 μE/m²s).
Samples were taken as indicated, the membranes isolated and
separated by SDS-PAGE and used for western blot.

time of the ELIP during recovery decreased progresively in leaves exposed
to lower light intensities during the photoinhibition process (Fig.3). The
reasons for the difference observed in the decay of ELIP in relation to the
photoinhbitory treatment are not yet understood.

SUMMARY AND CONCLUSIONS

1. High ligth treatment induced synthesis of the ELIP mRNA and protein in
green mature pea plants. This process correlated with the degree of photoinhibition obtained as a function of the light intensity and duration of
exposure to high light. Accumulation of ELIP mRNA reached a maximum after 3
hours from the beginning of high light treatment. The protein level
incresed during the whole period of illumination.

2. Increase of ELIP mRNA level was high light intensity dependent between
40-1500 μE/m²s and the saturation point was reached at 1500 μE/m²s.
Accumulation of protein was light intensity dependent in the range from
500-3000 μE/m²s.

3. ELIP mRNA was rapidly degraded after transferring the leaves to low
light conditions. The degradation of the protein during recovery was
related to the light intensity and extent of the high light treatment.

    The correlation between the high light induction of ELIP and the
photoinhibition process could indicate that this protein may play a role in
the process of protection from photoinhibition possibly related to the
xanthophyll cycle (Demmig et al.,1990).

ACKNOWLEDGEMENTS

This work was supported by a MINERVA grant awarded to I.A.

REFERENCES

Adamska, I., and Kloppstech, K., 1991, Evidence for the association of the early light-inducible protein (ELIP) of pea with photosystem II, Plant. Mol. Biol., 16:209-223.

Demming-Adams, B., 1990, Carotenoids and photoprotection in plants: A role for the xanthophyll zeaxanthin, Biochim. Biophysic. Acta, 1020:1-24.

Grimm, B., and Kloppstech, K., 1987, The early light-inducible proteins of barley. Characterization of two families óf 2-h-specific nuclear-coded chloroplast proteins, Eur. J. Biochem., 167:493-499.

Kloppstech, K., 1985, Diurnal and circadian rhythmicity in the expression of light-induced plant nuclear messenger RNAs, Planta, 165:502-506.

Lers, A., Levy, H., and Zamir, A., 1991, Co-regulation of an ELIP-like gene and ß-carotene biosynthesis in the alga Dunaliella bardawil, Plant Cell, in press.

Meyer, G., and Kloppstech, K., 1984, A rapidly light-induced chloroplast protein with a high turnover coded for by pea nuclear DNA, Eur. J. Biochem., 138:201-207.

Ohad, I., Kyle, D.J., and Hirschberg, J., 1985, Light-dependent degradation of the QB-protein in isolated pea thylakoids, EMBO J., 7:1655-1659.

Otto, B., Grimm, B., Ottersbach, P., and Kloppstech, K., 1988, Circadian control of the accumulation of mRNAs for light- and heat-inducible chloroplast proteins in pea (Pisum sativum L.), Plant Physiol., 88:21-25.

Prasil, O., Adir, N., and Ohad, I., 1991, Dynamics of photosystem II: Mechanisms of photoinhibition and recovery processes, in Topics in Photosynthesis, vol. 11, Barber, J., ed., in press.

Schuster, G., Even, D., Kloppstech, K., and Ohad, I., 1988, Evidence for protection by heat-shock proteins against photoinhibition during heat-shock, EMBO J., 7:1-6.

Tavladoraki, P., Kloppstech, K., and Argyroudi-Akoyunoglou, J., 1989, Circadian rhythm in the expression of the mRNA coding for the apoprotein of the light-harvesting complex of photosystem II, Plant Physiol., 90:665-672.

REGULATION BY PHYTOCHROME AND OXYGEN RADICALS OF

ASCORBATE SPECIFIC PEROXIDASE AND GLUTATHIONE REDUCTASE

H. Drumm-Herrel, B. Thomsen and H. Mohr

Biological Institute II,
University of Freiburg,
D-7800 Freiburg, Germany

INTRODUCTION

Three enzymes of the Halliwell-Asada-pathway (e.g. Nakano and Asada 1981, Anderson et al.,1983) known to abolish toxic ("reactive") oxygen species in illuminated chloroplasts (such as the superoxide radical,H_2O_2 and the highly toxic hydroxyl radical) were studied in the cotyledons of mustard seedlings (Sinapis alba L.). All three enzymes [superoxide dismutase (SOD, EC 1.15.1.1), ascorbate specific peroxidase (APO, EC 1.11.1.11) and glutathione reductase (GR, EC 1.6.4.2)] are phytochrome regulated.

GR as well as APO are present in two isoforms which can be separated by ion exchange chromatography (FPLC). The technique of in situ photooxidation of carotenoid-free plastids, made carotenoid-free by the herbizide Norflurazon (NF), was used to study the response of the isoforms under photooxidative stress.

RESULTS

1. Glutathione Reductase (GR)

GR from mustard seedling cotyledons exists in two iso-forms, cytosolic GR1 and plastidic GR2. Light mediated increase of GR level is phytochrome-controlled (Drumm and Mohr, 1973; Drumm-Herrel et al.,1989). If carotenoid-free seedlings grown so far in far-red light (FR) on a Norflurazon solution (NF, 10^- ^{5} M) are treated with photooxidative red light (R, 6.8 Wm^{-2}) for three h a rapid decrease in enzyme level is observed. Western blots (Fig.1) show that decrease of enzyme level during photooxidation correlates with decrease of GR2 protein. After transfer to darkness or FR a rapid recovery of enzyme level is observed only in FR (Drumm-Herrel et al.,1989). Decrease and recovery of total GR are only due to the decrease and reappearance of the plastidic isoform (GR2). Rapid recovery requires phytochrome action.

The cytosolic isoform (GR1) remains almost unchanged. Plastidic localization of GR2 was confirmed by isolating chloroplasts (Fig.2). GR2 activity coincides with NADP-GPD and

Regulation of Chloroplast Biogenesis
Edited by J.H. Argyroudi-Akoyunoglou, Plenum Press, New York, 1992

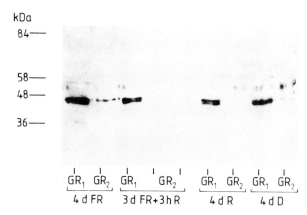

Fig.1. Western blot of GR1 and GR2 from mustard cotyledons
(separated by ion exchange chromatography) after separation
by SDS-PAGE. Seedlings were grown on Norflurazon (NF, 10^{-5}
M), programmes as indicated. Electroblotted nitrocellulose
membrane was treated with antibodies raised against spinach
GR. GR1 and GR2 were visualized with the enhanced
chemiluminescence (ECL) Western blotting system.

RUBISCO. Integrity of plastids was checked by the ferricyanide
assay. Increase of plastidic GR2 is to be attributed to
synthesis de novo as shown by immunotitration experiments
(Fig.3). Compared to the recovery kinetics of other plastidic
marker enzymes such as NADP-GPD and RUBISCO, reappearance is
much faster (see Schuster et al., 1988).

2. Ascorbate specific Peroxidase (APO)

Two other enzymes of the Halliwell-Asada-pathway, APO and
SOD, are phytochrome regulated as well. In the following we
consider only APO because of highly reproducible results (in
contrast to the less accurate SOD-test and the presence of
several isoforms).

The basic results on APO induction by phytochrome and possible involvement of oxygen radicals are shown in Fig.4. In
FR/NF-seedlings [phytochrome action via High Irradiance
Reaction (HIR), no carotenoids, small amount of chlorophyll
only, no photosynthesis] APO induction is the same as in
FR/H_2O-seedlings. No photooxidation takes place in FR due to
the very low absorption of FR by chlorophyll.

In R the herbizide has a strong <u>inducing effect.</u> The same
effect is also visible if FR/NF-seedlings are transferred to R
(Fig.4), conditions under which GR2 (and other plastidic
enzymes) are destroyed (see Fig.1). In NF-treated carotenoid-
free seedlings R leads to a strong photooxidative action. We
conclude that the increase of the APO level is caused by
reactive oxygen radicals.

Circumstantial evidence in favour of radicals as inducers
comes from experiments with quenchers. They indicate that in
the presence of α-tocopherol and p-benzochinone, quenching
electrochemical energy of radicals and other reactive molecules
produced during photostress, APO activity is <u>reduced</u> in R/NF-

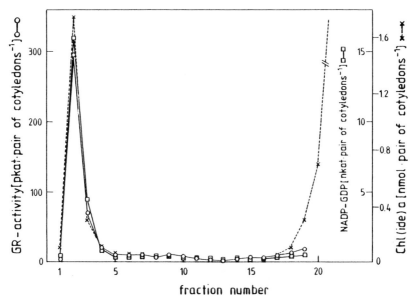

Fig.2. Distribution of Chl, GR and NADP-GPD in chloroplasts
isolated from 80 g of 4 d old R grown mustard seedlings on
a Percoll gradient. RUBISCO (not shown) coincides with
NADP-GPD.

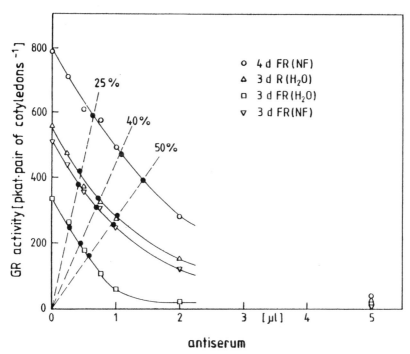

Fig.3. Immunotitration of GR2 from mustard cotyledon extracts.
The dashed lines show that the amount of antiserum required
to reduce enzyme activity by 25, 40 or 50 % is directly
proportional to the activity of the original extract (see
Brödenfeldt and Mohr, 1986).

seedlings. When R/NF-seedlings were grown under an elevated oxygen pressure - potentially producing more radicals - APO induction is even higher. In water controls a slight increase in R is also observed. Figure 5 demonstrates the presence of two isoforms (APO1 at 0.12 M and APO2 at 0.21 M NaCl). Only the major peak, APO2, is increased in R/NF-seedlings; APO1, the minor peak, remains unchanged.

Isolation of chloroplasts followed by fractionation of plastidic proteins by anion exchange chromatography demonstrates that only APO1, the minor peak, is the plastidic isoform of APO. The plastidic isoform is extremely resistant to photooxidative stress, even when isolated superetioplasts from FR/NF-seedlings were photooxidized in vitro (Tab.1).

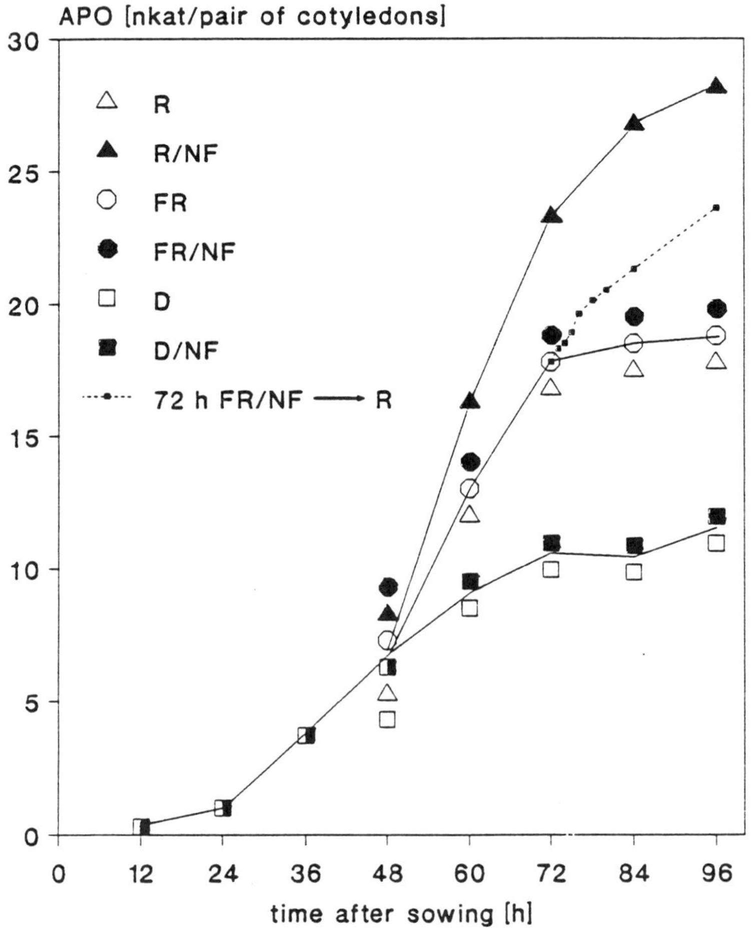

Fig.4. Time courses of APO activities isolated from mustard seedling cotyledons grown on water or on NF in darkness (D), R or FR. FR/NF-seedlings were treated with photooxidative R after 72 h in FR (⋯•⋯).

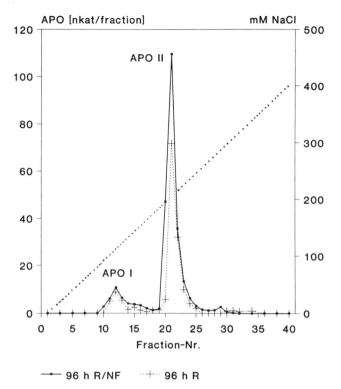

Fig.5. Elution profile of APO activity from mustard cotyledon
extracts. Seedlings were grown on water or on NF and kept
for 4 d in photooxidative R. Separation was carried out by
anion exchange chromatography (FPLC, MONO Q).

SUMMARY

    Glutathione reductase (GR) and ascorbate specific peroxi-
dase (APO), two enzymes of the Halliwell-Asada-pathway, both
consisting of two isoforms - a plastidic and an extraplastidic
one - both phytochrome regulated - behave completely
differently under photooxidative stress.
(1) The plastidic GR2 is extremely sensitive to photooxidation,
whereas the cytosolic isoform is not affected by photodynamic
action on the plastids. Recovery of GR2 level is fast and
phytochrome dependent.
(2) The plastidic isoform of APO (APO1) is completely resistant
to photooxidative stress in situ as well as in isolated
plastids while the level of the extraplastidic isoform (APO2)
is increased under these conditions.
(3) Experiments with quenchers and elevated oxygen levels
suggest that oxygen radicals from the photodamaged plastids
cause the increase of the extraplastidic APO2. Experiments
concerning localization of this isoform and the nature of the
signal that causes its increase are in progress.

Table 1. in vitro-photooxidation of plastids. They were
isolated from 4 d old FR/NF-seedlings. One half of
the carotenoid-free superetioplasts were photooxidized for 4 h with R whereas the other half
(controls) stayed in darkness.

Treatment	APO [nkat]	Chlorophyll [µg]
96 h FR/NF	119,6	65,3
96 h FR/NF + 4 h R	121,0	27,5
96 h FR/NF + 4 h D	120,0	61,5

REFERENCES

Anderson, J. W., Foyer, C. H., and Walker, D. A., 1983, Light-
dependent reduction of hydrogen peroxidase by intact
spinach chloroplasts, Biochim., Biophys., Acta, 724:69
Brödenfeldt, R., and Mohr, H., 1986, Use of immunotitration to
demonstrate phytochrome-mediated synthesis de novo of
chalcone synthase and phenylalanine ammonia lyase in
mustard seedling cotyledons, Z., Naturforsch.,41c:61
Drumm, H., and Mohr, H., 1973, Control by phytochrome of
glutathione reductase levels in the mustard seedling, Z.,
Naturforsch., 28c:559
Drumm-Herrel, H., Gerhäußer, U., and Mohr, H., 1989, Differential regulation by phytochrome of appearance of
plastidic and cytoplasmatic isoforms of glutathione reductase in mustard (Sinapis alba L.) cotyledons, Planta,
178:103
Nakano, Y., and Asada, K., 1981, Hydrogen peroxidase is
scavenged by ascorbate specific peroxidase in spinach
chloroplasts, Plant, Cell, Physiol., 22:867
Schuster, C., Oelmüller, R., Bergfeld, R., and Mohr, H., 1988,
Recovery of plastids from photooxidative damage :
Significance of a plastidic factor, Planta, 174:289

IMPORT OF NUCLEAR-CODED PROTEINS INTO CHLOROPLASTS AND MECHANISM OF CHLOROPLAST PROTEIN TRANSLOCATION

CHLOROPLAST-SPECIFIC IMPORT AND ROUTING OF PROTEINS

Peter Weisbeek and Douwe de Boer

Department of Molecular Cell Biology
University of Utrecht, P.O. Box 80.056
3508 TB Utrecht, The Netherlands

INTRODUCTION

Plant cells, like all other eukaryotic cells, contain various compartments, organelles, which serve to separate the different metabolic processes. All organelles are enclosed by one or two semipermeable membranes, constituted of lipids and proteins. Most of these proteins are involved in the import and export of a large variety of molecules.

One family of organelles, the plastids, is unique to plants; whereas most other organelles, like mitochondria and the endoplasmic reticulum, are also present in other eukaryotes. Plastids are surrounded by a two membrane envelope and contain a limited amount of DNA. Several types of plastids can be distinguished in the different tissues of a plant, but all originate in development from one form, the proplastid. The most well known type, the chloroplast, is present in green tissue and is the site were photosynthesis takes place. The coding capacity of the chloroplast genome is not sufficient to code for all the proteins that are present in the organelle. Chloroplasts are themselves also subdivided by membranes into three different compartments, the intermembrane space, the stroma and the thylakoid lumen. Therefore there are, together with the membranes, six different locations for proteins to reside. The two membranes of the envelope are separated from each other by the intermembrane space, but there are specific points, contact sites, where the two membranes are fused or held in close proximity (Cline et al., 1985b; Cremers et al., 1988).

Nuclear encoded chloroplast proteins are synthesized in the cytoplasm with an N- terminal extension, that contains the import signals. In the cytoplasm the precursor might interact with cytosolic factors during or shortly after its synthesis. After binding to receptor proteins in the outer membrane translocation across the envelope occurs. ATP is required for binding to the receptor and also during translocation. After translocation the chloroplast import signal is removed by a stromal processing peptidase and interaction with stromal factors might occur. The default import route to the stroma is only followed by a routing step towards one of the other locations in the chloroplast when specific routing signals are present in the protein. For membrane insertion signals are usually present in the mature protein. Whereas for thylakoid membrane translocation an increased N-terminal extension, composed of two domains, is present in front of the mature protein. The two domain targeting signal is removed from the precursor in two steps.

CHLOROPLAST TARGETING SIGNALS

Nuclear encoded chloroplast proteins are synthesized as higher molecular weight precursor proteins with an N-terminal extension, the chloroplast targeting signal (Dobberstein et al., 1977). This targeting signal is required for chloroplast import and sometimes also for routing within the chloroplast. The part of the targeting signal that is required for translocation across the envelope membranes is called a transit peptide (Chua and Schmidt, 1979). This part is removed by a stromal processing peptidase after import into the chloroplast (Robinson and Ellis, 1984). Thylakoid lumen proteins are synthesized with a targeting signal that contains a thylakoid transfer domain behind the transit peptide (Smeekens et al., 1986). This second domain is essential for thylakoid lumen targeting and it is removed from the protein by a thylakoidal processing peptidase, after translocation across the thylakoid membrane. A list of cloned and sequenced genes for these nuclear encoded chloroplast proteins is available (De Boer and Weisbeek, 1991), but more and more cloned genes are being published. From these sequencing data several characteristic features could be determined.

CHLOROPLAST ENVELOPE TRANSLOCATION

Although the presence of a transit peptide at the N-terminus of a precursor protein is sufficient to signal import into chloroplasts, several other requirements are necessary for the actual process. Cytosolic factors might be needed to assist in unfolding and translocation; receptor proteins, that recognize the transit peptide are usually present in the chloroplast envelope; energy is needed in the form of ATP; after translocation a stromal peptidase is needed and stromal factors might be necessary to assist in refolding and assembly.

Cytosolic factors

Proteins that enter the chloroplast, do so post-translationally. Therefore, they have to be transported through the cytoplasm towards the chloroplast. After expression of precursors in E. coli and subsequent purification, it has become possible to study the requirements for cytosolic factors for the import of these proteins into chloroplasts. Import into chloroplasts of the LHCP-II precursor was found to require at least two cytosolic factors (Waegemann et al., 1990). Although hsp70 stimulated the import of the E. coli purified protein, no association with a hsp70 protein was observed in a different study (Keegstra, personal communication). For mitochondrial import there is also evidence for the presence of additional cytosolic factors apart from the 70 kDa chaperone (Murakami and Mori, 1990). This additional factor might act by presenting the protein in an import competent state, maybe by exposing the transit peptide. For two other chloroplast precursors, ferredoxin and plastocyanin, that have been expressed in E. coli, no evidence exists for the requirement of cytosolic factors (Pilon et al., 1990; De Boer et al., in preparation). It might be possible that these small proteins are easily folded and unfolded and therefore do not require chaperones. Alternatively, it is possible that unfoldases, that were found in the chloroplast envelope, act on these proteins just before translocation.

Components of the translocation complex

The first step in the import of a precursor into the chloroplast is its binding to a proteinaceous receptor in the envelope membrane. The presence of receptor proteins in the chloroplast envelope was already recognized a few decades ago. It was found that mild protease treatment of intact chloroplasts reduced binding and import of proteins (Chua and Schmidt, 1978).

Binding and import could be separated from each other by destroying the ATP generating system of the chloroplast or by lowering the temperature (Cline et al., 1985a). Both treatments resulted in binding of precursors to the envelope in the absence of import. This binding assay made it possible to titrate the amount of binding sites. Saturation was reached between 1500 and 3500 molecules bound per chloroplast. Pretreatment of the chloroplasts with a protease reduced the number of binding sites to about 700, whereas the affinity for the precursor was not influenced. However, when chloroplasts were pretreated with the cysteine specific protein modifying agent NEM, the number of binding sites was unaltered, while the affinity for the precursor was reduced.

Import competition studies in our own laboratory, with the plastocyanin precursor purified after expression in E. coli and other precursors synthesized in a wheat germ system, showed that plastocyanin was able to reduce the import of the other precursor protein significantly (De Boer et al., in preparation). Plastocyanin, ferredoxin and the small subunit of rubisco seem to make use of the same import pathway. A similar conclusion is reached when competition studies were performed with synthetic transit peptides (Keegstra, personal communication). Competition studies for binding show that not only the translocation complex is identical, but also the amount of different receptors is limited. A limited number of different receptors was already expected, since chloroplast proteins could be imported into other plastids and vice versa (De Boer et al., 1988; Klösgen et al., 1989). It is not likely that specific receptors are present on plastids in tissues where the corresponding precursor normally is not present.

Besides receptor proteins there are also proteins needed for the actual translocation of precursor proteins across the envelope. These proteins probably form a channel through which the precursors are transported across the membranes. These channel forming proteins are expected to be localized in the contact sites between the inner and outer envelope membranes. The translocation complex in chloroplasts is less well characterized, so far only one protein has been identified by using anti-idiotypic antibodies against a synthetic transit peptide of the small subunit of rubisco (Pain et al., 1988). The gene for this 30 kDa protein is cloned and sequenced (Schnell et al., 1990). The protein is present in contact sites between the inner and outer chloroplast envelope. Antibodies against the 30 kDa protein inhibit import of precursor proteins into the chloroplast. There are, however, some arguments against a function in protein transport. The same DNA sequence of the 30 kDa protein is cloned and sequenced by another group and is claimed to be the gene for the phosphate translocator (Willey et al., 1991). More data are necessary to determine whether this gene codes for the phosphate translocator, a component of the translocation complex or a protein with a dual function.

Energetics of envelope translocation

Translocation of proteins across the chloroplast envelope requires the hydrolysis of ATP (Flügge and Hinz, 1986; Theg et al., 1989). A small amount of ATP is required for binding of precursor proteins to the outer membrane (Olsen et al., 1989). The amount of ATP is about five times lower than the total amount required for import of proteins into the chloroplast. Phosphorylation of a 51 kDa protein that was found to be correlated with protein import, also required only a limited amount of ATP. Since this phosphorylation site could be removed after protease treatment of the chloroplast, it might well be that the ATP necessary for binding is required for phosphorylation of this 51 kDa protein (Hinz and Flügge, 1988). This protein therefore is a good candidate for the receptor or for a subunit of the receptor complex. The low amount of ATP is probably utilized in the intermembrane space (Olsen et al., 1989).

Stromal processing

After translocation of precursor proteins across the envelope membranes, the transit peptide is specifically removed by the stromal processing peptidase (SPP). The presence of this stromal protease was first observed by Dobberstein et al., (1977) and was subsequently found to be localized in the chloroplast stroma (Smith and Ellis, 1979). The activity was partially purified and was found to consist of a complex of 180 kDa (Robinson and Ellis, 1984). The activity is only inhibited by metal-chelators like EDTA and ATP is not required for the activity.

Stromal factors

After translocation across the chloroplast envelope membranes, proteins usually associate with stromal factors before they assemble into protein complexes or before they are routed to a different location. The first stromal factor that was identified is a protein complex that binds the large subunit of rubisco (Barraclough and Ellis, 1980; Roy et al., 1982). Although the large subunit protein is synthesized in the stroma, it was found that many nuclear encoded proteins also associate with this 700 kDa complex after import into the chloroplast (Lubben et al., 1989).

The complex was found to consist of fourteen subunits, seven -subunits of 61 kDa and seven and was found to be highly homologous to the E. coli groEL protein (Hemmingsen et al., 1988). The groEL complex consists of 14 identical subunits, that are packed in two stacked rings of seven subunits each. The protein binds to newly synthesized proteins and prevents aggregation. Assembly of the bound protein into protein complexes is catalyzed by the groEL complex. The bound protein is released from the groEL complex after ATP-hydrolyses. GroEL-like proteins are found in prokaryotes, mitochondria (hsp60) and chloroplasts and this subclass of molecular chaperones has been termed chaperonins (Hemmingsen et al., 1988).

Association with chaperonins was observed after import of several nuclear encoded chloroplast proteins (Lubben et al., 1989), like the small subunit of rubisco, the of the F_1-ATPsynthase complex, glutamine synthetase and the LHCP-II. Moreover, also non-chloroplast proteins like chloramphenicol, acetyltransferase and found to associate with chaperonins after import into the chloroplast. No indication for association with a chaperonin was observed for ferredoxin, superoxide dismutase and plastocyanin.

Several other, mostly unknown, factors were found to be essential for assembly of proteins into protein complexes or into the thylakoid membrane. The assembly of rubisco is besides the large subunit binding protein dependent on a not fully identified stromal factor that seems to be homologous to the bacterial groES protein (Goloubinoff et al., 1989). The LHCP-II protein forms a 120 kDa complex with a stromal protein probably after its association with the chaperonin complex (Payan and Cline, 1991). Insertion into the membrane seems to be dependent on yet another stromal factor (Payan and Cline, 1991).

Non-stromal proteins need to contain information for routing to their final location. Proteins that are destined for the thylakoid lumen usually contain a cleavable thylakoid transfer domain in front of the mature protein Proteins destined for the thylakoid membrane normally contain information in the mature protein. Little is known about inner membrane proteins, but they seem to contain information in the mature for membrane insertion. There are no examples known of intermembrane space proteins. Therefore it is not known whether they import into the chloroplast and subsequently back into the intermembrane space, as has been suggested for several mitochondrial intermembrane space proteins (Hartl et al., 1989), or whether they only translocate across the outer membrane. All known nuclear encoded outermembrane proteins are synthesized without a transit peptide and they seem to enter the membrane directly, without translocation into the stroma.

Membrane insertion

There are only a few cloned genes available that code for chloroplast envelope proteins, consequently there is only a limited knowledge about protein insertion into the envelope membranes. The genes for a 6.7 kDa (Salomon et al., 1990) and a 14 kDa (Keegstra, personal communication) outer membrane protein, both with unknown function, have been cloned and sequenced. Both these nuclear encoded outer envelope membrane proteins are synthesized as native proteins without the presence of a cleavable targeting signal. ATP is not required for insertion of both proteins into the membrane. Pretreatment of the chloroplasts with a protease does not affect insertion. It is not known whether cytosolic factors are involved. The cryptic genes for two inner membrane proteins, a 37 kDa (Dreses-Werringloer et al., 1991) and a 30 kDa (Keegstra, personal communication) protein, have been cloned and sequenced, and their products have been used to study integration into the membrane. Both proteins are synthesized in the cytoplasm as precursor proteins with a cleavable N-terminal extension. ATP is necessary for arrival into the inner membrane and interaction with the chloroplast envelope does not occur when the chloroplasts are pretreated with a protease. Interaction with a receptor protein in the outer envelope is assumed from these data. It is not known whether the N-terminal extension acts as a normal transit peptide by translocating the protein into the stroma. If this turns out to be the case than the mature part of the protein has to integrate into the inner membrane from the stromal side (see Fig. 1).

The gene for one other envelope protein is cloned and sequenced. This 30 kDa protein is reported to be the phosphate translocator that is present in the inner membrane (Flügge et al., 1989; Willey et al., 1991), whereas an other group claims that it is the import receptor that is present in the contact sites.

Proteins that insert into the thylakoid membrane contain non-cleavable routing information in the protein. The exception to this rule is the chloroplast encoded protein cytochrome f (Willey et al., 1984). The main part of this protein is located at the lumen of the thylakoid membrane and as a consequence most of the protein has to translocate the membrane. The protein is synthesized in the stroma with a thylakoid transfer domain in front of the protein. A stop transfer sequence at the C- terminus of the protein prevents its complete translocation. The extension is removed from the protein by the thylakoidal processing peptidase. All other thylakoid membrane proteins that have been examined enter the membrane without a cleavable sequence. These thylakoid membrane proteins contain one or more membrane spanning domains that probably also signal membrane insertion. The

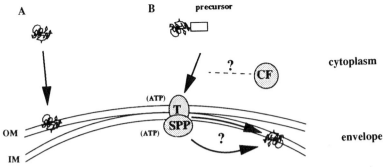

Figure 1. Model for the insertion of an outer membrane protein
(A) and an inner membrane protein (B) into the chloroplast
envelope. The outer membrane protein interacts with and
integrates into the outer membrane (OM) in the absence of ATP and
a proteinaceous receptor.

orientation of the proteins in the membrane is such that most of
the positive charges are located at stromal site of the membrane.
Insertion and orientation of proteins in the thylakoid membrane
is more thoroughly reviewed elsewhere (De Boer and Weisbeek,
1991).

Thylakoid membrane translocation

Translocation of proteins into the thylakoid lumen was first
studied with the precursor of plastocyanin (Smeekens et al.,
1986). After translocation of the protein into the stroma and
partial processing, the intermediate sized protein was imported
into the thylakoid lumen and processed to its mature size. This
two step import pathway (see Fig. 2) was also observed for the
16, 23 and 33 kDa subunits of the watersplitting complex (James
et al., 1989).
    The first step, the translocation across the envelope
membranes, is comparable to import of stromal proteins. The first
part of the targeting signal can be exchanged for a normal
transit peptide (Hageman et al., 1990). The requirements for the
second part, translocation across the thylakoid membrane, were
more difficult to determine. Fusions between the two domain
targeting signal and a passenger protein were usually not
successful in proper targeting of the passenger in vitro; an
intermediate sized protein was found in the stromal fraction
instead. This unsuccessful routing of passenger proteins was,
however, not observed in vivo (De Boer et al., 1991).

Thylakoidal processing

Thylakoid lumen proteins are processed to their mature sizes
by a processing peptidase that is present in the thylakoid
membrane (Hageman et al., 1986). The processing activity has been
partially purified and was found to be insensitive to several
types of protease inhibitors (Kirwin et al., 1988). The presence
of Triton X-100 was essential during purification to prevent
inactivation of the protease activity. The protease has a low pH
optimum of 6.5-7.0 and the active site of the protein is present
at the luminal side of the thylakoid membrane.
    The thylakoidal processing peptidase resembles the leader
peptidase of E. coli and the eukaryotic signal peptidase in the

endoplasmic reticulum with respect to enzymatic activity and membrane orientation (Halpin et al., 1989). The inner membrane protease I of yeast is also related to these three enzymes (Schneider et al., 1991). This probably means that the translocation machinery or at least the processing complex necessary for routing inside organelles is a remnant of the prokaryotic export device.

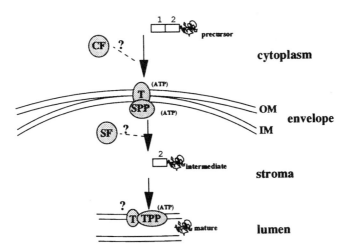

Figure 2. Model for the two step routing pathway of a thylakoid lumen protein. The precursor is synthesized with a two domain targeting signal and is expected to bind cytoplasmic factors (CF), before interaction with a proteinaceous translocation complex (T). After translocation the first domain of the targeting signal is removed by the stromal processing peptidase (SPP). The intermediate sized protein is supposed to interact with stromal factors. The second domain of the targeting signal is removed by the thylakoidal processing peptidase (TPP) after translocation into the lumen.

REFERENCES

Anderson, J.M., Hnilo, J., Larson, R., Okita, T.W., Morell, M., and Preiss, J. (1989). The encoded primary sequence of a rice seed ADP-glucose pyrophosphorylase subunit and its homology to the bacterial enzyme. J. Biol. Chem. 264, 12238-12242.

Barraclough, R. and Ellis, R.J. (1980). Protein synthesis in chloroplasts IX: assembly of newly-synthesized large subunits into ribulose bisphosphate carboxylase in isolated intact pea chloroplasts. Biochim. Biophys. Acta 608, 19-31.

Chua, N.-H. and Schmidt, G.W. (1978). Post-translational transport into intact chloroplasts of a precursor to the small subunit of ribulose-1,5-bisphosphate carboxylase. Proc. Natl. Acad. Sci. USA 75, 6110-6114.

Cline, K., Werner-Washburne, M., Lubben, T.H. and Keegstra, K. (1985a). Precursors to two nuclear-encoded chloroplast proteins bind to the outer envelope membrane before being imported into chloroplasts. J. Biol. Chem. 260, 3691-3696.

Cline, K., Keegstra, K., and Staehelin, L.A. (1985b). Freeze-fracture electron microscopic analysis of ultrarapidly frozen envelope membranes on intact chloroplasts and after purification. Protoplasma 125, 111-123.

Cremers, A.F.M., Voorhout, W.F., Van der Krift, T.P.,

Leunissen-Bijvelt, J.J.M., and Verkleij, A.J. (1988). Visualization of contact sites between outer and inner envelope membranes in isolated chloroplasts. Biochim. Biophys. Acta 933, 334-340.

De Boer, D., Cremers, F., Teertstra, R., Smits, L., Hille, J., Smeekens, S., and Weisbeek, P. (1988). In vivo import of plastocyanin and a fusion protein into developmentally different plastids of transgenic plants. EMBO J. 7, 2631-2636.

De Boer, A.D. and Weisbeek, P.J. (1991). Chloroplast Topogenesis: Protein import, sorting, and assembly. Biochim. Biophys. Acta, in press.

De Boer, D., Bakker, H., Lever, A., Bouma, T., Salentijn, E. and Weisbeek, P. (1991). Protein targeting towards the thylakoid lumen of chloroplasts: proper localization of fusion proteins is only observed in vivo. EMBO J., accepted for publication.

Dobberstein, B., Blobel, G., and Chua, N-H. (1977). In vitro synthesis and processing of a putative precursor for the small subunit of ribulose-1,5-bisphosphate carboxylase of Chlamydomonas reinhardtii. Proc. Natl. Acad. Sci. USA 74, 1082-1085.

Dreses-Werringloer, U., Fisher, K., Wachter, E., Link, T.A., and Flügge, U.I. (1991). cDNA sequence and deduced aminoacid sequence of the precursor of the 37-kDa inner envelope membrane polypeptide from spinach chloroplasts: its transit peptide contains an amphiphilic -helix as the only detectable structural element. Eur. J. Biochem. 195, 361-368.

Flügge, U.I. and Hinz, G. (1986). Energy dependence of protein translocation into chloroplasts. Eur. J. Biochem. 160, 563-570.

Függe, U.I., Fischer, K., Gross, A., Sebald, W., Lottspeich, F., and Eckerskorn, C. (1989). The triose phosphate-3-phosphoglycerate-phosphate translocator from spinach chloroplasts nucleotide sequence of a full-length complementary DNA clone and import of the in vitro synthesized precursor protein into chloroplasts. EMBO J. 8, 39-46.

Goloubinoff, P., Christeller, J.T., Gatenby, A.A., and Lorimer, G.H. (1989). Reconstitution of active dimeric ribulose bisphosphate carboxylase from an unfolded state depends on two chaperonin proteins and Mg-ATP. Nature 342, 884-889.

Hageman, J., Robinson, C., Smeekens, S., and Weisbeek, P. (1986). A thylakoid processing protease is required for complete maturation of the lumen protein plastocyanin. Nature 324, 567-569.

Hageman, J., Baecke, C., Ebskamp, M., Pilon, R., Smeekens, S., and Weisbeek, P. (1990). Protein import into and sorting inside the chloroplast are independent processes. Plant Cell 2, 479-494.

Halpin, C., Elderfield, P.D., James, H.E., Zimmermann, R., Dunbar, B., and Robinson, C. (1989). The reaction specificities of the thylakoidal processing peptidase and Escherichia coli leader peptidase are identical. EMBO J. 8, 3917-3921.

Hartl, F.-U., Pfanner, N., Nicholson, D.W., and Neupert, W. (1989). Mitochondrial protein import. Biochim. Biophys. Acta 988, 1-45.

Hemmingsen, S.M., Woolford, C., Van der Vies, S., Tilly, K., Dennis, D.T., Georgopoulos, C.P., Hendrix, R.W., and Ellis, R.J. (1988). Homologous plant and bacterial proteins chaperone oligomeric protein assembly. Nature 333, 330-334.

Hinz, G. and Flügge, U.I. (1988). Phosphorylation of a 51-kDa envelope membrane polypeptide involved in protein translocation into chloroplasts. Eur. J. Biochem. 175, 649-659.

James, H.E., Bartling, D., Musgrove, J.E., Kirwin, P.M., and Herrmann, R.G. (1989). Transport of proteins into chloroplasts import and maturation of precursors to the 33-kda 23-kda and 16-kda proteins of the photosynthetic oxygen-evolving complex. J. Biol. Chem. 264, 19573-19576.

Keegstra, K., Bauerle, C., Lubben, T., Smeekens, S., and Weisbeek, P. (1987). Targeting of proteins into chloroplasts. In "UCLA Symposia on Molecular and Cellular Biology New Series, Vol. 62. Plant Gene Systems and Their Biology" (J.L. Key and L. McIntosh. eds.) pp. 363-370. Alan R. Liss, Inc., New York, N.Y. USA.

Kirwin, P.M., Elderfield, P.D., Williams, R.S., and Robinson, C. (1988). Transport of proteins into chloroplasts organization orientation and lateral distribution of the plastocyanin processing peptidase in the thylakoid network. J. Biol. Chem. 263, 18128-18132.

Klösgen, R.B., Saedler, H., and Weil, J.-H. (1989). The amyloplast-targeting transit peptide of the waxy protein of maize also mediates protein transport in vitro into chloroplasts. Mol. Gen. Genet. 217, 155-161.

Lubben, T.H., Donaldson, G.K., Viitanen, P.V., and Gatenby, A.A. (1989). Several proteins imported into chloroplasts form stable complexes with the GroEL-related chloroplast molecular chaperone. Plant Cell 1, 1223-1230.

Murakami, H., Blobel, G., and Pain, D. (1990). Isolation and characterization of the gene for a yeast mitochondrial import receptor. Nature 347, 488-491.

Murakami, K. and Mori, M. (1990). Purified presequence binding factor (PBF) forms an import-competent complex with purified mitochondrial precursor protein. EMBO J. 9, 3201-3206.

Musgrove, J.E., Johnson, R.A., and Ellis, R.J. (1987). Association of the ribulosebisphosphate-carboxylase large-subunit binding protein into dissimilar subunits. Biochemistry 163, 529-534.

Olsen, L.J., Theg, S.M., Selman, B.R., and Keegstra, K. (1989). ATP is required for the binding of precursor proteins to chloroplasts. J. Biol. Chem. 264, 6724-6729.

Pain, D. and Blobel, G. (1987). Protein import into chloroplasts requires a chloroplast ATPase. Proc. Natl. Acad. Sci. USA 84, 3288-3292.

Pain, D., Kanwar, Y.S., and Blobel, G. (1988). Identification of a receptor for protein import into chloroplasts and its localization to envelope contact zones. Nature 331, 232-237.

Pain, D., Murakami, H., and Blobel, G. (1990). Identification of a receptor for protein import into mitochondria. Nature 347, 444-449.

Payan, L.A. and Cline, K. (1991). A stromal factor maintains the solubility and insertion competence of an imported thylakoid membrane protein. J. Cell Biol. 112, 603-613.

Pfanner, N. and Neupert, W. (1990). The mitochondrial protein import apparatus. Ann. Rev. Biochem. 59, 331-351.

Pilon, M., De Boer, A.D., Knols, S.L., Koppelman, M.H.G.M., Van der Graaf, R.M., De Kruijff, B., and Weisbeek, P.J. (1990). Expression in Escherichia coli and purification of a translocation-competent precursor of the chloroplast protein ferredoxin. J. Biol. Chem. 265, 3358-3361.

Robinson, C. and Ellis, R.J. (1984). Transport of proteins into chloroplasts partial purification of a chloroplast protease involved in the processing of imported precursor polypeptides. Eur. J. Biochem. 142, 337-342.

Roy, H., Bloom, M., Milos, P., and Monroe, M. (1982). Studies on the assembly of large subunits of ribulose bisphosphate carboxylase in isolated pea chloroplasts. J. Cell Biol. 94, 20-27.

Salomon, M., Fisher, K., Flügge, U.I., and Soll, J. (1990). Sequence analysis and protein import studies of an outer chloroplast envelope polypeptide. Proc. Natl. Acad. Sci. USA 87, 5778-5782.

Schneider, A., Behrens, M., Scherer, P., Pratje, E., Michaelis, G., and Schatz, G. (1991). Inner membrane protease I, an enzyme mediating intramitochondrial protein sorting in yeast. EMBO J. 10, 247-254.

Schnell, D.J., Blobel, G., and Pain, D. (1990). The chloroplast import receptor is an integral membrane protein of chloroplast envelope contact sites. J. Cell Biol. 111, 1825-1838.

Smeekens, S., Bauerle, C., Hageman, J., Keegstra, K., and Weisbeek, P. (1986). The role of the transit peptide in the routing of precursors toward different chloroplast compartments. Cell 46, 365-376.

Smeekens, S., Geerts, D., Bauerle, C., and Weisbeek, P. (1989). Essential function in chloroplast recognition of the ferredoxin transit peptide processing region. Mol. Gen. Genet. 216, 178-182.

Smith, S.M. and Ellis, R.J. (1979). Processing of small subunit precursor of ribulose bisphosphate carboxylase and its assembly into whole enzyme are stromal events. Nature 278, 662-664.

Theg, S.M., Bauerle, C., Olsen, L.J., Selman, B.R., and Keegstra, K. (1989). Internal ATP is the only energy requirement for the translocation of precursor proteins across chloroplastic membranes. J. Biol. Chem. 264, 6730-6736.

Von Heijne, G., Steppuhn, J., and Herrmann, R.G. (1989). Domain structure of mitochondrial and chloroplast targeting peptides. Eur. J. Biochem. 180, 535-545.

Von Heijne, G. and Nishikawa, K. (1991). Chloroplast transit peptides: the perfect random coil? FEBS Lett. 278, 1-3.

Waegemann, K., Paulsen, H., and Soll, J. (1990). Translocation of proteins into isolated chloroplasts requires cytosolic factors to obtain import competence. FEBS Lett. 261, 89-92.

Willey, D.L., Auffret, A.D., and Gray, J.C. (1984). Structure and topology of cytochrome f in pea chloroplast membranes. Cell 36, 555-562.

TARGETING OF PROTEINS INTO AND ACROSS

THE CHLOROPLASTIC ENVELOPE

Kenneth Keegstra, Hsou-min Li, Jerry Marshall, Jennifer Ostrom
and Sharyn Perry

Department of Botany
University of Wisconsin
Madison, WI 53706

INTRODUCTION

Chloroplasts are functionally complex organelles that perform a diverse array of metabolic processes in addition to their well known role in photosynthesis. Consistent with their functional complexity, chloroplasts are structurally complex organelles, possessing three different lipid bilayer membranes enclosing three different aqueous compartments (Fig. 1). Because of the limited coding capacity of the plastid genome, most of the proteins in each compartment are encoded by nuclear genes and synthesized in the cytoplasm (Fig. 1). Understanding how these precursor proteins are targeted from the cytoplasm to their proper location within chloroplasts is an intriguing problem that has received considerable attention in recent years. In this chapter, we will briefly summarize relevant work from other laboratories as well as some recent results from our laboratory, all of which are aimed at understanding the import and proper localization of cytoplasmically synthesized chloroplastic proteins.

TRANSPORT ACROSS THE ENVELOPE MEMBRANES

Several lines of evidence indicate that cytoplasmically synthesized proteins destined for the stroma, the thylakoid membrane and the thylakoid lumen share a common first step, i.e. transport across the envelope membranes. Studies with chimeric precursor proteins, in which transit peptides have been switched between the precursor for the light-harvesting chlorophyll a/b protein (prLHCP) and the precursor for the small subunit of ribulose-1,5-bisphosphate carboxylase (prSS), have demonstrated that the transit peptide of prLHCP functions as a stromal targeting domain (Hand et al., 1989; Lamppa, 1988). Moreover, the transit peptides of precursors destined for the thylakoid lumen contain two regions (Smeekens et al., 1986), one of which functions as a stromal targeting domain (Hageman et al., 1990). Thus, precursors destined for both the thylakoid membrane and the thylakoid lumen have transit peptides that function to transport proteins across the envelope membranes to the stromal space.

Regulation of Chloroplast Biogenesis
Edited by J.H. Argyroudi-Akoyunoglou, Plenum Press, New York, 1992

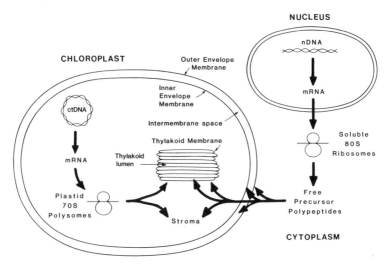

Figure 1. A schematic representation of the structural complexity of chloroplasts and the targeting of cytoplasmically synthesized proteins to the various compartments.

Binding of Precursors to the Chloroplastic Surface

Transport across the envelope membranes can be divided into at least two discrete stages. The first is the binding of precursor proteins to the surface of chloroplasts and the second is the translocation of bound precursors across the two envelope membranes (Keegstra *et al.*, 1989). Specific binding can be detected only if the translocation step is blocked. Translocation proceeds slowly or not at all below 10°C (Bauerle *et al.*, 1991), whereas binding occurs readily at low temperatures. Thus, low temperatures can be used to study specific binding. Another method is to limit the ATP concentration. It was originally thought that precursor binding did not have an energy requirement, but more recently it has been reported that precursor binding requires ATP or other nucleoside triphosphates (Olsen *et al.*, 1989). The ATP requirement for binding can be distinguished by several criteria from a separate ATP requirement for translocation (Theg *et al.*, 1989). One useful method is to limit the concentration of ATP because binding requires only low levels of ATP, whereas translocation requires much higher levels.

Bound precursors are stably associated with the envelope membranes and are no longer in equilibrium with free precursors (Olsen *et al.*, 1989). Indeed, it may be more appropriate to consider bound precursors as a transport intermediate blocked at an early stage in the translocation process. We have recently taken advantage of the stable association between precursors and the chloroplastic surface to localize the bound precursors (Ostrom and Keegstra, manuscript in preparation). Chloroplasts containing bound precursors were lysed and the envelope membranes isolated by sucrose density gradient centrifugation (Cline *et al.*, 1985). Surprisingly, the bound precursors were not found in the fractions containing outer envelope membranes. Rather, precursors were found in a region of the gradient containing mixed envelope membranes. Membranes in this region of the gradient had previously been demonstrated to contain putative contact sites (Cline *et al.*, 1985). These observations provide support for the popular

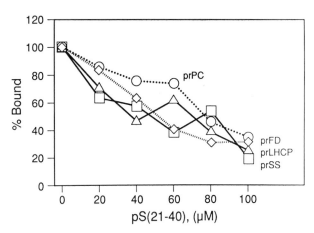

Figure 2. Inhibition of the binding of four different precursors by various concentrations of peptide pS(21-40), corresponding to residues 21-40 of the transit peptide of pea prSS.

hypothesis that protein transport occurs at special contact sites, where the two membranes of the chloroplastic envelope are tightly appressed.

One important question regarding precursor binding is the substrate specificity of the putative receptors involved in recognizing precursors. This is a particularly intriguing question in that the transit peptides of different precursors have no similarities in primary sequence that could serve as the motif for recognition by receptors (Keegstra *et al.*, 1989). Thus, the question arises whether a single receptor recognizes many different precursors or whether many different receptors are present in the envelope membranes. We have recently addressed this question by investigating the ability of synthetic peptides corresponding to regions of the transit peptide from one precursor to inhibit the binding of other precursors (Perry *et al.*, 1991). We found that a synthetic peptide corresponding to the middle region of the transit peptide from pea prSS could inhibit the binding of prSS to chloroplasts. The next step was to investigate whether this peptide could also inhibit the binding of other precursors that have no sequence similarity with this peptide. As shown in Fig. 2, the peptide could inhibit the binding of the precursor to ferredoxin, another stromal protein, the precursor to plastocyanin, a thylakoid lumen protein as well as the prLHCP, an integral thylakoid membrane protein. Thus, we conclude that precursors destined for all these different compartments interact with the same receptor on the chloroplastic surface.

Another important question regarding precursor binding is the identity of the putative receptor protein. Several different approaches have been utilized in attempts to identify a receptor protein. These efforts have led to the identification of at least two different proteins that may be components of the transport apparatus. Cornwell & Keegstra (1987) used a heterobifunctional, photoactivatable cross-linking reagent to identify a 66 kDa surface protein that was associated with bound prSS. Further work needs to be conducted to determine whether this protein is a receptor or some other component of the transport apparatus.

Pain *et al.* (1988) employed a very different strategy and came to a very different conclusion. They raised antibodies against antibodies that were directed against a synthetic peptide analog of the carboxyl-terminal 30 residues of the transit peptide of pea prSS. These anti-idiotypic antibodies mimic the transit peptide and presumably interact with the same binding site as the transit peptide. When used for immunoblotting the antibodies reacted with a 30 kDa protein of the envelope membranes. In subsequent work, Schnell *et al.* (1990), isolated a cDNA clone encoding this 30 kDa envelope membrane protein. Surprisingly, sequence analysis reveals that it is the same protein that others (Flügge *et al.*, 1989; Willey *et al.*, 1991) had identified as the phosphate translocator protein. Although it is possible that one protein could have two functions, it seems unlikely that one protein could be present in two locations. A precursor receptor should be in the outer envelope membrane, possibly at contact sites, whereas the phosphate translocator should be in the inner membrane. Further work will be needed to determine the precise location and function of this 30 kDa envelope polypeptide.

Translocation of Precursors across the Envelope Membranes

Under the proper conditions, i.e., with adequate temperatures and sufficient levels of ATP, precursor polypeptides are translocated across the two envelope membranes to the stromal space. Although mechanistic details of this translocation process are unclear, it is clear that translocation requires energy in the form of ATP (Flügge and Hinz, 1986; Pain and Blobel, 1987; Theg *et al.*, 1989). This ATP can be provided either via photophosphorylation or by adding exogenous ATP. As noted earlier, if ATP is provided exogenously, the levels of ATP needed are quite high, approximately 1 to 2 mM in the external solution. However, the ATP needed for protein translocation is utilized inside chloroplasts (Pain and Blobel, 1987; Theg *et al.*, 1989), so the concentration needed at the site of utilization is unknown. Protein transport across the envelope membranes does not require a membrane potential, as is the case for mitochondria (Neupert *et al.*, 1990).

Although the molecular mechanism of protein translocation is unknown, the most popular hypothesis posits that precursor proteins are "unfolded" and translocated through a proteinaceous pore present at contact sites (Neupert *et al.*, 1990). The evidence for such a hypothesis comes primarily from mitochondrial studies, but it seems reasonable that a similar mechanism might also apply to chloroplasts. The identification of the proteinaceous pore has not been accomplished in either mitochondria or chloroplasts. However, in mitochondrial systems, evidence has been presented that members of the hsp70 family of proteins are important, both in the cytoplasm and inside mitochondria, in "unfolding" and "refolding" the precursor protein. The best evidence comes from genetic studies with yeast, where a mutation in *ssc1*, a nuclear gene that encodes a mitochondrial hsp70, prevents protein transport into mitochondria (Kang *et al.*, 1990). It has been postulated that mitochondrial hsp70 interacts with the translocated protein as it emerges from the translocation apparatus, possibly providing some of the driving force for the vectorial translocation (Neupert *et al.*, 1990). Because the release of proteins from hsp70 requires ATP, this interaction may also account for the internal ATP requirement for protein translocation.

We recently identified at least two, and possibly three different hsp70 proteins in chloroplasts (Marshall *et al.*, 1990). Subsequently, a cDNA clone encoding one of the stromal hsp70 proteins has been isolated and sequenced (Marshall and Keegstra, manuscript in preparation). Sequence analysis shows that it has many similarities to the mitochondrial hsp70 proteins, but is most similar to an hsp70 from *Synecocystis* (Chitnis and Nelson, 1991). It has still not been established whether this, or any of the chloroplastic hsp70 proteins, have a role in protein import into chloroplasts.

Table 1. Comparison of the Import of Internal Proteins with
the Insertion of Outer Membrane Proteins

Feature	Internal Proteins	Outer Membrane Proteins
Larger Precursors	+	-
ATP Required	+	-
Thermolysin-sensitive Receptors Involved	+	-

TARGETING OF PROTEINS TO THE ENVELOPE MEMBRANES

In addition to the three compartments located inside chloroplasts (Fig. 1), there are three additional compartments that are part of the chloroplastic envelope. Although it is clear that most, if not all, envelope proteins are synthesized in the cytoplasm, little is know about the mechanisms that direct them to the envelope.

Targeting to the Outer Envelope Membrane

Recent work with two different proteins of the outer envelope membrane provide evidence that the targeting of proteins to this location is distinctly different than the pathway described above for transporting proteins across the two envelope membranes (Salomon et al., 1990; Li et al., 1991). A summary of some of the characteristics that distinguish the two processes is presented in Table I. One of the major differences is that all of the proteins transported across the envelope membranes that have been studied to date are synthesized as higher molecular weight precursors. The two examples of outer membrane proteins are synthesized without cleavable transit peptides. Other distinguishing features are that transport across the envelope membranes requires ATP, whereas the insertion of proteins into the outer envelope membrane does not. Transport of proteins across the envelope membranes is diminished or abolished by mild thermolysin pretreatment of chloroplasts, whereas insertion of proteins into the outer membrane is not. For precursors that are transported across the envelope membrane, the peptide fragment corresponding to the middle of prSS transit peptide inhibits their binding to the chloroplastic surface (Fig. 2), whereas this is not true for OM14, one of the outer membrane proteins (Li et al., 1991). Further work will be required to determine whether these conclusions apply more generally to the targeting of all outer membrane proteins.

Targeting to the Inner Envelope Membrane

The transport of cytoplasmically synthesized proteins to the inner envelope membrane is more similar to the transport of proteins across the envelope membrane than it is to the insertion of proteins into the outer envelope membrane. The transport of several inner membrane proteins has been studied (Flügge et al., 1989; Dreses-Werringloer et al., 1991; Li and Keegstra, unpublished observations), and in each case, the proteins are synthesized as higher molecular weight precursors. The import of these precursors requires energy and a thermolysin-sensitive component on the surface of chloroplasts. Although it has not yet been proven that inner membrane proteins interact with the same transport apparatus as other internal proteins, this conclusion seems likely. The question then arises as to how these proteins are directed to the inner envelope membrane. Two hypotheses can be considered, based on analogy with the targeting of proteins to the inner mitochondrial membrane. The first is the "stop-transfer" hypothesis (van Loon and Schatz, 1987). In this situation, the inner membrane proteins contain a stop-transfer signal that halts their translocation across the inner envelope membrane and functions to anchor the protein in the membrane. This hypothesis has been suggested for certain proteins of the inner mitochondrial membrane (Mahlke *et al.*, 1990). The second possibility is the "conservative sorting" hypothesis (Mahlke *et al.*, 1990). In this case, the inner membrane proteins are transported into the organelle, where the targeting signal is removed and then the processed protein is inserted into the inner envelope membrane. Although this pathway seems convoluted, it makes sense in terms of the postulated endosymbiotic origin of mitochondria and chloroplasts. Further work will be required to determine which of these hypotheses applies to proteins of the chloroplastic inner envelope membrane.

LITERATURE

Bauerle, C., Dorl, J.,and Keegstra, K., 1991, Kinetic analysis of the transport of thylakoid lumenal proteins in experiments using intact chloroplasts, J.Biol.Chem., 266:5884.

Chitnis, P.R.,and Nelson, N., 1991, Molecular cloning of the genes encoding two chaperone proteins of the cyanobacterium *Synechocystis* sp.PCC 6803, J. Biol. Chem., 266:58.

Cline, K., Keegstra, K.,and Staehelin, L.A., 1985, Freeze-fracture electron microscopic analysis of ultrarapidly frozen envelope membranes on intact chloroplasts and after purification, Protoplasma, 125:111.

Cornwell, K.L.,and Keegstra, K., 1987, Evidence that a chloroplast surface protein is associated with a specific binding site for the precursor to the small subunit of ribulose-1,5-bisphosphate carboxylase, Plant Physiol., 85:780.

Dreses-Werringloer, U., Fischer, K., Wachter, E., Link, T.A.,and Flügge, U.-I., 1991, cDNA sequence and deduced amino acid sequence of the precursor of the 37-kDa inner envelope membrane polypeptide from spinach chloroplasts--Its transit peptide contains an amphiphilic α-helix as the only detectable structural element, Eur.J.Biochem., 195:361.

Flügge, U.I.,and Hinz, G., 1986, Energy dependence of protein translocation into chloroplasts, Eur. J. Biochem., 160:563.

Flügge, U.I., Fischer, K., Gross, A., Sebald, W., Lottspeich, F.,and Eckerskorn, C., 1989, The triose phosphate-3-phosphoglycerate-phosphate translocator from spinach chloroplasts: nucleotide sequence of a full length cDNA clone and import of the *in vitro* synthesized precursor protein into chloroplasts, EMBO J., 8:39.

Hageman, J., Baecke, C., Ebskamp, M., Pilon, R., Smeekens, S.,and Weisbeek, P., 1990, Protein import into and sorting inside the chloroplast are independent processes, Plant Cell, 2:479.

Hand, J.M., Szabo, L.J., Vasconcelos, A.C.,and Cashmore, A.R., 1989, The transit peptide of a chloroplast thylakoid membrane protein is functionally equivalent to a stromal targeting sequence, EMBO J., 8:3195.

Kang, P.-J., Ostermann, J., Shilling, J., Neupert, W., Craig, E.A.,and Pfanner, N., 1990, Requirement for hsp70 in the mitochondrial matrix for translocation and folding of precursor proteins, Nature, 348:137.

Keegstra, K., Olsen, L.J.,and Theg, S.M., 1989, Chloroplastic precursors and their transport across the envelope membranes, Annu. Rev. Plant Physiol. Plant Mol. Biol., 40:471.

Lamppa, G.K., 1988, The chlorophyll a/b binding protein inserts into the thylakoids independent of its cognate transit peptide, J.Biol.Chem., 263:14996.

Li, H-m., Moore, T., and Keegstra, K., 1991, Targeting of proteins to the outer envelope membrane uses a different pathway than transport into chloroplasts, Plant Cell, 3:in press.

Mahlke, K., Pfanner, N., Martin, J., Horwich, A.L., Hartl, F.-U.,and Neupert, W., 1990, Sorting pathways of mitochondrial inner membrane proteins, Eur.J.Biochem., 192:551.

Marshall, J.S., DeRocher, A.E., Keegstra, K.,and Vierling, E., 1990, Identification of heat shock protein hsp70 homologues in chloroplasts, Proc. Natl. Acad. Sci. USA, 87:374.

Neupert, W., Hartl, F.-U., Craig, E.A.,and Pfanner, N., 1990, How do polypeptides cross the mitochondrial membranes, Cell, 63:447.

Olsen, L.J., Theg, S.M., Selman, B.R.,and Keegstra, K., 1989, ATP is required for the binding of precursor proteins to chloroplasts, J.Biol.Chem., 264:6724.

Pain, D.,and Blobel, G., 1987, Protein import into chloroplasts requires a chloroplast ATPase, Proc. Natl. Acad. Sci. USA, 84:3288.

Pain, D., Kanwar, Y.S.,and Blobel, G., 1988, Identification of a receptor for protein import into chloroplasts and its localization to envelope contact zones, Nature, 331:232.

Perry, S.E., Buvinger, W.E., Bennett, J., and Keegstra, K., 1991, Synthetic analogues of a transit peptide inhibit binding or translocation of chloroplastic precursor proteins, J. Biol. Chem., 266:11882.

Salomon, M., Fischer, K., Flügge, U.-I.,and Soll, J., 1990, Sequence analysis and protein import studies of an outer chloroplast envelope polypeptide, Proc. Natl. Acad. Sci. USA, 87:5778.

Schnell, D.J., Blobel, G.,and Pain, D., 1990, The chloroplast import receptor is an integral membrane protein of chloroplast envelope contact sites, J. Cell Biol., 111:1825.

Schnell, D.J., Blobel, G.,and Pain, D., 1991, Signal peptide analogs derived from two chloroplast precursors interact with the signal recognition system of the chloroplast envelope, J. Biol. Chem., 266:3335.

Smeekens, S., Bauerle, C., Hageman, J., Keegstra, K.,and Weisbeek, P., 1986, The role of the transit peptide in the routing of precursors toward different chloroplast compartments, Cell, 46:365.

Theg, S.M., Bauerle, C., Olsen, L.J., Selman, B.R.,and Keegstra, K., 1989, Internal ATP is the only energy requirement for the translocation of precursor proteins across chloroplastic membranes, J. Biol. Chem., 264:6730.

van Loon, A.P.G.M.,and Schatz, G., 1987, Transport of proteins to the mitochondrial intermembrane space: the 'sorting' domain of the cytochrome c_1 presequence is a stop-transfer sequence specific for the mitochondrial inner membrane, EMBO J., 6:2441.

Willey, D.L., Fischer, K., Wachter, E., Link, T.A.,and Flügge, U.I., 1991, Molecular cloning and structural analysis of the phosphate translocator from pea chloroplasts and its comparison to the spinach phosphate translocator, Planta, 183:451.

IMPORT AND PROCESSING OF THE MAJOR

LIGHT-HARVESTING CHLOROPHYLL A/B BINDING PROTEIN OF PSII

Gayle Lamppa, Mark Abad, Steven Clark, John Oblong

Molecular Genetics and Cell Biology
University of Chicago
920 E. 58th Street
Chicago, Il 60637 USA

INTRODUCTION

An essential step in the pathway of protein import into the chloroplast is the proteolytic removal of the N-terminal extension, the transit peptide. We have demonstrated previously that upon import wheat preLHCP is cleaved at two sites, referred to as the primary and secondary sites, which, respectively, give rise to two peptides of approximately 26 and 25 kD (Lamppa and Abad, 1987; Clark et al., 1990). We have recently identified a soluble processing enzyme of ~200 kD that cleaves preLHCP in an organelle-free assay, yielding only the 25 kD peptide (Abad et al., 1989). In the present study, the specificity of the processing enzyme was investigated. Details of this study have recently been presented (Abad et al., 1991; Clark et al., 1991). To expedite our analysis the precursor was synthesized in Escherichia coli. Three questions were addressed: 1) where within preLHCP does the processing enzyme cleave? 2) does the same enzyme that cleaves preLHCP at the secondary site in the organelle-free assay also cleave other precursors, i.e. is it a general processing enzyme? and 3) what are the structural determinants within preLHCP that are critical for cleavage at the primary and secondary sites?

RESULTS AND DISCUSSION

Processing of preLHCP synthesized in E. coli in an organelle-free reaction and during import into chloroplasts

PreLHCP encoded by the wheat Cab-1 gene (Lamppa et al., 1985) was synthesized in E. coli using a T7-directed expression system (Studier et al., 1990). At least 95% of the expressed preLHCP was found in inclusion-like bodies, which could be solubilized, after washing with 0.5% Triton, in 6 M urea. We examined if the precursor recovered from E. coli would be recognized in the organelle-free processing assay. As shown in Figure 1, and described earlier (Abad et al., 1991), the precursor was cleaved in a highly specific reaction, producing only a single peptide. Processing was inefficient, however, suggesting either that urea inhibited the reaction or the precursor was in a conformation not readily accessible to the processing enzyme. The precursor was recovered from E. coli in 8 M urea to promote complete denaturation and then used directly in an organelle-free reaction, or gradually dialyzed into 20 mM Tris, pH 8. The precursor added to the reaction from 8 M urea (final concentration 1.6 M) was not cleaved, but gradual removal of the urea generated a substrate with a conformation that was recognized in the organelle-free assay.

Regulation of Chloroplast Biogenesis
Edited by J.H. Argyroudi-Akoyunoglou, Plenum Press, New York, 1992

Radiolabelled preLHCP made in E. coli was introduced into a standard import assay. As shown in Figure 2 (right), preLHCP was found in the total membrane fraction and two new peptides of 25 and 26 kD appeared which were resistant to thermolysin treatment of the chloroplasts, providing evidence that they were within the organelle. A considerable fraction of the precursor remained on the exterior of the organelle, which could be due to a block in translocation across the envelope, or nonspecific binding. No radiolabeled protein was present in the soluble phase. We conclude that preLHCP made in E. coli can be imported and proteolytically processed at two sites, whereas only the secondary site is accessible in the organelle-free assay. These results reconstitute our earlier observations using preLHCP synthesized in a reticulocyte lysate (Lamppa et al., 1987; Abad et al., 1989; Clark et al., 1990). The availability of the E. coli-synthesized precursor should now make it possible to more clearly define the basic requirements for individual steps in the pathway of preLHCP import into the chloroplast.

Identification of the site cleaved in the organelle-free assay

The 25 kD produced during organelle-free processing was separated by SDS-PAGE, and then transferred to PVDF membrane for N-terminal sequence

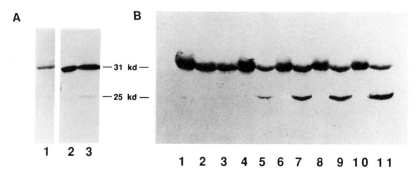

Fig. 1. Organelle-free processing of preLHCP recovered from E. coli. A. The protein was analyzed by SDS-PAGE (lane 1). LHCP-specific antisera in Western blot analysis identified a 31 kD protein (lane 2) recovered in 6 M urea, which was cleaved (lane 3). B. preLHCP was resuspended in 8 M urea and used in an organelle-free assay at a final concentrations of 1.6, 0.9, 0.6, 0.3, and 0 M urea (lanes 1, 4, 6, 8, 10). The products (lanes 2, 5, 7, 9 and 11) were analyzed by Western blotting. The 0 M urea sample of preLHCP was also added to a reaction containing 1.6 M urea (lane 3).

analysis by Edman degradation. The sequence AKQVSSS was obtained, which begins in wheat preLHCP at residue 41, numbered from the initiator methionine. The N-terminus of the major LHCP species of pea thylakoids begins with [K,R]SATTKK (Mullet, 1983). By comparing this sequence with wheat LHCP, we inferred that the N-terminal sequence of the major endogenous wheat LHCP species starts with RKTAAKAKQ. Thus, processing of preLHCP between residues 40 and 41 at the secondary site removes not only the transit peptide, but also a basic hexapeptide RKTAAK normally thought be to present on all LHCP molecules of Type I associated with PSII. This domain has been implicated in thylakoid stacking (Mullet, 1983), and contains a threonine residue found to be a preferred site for phosphorylation (Mullet, 1983; Michel and Bennet, 1989). We predict that in vivo there are at least two forms of LHCP with distinct N-termini produced by selective cleavage of a single precursor polypeptide encoded by a Type I gene.

Organelle-free processing of the E. coli synthesized precursor liberated a single peptide. To determine if this peptide co-migrated ith either of the two products (25 and 26 kD) typically observed upon transport of preLHCP into the chloroplast, radiolabeled preLHCP was synthesized in E. coli, and introduced into an organelle-free reaction. The products of the reaction were analyzed by SDS-PAGE alongside the products of an import reaction using preLHCP synthesized in a reticulocyte lysate (Figure 2, left). When processed in the organelle-free assay, the radiolabeled E. coli precursor gave rise to one peptide that migrated at the same position as the 25 kD peptide found upon import of preLHCP made in the reticulocyte lysate.

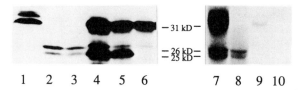

1 2 3 4 5 6 7 8 9 10

Figure 2. Organelle-free processing and chloroplast import of preLHCP made in E. coli. PreLHCP was synthesized in a reticulocyte lysate and the products (lane 1) were imported into chloroplasts. Lanes 2 and 3: protein from chloroplasts untreated and treated with thermolysin before lysis. Radiolabeled E. coli synthesized preLHCP (lane 6) was recovered in 8 M urea, dialyzed to 0 M urea and both stocks were used in an organelle-free reaction and the products analyzed (lanes 4 and 5). PreLHCP made in E. coli was used in an import reaction and the membrane (lanes 7 and 8) and soluble fractions (lanes 9 and 10) were analyzed. Lanes 8 and 10: chloroplasts were treated with thermolysin before lysis.

Substrate specificity of the processing enzyme

To determine whether the soluble enzyme that cleaves preLHCP to the 25 kD form is related to the stromal endopeptidase that processes other precursors, we partially purified the enzyme (see Abad et al., 1991). Five other precursors were tested to see if they would be processed by the same peak of activity recovered from a Superose 6 sizing column, the last step in the purification scheme. At this stage the processing enzyme eluted as an approximately 200 kD species, and only 25 proteins remained detectable by silver-staining of an SDS-acrylamide gel. cDNAs encoding the precursors for the small subunit of Rubisco (S), Rubisco activase (RA), plastocyanin (PC), hsp 21, and acyl carrier protein (ACP) were used to generate in vitro translation products for organelle-free processing reactions. We have found that the identical peak of activity, eluting from the Superose 6 column, that cleaves preLHCP also processed these other precursors to their mature forms, suggesting that they are substrates for a common protease. Purification of the processing enyzme to homogeneity will unequivocally establish this point.

A basic residue is essential for primary site cleavage of preLHCP

Transit peptides differ considerably even for the same protein between species, but sequence comparisons have revealed several potentially critical features for processing (Gavel and von Heijne, 1990). For example, most transit peptides (80%) have an arginine or lysine positioned between -1 and -7 relative to the site cleaved. In light of the prevalence of basic amino acids in this region we have investigated the importance of a basic residue amino-proximal to the primary cleavage site of preLHCP. Pea preLHCP contains an arginine at position -4; there is also a glutamic acid at position -9 which was changed in order to determine the importance of a balanced charge in this domain. Substitution of leucine for glutamic acid had no effect on processing of the modified precursor, pea:preLHCP[l], during import into chloroplasts. However, when valine was substituted for the arginine, creating pea:preLHCP[v], processing at the primary site was inhibited. Only 20% of the imported protein was cleaved to yield the 26 kD peptide which usually accounts for 85% of the radiolabeled mature protein (Figure 3, compare lanes 2 and 8). Processing at the secondary site of the precursor was not inhibited, and may have been slightly enhanced. There was also an accumulation of a high molecular mass intermediate which was thermolysin resistant. A time course of import showed that this intermediate is relatively stable (Figure 3, right panel, lanes 13-16).

The arginine at position -4 was also changed to lysine (see

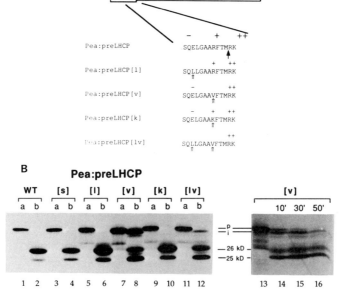

Figure 3. A basic residue is required for efficient primary site cleavage of preLHCP. A. Schematic representation of each precursor polypeptide (TP, transit peptide). Directly below are the amino acid sequences of the primary processing site regions coded for by the mutant constructs. The letters in brackets identify each mutant. B. The translation products for wildtype (WT) and mutant precursors (lanes a) were incubated with chloroplasts for 30 min., and then the membrane fractions of thermolysin-treated organelles (lanes b) were analyzed by SDS-PAGE. Translation products for pea:preLHCP[v] (lane 13) were incubated in a time course experiment for 10 (lane 14), 30 (lane 15) and 50 (lane 16) minutes. p, precursor; i, intermediate.

pea:preLHCP[k]) to determine if a basic residue, or arginine per se, is critical for primary site cleavage. This substitution had no affect on primary site recognition upon import (Figure 3). To maintain the net charge of the transit peptide (+2), a double mutant, pea:preLHCP[lv], was constructed in which leucine replaced glutamic acid and valine replaced arginine. This mutant precursor also showed an inhibition of primary site processing; only 40% of the imported protein was cleaved to the 26 kD form (Figure 3).

Similarly, changes in wheat:preLHCP were made and the modified precursors were employed in import reactions. An arginine at position -4 was replaced with leucine, and an aspartic acid at -6 with asparagine. The latter had no affect on processing, but the loss of the arginine again lead to an 80% reduction in the level of the 26 kD peptide produced (see Clark and Lamppa, 1991). Taken together, our analyses demonstrate that an amino-proximal basic residue is necessary for primary site cleavage of preLHCP.

The hypothesis that a basic residue is a general requirement for transit peptide removal was tested by extending our analysis to preS and preRA. We find that preS and preRA do not require a basic amino acid within seven amino acids of the cleavage site for maturation.

Determinants for secondary site processing

Both the wheat and pea precursors give rise to two peptides upon import, but the relative amounts of the 26 and 25 kD mature forms consistently differ. Pea:preLHCP is cleaved predominantly at the primary site, giving a 26/25 kD peptide ratio of about 5:1, while wheat:preLHCP is cleaved almost equally at both sites to produce a 26/25 kD peptide ratio of 1.5:1. The pea and wheat precursors contain two regions of sequence divergence: the transit peptide and the first 10 residues of the "mature" protein. Hybrids precursors were constructed in which the transit peptides were exchanged to investigate the relative importance of these two regions in directing selective cleavage at the primary and secondary sites. The hybrid whe/pea:preLHCP contained the wheat transit peptide joined to the pea mature protein, and in pea/whe:preLHCP the pea transit peptide was joined to the wheat mature protein. The results of incubating the hybrid precursors with intact chloroplasts are presented in Figure 4. In both cases, processing was determined by the nature of the mature protein and not the transit peptide. That is, whe/pea:preLHCP produced a 26/25 kD peptide ratio of 5.2:1, almost identical to the wildtype pea precursor. The hybrid pea/whe:preLHCP produced a 26/25 kD peptide ratio of 1.6:1, characteristic of the wildtype wheat precursor. In summary, the determinants for utilization of the secondary site reside carboxy to the primary site in the N-terminus of "mature" LHCP.

An AKA motif promotes efficient secondary site processing

The secondary processing site of wheat preLHCP occurs between lys40 and ala41 within the sequence AKAK (underlined), as described above. This sequence is not found at the N-terminus of pea LHCP which has a TTKK in the same position. We converted the three different residues, TTK, in the pea substrate to AKA (pea:preLHCP[aka]), and conversely, changed the AKA to TTK in the wheat precursor (whe:preLHCP[ttk]) to investigate if this would be sufficient to alter secondary site cleavage. When pea:preLHCP[aka] was used in an import reaction, it was processed efficiently at the secondary site to yield a 26/25 kD ratio of 1.2:1. In sharp contrast, whe:preLHCP[ttk] showed a complete loss of secondary site cleavage. In addition, when whe:preLHCP[ttk] was used in an organelle-free reaction it was not processed to any detectable level (Figure 4).

These results indicate that a limited number of residues can confer specificity for secondary site processing. Interestingly, the sequence KAK (underlined below) is found in two-thirds of the precursors in the Type I class, which all share the sequence motif RKT(V/A)(T/A)KAK at the N-terminus of "mature" LHCP (Demmin et al., 1989). We propose that precursors with the same sequence motif as the wheat precursor have the potential to be cleaved at two sites during import. Based on the observation that a 4 amino acid substitution at the transit peptide-mature protein junction can abolish secondary site processing indicates, however, that the presence of the AKA motif alone is not sufficient for cleavage but must be localized within the correct structural context, i.e. conformation of the precursor.

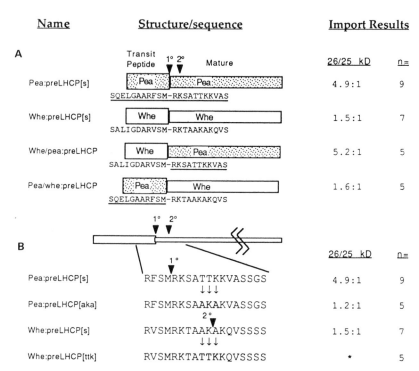

Figure 4. Analysis of cleavage at the primary and secondary sites of preLHCP from pea and wheat. The name of each precursor is indicated on the left, and the structure is shown diagrammatically with the amino acids at the transit peptide-mature protein junction given immediately below. The hyphen indicates the cleavage site. The ratio of the 26 and 25 kD peptides observed during import is given at right (26/25 kD), averaged over *n* experiments. A. Analysis of import using the hybrid precursors. B. Analysis using the secondary sites mutants. The asterisk denotes the fact that whe:preLHCP[ttk] produced no detectable 25 kD peptide during import.

SUMMARY

It has been demonstrated that preLHCP has two independently utilized cleavage sites separated by 6 amino acids that can be used to remove the transit peptide. Cleavage at the secondary site produces an alternate form of mature LHCP lacking a basic N-terminal domain. We propose that in vivo, besides being encoded by a multigene family, there is another strategy for generating LHCP heterogeneity.

ACKNOWLEDGEMENTS

Figures 1 and 2 and portions of related text were reproduced from
Plant Physiology (Abad, Oblong, and Lamppa, 1991) with the permission of
the American Society of Plant Physiology. Figures 3 and 4 with
accompanying text were reproduced from the Journal of Cell Biology (Clark
and Lamppa, 1991) by copyright permission of the Rockefeller University
Press. This work was supported by a National Institutes of Health Grant
GM36419 to G. K. L.

REFERENCES

Abad, M.S., Clark, S.E., and Lamppa, G.K., 1989, Properties of a
    chloroplast enzyme that cleaves the chlorophyll a/b binding protein
    precursor. Plant Physiol. 90: 117-124.
Abad, M. S., Oblong, J., E. and Lamppa, G. K., 1991, A soluble chloroplast
    enzyme cleaves preLHCP made in E. coli to a mature form backing a
    basic N-terminal domain. Plant Physiol. 96: 1221-1228.
Clark, S.E., Abad, M.S., and Lamppa, G.K., 1989, Mutations at the transit
    peptide-mature protein junction separate two cleavage events during
    chloroplast import of the chlorophyll a/b-binding protein. J. Biol.
    Chem. 264: 17544-17550.
Clark, S. E., Oblong, J. E., and Lamppa, G. K., 1990, Loss of efficient
    import and thylakoid insertion due to N- and C-terminal deletions in
    the light-harvesting chlorophyll a/b binding protein. Plant Cell 2:
    173-184.
Clark, S. E. and Lamppa, G. K., 1991, Determinants for cleavage of the
    chlorophyll a/b binding protein precursor: a requirement for a basic
    reside that is not universal for chloroplast imported proteins. J.
    Cell Biol. 114: in press.
Demmin, D. S., Stockinger, E. J., Change, Y. C., and Walling, L. L.,
    1989, Phylogenetic relationships between the chloropohyll a/b
    binding protein 9CAB) multigene family: an intra-interspecies study.
    J. Mol. Evol. 29: 266-279.
Gavel, Y. and von Heijne, G., 1990, A conserved cleavage-site motif in
    chloroplast transit peptides. FEBS Lett. 261: 455-458.
Lamppa, G.K., Morelli, G., and Chua, N.-H., 1985, Structure and
    developmental regulation of a wheat gene encoding the major
    chlorophyll a/b binding polypeptide. Mol. Cell. Biol. 5, 1370-1378.
Lamppa, G.K. and Abad, M.S., 1987, Processing of a wheat light-harvesting
    chlorophyll a/b protein precursor by a soluble enzyme from higher
    plant chloroplasts. J. Cell Biol. 105: 2641-2648.
Michel, H.P. and Bennett, J., 1989, Use of synthetic peptides to study the
    substrate specificity of a thylakoid protein kinase. FEBS Lett. 254:
    165-170.
Mullet, J.E., 1983, The amino acid sequence of the polypeptide segment
    which regulates membrane adhesion (grana stacking) in chloroplasts.
    J. Biol. Chem. 258: 9941-9948.
Studier, R. W., Rosenberg, A. H. Dunn, J.J., and Dubendorff, J. W., 1990,
    Use of T7 RNApolymerase to idrect expressionof cloned genes. Methods
    Enzymol. 185: 60-89.

PROCESSING OF THE LIGHT-HARVESTING CHLOROPHYLL a/b PROTEIN II PRECURSOR - POSSIBLE INTERMEDIATE STEPS EXAMINED BY USING TRUNCATED PRECURSORS

Amos Sommer, Emma Ne'eman and Eitan Harel

Botany Department, The Hebrew University, Jerusalem 91904
Israel

INTRODUCTION

Light-harvesting chlorophyll a/b protein II (LHCP) is encoded in the nucleus and synthesized in the cytoplasm as a precursor (pLHCP). It is imported into chloroplasts, processed to the mature form and assembled in the light-harvesting complex of PS II[1]. Processing was reported to occur in one step in the stroma[2,3]. However, the precursor has been observed in thylakoids[4] and processing was reported to take place there[5]. A truncated precursor was transiently observed during import of pLHCP and was suggested to be an envelope-bound processing intermediate[6]. We have observed two truncated forms of pLHCP, intermediate in size between the precursor and mature proteins. These two forms were observed during import by intact plastids or after incubation of the precursor with plastid lysates or extracts of lumen proteins. The targeting and intra-organellar sorting of nuclear-encoded plastidial[7] and mitochondrial[8] proteins often occurs in several steps, e.g. lumen proteins[9] and Rubisco small subunit[10]. It is assumed that the transit peptide can carry composite information that determines targeting to the organelle or to a particular compartment in it[11,12]. We examined the possibility that the two truncated forms of pLHCP observed, are intermediates in its import and processing.

MATERIALS AND METHODS

AB30 gene from Lemna gibba was used in all experiments[13]. Most of the methods and techniques employed - DNA cloning, in vitro transcription and translation, isolation of plastids and sub-organellar fractions, import of pLHCP, insertion into thylakoids, protein fractionation etc., were performed essentially as previously described[4,5,13,14]. Intact plastids were collected from Percoll gradients. Envelopes, thylakoids and stromal proteins were isolated from washed, intact plastids following import of pLHCP, by gently bursting the organelles and immediately fractionating on discontinuous sucrose gradients[15]. Lumen proteins were isolated after disrupting washed thylakoids with a Yeda Press and keeping the 250,000 g supernatant. Canavanyl-containing pLHCP was prepared by in vitro translation of AB30 mRNA in the presence of L-canavanine instead of L-arginine. The deletion mutants Da and Db were prepared by oligonucleotide-directed mutagenesis[16]. Mutants were isolated by colony hybridization and verified by sequencing.

RESULTS AND DISCUSSION

Small amounts of the two putative intermediate forms, designated Ia and Ib, were observed in thylakoids after import of pLHCP by intact plastids (Fig. 1A). Ia was observed more consistently, in plastids from a variety of plants. Both Ia and Ib were observed after treatment of intact plastids with thermolysin at the end of import and must therefore reside in thylakoids or inner envelope membranes. When import was slowed down by the addition of apyrase and transfer to darkness, Ia and Ib became more evident (Fig. 1A). Ia is shorter than pLHCP by about 0.5 kD and corresponds apparently to the thylakoid-bound precursor reported earlier[4,5]. It showed a precursor-product relationship with mature LHCP (Fig. 2) and resembled in its behaviour the putative intermediate observed by Kohorn and Tobin[6] except for the sensitivity of the latter to thermolysin. Ib is shorter than pLHCP by a further 1.5 kD. It was observed alongside Ia after incubation of pLHCP with plastid lysates and was produced by a lumen protease (Fig. 1B) which could be extracted from thylakoids and appeared to be located on their lumenal side. The protease showed optimal activity around pH 6.0, required Ca or Mg ions and was inhibited by DTT, EGTA and citrate. Isolated thylakoids were able to convert Ia and Ib to mature LHCP to a limited extent (Harel et al. submitted).

An attempt to cause accumulation of Ia and Ib through interfering with the ultimate processing step was unsuccessful. The amino acid sequence around the processing site is rich in Arg and Lys residues (Fig. 4), a feature conserved in all the sequences reported for pLHCP II[1]. Replacing Arg residues with canavanyl resulted in progressive enhancement of pLHCP processing, rather than its inhibition, when the pH was raised from 7.0 to 9.0 (Fig. 3). This could be ascribed to the increased ionization of the guanidinoxy group of canavanine (pK=7.4) at these pH values, compared to that of the guanido group of arginine (pK=12.5). Thus, the positively charged residues appear to be important for the processing of pLHCP.

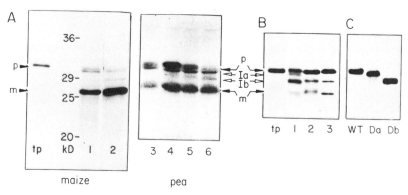

Figure 1. Possible intermediates in the processing of pLHCP by isolated plastids and sub-organellar fractions. A - Thylakoids from etiochloroplasts which were incubated with [^{35}S]-pLHCP for ten (1) or thirty min (2), then treated with thermolysin before fractionation, or which were incubated in light for 5 min, then transferred to darkness in the presence of apyrase for 0, 5, 10 and 20 min (3-6). B - pLHCP incubated for 20 min with lysates of maize (1) or pea plastids (2), or with an extract of lumen proteins (3). C - Translation products of Lemna AB30 and its deletion mutants. tp - translation products. Precursor (p) and mature (m) forms of LHCP are indicated.

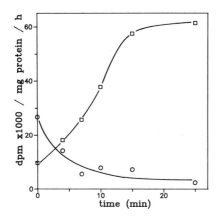

Figure 2. Disappearance of the truncated form, Ia, of pLHCP (o) from thylakoids and accumulation of LHCP (◘). Maize plastids were incubated with
[³⁵S]-pLHCP for 5 min, then an ATP-scavenging system and 0.2 μM nigericin
were added to stop further import. The plastids were resuspended in a
medium lacking pLHCP and MgATP but containing nigericin, and further incubated for the periods of time indicated.

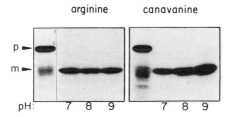

Figure 3. pH dependence of the processing of normal (Arg-containing) and
canavanine-containing [³⁵S]-pLHCP by pea chloroplasts. The left lane in each
group shows the translation product. p – precursor; m – mature LHCP.

Figure 4. Amino acid sequence of the transit peptide of AB30 pLHCP from
Lemna gibba. The residues deleted in preparing Da and Db and the positively
charged ones around the ultimate processing site (arrow) are pointed out.

We prepared two N-terminal deletions in the transit peptide of LHCP to examine the possibility that Ia and Ib are processing intermediates. In the absence of information from microsequencing, the mutants were designed to approximate Ia and Ib in size, by deleting four amino acids in Da and 15 in Db, leaving the first Met and Ala intact to allow initiation of translation (Fig. 4). In addition we assumed that intermediate processing is likely to take place in conserved regions of the transit peptide[12]. The in vitro translation products of Da and Db are shown in Fig. 1C. The import of Da and Db into intact chloroplasts was compared to that of pLHCP. The two truncated precursors were bound to envelope membranes, imported into the organelle and integrated in thylakoids. Da underwent processing to form mature LHCP, albeit at a much slower rate than pLHCP (Fig. 5). Neither the truncated forms, nor pLHCP or LHCP were observed in the stroma after 30 min import (Fig. 5). The small amounts of mature LHCP observed in the envelope fraction

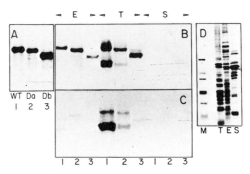

Figure 5. Distribution of native pLHCP and its deletion mutants among sub-chloroplast fractions following import into intact pea plastids. A – Translation products of AB30 (WT) and its mutants. B and C – Envelopes (E), thylakoids (T) and stromal proteins (S) from control [B] and thermolysin-treated [C] intact plastids. D – Coomassie-Blue staining of the three fractions. M – molecular weight markers: 66, 45, 36, 29, 20 and 14.2 kD.

resulted probably from contamination by thylakoid material. No contamination by envelopes was apparent in the thylakoid fraction (Fig. 5D). The thylakoid-bound Ia, Da and Db were partially resistant to thermolysin. This suggests that import and insertion, and not merely binding to envelopes, had taken place. pLHCP Da and Db were inserted into isolated thylakoids in a process dependent on MgATP and stromal proteins[14,17]. All three were stable to washing with 2 M NaBr. However, preincubation with lumen proteins markedly increased stable insertion of all three precursor forms into thylakoids (Fig. 6). None of the precursor forms underwent further processing by thylakoids alone or after the addition of stromal or lumen proteins, irrespective of whether thylakoids were incubated first with the precursor and stromal proteins then with lumen proteins, or first with the latters and then with stromal proteins (Fig. 7). Preparations of lumen proteins converted pLHCP to a form which co-migrated with mature LHCP on SDS-PAGE. Ia was discernible alongside LHCP. Da and particularly Db were also converted

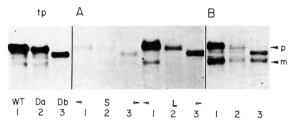

Figure 6. Enhancement of thylakoid insertion and processing of pLHCP and its deletion mutants by an extract of lumen proteins. A - The three precursor forms were preincubated with stromal (S) or lumen (L) proteins for 20 min, then isolated thylakoids were added for a further 30 min. The thylakoid membranes were washed with 2 M NaBr before fractionation of their proteins. tp - translation products, B - the three precursor forms were incubated with an extract of lumen proteins for 30 min.

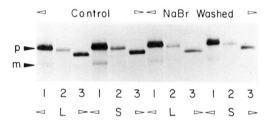

Figure 7. No further processing occurred after insertion of the three precursor forms into thylakoids. Thylakoids were incubated with the three precursor forms and lumen (L) or stromal (S) proteins for 20 min. Stromal or lumen proteins preparations respectively were then added for a further 30 min. Thylakoids were washed with buffer (control) or 2 M NaBr before fractionation of their proteins. p - precursor; m - mature form of LHCP; 1 - pLHCP, 2 - Da, 3 - Db.

to LHCP (Fig. 6B). However the LHCP produced was not inserted into thylakoids in a manner stable to washing with NaBr.

It appears that targeting to, import into plastids and sub-organellar sorting of pLHCP are not dependent on the integrity of the transit peptide to a considerable extent. Deletion of nearly half of its amino terminal part still enabled binding to the envelope, import and insertion to take place. Processing was markedly affected by deletion (Fig. 5), though the deletion of amino acid residues 3-6 still permitted some processing to take place. It is interesting to note that even small deletions in the N-terminal part of the transit peptide of Rubisco small subunit markedly affected import, but not processing[18].

The partial processing of pLHCP by sub-organellar fractions could result from non-specific proteolytic degradation. However, several observations would tend to discredit such an interpretation: a) The truncated forms observed were always intermediate in size between the precursor and mature forms. b) Lumen preparations from a variety of plants (pea, maize, lemna, barley) cleaved AB30 pLHCP to give a similar array of products. c) Precursors of other integral thylakoid proteins were processed by the lumen preparations to the mature forms (Harel et al., unpublished observations). d) The Ia and Ib forming activities were strongly dependent on the developmental stage of the plastid and corresponded to their ability to process and assemble pLHCP (Harel et al., unpublished). The apparent conversion of pLHCP, Da and Db to LHCP by lumen preparations suggests that this activity is quite specific and may be related to the authentic processing system. It also suggests that Ia and Ib indeed result from cleavages in the transit peptide of LHCP. Possible intermediates in the processing of pLHCP have been reported in two cases[6,19]. The sensitivity to thermolysin of the Ia-like putative intermediate mentioned earlier[6] could result from its formation in envelopes, prior to the migration to thylakoids. It is also possible that envelopes of plastids from various sources differ in their sensitivity to thermolysin. Two processing products were described in pea plastids during import of a wheat pLHCP[19]. The phenomenon was ascribed to the multiplicity of mature LHCP II in vivo. However, it is possible that the 26 kD form is related to Ib.

It appears that processing of pLHCP is more complex than formerly thought and may occur in more than one step. Our attempt to prepare truncated forms which will imitate putative intermediates in the processing of pLHCP failed, insofar that isolated envelopes or thylakoids did not process Da or Db to LHCP. However, since thylakoids contain extractable activity which will cause such processing (Fig. 6B) it appears that stable insertion of the putative intermediates is not enough to allow further processing. Additional factors must be involved. Alternatively, it is possible that the sequences of Da and Db are not identical to those of Ia and Ib and hence the lack of further processing. This could be examined by accumulating enough Ia and Ib to determine their exact amino terminal sequence. 'Correct' intermediates could then be prepared and tested

Acknowledgements: This research was supported by the Basic Research Foundation administered by the Israel Academy of Sciences and Humanities. We are grateful to Dr. E.M. Tobin and Dr. J.P. Thornber (UCLA) for the AB30 gene and many stimulating discussions and to Miss Hamutal Mazrier for technical help.

REFERENCES

1. P. R. Chitnis and J. P. Thornber, The major Light-harvesting complex of photosystem II: Aspects of its molecular and cell biology, Photosynth. Res. 16:41 (1988).
2. G. K. Lamppa and M. S. Abad, Processing of a wheat light-harvesting chlorophyll a/b protein precursor by a soluble enzyme from higher plant chloroplasts, J. Cell Biol. 105:2641 (1987).
3. J. M. Hand, L. J. Szabo, A. C. Vasconcelos, and A. R. Cashmore, The transit peptide of a chloroplast tylakoid membrane protein is functionally equivalent to a stromal-targeting sequence, EMBO J. 8:3195 (1989).
4. P. R. Chitnis, E. Harel, B. D. Kohorn, E. M. Tobin, and J. P. Thornber, Assembly of the precursor and processed Light-harvesting Chlorophyll a/b protein of Lemna into the Light-harvesting Complex II of barley etiochloroplasts, J. Cell Biol. 102:982 (1986).

5. P. R. Chitnis,   D. T. Morishige, R. Nechushtai and J. P. Thornber Assembly of the barley Light-harvesting Chlorophyll a/b proteins in barley etiochloroplasts involves processing of the precursor on thylakoids, _Plant Mol. Biol._ 11:95 (1988).

6. B. D. Kohorn and E. M. Tobin, A hydrophobic carboxy-proximal region of LHCP, mediates insertion into thylakoid membranes, _Plant Cell_ 1:159 (1989).

7. M. L. Mishkind and S. E. Scioly, Recent developments in chloroplast protein transport, _Photosynth. Res._ 19:153 (1988).

8. N. Pfanner and W. Neupert, The mitochondrial protein import apparatus, _Ann. Rev. Biochem._ 59:331 (1990).

9. S. Smeekens and P. Weisbeek, Protein transport towards the thylakoid lumen: post-translational translocation in tandem, _Photosynth. Res._ 16:177 (1988).

10. C. Robinson and J. R. Ellis, Transport of proteins into chloroplasts: the precursor of small subunit of ribulose bisphosphate carboxylase is processed to the mature size in two steps, _Eur. J. Biochem._ 142:343 (1984).

11. A. Colman and C. Robinson, Protein import into organelles: Hierarchial targeting signals, _Cell_ 46:321 (1986).

12. G. A. Karlin-Neumann and E. M. Tobin, Transit peptides of a nuclear-encoded chloroplast proteins, share a common framework, _EMBO J._ 5:9 (1986).

13. B. D. Kohorn, E. Harel, P. R. Chitnis, J. P. Thornber and E. M. Tobin, Functional and mutational analysis of the Light-harvesting Chlorophyll a/b protein of thylakoid membranes, _J. Cell Biol._ 102:972 (1986).

14. P. R. Chitnis, R. Nechushtai and J. P. Thornber, Insertion of the precursor of the Light-harvesting Chlorophyll a/b protein into the thylakoids requires the presence of a developmentally regulated stromal factor, _Plant Mol. Biol._ 10:3 (1987).

15. K. Cline, J. Andrews, B. Mersey, E. H. Newcomb and K. Keegstra, Separation and characterization of inner and outer envelope membranes of pea chloroplasts, _Proc. Natl. Acad. Sci. USA_ 78:3595 (1981).

16. M. J. Zoller and M. Smith, Oligonucleotide-directed mutagenesis: A simple method using two oligonucleotide primers and a single-stranded DNA template, _Methods Enzymol._ 154:329 (1987).

17. K. Cline, Import of proteins into chloroplasts, membrane integration of a thylakoid precursor protein reconstituted in chloroplast lysates, _J. Biol. Chem._ 261:14804 (1986).

18. B. Reiss, C. C. Wasmann, J. Schell and H. J. Bohnert, Effect of mutation on the binding and translocation functions of a chloroplast transit peptide, _Proc. Natl. Acad. Sci. USA_ 86:886 (1989).

19. S. E. Clark, M. S. Abad and G. K. Lamppa, Mutations at the transit peptide-mature protein junction separate two cleavage events during chloroplast import of the chlorophyll a/b-binding protein, _J. Biol. Chem._ 264:17544 (1989).

PROTEIN TARGETING TO THE THYLAKOID LUMEN

Ruth M. Mould and Colin Robinson

Department of Biological Sciences
University of Warwick
Coventry,   CV4 7AL

INTRODUCTION

Chloroplast biogenesis is a complex process involving the activities of two genetic systems. Although some chloroplast proteins are encoded by the chloroplast genome and synthesised within the organelle, the majority are nuclear-encoded and synthesised on cytosolic ribosomes. Nuclear-encoded chloroplast proteins are transported into the organelle post-translationally; this process requires specific targetting to the organelle and a mechanism for 'sorting' proteins within the organelle so that they become located in the correct compartment.

Higher plant chloroplasts are structurally particularly complex, consisting of three distinct membrane systems (the outer- and inner-envelope membranes and the thylakoid membrane network) and three distinct soluble phases (the inter-envelope space, the stroma and the thylakoid lumen).

The biogenesis of nuclear-encoded thylakoid lumen proteins is interesting since these hydrophilic proteins must cross all three chloroplast membranes to reach their site of function. The main body of work concerning import of thylakoid lumen proteins has centred on plastocyanin and three extrinsic photosystem II proteins; the 33kDa, 23kDa and 16kDa components of the oxygen-evolving complex (Andersson, 1986; Murato and Miyao, 1985). Figure 1 illustrates the import pathway of a nuclear-encoded thylakoid lumen protein. The protein is synthesised in the cytosol as a larger precursor form with an aminoterminal extension consisting of two domains. The first domain contains targetting information for translocation across the envelope membranes and into the stroma. This 'envelope transfer' domain is cleaved by stromal processing peptidase (SPP) to yield an intermediate-size form of the protein. The second domain directs transport of the intermediate across the thylakoid membrane and is cleaved off by thylakoidal processing peptidase (TPP) to yield the mature-sized protein (Hageman *et al.*, 1986; Smeekens *et al.*, 1986; James *et al.*, 1989).

Regulation of Chloroplast Biogenesis
Edited by J.H. Argyroudi-Akoyunoglou, Plenum Press, New York, 1992

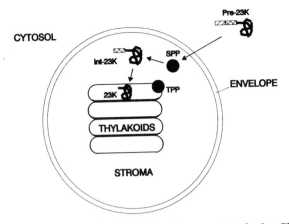

Figure 1. Import Pathway of the 23kDa Component of the Photosystem II
Oxygen-Evolving Complex.
Pre-23K, int-23K and 23K are the precursor, intermediate and mature
forms of the protein respectively. SPP = Stromal processing peptidase.
TPP = Thylakoidal processing peptidase.

ENERGY REQUIREMENTS FOR PROTEIN TRANSPORT ACROSS THE THYLAKOID MEMBRANE

Until recently, little was known about the mechanism of protein
transport across the thylakoid membrane. The standard import assay
utilised isolated intact chloroplasts and therefore the thylakoid-
transfer step could not be studied in isolation from the envelope-
transfer step. However, Kirwin et al. (1989) demonstrated import of the
33K into isolated thylakoids in the presence of stromal extract and ATP.
This import assay has been optimised further in order to study the
thylakoid-transfer step of import (Mould et al., in press). We found
that light has an essential role in the efficient import of 23K and 33K
into isolated thylakoids. Figure 2 shows light-dependent import of 23K
into isolated thylakoids. Incubation of the precursor with isolated
thylakoids and stromal extract yields the intermediate form (due to
cleavage by SPP in the stroma) both in the light (lane 2) and in the
dark (lane 3). The mature form is only generated in the light, and this
is located within the thylakoid lumen since it is resistant to protease
treatment (lane 4).

The stimulatory effect of light on import was much greater than
that achieved by the addition of ATP and incubation in the dark (results
not shown). These results suggested a role for the protonmotive force
(pmf), generated across the thylakoid membrane in the light, in
promoting protein transport across this membrane. To test this
possibility, import assays were carried out in the presence of a
combination of the electron transport inhibitors DCMU
(dichlorophenyldimethylurea) and methyl viologen, which prevent the
formation of a pmf. The pmf has an electrical component and a proton
graident component which were selectively collapsed by the addition of
valinomycin and nigericin respectively. Figure 3 shows the effect on
23K import of collapsing the pmf or each of its components. The
intermediate form is generated in all incubations (lanes 1-4) due to the
presence of SPP but this is susceptible to protease digestion (lanes 5-
8) and is therefore on the outside of the thylakoid membrane. In the
light, the mature form is also generated (lane 1) and this is protease
protected (lane 5) showing that it has been imported into the thylakoid

160

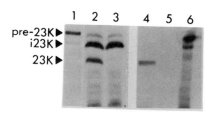

Figure 2.   Light-Dependent Import of 23K by Isolated Thylakoids.
Lanes 1 and 6, radiolabelled translation product (import substrate). Pre-23K incubated with stromal extract and isolated thylakoids in the light (lane 2) and in the dark (lane 3). Lanes 4 and 5 are as lanes 2 and 3 respectively except that after import incubations, samples were protease treated. Pre-23K, i23K and 23K denote precursor, intermediate and mature forms respectively.

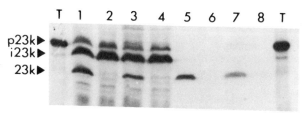

Figure 3.   Effect of Protonmotive Force on 23K Import into Isolated Thylakoids.
Lane T, translation product (import substrate). Pre-23K incubated with stromal extract and isolated thylakoids in the light (lane 1) and with DCMU/methyl viologen (lanes 20, valinomycin (lane 3) or nigericin (lane 4). Lanes 5-8 as lanes 1-4 respectively, except that import incubations were followed by protease treatment.

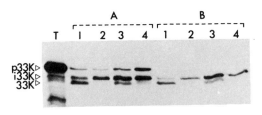

Figure 4. Effect of Protonmotive Force on 33K Import into Isolated Chloroplasts.
Lane T, radiolablled translation product (import substrate). Pre-33K incubated with isolated chloroplasts and ATP in the light (lane 1) and DCMU/methyl viologen (lane 2), valinomycin (lane 3) or nigericin (lane 4). After import incubations, samples denoted 'B' were protease treated.

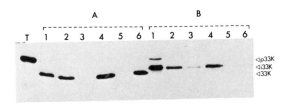

Figure 5. Localisation of Imported 33K Forms Within the Chloroplast.
Lane T, radiolabelled translation product (import substrate). Pre-33K added to chloroplasts preincubated with ATP and then incubated in the light (lanes denoted 'A') and in the presence of DCMU/methyl viologen (lanes denoted 'B'). Total import assays (lane 1), protease treated total import assays (lane 2), stromal and thylakoid fractions (lanes 3 and 4 respectively) and protease treated stromal and thylakoid fractions (lanes 5 and 6 respectively).

lumen. If a pmf is not generated (lanes 2 and 6), no mature form is detected, showing that import has been completely inhibited. Dissipation of the electrical component of the pmf (lanes 3 and 7) leads to slight inhibition of import, whereas dissipation of the proton gradient completely inhibits import. Identical results are obtained when using pre-33K as the import susbstrate.

These import conditions were repeated using intact chloroplasts, instead of thylakoids and stromal extract, and the results from the isolated thylakoid system were confirmed as shown in Figure 4. Isolated chloroplasts were preincubated with ATP before addition of inhibitors and the import substrate since ATP is required within the chloroplast for efficient import into the stroma (Theg et al., 1989). In the light, intermediate and mature forms are generated within the chloroplast (lanes 1A and 1B). If the formation of a pmf is prevented, the intermediate, but not the mature form, is generated (lanes 2A and 2B). If the electrical component of the pmf is dissipated, both the intermediate and mature forms are generated (lanes 3A and 3B), whereas if the proton gradient is dissipated, only the intermediate form is produced (lanes 4A and 4B).

A disadvantage of the intact chloroplast import assay is that protease protection demonstrates a location within the chloroplast, but to determine the precise suborganellar location of imported proteins, the chloroplast must be fractionated, as in Figure 5. In the light, virtually all the substrate is converted to the mature form (lane 1A), whereas if the formation of a pmf is prevented, precursor is processed only to the intermediate form (lane 1B). After protease treatment (lanes 2A and 2B), the chloroplasts are lysed and divided into thylakoid and stromal fractions. In the light, all the mature form is associated with the thylakoid fraction (lane 4A) and is located within the thylakoid lumen since it is protease protected (lane 6A). If a pmf is prevented from forming in the light, some intermediate is in the stromal fraction, but the majority is associated with the thylakoid fraction (lanes 3B and 4B respectively). If these fractions are protease treated, none of the intermediate is protease protected and therefore has not been imported into the thylakoid lumen. Similar results are obtained when using pre-23K as the import substrate.

SUMMARY

We conclude from these results that transport of the 23K and 33K across the thylakiod membrane requires a protonmotive force. The dominant component, in terms of driving protein transport, is the proton gradient and not the electrical potential. Interestingly, Theg *et al*. (1989) have reported that the transport of the small electron carrier protein plastocyanin, does not require a pmf for efficient import into the thylakoid lumen. We have used pre-plastocyanin as an import substrate in our import assays and our results are in agreement. It remains to be determined exactly how a pmf drives transport of the 23K and 33K across the thylakoid membrane.

REFERENCES

Andersson, B., 1986, Proteins participating in photosynthetic water oxidation, in: "Encyclopedia of plant physiology 19", Staehelim, L.A., Arntzen, C.J., eds., Springer-Verlag, Berlin.

Hageman, J., Robinson, C., Smeeken, S. and Weisbeck, P., 1986, A thylakoid processing peptidase is required for complete maturation of the lumen protein plastocyanin, Nature, 324: 567-569.

James, H.E., Bartling, D., Musgrove, J.E., Kirwin, P.M., Herrmann, R.G. and Robinson, C., 1989, Transport of proteins into chloroplasts. Import and maturation of precursors to the 33, 23 and 16kDa proteins of the oxygen-evolving complex, J. Biol. Chem., 264, 19573-19576.

Kirwin, P.M., Meadows, J.W., Shackleton, J.B., Musgrove, J.E., Elderfield, P.D., Hay, N.A., Mould, R.M. and Robinson, C. 1989, ATP-dependent import of a lumenal protein by isolated thylakoid vesicles, EMBO J., 8: 3917-3921.

Mould, R.M., Shackleton, J.B. and Robinson, C., 1991, Transport of proteins into chloroplasts. Requirements for the efficient import of two lumenal oxygen-evolving complex proteins into isolated thylakoids, J. Biol. Chem., in press.

Murata, N. and Miyao, M., 1985, Extrinsic membrane proteins in the photosynthetic oxygen-evolving complex, Trends in Biochem. Sci., 10: 122-124.

Smeeken, S., Kauelle, C., Hageman, J., Keegstra, K. and Weisbeck, P. 1986, The role of the transit peptide in the routing of precursors toward different chloroplast compartments, Cell, 46: 365-375.

Theg, S.M., Bauerle, C., Olsen, L.J., Selman, B.R. and Keegstra, K. 1989, Internal ATP is the only energy requirement for the translocation of precursor proteins across chloroplastic membranes, J. Biol. Chem., 264: 6730-6736.

BIOSYNTHESIS AND ORIGIN OF CHLOROPLAST PIGMENTS

PIGMENTS OF THE PLASTID ENVELOPE MEMBRANES

Jacques JOYARD, Maryse A. BLOCK, Bernard PINEAU* and Roland DOUCE

Laboratoire de Physiologie Cellulaire Végétale, URA CNRS 576
Département de Biologie Moléculaire et Structurale
Centre d'Etudes Nucléaires de Grenoble
et Université Joseph Fourier
85 X, F-38041 Grenoble-cedex, France

INTRODUCTION

One of the most characteristic features of a plant cell is the existence of plastids representing a whole family of interrelated organelles
(Kirk & Tilney-Bassett, 1978). Meristematic cells contain small undifferentiated plastids or proplastids, which have little internal structure
apart from a few flattened sacs which occasionally have continuity with
the inner envelope. They ensure the continuance of plastids within a species from generation to generation and are capable of considerable structural and metabolic transformations which give rise to more mature plastids. During root cell differentiation, proplastids develop mostly into
amyloplasts, which are storage plastids containing large starch grains.
During leaf cell differentiation in the presence of light, proplastids develop into chloroplasts, which are the site of photosynthesis, but in the
absence of light, they develop into etioplasts, containing a crystal-like
structure called the prolamellar body. Etioplasts can then be transformed
into chloroplasts upon illumination. Carotenoid-rich plastids, or chromoplasts, can develop from proplastids or chloroplasts during flower development, fruit ripening or senescence. Therefore, plastid differentiation
within plant cells is both light- and tissue-specific, and mechanisms
should exist that control nuclear and plastid gene expression.

Despite their wide size range and their considerable structural and
physiological diversity, all plastids are bounded by an envelope which
separates spatially and temporally the plastid compartment from the surrounding cytosol. During all the interconversions involving non-green
plastids and chloroplasts, dramatic changes occur within the organelles,
with the development or the regression of the internal membrane systems
(thylakoids, prolamellar bodies, ...) and of the stromal enzymes (for instance of Rubisco); but at all stages of these transformations, the two envelope membranes remain apparently identical (Douce & Joyard, 1979, 1990,
1991). Because of this apparent continuity, one can question how the envelope composition and enzymatic equipment are regulated. In fact, we know
almost nothing about this problem.

In contrast, knowledge of the structure, functions and biochemical
properties of envelope membranes has increased greatly in the last few
years (Douce & Joyard, 1979, 1990, 1991, Joyard *et al*, 1991) through the
refinement of procedures to purify membrane fractions from different

---

* *Laboratoire de Biochimie Fonctionnelle des Membranes Végétales, UPR 39,
CNRS, 91190, Gif sur Yvette, France.*

Regulation of Chloroplast Biogenesis
Edited by J.H. Argyroudi-Akoyunoglou, Plenum Press, New York, 1992

plastid types (Douce *et al*, 1973; Cline *et al*, 1981; Block *et al*, 1983a,b; Douce & Joyard, 1990). Two major functions have been demonstrated in envelope membranes. First, the inner envelope membrane contains specific translocators and therefore regulates metabolite transport between the cytosol and the plastid stroma, especially during photosynthesis. The second major function of the plastid envelope is related to plastid biogenesis, since the envelope is involved in the biosynthesis of specific plastid constituents (glycerolipids, pigments or prenylquinones) and in the transport and targetting of plastid proteins that are coded for by nuclear DNA and synthesized on cytosolic ribosomes.

Among typical constituents of plastid membranes, light-absorbing pigments (carotenoids, chlorophyll) are well-known for their role in thylakoids for photosynthesis. In contrast with thylakoids, envelope membranes are yellow, due to the presence of carotenoids and the absence of chlorophyll (Figure 1). In this article, we wish to focus our attention on the pigments of plastid envelope membranes. We wish also to discuss the possible role of envelope membranes in the biosynthesis of these compounds.

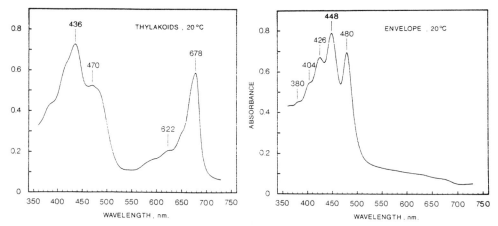

Figure 1. Absorption spectra of isolated thylakoids and envelope membranes from spinach chloroplasts.

CAROTENOIDS AS GENUINE CONSTITUENTS OF ENVELOPE MEMBRANES

Envelope membranes from all plastid analyzed so far contain carotenoids (Douce and Joyard, 1990), despite claims that there is no evidence for this (Britton, 1988). Chloroplasts and from most non-green plastids have an almost identical absorption spectrum (Figure 2) and have a very similar pigment composition (Figure 3). In all envelope membranes, violaxanthin is the major carotenoid whereas thylakoids are richer in ß-carotene (Jeffrey *et al*, 1974). The xanthophyll to ß-carotene ratio is about 6 in chloroplast envelope membranes, but only 3 in thylakoids. Both chloroplast envelope membranes contain carotenoids, the outer membrane being slightly richer in neoxanthin (Block *et al*, 1983b), but it is not yet known whether carotenoids are present as protein-pigment complexes, as in thylakoids. When present in protein-pigment complexes, carotenoids are not usually covalently bound to proteins, thus the procedures used to separate envelope membrane proteins involving detergents are not mild enough to preserve the integrity of such complexes, if they do occur in envelope membranes. Therefore, carotenoids are generally released as "free" carotenoids during membrane solubilization. This strongly limits our understanding of the functions and physiological significance of envelope carotenoids, although a role (as a singlet-oxygen quencher) in preventing (photo)oxidative damage cannot be excluded.

The comparison with cyanobacteria is most interesting. Several groups have purified the cytoplasmic membranes from cyanobacteria and have demonstrated that they are yellow, due to the lack of chlorophyll and the presence of carotenoids. Zeaxanthin is the major carotenoid in the cytoplasmic membrane of *Anacystis nidulans*, whereas thylakoids contain mostly ß-carotene (Omata & Murata, 1983). Therefore, the xanthophyll to ß-carotene ratio is much higher in the cytoplasmic membrane than in the thylakoids, like in the chloroplast (see above). The major difference with chloroplasts is the lack of violaxanthin in cyanobacteria. Violaxanthin differs from zeaxanthin in the presence of epoxide cycles, and cyanobacteria probably lack the enzyme(s) involved in zeaxanthin epoxidation.

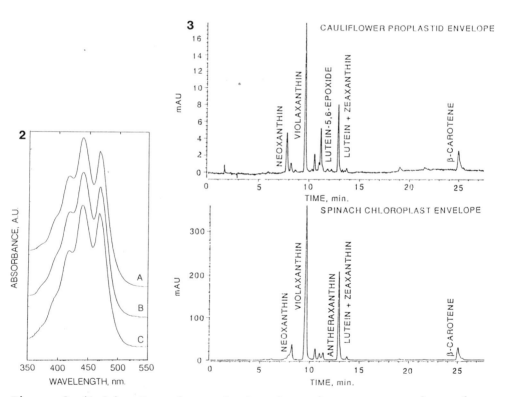

Figure 2 (left). Comparison of the absorption spectrum of envelope membranes from spinach chloroplasts (A), cauliflower proplastids (B) and potato tuber amyloplasts (C).

Figure 3 (right). Comparison of the carotenoid composition of envelope membranes from spinach chloroplasts and cauliflower proplastids. The envelope carotenoids were extracted and analyzed by HPLC as described by Barry and Pallett (1990). These experiments were done in collaboration with P. Barry and K. Pallett (Rhône-Poulenc Agrochemicals).

ARE PLASTID MEMBRANES INVOLVED IN CAROTENOID BIOSYNTHESIS ?

Jones and Porter (1986) and Britton (1988) have summarized our present knowledge on the different enzymes involved in carotenoid biosynthesis. These studies largely used chromoplasts, since these contain large amounts of carotenoids. In chromoplasts, carotenoids are concentrated in a concentrically stacked multilayer (about 20 layers in *Narcissus pseudonarcissus*) of lipid-rich membranes deriving from the inner envelope membrane (Gunning & Steer, 1975). Herbicides which block carotenoid biosynthesis and/or desaturation, such as norflurazon, amitrole, etc..., proved to be very useful tools in understanding carotenogenesis in plastids (Barry & Pallett, 1990). The first C40 carotenoid synthesized is 15-cis-phytoene, which is formed from geranylgeranyl pyrophosphate and is the precursor for all desaturated and oxygenated carotenoids. It is not known, however, at which stage pathways for carotene and the different xanthophylls diverge and how the formation of the different carotenoids is regulated (Britton, 1988). The localization of carotenoid synthesis within chloroplasts is not clearly established. It is generally assumed that thylakoids, and not envelope membranes, play the central role in carotenogenesis (Britton, 1888). However, since (a) carotenoid synthesis occurs in all plastids, (b) all the last steps in carotenoid biosynthesis involve membrane-bound enzymes and (c) carotenoids are present in given proportions in all plastid envelope membranes (which are almost the only membrane in some plastids such as proplastids and amyloplasts), the participation of envelope membranes from all plastid types in carotenogenesis cannot be excluded. This hypothesis is supported by several lines of evidence. First, chromoplast membranes, assumed to derive from the inner envelope membrane, are very active in carotenoid biosynthesis. For instance, *in vitro* phytoene desaturation in chromoplast membranes employs molecular oxygen as the terminal acceptor, but quinone compounds were able to replace molecular oxygen (Mayer *et al*, 1990). In addition, phytoene synthase and desaturase (Lütke-Brinkhaus *et al*, 1982) and zeaxanthine epoxidase (Costes *et al*, 1979) activities were demonstrated in envelope membranes from mature spinach chloroplasts. Recently, Laferrière *et al* (in preparation) have provided further evidence for a role of chloroplast envelope membranes in carotenoid biosynthesis: that envelope membranes from spinach chloroplasts contain the enzymes (desaturase(s) and cyclase) necessary for the conversion of phytoene into ß-carotene.

Despite the crucial roles of carotenoids in higher plants, such as photooxidative protection and participation in light-harvesting protein-pigment complexes, few carotenoid biosynthetic enzymes have been characterized and at least partially purified (Schmidt *et al*, 1989), only one gene involved in carotenoid biosynthesis has been cloned (Buckner *et al*, 1990) and the molecular mechanisms regulating carotenogenesis have not yet been elucidated. As discussed by Taylor (1989), the pigmentation of plastids makes it easy to find mutants that block their development. Unfortunately, since carotenoids protect plastids against photooxidative damage, carotenoid deficiencies will cause a wide range of pleiotropic effects (Taylor, 1989). It is therefore difficult to analyze higher plant mutants devoid of carotenoids: exposure to light will kill the mutant plant (bleaching compounds are powerful herbicides) or at least will significantly affect its development (for instance, they grow only in limited light conditions, Taylor, 1989). Despite this major limitation, several albino mutants have been characterized (Taylor, 1989). For instance, maize mutants defective in carotenoid accumulation in the endosperm of their kernels have been analyzed at the genetic level: some of these mutants are devoid of ß-carotene but they do accumulate precursors of ß-carotene in both their leaves and endosperm (Buckner *et al*, 1990). The y1 locus of maize, a gene involved in regulation of ß-carotene biosynthesis, was recently isolated by transposon tagging and cloning (Buckner *et al*, 1990). It is now possible to determine whether the *y1* gene product regulates the carotenoid biosynthetic pathway, or if it is essential for the assembly or stability of a carotenoid biosynthetic enzyme complex (Buckner *et al*, 1990). Genetic studies on carotenogenesis have also been carried out using photosynthesizing prokaryotes. However, although the enzymes for carotenoid biosynthesis are probably highly conserved beyond species (antibodies raised against phytoene desaturase cross react with the enzyme from cyanobacteria and higher plants (Schmidt *et al*, 1989), the genes coding for these enzymes apparently do not share the same nucleotide sequence among the different species analyzed (Misawa *et al*, 1990). Therefore, although these systems could be valuable models for understanding carotenoid biosynthesis and its mechanisms, it is not yet clear whether this could be useful for characterizing higher plant genes and their regulation.

ENVELOPE MEMBRANES ARE DEVOID OF CHLOROPHYLL, BUT CONTAIN CHLOROPHYLL PRE- CURSORS

Although devoid of the most conspicuous plastid pigment (chlorophyll) envelope membranes from mature spinach chloroplasts contain low amounts of protochlorophyllide and chlorophyllide (Pineau *et al*, 1986). Figure 4 demonstrates that envelope and thylakoids have very different fluorescence emission spectrum, and therefore that envelope protochlorophyllide and chlorophyllide are not due to thylakoid contamination. Protochlorophyllide is a well-known constituent of etioplast prolamellar body, therefore its presence in etioplast envelope membranes was considered as a contamination by vesicles deriving from the prolamellar body.

Interestingly, Hinterstoisser *et al* (1988) and Peschek *et al* (1989) have demonstrated that chlorophyll-free cytoplasmic membranes from the cyanobacterium *Anacystis nidulans* contain significant amounts of protochlorophyllide and chlorophyllide.

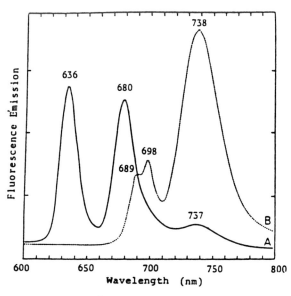

Figure 4. Low temperature (77°K) fluorescence emission spectra of isolated envelope membranes (A) and thylakoids (B) from spinach chloroplasts. Excitation light: 440 nm. Experimental conditions, see Pineau et al (1986).

169

ENVELOPE MEMBRANES CONTAIN PROTOCHLOROPHYLLIDE REDUCTASE

As shown in Figure 5, incubation of chloroplast envelope membranes under weak light in the presence of NADPH induced a progressive decrease in the level of fluorescence at 636 nm (attributed to protochlorophyllide) together with a parallel increase in fluorescence at 680 nm (attributed to chlorophyllide) due to the action of a protochlorophyllide reductase (Pineau *et al*, 1986; Joyard *et al*, 1990). In etioplasts, this enzyme is concentrated in much larger amounts in the prolamellar body, and is barely detectable in envelope membranes (Shaw *et al*, 1985). In mature spinach chloroplasts, a 37,000 Da-polypeptide from envelope membranes was immunodecorated by specific antibodies raised against oat protochlorophyllide reductase (anti-PCR) and no reaction was observed with thylakoids or stroma proteins (Joyard *et al*, 1990). In fact, we have demonstrated that envelope protochlorophyllide reductase was not the major envelope E37 polypeptide: (a) protochlorophyllide reductase can be separated from the major E37 polypeptide by two-dimensional polyacrylamide gel electrophoresis as shown in Figure 4 (Joyard *et al*, 1990), (b) in contrast to protochlorophyllide reductase, E37 is a basic envelope protein and can be purified from envelope membranes by MonoS chromatography (Block *et al*, 1991) and (c) cDNA corresponding to E37 (Dreses-Werringloer *et al*, 1991) and protochlorophyllide reductase (Darrah *et al*, 1990) were cloned and sequenced and do not present any homology. In fact, protochlorophyllide reductase is only a very minor envelope protein (Joyard *et al*, 1990). Finally, we have demonstrated that protochlorophyllide reductase from mature spinach chloroplasts is present on the cytosolic side of the outer envelope membrane, because *(a)* this polypeptide was susceptible to thermolysin digestion of isolated intact chloroplasts and *(b)* anti-PCR induced agglutination of isolated intact chloroplasts (Joyard *et al*, 1990).

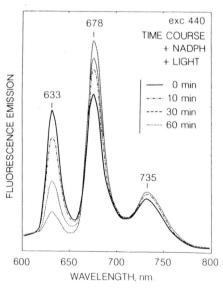

Figure 5. Evolution of the fluorescence emissions during light incubation, in the presence of NADPH, of envelope membranes isolated from intact spinach chloroplasts. The experimental conditions are described by Joyard et al (1990).

Protochlorophyllide reductase is coded for by nuclear DNA and is synthesized, in oat, as a 41 kDa precursor (Darrah et al, 1990). This compares with a doublet of Mr 35,000 and 37,000 as the mature polypeptides in 6-days old etiolated oats (Darrah et al, 1990). In our hands, the envelope polypeptide from mature spinach chloroplasts always migrates as a single polypeptide. cDNA clones for protochlorophyllide reductase were used as a probe for investigating some aspects of protochlorophyllide reductase formation (Apel et al, 1983; Harpser & Apel, 1985; Darrah et al, 1990) and to determine its nucleotide sequence (Darrah et al, 1990). The protochlorophyllide reductase contain a relatively high content of basic amino acids, with arginine and lysine accounting for more than 12 % of the total (Darrah et al, 1990). However, it is not yet known how many genes code for protochlorophyllide reductase and whether the presence of multiple genes could be responsible for the existence of the different isoenzymes (Ikeuchi & Murakami, 1982) present in different membrane systems (outer envelope membrane, prolamellar body, etc...). This also raises the question of whether a single protochlorophyllide reductase precursor could be transferred to different membranes or whether different precursors with distinct transit peptides could be synthesized in a plant cell.

The different localization of protochlorophyllide reductase in etioplasts and chloroplasts also raises the question of the physiological significance of the envelope enzyme. One can question whether the presence of protochlorophyllide reductase on the envelope membrane is indeed relevant to the participation of this membrane system in chlorophyll biosynthesis. We should first keep in mind that mature chloroplasts need chlorophyll to replace the molecules which are continuously destroyed by photooxidation. The sequence of the reactions involved in chlorophyll biosynthesis has been thoroughly studied (for a review, see Castefranco & Beale, 1983), but little is known about compartmentation of the enzymes involved within chloroplasts. The first steps in chlorophyll biosynthesis, namely those from (2)5-aminolevulinic acid to protoporphyrinogen IX, occur in the soluble phase of chloroplasts, but all the subsequent steps in protoporphyrinogen IX transformation are catalyzed by membrane-bound enzymes (Castefranco & Beale, 1983). The final step of chlorophyll formation involves addition of the phytoyl moiety and is specifically associated with thylakoids (Block et al, 1980). At least some of the enzymes involved in the transformation of protoporphyrinogen IX to chlorophyllide could be associated with envelope membranes (see above). This hypothesis is supported by several observations. Matringe et al (in preparation) have demonstrated that the enzyme protoporphyrinogen oxidase, involved in the conversion of protoporphyrinogen IX to protoporphyrin IX, was present in envelope membranes. Interestingly, this enzyme is the target for diphenylether herbicides which bind specifically to the envelope membranes. Protoporphyrin IX formed by protoporphyrinogen oxidase is the substrate for Mg-chelatase, Fuesler et al (1984) have shown that in intact chloroplasts, this enzyme was accessible to molecules unable to pass through the inner envelope membrane, thus suggesting that Mg-chelatase could be present in the envelope membranes. Furthermore, Johanningmeier and Howell (1984) have shown that the level of Mg-protoporphyrin methyl esters can regulate the accumulation of light-induced cytosolic mRNA for light-harvesting chlorophyll a/b binding protein. Consequently, the authors postulated that such intermediates in chlorophyll synthesis could be located in the chloroplast envelope.

Finally, we know almost nothing on the regulation of gene expression for enzymes involved in chlorophyll biosynthesis. Chlorophyll-free chromoplasts from daffodil contain all the enzymes involved in the conversion of (2)5-aminolevulinic acid to Mg-protoporphyrin IX monomethyl ester (Lützow & Kleinig, 1981) and the last enzyme of the pathway, chlorophyll synthase, involved in the transfer of the geranylgeranyl moiety to exogenous chlorophyllide a (Kreuz & Kleinig, 1981), but the formation of the isocyclic chlorophyll ring was not observed (Lützow & Kleinig, 1990). Thus, the genes for most of the membrane-bound enzymes involved in chlorophyll biosynthesis were not expressed in chromoplasts, in contrast with those for soluble enzymes. Albino mutants that are devoid of chlorophyll, are potentially interesting tools for studies of gene expression and regulation in chlorophyll synthesis. In fact, in most of these mutants, the primary block is in carotenoid biosynthesis whereas chlorophyll is synthesized, but then, due to overwhelming photooxidative damage, chlorophyll is destroyed and fails to re-accumulate.

CONCLUSION

The presence of several enzymes involved in carotenoid and chlorophyll biosynthesis on chloroplast envelope membranes provide further evidence for a major role of this membrane system in plastid biogenesis (Douce & Joyard, 1990; Joyard et al, 1991). However, the other plastid compartments (stroma, thylakoids) also play a role in the synthesis of carotenoids or of chlorophyll. Therefore, the compartimentation of these biosynthetic pathways is probably much more complex than usually assumed. In addition, thylakoids contain the largest amount of these pigments. Therefore, transport of the molecules from their site of synthesis to their site of synthesis to their site of accumulation should take place during development, but also in mature chloroplasts. The mechanisms involved are still unknown and both the biochemical and molecular details related to the dynamic of the transfer of lipid monomer between envelope and thylakoids will undoubtely emerge from future studies.

REFERENCES

Apel, K., Gollmer, I. & Batschauer, A. (1983) J. Cell. Biochem. 23, 181-189.

Barry, P. & Pallett, K.E. (1990) Z. Naturforsch. 45, 492-497.

Bartley, G.E., Schmidhauser, T.J., Yanofsky, C. & Scolnik, P.A. (1990) J. Biol. Chem. 265, 16020-16024.

Block, M.A., Joyard, J. & Douce, R. (1980) Biochim. Biophys. Acta 631, 210-219.

Block, M.A., Joyard, J. & Douce, R. (1991) FEBS Lett. in press

Block, M.A., Dorne, A.-J., Joyard, J. & Douce, R. (1983) J. Biol. Chem. 258, 13273-13280.

Block, M.A., Dorne, A.-J., Joyard, J. & Douce, R. (1983) J. Biol. Chem. 258, 13281-13286.

Britton, G. (1988) In Plant Pigments, (T.W. Goodwin, ed.), pp. 133-182, Academic Press,

Buckner, B., Kelson, T.L. & Roberston, D.S. (1990) The Plant Cell 2, 867-876.

Castelfranco, P.A. & Beale, S.I. (1983) Annu. Rev. Plant Physiol. 34, 241-278.

Cline, K., Andrews, J., Mersey, B., Newcomb, E.H. & Keegstra, K. (1981) Proc. Natl. Acad. Sci. USA 78, 3595-3599.

Costes, C., Burghoffer, C., Joyard, J., Block, M.A. & Douce, R. (1979) FEBS Lett. 103, 17-21.

Darrah, P.M., Kay, S.A., Teakle, G.R. & Griffiths, W.T. (1990) Biochem. J. 265, 789-798.

Douce, R. & Joyard, J. (1979) Adv. Bot. Res. 7, 1-116.

Douce, R. & Joyard, J. (1990) Annu. Rev. Cell Biol. 6, 173-216.

Douce, R. & Joyard, J. (1991) In Cell Culture and Somatic Cell Genetics of Plants, vol. 7A, The Molecular Biology of Plastids and Mitochondria (Bogorad, L. & Vasil, I.K., eds.), pp. 217-256, Academic Press, New York.

Douce, R., Holtz, R.B. & Benson, A.A. (1973) *J. Biol. Chem.* 248, 7215-7222.

Dreses-Werringloer, U., Fisher, K., Wachter E., Link, T.A. & Flügge, U-I (1991) Eur. J. Biochem. 195, 361-368.

Fuesler, T.P., Wong, Y.S. & Castelfranco, P.A. (1984) *Plant Physiol.* 75, 662-664.

Gunning, B.E.S. & Steer, M.W. (1975) *Ultrastructure and the Biology of Plant Cell*, Edward Arnold, London.

Harpser, M. & Apel, K. (1985) *Physiol. Plant.* 64, 147-152.

Hinterstoisser, B., Missbichler, A., Pineau, B. & Peschek, G. (1988) *Biochem. Biophys. Res. Comm.* 154, 839-846.

Ikeuchi, M. & Murakami, S. (1982) *Plant Cell Physiol.* 23, 575-583.

Jeffrey, S.W., Douce, R. & Benson, A.A. (1974) Proc. Natl. Acad. Sci. USA 71, 807-810.

Johanningmeier, U. & Howell, S.H. (1984) *J. Biol. Chem.* 259, 13541-13549.

Jones, B.L. & Porter, J.W. (1986) *CRC Crit. Rev. Plant Sci.* 3, 295-324

Joyard, J., Block, M.A., Pineau, B., Albrieux, C. & Douce, R. (1990) *J. Biol. Chem.* 265, 21820-21827.

Joyard, J., Block, M.A. & Douce, R. (1991) *Eur. J. Biochem.* in press.

Kirk, J.T.O. & Tilney-Bassett, R.A.E. (1978) *The Plastids. Their Chemistry, Structure, Growth and Inheritance.* 2nd ed., Elsevier, Amsterdam.

Kreuz, K. & Kleinig, H. (1981) *Plant Cell Rep.* 1, 40-42.

Lütke-Brinkhaus, F., Liedvogel, B., Kreuz, K. & Kleinig, H. (1982) *Planta* 156, 176-180.

Lützow, M. & Kleinig, H. (1990) *Arch. Biochem. Biophys.* 277, 94-100.

Matringe, M., Camadro, J.M., Block, M.A., Joyard, J., Scalla, R., Labbe, P., & Douce, R. (in preparation)

Mayer, M.P., Beyer, P. & Kleinig, H. (1990) Eur. J. Biochem. 191, 359-363.

Misawa, N., Nakagawa, M., Kobayashi, K., Yamano, S., Izawa, Y., Nakamura, K. & Harashima, K. (1990) *J. Bacteriol.* 172, 6704-6712.

Omata, T. & Murata, N. (1983) *Plant Cell Physiol.* 24, 1101-1112.

Peschek, G., Hinterstoisser, B., Wastyn, M., Kuntner, O., Pineau, B., Missbichler, A. & Lang, J. (1989) *J. Biol. Chem.* 264, 11827-11832.

Pineau, B., Dubertret, G., Joyard, J. & Douce, R. (1986) *J. Biol. Chem.* 261, 9210-9215.

Schmidt, A., Sandmann, G., Armstrong, G.A., Hearst, J.E. & Böger, P. (1989) *Eur. J. Biochem.* 184, 375-378.

Shaw, P., Henwood, J., Oliver, R. & Griffiths, T. (1985) *Eur. J. Cell Biol.* 39, 50-55.

Taylor, W.C. (1989) *Annu. Rev. Plant Physiol. Plant Mol. Biol.* 40, 211-233.

NEW ASPECTS OF THE INTERMEDIATES, CATALYTIC COMPONENTS AND THE REGULATION OF THE C_5-PATHWAY TO CHLOROPHYLL

D. Dörnemann

FB Biologie/Botanik
Philipps-Universität Marburg
Lahnberge, D-W3550 Marburg (Germany)

Introduction

Tetrapyrroles occur in all kinds of living cells as hemes, cytochromes, prosthetic groups of various enzymes and chlorophylls. Among tetrapyrrols chlorophylls represent the quantitatively biggest group. About 1.6×10^9 t of this compound are biosynthesized per year by plants, algae and cyanobacteria[1].

The first committed intermediate of all tetrapyrrols is 5-aminolevulinate (ALA) which can be formed via two different pathways, the Shemin-pathway[2] or the C_5-pathway[3]. Most of the phototrophic organisms synthesize ALA via the C_5-pathway, using the intact C_5-skeleton of glutamate or 2-oxoglutarate as first substrate. Chlorophylls in green algae and higher plants are exclusively formed via the C_5-pathway[4,5]. Also in other organisms with oxygenic photosynthesis like cyanobacteria[6] and red algae[7] chlorophylls are synthesized via this pathway.

However, the C_5-pathway to ALA also occurs in a great variety of other organisms. Thus, green sulfur bacteria (Chlorobium vibriforme)[8] and green non-sulfur bacteria (Chloroflexus aurantiacus[8]) form ALA via this pathway. Even species like the gram-positive green bacterium Heliospirillum[8], the purple sulfur bacterium Chromatium vinosum[8,9] and Desulfovibrio desulfuricans[8] synthesize ALA via the C_5-pathway. The only exception in the group of photosynthetically active organisms are purple non-sulfur bacteria like Rhodobacter spheroides[8] and Rhodospirillum rubrum[8], which only show ALA-synthetase activity. Furthermore, even non-photosynthetic bacteria like Clostridium thermoaceticum[10], Methanobacterium thermoautotrophicum[11], Escherichia coli[12,13] and Bacillus subtilis[12] use the C_5-pathway for the biosynthesis of ALA. A compilation of the above presented data and references is given in Tab. 1.

Results and discussion

The first step of the C_5-pathway is the ligation of glutamate to $tRNA^{glu}$ by an ATP dependent glutamyl-tRNA-synthetase. Concerning chlorophyll biosynthesis this was demonstrated for quite a number of organisms, e.g. barley[27], the green algae Scenedesmus[17], Chlorella[18] and Chlamydomonas[19], the red alga Cyanidium[20] and the cyanobacterium Synechocystis 6803[28]. Experimental proof for the participation of a tRNA was always given by treatment of a soluble enzyme preparation, capable to convert glutamate to ALA, with RNAse, leading to a drastical decrease of label from glutamate in ALA. Beside others[27] the $tRNA^{glu}$ of barley was sequenced and shown to be a chloroplast $tRNA^{29}$ bearing the UUC glutamate anticodon necessary for ALA biosynthesis. From Scenedesmus two $tRNA^{glu}$ could be isolated by HPLC. Two of them could also be loaded with glutamine, which is in agreement with results from barley[30].

Regulation of Chloroplast Biogenesis
Edited by J.H. Argyroudi-Akoyunoglou, Plenum Press, New York, 1992

Tab. 1. Distribution of the two pathways to ALA among different organisms. References and organisms represent only a selection of studied objects. For recent reviews see 8, 24, 25, 26

Biological Material	C_5-pathway	Shemin-pathway	Ref.
Higher Plants *Hordeum vulgare* *Zea mais*	+ +	− −	3,14,15 5
Green algae *Scenedesmus obliquus* *Chlorella vulgaris* *Chlamydomonas reinhardtii*	+ + +	+ − −	16, 17 18 19
Red algae *Cyanidium caldarium*	+	−	7, 20
Euglenophyta *Euglena gracilis*	+	+	21
Cyanobacteria *Anabaena variabilis* *Synechococcus* PCC 6301	+ +	− −	6 13
Green sulfur and non-sulfur bacteria *Chlorobium vibrioforme* *Chloroflexus aurantiacus*	+ +	− −	8 8
Gram-positive green bacteria *Heliospirillum*	+	−	8
Purple sulfur bacteria *Chromatium vinosum* *Desulfovibrio desulfuricans*	+ +	− −	8, 9 8
Purple non-sulfur bacteria *Rhodobacter sphaeroides* *Rhodospirillum rubrum* *Rhizobium sesbaniae*	− (+) (+)	+ + +	8 8 8
Clostridium thermoaceticum	+	−	10
Archaebacteria *Methanobacterium thermoautotrophicum*	+	−	11
Enterobacteria *Escherichia coli*	+	−	12, 13
Bacilli *Bacillus subtilis*	+	−	12
Other bacteria families	n.d.	n.d.	
Fungi	−	+	8, 22
Animals	−	+	23

The enzyme that ligates glutamate to tRNA[glu] is an ATP-dependent glutamyl-tRNA-synthetase. From barley, *Chlorella* and *Synechocystis* the enzymes could be purified and their molecular weights were determined to be 54, 73 and 63 kDa, respectively[26]. From *Scenedesmus* (unpublished) and also from *Cyanidium*[20] two tRNA-ligases could be isolated, the latter not beeing further characterized. However, the enzymes from *Scenedesmus*, one binding to Blue Sepharose, the other to Red Agarose, were highly purified. Their molecular weight was determined to be 83 kDa. Both reacted after Western-blotting with an antibody raised against the ligase from barley. For further details see Vothknecht *et al.*, this issue.

The next enzyme in the reaction sequence is the NAD(P)H-dependent glutamyl-tRNA reductase (dehydrogenase). This protein reduces the glutamyl-tRNA complex to glutamate-1-semialdehyde (G-1-SA). The reductase seems to be the most labile enzyme of the C_5-pathway and attempts to purify the protein were, up to now, only successfull with the green alga *Chlamydomonas reinhardtii*[31]. However, apparent homogenity was not achieved. Purification was performed employing six different chromatographic separations. The apparent molecular mass was determined to be 130 ± 5 kDa by SDS-gel electrophoresis. This is contradictory to results of Wang *et al.*[32] who also partially purified the protein and found a molecular mass of 51 kDa for the same enzyme. Under non-denaturing conditions (rate zonal sedimentation on glycerol gradients) the protein showed the same molecular weight.

From this it is concluded that the active protein is a monomer. Furthermore, Chen *et al.*[31] report that the reductase forms a stable complex with tRNA[glu] in the absence of NADPH when reduction of glutamate cannot take place. No GTP or another protein is needed to form the complex, which is again contradictory to findings of Kannangara *et al.*[27].

The further steps from G-1-SA to ALA are still a matter of controversy. Grimm *et al.*[13,33] postulate two different mechanisms for both barley and *Synechoccocus*. One mechanism involves head to tail dimerization of two G-1-SA molecules to form a double Schiff-base which is subsequently hydrolyzed with a reciprocal exchange of the amino groups. The second mechanism proposes first the formation of 4,5-diaminovalerate from G-1-SA. The amino group is donated by the enzyme or the cofactor pyridoxamine-phosphate. The amino group at C_5 is then accepted by pyridoxal-phosphate forming ALA and pyridoxamine-phosphate. 4,5-diaminovalerate as intermediate between G-1-SA and ALA is supported by Houghton *et al.*[20] who found a yet unidentified intermediate during the conversion of glutamate to ALA in the red alga *Cyanidium caldarium*. They suggested that this compound could be 4,5-diaminovalerate. For the green alga *Scenedesmus* and barley we could, however, show that 4,5-dioxovalerate (DOVA) accumulates, when ALA dehydratase is inhibited by levulinic acid. The ratio of ALA and DOVA was determined to be 180 and 250, respectively.

The role of DOVA as intermediate in the formation of ALA from G-1-SA is further supported by the fact that on the one hand the label in ALA from ^{14}C-glutamate decreases to about 50%, when unlabelled DOVA is added to the enzyme assay[17]. On the other hand labelled G-1-SA is accumulated, when 1 mM unlabelled DOVA is present in the assay, indicating that a competition of labelled and unlabelled external DOVA occurs[17]. G-1-SA accumulation can also be verified by the inhibition of the corresponding G-1-SA aminotransferase by gabaculine, whereas the following step, the transamination of DOVA to ALA is inhibited by gabaculine to about only 60%. Concerning the conversion of G-1-SA to ALA it is thus proposed that a pyridoxal-phosphate/pyridoxamine-phosphate shuttle with DOVA as intermediate takes place. The shuttle is shown in Fig.1.

Taking into account all data about the conversion of glutamate to ALA the following reaction sequence is proposed for the C_5-pathway: glutamate is activated by its ligation to tRNA[glu] by glutamyl-tRNA-ligase. The glutamyl-tRNA[glu]-complex is subsequently reduced by a NAD(P)H-dependent reductase to G-1-SA, which is then converted to ALA via DOVA. The proposed reaction sequence[17] is shown in Fig.2.

Further support for the above described reaction sequence and the proposed pyridoxamine/pyridoxal-phosphate shuttle is given by the fact that an enzyme preparation from *Scenedesmus*, purified by affinity chromatography[17], followed by FPLC on MonoQ

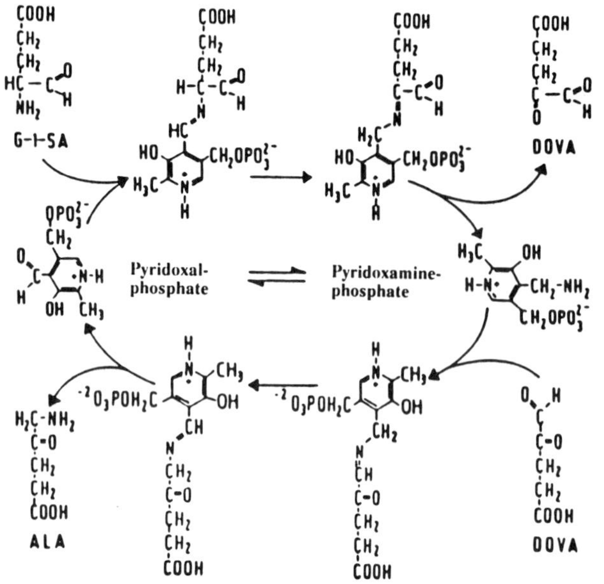

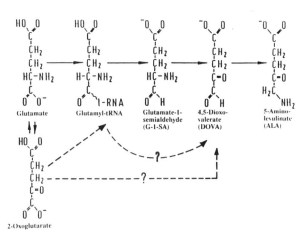

Figure 1. The postulated pyridoxamine-/pyridoxalphosphate shuttle for the conversion of G-1-SA *via* DOVA to ALA by a single protein.

Figure 2. The proposed reaction sequence of the C_5-pathway.

showed both, G-1-SA-aminotransferase and DOVA-transaminase activity in the same protein fraction. Attempts to separate both enzyme activities were yet unsuccessful and it will have to be shown in the near future by partial sequencing of the homogenous protein, that both activities are located in the same protein. G-1-SA-aminotransferases have by now been purified and characterized from barley[33] and *Chlamydomonas*[34], however, were these proteins not yet tested for their ability to convert DOVA to ALA. Thus it cannot be ruled out that these proteins have both enzyme activities. Also the fact that the structural gene of G-1-SA-aminotransferase from *Synechoccocus* PCC 6301 has been sequenced[13] does not exclude the possibility that at least in *Scenedesmus* both enzyme activities are performed by a single protein.

Little is known about the regulation of chlorophyll biosynthesis. Angiosperms and some mutants of algae have a light-dependent protochlorophyllide reductase. It should thus be expected that in darkness protochlorophyllide (Pchlide) accumulates to a certain degree in these organisms, however, the pool size of this compound remains very low. When 4 mM ALA is added to cells of the light-dependent greening mutant C-2A' of *Scenedesmus* in darkness, the amount of Pchlide increases up to 20-fold. This points out that a regulatory step upstream of ALA biosynthesis must be present preventing ALA formation and thus Pchlide accumulation under physiological conditions during darkness. For the *Scenedesmus* mutant[35] and barley (unpublished data) it could be shown that Pchlide directly acts on glutamyl-tRNA-ligase preventing the ligation of glutamate to tRNA^{glu} and thus the first step of ALA biosynthesis. The K_i-value for Pchlide was determined to be 1.25 μM. That Pchlide acts on the ligase and not on the tRNA^{glu} was shown by the fact that with increasing ligase amounts at constant tRNA and Pchlide concentrations enzyme activity could be restored. Fig.3 shows a scheme for the dark regulation of chlorophyll biosynthesis in light-dependent greening organisms.

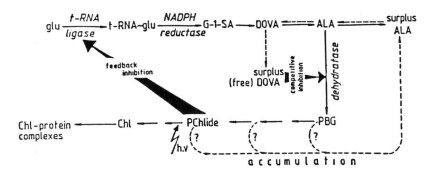

Figure 3. Regulation scheme of chlorophyll biosynthesis in darkness and under light conditions. The dashed lines indicate possible accumulations of intermediates, finally causing free (not enzyme-bound) DOVA as competitive inhibitor of ALA-dehydratase.

When DOVA, the immediate precursor of ALA, is added to greening cultures of *Scenedesmus* mutant C-2A' it should be expected that chlorophyll biosynthesis is enhanced or at least should remain constant. In darkness, in analogy to the addition of ALA, Pchlide accumulation should occur. However, it was observed that as well in the light as in darkness the formation of chlorophyll and Pchlide, respectively, decreased. In the light during the first 8 hours, when chlorophyll formation rate is maximum, DOVA had only little influence, which means, that at high chlorophyll formation rates the external substrate DOVA is consumed. When chlorophyll biosynthesis lowers the internal DOVA concentration becomes higher and acts inhibitory on chlorophyll formation. This can also happen when the C_5-pathway for some reason runs too fast or when the amount of chlorophyll is sufficient for the formation of all Chl-protein complexes. From the fact, that DOVA is a substrate analog of levulinic acid, which is a strong competitive inhibitor of ALA-dehydratase ist was concluded that DOVA at higher internal concentrations acts as a competitive inhibitor of ALA-dehydratase. Actually we could show that in crude enzyme preparations containing the enzyme, porphobilinogen and uroporphyrin formation from added ALA was inhibited depending on DOVA concentration. The K_i-value for DOVA was determined to be 60 ± 5 μM[36]. Via this mechanism the regulation of chlorophyll biosynthesis in the light could be facilitated. From the presented data a general regulation scheme for chlorophyll biosynthesis under light and dark conditions was derived (Fig.3).

From the fact that in earlier experiments 2-oxoglutarate was more effectively incorporated into ALA than glutamate the question arose whether this compound participates directly or via glutamate in the C_5-pathway. In preliminary experiments we noted that 2-oxoglutarate or a derivative can be bound to tRNA[glu]. It could be excluded by the addition of aminooxyacetate that prior to the ligation of the compound to the tRNA glutamate was formed. In forthcoming experiments we plan to reveal the real nature of the bound residue by recleavage of the complex and successive identification of the product by HPLC. Furthermore, it is intended to elucidate whether this tRNA-complex is reduced to G-1-SA or is, which is more likely, directly converted to DOVA. This would mean that an alternative reaction sequence exists for the C_5-pathway.

References

1. G.A.F. Hendry, J.D. Houghton and S.B. Brown, The degradation of chlorophyll - a biological enigma, New Phytol. 107: 255 (1987).
2. G. Kikuchi, A. Kumar, P. Talmage and D. Shemin, The enzymatic synthesis of δ-aminolevulinic acid, J. Biol. Chem. 233: 1214 (1958).
3. S.I. Beale and P.A. Castelfranco, The biosynthesis of δ-aminolevulinic acid in higher plants II. Formation of ^{14}C-δ-aminolevulinic acid from labelled precursors in greening plant tissues, Plant Physiol 53: 297 (1974).
4. T. Oh-hama, H. Seto, N. Otake and S. Miyachi, ^{13}C-NMR-evidence for the pathway of chlorophyll biosynthesis in green algae, Biochem. Biophys. Res. Commun. 105: 647 (1982).
5. R.J. Porra, O. Klein and P.E. Wright, The proof by ^{13}C-NMR-spectroscopy of the predominance of the C_5-pathway over the Shemin-pathway in chlorophyll biosynthesis in higher plants and of the formation of the methylester group of the chlorophyll from glycine, Eur. J. Biochem. 130: 509 (1983).
6. Y.J. Avissar, Biosynthesis of 5-aminolevulinate from glutamate in *Anabaena variabilis*, Biochim. Biophys. Acta 613: 220 (1980).
7. J.D. Weinstein and S.J. Beale, Biosynthesis of protoheme and heme a precursors solely from glutamate in the unicellular red alga *Cyanidium caldarium*, Plant Physiol. 74: 146 (1984).
8. Y.J. Avissar, J.G. Ormerod and S.I. Beale, Distribution of δ-aminolevulinic acid biosynthetic pathways among phototrophic bacterial groups, Arch. Microbiol. 151: 513 (1989).
9. T. Oh-hama, H. Seto and S. Miyachi, ^{13}C-NMR evidence of bacteriochlorophyll a formation by the C_5-pathway in *Chromatium*, Arch. Biochem. Biophys. 246: 192 (1986).
10. T. Oh-hama, N.H. Stotowich and A.I. Scott, 5-Aminolevulinic acid formation from glutamate via the C_5-pathway in *Clostridium thermoaceticum*, FEBS Lett. 228: 89 (1988).
11. H.C. Friedman, R.K. Thauer, S.P. Gough and C.G. Kannangara, δ-Aminolevulinic acid formation in the archaebacterium *Methanobacterium thermoautotrophicum* requires tRNA[glu], Carlsberg Res. Commun. 52: 363 (1987).
12. G.P. O'Neill, M.-W. Chen and D. Söll, δ-Aminolevulinic acid biosynthesis in *Escherichia coli* and *Bacillus subtilis* involves formation of glutamyl-tRNA, FEMS Microbiol. Lett. 60: 255 (1989).
13. B. Grimm, A. Bull and V. Breu, Structural genes of glutamate-1-semialdehyde aminotransferase for porphyrin synthesis in a cyanobacterium and *Escherichia coli*, Mol. Gen. Genet. 255: 1 (1991).
14. C.G. Kannangara and S.P. Gough, Biosynthesis of δ-aminolevulinate in greening barley leaves. Glutamate-1-semialdehyde aminotransferase, Carlsberg Res. Commun. 43: 185 (1978).
15. W.Y. Wang, S.P. Gough and C.G. Kannangara, Biosynthesis of δ-aminolevulinate in greening barley leaves. IV. Isolation of three soluble enzymes required for the conversion of glutamate to δ-aminolevulinate, Carlsberg Res. Commun. 46: 243 (1981).
16. O. Klein and H. Senger, Two biosynthestic pathways to δ-aminolevulinic acid in a pigment mutant of the green alga *Scenedesmus obliquus*, Plant Physiol. 62: 10 (1978).
17. V. Breu and D. Dörnemann, Formation of 5-aminolevulinate via glutamate-1-semialdehyde and 4,5-dioxovalerate with participation of an RNA component in *Scenedesmus obliquus* mutant C-2A', Biochim. Biophys. Acta 967: 135 (1988).

18. J.D. Weinstein and S.I. Beale, RNA is required for enzymatic conversion of glutamate to δ-aminolevulinate by extracts of *Chlorella vulgaris*, Arch. Biochem. Biophys. 239: 87 (1985).

19. D.-D. Huang and W.-Y. Wang, Chlorophyll synthesis in *Chlamydomonas* starts with the formation of glutamyl-tRNA, J. Biol. Chem. 261: 13451 (1986).

20. J.D. Houghton, S.B. Brown, S.P. Gough and C.G. Kannangara, Biosynthesis of δ-aminolevulinate in *Cyanidium caldarium*: Characterization of tRNA^{glu}, ligase, dehydrogenase and glutamate-1-semialdehyde aminotransferase, Carlsberg Res. Commun. 54: 131 (1989).

21. J.D. Weinstein and S.I. Beale, Separate physiological roles and subcellular compartments for two tetrapyrrole biosynthetic pathways in *Euglena gracilis*, J. Biol. Chem. 258: 6799 (1983).

22. R.J. Porra, R. Barnes and O.T.G. Jones, The level and subcellular distribution of 5-aminolevulinate synthetase activity in semi-anaerobic yeast, Hoppe-Seyler's Z. Physiol. Chem. 353: 1365 (1972).

23. D. Shemin, C.-S. Russell and T. Abramsky, The succinate-glycine cycle. I. Mechanism of pyrrole synthesis, J. Biol. Chem. 215: 613 (1955).

24. W. Rüdiger and S. Schoch, Chlorophylls, in: "Plant Pigments", T.W. Goodwin, ed., Academic Press, London, San Diego (1988).

25. S.I. Beale and J.D. Weinstein, Tetrapyrrole metabolism in photosynthetic organisms, in: "Biosynthesis of heme and chlorophylls", H.A. Dailey, ed., McGraw-Hill, New York (1989).

26. S.I. Beale, Biosynthesis of the tetrapyrrole pigment precursor, δ-aminolevulinic acid, from glutamate, Plant Physiol. 93: 1273 (1990).

27. C.G. Kannangara, S.P. Gough, P. Bruyant, J.K. Hoober, A. Kahn and D. v.Wettstein, tRNA^{glu} as a cofactor in δ-aminolevulinate biosynthesis: steps that regulate chlorophyll synthesis, TIBS 13: 139 (1988).

28. S. Rieble and S.I. Beale, Enzymatic transformation of glutamate to δ-aminolevulinic acid by soluble extracts of *Synechocystis* sp. 6803 and other oxygenic procaryotes, J. Biol. Chem. 263: 8864 (1988).

29. A. Schön, G. Krupp, S.P. Gough, S. Berry-Lowe and C.G. Kannangara, The RNA required in the first step of chlorophyll biosynthesis is a chloroplast glutamate tRNA, Nature 322: 281 (1986).

30. A. Schön, C. Kannangara, S.P. Gough and D. Söll, Protein biosynthesis in organelles requires misacylation of tRNA, Nature 331: 187 (1988).

31. M.W. Chen, D. Jahn, G.P. O'Neill and D. Söll, Purification of the glutamyl-tRNA reductase from *Chlamydomonas reinhardtii* involved in the δ-aminolevulinic acid formation during chlorophyll biosynthesis, J. Biol. Chem. 265: 4058 (1990).

32. W.-Y. Wang, D.-D. Huang, T.-E. Chang, D. Stachon and B. Wegmann, Regulation of chlorophyll biosynthesis. Genetics and biochemistry of δ-aminolevulinate synthesis, in: "Progress in photosynthesis research", J. Bigginsed., Martinus Nijhoff Publinshers, Dordrecht (1987).

33. B. Grimm, A. Bull, K.G. Welinder, S.P. Gough and C.G. Kannangara, Purification of the glutamate-1-semialdehyde aminotransferase of barley and *Synechococcus*, Carlsberg Res. Commun. 54: 67 (1989).

34. D. Jahn, M.W. Chen and D. Söll, Purification and functional characterization of glutamyl-1-semialdehyd aminotransferase from *Chlamydomonas reinhardtii*, J. Biol. Chem. 266: 1611 (1991).

35. D. Dörnemann, K. Kotzabasis, P. Richter, V. Breu and H. Senger, The regulation of chlorophyll biosynthesis by the action of protochlorophyllide on ^{glu}tRNA-ligase, Bot. Acta 102: 112 (1989).

36. K. Kotzabasis, V. Breu and D. Dörnemann, The inhibitory effect of 4,5-dioxovalerate on 5-aminolevulinate dehydratase and its implication in the regulation of light-dependent chlorophyll formation in pigment mutant C-2A' of *Scenedesmus obliquus*, Biochim. Biophys. Acta 977: 309 (1989).

LAST STEPS IN CHLOROPHYLL BIOSYNTHESIS: ESTERIFICATION AND INSERTION

INTO THE MEMBRANE

W. Rüdiger

Botanisches Institut der Universität
Menzinger Str. 67, D-8000 München 19

INTRODUCTION

Chloroplast biogenesis requires coordination of many biosynthetic pathways. Coordinate accumulation of nuclear encoded and plastid encoded proteins is only one aspect. Such coordination exists also in the area of chloroplast pigments. An example is chlorophyll biosynthesis and accumulation: Chlorophyll precursor molecules are derived not only from the porphyrin pathway but also from the isoprenoid pathway. For stabilization of the pigment, plastid and nuclear encoded chlorophyll apoproteins are required. Assembly of chlorophyll protein complexes depends also on other cofactors, e.g. carotenoids. I wish to concentrate upon the esterification of chlorophyllide at first which connects the porphyrin and the isoprenoid pathways. In the second part of my contribution, the present knowledge on interaction of chlorophyll and chlorophyll precursors with apoprotein formation and assembly will be summarized.

PROPERTIES OF CHLOROPHYL SYNTHETASE

Chlorophyll synthetase activity was detected in 1980 in crude membrane fractions from oat etioplasts (Rüdiger et al., 1980). The activity was subsequently also described to occur in the stroma fraction of chromoplasts from daffodil (Kreuz and Kleinig, 1981) and in membrane fractions from spinach chloroplasts (Soll et al., 1983), paprika chloroplasts and chromoplasts (Dogbo et al., 1984). In mature chloroplasts, the enzyme was found only in the thylakoid membrane but not in stroma or envelope (Soll et al., 1983). In etioplasts, both prolamellar body (PLB) and prothylakoid (PT) fractions contain chlorophyll synthetase activity (Lütz et al., 1981; Lindsten et al., 1980). The activity is latent in "native" PLB's and only apparent after dissociation of the "paracrystalline" structure (Lindsten et al., 1990). It is very probable (but not proven so far) that the enzyme migrates upon illumination of etiolated leaves, which leads to dissociation of PLB's, into the PT's. This would explain the localization exclusively in thylakoid membranes of mature chloroplasts (Soll et al., 1983).

The enzyme catalyzes the reaction of chlorophyllide (Chlide) with phytylpyrophosphate (PhyPP) or geranylgeranylpyrophosphate (GGPP) under formation of the respective esterified chlorophyll (Chl) and release of inorganic pyrophosphate (PP_i).

$$Chlide \ + \ GGPP \ \rightarrow \ Chl(GG) \ + \ PP_i$$

$$Chlide \ + \ PhyPP \ \rightarrow \ Chl(Phy) \ + \ PP_i$$

In the first crude preparations, GGPP could be substituted by GGMP or GG plus ATP (Rüdiger et al., 1980). Later stroma-free preparations required the pyrophosphate proper. The isoprenoid kinase which can form GGPP from GGMP or GG plus ATP has been localized in the stroma fraction (Soll et al., 1983). Various C_{20} isoprenoids are accepted nearly equally well; etioplast preparations have only a slight preference for GGPP over PhyPP (Rüdiger et al., 1980) whereas chloroplast preparations have only a slight preference for PhyPP over GGPP (Soll et al., 1983). This seemingly slight difference is reflected in different findings in etiolated and green tissue: The first esterified Chl in etiolated seedlings (after phototransformation of Protochlide is Chl(GG) (Schoch et al., 1977) whereas the first esterified Chl in green seedlings is Chl(Phy) (Rüdiger, 1987). The end product is Chl(Phy) in both cases: according to investigated kinetics, Chl(GG) is hydrogenated stepwise to yield Chl(Phy) in etiolated seedlings (Schoch et al., 1977; Schoch, 1978). Two different pathways have been proposed for green and etiolated tissues.

Green
tissues:

$$GGPP \ \xrightarrow{NADPH} \ PhyPP \ + \ Chlide \ \rightarrow \ Chl(Phy)$$

Etiolated
tissues:

$$GGPP \ + \ Chlide \ \rightarrow \ Chl(GG) \ \xrightarrow{NADPH} \ Chl(Phy)$$

The Michaelis constant (K_M) has been determined for PhyPP and the enzyme in etiolated membranes to be about 1×10^{-5} M (Gronau, 1983). This value indicates a high affinity of chlorophyll synthetase for this substrate.

The tetrapyrrole substrate, Chlide a, can either be formed by photoconversion of Protochlide in the membrane or added as such in solution. In the latter case, the reaction can be carried out in the dark and with modified Chlides. Variation of the structure revealed the following requirements for a suitable substrate of chlorophyll synthetase (see Fig. 1): The propionic side chain of ring D must be above the tetrapyrrole plain in order to be esterified. Protochlide which has the propionic acid side chain in the tetrapyrrole plain is not a substrate (Rüdiger et al., 1980). Ring B must not be hydrogenated as in bacteriochlorophyllide but the formyl group at ring B (as in Chlide b) is accepted as well as the methyl group (as in Chlide a) (Benz and Rüdiger, 1981). Side chains have also been modified at ring A and at the isocyclic ring without loss of substrate properties (Benz and Rüdiger, 1981). A central metal ion is required. This can be either magnesium or zinc; nickel, copper or cobaltum as central metal ion are not accepted by the enzyme (Helfrich and Rüdiger, unpublished). It is interesting to note that only those metals (Mg, Zn) which form pentacoordinate complexes are accepted by the enzyme. It may be that the 5th ligand in the enzyme-substrate complex is an amino acid side chain of the active center of chlorophyll synthetase.

Addition of exogenous Chlide to PhyPP and a membrane fraction containing chlorophyll synthetase gives different results according to the reaction conditions (Fig. 2). At low concentrations of exogenous Chlide (up to about 6 nmol Chlide per sample), a constant percentage of esterification (about 70-80%) is found. In the standard volume of phosphate buffer (2 ml), higher concentrations of Chlide yield typical substrate inhibition. This can be avoided when the buffer volume is increased (e.g. to 6 ml, see Fig. 2). With the standard membrane fraction (containing 2 nmol Protochlide), a constant upper amount of about 4 nmol esterified Chl is obtained under these conditions. The fact, that only a constant amount of Chl is formed in

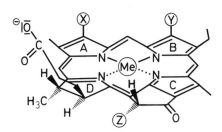

Fig. 1. Structure of substrates for chlorophyll synthetase, X, Y and Z are side chains which can be modified without loosing substrate properties. Me = Mg or Zn.

a standard membrane sample can be interpreted as product inhibition of the enzyme. Chl is apparently not released from the enzyme because it is insoluble in the applied phosphate buffer. The upper limit of esterification can be extended to 10 nmol Chl or higher, however, if the supernatant of an etioplast membrane suspension is used instead of the phosphate buffer. This membrane suspension had briefly been heated to 80°C in order to inactivate any enzyme activity and to extract heat-stable compounds.

Such compounds, probably polar lipids, are apparently able to act as "carrier" for Chl and by this means liberate the product site of chlorophyll synthetase. In vivo, Chl is probably released into the lipid phase of the membrane before it is inserted into Chl-binding proteins. Product inhibition may come into play when not enough Chl-binding apoproteins are formed. Concentration of Chl in the lipid phase of the membrane might become high and inhibitory to Chl synthetase in this case. Such type of regulation can be considered an "archaic" type of metabolic regulation, we found it also for synthesis of GGPP from isopentenylpyrophosphate (Steiger et al., 1985).

185

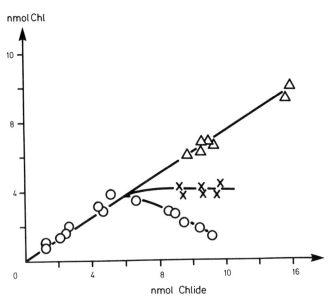

Fig. 2. Esterification of Chlide a with an etioplast membrane fraction.
Membrane samples of lysed oat etioplasts containing 2 nmol Protochlide each were incubated with PhyPP and various amounts of Chlide
a for 60 min. Membrane samples suspended in:
o-o   2 ml 50 mM phosphate buffer, pH 7.5;
x-x   6 ml phosphate buffer;
Δ-Δ   6 ml supernatant of heated etioplast membrane suspension
          (after Gronau, 1983)

In the case of Chl, it may be essential for survival of plants in bright
sunlight: Chl and Chl precursors which are not protein-bound can cause photodynamic damage (Rebeiz et al., 1984; Kittsteiner et al., 1991a).

INTERACTION OF CHLOROPHYLL BIOSYNTHESIS WITH OTHER BIOSYNTHETIC PATHWAYS

Coordination of several biosynthetic pathways during chloroplast biogenesis has so far been described primarily for accumulation of single compounds rather than for rates of their biosynthesis. A general idea is that
single compounds are taken in stoichiometric amounts for the assembly process and that any compound synthesized in excess will be degraded. Such a
mechanism works economically only if the pool sizes of single compounds are
strictly regulated. Otherwise massive synthesis and degradation of compounds would vaste too much energy.

We have described such kind of coordination between light-dependent
synthesis of chlorophyllide (Chlide) and light-independent synthesis of
geranylgeranylpyrophosphate (GGPP) (Benz et al. 1983 a,b; Steiger et al.,
1985):

In the dark, no Chlide is formed in Angiosperms. The GGPP pool is
saturated; the pool size is limited by feed-back inhibition of its own synthesis. The same is true for the Protochlide pool (Dörnemann et al., 1989).

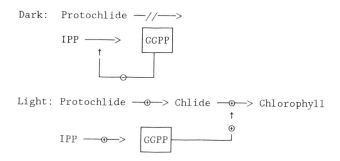

Dark: Protochlide —//—>

Light: Protochlide —⊕—> Chlide —⊕—> Chlorophyll

In the light, Chlide is formed from Protochlide. The GGPP pool is immediately emptied due to consumption in the esterification reaction. It is slowly built up again when esterification of Chlide is completed. When the
esterification ceases, the GGPP pool has about the same size in the light
as in the dark (Benz et al., 1983b)

A similar mechanism has been described for coordination between carotenoid and Chl accumulation (Frosch and Mohr, 1980). According to this proposal, carotenoids are taken from the free carotenoid pool for formation of
carotenoid/chlorophyll-protein complexes ("pull" regulation). The negative
feedback of free carotenoids to their own synthesis must differentiate between single carotenoids: Chl formation changes the carotenoid pattern,
ß-carotene showing the strongest relative increase. It is not yet known on
which step this differential regulation takes place. It is not regulated
via phytochrome: Pfr formation leads to a general increase in carotenoids
but is unspecific with regard to the carotenoid pattern (coarse "push"
control).

The specificity for amount and nature of pigments in the assembly process resides in the apoproteins. This can be shown by reconstitution of a
pigment-protein complex in vitro even with a single apoprotein (LHCP) produced in E.coli by overexpression (Paulsen et al., 1990). This phenomenon
will be described in more detail in the chapter of H. Paulsen. I wish to
mention only two aspects here: (1) Esterification of Chlide is a precondition for reconstitution; Chlide is not accepted by the apoprotein. This was
not obvious before the experiment since cleavage of the phytyl residues of
Chls in the LHPcomplex with chlorophyllase still left a functional pigment-protein complex, i.e. it did not affect the efficiency of energy
transfer between Chl-(ide)s (Schoch and Brown, 1987). (2) Reconstitution
with the required pigments (Chl a and b, 2-3 xanthophylls) in unfavorable
ratios proceeds only to the extent of the "minimum factor". This means that
all compounds which are present in higher amounts are left over. One may
assume that excess pigments are left in vivo in the lipid phase of the
membrane and can exert there negative feed-back control.

COORDINATION OF CHLOROPHYLL AND APOPROTEIN SYNTHESIS

For stabilization and insertion into the membrane, Chls have generally
to be bound to apoproteins. Vice versa, several apoproteins have been re-

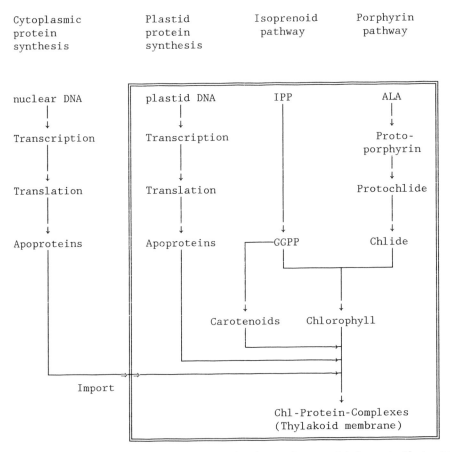

Fig. 3. Correlation of several biosynthetic pathways which contribute to formation of Chl-protein complexes. The plastid compartment is enciled by a double line. All pathways are regulated by Chl synthesis: transcription of nuclear genes probably by Chl precursors, translation of plastid-encoded proteins and GGPP synthesis by esterification of Chlide.

ported to be stabilized by insertion of Chl (Apel and Kloppstech, 1980; Bennett, 1981; Mullet et al., 1990). I consider stabilization (i.e. avoidence of proteolysis) as fine tuning rather than as coarse regulation; otherwise uninhibited protein synthesis in the absence of Chl would lead to uninhibited proteolysis and hence to enormous amounts of vasted bioenergy. Negative feed-back control of apoproteins upon their own biosynthesis has not been reported. We asked therefore whether Chl itself or its precursors could have a regulatory function in apoprotein synthesis.

It has been reported that certain Chl precursors (especially magnesium protoporphyrin monomethylester, MgPMe) inhibit specifically cab gene transcription in <u>Chlamydomomas</u> (Johanningmeier and Howell, 1984; Johanningmeier, 1988). Such an effect had not been described for higher plants. We tested this possibility with cress (<u>Lepidium sativum</u> L.) seedlings. The application of metal chelators which had successfully been applied to <u>Chlamydomonas</u> proved to be useless for this problem: Although Chl precursors were accumulated (Kittsteiner et al., 1991a) no specific effect upon transcription of the cab gene(s) could be demonstrated since a general inhibition of nuclear transcription occurred in cress seedlings (Kittsteiner et al., 1991b). Incubation with 5-aminolevulinate (ALA) decreased the light-dependent cab gene expression in 4 days old but not in 2 days old cress seedlings (Kittsteiner et al., 1991b). Pigment analysis revealed that the amount of MgPMe was about equal in younger and older cress seedlings after ALA feeding but that other precursors (e.g. protoporphyrin) were accumulated in addition only in younger seedlings (Oster et al., 1991). If Chl precursors were to regulate cab gene transcription in cress seedlings, the signal should be the ratio MgPMe to other porphyrins rather than the concentration of MgPMe alone.

Light-dependent accumulation of plastid-encoded apoproteins does not occur on the level of transcription: Transcripts are present in dark-grown plants (Herrmann et al., 1985; Klein and Mullet, 1986; Kreuz et al., 1986) but most translation products appear only in the light together with Chl. The direct correlation of Chl synthesis (i.e. esterification of Chlide) with translation of plastid-encoded apoproteins was demonstrated in an <u>in vitro</u> system (Eichacker et al., 1990). More details on the system and more recent results will be delt with in the chapter of L. Eichacker. An important aspect was the evidence that Chl synthesis triggers translation of apoproteins independent of light. The coordination between Chl synthesis and translation of apoproteins might therefore also work in those organisms in which Chl is synthesized in the dark.

In summary, the last steps of Chl biosynthesis imply important regulatory steps (Fig. 3). Whereas light regulation via phytochrome is probably restricted to early steps of Chl biosynthesis and to transcription of nuclear genes, light-regulation via NADPH:Protochlide oxidoreductase has an indirect influence upon GGPP synthesis and translation of plastid-encoded apoproteins. This influence is realized via chlorophyll synthetase. The possible correlation between Chl synthesis and carotenoid synthesis on the one hand and transcription of nuclear genes on the other hand remains to be investigated in more detail.

ACKNOWLEDGEMENT

I wish to thank Mrs. R. Dennig for careful preparation of the manuscript. The work in my laboratory was supported by the Deutsche Forschungsgemeinschaft, Bonn, and the Fonds der Chemischen Industrie, Frankfurt.

REFERENCES

Apel, K., and Kloppstech, K., 1980, <u>Planta</u> 150:426-430.

Bennett, J., 1981, <u>Eur.J.Biochem.</u> 118:61-70.

Benz, J., Fischer, I., and Rüdiger, W., 1983a, <u>Phytochemistry</u> 22:2801-2804.

Benz, J., Haser, A., and Rüdiger, W., 1983b, <u>Plant Physiol.</u> 69:112-116.

Benz, J., and Rüdiger, W., 1981, <u>Z.Naturforsch.</u> 36c:51-57.

Dogbo, O. Bardat, F., and Camara, B., 1984, <u>Physiol.Veg.</u> 22:75-.

Dörnemann, D., Kotzabasis, K., Richter, P., Breu, V., and Senger, H., 1989, <u>Botanica Acta</u> 102:112-115.

Eichacker, L., Soll, J., Lauterbach, P., Rüdiger, W., Klein, R.R., and Mullet, J.E., 1990, <u>J.Biol.Chem.</u> 23:13566-13571.

Frosch, S., and Mohr, H., 1980, <u>Planta</u> 148:279-286.

Gronau, G., 1983, Dissertation Univ. München.

Herrmann, R.G., Westhoff, P., Alt, J. Tittgen, J. and Nelson, N. (1985) <u>in</u> "Molecular Form and Function of the Plant Genomes", L. van Vloten-Doting, G.S.P. Groot and T.C. Hall eds., pp. 233-256, Plenum Publishing Corp., Amsterdam.

Johanningmeier, U., 1988, <u>Eur.J.Biochem.</u> 177:417-424.

Johanningmeier, U., and Howell, St.H., 1984, <u>J.Biol.Chem.</u> 259:13541-13549.

Kittsteiner, U., Brunner, H., and Rüdiger, W., 1991b, <u>Physiol.Plant.</u> 81:190-196.

Kittsteiner, U., Mostowska, A., and Rüdiger, W., 1991a, <u>Physiol.Plant.</u> 81:139-147.

Klein, R.R., and Mullet, J.E., 1986, <u>J.Biol.Chem.</u> 261:11138-11145.

Kreuz, K., and Kleinig, H., 1981. <u>Plant Cell Rep.</u> 1:40-42.

Kreuz, K., Dehesh, K. and Apel, K., 1986, <u>Eur.J.Biochem.</u> 159:459-467.

Lindsten, A., Welch, C.J., Schoch, S., Ryberg, M., Rüdiger, W., and Sundqvist, C., 1990, <u>Physiol.Plant</u> 80:277-285.

Lütz, C., Benz, J., and Rüdiger, W., 1981, <u>Z.Naturforsch.</u> 36c:58-61.

Mullet, J.E., Klein, G.P., and Klein R.R., 1990, <u>Proc.Natl.Acad.Sci. USA</u> 87:4038-4042.

Oster, U., Blos, I., and Rüdiger, W., 1991, <u>Z.Naturforsch.</u> in press.

Paulsen, H., Rümler,U., and Rüdiger, W., 1990, <u>Planta</u> 181:204-211.

Rebeiz, C.A., Montazer-Zouhoor, A., Hopen, H.J., and Wu, S.M., 1984, <u>Microb.Technol.</u> 6:390-401.

Rüdiger W. (1987), <u>in</u>: "Progress in Photosynthetic Research". (J. Biggins, ed.) Vol. 4, pp. 461-467, Martin Nijhoff Publ., Dordrecht, Boston, Lancaster.

Rüdiger, W., Benz, J., and Guthoff, C., 1980, <u>Eur.J.Biochem.</u> 190:193-200.

Schoch, S., 1978, <u>Z.Naturforsch.</u> 33c:712-714.

Schoch, S., Lempert, U., and Rüdiger, W., 1977, <u>Z.Pflanzenphysiol.</u> 83: 427-436.

Schoch, S., and Brown, J., 1987, <u>J.Plant Physiol.</u> 126:483-495.

Soll, J., Schultz, G., Rüdiger, W., and Benz, J., 1983, <u>Plant Physiol.</u> 71: 849-854.

Steiger, A., Mitzka-Schnabel, U., Rau, W., Soll, J., and Rüdiger, W., 1985, <u>Phytochemistry</u> 24:739-743.

CONTROL OF CAROTENOID SYNTHESIS AND ASSEMBLY

OF PS-II BY LIGHT IN MUTANT C-6D OF *SCENEDESMUS**

Klaus Humbeck[1] and Susanne Römer[2]

[1]Institut für Allgemeine Botanik, Ohnhorststr. 18,
D-2000 Hamburg 52, Germany
[2]Institut de Biologie Moleculaire des Plantes du
CNRS, Universite Louis Pasteur, 12 rue du General
Zimmer, 67084, Strasbourg, France

INTRODUCTION

Besides the chlorophylls carotenoids also play an important role in the photosynthetic process. The two main functions of carotenoids in plants are light-harvesting and protection against harmful effects of excess light energy (Larkum and Barrett, 1983; Siefermann-Harms, 1985). Other functions as components of the electron transport chain and structural elements of the photosynthetic apparatus are discussed (Mathis, 1987; Senger and Strassberger, 1978). The carotenogenic pathway comprises the formation of C-40 tetraterpenoids starting from mevalonate. Later steps involve dehydrogenase, isomerase and cyclase activities leading to the cyclic carotenes and oxidation-reactions which yield the various xanthophylls.
Normally, all photosynthetic organisms are able to synthesize certain amounts of carotenoids during growth in darkness. But, in addition to this, parallel to light-induced chloroplast development light induces new synthesis of carotenoids and changes in the carotenoid composition in many plants (Rau, 1980; Rau and Schrott, 1987).
In the present study we pose two questions: i) how does light regulate carotenoid biosynthesis and ii) which role do carotenoids play in light-induced development of photosystem II.

MATERIAL AND METHODS

Organism and growth: *Scenedesmus obliquus* mutant C-6D which was induced by X-ray irradiation (Bishop, 1982) was employed. The culture conditions for heterotrophic growth were the same as described by Senger and Bishop (1972).

---

*Dedicated to Professor Dr. Horst Senger on the occasion of his 60th birthday

In some cases nicotine ($5*10^{-3}$ M) was added to the cultures 30 min before onset of illumination.

Quantitative and qualitative analysis of pigments: High performance liquid chromatography (HPLC) was used as described in Humbeck (1990). Pigments were identified by comparison of spectral characteristics with literature values, co-chromatography with authentic standards and analysis of peaks after chemical treatments (Humbeck, 1989).

Photosynthetic activity: Photosystem II activity was measured as Hill-reaction with p-benzoquinone following the method of Mell and Senger (1978).

Polyacrylamide gel electrophoresis: Pigment-protein complexes were isolated according to Römer et al. (1990). For PAGE, samples were prepared by solubilization at 4°C in a solution containing 0.3 M Tris-HCl (pH 8.8), 13% (v/v) glycerol and Triton X-100 and lithiumdodecylsulfate (LDS) to give final Triton X-100:LDS:protein proportions of 1.8:0.6:1 (by weight). The acrylamide was 4% (w/v) in the stacking gel and 7.5-15% (w/v) in the resolving gel. The distribution of pigment-protein complexes on the gel was determined by scanning at 670 nm in a spectrophotometer.

RESULTS AND DISCUSSION

Light regulation of carotenoid biosynthesis

Mutant C-6D of the unicellular green alga *Scenedesmus obliquus* is able to synthesize chlorophyll a during heterotrophic growth in the dark. Analysis of the carotenoid composition by HPLC revealed that acyclic carotenes accumulate in these cells (Table 1). The following intermediates could be identified: phytoene, ʓ -carotene, neurosporene and lycopene. In the dark no cyclic carotenes and xanthophylls are formed (Table 1). After transfer to light the acyclic intermediates disappear and a normal set of carotenes (α - and β -carotene) and xanthophylls (neoxanthin, violaxanthin, lutein, zeaxanthin, antheraxanthin and trihydroxy-α-carotene) is synthesized. These results clearly indicate that

Table 1. Pigment content of dark-grown and 3h illuminated cells of mutant C-6D of *Scenedesmus*. Analysis was performed by HPLC.

acyclic carotenes			cyclic carotenes			chlorophylls		
	dark	light		dark	light		dark	light
phytoene	+	-	α -carotene	-	+	chlorophyll a	+	+
ʓ-carotene	+	-	ß-carotene	-	+	chlorophyll b	-	+
neurosporene	+	-	neoxanthin	-	+			
lycopene	+	-	violaxanthin	-	+			
			lutein	-	+			
			zeaxanthin	-	+			
			antheraxanthin	-	+			
			trihydroxy-α-carotene	-	+			

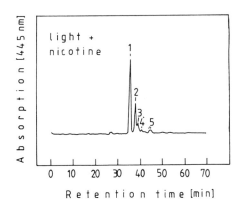

Figure 1.   HPLC-chromatogram   of   total   pigment   extract of
mutant C-6D of *Scenedesmus*. Cells were illuminated for 3h
in the presence of nicotine ($5*10^{-3}$ M).

Table 2. Absorption maxima of eluted pigments shown in Fig. 1
Solvents: chlorophylls - acetone, carotenes - hexane.

Peak	Pigment	Absorption maxima (nm)
1	chlorophyll a	661.2, 428.9
2	all-trans lycopene	501.7, 470.6, 443.9
3	chlorophyll a'	663.2, 430.7
4	prolycopene	438.5
5	neurosporene	460.0, 431.8, 410.0
	⟍ -carotene	423.3, 399.1, 379.0

light acts on the level of the cyclase-reaction in which the
last intermediate lycopene is transformed to carotene.
In order to investigate this step in more detail the reaction
was inhibited by nicotine ($5*10^{-3}$ M) in the dark. Then the
cells were illuminated for 3h. Pigments were extracted and
separated by HPLC. The resulting chromatogram (Fig. 1) shows
that   no   carotenes   and   xanthophylls were
synthesized under these conditions. Besides chlorophyll a
(peak 1) and chlorophyll a' (peak 3) low amounts of the
acyclic intermediates prolycopene (peak 4) and ⟍ -carotene and
neurosporene (both peak 5) which are also present in dark-
grown cells could be detected. But one major peak (peak 2)
which was not observed in dark-grown and illuminated cells
appeared. This pigment was identified to be all-trans
lycopene.The absorption maxima of the eluted pigments are
listed in Table 2.
The spectra of the cis-isomer prolycopene which accumulates in
the dark and of all-trans lycopene formed during illumination
in the presence of nicotine are shown in Figure 2. Whereas
prolycopene has one major peak at 438.5 nm in hexane the all-
trans isomer exhibits three maxima at 443.9, 470.6 and 501.7
nm.
Our data indicate a light-induced isomerization of prolycopene
to all-trans lycopene in mutant C-6D. Since the spectral
characteristics of these isomers differ drastically (Fig. 2)
the photoisomerization can be followed <u>in vivo</u> as an increase
in absorption at 520 nm. Actinic light of various

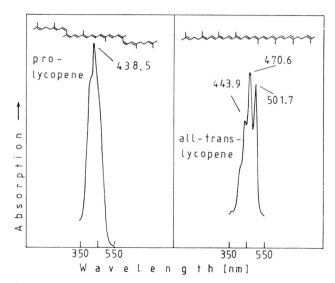

Figure 2. Absorption spectra of prolycopene and all-trans lycopene. Both pigments were solved in hexane.

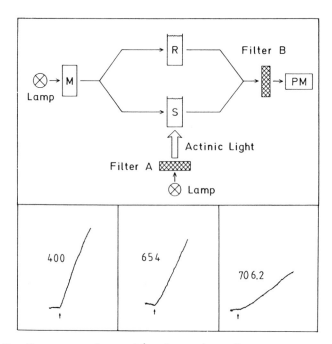

Figure 3. Upper part: optical system for measurement of light-induced absorption changes. M = monochromator (supplies light of 521 nm), R = reference, S = sample, PM = photomultiplier. Filter A = actinic light, filter B = DIL 521, Schott Mainz.
Lower part: typical tracings of light-induced increase in absorption.

wavelengths and intensities was provided by cross-illumination
to one aliquot (sample) and the change in absorption at 520 nm
was detected against another aliquot (reference) kept in the
dark (Fig. 3). In the lower part of Figure 3 some typical
experiments are shown. Actinic light of 400, 654 and 706.2 nm
had the same photon fluence rate. The slope of these tracings
represents a value for the rate of absorption change and
therefore for the rate of photoisomerization of prolycopene to
all-trans lycopene.
In Table 3 these values are listed for actinic blue (400,
438.7 and 467 nm) and red (672 and 706.2 nm) light. The data
prove that both parts of the spectrum are effective in
phototransformation. Since lycopene itself does not absorb red
light another pigment system must be involved in cis-trans
isomerization of lycopene. The very high value at 672 nm
(0.217 $\triangle$E/h) indicates that chlorophylls might act in this way
in mutant C-6D.
Our results correlate with investigations from Claes (1966),
Powls and Britton (1977) and Schiff et al. (1982) who also
found accumulation of cis-isomers of acyclic intermediates

Table 3. Rate of photoisomerization using different wave-
lenghts for actinic light. The rate of absorption change
was detected photometrically as described in Figure 3. The
actinic light had always a photon fluence rate of
12 μmol m^{-2} s^{-1}.

wavelenght (nm)	rate of isomerization ($\triangle$E/h)
400.0	0.150
438.7	0.168
467.0	0.032
672.0	0.217
706.2	0.040

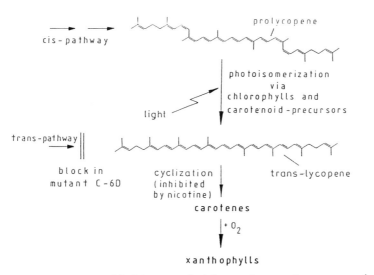

Figure 4. Model of light-regulation of carotenogenesis in
mutant C-6D of *Scenedesmus obliquus*.

during dark-growth of algal mutants. Furthermore, Beyer et al. (1989) reported a photoisomerization of ℑ -carotene and a subsequent isomerization of lycopene during carotenogenesis in daffodil chromoplasts. One author (Claes, 1966) investigated the spectral effectiveness of carotenoid biosynthesis and could also show that both, blue and red light, are effective in carotenogenesis.

Based on our results we propose a model for light-regulation of carotenoid biosynthesis in mutant C-6D (Fig. 4). In mutant C-6D prolycopene accumulates in the dark. Obviously, this isomer can not be transformed to carotene by the cyclase. Only after light-induced isomerization to all-trans lycopene the cyclase reaction takes place leading to carotenes and xanthophylls. As proposed by Ernst and Sandmann (1988) a mutation at the level of phytoen desaturase might cause an accumulation of only cis-isomeric forms. Characterization of this enzyme will be the subject of further experiments.

Assembly of photosystem II

As outlined above in mutant C-6D carotenoid biosynthesis is strictly light dependent whereas chlorophyll formation is not. These cells represent an ideal tool to study regulatory interactions between carotenoid biosynthesis and the formation of photosynthetic units during chloroplast development. During growth in darkness mutant C-6D is able to form an active photosystem I as proven by high activity and the presence of pigment-protein complex CP I (Römer et al., 1990). But no photosystem II activity and no photosystem II complex CPa could be detected in these cells (Fig. 5). After transfer to light both, photosystem II activity and CPa, appear with a fast kinetics (Fig. 5).

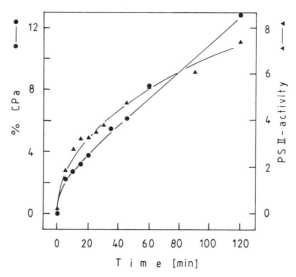

Figure 5. Changes in CPa-content and photosystem II activity in dark grown cells of mutant C-6D of *Scenedesmus* after transfer to light. Polyacrylamide gel electrophoresis and measurement of PS II activity were as described in Materials and Methods.

Neither chlorophyll a-synthesis nor the formation of the apoproteins of CPa which both take place during growth in darkness (Humbeck et al., 1990) can be the limiting factor regulating the assembly of active photosystem II units. But the light-dependent formation of carotenoids might be a good candidate. In order to test this hypothesis carotenoid synthesis was inhibited by nicotine and xanthophyll synthesis was prevented by the exclusion of oxygen. After 3h of illumination pigment-protein complexes were separated by polyacrylamide gel electrophoresis to detect the photosystem II-complex CPa. Photosystem II activity was measured as Hill-reaction. In control cells which synthesized carotenes and xanthophylls CPa and photosystem II activity could be detected (Table 4). The addition of nicotine prevents the formation of active photosystem II units. During illumination under anaerobic conditions also no active photosystem II complex is assembled. Here, carotenes are synthesized but xanthophylls are lacking. Only subsequent transfer to an aerobic environment allows new synthesis of xanthophylls and a parallel development of a functioning photosystem II. These data clearly indicate that carotenoids (especially xanthophylls) are essential components for the assembly of an active photosystem II-unit. Similar results have been reported by Lebedev et al. (1988) who treated barley seedlings with norfluorazone. The exact mechanism by which carotenoids act during the assembly-process is not yet clear. This has to be clarified in future experiments.

Table 4. CPa-content, photosystem II activity and presence of carotenes and xanthophylls in cells of mutant C-6D. Cells were illuminated for 3h (control). Nicotine was added with a final concentration of $5*10^{-3}$ M. Oxygen was excluded by bubbling the cell suspension with nitrogen and the cells were either analyzed at once or after subsequent 3h of illumination in the presence of oxygen.

	CPa	PS II-activity	carotenes	xanthophylls
control	+	+	+	+
+nicotine	-	-	-	-
- O_2	-	-	+	-
- O_2 → + O_2	+	+	+	+

REFERENCES

Beyer, P., Mayer, M. and Kleinig, H., 1989, Molecular oxygen and the state of geometric isomerism of intermediates are essential in the carotene desaturation and cyclization reactions in daffodil chromoplasts, Eur. J. Biochem., 184:141.
Bishop, N.I., 1982, Isolation of mutants of Scenedesmus obliquus defective in photosynthesis, in: "Methods in Chloroplast Molecular Biology", M. Edelman, R.B. Hallick, N.-H. Chua, eds., Elsevier, Amsterdam, pp. 51.
Claes, H., 1966, Maximal effectiveness of 670 mμ in the light dependent carotenoid synthesis in Chlorella vulgaris, Photochem. Photobiol. 5:515.

Ernst, S. and Sandmann, G., 1988, Poly-cis carotene pathway in the *Scenedesmus* mutant C-6D, Arch. Microbiol. 150:590.

Humbeck, K., 1989, Light dependent carotenoid biosynthesis in mutant C-6D of *Scenedesmus obliquus*, Photochem. Photobiol. 51:113.

Humbeck, K., 1990, Photoisomerization of lycopene during carotenogenesis in mutant C-6D of *Scenedesmus obliquus*, Planta 182:204.

Humbeck, K., Römer, S. and Senger, H., 1990, Light-dependent assembly of the components of photosystem II core complex CPa in mutant C-6D of *Scenedesmus obliquus*, J. Plant Physiol. 136:569.

Larkum, A.W.D. and Barrett, J., Light-harvesting processes in algae, 1983, Adv. Bot. Res. 10:1.

Lebedev, N.N., Pakshina, E.V., Bolychevtseva, Y.V. and Karapetyan, N.V., 1988, Fluorescence characterization of chlorophyll-proteins in barley seedlings grown with the herbicide norlurazon under low irradiance, Photosynthetica 22:371.

Mathis, P., 1987, Primary reactions of photosynthesis: discussion of current issues, in: "Progress in Photosynthesis Research", Vol. 1, J. Biggins, ed., Martinus Nijhoff, Dordrecht, pp. 151.

Mell, V. and Senger, H., Photochemical activities, pigment distribution and photosynthetic unit size of subchloroplast particles isolated from synchronized cells of *Scenedesmus obliquus*, 1978, Planta 143:315.

Powls, R. and Britton, G., 1977, The role of isomers of phytoene, phytofluene and $\mathcal{J}$-carotene in carotenoid biosynthesis by a mutant strain of *Scenedesmus obliquus*, Arch. Microbiol. 115:175.

Rau, W., 1980, Photoregulation of carotenoid biosynthesis: an example of photomorphogenesis, in: "Pigments in Plants", F.C. Czygan, ed., Fischer, Stuttgart, pp.80.

Rau, W. and Schrott, E.L., 1987, Blue light control of pigment biosynthesis, in: "Blue Light Responses: Phenomena and Occurance in Plants and Microorganisms", H. Senger, ed., CRC Press, Boca Raton, pp.43.

Römer, S., Humbeck, K. and Senger, H., 1990, Relationship between biosynthesis of carotenoids and increasing complexity of photosystem I in mutant C-6D of *Scenedesmus obliquus*, Planta 182:216.

Schiff, J.A., Cunningham, F.X. and Green, M.S., 1982, Carotenoids in relation to chloroplasts and other organelles, in: IUPAC, Carotenoid Chemistry and Biochemistry, G. Britton and T.W. Goodwin, eds., Pergamon Press, Oxford, pp.329.

Senger, H. and Bishop, N.I., 1972, The development of structure and function in chloroplasts of greening mutants of *Scenedesmus*. I Formation of chlorophylls, Plant Cell Physiol. 13:633.

Senger, H. and Straßberger, G., 1978, Development of photosystems in greening algae, in: "Chloroplast Development", G. Akoyunoglou, ed., Elsevier, Amsterdam, pp. 367.

Siefermann-Harms, D., 1985, Carotenoids in photosynthesis. I Location in photosynthetic membranes and light-harvesting function, Biochim. Biophys. Acta 811:325.

PURIFICATION AND CHARACTERIZATION OF TWO GLUTAMYL-tRNA LIGASE

FRACTIONS FROM THE UNICELLULAR GREEN ALGA *SCENEDESMUS OBLIQUUS*,

MUTANT C-2A'

U. Vothknecht., D. Dörnemann and H. Senger

FB Biologie/Botanik, Philipps-Universität
Marburg, Lahnberge
D-3550 Marburg (Germany)

Introduction

Two different pathways are known for 5-aminolevulinic acid (ALA) biosynthesis, the first committed intermediate in the synthesis of all tetrapyrrols (Shemin and Russel, 1953; Beale and Castelfranco, 1973). In plants and algae ALA as the precursor of chlorophyll biosynthesis is synthesized via the C_5-pathway from the intact five-carbon chain of glutamate or 2-oxoglutarate, respectively (for review see: Rüdiger and Schoch, 1988; Beale, 1990).

The first step of the C_5-pathway is the ligation of glutamate to a $tRNA^{glu}$ by glutamate-tRNA ligase (Schön et al., 1986). This step requires ATP, tRNA and Mg^{2+}. The activated glutamate is subsequently reduced by a NADPH-dependent dehydrogenase to glutamate-1-semialdehyde (GSA). The next steps of the C_5-pathway are still discussed controversly. Kannangara and Gough (1978) proposed a direct conversion of GSA to ALA by GSA-aminotransferase. A second two-step mechanism includes the deamination of GSA to dioxovalerate (DOVA) by GSA-aminotransferase, followed by the transamination of DOVA to ALA by DOVA-transaminase (Kah et al., 1988; Rüdiger and Schoch, 1988). A third mechanism proposed by Hoober et al. (1988) postulates the transamination of GSA to diaminovalerate (DAVA) followed by the deamination of DAVA to ALA.

In *Scenedesmus obliquus* two ligases have been separated during chromatography purification. In the current contribution a further purification and characterization of the ligase(s), an essential enzyme in the tetrapyrrol biosynthesis is described.

Results

Three day old cultures of the yellow pigment mutant C-2A' of *Scenedesmus obliquus*, grown heterotrophically in darkness, were exposed to light for 8 hours. Afterwards the cells were harvested by centrifugation, washed and resuspended in cold extraction buffer containing 0.1 M tricine (pH 9), 0.3 M glycerol, 25 mM $MgCl_2$ and 1 mM DTT and homogenized in a Vibrogen cell mill (Senger and Mell, 1977).

The 300,000 g supernatant was fractionated by gel filtration and serial chromatography on S-300, Blue Sepharose, Red Agarose and chlorophyllin Sepharose as previously described by Wang et al. (1981). The method was adapted to *Scenedesmus* by Breu and Dörnemann (1988). Blue Sepharose and Red Agarose binding proteins were eluded with 1 M NaCl in extraction buffer. The eluates of both columns were desalted by gel filtration on Sephadex G-50 and concentrated using an Amicon pressure dialysis cell with a PM 30 membrane. tRNA, the substrate of the activity test, was isolated from the chlorophyllin Sepharose eluate.

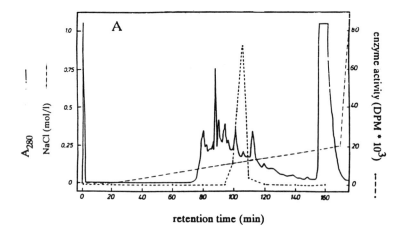

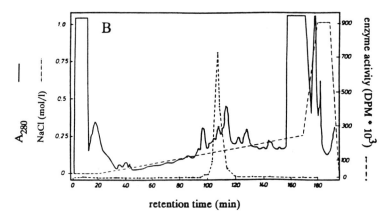

Fig 1. Elution profiles of the separation of Blue Sepharose and Red Agarose eluates by anion exchange chromatography on Mono Q. Binding protein were eluated with a continous NaCl gradient. All fractions were tested of glutamyl-tRNA ligase activity.

Ligase activity was determined in 0.5-1 ml assays containing 100 μg of protein, 100 μg RNA, 0.1 M tricine (pH 7.9), 0.3 M glycerol, 25 mM MgCl$_2$, 1 mM DTT and 1 mM ATP. Incubation time was 20 min at 28 °C. The reaction was started by adding 5 μCi of L-[U-^{3}H]-glutamate and stopped by adding 1/10 TCA. The reaction mixture was collected on fiberglass filters (Schleicher & Schüll, G25) and washed twice with icecold 10% TCA and ethanol. The TCA-insoluble radioactivity on the filter was measured by liquid scintillation counting. The protein fractions from both, Blue Sepharose and Red Agarose, showed glutamate-tRNA ligase activity. This is in accordance with the results from *Cyanidium caldarium* (Houghton et al., 1989). The ligation requires tRNA and ATP.

In isolated chloroplasts from barley and *Chlamydomonas* only one ligase could be found (Bryant and Kannangara, 1986; Chen et al., 1990; Chang et al., 1990). In order to find out whether the two ligase fractions are isoenzymes, or arbitrary fractions of one of the same enzyme, isolation and characterization of both ligase fractions were carried out. It was investigated whether the two ligase fractions can be distinguished by their physiological reactions or by their enzymatic properties.

Therefore the protein fractions from Blue Sepharose and Red Agarose were further purified by anion exchange chromatography on Mono Q HR 10/10. The column was equilibrated with buffer containing 20 mM tricine (pH 9), 0.3 M glycerol, 10 mM $MgCl_2$ and 1 mM DTT. Bound proteins were eluded by a continous salt gradient from 0 to 250 mM NaCl. The elution profile of both Blue Sepharose and Red Agarose binding fractions are shown in Fig 1a and b. The activity of both ligase fraction eluded from the column as a single peak at a NaCl concentration of 150 mM. The active fractions were pooled, desalted and referred to as Mono Q BS and Mono Q RA respectively.

For characterization of the Mono Q BS and Mono Q RA fractions, their K_m-values for glutamate were determined. 3 μg protein from both fractions were incubated with varying glutamate concentrations while all the other components of the assay were kept identically. To make sure that the RNA-amount was not a limiting factor, RNA saturation was measured before (Data not shown). The K_m-value was examined in a Lineweaver-Burke-plot of the kinetics. The plots for both fractions are shown in Fig 2a and b. From the plot a K_m-value of 2.5 ± 0.4 μM glutamate was determined for Mono Q BS. The Mono Q RA ligase has a K_m-value of 2.2 ± 0.3 μM glutamate.

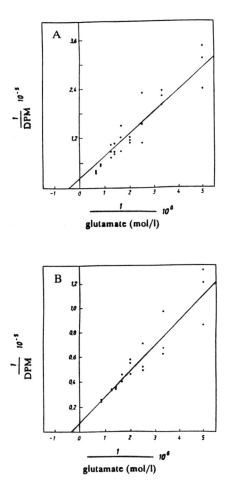

Fig 2. Determination of the K_m-values for glutamate for Mono Q BS and Mono Q RA. The measurements were carried out in an assay with varying glutamate concentrations. The K_m-values were examined using a Lineweaver-Burke-plot.

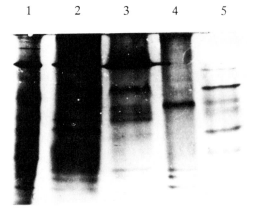

Fig.3. Samples of all purification steps of the ligase fractions from *Scenedesmus obliquus*, separated by non-denaturing gelelectrophoresis.
1: cell-free extract
2: BS eluate
3: RA eluate
4: Mono Q eluate BS
5: Mono Q eluate RA

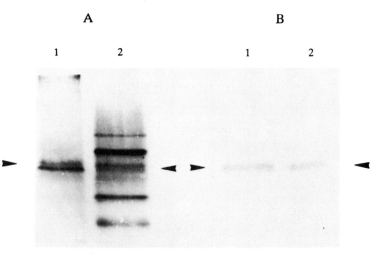

Fig 4. Comparison of the separation of the Mono Q eluats from the *Scenedesmus obliquus* ligase fractions by non-denaturing gel electrophoresis (A) with the Western blot analysis (B).
1: Mono Q BS eluate
2: Mono Q RA eluate

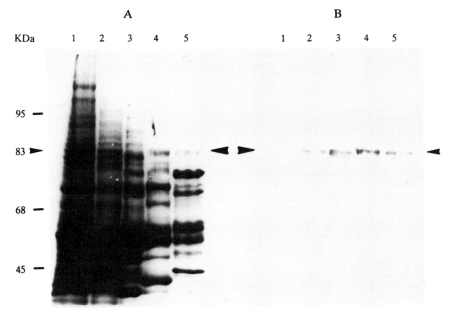

Fig 5. Samples of all purification steps of the *Scenedesmus obliquus* ligase fractions were separated by SDS-page. The Coomassie stained gel (A) was compared with the Western blot analysis (B).
1: cell free extract
2: BS eluate
3: RA eluate
4: Mono Q eluate BS
5: Mono Q eluate RA

Samples of all purification steps (50 µg protein) were separated by non-denaturing gel electrophoresis to show the enrichment of the ligases. The non-denaturing gelelectrophoresis was carried out by the modified method of Davis (1968) using flat gels with 6 % polyacrylamide. A gel stained with Coomassie Blue is shown in Fig 3. In the Mono Q BS and Mono Q RA fraction 5 or 6 protein bands are detectable, respectively. This indicates that both fractions are highly purified. However, further purification steps are necessary to achieve apparent homogenity of the proteins.

To identify the ligases, separated on the gel, Western blot analysis was performed as described by Tobwin (1979) and immunostained using antibodies conjugated with alkaline phosphatase. For the identification it was tried whether the ligases react with an antibody raised against the barley chloroplast glutamate-tRNA ligase (Bryant and Kannangara, 1986). The antibody was a generous gift from Dr. C.G. Kannangara, Copenhagen. A single reaction in identical position for both ligases is shown in Fig 4.

To determine the molecular mass of the ligases, SDS-gel electrophoresis was carried out using flat gels with 8% polyacrylamide. The protein bands were stained with Coomassie Blue.The enriched protein fractions were separated and the ligases were identified by Western blot analysis (Fig 5). A single band in identical position for both ligase fractions appeared with a molecular mass of 83 kDa.

As a summary the two ligase fractions show no differences in the measured enzymatic properties. They are only distinguishable by their different affinity to Blue Sepharose. The question whether they are iso-enzymes or derive from different compartments of the cell cannot be answered at the current stage.

Acknowledgments

The authors wish to thank Dr. C.G. Kannangara and colleagues for providing the barley glutamyl-tRNA ligase antibody. Furthermore we want to thank Ms. K. Bölte and Ms. B. Böhm for their skillful technical assistance. This research was supported by grants of the Deutsche Forschungsgemeinschaft.

References

Beale, S.I. 1990, Biosynthesis of tetrapyrrole pigment precursor, δ-aminolevulinic acid, from glutamate., Plant Physiol., 93:1273

Beale, S.I. and Castelfranco, P.A. 1973, ^{14}C-incoporation from exogeneous compounds into 5-aminolevulinic acid by greening cucumber cotyledons., Biochem. Biophys. Res. Commun. 52:143

Breu, V. and Dörnemann, D. 1988, Formation of 5-aminolevulinate via glutamate-1-semialdehyde and 4,5-dioxovalerate with participation of an RNA component in *Scenedesmus obliquus* mutant C-2A'., Biochem. Biophys. Acta 967:135

Bryant, P. and Kannangara, C.G. 1987, Biosynthesis of δ-aminolevulinate in greening barley leaves. VIII. Purification and characterization of the glutamate-tRNA-ligase., Carlsberg Res. Commun. 52:99

Chang, T.-E., Wegmann, B. and Wang, W.-Y. 1990, Purification and characterization of glutamyl-tRNA synthetase., Plant Physiol. 93:1641

Chen, M.-W., Jahn, K., Schön, A., O'Neill, G.P. and Söll, D. 1990, Purification and characterization of *Chlamydomonas reinhardtii* chloroplast glutamyl-tRNA synthetase, a natural misacylating enzyme., J. of Biological Chemistry 265:4054

Davis, B.J. 1964, Disc electrophoresis-II: method and application to human serum proteins., An. N.Y. Acad. Sci. 121:404

Hoober, J.K., Kahn, A., Ash, D.E., Gough, S.P. and Kannangara, C.G. 1989, Biosynthesis of δ-aminolevulinate in greening barlay leaves. XI. Structure of the substrate, mode of gabaculine inhibition and catalytic mechanisms of glutamate-1-semialdehyde aminotransferase., Carlsberg Res. Commun. 53:11

Houghton, J.D., Brown, S.B., Gough, S.P. and Kannangara, C.G. 1989, Biosynthesis of δ-aminmolevulinate in *Cyanidium caldarium*: Characterisation of tRNA^{glu}, Ligase, Dehydrogenase and Glutamate-1-semialdehyde-aminotransferase., Carlsberg Res. Commun. 54:131

Kah, A., Breu, V. and Dörnemann, D. 1988, Quantitative determination of 4,5-dioxovaleric acid as a metabolite in *Scenedesmus obliquus* after complete separation from δ-aminolevulinic acid., Biochem. Biophys. Acta 964:62

Kannangara, C.S. and Gough, S.P. 1978, Biosynthesis of δ-aminolevulinate in greening barley leaves: glutamate-1-semialdehyde aminotransferase., Carlsberg Res. Commun. 43:185

Rüdiger, W. and Schoch, S. 1988, Chlorophylls, in: Plant Pigments ,T.W. Goodwin, ed., Acad. Press, London, San Diego

Schön, A., Krupp, G., Gough, S.P., Berry-Lowe, S., Kannangara, C.G. and Söll, D. 1986, The RNA required in the first step of chlorophyll biosynthesis is a chloroplast glutamate tRNA., Nature 322:281

Senger, H. and Mell, V. 1977, Preparation of photosynthetically active particles from synchronized cultures of unicellular algae., in: Methods in Cell Biology XV, Prescott, ed., Acad. Press, New York and London

Shemin, D. and Russel, C.S. 1953, δ-Aminolevulinic acid, its role in the biosynthesis of porphyrins and purines., J. Am. Chem. Soc. 75:4837

Tobwin, H., Staehelin, T. and Gordon, S. 1979, Electrophoretic transfer of proteins from polyacrylamid gels to nitrocellulose sheets. Procedure and some applications., Proc. Natl. Acad. Sci., 76:4350

Wang, W., Gough, S.P. and Kannangara, C.G. 1981, Biosynthesis of δ-aminolevulinate in greening barley leaves. IV. Isolation of three soluble enzymes required for the conversion of glutamate to δ-aminolevulinate., Carlsberg Res. Commun. 46:243

THE PHOTOREDUCTION OF PROTOCHLOROPHYLL(IDE) IN *SCENEDESMUS* AND

BARLEY (*HORDEUM VULGARE*)

R. Knaust, B. Seyfried, K. Kotzabasis, H. Senger

Fachbereich Biologie/Botanik
Philipps-Universität Marburg
Lahnberge, 3550 Marburg (Germany)

Dark grown seedlings of angiosperms and some dark grown pigment mutants of algae accumulate protochlorophyllide (PChlide) and also small anounts of protochlorophyll (PChl) but contain no chlorophyll (Chl) or only traces.

It has been shown that the Chl-precursors PChlide and PChl both exist in the monovinyl-form (MV) with only one vinyl-group at C3 and in the divinyl-form (DV) with an additional vinyl-residue at C8 (Belanger and Rebeiz, 1980a; 1980b).

The enzyme catalyzing the photoconversion of PChlide to chlorophyllide (Chlide), NADPH-Protochlorophyllide oxidoreductase (PChlide reductase) (Griffiths, 1978), needs NADPH as cofactor for this reaction. The reduction is mediated by light. In angiosperms, also Chlide occurs in the MV- and the DV-form which indicates that MV- and DV-PChlide both serve as a substrate for the PChlide reductase (Rebeiz et al., 1983; Rebeiz et al., 1986). Other authors, however, postulate that first DV-PChlide is reduced to MV-PChlide and then subsequently photoconverted to Chlide (Castelfranco and Beale, 1983). PChlide acts as the main substrate for the PChlide reductase, whereas the ability of PChlide reductase to reduce PChl, the esterified form of PChlide, is still a matter of debate. Some authors found a photoreduction of PChl (Kotzabasis et al., 1989; Liljenberg, 1974) whereas others deny the existence of this reaction for higher plants (Griffiths, 1979).

In this article some aspects of the last steps of Chl-biosynthesis will be discussed for the pigment-mutant C-2A' of *Scenedesmus obliquus* and for dark-grown barley (*Hordeum vulgare*).

1. EVIDENCE FOR THE PRESENCE OF MV AND DV FORMS OF PCHL(IDE) IN *SCENEDESMUS* C-2A'

Pigment mutant C-2A' of *Scenedesmus obliquus* was grown heterotrophically in the presence of 5-aminolevulinic acid (ALA) of a final concentration of 1 mM. After 20 h cells were harvested by centrifugation. Cells were treated in different ways for the isolation of either PChlide or PChl.

Isolation of PChlide and PChl

Pelleted cells were first suspended in acetone/ 0.05 M NH_4OH (9:1) and then disrupted in a Vibrogen cell mill (Senger and Mell, 1977). The pigment extract was centrifuged for 5 min at 1400 x g to remove unbroken cells and cell fragments. The supernatant was then extracted three times with petroleumether. The nonpolar petroleumether-phase contained the carotenoids, Chl and PChl, whereas the more polar PChlides stayed in the

Table 1. Spectroscopic properties of MV- and DV- PChlide and -PChl:
Absorbance maxima, fluorescence emission maxima (EM) and fluorescence
excitation maxima (EX)

	Absorbance [nm]	Fluorescence EM [nm]	EX [nm]	Reference
MV-PChlide				
Cucumis	431/535/571/623	–	432	Belanger and Rebeiz (1980a)
Scenedesmus	433/ 575/624	630	428	this study
C-2A'	431/ 575/628	632	429	Kotzabasis and Senger (1986)
MV-PChl				
Cucumis	432/530/568/620	–	432	Belanger and Rebeiz (1980)
Cucurbita	432/535/570/622	630/ 685	–	Houssier and Sauer (1969)
Cucurbita	432/535/571/623	–	–	Jones (1966)
Scenedesmus C-2A'	432/ 572/624	630	428	this study
DV-PChlide				
Cucumis	436/537/574/625	–	438	Belanger and Rebeiz (1980)
Scenedesmus	438/ 575/624	630	433	this study
C-2A'	438/ 578/629	633	437	Kotzabasis and Senger (1986)
DV-PChl				
Cucumis	438/530/570/622	–	438	Belanger and Rebeiz (1980)
Cucurbita	438/536/572/622	630/ 685	–	Houssier and Sauer (1969)
Cucurbita	438/537/574/624	–	–	Jones (1966)
Scenedesmus C-2A'	438/ 575/624	630	433	this study

acetone-phase. Separation of MV- and DV-PChlide was carried out by an HPLC-method modified after Shioi and Beale (1987) using a polyethylene-column. An isocratic system of 80% acetone/distilled water was used as the eluent at a flow rate of 0.4 ml/min (Kotzabasis et al, 1990).

For the isolation of PChl cells of *Scenedemus* C-2A' were extracted with hot methanol. Then the pigments were transferred into acetone and a fraction containing PChl and Chl was separated from carotenoids and PChlides on FRACTOGEL-TSK-DEAE 650 (Merck, Darmstadt, FRG) following Omata and Murata (1980; 1983). The separation of MV- and DV-PChl was performed by HPLC analogous to the separation of the PChlides using an isocratic system of 95% acetone/distilled water at a flow rate of 1 ml/min.

In both cases separation of the pigments on HPLC yielded pure MV- and DV-PChlide and -PChl. Spectroscopic data of absorption and fluorescence spectroscopy correspond to data from literature (see table 1.).

For PChlide, Plasma desorption mass spectrometry (PDMS) showed that the two PChlides were indeed the MV- and the DV-form of the pigment. The mass spectra of MV- and DV-PChlide are shown in Fig.1a+b. The peak at 610 mass units for DV- and at 612 mass units for MV-PChlide are identical with theoretically calculated values.
The spectroscopic data of the two PChl-forms showed, as expected, the same maxima as MV-/DV-PChlide, respectively, and correspond to data from the literature.

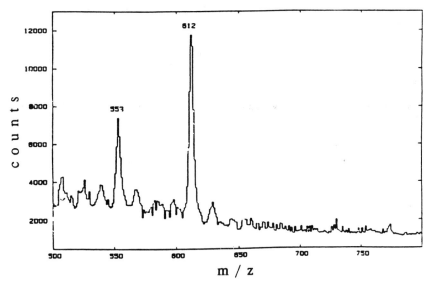

Fig.1a. PDMS mass spectrum of MV-PChlide

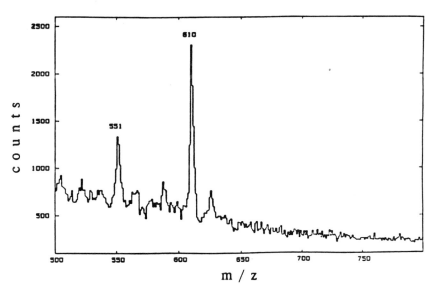

Fig.1b. PDMS mass spectrum of DV-PChlide

2. THE *IN VITRO*-PHOTOCONVERSION OF MV- AND DV-PCHLIDE WITH AN ENRICHED FRACTION OF PCHLIDE REDUCTASE FROM *SCENEDESMUS OBLIQUUS* MUTANT C-2A'

The question whether both forms of PChlide are able to serve as a substrate for the PChlide reductase from dark-grown *Scenedesmus* is of general interest. Therefore enzyme-assays with an enriched PChlide reductase-fraction and MV- or DV-PChlide were performed.

For the assay 1 ml of an enriched PChlide reductase-fraction, pH 9, and 1 ml of tricine-buffer, pH 7, was supplemented with MV- or DV-PChlide (final concentration 50-100 nM). Fluorescence emission spectra were taken, then 1 mM NADPH was added and the fractions were exposed for 7 sec to white light. After illumination fluorescence emission spectra were taken again (Fig.2). With both substrates, MV- and DV-PChlide, the spectra showed an increase of Chlide and a decrease of PChlide after illumination. Thus, both PChlide forms were accepted by PChlide reductase and transformed to Chlide.

Fluorescence excitation spectra of the dark controls and the illuminated fractions showed different excitation maxima (MV-PChlide: 436 nm, MV-Chlide 433 nm, and DV-PChlide: 442 nm, DV-Chlide: 441 nm). The difference between the maxima for fluorescence excitation of the MV- and DV-forms was about 8 nm. Two different Chlide-species were formed in this test which obviously have to be interpreted as MV- and DV-Chlide.

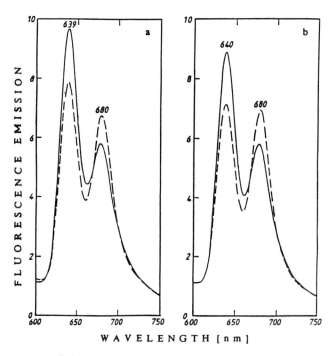

Fig. 2. Fluorescence emission spectra of *in vitro* enzyme assays of PChlide reductase with either MV- (a) or DV-PChlide (b).
The assay first contained enzyme and substrate (———). Then, NADPH was added and the assay was illuminated for 7 s (---). (Ex 435 nm)

3. THE *IN VIVO*-PHOTOCONVERSION OF PCHL IN ETIOLATED BARLEY

The question whether PChl can be photoreduced to Chl in analogy to the PChlide/Chlide-photoconversion is yet not answered. For *Scenedesmus* there is evidence that PChl is photoconvertible (Kotzabasis et al., 1989). A coherence was found between the photoreduction of PChl to Chl and the development of the reaction centers of the photosynthetic apparatus in *Scenedesmus* C-2A' (Kotzabasis et al., 1991). This explains the physiological role of PChl. For higher plants the photoconversion of PChl was denied (Griffiths, 1980).

To clarify this controversy, five day old dark grown barley seedlings were harvested under a dim green safelight, illuminated with one flash of white light, then immediately frozen in liquid nitrogen and extracted with hot methanol. The pigment extract and the extract of a dark-control treated in the same manner as the light-sample were transferred into acetone and separated on thin layer chromatography according to Kotzabasis et al. (1989). The PChl- and Chl-containing bands were scraped off and fluorescence emission spectra in acetone were taken (Fig.3). The decrease of the PChl-maximum at 630 nm and the corresponding increase of the Chl-maximum at 660 nm in the illuminated sample compared with the dark control show that PChl can be photoconverted to Chl in barley too. A reduction of PChlide to Chlide and the following phytylation of Chlide to Chl starts immediately after exposure of the etiolated plant material to light. The short time from illumination of the sample to the stop of all reactions by liquid nitrogen however is not sufficient for the slow phytylation of Chlide to Chl.

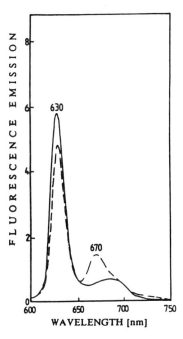

Fig. 3. *In vivo*-photoconversion of PChl in barley (*Hordeum vulgare*)
Fluorescence emission spectra of the PChl/Chl-fraction of five day old dark-grown barley seedlings separated by TLC according to Kotzabasis et al. (1989).
Dark control (——) and extract from seedlings that were flashed one time and immediately frozen in liquid nitrogen after illumination (---). (Ex 432 nm)

Thus it is concluded that PChl is photoconverted to Chl even in higher plants and can be used for the fast formation of reaction centers of the photosynthetic apparatus immediately after exposure to light.

Acknowledgment

This work was supported by a grant of the Deutsche Forschungsgemeinschaft. We want to thank Ms. K. Bölte and Ms. B. Böhm for their technical assistance.

REFERENCES

Belanger, F. C., and Rebeiz, C. A., 1980a, Chloroplast biogenesis: Detection of divinyl protochlorophyllide in higher plants, J. Biol. Chem., 225:1266-1272.

Belanger, F. C., and Rebeiz, C. A. 1980c, Chloroplast biogenesis: Detection of protochlorophyllide ester in higher plants, Biochemistry, 19:4875-4883.

Castelfranco, P. A. and Beale, S. I., 1983, Chlorophyll biosynthesis: Recent advances and areas of current interest, Ann. Rev. Plant. Physiol., 34:241-278.

Griffiths, W. T., 1978, Reconstitution of Chlorophyllide formation by isolated etioplast membranes, Biochem. J., 174:681-692.

Griffiths, W. T., 1980, Substrate-specifity studies on protochlorophyllide reductase in barley (Hordeum vulgare) etioplast membranes, Biochem. J., 186:267-278.

Houssier, C., and Sauer, K., 1969a, Optical properties of the protochlorophyll pigments. 1. Isolation, Characterization and IR-spectra, Biochim. Biophys. Acta, 172:476-491.

Houssier, C., and Sauer, K., 1969b, Optical properties of the protochlorophyll pigments. 2. Electronic absorption, fluorescence and circular dichroism spectra, Biochim. Biophys. Acta, 172:492-502.

Kotzabasis, K., Humbeck, K., and Senger, H., 1991; Incorporation of photoreduced protochlorophyll into reaction centers, J. Photochem. Photobiol., B:Biol., 8:255-262.

Kotzabasis, K., Schüring, M.-P., and Senger, H., 1989, Occurence of Protochlorophyll and its phototransformation to Chlorophyll in mutant C-2A' of Scenedesmus obliquus, Physiol. Plant., 75:221-226.

Kotzabasis, K., Senge, M., Seyfried, B., and Senger, H., 1990, Aggregation of monovinyl- and divinyl protochlorophyllide in organic solvents, Photochem. Photobiol., 52:95-101.

Omata, T., and Murata, N., 1980, A rapid and efficient method to prepare Chlorophyll a and b from leaves, Photochem. and Photobiol., 31:183-185.

Omata, T., and Murata, N., 1983, Preparation of Chlorophyll a, Chlorophyll b and Bacteriochlorophyll a by column chromatography with DEAE Sepharose CL-6B and Sepharose CL-6B, Plant and Cell Physiol., 24:1093-1100.

Rebeiz, C. A., Tripathy, B. C., Wu, S.-M., Montazer-Zouhoor, A., and Carey, E. C., 1986, Chloroplast biogenesis 52: Demonstration in toto of monovinyl and divinyl monocarboxylic Chlorophyll biosynthetic routes in higher plants, in: "Regulation of Chloroplast differentiation", G. Akoyunoglou, H. Senger, eds., Alan R. Liss. Inc. New York.

Rebeiz, C. A., Wu, S.-M., Kuhadja, M., Daniell, H., and Perkins, E. Biosynthetic routes and Chlorophyll a chemical heterogeneity in plants, Mol. Cell. Biochem., 57:97-125.

Senger, H., and Mell, V., 1977, Preparation of photosynthetically active particles from synchronized cultures of unicellular green algae, Methods Cell. Biol., 15:201-210.

Shioi, Y., and Beale, S. I.,1987, Polyethylene-based high-performance liquid chromatography of chloroplast pigments: Resolution of mono- and divinyl Chlorophyllides and other pigment mixtures, Analytical Biochemistry, 162:493-499.

THE REGULATION OF PROTOCHLOROPHYLL SYNTHESIS AND ITS PHYSIOLOGICAL ROLE

Kiriakos Kotzabasis[1], Klaus Humbeck[2]
and Horst Senger[2]

[1]Department of Biology, University of Crete,
P.O. Box 1470, 711 10 Heraklio, Crete, Greece
[2]Fachbereich Biologie/Botanik, Philipps-Universität
Marburg, Lahnberge, D-3550 Marburg, Deutschland

INTRODUCTION

It is well established that the last two steps of chlorophyll (Chl) biosynthesis are a fast photoreduction of protochlorophyllide (PChlide) to chlorophyllide (Chlide) [1,2] and a rather slow esterification of Chlide to Chl [3]. The fast reduction of PChlide to Chlide is light dependent in angiosperms and some mutants of algae [4,5]. Besides the generally occurring PChlide also the occurrence of protochlorophyll (PChl), i.e. the already esterified PChlide, has been frequently reported [6-13]. PChl which was found in angiosperms, could not be photoreduced to Chl [14,15], whereas in Euglena gracilis [8] and in a mutant strain of Chlorela regularis [9] PChl could be photoreduced to Chl.

In this contribution we try to assess the role of PChl in Chl biosynthesis and its physiological role in the assembly of the photosynthetic apparatus in the pigment mutant C-2A' of Scenedesmus obliquus.

MATERIALS AND METHODS

Pigment mutant C-2A' of Scenedesmus obliquus [16] was used in this study. This mutant was grown heterotrophically in the dark as described by [17]. Pigments were extracted with hot methanol and the residue transferred into acetone. PChl was isolated, using a combination of DEAE-chromatography and a TLC-system according to [18]. The PChl/Chl and the PChlide/Chlide fractions were separated with a TLC-system as described by [18]. For the preparation of membrane fraction from C-2A' mutant cells of Scenedesmus obliquus the method of [19] is used. The separation of thylakoid membrane proteins by polyacrylamide gel electrophoresis (PAGE) and the radioactive labelling of polypeptides was according to [20]. For fluorescence measurements we used the method of [21].

RESULTS AND DISCUSSION

Pigment mutant C-2A' of <u>Scenedesmus</u> <u>obliquus</u> synthesizes small amounts of PChlide and PChl when grown heterotrophically in darkness. PChl was isolated and characterized with absorption, fluorescence emission and fluorescence excitation spectra. The peaks in the absorption spectrum (624 and 432 nm in acetone) were identical with those reported by [22] and [11]. The ratio of the blue to red maximum of 7.2 corresponded to the value of 7.3 reported by [12]. The fluorescence emission and excitation maxima (F629/E429) corresponded to the emission and excitation peaks (F631/E429) of monovinyl-PChlide [23].

Immediately after irradiation (2 min 652 nm, 3 Wm^{-2}) the pigments were extracted and separated into PChl/Chl and PChlide/Chlide fraction. Fluorescence emission spectrum of the PChl/Chl fraction compared with the corresponding fraction of the dark control, showed that PChl is photoconvertible to Chl (Fig. 1). Kinetics of PChl and PChlide photoreduction after irradiation (white light, $20Wm^{-2}$) followed the same pattern. After 5 min light ca 60% of the non reduced precursor forms were reduced. This is a good indication that in the C-2A' mutant of <u>Scenedesmus</u> <u>obliquus</u>, PChl is photoconvertible to Chl in the same way as PChlide to Chlide.

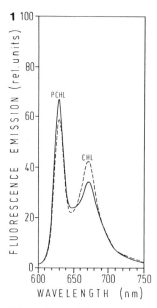

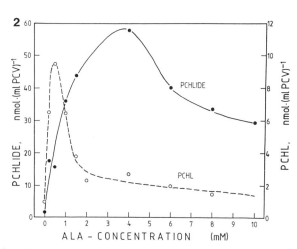

Fig. 1. Fluorescence emission spectra of PChl/Chl fractions in acetone extracts of the dark control (solid line) and after 2 min of irradiation (dotted line). Fluorescence was excited by 435nm.

Fig. 2. Concentrations of PChlide and PChl in the pigment mutant C-2A' of <u>Scenedesmus</u> <u>obliquus</u> incubated with various ALA concentrations in the dark for 20h.

212

Investigations of the influence of 5-aminolaevulinic acid (ALA) on the accumulation of PChlide and PChl, demonstrate that the addition of ALA causes an increase in the amounts of PChlide and PChl, confirming that ALA is the limiting factor for PChlide and PChl biosynthesis in darkness. Moreover, the results suggest that the concentration of ALA regulates the ratio of PChlide to PChl in different ways. The PChl level reaches a sharp maximum at an ALA concentration of 0.5 mM, representing a tenfold increase over the control level (Fig. 2). The PChlide level shows a broader maximum at 4mM ALA. At this ALA concentration PChlide increases about 30fold (Fig. 2). The ratio of PChlide to PChl is low in the darkness and changes from 2 at 0.5mM ALA to 22 at 4mM ALA. Kinetics of PChlide and PChl formation during incubation with ALA show, that at high concentration of ALA (5mM) PChlide reaches a maximum after 16h while a small increase of PChl is noticed after 24h. At the low concentration of ALA (0.5mM) PChlide increases steadily, whereas the PChl amount reaches a sharp maximum after 6h incubation and declines thereafter (Fig. 3).

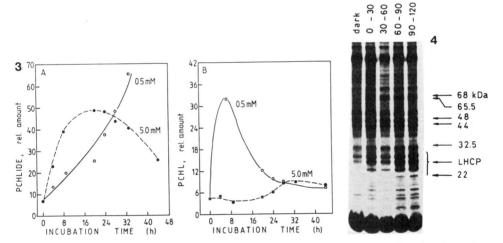

Fig. 3. Time course of PChlide (A) and PChl (B) formation in the dark grown cells in the presence of 0.5 and 5mM ALA respectively.

Fig. 4. Fluorography of thylakoid membrane polypeptides after labelling with ^{35}S for various time periods during the first 2h of illumination of mutant C-2A' of <u>Scenedesmus</u> <u>obliquus</u>. ^{35}S (250µCi per 100ml) was added to the cultures as sodium sulphate during dark growth, at onset of illumination (0-30) and after 0.5h (30-60), 1h (60-90) and 1.5h (90-120) of illumination. After 30 min of incubation the polypeptides (250µg/slot) were separated by PAGE. The gels were impregnated with 1M sodium salicylate, dried and exposed to X-ray film for 14 days.

Table 1. Concentration of PChl and PChlide and relative amounts of PSII reaction centres in cells incubated for 20h in the dark with various concentrations of ALA.

	ALA-concentration [mM]		
	0	0.5	5.0
PChl [nmol (ml PCV)$^{-1}$]	0.8	10	2
PChlide [nmol (ml PCV)$^{-1}$]	2	15	50
PSII reaction centres (relative values)	6	17	1.5

First the influence of ALA on the ratio of PChlide to PChl in comparison to the fact that the Chl formation via PChl involves only a fast photoreduction, and second the fact that Chl formation via PChlide includes the fast photoconversion to Chlide and a rather slow esterification to Chl, indicate that the rapidly photoconverted Chl via PChl immediately after irradiation, is inserted into photosynthetic reaction centres.

Parallel to Chl formation, the development of the photosynthetic apparatus takes place. To analyse the biosynthesis of the apoproteins of reaction centres and light-harvesting complexes, cells were incubated with ^{35}S-labelled sodium sulphate for various periods during the first 2h of light induced chloroplast development. The thylakoid membrane polypeptides were separated by PAGE and the incorporation of radioactivity in the newly synthesized proteins was monitored by fluorography. The dark control cells already show some incorporation into the polypeptides of the core of PSI (68 and 65.5 KD) and PSII (48, 44 and 32.5 KD) [24]. The polypeptides of PSI and PSII reaction centres are synthesized with high rates immediately on exposure to light (Fig. 4). The biosynthesis of the apoproteins of the light-harvesting complexes is much slower. They run on the gel between 30 and 20 KD [25]. The delay in formation (about 1h) is most evident for the band at 22 KD, representing one major integral apoprotein of the light-harvesting complex. The coincidence of the fast photoconversion of PChl to Chl with the appearance of the reaction centre proteins, suggests that newly synthesized Chl via PChl is immediately inserted into the reaction centres. The slower Chl formation via PChlide serves for the synthesis of light-harvesting pigment-protein complexes. Mutant cultures after incubation with 0.5 and 5mM ALA or without ALA for 20h in the dark, were supplemented with laevulinic acid (10mM) in order to prevent further formation of Chl precursors and were irradiated for 15 min with white light (20 Wm^{-2}). Results were recorded fluorimetrically at low temperature. The dark control and the 5mM ALA sample have low amounts of PChl (0.8 and 2nmol.mlPCV^{-1}, respectively) and showed only a peak at 688 and 686 nm respectively. Only the 0.5mM ALA sample, which has considerable amounts of PChl (10nmol.mlPCV^{-1}) presents two separate maxima at 686 and 696nm. The 696nm peak corresponds to the core complex of PSII. The formation of reaction centres was also measured in aliquots of the samples used for the previous experiment by fluorescence induction with similar results (Tab. 1). The

Chl, synthesized directly after instant photoreduction of PChl, is used for the rapid formation of reaction centres. Because of the fact that the cells incubated with 5mM ALA have less PSII reaction centres than those without ALA we can assume that the small amount of PChlide and PChl in the cells in the absence of ALA can be photoreduced by protochlorophyllide oxidoreductase (PCR), whereas in the cells with a high concentration of PChlide the sites of the PCR will be mainly occupied by PChlide.

All these results established the occurrence and explain the physiological role of PChl in the mutant C-2A' of Scenedesmus obliquus.

REFERENCES

1. W.T. Griffiths, 1975, Biochem. J. 152:623-635.
2. W.T. Griffiths, 1978, Biochem. J. 174:681-692.
3. W. Rudiger, J. Benz and C. Guthoff, 1980, Eur. J. Biochem. 109:193-200.
4. T. Oh-hama and E. Hase, 1980, Plant Cell Physiol. 21: 1263-1272.
5. H. Senger and G. Brinkmann, 1986, Physiol. Plant. 68:119-124.
6. B.Singh, 1953, Phytomorphology 3: 224-239.
7. O.T.G. Jones, 1966, Biochem. J. 101: 153-160.
8. C. Cohen and J.A. Schiff, 1976, Plant Cell Physiol. 24: 555-567.
9. T. Sasa and K. Sugahara, 1976, Plant Cell Physiol. 17: 273-279.
10. H.A. Lancer, C.E. Cohen and J.A. Schiff, 1976, Plant Physiol. 57: 369-374.
11. C. Sundqvist, H. Ryberg, B. Boddi and F. Lang, 1980, Physiol. Plant. 48: 297-301.
12. C. Sundqvist and H. Ryberg, 1983, Physiol. Plant. 59: 42-45.
13. H.I. Virgin, 1984, Physiol. Plant. 61: 303-307.
14. W.T. Griffiths, 1980, Biochem. J. 186: 267-278.
15. Y. Shioi and T. Sasa, 1983, Plant Cell Physiol. 24: 835-840.
16. N.I. Bishop, 1971, Methods Enzymol. 23: 130-143.
17. N.I. Bishop and J. Wong, 1971, Biochim. Biophys. Acta 234: 433-445.
18. K. Kotzabasis, M.P. Schuring and H. Senger, 1989, Physiol. Plant. 75: 221-226.
19. H. Senger and V. Mell, 1977, In: Methods in Cell Biology (D.M. Prescott, ed.), Vol. XV, pp. 201-218. Academic Press Inc., New York ISBN 0-12-56115-X.
20. K. Kotzabasis, K. Humbeck and H. Senger (1991), J. Photochem. Photobiol. B8: 225-262.
21. S. Malkin, P.A. Armond, H.A. Mooney and D.C. Fork, 1981, Plant Physiol. 67: 570-579.
22. W.W. Svec, 1978, In: The Porphyrins. Chlorophylls and Bacteriochlorophylls (D. Dolphin, ed.) Vol. V, pp. 341-349. ISBN 0-12-220107-8.
23. K. Kotzabasis and H. Senger, 1986, Z. Naturforsch. 41c: 1001-1003.
24. B.R. Green, 1988, Photosynthes. Res. 15: 3-32.
25. P. Delepelaire and N.H. Chua, 1981, J. Biol. Chem. 256: 9300-9307.

PIGMENT-PROTEIN COMPLEXES OF CHLOROPHYLL PRECURSORS

M. Ryberg, N. Artus, B. Böddi, A. Lindsten,
B. Wiktorsson, and C. Sundqvist

Department of Plant Physiology,
University of Göteborg, Carl Skottsbergs Gata 22
S-413 19 Göteborg, Sweden

INTRODUCTION

In this paper we describe the early light-induced changes of the organization of the chlorophyllous pigments as etioplasts of dark-grown wheat leaves (Triticum aestivum) develop to chloroplasts. Based on results of both in vitro and in vivo experiments, we present our view on how changes in the state of protochlorophyllide (PChlide) and chlorophyllide (Chlide), and of NADPH-PChlide oxidoreductase (PChlide-reductase) are related to the transformation of prolamellar bodies (PLBs), and how the early stages of greening may influence plastid activities, like esterification of Chlide and translation of plastid-encoded proteins. The paper is focused on recent results, but some background may be appropriate.

PChlide is present in dark-grown leaves in several spectral forms. The relative amounts depend on e.g the developmental stage of the plastids. Phototransformable PChlide657 (fluorescence maximum at 657 nm) is the dominating form of PLBs (Ryberg and Sundqvist, 1982), and is generally considered as PChlide complexed with PChlide-reductase and NADPH. Non-phototransformable PChlide633 is the dominating form of prothylakoids (PTs), and is considered as not associated to PChlide-reductase (Oliver and Griffiths, 1982). Phototransformation of PChlide657 results in a long-wavelength fluorescing Chlide, which undergoes a spectral shift, the Shibata-shift, to a short-wavelength Chlide within ca 30 min (Shibata, 1957). The cause for the Shibata shift has been ascribed the esterification of Chlide (Akoyunoglou and Michalopoulos, 1971) or the dissociation of Chlide from the PChlide-reductase (Brodersen, 1976), but is still a matter of debate.

Immuno-electron microscopical studies showed that PChlide-reductase is localized almost exclusively in the PLBs of 6-day-old dark-grown wheat leaves (Ryberg and Dehesh, 1986). Illumination of the leaves resulted in an apparent translocation of PChlide-reductase from the transforming PLBs to the developing

Regulation of Chloroplast Biogenesis
Edited by J.H. Argyroudi-Akoyunoglou, Plenum Press, New York, 1992

thylakoids. The time course for this redistribution of the protein coincided with the Shibata shift. No drastic decline in the amount of PChlide-reductase during the first 30 min of irradiation was observed. These results are in contrast to the observation of a rapid decline in PChlide-reductase reported by Shaw et al. (1985).

PIGMENT COMPLEXES

Circular dichroism (CD) spectroscopy of isolated PLBs (Ryberg and Sundqvist, 1982) was used to analyse interaction between the pigment molecules. PLBs at three different conditions were compared; unirradiated, immediately after 3 flashes of white light, and 40 min later (Böddi et al., 1990). Unirradiated PLBs excibited a strong CD-signal at 655 nm, an indication of PChlide-PChlide interaction. After the flash irradiation the band shifted to 685 nm, still showing chromophore-chromophore interaction. During the following dark period a decreasing CD-signal was observed and reflects a decrease in the interaction between the Chlide molecules. This coincides with the time period of a spectral blue-shift similar to the Shibata shift of intact leaves. The results were compared with the CD-signals of isolated PLBs, treated with the non-ionic detergent octylglucoside. Increasing concentrations of the detergent resulted in decreasing CD-signals. Thus, the solubilization reduced the interaction between the PChlide molecules. Solubilization of the PLBs resulted also in a spectral shift of PChlide to shorter wavelengths. Based on these results we conclude that *a breakup of the pigment interactions correlate with the spectral shifts to shorter wavelengths.*

PROTEIN COMPLEXES

Crosslinking agents were used to analyse the size of PChlide-reductase complexes of PLBs. Isolated PLBs were treated with the zero-crosslinkers 1-ethyl-3-[3-(dimethylamino)propyl] carbodiimide (EDC) or 2-iminothiolane (Sundqvist et al., this Proceedings). In this way only proteins closely situated in the membranes were crosslinked. The treatment resulted in a strong (covalent) binding which is not broken by SDS. SDS-PAGE and immuno-blots showed the presence of the 36 kD polypeptide of PChlide-reductase and a double-band at ca 56 kD, probably the CF1 of chloroplast ATPase, in the non-crosslinked control. Only minor bands of higher molecular weights were observed. By increasing concentrations of the crosslinkers the amount of the PChlide-reductase monomer decreased and new bands of higher molecular weights appeared. High amounts of protein did not even penetrate into the gel or stopped at the interface between the stacking gel and the separation gel. Immunoblots of 2-dimensional SDS-PAGE showed that these were mainly polymers of PChlide-reductase. After the first dimension the gel strip was treated with mercaptoethanol to break the crosslinking. The second step revealed that mainly PChlide-reductase molecules were crosslinked to each other, and thus situated close to each other in the PLB membranes. The apparent molecular weight in the first gel at ca 72 kD and higher indicates that PChlide-reductase is present as dimers and larger aggregates in the PLBs.

As mentioned above, octylglucoside caused a spectral shift of PChlide of isolated PLBs to shorter wavelengths, and CD spectra indicated that this shift was correlated with a breakup of the pigment complexes. Fluorescence spectra of octyl-glucoside-treated isolated PLBs with or without crosslinking agents showed that a crosslinking of the PChlide-reductase partially prevented such a shift caused by the detergent. Based on these observations we conclude that *PChlide-reductase is present as large aggregates in PLBs, and there is a correlation between the spectral shifts to shorter wavelengths and a breakup of the PChlide-reductase aggregates.* Taken together, these results indicate that *PChlide and PChlide-reductase in highly regular PLBs are present as aggregates of pigment-protein complexes.*

PIGMENT-PROTEIN ASSOCIATIONS

From previous studies we know that isolated PLBs can retain the highly regular structure despite a transformation of PChlide to Chlide, provided the Chlide is preserved in the long-wavelength form (Ryberg and Sundqvist, 1988). The transformation of the PLB structure in vitro coincided with a spectral shift of Chlide to shorter wavelengths. From the results and conclusions presented so far it follows, that the structural transformation of PLBs coincides with a breakup of the Chlide aggregates as well as of the PChlide-reductase aggregates. Does this mean that the association of Chlide with the reductase also is broken?

In an approach to analyze the association of PChlide or Chlide with PChlide-reductase at various stages of structural transformation of PLBs, isolated PLBs were treated with octyl-glucoside, and the solubilized components were subjected to preparative isoelectric focusing (IEF; Wiktorsson et al., this proceedings). The fluorescence emission maxima of the pigments before the detergent treatment were used as criteria for the state of the pigments and the structure of the PLBs. Solubili-zation at 3 different stages were performed; unirradiated PLBs containing PChlide657, irradiated PLBs containing long-wave-length Chlide, and irradiated PLBs where the Chlide had shifted to shorter wavelengths. The IEF was run in the presence of sucrose and NADPH, as these components prevent an artificial dissociation of the pigment-protein complexes. After completion of the IEF, the resulting 20 fractions were analysed for PChlide and Chlide by fluorescence spectroscopy, and for proteins by SDS-PAGE and immuno-blots. PChlide of the unirra-diated PLBs was heavily enriched in one fraction, at a pH of ca 5. Similarly, Chlide of the irradiated PLBs were enriched in a few fractions, both when long-wavelength Chlide- and short-wavelength Chlide-containing PLBs were solubilized. SDS-PAGE showed that PChlide-reductase was enriched in the same frac-tions as the PChlide or Chlide, independently of the pretreat-ment of the PLBs. The activity of PChlide-reductase was well preserved. More than 75% of the PChlide of unirradiated PLBs was transformed to Chlide by flash irradiation, which is taken as an indication of PChlide being associated to the enzyme. Based on the similar distribution of the pigments and the enzyme in all three cases we conclude that *Chlide of isolated PLBs is associated with PChlide-reductase also after the spectral shift to shorter wavelengths, and PChlide - PChlide-reductase complexes as well as Chlide - PChlide-reductase complexes focus at a pH of ca 5.*

The presence of several isoforms of PChlide-reductase in dark-grown leaves was reported (Ikeuchi and Murakami, 1982; Dehesh et al., 1986). Two-dimensional electrophoresis showed at least 4 isoforms, focusing at pH 8-9. (A calculation of the theoretical pI of PChlide-reductase, based on the amino acid sequence of a cDNA for the protein, gives a pI value of ca 6.2; Schulz et al., 1989). In 16 h illuminated dark-grown leaves only one isoform was seen. This could in part be explained by a light-induced proteolytic break-down of PChlide-reductase after the depletion of the substrate PChlide (Kay and Griffiths, 1983; Häuser et al., 1984). The remaining isoform may then be the PChlide-reductase active in green leaves. This remains however to be proven. In this connection it is interesting to mention the report by Covello et al. (1987), that PChlide-reductase of dark-grown leaves is a phosphorylated protein. Could the presence of several apparently different isoforms in the 2-D gels be explained by varying degrees of phosphorylation? ATP was shown to have a positive effect on the phototransformation of PChlide (Horton and Leech, 1972), and we found that the addition of ATP to isolated etioplast membranes preserved or even increased the ratio PChlide657/PChlide633, especially at low pH (Grevby et al., 1987). Thus the presence of large aggregates of PChlide and PChlide-reductase is favoured by the addition of ATP. It is challenging to suggest that the state of the pigment-enzyme complexes depend on the degree of phosphorylation of PChlide-reductase.

ESTERIFICATION OF CHLOROPHYLLIDE

As mentioned in the introduction, the Shibata shift has been suggested to be due to the dissociation of Chlide from the PChlide-reductase or to the esterification of Chlide. From our in vitro experiments we have now seen that a spectral shift, similar to the Shibata shift of intact leaves, can take place without the dissociation of Chlide and PChlide-reductase. We have analysed the esterification capacity of isolated etioplast membrane fractions (Lindsten et al., 1990). (So far, the esterifying enzyme, chlorophyll synthetase, has not been identified). Esterification of Chlide took place in a highly purified PLB- as well as in a PT fraction. Thus Chl synthetase seems to be present in both types of membranes. However, when endogenous Chlide was used as the substrate, and the PLBs were irradiated under conditions that preserved the long-wavelength form of Chlide, no esterification activity was seen. Under such conditions large aggregates of Chlide and PChlide-reductase are still present, and the PLBs retain their highly regular structure. We concluded that the lack of esterification was due to either, that Chl synthetase was in an inactive state, or that Chlide was not available for esterification to occur. We have continued our studies on esterification during greening by analysing the activity in plastid membrane fractions isolated after varying times of preirradiation (0, 5, 10, and 20 min) of the leaves. In this way, the plastids contain PLBs at varying degrees of transformation, as well as thylakoids at increasing stages of development. The plastids were fractionated as described previously (Lindsten et al., 1988). Analyses of the total amounts of pigments and proteins showed that there was a successive increase in pigment- and protein contents in the "PT" fraction and a decrease in the "PLB" fraction with preirradiation time. At 5 min, the "PLB" fraction was dominated by long-wavelength Chlide, and the "PT" fraction by short-wavelength Chlide. With prolonged preirradiation, the relative amounts of short-wavelength Chlide increased in the "PT" fraction. A slight increase in the specific esterification activity in the "PT" fraction, as well as an increase in total esterification activity in the "PT" fraction, was observed. SDS-PAGE indicated a successive increase of PChlide-reductase in the "PT" fraction, but no other obvious changes in the polypeptide pattern were seen (Lindsten et al., in preparation). Thus, the esterifying enzyme remains unidentified.

If we go back to in vivo studies and look at the transformation of PLBs, the Shibata shift, and the esterification of Chlide, we can see that these events are correlated with each other, apparently in a similar way as in vitro. Does Chlide, PChlide-reductase and Chl synthetase relocalize simultaneously from the PLBs to the PTs during irradiation? In an attempt to answer at least part of the question, we performed a short-term (over-night) treatment of dark-grown wheat leaves with clomazone or amiprophos-methyl (APM; Artus et al., 1991). This short-term incubation did not significantly alter the carotenoid contents (and probably not the amount of the alcohol substrate for esterification of Chlide. The synthesis of terpenoids are inhibited at the C15 stage; Duke et al., 1991). The light-induced structural transformation of the PLBs was significantly delayed by clomazone. As could be expected, the inhibition of PLB-transformation also correlated with a lack of PChlide-reductase relocation, as indicated by immuno-electron-microscopical studies. In contrast to control leaves, the distribution of PChlide-reductase remained as in etioplasts in the irradiated APM- or clomazone treated leaves. (By 60 min light the amount of PChlide-reductase was drastically reduced, but still within the PLB-area, in the clomazone treated leaves). Fluorescence spectra showed a clearly delayed Shibata shift in the APM- or clomazone treated leaves. This also correlated with a delayed esterification of Chlide. We believe that this was due to the lack of PLB-transformation and the absence of available Chlide (or inactive Chl synthetase. APM or clomazone did not, however, inhibit the activity of Chl synthetase in vitro under conditions where esterification was possible in the control). We conclude that *chlorophyll synthetase is present but inactive in well preserved PLBs, and esterification of Chlide is preceded by the breakup of the large Chlide - PChlide-reductase aggregates, seen as the Shibata shift, and is also correlated with the relocation of PChlide-reductase.*

Further indications for the validity of the described hypothesis on the series of events during greening of etiolated leaves, are the results obtained by the use of 8-hydroxy-quinoline. The reformation of PLBs, after returning irradiated dark-grown leaves to darkness, was inhibited by a pretreatment with this metal chelator, despite the presence of PChlide-reductase (Ryberg and Sundqvist, 1991). Immuno-electron micro-scopy indicated presence of large amounts of the protein in the prothylakoids, in the same way as in control leaves before returning the leaves to darkness. The results may be explained by the lack of reformed PChlide (8-hydroxyquinoline inhibits chlorophyll synthesis at the step after Mg-protoporphyrin (ME)). More interesting in this connection is that the esterification of Chlide was inhibited. It is reasonable to assume that PChlide-reductase was still associated with Chlide, possibly co-transported with PChlide-reductase to the prothylakoids. These leaves did not perform a normal Shibata shift, an indication that the large pigment-protein complexes present in the PLBs do not have to be broken up to single pigment-protein complexes in order to be translocated to the PTs. In vivo, esterification may occur immediately after the breakup of the large pigment-protein aggregates, and may thus easily be interpreted as the cause for the Shibata shift.

The detailed analyses of the light-induced changes during etioplast to chloroplast development are interesting in view of the regulatory role of light for the synthesis of chloroplast proteins. Recently it was shown by Eichacker et al. (1990), that the esterification of Chlide induced an accumulation of P700 apoproteins in lysed etioplasts. Maybe this light-induced translation is triggered by the transformation of the PLBs, which leads to the relocation of PLB components, e.g. PChlide-reductase, and to the release of the Chlide from the PChlide-reductase, rather than by the esterification process as such?

Results obtained by in vitro experiments can certainly not explain all the events in the intact leaf. The changes observed to take place in isolated PLBs can be influenced by factors outside the PLBs. This is shown by an enhancement of the spectral shift of Chlide to shorter wavelengths by the addition of a crude stroma fraction (Sundqvist et al., unpublished). Despite the limitations, in vitro experiments can help interpreting the events observed in vivo.

ACKNOWLEDGEMENTS

This work was supported by the Swedish Natural Science Research Council.

REFERENCES

Artus, N.N., Ryberg, M., Lindsten, A., Ryberg, H., and Sundqvist, C., 1991, The Shibata shift and the transformation of etioplasts to chloroplasts in wheat with clomazone (FMC 57020) and amiprophos-methyl (Tokunol M), Plant Physiol., in press.

Akoyunoglou, G., and Michalopoulos, G., 1971, The relation between the phytylation and the 682 to 672 nm shift in vivo of chlorophyll a, Physiol. Plant., 25:324.

Brodersen, P., 1976, Factors affecting the photoconversion of protochlorophyllide to chlorophyllide in etioplast membranes isolated from barley, Photosynthetica, 10:33.

Böddi, B., Lindsten, A., Ryberg, M., and Sundqvist, C., 1990, Phototransformation of aggregated forms of protochlorophyllide in isolated etioplast inner membranes, Photochem. Photobiol., 52:83.

Covello, P.S., Webber, A.N., Danko, S.J., Markwell, J.P., and Baker, N.A., 1987, Phosphorylation of thylakoid proteins during chloroplast biogenesis in greening etiolated and light-grown wheat leaves, Photos. Res., 12:243.

Dehesh, K., van Cleve, B., Ryberg, M., and Apel, K., 1986, Light-induced changes in the distribution of the 36000-M$_r$ polypeptide of NADPH-protochlorophyllide oxidoreductase within different cellular compartments of barley (Hordeum vulgare L.), Planta, 169:172. et al

Duke, S.O., Paul, R.N., Becerril, J.M. Schmidt, J.H., 1991, Clomazone causes accumulation of sesquiterpenoids in cotton (Gossypium hirsutum L.), Weed Sci., in press.

Eichacker, L.A., Soll, J., Lauterbach, P., Rüdiger, W., Klein, R.R., and Mullet, J.E., 1990, In-vitro synthesis of chlorophyll a in the dark triggers accumulation of chlorophyll a apoproteins in barley etioplasts, J. Biol. Chem., 265:13566.

Grevby, C., Ryberg, M., and Sundqvist, C., 1987, Transformation of photoactive protochlorophyllide in isolated prolamellar bodies of wheat (Triticum aestivum) exposed to low pH and ATP, Physiol. Plant., 70:155.

Häuser, I., Dehesh, K., and Apel, K., 1984, The proteolytic degradation in vitro of the NADPH-protochlorophyllide oxidoreductase of barley (Hordeum vulgare L.), Arch. Biochem. Biophys., 228:577.

Horton, P., and Leech, R.M., 1972, The effect of ATP on the photoconversion of protochlorophyllide in isolated etioplasts, FEBS Lett. 26:277.

Ikeuchi, M., and Murakami, S., 1982, Behavior of the 36,000-dalton protein in the internal membranes of squash etioplasts during greening, Plant Cell Physiol., 23:575.

Kay, S.A., and Griffiths, W.T., 1983, Light-induced breakdown of NADPH-protochlorophyllide oxidoreductase in vitro, Plant Physiol., 72:229.

Lindsten, A., Ryberg, M., and Sundqvist, C., 1988, The polypeptide composition of highly purified prolamellar bodies and prothylakoids from wheat (Triticum aestivum) as revealed by silver staining, Physiol. Plant., 72:167.

Lindsten, A., Welch, C.J., Schoch, S., Ryberg, M., Rüdiger, W., and Sundqvist, C., 1990, Chlorophyll synthetase is latent in well preserved prolamellar bodies of etiolated wheat, Physiol. Plant., 80:277.

Oliver, R.P., and Griffiths, W.T., 1982, Pigment-protein complexes of illuminated etiolated leaves, Plant Physiol., 70:1019.

Ryberg, M., and Dehesh, K., 1986, Localization of NADPH-protochlorophyllide oxidoreductase in dark-grown wheat (Triticum aestivum) by immuno-electron microscopy before and after transformation of the prolamellar bodies, Physiol. Plant., 66:616.

Ryberg, M., and Sundqvist, C., 1982, Characterization of prolamellar bodies and prothylakoids fractionated from wheat etioplasts, Physiol. Plant., 56:125.

Ryberg, M., and Sundqvist, C., 1988, The regular ultrastructure of isolated prolamellar bodies depends on the presence of membrane-bound NADPH-protochlorophyllide oxidoreductase, Physiol. Plant., 73:218.

Ryberg, M., and Sundqvist, C., 1991, Structural and functional significance of pigment-protein complexes of chlorophyll precursors, in:" Chlorophylls," H. Scheer, ed., CRC Press, Boca Raton, ISBN 0-8493-6842-1.

Schulz, R., Steinmüller, K., Klaas, M., Forreiter, C., Rasmussen, S., Hiller, C., and Apel, K., 1989, Nucleotide sequence of a cDNA coding for the NADPH-protochlorophyllide oxidoreductase (PCR) of barley (Hordeum vulgare L.) and its expression in Escherichia coli, Mol. Gen. Genet., 217:355.

Shaw, P., Henwood, J., Oliver, R., and Griffiths, T., 1985, Immunogold localization of protochlorophyllide oxidoreductase in barley etioplasts, Eur. J. Cell Biol., 39:50.

Shibata, K., 1957, Spectroscopic studies of chlorophyll formation in intact leaves, J. Biochem., 44:147.

Sundqvist, C., Wiktorsson, B., Zhong, L.B., Böddi, B., and Ryberg, M., Cross-linking of NADPH-protophyllide oxidoreductase in isolated prolamellar bodies, this Proceedings.

Wiktorsson, B., Ryberg, M., and Sundqvist, C., IEF of solubilized active protochlorophyllide reductase, this Proceedings.

CROSS-LINKING OF NADPH-PROTOCHLOROPHYLLIDE OXIDOREDUCTASE IN

ISOLATED PROLAMELLAR BODIES

Christer Sundqvist, Bengt Wiktorsson, Zhong Lin
Bang, Bela Böddi[1] and Margareta Ryberg

Dept. of Plant Physiology, University of
Göteborg, Carl Skottsbergs Gata 22, S-413 19
Göteborg, Sweden, [1]Dept. of Plant Physiology,
Eötvös Lorand University, Múzeum krt 4/A,
H-1088 Budapest Hungary

INTRODUCTION

In dark-grown leaves the plastid development is arrested at the etioplast level with the formation of typical prolamellar bodies (PLBs) and prothylakoids (PTs). The dominating protein of isolated PLBs is NADPH-protochlorophyllide oxidoreductase (PChlide-reductase). The PLBs also contain most of the phototransformable PChlide (Lindsten et al., 1988; Ikeuchi and Murakami, 1983). Immunocytochemical labelling of ultrathin sections of dark-grown wheat leaves confirmed that the PChlide-reductase is localized in the PLBs (Ryberg and Dehesh, 1986).

During irradiation PChlide is transformed to chlorophyllide (Chlide) which then is esterified to chlorophyll. During this process several spectral alterations occur, one of which is referred to as the Shibata shift. The phototransformable PChlide has been described as an aggregate of PChlide and PChlide-reductase molecules (Böddi et al., 1989, 1990). The use of cross-linking agents makes it possible to analyze the aggregational state of the proteins. By the cross-linking technique it is also possible to estimate the proximity between a certain protein and its neighbouring proteins.

MATERIALS AND METHODS

Prolamellar bodies were isolated from dark-grown 6.5-day-old leaves of Triticum aestivum (L) cv. Kosack in a bottom loaded sucrose gradient as previously described (Ryberg and Sundqvist, 1982). The PLBs were collected and stored frozen until use. The cross-linking with the zero-length cross-linker 1-ethyl-3-[3-(dimethylamino)propyl]carbodiimide (EDC) was made according to Wynn and Malkin (1988). The pelleted PLBs were resuspended in 5 mM magnesium chloride, 20 mM MOPS, pH 6.5, and 3 mM EDC. After incubation for 30 min at 25° C the cross-linked PLBs were solubilized in SDS buffer and subjected to PAGE in the PhastSystem (Pharmacia LKB Biotechnology) in a 4-15% gradient gel. Cross-linking with 2-iminothiolane was made mainly according to Traut et al. (1989). After incubation in 12 mM 2-iminothiolane for 2.5 h in darkness at 25°C, the cross-linking was induced by adding hydrogen peroxide to a final concentration of 40 mM. The suspension was made 40 mM with iodoacetamide and incubated for 30 min at 25°C. The PLBs were dissolved in SDS buffer without mercaptoethanol. Samples were subjected to PAGE in the PhastSystem in a 8-

25% gradient gel. The Coomassie stained gels were scanned with an image-analyser. Solubilization of PLBs was performed with octylglucoside in isolation medium (Ryberg and Sundqvist, 1982).

Fluorescence emission spectra were measured at 77 K with an SLM 8000C spectrofluorometer according to Lindsten et al. (1988). Smoothed and corrected fluorescence spectra were resolved into Gaussian components by the SPSERV V1.2 program. (Böddi et al., 1990).

RESULTS

After cross-linking of the isolated PLBs with 2-iminothiolane the monomeric subunit of the PChlide-reductase was the most prominent protein band, with a molecular weight of ca 36 kD (Fig. 1). It showed a slightly slower migration in the gel after cross-linking. A number of high molecular weight bands were present in the cross-linked samples as compared to the non cross-linked PLBs (Ref). A dimer of the PChlide-reductase was found around 70 kD (relative migration distance, rmd = 0.45). The relation to the PChlide-reductase was confirmed by a positive reaction with antibodies against PChlide-reductase on western blots (results not shown). Some proteins were also found at the application spot (App) and at the interface between the stacking gel and the separation gel (Sep). These proteins reacted with antibodies against PChlide-reductase. Minor bands with the rmd 0.31, 0.34, 0.38 and 0.42, respectively, were also seen. These can correspond to a trimer of CF1, a tetramer of PChlide reductase, a dimer of CF1, and a complex of CF1 and PChlide-reductase, respectively.

A similar pattern was found when EDC was used as the cross-linker (Fig. 2). The PChlide-reductase dimer and the large aggregates at the application spot and at the interface between the stacking gel and separation gel increased with increasing concentrations of the cross-linker. A succesive decrease of the CF1 band (rmd 0.65-0.70) was also evident.

Cross-linking of the proteins with EDC had only minor effects on the fluorescence emission spectra. There was a small relative increase of the 633-634 nm fluorescence. However, the cross-linking increased the fluorescence yield, which was also obvious after

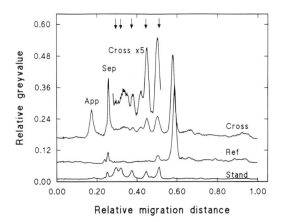

Fig. 1. Image analyser scanning of SDS-polyacrylamide gels after electrophoresis of 2-iminothiolane cross-linked proteins from PLBs. Standard proteins (Stand and arrows) have molecular weights of 212, 170, 116, 76 and 53 kD (left to right). The application point (App) and the interface between the stacking gel and separation gel (Sep) are indicated.

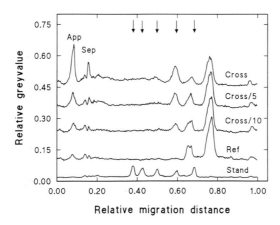

Fig. 2. Image analyser scanning of SDS-polyacrylamide gels after electrophoresis of EDC cross-linked proteins from PLBs. EDC was used as a 3 mM solution (Cross) or diluted 5 (Cross/5) or 10 (Cross/10) times. Otherwise denominations as for Fig. 1.

irradiation (Fig. 3). A prolonged incubation (320 min) in cross-linking solution reduced the degree of phototransformation by 50%, as compared to 30% for the reference. After irradiation the Chlide emission maximum was at about 690 nm (Fig. 3). Gaussian resolution of the fluorescence emission curves showed that the 678 nm component was more prominent in the reference sample than in the cross-linked sample. During a subsequent dark period the 678 nm component increased and the 691 nm component decreased, which was seen as a blue shift (Shibata shift) of the fluorescence peak. In the cross-linked preparations, this blue-shift was retarded (Fig. 4).

As an effect of octylglucoside-solubilization, the 633-635 nm emission band increased and the main peak shifted from 657 nm to 645 nm (Fig. 5). At high concentrations of octyl

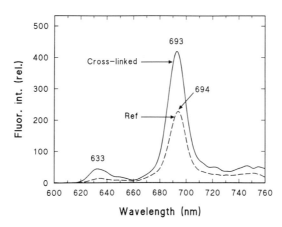

Fig. 3. Low-temperature fluorescence emission spectra of isolated PLBs irradiated with 3 flashes of white light and of PLBs treated with the cross-linker EDC and then irradiated.

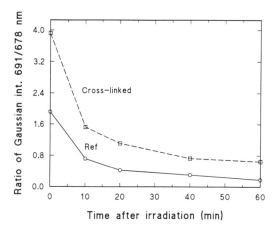

Fig. 4. Ratio of Gaussian intensities 691/678 nm during the time
period of the Shibata shift. Low-temperature fluorescence
emission spectra were measured and resolved into Gaussian
components. The sample marked Cross-linked was treated
with the cross-linker EDC before irradiation.

glucoside, the emission band at 635 nm became the dominant one. Higher detergent concentrations were required for the spectral shifts to occur in cross-linked preparations (Fig. 6).

DISCUSSION

PChlide was early suggested to be present in the dark-grown leaves as dimers (Schulz and Sauer, 1972). The presence of large aggregates has been suggested from spectroscopic measurements (Sironval, 1972; Böddi et al., 1989, 1990). The theoretical consideration of energy transfer and photoreduction kinetics (Sironval, 1972), and the effects of detergents on

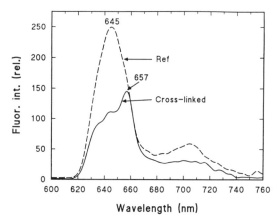

Fig. 5. Low-temperature fluorescence emission spectra of isolated
PLBs treated with 8 mM octylglucoside (Ref) or treated
with the cross-linker EDC before addition of
octylglucoside.

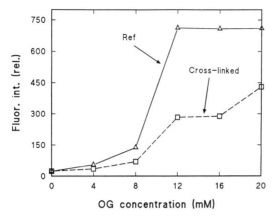

Fig. 6. Fluorescence intensity at 633 nm during treatment of PLBs and cross-linked PLBs with octylglucoside. The detergent was added stepwise to increase the concentration. The fluorescence emission was measured at -196° C.

the flourescence and CD signal of isolated PLBs indicate the presence of large aggregates also of PChlide-reductase in the PLB. There is, however, no experimental evidence supporting the existence of such aggregates.

The bifunctional EDC and 2-imminothiolane protein cross-linkers provide tools for analysing the quartenary structure of proteins. The apparent molecular weight of the PChlide-reductase monomer of barley is 36 kD (Apel et al., 1980). The results from the cross-linking of PLB proteins showed that the PChlide-reductase can be cross-linked into large aggregates. An evident increase in a 70 kD protein in the cross-linked samples indicated the presence of a PChlide-reductase dimer, but also tetramers of the protein were present. The presence of even larger aggregates in the cross-linked samples was obvious, as high amounts of proteins did not penetrate into the gels. The results thus indicate that the PChlide-reductase molecules are situated close together in the PLB structure. Further some of the PChlide-reductase is localized in the vicinity of the CF1 protein. The fact that the results were similar when two different cross-linkers were used strengthens this conclusion.

The cross-linking had only minor effects on the spectral properties PChlide and on the PChlide-reductase activity. This indicates that the PChlide-reductase is kept more or less in its native state during the experimental procedure. The increased fluorescence yield indicates a higher mobility of the chromophore after cross-linking with EDC, which in turn indicates that EDC in some way slightly interfere with the arrangements of the PChlide-reductase molecules or the binding of the PChlide to the active site of the PChlide-reductase molecules.

PLBs treated with octylglucoside loose their highly regular structure and planar membranes are formed (Grevby et al., 1989). Possible aggregates of the reductase can then be relieved of the constraints of the membranes and may undergo a relaxation process. After the cross-linking with EDC this process is obviously delayed. It is also of interest that the spectral blue-shifts, similar to the Shibata shift, is not stopped after the cross-linking but only delayed. This indicates that the Shibata shift can not reflect a dissociation of PChlide-reductase aggregates to monomers or a release of Chlide but instead may reflect a conformational change preceeding these events.

REFERENCES

Apel, K., Santel, H-J., Redlinger, T. E. and Falk, H., 1980, The protochlorophyllide holochrome of barley (Hordeum vulgare L.) Isolation and characterisation of the NADPH:protochlorophyllide oxidoreductase, Eur. J. Biochem., 111:251.

Böddi, B. Lindsten, A., Ryberg, M. and Sundqvist, C. 1989, The aggregational state of protochlorophyllide in isolated prolamellar bodies, Physiol. Plant., 76:135.

Böddi, B. Lindsten, A., Ryberg, M. and Sundqvist, C., 1990, Phototransformation of aggregated forms of protochlorophyllide in isolated etioplast inner membranes, Photochemistry and Photobiology, 52:83.

Grevby, C., Engdahl, S., Ryberg, M. and Sundqvist, C., 1989, Binding properties of protochlorophyllide oxidoreductase to isolated immobilized prolamellar bodies as revealed by detergent and ion treatments, Physiol. Plant., 77:493.

Ikeuchi, M. and Murakami, S., 1983, Separation and characterization of prolamellar bodies and prothylakoids from squash etioplasts, Plant Cell Physiol., 24:71.

Lindsten, A., Ryberg, M. and Sundqvist, C., 1988, The polypeptide composition of highly purified prolamellar bodies and prothylakoids from wheat (Triticum aestivum) as revealed by silver staining, Physiol. Plant., 72:167.

Ryberg, M. and Sundqvist, C., 1982, Characterization of prolamellar bodies and prothylakoids fractionated from wheat etioplasts, Physiol. Plant., 56:125.

Ryberg, M. and Dehesh, K., 1986, Localization of NADPH-protochlorophyllide oxidoreductase in dark-grown wheat (Triticum aestivum) by immuno-electron microscopy before and after transformation of the prolamellar bodies, Physiol. Plant., 66:616.

Schultz, A. and Sauer, K., 1972, Circular dichroism and fluorescence changes accompanying the protochlorophyllide to chlorophyllide transformation in greening leaves and holochrome preparations, Biochim. Biophys. Acta, 267:320.

Sironval, C., 1972, The reduction of protochlorophyllide into chlorophyllide VI. Calculation of the size of the transfer unit and the initial quantum yield of the reduction in vivo, Photosynthetica, 6:375.

Traut, R. R., Casiano, C. and Zecherle, N., 1989, Crosslinking of protein subunits and ligands by the introduction of disulphide bonds In: Protein structure a practical approach, T. E. Creighton ed., IRL Press, Oxford p. 101.

Wynn, R. M. and Malkin, R., 1988, Interaction of plastocyanin with photosystem I: A chemical cross-linking study of the polypeptide that binds plastocyanin, Biochemistry, 27:5863.

IEF OF SOLUBILIZED ACTIVE PROTOCHLOROPHYLLIDE REDUCTASE

B. Wiktorsson, M. Ryberg and C. Sundqvist

Department of Plant Physiology,
University of Göteborg, Carl Skottsbergs Gata 22
S-413 19 Göteborg, Sweden

INTRODUCTION

NADPH-protochlorophyllide oxidoreductase (PChlide-reductase) catalyses the light-dependent reduction of proto-chlorophyllide (PChlide) to chlorophyllide (Chlide) in higher plants. The major part of the PChlide which accumulates in dark-grown leaves occurs complexed to PChlide-reductase, with the characteristic fluorescence emission maximum at 657 nm. This form is often referred to as phototransformable PChlide, and is typical of prolamellar bodies (PLBs) with a highly regular ultrastructure. When PChlide is transformed to Chlide a series of spectral shifts occur, e.g. the Shibata shift, which reflects changes in the aggregational state of the pigment-protein complexes. PChlide-reductase is a membrane associated protein in highly regular PLBs. The amino-acid sequence does not, however, implicate any membrane spanning region of the protein (Schulz et al. 1989, Benli et al. 1991). One of the aims with this study is to determine the relation between the spectral shifts that Chlide goes through shortly after its formation from PChlide and the association of Chlide with PChlide-reductase. We have used preparative isoelectric focusing (IEF) to purify photoactive PChlide, i.e. PChlide complexed with PChlide-reductase, from octylglucoside-solubilised isolated PLBs. The association of Chlide with PChlide-reductase in irradiated PLBs was also analyzed.

MATERIALS AND METHODS

PLBs were isolated from dark-grown wheat (Triticum aestivum) as previously described (Ryberg & Sundqvist 1982). Non-irradiated PLBs or irradiated PLBs containing the 697 or the 684 nm fluorescing form of Chlide, representing different aggregational states of the pigment molecules, were treated with 1-O-n-octyl-ß-D-glucopyranoside (octylglucoside) in 2mM Tris-maleate, pH 8.0, in the presence of NADPH. The solubilized material was subjected to IEF in a Rotofor Preparative IEF Cell (Biorad). The IEF medium was 2 mM Tris-maleate, pH 8,

Regulation of Chloroplast Biogenesis
Edited by J.H. Argyroudi-Akoyunoglou, Plenum Press, New York, 1992

Table 1. Relative distribution of the recovered PChlide and Chlide from non-irradiated and irradiated PLBs, after completion of the IEF. The more basic fractions (not shown) contained no or only trace amounts of pigments.

Fraction	Non-irradiated		Irradiated, containing Chlide F697			Irradiated, containing Chlide F684		
	pH	PChlide	pH	PChlide	Chlide	pH	PChlide	Chlide
1	3.0	1.5	1.9	0.14	0.99	2.3	0	0.9
2	3.7	0.4	3.0	0.07	1.8	2.9	0	0.8
3	4.4	0.4	3.0	22.7	15.6	3.8	0	2.9
4	4.6	5.6	4.2	68.2	70.8	4.1	28.6	15.7
5	5.2	89.1	4.7	7.8	8.3	4.5	61.2	63.9
6	5.6	·2.2	5.2	0.5	1.0	5.0	10.2	10.5
7	5.8	0.2	5.8	0.2	0.6	5.3	0	0.4

1% Biolyte 3/10 (Biorad), 0.5 M sucrose, and 0.3 mM NADPH. The resulting 20 IEF fractions were collected, and the pHs were measured and adjusted to 8. All fractions were analyzed by fluorescence- and absorption spectroscopy, SDS-PAGE, and immuno-blots to detect PChlide-reductase. The phototransformation of PChlide was analyzed by flash irradiation; PChlide and Chlide contents were calculated from acetone extracts.

RESULTS AND DISCUSSION

The detergent treatment resulted in absorbance- and fluorescence emission maxima at shorter wavelengths, as compared to that of non-solubilized PLBs. This indicates a disaggregation of the large native aggregates of the pigment-enzyme complexes (Böddi et al. 1990).

When non-irradiated PLBs were solubilized and subjected to IEF, 89 % of the recovered PChlide was enriched in one fraction, at a pH of 5.2 (Table 1). The fluorescence emission spectrum (- 196°C) of the enriched fraction showed a maximum at 645 nm (Table 2) and a small shoulder at 657 nm. Flash irradiation caused a 75 % phototransformation of the PChlide to Chlide. The fluorescence maximum of the Chlide was at 679 nm. PLBs which were not treated with the detergent showed a fluorescence maximum at 657 nm before and at 697 nm after the flash irradiation (Table 2).

When PLBs which were irradiated to get a long-wavelength fluorescing Chlide (fluorescence emission maximum at 697 nm) were treated with octylglucoside, the maximum shifted to 679 nm during the solubilisation. This spectrum is very similar to that of the flash irradiated PChlide-enriched fraction from the non-irradiated experiment. The IEF resulted in an enrichment of the pigments to a few fractions (Table 1); 71 % of the Chlide and 68 % of the PChlide was found in fraction 4, pH 4.2.

The fluorescence maximum of the PLBs irradiated to get a short-wavelength form of Chlide (fluorescence emission maximum at 684 nm) shifted to 678 nm by the detergent treatment. The spectrum after solubilisation is similar to that of the long-

Table 2. Maxima of the absorbance (A) and fluorescence (F) spectra at different stages of the experiments. The wavelengths are in nm and brackets indicate shoulders in the spectra.

	Non-irradiated		Irradiated, Chlide F697		Irradiated, Chlide F684	
	A	F	A	F	A	F
Before solubilisation	650	657	684	697	678	684
After solubilisation	640	644	673	679 (693)	673	678
Pigment enriched fractions after IEF	640	645 (657)	673	679	673	679

wavelength form of Chlide after solubilisation. As in the previous experiments, the IEF resulted in an enrichment of the pigments to a few fractions (Table 1); 64 % of the Chlide and 61 % of the PChlide were found in fraction 5, pH 4.5.

The protein pattern of the IEF-separated non-irradiated PLBs is shown in figure 1A. A 36 kD polypeptide was heavily enriched in fraction 5. Immunodetection showed that the 36 kD polypeptide is the PChlide-reductase (Fig. 1B). The three bands with molecular weights > 75kD, and which reacted with the PChlide-reductase-specific antiserum, are likely polymers of the enzyme. No polypeptides with molecular weights lower than 36 kD showed any reaction with the antiserum. Thus we conclude that there were no break-down products of the PChlide-reductase.

Both types of irradiated experiments showed very similar relation between pigment distribution and protein pattern as

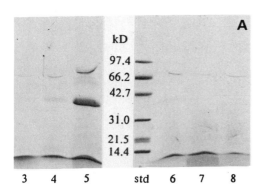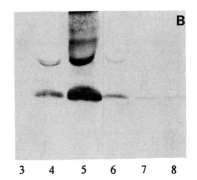

Fig. 1. A, SDS-PAGE of the pigment- and protein enriched fractions after IEF of solubilized non-irradiated PLBs. B, Immunoblot of the same fractions, showing reaction with PChlide-reductase antibodies.

the non-irradiated did (see above). The pigments and the PChlide-reductase were in each case enriched in the same fractions, with a pH lower than 5. The fact that in all three types of experiment both the enzyme and the pigments were enriched in one and the same fraction, and the fact that the PChlide was possible to transform to Chlide with flash-illumination, indicate that PChlide as well as Chlide were closely associated with the enzyme, independently of the pretreatment of the PLBs.

The shift of the absorbance- and fluorescence maxima to shorter wavelengths in intact leaves, the Shibata shift, has been ascribed the dissociation of the Chlide from the enzyme (Brodersen 1976) or the esterification of Chlide (Akoyunoglou and Michalopoulos 1971). More recent results from in vitro experiments indicate that such a shift may reflect a dissociation of large aggregates of Chlide and PChlide-reductase to smaller units (Böddi et al. 1990). The results presented here indicate that a spectral shift to shorter wavelengths, similar to the Shibata shift of intact leaves, can take place without a dissociation of the Chlide and the enzyme protein. The pIs of the pigment-PChlide-reductase complexes differed from the reported pIs of the pigment-free PChlide-reductase polypeptides (8.1-9.2, Ikeuchi & Murakami 1982; 6.2, Schulz et al. 1989). One reason for this could be that the binding of the pigments to the enzyme changes its conformational state and surface properties compared to the pigment-free polypeptide.

REFERENCES

Akoyunoglou, G., and Michalopoulos, G. 1971, The relation between the phytylation and the 682 to 672 nm shift in vivo of chlorophyll a, Physiol. Plant., 25:324.

Benli, M., Schulz, R., and Apel, K. 1991, Effect of light on the NADPH-protochlorophyllide oxidoreductase of Arabidopsis thaliana, Plant. Mol. Biol., 16:615.

Böddi, B., Lindsten, A., Ryberg, M. and Sundqvist, C. 1990, Phototransformation of aggregated forms of proto-chlorophyllide in isolated etioplast inner membranes, Photochem. Photobiol., 52:83.

Brodersen, P.1976, Factors affecting the photoconversion of protochlorophyllide to chlorophyllide in etioplast membranes isolated from barley, Photosynthetica, 10:33.

Ikeuchi,M. and Murakami,S. 1982, Behaviour of the 36000-dalton protein in the internal membranes of squash etioplasts during greening, Plant. Cell Physiol., 23:575.

Schulz, R., Steinmüller, K., Klaas, M., Forreiter, C., Rasmussen, S., Hiller, C. and Apel, K. 1989, Nucleotide sequence of a cDNA coding for the NADPH-protochloro-phyllide oxidoreductase (PCR) of barley (Hordeum vulgare L.) and its expression in Escherichia coli, Mol. Gen. Genet., 217:355.

Ryberg, M. and Sundqvist C. 1982, Characterization of prolamellar bodies and prothylakoids fractionated from wheat etioplasts, Physiol. Plant., 56:125.

EFFECT OF PHOSPHONIC ACID ESTERS ON PHOTOCONVERSION OF PROTOCHLOROPHYLLIDE

IN BARLEY ETIOPLASTS

Fabrice Franck[1], Olga Górnicka[2] and Kazimierz Strzałka[2]

1 - Lab. Photobiologie, B22, Liege University, Sart-Tilman
    B-4000 Liege, Belgium
2 - Inst. Molecular Biology, Jagiellonian University
    Al. Mickiewicza 3, 31-120 Kraków, Poland

INTRODUCTION

Photoactive protochlorophyllide (pchlide) exists in etioplasts in the form of an active complex with the enzyme and NADPH, which has 77 K fluorescence maximum at 657 nm. In this complex pchlide is reduced into chlorophyllide (chlide) as soon as light is turned on. Part of the pchlide pool cannot be photoconverted because either the pigment or NADPH is not bound to the enzyme. This inactive pchlide has 77 K fluorescence maximum at 633 nm. After the initial photoconversion of the active complex, some inactive pchlide is transformed into new photoactive one [Oliver and Griffiths, 1982]. The photoactive pchlide is located mainly in the prolamellar body [Lindsten et al., 1988], where it is thought to form aggregates [Böddi et al., 1989]. The relationship between the ordered structure of the prolamellar body and the occurrence of the photoactive pchlide has not yet been established, although some data suggest that such relationship may exist [Klein and Schiff, 1972; Grevby et al., 1989].

In this paper we report on the effect of phosphonic acid dialkyl esters (PAE) on the occurrence of the photoactive pchlide complex. PAE have the general formula of $RP(O)(OR')_2$ where R' stands for ethyl group and R is hydrocarbon chain differing in length (6 carbons in PAE-6 and 12 carbons in PAE-12). Due to the presence of a long hydrophobic chain, PAE are easily incorporated into membranes. They are used as herbicides and defoliants [Kochmann et al., 1973] and have been found to inhibit growth and propagation of Chlorella vulgaris [Klose et al., 1984]. They inhibit PSI and PSII activities in isolated thylakoid membranes, and cause changes in membrane fluidity [Górnicka and Strzałka, 1990]. As shown in this paper, PAE have large effects on the stability of the photoactive pchlide complex and its regeneration in darkness.

MATERIALS AND METHODS

Etiolated barley seedlings were grown and etioplasts were isolated as described previously by Franck and Schmid [1984] except that bovine serum albumin was omitted in the isolation medium. PAE of the highest purity were synthesized in the laboratories of Prof. Grossmann (Dresden) and Dr. Haage (Berlin) and kindly provided by Prof. Klose from Univ. of Leipzig.

Regulation of Chloroplast Biogenesis
Edited by J.H. Argyroudi-Akoyunoglou, Plenum Press, New York, 1992

Respective amounts of PAE dissolved in methanol were evaporated under
nitrogen stream to form a thin film on the side walls of a conical tube.
Then an 0.9 ml aliquot of etioplast membrane suspension was added and
vortexed at room temperature for 2 min at maximum speed. An aliquot of 100
μl was taken and frozen in liquid nitrogen on a special holder for
spectroscopic measurements. The remaining sample was illuminated by single
saturating 1 ms white flash and after 30 s another 100 μl aliquot was
immediately frozen on a second holder. Then NDPH was added (final conc. 1
mM) and after another 2 min third 100 μl aliquot was withdrawn and frozen
on a holder. Fluorescence spectra were measured at the excitation light of
436 nm using the instrumentation described by Sironval et al. [1968]. The
pchlide amount in the etioplast membrane preparation was measured after
extraction with 80% acetone using molar extinction coefficient of 30.41 x
$Mol^{-1} x\ cm^{-1}$, at 626 nm.

RESULTS

A typical control experiment is described in Fig. 1 A. Etioplast
membranes frozen in darkness exhibited two main 77 K fluorescence bands at
633 and 657 nm (Fig. 1 A, spectrum 1), corresponding respectively to the
inactive pchlide and to the photoactive, membrane bound ternary complex
NADPH-pchlide-reductase. Upon illumination at room temperature by a
saturating light flash, the 657 nm band was replaced by a chlide band at 688
nm as a result of pchlide photoreduction (Fig. 1 A, spectrum 2). The chlide
band was very broad and clearly showed a shoulder around 678 nm. This was

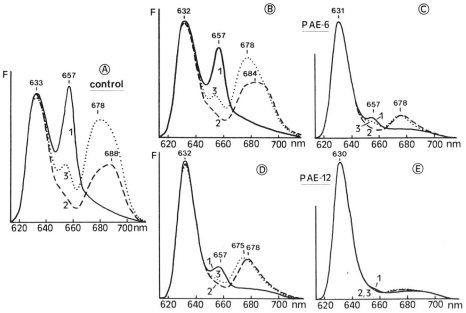

Fig. 1. 77 K fluorescence spectra of etioplast membranes in
control (A) and after treatment with two different
concentrations (8 and 160 μMole/μMole pchlide) of PAE–6
(B and C) and PAE–12 (D and E). 1 – non-illuminated
sample, 2 – sample illuminated by one flash and kept in
darkness for 30 s, 3 – same as 2 after addition 1mM
NADPH and incubation for further 2 min in darkness. Each
set of spectra was normalized to the value of peak at
around 630–633 nm.

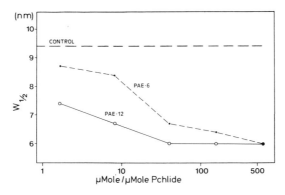

Fig. 2. Effect of PAE-6 and PAE-12 on half-band width of the
inactive pchlide emission (only the short wavelength part
of the band was measured).

due to the fact that the sample was not frozen immediately but 30 s after
the flash. During that time, a blue shift from 688 to 678 nm began to occur,
corresponding to the release of part of the chlide from the apoprotein. This
shift, which may be considered as an equivalent to the Shibata shift
[Shibata, 1957], was completed within a few min at room temperature. If an
aliquot of the illuminated suspension was supplied with 1 mM NADPH and was
then kept for a further 2 min in darkness before being frozen, the Shibata
shift had then proceeded almost completely and the chlide maximum was
shifted to 679 nm (Fig. 1 A, spectrum 3). In the same sample, the 657 nm
band had regenerated to the extent of about 30% relatively to the "dark"
sample. This regeneration was dependent on NADPH, whereas the chlide shift
was not influenced by the presence of the nucleotide (data not shown).

The whole experiment described above was repeated in the presence of
various amounts of PAE-6 or PAE-12 in order to observe the effect of these
compounds on the amount of photoactive pchlide, on the rate of the Shibata
shift and on the extent of the NADPH-stimulated regeneration of the
photoactive pchlide. Fig. 1 B to E shows the fluorescence spectra obtained
at two particular concentrations of the two esters.

Comparison of the fluorescence spectra of membranes frozen in darkness
after the addition of PAE-6 or PAE-12 at concentrations of 8 or 160
μMole/μMole pchlide shows that both compounds progressively reduced the
relative amount of the photoactive pchlide emitting at 657 nm (Fig. 1, B to
E, spectra 1). This effect was observed at lower concentrations of PAE-12
than of PAE-6. Comparison of the spectra obtained 30 s after the flash in
the case where some photoactive pchlide was still present (Fig. 1, B to D,
spectra 2) reveals that the chlide maximum was blue-shifted upon the
addition of the two esters. This suggests that the Shibata shift was
accelerated in the presence of PAE-6 or PAE-12. This interpretation is
confirmed by the third series of spectra recorded after a further 2 min in
darkness in the presence of NADPH. In these spectra the chlide band also
exhibited a shift towards shorter wavelengths in comparison to the control
sample. At equal concentrations, PAE-12 was more efficient than PAE-6 in
accelerating the Shibata shift. 77 K absorbance measurements showed that the
628 nm band corresponding to inactive pchlide, increased in parallel with
the decrease of the 650 nm band, corresponding to photoactive pchlide (data
not shown).

The shape of the emission band of the inactive pchlide was not constant
upon treatment with the inhibitors. Together with the loss of the 657 nm

237

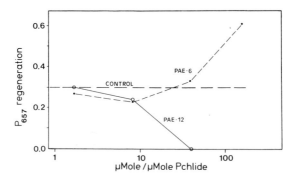

Fig. 3. Effect of PAE-6 and PAE-12 on the relative extent of
photoactive pchlide regeneration expressed as the ratio
of fluorescence intensity at 657 nm of illuminated sample
treated with NADPH to the fluorescence intensity at
657 nm of non-illuminated sample.

band, a shift of the inactive pchlide band from 633 nm to 630 nm occurred,
as well as a significant decrease of its bandwidth. In order to minimize the
error due to overlapping, bandwidth estimations were made by calculating the
half-bandwidth of the short-wavelength side of the inactive pchlide band.
Its decrease upon PAE-6 or PAE-12 addition is shown in Fig. 2 as a function
of concentration. PAE-12 was also more efficient in this respect.

Another type of information that was deduced from fluorescence spectra
concerns the regeneration of the photoactive pchlide after the flash in the
presence of NADPH. The regeneration efficiency was expressed as the
amplitude of the 657 nm band after NADPH addition relatively to its
amplitude in the non-illuminated sample. It thus represents the proportion
of 657 nm fluorescence which could be restored within 2 min when NADPH was
added after the flash. The evolution of this proportion is shown in Fig. 3
as a function of PAE-6 or PAE-12 concentration. A two-phase effect was
observed here. At low concentrations, the regeneration was slightly
inhibited by both compounds. At higher concentrations, it was markedly
enhanced by PAE-6 whereas it was completely abolished by PAE-12.

DISCUSSION

The main observation that can be drawn from the results obtained is
that both PAE interact with etioplast membranes in a way which results in
the transformation of the active form of pchlide into the inactive one.
According to Oliver and Griffiths [1982], the active form of pchlide
consists of a complex between the enzyme protochlorophyllide reductase (PCR)
and its substrates; pchlide and NADPH. The obtained results could be
explained if we assume that the esters cause the dissociation of this
complex, either by loss of pchlide, NADPH or both compounds. In either case,
a shift of the fluorescence from 657 towards 630 nm is expected.

It is widely accepted that the majority, or all of active pchlide and
the enzyme reductase are located in the prolamellar body [Lindsten et al.,
1988]. Therefore the effect of PAE-6 and PAE-12 may be discussed in terms of
their interactions with prolamellar body structure. Both molecules are
amphipatic and have a conical shape which results in their affinity for
curved lipid layers. Such layers are abundant in prolamellar bodies
[Murakami et al., 1985] which are essentially composed of MGDG and DGDG (52
and 32 mol%, respectively [Selstam and Sandelius, 1984]) and the PCR. The

perturbation of the regular lipid structure by PAE might affect the conformation of the enzyme structure in such a way that it is no more able to stabilize NADPH or pchlide. It is also possible that a multimeric structure as proposed by Böddi et al, [1989] is lost under treatment by the esters. However, we cannot also exclude the possibility of direct interactions between the enzyme and the esters. This may again lead to conformational changes of PCR which would result in the liberation of the pigment and/or NADPH. Alternatively the esters may successfully compete with pchlide at its binding site.

As to the fate of the liberated pigment, from the shift in the emission band of the inactive form towards shorter wavelength, we may conclude that the liberated pigment is in a different environment than the native inactive pchlide. The observed blue shift and narrowing of the band suggest a more mobile environment. This is also in agreement with increased membrane fluidity caused by both esters as shown by spin probe technique [Górnicka and Strzałka, 1990].

The chlorophyllide Shibata shift has been interpreted by Oliver and Griffiths [1982] as reflecting the release of the product chlide from the enzyme. If pchlide is efficiently released from its apoprotein by PAE treatment, one can assume that the same treatment would also favour the release of chlide after photoreduction. This is indeed observed as an acceleration of the Shibata shift as a function of the ester concentration. The two esters exhibit different efficiencies in affecting the processes described above, the more active being PAE-12. Since the only difference is a longer hydrophobic part of the molecule for PAE-12, we must conclude that the observed differences are due to a deeper penetration of PAE-12 tail into the hydrophobic part of the membranes.

It is known from in vivo experiments that regeneration of photoconvertible form of pchlide is greatly dependent on the excess of pchlide in the medium [Oliver and Griffiths, 1982]. If we assume that all inactive pchlide is initially present in prothylakoids (as suggested by Lindsten et al. [1988]), then the transport of pchlide from prothylakoids to prolamellar bodies would be the rate-limiting step in the regeneration of the photoactive complex in the untreated system. Upon the action of esters the observed rate of regeneration may be the result of two competing processes: on the one hand the release of pchlide from the protein would diminish the possibility of regeneration, on the other hand the excess of free pigment formed through the above process would favour the regeneration of the few intact complexes that still remained photoactive. At low concentrations of the esters the inhibition would prevail. Above a certain level of pchlide available the enhancement would be the dominating effect. This behaviour was indeed observed with PAE-6. With PAE-12 however, pchlide regeneration was totally blocked in the situation where the ratio of inactive to active pchlide was high. This can only be explained if PAE-12 interacts directly with the binding site of pchlide whereas PAE-6 only indirectly decreases its affinity towards the pigment.

ACKNOWLEDGEMENTS

This research was supported in part by Polish Ministry of Education grant CPBP 05.02 awarded to O. G. and K. S. in Poland. F. Franck is research associate of the National Fund of Scientific Research of Belgium

REFERENCES

Böddi, B., Lindsten, A., Ryberg, M., and Sunqvist, C., 1989, On the aggregational states of protochlorophyllide and its protein complexes in wheat etioplasts, Physiol. Plant., 76:135.

Franck, F., and Schmid, G. H., 1984, Flash pattern of oxygen evolution in greening etioplasts of oat, Z. Naturforsch., 39c:1091.

Górnicka, O., and Strzałka, K., 1990, Effect of phosphonic acid ester PAE-6 on electron transport in thylakoid membranes of wheat and broad bean, in: "Current Research in Photosynthesis," M. Baltscheffsky, ed., Kluwer Acad. Publ., Dordrecht / Boston / London.

Grevby, C., Engdahl, S., Ryberg, M., and Sundqvist, C., 1989, Binding properties of NADPH-protochlorophyllide oxidoreductase as revealed by detergent and ion treatments of isolated and immobilized prolamellar bodies, Physiol. Plant., 77:493.

Kahn, A., Boardman, K., and Thorne, S. W., 1970, Energy transfer between protochlorophyllide molecules: evidence for multiple chromophores in the photoactive protochlorophyllide-protein complex in vivo and in vitro, J. Molec. Biol., 48:85.

Kahn, A., and Nielsen, O. F., 1974, Photoconvertible protochlorophyll(ide) 635/650 in vivo: a single species or two species in dynamic equilibrium?, Biochim. Biophys. Acta, 333:409.

Klein, S., and Schiff, J. A., 1972, The correlated appearance of prolamellar bodies, protochlorophyll(ide) species, and the Shibata shift during development of bean etioplasts in the dark, Plant Physiol., 49:619.

Klose, G., Hentschel, M., Bayerl, T., and Strobel, U., 1984, The motional behaviour of phosphonic acid dialkyl esters in model and microsomal membranes and their influence on the ion permeability, in: "Proc.of the Seventh School on Biophysics of Membrane Transport," B. Tomicki, J. Kuczera and S. Przestalski, eds., Publ. Dep. of the Agricultur. Univ., Wrocław.

Kochmann, W., Gunther, E., Lottge, W., and Kramer, W., 1973, Aminophosphon-säureester mit herbizider Aktivität, Chem. Techn., 25:434.

Lindsten, A., Ryberg, M., and Sundqvist, C., 1988, The polypeptide composition of highly purified prolamellar bodies and prothylakoids from wheat (Triticum aestivum) as revealed by silver staining, Physiol. Plant., 72:167.

Murakami, S., Yamada, N., Nagano, M., and Osumi, M., 1985, Three-dimentional structure of the prolamellar body in squash etioplasts, Protoplasma, 128:147.

Oliver, R. P., and Griffiths, W. T., 1982, Pigment-protein complexes of illuminated etiolated leaves, Plant Physiol., 70:1019.

Selstam, E., and Sandelius, A. S., 1984, A comparison between prolamellar bodies and prothylakoid membranes of etioplasts of dark-grown wheat concerning lipid and polypeptide composition, Plant Physiol., 76:1036.

Shibata, K., 1957, Spectroscopic studies on chlorophyll formation in intact leaves, J. Biochem., 44:147.

Sironval, C., Brouers, M., Michel, J. M., and Kuiper, Y., 1968, The reduction of protochlorophyllide into chlorophyllide. I. The kinetics of the P_{567-47} — $P_{688-676}$ phototransformation, Photosynthetica, 2:268.

ON THE FORMATION OF CHLOROPHYLLIDE AFTER PHOTOTRANSFORMATION

OF PROTOCHLOROPHYLLIDE IN 2-DAY OLD BEAN LEAVES: COMPARISON

WITH ETIOLATED LEAVES

B. Schoefs[1], M. Bertrand[2], and F. Franck[1]

[1]Photobiology Laboratory, University of Liege,
4000 Liege, Belgium
[2]Institut National des Techniques de la Mer, Digue
de Collignon, 50110 Tourlaville, France (corresponding author)

INTRODUCTION

Dark-grown angiosperms accumulate a well known chlorophyll precursor, protochlorophyllide (Pide), in the prolamellar
body and in prothylakoids (Granick and Gassman, 1970;
Brodersen, 1976; Lindsten et al., 1988). In vitro Pide occurs
in two states: one is a non-photoreducible form noted P633 and
the other one is a reducible form noted P657 (Sironval et al.,
1968; Griffiths, 1980; Griffiths, 1981). P657 is usually
described as a ternary complex between Pide, NADPH and NADPH
protochlorophyllide oxidoreductase (PCR, EC 1.6.99.1) (Oliver
and Griffiths, 1982; Griffiths et al., 1984; Ryberg and
Sundqvist, 1988). When an etiolated leaf is illuminated by a
short flash, PCR reduces P657 in a chlorophyllide form
fluorescing at 688 nm (C688) as described by Sironval et al.,
(1968); Goedheer and Verhulsdonk (1970); Shoefs and Franck
(1990). When the sample is maintained in the dark after an
illumination, the maximum of chlorophyllide emission shows the
well known rapid and Shibata shifts (Shibata, 1957; Sironval
et al., 1967; Sironval et al., 1968).

In contrast to etiolated leaves, 2-d old leaves only
contain proplastids (amoeboid stage II in the nomenclature
established by Whatley (1974) and Khandakar and Bradbeer
(1989) which only have short stromatal membranes (Klein and
Schiff, 1972; Whatley, 1974; Khandakar and Bradbeer, 1989). We
have previously shown that 2-d old leaves do however synthesize active Pide in darkness and that the pigment can be
photoreduced by a continuous He-Ne laser illumination
(Schoefs, 1990; Schoefs and Franck, 1990). We present here
results on the formation of the different chlrophyllide
species in 2-d old bean leaves.

MATERIAL AND METHODS

Bean culture and devices used to irradiate bean leaves by a flash and by a continuous illumination were described by Jouy (1982) and Schoefs (1990). The apparatus used to record the 77 K fluorescence spectra and room temperature 690 nm fluorescence kinetics were described by Sironval et al., (1968) and Jouy and Sironval (1979) respectively.

RESULTS

Figure 1 shows 77 K fluorescence spectra of 2-d old and 10-d old non-illuminated leaves. The spectra present bands at 630 and 657 nm traducing the presence of non-reducible and reducible Pide (P633 and P657) respectively. The relative amplitude of the 657 nm band is much weaker in 2-d old leaves than in 10-d old ones.

In 2-d old leaves, a short white flash triggers the photoreduction of P657 in two chlorophyllide species fluorescing at 678 and 690 nm (C678 and C690 respectively). The two bands have almost the same amplitude (Fig. 2a). When leaves are placed in darkness after the flash for only some seconds, the 678 and 690 nm bands progressively disappear whereas a new band at 675 nm increases. A shoulder at about 696 nm shows that only a small portion of the chlorophyllide pool undergoes the same rapid shift as in older leaves (Fig. 2b).

The illumination also induces the reduction of P657 to C690 nm (Fig. 3a) in 10-d old etiolated leaves. When they are placed in darkness after the flash, the fluorescence first shifts to 696 nm (Fig. 3b; Rapid shift) and then back to 684 nm (Shibata shift; Fig. 3c).

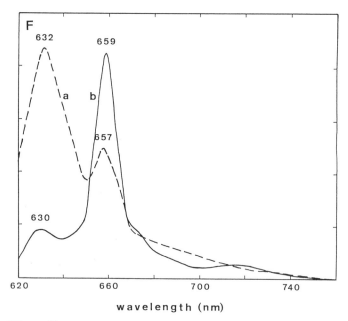

Fig. 1.   77 K fluorescence spectra of non-illuminated 2-d and
          10-d old bean leaves.

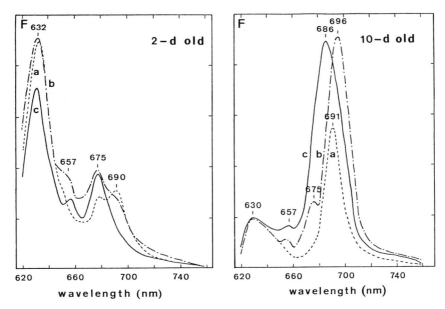

Fig. 2. 77 K fluorescence spectra of 2-d old non-illuminated bean leaves illuminated with a short white flash and then maintained in darkness for (a) 0 s, (b) 40 s and (c) 30 min.

Fig. 3. 77 K fluorescence spectra of 10-d old etiolated leaves placed, first flashed (a) and maintained in darkness for (b) 40 s and (c) 30 min before freezing.

Because the 696 nm emitting chlorophyllide (P696) quenches the fluorescence of other pigments at room temperature (Gassman et al., 1968; Jouy and Sironval, 1979), we have recorded room temperature fluorescence kinetics at 690 nm in order to confirm the occurrence of a minor rapid shift towards P696 in 2-d old leaves (Fig. 4). After an initial jump corresponding to the shutter aperture (0 -> 0 transition), the 690 nm fluorescence increases till a maximum (0 -> B transition) and decreases afterwards (B -> S transition). The "0 -> B transition" comes from the photoreduction of P657 whereas the "B -> S transition" is due to the formation of P696 (Jouy and Sironval, 1979).

DISCUSSION

We show here by spectrofluorimetry that very young leaves contain the active protochlorophyllide fluorescing at 657 nm. This result contrasts with those presented by Klein and Schiff (1972) who did not find in bean proplastids the absorbance band corresponding to this protochlorophyllide. Because proplastids only contain some stromatal membranes, this result demonstrates that P657 can be formed in the absence of prolamellar body.

243

2-d old leaves form two types of chlorophyllide fluorescing at 678 and 690 nm when they are illuminated whereas etiolated leaves form mainly chlorophyllide fluorescing at 690 nm. When 10-d old etiolated leaves are maintained in the dark after photoreduction, the chlorophyllide fluorescence bands first show the rapid shift to 696 nm (Sironval et al., 1967; Sironval et al., 1968) and then the Shibata shift back to 684 nm (Shibata, 1957). In 2-d old leaves, only a small proportion of chlorophyllide undergoes these two shifts. This result confirms those presented by Klein and Schiff (1972). In 2-d old leaves, the 678 and 690 nm fluorescence maxima rapidly shift to 675 nm, a wavelength close to that of chlorophyllide fluorescence in organic solvents (Rebeiz et al., 1973).

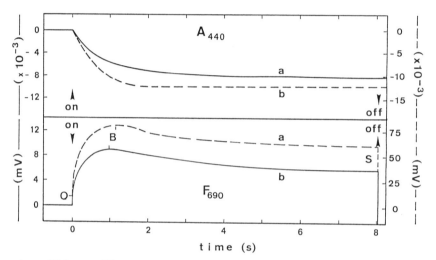

Fig. 4. 690 nm fluorescence kinetics recorded during an actinic and continuous He-Ne laser illumination at room temperature of 2-d old (A) and 10-d old (B) bean leaves.

This result suggests that in 2-d old leaves a large amount of the chlorophyllide formed, as a result of protochlorophyllide photoreduction, is quickly released from the active site of PCR. However some chlorophyllide molecules perform the shifts.

Rapid liberation of Cide supports its rapid binding to polypeptides of the photosynthetic apparatus (Bertrand, 1988; Bertrand et al., 1988; Schoefs, 1990).

ACKNOWLEDGMENTS

B. Schoefs thanks the FEBS for his Youth Travel Fund. The authors are grateful to the FNRS for its help.

REFERENCES

Bertrand, M., 1988, Ph.D. Thesis, University of Paris VI
    (France).
Bertrand, M., Bereza, B. and Dujardin, E., 1988, Z.
    Naturforsch., 43c: 443.
Brodersen, P., 1976, Photosynthetica, 10:33.
Dujardin, E., 1984a, Israel J. Bot., 33:83.
Dujardin, E., 1984b, In: Protochlorophyllide Reduction and
    Greening (Sironval, C. and Brouers, M., eds), Martinus
    Nijhoff Publ., pp. 87.
El Hamouri, B. and Sironval, C., 1979, FEBS Lett., 103:345.
Gassman, M.S., Granick, S. and Mauzerall, D., 1968, Biochem.
    Biophys. Res. Commun., 32:295.
Goedheer, J.C. and Verhulsdonk, C.A.H., 1970, Biochem.
    Biophys. Res. Commun., 39:260.
Granick, S. and Gassman, M., 1970, Plant Physiol., 45:201.
Griffiths, W.T., 1980, Biochem. J., 186:167.
Griffiths, W.T., 1981, In: Photosynthesis. (Akoyunoglou, G.,
    ed) Vol. 5, Balaban Int. Serv., pp. 65.
Griffiths, W.T., Oliver, R.P. and Kay, S.A., 1984, In:
    Progress in Photosynthesis Research. (Biggins, J., ed).
    Martinus Nijhoff Publ., Vol. IV, pp. 469.
Jouy, M., 1982, Photosynthetica, 16:123.
Jouy, M., and Sironval, C., 1979, Planta, 147:127.
Khandakar, K. and Bradbeer, J.W., 1989, Cytologia, 54:409.
Klein, S. and Schiff, J.A., 1972, Plant Physiol., 49:619.
Lindsten, A., Ryberg, M. and Sundqvist, C., 1988, Physiol.
    Plant., 72:167.
Litvin, F.F. and Belyaeva, O.B., 1971, Photosynthetica, 5:200.
Oliver, R.P. and Griffiths, W.T., 1982, Plant Physiol., 70:
    1019.
Rebeiz, C.A., Crane, J.C., Nishijimura, C. and Rebeiz, C.C.,
    1973, Plant Physiol., 51:660.
Ryberg, M. and Sundqvist, C., 1988, Physiol. Plant., 73:218.
Schoefs, B., 1990, Memoire de licence, University of Liege,
    Belgium.
Schoefs, B., and Franck, F., 1990, In: Current Research in
    Photosynthesis (Baltscheffsky, M., ed) Vol. III, pp. 755.
Shibata, K., 1957, Biochem. J., 44:147.
Sironval, C., 1981, In: Photosynthesis. (Akoyunoglou, G.,
    ed), Vol. 5, Balaban Int. Serv., pp. 3.
Sironval, C., Kuyper, Y., Michel, J.M. and Brouers, M.,
    1967, Biofizika, 5:43.
Sironval, C., Brouers, M., Michel, J.M. and Kuiper, Y., 1968,
    Photosynthetica, 2:268.
Whatley, J.M., 1974, New Phytol., 73:1097.

CHLOROPHYLL SYNTHESIS AND DEGRADATION IN GERMINATING BARLEY SEEDLINGS

Jane Walmsley and Heather Adamson

School of Biological Sciences,

Macquarie University, North Ryde, N.S.W. 2109, Australia

ABSTRACT

The effect of gabaculine (GAB) on dark chlorophyll (Chl) accumulation in diurnally-grown 3.5-day-old barley seedlings was investigated under conditions in which Chl b was rapidly degrading in darkness. GAB inhibited accumulation of Chl a and promoted loss of Chl b with the net result that Chl levels declined in darkness in the presence of GAB. The inhibitory effect of GAB in darkness is evidence for the existence of light-independent Chl biosynthesis during very early leaf development.

INTRODUCTION

Massive destruction of Chl b is well known to occur in greening etiolated tissue exposed to light for a short time and then returned to darkness. For example, Kupke and Huntingdon [1] noted substantial loss of Chl b in dark-grown bean leaves, greened for several hours and then returned to darkness. Tanaka and Tsuji [2] reported an 80% decline in Chl b of etiolated cucumber cotyledons, illuminated for 4 h and darkened for 24 h. Popov and Dilova [3] previously reported similar findings with 6-day-old etiolated barley greened for 6 h and transferred to darkness for 24 h. Much less attention has been paid to the observations of Kupke and Huntingdon [1], Tanaka and Tsuji [2] and

Regulation of Chloroplast Biogenesis
Edited by J.H. Argyroudi-Akoyunoglou, Plenum Press, New York, 1992

Popov and Dilova [3] that the loss of Chl b in darkness was accompanied by a significant increase in the level of Chl a.   We have observed a similar net gain in Chl a in germinating light-grown barley seedlings transferred to darkness under conditions where Chl b is rapidly degrading. Logically, there can be only two possible explanations for this increase. Either, (1) Chl a is being synthesised *de novo* or, (2) Chl a is being formed from Chl b. In the experiments described here, gabaculine, a powerful inhibitor of ALA, and hence Chl, synthesis [4-8] was used to distinguish between these possibilities in germinating barley seedlings.

MATERIALS AND METHODS

Barley seeds (*Hordeum vulgare* L. cv Clipper) were germinated and grown for 3-4 days in moist vermiculite in a controlled environment cabinet (16-23 °C, 11 h day/13 h night, 350-400 μE m^{-2} sec^{-1}). When seedlings were about 3.5 days old and less than 2 cm long, plants with leaf lengths close to the population mean were sorted from the rest as previously described [9, 10]. Seedlings were selected for uniformity of length to provide starting material of the same developmental as well as chronological age for all treatments. Matched intact seedlings (20 plants per replicate) were randomly assigned to experimental and control treatments. Plants were transferred to darkness with cut roots immersed in 10 mM potassium phosphate buffer, pH 6.8 with and without 150 μM GAB. Chl content of leaves was determined at T_o and after 17 h dark treatment with or without 150 μM GAB.

Pigment analysis: Pigments were extracted in 85% (v/v) acetone and Chl a and b were determined spectrophotometrically as previously described from extract absorbance values at 663, 645 and 626 nm. [10]. Chl a, b and total Chl were expressed on a leaf basis (μg leaf^{-1}).

RESULTS AND DISCUSSION

When very young barley seedlings germinated and grown for 3.5 days (1.7 ± 0.2 cm) were transferred from light to darkness, total leaf Chl content remained constant (Fig. 1). However there was a very marked increase in the Chl a:b ratio as a result of small but significant increases (10 %) in Chl a and substantial losses (>50 %) of Chl b. Similar results have been reported in the literature for barley and other angiosperms [1-3, 11]. For example, Popov and Dilova [3] noted an 8% increase in Chl a and a 10 % loss in Chl b when dark-grown 6-day-old barley seedling illuminated for 6 h were transferred to darkness for 20 h. Kupke and Huntingdon [1] observed more than 50 % decline in Chl b while Chl a increased by 17 % in dark-grown greening bean leaves transferred to darkness for 24 h. Similar observations were reported by Tanaka and Tsuji [2] for greening dark-grown cucumber cotyledons transferred to darkness.   Table 2 compares the percentage changes of Chl a and Chl b observed, in our experiments, for light-grown 3.5-day-old barley seedlings transferred to darkness with previously published data describing significant increases of Chl a in darkness while Chl b is rapidly degrading [1-3, 11].

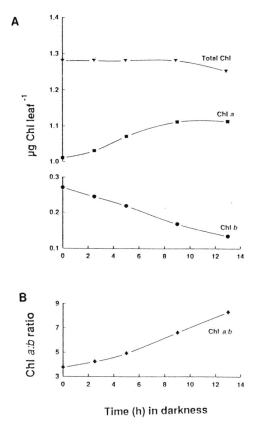

Fig 1. Chl a,ˈb, total Chl content and Chl a:b ratios in the first leaf of 3-3.5-day-old barley seedlings at time of transfer to darkness (T_o) and during 17 h dark treatment in the absence of 150 µM GAB. Mean leaf length at T_o was 1.7 ± 0.2 cm.

Table 1. Chl a, b and fresh weight of the first leaf of intact 3.5-day-old light-grown barley seedlings at time of transfer to darkness (T_o) and after 17 h in the presence or absence of 150 µM GAB. Each result is the mean of 9 replicates each containing 20 plants matched for size (1.9 ± 0.2 cm) at T_o. P indicates the probability that the differences between T_o and after 17 h darkness could have arisen by chance. P* indicates the probability that difference between dark treatments ($\pm$ GAB) could have arisen by chance.

Treatment	F.W. g	Chl a (µg leaf⁻¹)	Chl b (µg leaf⁻¹)	Total Chl (µg leaf⁻¹)	Chl a:b	Chl Conc. (µg g⁻¹ F.W.)
Initial T_o	0.0064±0.0001	3.38±0.02	0.93±0.01	4.31±0.02	3.62±0.06	671±13
After 17 h D − GAB	0.0121±0.0003	3.71±0.05	0.79±0.02	4.49±0.06	4.73±0.08	373±12
P	<0.01	<0.01	<0.01	0.05	<0.01	<0.01
After 17 h D + 150 µM GAB	0.0122±0.0002	3.43±0.07	0.69±0.01	4.13±0.08	5.00±0.08	339±9
P	<0.01	n.s.	<0.01	<0.05	<0.01	<0.01
P*	n.s.	<0.01	<0.01	<0.01	<0.01	<0.05

Table 2. Comparison of the percentage changes in the level of Chl *a, b* and total Chl in leaves of light-grown 3.5-day-old barley seedlings transferred to darkness for 17 h (data from Fig. 1) with those reported for dark-grown illuminated plants and light-grown plants also measured after transfer from light to dark.

	Chl *a*	Chl b	Total
Kupke and Huntingdon, 1963 [1], Bean	17.3 %	-54.4 %	0
Popov and Dilova, 1969 [3], Barley	8 %	-10.0 %	4 %
Wieckowski and Ficek, 1969 [11], Bean	12.6 %	-16.8 %	4.4 %
Tanaka and Tsuji, 1981 [2], Cucumber	16 %	-80 %	-4.5 %
Walmsley and Adamson, 1991 (Fig.1)	11 %	-52%	4.2 %

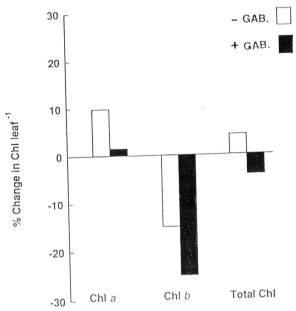

Fig. 2. Percentage change in Chl *a, b* and total Chl in the first leaf of 3.5-day-old germinating barley seedlings transferred to darkness for 17 h in the presence and absence of GAB. Data derived from Table 1.

Increases of Chl *a*, observed in our experiments, were eliminated by 150 µM GAB which also promoted loss of Chl *b* and a net loss of Chl overall (Fig. 2.). We attribute the decrease in Chl *b* to inhibition of Chl synthesis. In general, the evidence for synthesis of small amounts of Chl *a* in darkness in angiosperms (as shown in Table 2.) has tended to be disregarded. However, the findings are all reproducible (unpublished) and, in the case of barley, consistent with other evidence from labelling [12, 13] and ultrastructure [14] studies. The fact that GAB, a specific inhibitor of ALA synthesis [15] eliminates the increase in Chl *a* and promotes the loss of Chl *b* in darkness (Table 1) adds weight to our claim that very young barley seedlings have a functioning light-independent Chl biosynthetic pathway. In this context is worth noting that GAB has absolutely no effect on leaf expansion in darkness. This means that the reduction in Chl concentration (μg Chl g FW^{-1}), which occurs in the presence of GAB, can only be attributed to inhibition of Chl synthesis.

In summary, our results are entirely consistent with *de novo* synthesis of Chl *a*. They provide no support for the suggestion that Chl *a* might be produced from Chl *b* [1, 11]. We believe our observations and those of earlier workers are important because they demonstrate the potential for *de novo* synthesis of Chl in darkness under conditions where total Chl content is either constant or decreasing. It is thus conceivable that *de novo* synthesis of Chl *a* when Chl *b* is rapidly degrading could account for the observations of Tanaka and Tsuji [2] that, "CP1 accumulated with a concomitant decrease in LHCP" and "that the chlorophyll of CP1 came from chlorophyll that was bound to other proteins". If light-independent Chl *a* synthesis is occurring then it may be unnecessary to postulate Chl redistribution between proteins during the early phases of greening.

References

1. D.W. Kupke and J.L. Huntingdon, Chlorophyll *a* appearance in the dark in higher plants. Science, 140 (1963) 49-51.
2. A.Tanaka,. and H. Tsuji, Changes in the chlorophyll *a* and *b* content in dark-incubated cotyledons excised from illuminated cucumber seedlings.- Plant Physiol. 68: (1981) 567-570.
3. K. Popov and S. Dilova, On the dark synthesis and stabilization of chlorophyll, in: H. Metzner (Ed), Progress in Photosynthetic Research, Vol. 2, Inst.Union Biol. Sci., Tübingen, FRG., 1969, pp. 606-609.
4. D.P.Caiger, S.A. Pearson, A.J. Smith and L.J. Rogers, Differential effects of gabaculin and laevulinic acid on protochlorophyllide regeneration. Plant, Cell and Environment, 9 (1986) 495-499.
5. J.L.Corriveau and S.I.Beale, Influence of gabaculine on growth, chlorophyll synthesis, and 5-aminolevulinic acid synthase activity in *Euglena gracilis*. Plant Sci.,45 (1986) 9-17.
6. D.H. Flint, Gabaculine inhibits δ-aminolevulinic acid in chloroplasts. Plant Physiol., 75S (1984) 170.
7. C.M. Hill, S.A. Pearson, A.J. Smith, and L.J. Rogers, Inhibition of chlorophyll synthesis in *Hordeum vulgare* by 3- amino 2,3-dihydrobenzoic acid (gabaculin). Biosci. Rep., 5 (1985) 775-781.
8. J.Walmsley and H. Adamson, Gabaculine inhibition of chlorophyll synthesis in the light and in darkness in intact barley (*Hordeum vulgare*). Plant Sci. 68 (1990) 65-70.
9. H. Adamson, Evidence for a light-independent protochlorophyllide reductase in green barley leaves, in: G. Akoyunoglou, A.E. Evangelopoulos, J.Georgatsos, G. Palaiolgos, A. Trakatellis and C. Tsiganos, (Eds), Progress in Clinical and Biological Research; Cell Function and Differentiation Vol. 102B, Alan R. Liss, Inc., New York. 1982, pp. 33-41. ISBN 0-8451-0166-8.
10. J. Walmsley and H. Adamson, Chlorophyll accumulation and breakdown in light-grown barley transferred to darkness : effect of seedling age. Physiol. Plant. 77 (1989) 312-219.
11. S. Wieckowski and S. Ficek, On the chlorophyll metabolism of green seedlings in darkness, in: Bull. Acad. Polon. Sci. Vol. XVIII (1970) 47-51.

12. N. Packer and H. Adamson, Incorporation of 5-amino- levulinic acid into chlorophyll in darkness in barley. Physiol. Plant., 68 (1986) 222-230.

13. N.Packer, H. Adamson, and J. Walmsley, Comparison of chlorophyll accumulation and 14-C ALA incorporation into chlorophyll in the dark and light in green barley , in: J.Biggins (Ed.), Progress in Photosynthetic Research, Vol. IV, Martinus Nijhoff/Dr W. Junk, Dordrecht, The Netherlands, 1987, pp. 487-490.

14. H. Adamson, W.T. Griffiths, N. Packer and M. Sutherland, Light-independent accumulation of chlorophyll *a* and *b* and protochlorophyllide in green barley *(Hordeum vulgare)*. Physiol. Plant., 64(1985) 345-352.

15. R.R. Rando, Mechanism of the irreversible inhibition of γ-aminobutyric acid-α-ketoglutaric acid transaminase by the neurotoxin gabaculine. Biochemistry, 16 (1977) 4604-4610.

SUSCEPTIBILITY OF DARK CHLOROPHYLL SYNTHESIS IN BARLEY AND PINE SEEDLINGS TO INHIBITION BY GABACULINE

Ke-Li Ou, Jane Walmsley and Heather Adamson

School of Biological Sciences, Macquarie University, North Ryde, 2109, Australia

ABSTRACT

Dark chlorophyll accumulation in pine (*Pinus pinea* and *Pinus nigra*) and barley (*Hordeum vulgare*) seedlings was inhibited by gabaculine (3-amino 2,3-dihydrobenzoic acid) (GAB). The inhibitory effect of GAB was overcome by the addition of 5-aminolevulinic acid (ALA). Since GAB specifically interferes with ALA synthesis via the C5 (but not Shemin) pathway we conclude that ALA for light-independent chlorophyll synthesis in angiosperms and gymnosperms is formed via this route.

INTRODUCTION

Dark chlorophyll synthesis occurs in all major plant groups[1] but differences have evolved with respect to its regulation in seed plants. In angiosperms, chlorophyll synthesis and assembly into thylakoids only occur in darkness in plants which have been exposed to light during development[2,3,4,5]. In gymnosperms and lower plants exposure to light is not necessary to initiate these processes[6]. Two pathways of chlorophyll biosynthesis have been postulated for gymnosperms[7,8]: (i) a light-dependent pathway, rate limited by ALA formation and requiring light for the reduction of protochlorophyllide to chlorophyllide and (ii) a light-independent pathway, *not* rate limited by ALA and *not* requiring light for protochlorophyllide reduction. Separate light and dark pathways could equally well be postulated for angiosperms.

There are two main routes for the formation of ALA: (i) the C5 pathway from glutamate involving a reaction sequence catalysed by aminotransferase and a specific tRNA molecule and (ii) the Shemin pathway involving the condensation of succinyl Co-A and glycine, catalysed by ALA synthase[9]. The former is inhibited by GAB[10] the latter is not[11]. It is generally agreed that ALA for chlorophyll synthesis via the light-dependent pathway in angiosperms is formed via the C5 pathway[12]. There is also evidence that ALA synthesis for chlorophyll in light-grown pine seedlings is by the C5 pathway[8].

The origin of ALA for light-independent chlorophyll synthesis is more contentious. Jelić and Bogdanović[8] have argued that since gabaculine inhibited chlorophyll synthesis in pine seedlings in light but not in darkness in their experiments, ALA destined for incorporation into chlorophyll via the light-independent pathway in gymnosperms is synthesised via the GAB-insensitive ALA synthase / Shemin pathway. On the other hand we have observed that GAB inhibits dark chlorophyll synthesis in 5-6 day old glasshouse grown barley seedlings and that the inhibitory effect of GAB is eliminated by ALA[13]. Taken at face value, these findings suggest that ALA for light-independent chlorophyll synthesis in gymnosperms and angiosperms is synthesised via different routes. However, a negative result with GAB in darkness might simply mean that the endogenous concentration of inhibitor is too low or the rate of ALA synthesis too slow for an effect to be observed.

MATERIALS AND METHODS

Excised cotyledon experiments: *Pinus pinea* seedlings were germinated and grown in vermiculite in darkness (25 $^{\circ}$C) for 12 days.Two series of matched samples were obtained by: (1) harvesting and pooling 10 cotyledons per seedling from about 100 seedlings; (2) selecting for each replicate pair, 10 cotyledons and dividing each transversely into 2 equal parts; (3) determining the initial fresh weight of each group of 10 half cotyledons (4) assigning one group to control (T_0/initial) and the other to treatment (T_D/dark treatment). Excised cotyledons were floated on 5 mM K-phosphate buffer (pH 6.8) with and without GAB and ALA. Details of concentrations and the number of replicates per treatment, each with matched To are given on the Figures.

Experiments with intact seedlings: *Pinus nigra* seedlings were germinated and grown in darkness as above for 9 days. Seedlings of similar height were then harvested and placed in small beakers with sufficient treatment solution to cover their roots or dissected (roots and testa removed) to provide T_0 samples. *Hordeum vulgare* (barley) seedlings were germinated and grown in a glasshouse under natural light conditions for 5 -6 days, harvested, rigourously matched for first leaf size, transferred to darkness and treated as previously described [13]. Chlorophylls were extracted in 85% (v/v) acetone and measured by absorbance at 663, 645 and 626 nm[13]. Results were expressed as chlorophyll per gram *initial* fresh weight (excised *Pinus pinea* cotyledons), chlorophyll per seedling (*Pinus nigra*) or chlorophyll per leaf (barley). All dark treatments and manipulation were carried out in total darkness or under a dim green safe light (40 W green Osram globe, Ilford filter 909).

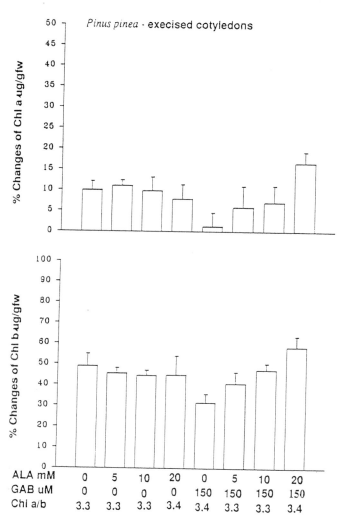

Fig.1 % changes in Chl a and b in excised cotyledons of dark-grown *Pinus pinea* in the presence and absence of GAB and ALA. n=3 replicates per treatment, 10 half cotyledons per sample.

RESULTS AND DISCUSSIONS

As shown in Fig.1 there was a small (10%) butsignificant increase in chlorophyll a (per gram initial fresh weight) in excised *Pinus pinea* cotyledons after 24 hours dark treatment and a proportionately larger increase in chlorophyll b (50%). GAB virtually eliminated the former and reduced the latter. In both cases, the inhibitory effect of GAB was overcome by the addition of ALA in a dose-related manner. ALA (5 - 20 mM) however did not promote the synthesis of either chlorophyll a or b.

Similar results were obtained with *Pinus nigra* (Fig. 2) and barley (Fig. 3). The chlorophyll content (a+b) of the intact pine seedlings doubled during the experiment. The increase was significantly less in the presence of 150 uM GAB and the inhibitory effect of this compound was relieved by 10 mM ALA. In the case of barley (the details of which have been presented elsewhere)[13] GAB converted a small but significant net gain in chlorophyll per leaf to a slight loss. As with pine, the inhibitory effect of GAB on barley was overcome by the addition of ALA.

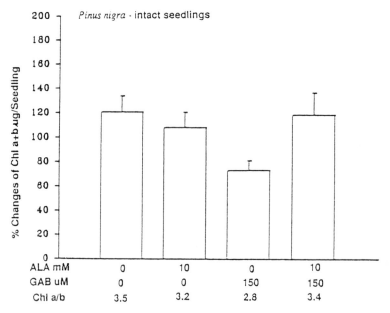

Fig.2 % changes in total Chl in dark-grown *Pinus nigra* seedlings in the presence and absence of GAB and ALA. n=3 replicates per treatment, 7 plants per sample.

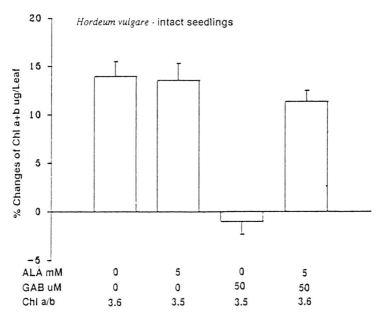

Fig.3 % changes in total Chl per leaf in 5-6 day old barley seedlings transferred to darkness in the presence and absence of GAB and ALA. n=6 replicates per treatment, 10 plants per sample.

These results strongly suggest that ALA destined for dark chlorophyll synthesis in angiosperms and gymnosperms is synthesised via the C5 pathway.

Since ALA for light-dependent chlorophyll synthesis is also synthesised via the C5 pathway in angiosperms, gymnosperms and lower plants[8,11,12] it seems likely that light-dependent and light-independent chlorophyll biosynthetic routes are the same, at least to protochlorophyllide.

We attribute the failure of Jelic and Bogdanovic[8] to observe GAB inhibition of dark chlorophyll synthesis in *Pinus nigra* to the extremely low rate of chlorophyll accumulation in their experiments.

REFERENCES

1. P.A. Castelfranco and S.I. Beale, Chlorophyll biosynthesis: recent advances and areas of current interest, in: Ann. Rev. Physiol. Plant, 34: 241-278 (1983).
2. K. Popov, and S. Dilova, On the dark synthesis and stabilization of chlorophyll, in: Progress in Photosynthesis Research, H. Metzner ed. 2: 606-

609, Inst. Union Biol. Sci., Tubingen, F.R.G (1969).

3. H. Adamson, R.G. Hiller and M. Vesk, Chloroplast development and the synthesis of Chl a and Chl b and Chl-protein complexes I and II in the dark in *Tradescantia albiflora* (Kunth), in: Planta, 150:269-274 (1980).

4. H. Adamson, N. Packer and J. Gregory, Chloroplast development and the synthesis of chlorophyll and protochlorophyllide in Zostera transferred to darkness, in: Planta, 165:469-476 (1985).

5. H. Adamson, T. Griffiths, N. Packer and M. Sutherland, Light-independent accumulation of chlorophyll a and b and protochlorophyllide in green barley (Hordeum vulgare), in: Physiol Plant, 64:345-352 (1985).

6. J.T.O. Kirk, and R.A.E. Tilney-Bassett, The plastids. -Elsevier, North Holland Biomedical Press, Amsterdam New York Oxford (1978).

7. M.R. Michel-Wolwertz, Chlorophyll formation in cotyledons of *Pinus jeffreyi* during germination in the dark. -occasional accumulation of protochlorophyll(ide)s forms. in: Plant Science Letters, 8:125-134 (1977).

8. G. Jelić and M. Bogdanović, Gabaculine effect on chlorophyll accumulation in Pinus nigra seedlings, in: Current Research in Photosynthesis. M. Baltscheffsky ed. Kluwer Academic Publishers. 3:683-686 (1990).

9. S.I. Beale and J.D. Weinstein, Biosynthesis of 5-aminolevulinic acid in phototrophic organisms, in: Chlorophylls. H. Scheer ed. CRC Press, Inc. 385-406 (1991).

10. R.R. Rando, Mechanism of the irreversible inhibition of γ-aminobutyric acid-a-ketoglutaric acid transaminase by the neurotoxin gabaculine. in: Biochemistry. 16:4604-4610, (1977).

11. J.L. Corriveau and S.I. Beale, Influence of gabaculine on growth, chlorophyll synthesis, and 5-aminolevulinic acid synthase activity in *Euglena gracilis*, in: Plant Sci, 45:9-17 (1986).

12. R.J. Porra, O. Klein and P.E. Wright, The proof by 13C-NMR spectroscopy of the predominance of the C5 pathway over the Shemin pathway in chlorophyll biosynthesis in higher plants and of the formation of the methyl ester group of chlorophyll from glycine, in: Eur. J. Biochem, 130:509-516 (1983).

13. J. Walmsley and H. Adamson, Gabaculine inhibition of chlorophyll synthesis in light and darkness in intact barley (Hordeum vulgare) seedlings, in: Plant Science, 68:65-70 (1990).

CHLOROPHYLL BIOSYNTHESIS: SPECTROFLUOROMETRIC ESTIMATION OF

PROTOPORPHYRIN IX, Mg-PROTOPORPHYRIN AND PROTOCHLOROPHYLLIDE[1]

Pinki Hukmani and Baishnab Charan Tripathy[*]

School of Life Sciences, Jawaharlal Nehru University
New Delhi 110067, India

The chlorophyll (chl) molecules are synthesized from 5 aminolevulinic acid (ALA) via various intermediates. It is already demonstrated that chl biosynthesis takes place via monovinyl and divinyl monocarboxylic routes (2,4-6). To elucidate the chl biosynthetic pathways and to monitor the amount of over-accumulated porphyrins to induce photodynamic damage to the plants (1,3), the various pools of intermediates of chl biosynthetic pathway like protoporphyrin IX (Proto IX), mg-protoporphyrin monoester (MPE) or protochlorophyllide (Pchlide) etc. need to be quantified (4). In small amount of tissues and especially in isolated etioplasts or chloroplasts the pools of above intermediates are too small to be estimated spectrophotometrically. As Proto IX, MPE and Pchlide fluoresce, and spectrofluorometry is a very sensitive tool, the fluorescence of these compounds can be measured for their quantification. However, as these components especially Proto IX, Pchlide and Chlide are having overlapping fluorescence spectra, it would be incorrect to quantify the above components from their mixtures by measuring fluorescence amplitudes at their respective peaks. Therefore it is essential to correct the fluorescence amplitude of the compound to be quantified (say Proto IX) for the contribution due to other fluorescing compounds (say Pchlide and Chlide) at the measured wavelength.

The present investigation aims at quantitatively estimating Proto IX, MPE, Pchlide, the tetrapyrrole intermediates of chl metabolism by spectrofluorometry.

MATERIALS AND METHODS

Plant Material: Barley (Hordeum Vulgare L.cv IB65) plants were grown on germination paper.
Source of MPE, Pchlide and Chlide: These tetrapyrroles were prepared from barley leaves which were induced to accumulate massive amounts of these compounds as described previously (5).
Spectrophotometry : Absorption spectra were recorded on shimadzu UV-260 spectrophotometer. For calculation of concentrations of pure samples of Proto IX, MPE and Pchlide in ether, the extinction coefficients of $1.5 \times 10^5 M^{-1} cm^{-1}$ (404 nm), $1.82 \times 10^4 M^{-1} cm^{-1}$ (589 nm) and $3.56 \times 10^4 M^{-1} cm^{-1}$ (624 nm) were used respectively. Diethyl ether was used as a solvent. The ether was dried under N_2 gas and required amount of hexane extracted acetone residue (HEAR) was added to redissolve the pigments. HEAR was used as solvent since Proto IX, MPE and Pchlide are quantitated spectrofluometricaly in HEAR. Chlide (a+b) pool was also dissolved in HEAR and prepared from the extinction coefficients (in 80% acetone) of $7.69 \times 10^4 M^{-1} cm^{-1}$ (663nm) for Chlide a and $4.45 \times 10^4 M^{-1} cm^{-1}$ (645nm) for Chlide b.
Spectrofluorometry: Fluorescence spectra were recorded in ratio mode in a computer driven SLM AMINCO 8000C spectrofluorometer and were corrected for photomultiplier tube sensitivity. Rhodamine B was used in the reference channel as a quantum counter. The sample channel photomultiplier tube was cooled to -20^0C by a thermo-electric cooling device to reduce the noise level. A tetraphenylbutadiene (TPD) block was used to adjust the voltage in both the channels (i.e., sample as well as reference channel) to 20,000 counts per second at excitation and emission wavelengths of 348nm and 422nm respectively. The emssion spectra were recorded at excitation and emission bandwidths of 4nm. The emission spectra of

----

This work was supported by a grant (DST/SP/SO/A44-88) from Department of Science and Technology, Government of India, to BCT. * To whom correspondence should be addressed.

pigments dissloved in HEAR were recorded from 580-700 nm. The data were stored in a micro computer (IBM PS 30) diskettes.

Chemicals: Protoporphyrin IX was obtained from the Porphyrin Products, Logan, Utah.

RESULTS

Calculation of net fluorescence emission amplitudes of Proto IX and Pchlide in mixtures of Proto IX, Pchlide & Chlide

In HEAR (293K) Proto IX has the emission maximum at 633 nm, when excited at 400 nm (E400 F633). Under identical conditions Pchlide and Chlide when excited at 440 nm, fluoresce at 638 nm (E440 F638) and 675 nm (E440 F675) respectively (Fig. 1). As Chlide includes two components ie Chlide a and Chlide b, their individual concentrations were not determined from the fluorescence emission spectra. The net fluorescence emission amplitude due to Proto IX (E400 F633) and Pchlide (E440 F638) in the mixture of Proto IX, Pchlide and Chlide calculated from pure samples are given in equations [1] & [2] and described in the appendix.

Proto IX (E400 F633) = 1.021 (E400 F633) - 0.4268 (E440 F638) - 0.0184 (E440 F675)   [1]

Pchlide (E440 F638) = 1.03 (E440 F638) - 0.0458 (E400 F633) - 0.0358 (E440 F675)   [2]

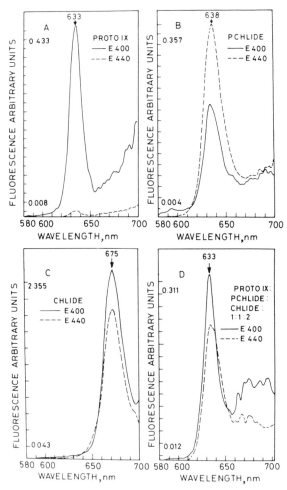

Figure 1. The room temperature (293k) fluorescence emission spectra of various tetrapyrroles in HEAR excited at 400nm (E400) and 440nm (E440). A) Pure sample of Proto IX, (B) Pure sample of Pchlide, C) Pure sample of Chlide (D) Mixture of Proto IX (50nM), Pchlide (50nM) and Chlide (100nm). The fluorescence spectra were correted for phototomultiplier tube sensitivity. They were recorded in ratio mode at excitation and emission slit widths of 4nm.

260

Conversion of Proto IX and Pchlide emission amplitudes to Proto IX and Pchlide concentrations
Eqs [1] and [2] allow calculation of emission amplitudes of Proto IX and Pchlide in pure samples as well as in mixtures of these two tetrapyrroles in the presence of a third interfering compound Chlide. Once this is calculated, it can be readily converted to authentic Proto IX and Pchlide concentration by reference to stadard calibration curves which is constructed (a) by either having various concentrations of pure Proto IX and Pchlide or mixing known proportion of Proto IX and Pchlide in the presence of Chlide, (b) by recording E400 and E440 emission spectra on each sample (c) by calculating Proto IX (E400 F633) & Pchlide (E440 F638) deconvoluted emission amplitude for each mixture with the help of equation [1] and [2] and finally by plotting Proto IX or Pchlide concentration on abscissa against calculated net deconvoluted amplitudes of Proto IX and Pchlide respectively on ordinate. Figure 1 shows the fluoresecence emission spectra (E400 and E440) of pure smaples of Proto IX, Pchlide and Chlide and mixture of Proto: Pchlide: Chlide at a molar ratio of 1:1:2:

Calculation of amounts of Proto IX and Pchlide in mixtures of Proto IX, Pchlide and Chlide:
Eqs [1] & [2] allow calculation of emission amplitudes of Proto IX & Pchlide in mixtures of Proto IX, Pchlide & Chlide pigments and these are converted to concentrations by reference to a standard celibration curve. The latter is constructed exactly as decribed before. The various mixtures of Proto IX and Pchlide were adjusted to total of 100 pmol/ml and in all the tetrapyrrole mixtures, the ratio of Proto IX + Pchlide : Chlide was maintained at a molar ratio of 1:1 The curve which related the concentration (on abscissa) to the calculated deconvoluted emission amplitudes (on ordinate) was a straight line that passed through the origin and it exhibited a slope of 0.0037 for Proto IX (r=0.980)and 0.0038 for Pchlide (r=0.992) (Fig 2 A). These slopes closely matched those derived for pure samples ie: 0.0037 for Proto IX and 0.0039 for Pchilde (Fig 2B).

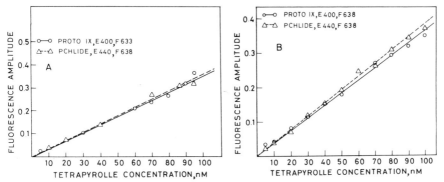

Figure 2. The standard calibration curves for Proto IX (E400 F633) and Pchlide (E440 F638) in A) Mixtures of Proto IX, Pchlide and Chlide and, B) Pure samples of Proto IX or Pchlide.

Table 1

RELIABILITY OF EQUATION 1 FOR PROTO IX IN THE MIXTURE OF PROTO IX, PCHLIDE AND CHLIDE

Amt of proto added (pmol/ml)	Amt of Proto calculated (pmol/ml)	% error between amt of Proto added and calculated	mean % error ± SD
5	5.39	7.80	
30	29.86	-0.46	
60	58.03	-3.28	
70	65.50	-6.43	-0.73 ± 5.5
80	73.38	-8.27	
90	89.64	-0.40	
95	100.64	5.93	

The deconvuluted fluorescence emission amplitude [E400 F633] was used to determine the concentration of Proto IX. The concentration of Chlide was 100 nM and the concentration of Pchlide was so adjusted that in each experiment Proto IX + Pchlide was 100 nM.

Table 2

RELIABILITY OF EQUATION 2 FOR PCHLIDE IN THE MIXTURE OF PROTO IX, PCHLIDE AND CHLIDE

Amt of Pchlide added (pmol/ml)	Amount of Pchlide calculated (pmol/ml)	% error between Pchlide added and calculated	mean % error ± SD
10	9.55	-4.50	
20	21.19	5.95	
30	31.28	4.26	
40	38.05	-4.87	-0.575 ± 5.589
70	74.94	7.05	
90	85.69	-4.79	
95	88.22	-7.13	

The deconvuluted fluorescence emission amplitude [E440 F638] was used to determine the concentration of Pchlide. The concentration of Chlide was always 100 nM and the concentration of Proto IX was so adjusted that Pchlide + Proto IX was 100 nM.

Table 3

RELIABILITY OF STANDARD CURVE WHICH WAS CONSTRUCTED FROM AUTHENTIC CONCENTRATIONS OF THE PURE SAMPLES OF MPE

Amt of MPE added (pmol/ml)	Amt of MPE calculated (pmol/ml)	%error between amt of MPE added & calculated	Mean % error ± SD
10	11.33	13.30	
20	21.66	8.30	
30	31.89	6.30	
40	41.69	4.22	
50	47.63	-4.74	2.96 ± 5.5098
60	62.34	3.90	
70	70.12	0.17	
80	78.33	-2.08	
90	87.59	-2.67	

The integration values from 580-610nm [E420] were used to determine the concentration of MPE.

The concentrations of Proto IX and Pchlide was calculated using the slopes of the calibration curves. The reliability of eq.1 in determining Proto IX in mixture of Proto, Pchlide and Chlide was - 0.73 ± 5.5 (Table 1) and same for Pchlide in the above mixture was -0.575 ± 5.589 (Table 2).

Calculation of Net Fluorescence of MPE

In HEAR (293K) the soret band of MPE, excited at 420 nm has the peak around 592 nm and this soret fluorescence band extends from 580 nm to 610 nm. The other components like Proto IX and Pchlide present in HEAR do not substantially fluoresce in the above wavelength range. Therefore the fluorescence emanating from the MPE was integrated from 580 nm to 610 nm. The integration values of different concentrations of pure samples of MPE was converted to the concentrations by reference to a standard calibration curve. The latter was constructed (a) by preparing various concentrations of MPE (b) by recording one 293K emission spectra (E420) on each concentration (c) by calculating the area under the fluorescence emission band between 580-610nm (d) by plotting the concentration on abscissa against the calculated area on the ordinate. Such a plot yielded a straight curve that passed through the orgin and exhibited a slope of -0.73 and a coefficient of correlation of 0.998. The reliability of this calibration curve in determining MPE was 2.96 ± 5.5098 (Table 3).

DISCUSSION

The present investigation demonstrates the sensitivity and resolution of room temperature fluorescence spectroscopy as an analytical tool for quantitative estimation of tetrapyrroles. It is often extremely difficult to seperate minute qunatities of pigments chromatographically which sometimes ends up in either loss or destruction of the pigments. Therefore, the elimination of the need of seperation and purification of various tetrapyrroles prior to quantitative analysis of pigments is an unique feature of the above procedure. Besides convenience, rapidity and accurate quantitation of minute (2-3 picomoles) quantities of pigments; elimination of a) analytical uncertainities due to recovery losses caused by chromatography and b) the problems associated with incomplete extraction of pigments from 80% acetone solvent to appropriate experimental medium are the unique advantages of the spectrofluorometric method of analysis of pigments.

APPENDIX

When three fluorescent compounds X, Y, and Z having different but overlapping fluorescence excitation and emission properties occur together in a particular sample, the net fluorescence signals generated by compound X can be seperated from the fluorescence signals generated by other two Y and Z compounds by using three unknown simultaneous equations. Let a and b represent any two fluoresence excitation or emission wavelengths of compound X, c and d represent the same for compound Y and g and h for the third interfering tetrapyrrole Z. Deconvoluted net fluorescence signals X $(E_a F_b)$, Y $(E_c F_d)$ & Z $(E_g F_h)$ generated by compounds X,Y and Z respectively at designated wavelength a,b,c,d,g and h can be determined using eqs. 3-5.

$$X\ (E_a\ F_b) = \frac{(E_a\ F_b) - C_1\ (E_c\ F_d) - C_2\ (E_g\ F_h)}{C_3} \quad [3] \qquad Y\ (E_c\ F_d) = \frac{(E_c\ F_d) - C'_1\ (E_a\ F_d) - C'_2\ (E_g\ F_h)}{C'_3} \quad [4]$$

$$Z\ (E_g\ F_h) = \frac{(E_g\ F_h) - C''_1\ (E_a\ F_b) - C''_2\ (E_c\ F_d)}{C''_3} \quad [5]$$

where

$$k_1 = \frac{X(E_c\ F_d)}{X(E_a\ F_b)} \quad k_2 = \frac{Y(E_c\ F_d)}{Y(E_a\ F_b)} \quad k_3 = \frac{Z\ (E_c\ F_d)}{Z\ (E_a\ F_b)} \quad k_4 = \frac{X(E_g\ F_h)}{X(E_a\ F_b)} \quad k_5 = \frac{Y(E_g\ F_h)}{Y(E_a\ F_b)} \quad k_6 = \frac{Z\ (E_g\ F_h)}{Z\ (E_a\ F_b)}$$

$$k'_1 = \frac{Y(E_a\ F_b)}{Y(E_c\ F_d)} \quad k'_2 = \frac{X(E_a\ F_b)}{X(E_c\ F_d)} \quad k'_3 = \frac{Z\ (E_a\ F_b)}{Z\ (E_c\ F_d)} \quad k'_4 = \frac{Y(E_g\ F_h)}{Y(E_c\ F_d)} \quad k'_5 = \frac{X(E_g\ F_h)}{X(E_c\ F_d)} \quad k'_6 = \frac{Z\ (E_g\ F_h)}{Z\ (E_c\ F_d)}$$

$$k''_1 = \frac{Z(E_a\ F_b)}{Z(E_g\ F_h)} \quad k''_2 = \frac{X(E_a\ F_b)}{X(E_g\ F_h)} \quad k''_3 = \frac{Y\ (E_a\ F_b)}{Y\ (E_g\ F_h)} \quad k''_4 = \frac{Z(E_c\ F_d)}{Z(E_g\ F_h)} \quad k''_5 = \frac{X(E_c\ F_d)}{X(E_g\ F_h)} \quad k''_6 = \frac{Y\ (E_c\ F_d)}{Y\ (E_g\ F_h)}$$

$$K_1 = 1 - \frac{k_1}{k_2}, \quad K_2 = 1 - \frac{k_3}{k_2}, \quad K_3 = k_6 - \frac{k_3 k_5}{k_2}, \quad K_4 = \frac{k_5}{k_2}, \quad K_5 = \frac{k_1.k_5}{k_2} - k_4, \quad K'_1 = 1 - \frac{k'_1}{k'_2},$$

$$K'_2 = 1 - \frac{k'_3}{k'_2}, \quad K'_3 = k'_6 - \frac{k'_3.k'_5}{k'_2}, \quad K'_4 = \frac{k'_5}{k'_2}, \quad K'_5 = \frac{k'_1.k'_5}{k'_1.k'_5} - k'_4, \quad K''_1 = 1 - \frac{k''_1}{k''_2}, \quad K''_2 = 1 - \frac{k''_3}{k''_2},$$

$$K''_3 = k''_6 - \frac{k''_3.k''_5}{k''_2}, \quad K''_4 = \frac{k''_5}{k''_2}, \quad K''_5 = \frac{k''_1.k''_5}{k''_2} - k''_4$$

$$C_1 = \frac{1}{k_2} \cdot \frac{K_2.K_4}{K_3}, \quad C_2 = \frac{K_2}{K_3}, \quad C_3 = K_1 + \frac{K_2.K_5}{K_3}$$

$$C'_1 = \frac{1}{k'_2} \cdot \frac{K'_2.K'_4}{K'_3}, \quad C'_2 = \frac{K'_2}{K'_3}, \quad C'_3 = K'_1 + \frac{K'_2.K'_5}{K'_3},$$

$$C''_1 = \frac{1}{k''_2} \cdot \frac{K''_2.K''_4}{k''_3}, \quad C''_2 = \frac{K''_2}{K''_3}, \quad C''_3 = K''_1 + \frac{K''_2.K''_5}{K''_3}$$

In this context (E_aF_b), (E_cF_d) and (E_gF_h) represent the fluorescence excitation or fluorescence emission amplitudes of X,Y and Z mixture, at the E_a or F_b, or E_c or F_d and E_g or F_h wavelengths respectively.

Calculation of amounts of Proto IX and Pchlide in the mixture of Proto IX, Pchlide and Chlide

The fluorescence due to proto IX and Pchlide can be determined from their emission spectra and calculated by appropriate adaptation of equations [3] and [4]. In this case $[E_aF_b]$ refers to [E440 F633] and is adopted for Proto IX, $[E_cF_d]$ represents [E440 F638] and is adopted for Pchlide, $[E_gF_h]$ is adopted for Chlide and represents [E440 F675]. With aforementioned assignments the Eqs for Proto IX and Pchlide are

$$\text{Proto IX (E400 F633} = \frac{\text{E400 F633} - C_1 \text{ (E440 F638)} - C2 \text{ (E440 F675)}}{C_3} \qquad ....[6]$$

$$\text{Pchlide (E440 F638)} = \frac{\text{E440 F638} - C'_1 \text{ (E400 F633)} - C'_2 \text{ (E440 F675)}}{C'_3} \qquad ....[7]$$

$$k_1 = \frac{\text{ProtoIX (E440 F638)}}{\text{Proto IX (E400 F633)}} \qquad k_2 = \frac{\text{Pchlide (E440 F638)}}{\text{Pchlide (E400 F633)}} \qquad k_3 = \frac{\text{Chlide (E440 F638)}}{\text{Chlide (E400 F633)}}$$

$$k_4 = \frac{\text{Proto IX (E440 F675)}}{\text{Proto IX (E400 F633)}} \qquad k_5 = \frac{\text{Pchlide (E440 F675)}}{\text{Pchlide (E400 F633)}} \qquad k_6 = \frac{\text{Chlide (E400 F675)}}{\text{Chlide (E400 F633)}}$$

$$k'_1 = \frac{\text{Pchlide (E400 F633)}}{\text{Pchlide (E440 F638)}} \qquad k'_2 = \frac{\text{Proto IX (E400 F633)}}{\text{Proto IX (E440 F638)}} \qquad k'_3 = \frac{\text{Chlide (E400 F633)}}{\text{Chlide (E440 F638)}}$$

$$k'_4 = \frac{\text{Pchlide (E440 F675)}}{\text{Pchlide (E440 F638)}} \qquad k'_5 = \frac{\text{Proto IX (E440 F675)}}{\text{Proto IX (E440 F638)}} \qquad k'_6 = \frac{\text{Chlide (E440 F675)}}{\text{Chlide (E440 F638)}}$$

The numerical Values of above constants were calculated and was utilized in solving E_{qs} (6) and (7) for determination of fluorescence of Proto IX and Pchlide respectively .

REFERENCES

1. N. Chakraborty and B. C. Tripathy, Involvement of singler oxygen in 5-aminolevulinic acid induced photodynamic damage of cucumber (Cucumis sativus L.) chloroplasts. Plant Physiol. (In Press).
2. W. Rudiger and S. Schoch, Chlorophylls in "Plant Pigments," T. W. Goodwin ed., Academic Press, London (1988).
3. B. C. Tripathy and N. Chakraborty, 5-aminolevulinic acid induced photodynamic damage of the photosynthetic electron transport chain of cucumber (Cucumis sativus L.) cotyledons, Plant Physiol. 96: 761-767 (1991).
4. B. C. Tripathy and C. A. Rebeiz, Chloroplast biogenesis: Quantitative estimation of monovinyl and divinyl monocarboxylic routes of chlorophyll biosynthesis in higher plants, Analytical Biochem. 149:43-61 (1985).
5. B. C. Tripathy and C. A. Rebeiz, Chloroplast Biogenesis: Demonstration of monovinyl and divinyl monocarboxylic routes of chlorophyll biosynthesis in higher plants, J. Biol. Chem. 261 : 13356-13564 (1986)
6. B. C. Tripathy and C. A. Rebeiz, Chloroplast biogenesis 60: Conversion of divinyl protochlorophyllide to monovinyl protochlorophyllide in greening barley, a dark monovinyl/light divinyl plant species, Plant Physiol. 87 : 89-94 (1988).

PHOTOSYNTHETIC UNIT ASSEMBLY AND ITS DYNAMIC
REGULATION

IN VITRO SYNTHESIS OF CHLOROPHYLL A REGULATES TRANSLATION OF CHLOROPHYLL A–APOPROTEINS IN BARLEY ETIOPLASTS

Lutz A. Eichacker

Botanisches Institut der Universität München
Menzinger Str. 67, D-8000 München 19
Tel. 089/1792-227, Fax 089/171683

The Signal and its Molecular Transmitter

The development of photosynthetic plastids from proplastids requires time-resolved coordination of a large number of
nuclear and plastid genes. A central question, therefore, concerns the nature and origin of signals and molecular transmit-
ters that initiate and control this process. A key signal in
the control of developmental processes in higher plants is
light which acts as an environmental control factor regulating
the expression of some nuclear genes for plastid-localized
proteins on such diverse levels as transcription[1], translation[2] and protein stability[3]. A second level of light regulation involves the coordination of nuclear and plastid-
apoprotein synthesis and cofactor (e.g. chlorophyll, carotenoid, heme, quinone, iron and manganese) accumulation during
plastid biogenesis.

In monocots such as barley, early primary leaf and plastid
development proceeds uninhibited in the absence of light[4].
During this developmental phase proplastids develop into etioplasts. Etioplasts do not synthesize chlorophyll (Chl) and
neither accumulate plastid-encoded Chl $\underline{a}$-binding apoproteins[5,6] nor nuclear encoded Chl $\underline{a/b}$-binding apoproteins[7,8]
although they accumulate protochlorophyllide (PChlide), a Chl
precursor. When plants are illuminated, PChlide is reduced to
chlorophyllide (Chlide) by protochlorophyllide-oxidoreductase[9]
within the plastid in a light and NADPH-dependent reaction.
Chlide is subsequently esterified with geranylgeranylpyrophosphate (GGPP) or phytylpyrophosphate (PhPP) in a light-
independent step to form Chl $\underline{a}$[10,11] (Fig. 1B and 1C).

During the first hour of greening *in vivo* the synthesis of
Chl $\underline{a}$ is paralleled by the accumulation of plastid encoded
Chl $\underline{a}$-binding apoproteins P700, CP47, CP43, D2 and D1
(Fig. 1A). To directly show the induction of apoprotein
accumulation by cofactor synthesis, the potential elicitor
activities of PChlide, Chlide $\underline{a}$, Chl $\underline{a}$ and PhPP had to be
separated from the influence of light. On the basis of isola-
ted and lysed etioplasts light-independent synthesis of Chl $\underline{a}$
from exogenously added Chlide and PhPP proved[13] to be the
molecular transmitter required *in vitro* to trigger the
accumulation of plastid encoded Chl $\underline{a}$-binding apoproteins

Tab. 1. Accumulation of Chl a-apoproteins by synthesis of Chl a in lysed etioplast.

Light[a]	(29 W/m^2)	−	−	+	+
Chlide a	(5.3 μM)	−	+	−	−
PhPP	(26.4 μM)	+	+	−	+
Chl a[b]	(pmoles/10^7 plastids)	2	180	5	63
P700[c]	(radioactivity	1	11.12	1	16.47
CP47	in arbitrary	1	31.18	1	39.25
CP43	units)	1	16.06	1	17.23

[a]Light treatment were given as described[13].
[b]Chl a was quantitated by using HPLC[13].
[c]Accumulation of Chl a-apoproteins P700, CP47 and CP43 is expressed relative to their radiolabeling in control etioplasts. Apoprotein radiolabeling was quantitated by laser-scanning densitometry of fluorograms. Lysed etioplasts were pulse-labeled with [^{35}S]-methionine for 15 min and radiolabel chased with L-methionine for an additional 15 min.

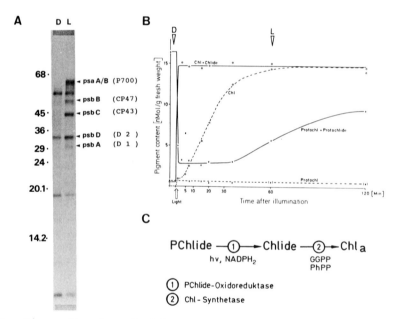

Fig. 1. Biosynthesis of Chl a-apoproteins and Chl a after illumination of etiolated barley seedlings.

    (A) *In vitro* synthesis of Chl a-apoproteins P700, CP47, CP43, D2 and D1, genproducts of psaA/B, psbB, psbC, psbD and psbA, after illumination of etiolated barley seedlings with white light for 1h (D, Dark; L, Light).

    (B) *In vivo* synthesis of Chl a between 0 and 120 min. after illumination of etiolated oat seedlings with white light for 1 min (D, Dark; L, 1h after illumination) (redrawn from Schoch et al.[12])

    (C) Components for photoreduction (1) and esterification (2) in etioplasts from angiosperms.

P700, CP47 and CP43 (Tab. 1). Interestingly, exogenous Chl _a_ added to the translation assays did not elicit the appearance of the Chl _a_-apoproteins. The exact reasons for this are unknown so far.

Regulation of Chl a-Apoprotein Accumulation

In principal, accumulation of Chl _a_-apoproteins could be due to an increased rate of protein synthesis or to Chl-dependent stabilization of _de novo_-synthesized apoproteins against proteolysis. In order to compare the stability of Chl _a_-apoproteins in the absence and presence of Chl, the kinetics of [^{35}S]-methionine incorporation into full-length Chl _a_-apoproteins were investigated during pulse-chase experiments. Lysed etioplasts were pulse-labeled for 5 to 15 min and subsequently chased for 0 to 30 min with unlabeled L-methionine (9 mM final concentration). The incubations were performed with GGPP in the dark either without photoconversion of PChlide or after short (15s) irradiation of lysed etioplasts with red light[13]. The quantitation of radiolabeled apoproteins revealed that degradation of P700, CP47, CP43 and D2 occures in the absence and presence of Chl synthesis. Under all conditions, the degradation rate was larger with Chl synthesis than the rate without Chl synthesis (Fig. 2A and 2B). Interestingly, radioactivity increased during the chase for P700, CP47, CP43 and D2 only if samples were pulse-labeled for 5 min in the presence of Chl synthesis. After a 10 min pulse such an increase in radioactivity was only observed for CP47 whereas an immediate degradation during the chase was found for CP43 and D2 in the same sample. The differential behavior of CP47 in contrast to CP43 and D2, the stability of P700 and similar degradation rates of D2 in the absence and presence of Chl argue against the possibility of an ineffective chase.

The most probable situation is readout of paused translation intermediates of low molecular weight or initiation of new peptide chains during the pulse which are then translated to full-length and released during the chase. (Fig. 2A and 2B). The results further reveal that the highest relative degradation (in percent of synthesis) is observed for CP43 followed by CP47 and D2. The relative stability of P700 is evident. The relative degradation without Chl synthesis is only 4-6% higher than the relative degradation with Chl synthesis. In summary, the data indicate a similar percentage of proteolysis in the absence and presence of Chl synthesis. A higher rate of synthesis induces a correspondingly higher rate of proteolysis. Thus, Chl _a_-apoprotein accumulation cannot be regulated merely via stabilization of nascent apoprotein chains by synthesis of Chl _a_. Regulation by Chl synthesis must also include increased rates of apoprotein synthesis.

Inhibition of Translation Initiation

Translation can generally be regulated at various steps, e.g. during initiation, elongation or termination. From transcript levels of Chl _a_-apoproteins[6] and from polysome profiles of membrane-associated Chl _a_-apoproteins P700 and D1 which did not change[14] upon illumination of 4.5 day-old dark-grown barley leaves for 1h with white light, it was concluded that regulation of Chl _a_-apoprotein translation operates during translation elongation or termination. To select against the impact of non-functional polysomes and to differentiate

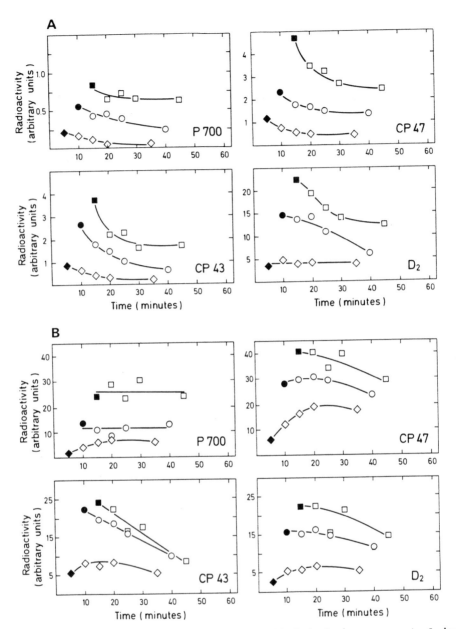

Fig. 2. Increase and stability of radiolabel incorporated into
full-length Chl a-apoproteins

Etioplasts were lysed in the reaction mixture and
incubated in the dark either under non-inductive (A)
or inductive (B) conditions for Chl synthesis. Apoproteins were pulse-labeled for 5 (◆), 10 (●) or 15
(■) min (closed symbols) with [^{35}S]-methionine and
then chased for additional 5, 10, 15 and 30 min (open
symbols) with unlabeled methionine. Radiolabel
stability of P700, CP47, CP43, and D2 was determined
by twodimensional laser densitometry of fluorograms
(not shown). Membrane samples corresponding to all
shown timepoints were loaded on one SDS-PAGE on an
equal plastid-number basis.

between a resuming of translation from paused translation intermediates and initiation of Chl apoprotein mRNA after induction of Chl synthesis, translation initiation was inhibited. The effect of aurintricarboxylic acid (ATA)[15] and kasugamycin (KSG)[16] was tested in the *in vitro* translation system[13]. ATA (at 10^{-5}M) specifically blocks the binding of mRNA to the ribosomal subunit, probably by preventing the binding of protein S1 to the 30S subunit. At higher concentrations (10^{-4}M) ATA inhibits various functions including steps during elongation. Kasugamycin prevents formyl methionine-tRNA binding to the 30S subunit[17].

When initiation inhibitors were added to the *in vitro* system at the beginning of translation they resulted in reduced incorporation of $[^{35}S]$-methionine into membrane proteins (Fig. 3A). The Chl-induced increase in Chl a-apoprotein specific radioactivity was inhibited by increasing concentrations of KSG (5 to 100 μM) to about 40-60%. The concentration-dependent inhibition by ATA showed two phases: between 6.25 and 25 μM ATA radiolabeling was reduced to 30-50% whereas between 50 and 100 μM ATA 50-95% inhibition of

A   **B**

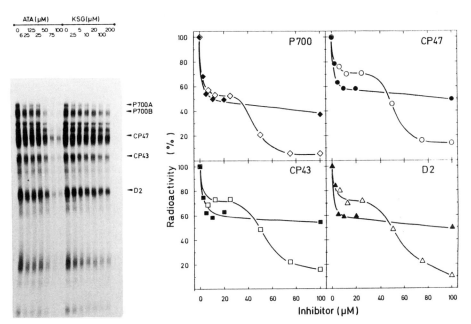

Fig. 3. Inhibition of Chl a-apoprotein synthesis by aurintri-
        carboxylic acid (ATA) and kasugamycin (KSG).

(A) Etioplasts were lysed in translation assays in
    the presence of various concentrations (0-100 μM)
    ATA or (0-200 μM) KSG. Incorporation of $[^{35}S]$-
    methionine into membrane proteins was analyzed by
    SDS-PAGE and fluorography after pulse-labeling
    for 30 min under inductive conditions for Chl
    synthesis.

(B) Quantification of radiolabeled Chl a-apoproteins
    P700, CP47, CP43 and D2 was achieved by twodimen-
    sional laser densitometry of the fluorogram
    Fig. 3A. Open symbols correspond to values for
    ATA, closed symbols to values for KSG.

translation was observed (Fig. 3B). This clearly demonstrates that Chl-dependent accumulation of Chl a-apoproteins in vitro results to about 50-60% from elongation of preinitiated and possibly paused translation intermediates and to about 40-50% from de novo initiation of apoprotein translation.

Summary

1. Light-dependent synthesis of Chl a-apoproteins is signaled in vitro via light-independent synthesis of Chl a.
2. Synthesis of Chl a does not alter the relative stability of de novo synthesized Chl a-apoproteins.
3. Accumulation of Chl a-apoproteins results in vitro from elongation of preinitiated and possibly paused translation intermediates (50-60%) and from de novo initiation of apoprotein translation (40-50%).

References

1. R. Fluhr, F. Kuhlemeier, F. Nagy, and N-H. Chua, Science 232: 1106-1112 (1986).
2. J. P. Slovin, and E. M. Tobin, Planta 154: 465-472 (1982).
3. J. Bennett, Eur. J. Biochem. 118: 61-70 (1981).
4. D. Robertson, and W. McM. Laetsch, Plant Physiol. 54: 148-159 (1974).
5. R. G. Herrmann, P. Westhoff, J. Alt, J. Tittgen, and N. Nelson, In "Molecular Form and Function of the Plant Genome" (L. van Vloten-Doting, G. S. P. Groot, T. C. Hall Eds.), pp. 233-256, Plenum, Amsterdam (1985)
6. R. R. Klein, and J. E. Mullet, J. Biol. Chem. 261: 11138-11145 (1986).
7. K. Apel, Eur.J.Biochem. 97: 183-188 (1979).
8. J. Bennett, G. I. Jenkins, and M. R. Hartley, J. Cell Biochem. 25: 1-13 (1984).
9. K. Apel, H. Santel, T. E. Redlinger, and H. Falk, Eur. J. Biochem. 111: 251-258 (1980)
10. W. Rüdiger, J. Benz, and C. Guthoff, Eur. J. Biochem. 109: 193-200 (1980).
11. J. Soll, G. Schultz, W. Rüdiger, and J. Benz, Plant Physiol. 71: 849-854 (1983).
12. S. Schoch, U. Lempert, and W. Rüdiger, Z.Pflanzenphysiol. 83: 427-436 (1977)
13. L. A. Eichacker, J. Soll, P. Lauterbach, W. Rüdiger, R. R. Klein, and J. E. Mullet, J. Biol. Chem. 265: 13566-13571 (1990)
14. R. R. Klein, H. S. Mason, and J. E. Mullet, J. Cell Biol. 106: 289-301 (1988).
15. A. P. Grollman, and M. L. Stewart, Proc. Nat. Acad. Sci. 61: 719-724 (1968)
16. A. Okuyama, N. Machiyama, T. Kinoshita, and N. Tanaka, Biochem. Biophys. Res. Comm. 43: 196-199 (1971).
17. K. H. Nierhaus, and H. G. Wittmann, Naturwiss. 67: 234-250 (1980).

IN VITRO SYNTHESIS AND MEMBRANE INTEGRATION OF THE

CHLOROPLAST ENCODED D-2 PROTEIN OF PHOTOSYSTEM II

Andreas Friemann, Hans Jürgen Schwarz and Wolfgang Hachtel

Botanisches Institut
Universität Bonn
D-5300 Bonn, F. R. Germany

SUMMARY

    Cell-free reconstitution systems were used to study in vitro synthesis
and membrane integration of the D-2 protein of photosystem II encoded in the
chloroplast genome. The D-2 protein was synthesized in vitro from a tobacco
psbD gene. A large amount of labelled D-2 was incorporated into broad bean
thylakoid membranes that were added to the translation assays either before
protein synthesis was started or after translation was finished. Most of the
incorporated protein was resistant to a wash with 2 M NaBr or 0.1 M NaOH.
The extent of membrane integration of newly formed D-2 was affected neither
by using partially extracted thylakoids nor by the addition of a stroma
fraction to the expression systems. These observations do not suggest the
existence of plastid specific factors involved in the integration of an in
vitro synthesized chloroplast encoded thylakoid protein. Possibly, factors
in the E. coli and rabbit reticulocyte lysate used for in vitro translation
were responsible for preserving an integration-competent conformation of the
newly synthesized protein.

INTRODUCTION

    A number of the intrinsic thylakoid membrane proteins are encoded by
chloroplast genes (Herrmann et al., 1985; Shinozaki et al., 1986) and are
synthesized on chloroplast ribosomes (Ellis, 1977; Hachtel, 1987). At least
some of these proteins are made by polysomes attached to thylakoids. There
are, however, only a few experimental results supporting the idea that the
insertion of these proteins into the thylakoid membranes may be cotranslationally (reviewed by Jagendorf and Michaels, 1990).

    Cotranslational integration of membrane proteins into the endoplasmic
reticulum of eukaryotic cells involves rather elaborate mechanisms (reviewed
by Wickner and Lodish, 1985). In bacteria, some membrane proteins are inserted cotranslationally, others enter the membrane after translation is
finished (Wickner and Lodish, 1985). Stabilization of the newly formed polypeptide in an unfolded conformation has been shown to be a prerequisite to
permit a posttranslational insertion (Meyer, 1988).

    An obvious question is whether complex mechanisms similar to those in
the cytoplasm of eukaryotic cells also occur in chloroplasts. An alternative

possibility would be systems similar to those found in bacteria. Solving this question seems only feasible if all essential components involved in the recognition and integration of membrane proteins can be identified. Most of our knowledge about transfer of cytosolically synthesized proteins into membranes comes from reconstitution experiments that have been carried out in cell-free systems (Meyer, 1988). These systems have consisted of mRNA encoding a membrane protein, a lysate capable of translating this message, and a fraction enriched in the membrane into which the protein is to be transported. An essential feature of cell-free assays is to synthesize the protein in a heterologous lysate from one species that lacks the components required for integration of the protein into the membrane of another species.

We describe here reconstitution experiments to perform synthesis and membrane integration of the D-2 protein of photosystem II in vitro. Heterologous prokaryotic and eukaryotic systems were used for transcription and translation of the chloroplast gene psbD encoding the D-2 protein, and thylakoid membranes were added to study membrane integration of the newly forming polypeptides.

MATERIALS AND METHODS

In Vitro Transcription and Translation

DNA fragments containing psbD were subcloned either into the expression vector pKK233-2 or into the dual promoter plasmid pSPT18 (Pharmacia, Freiburg, FRG). The single NcoI site of pKK233-2 was used for subcloning to avoid formation of a fusion protein from the recombinant plasmid. Cloning of recombinant plasmids in the E. coli strain JM103, preparation of plasmid DNA, enzymatic modification and agarose gel electrophoresis of DNA followed standard procedures (Sambrook et al. 1989). The plasmid derived from pK233-2 was transcribed and translated in an E. coli S30 lysate (Zubay, 1973) that was prepared as described by Pratt (1984) using the E. coli strain MRE-600. Transcription of DNA cloned in the pSPT18 derived plasmid was performed according to Krieg and Melton (1988) using SP6 RNA-polymerase (Atlanta, Heidelberg, FRG). RNA obtained by in vitro transcription was translated in a rabbit reticulocyte lysate (Boehringer, Mannheim) in the presence of 35-S-methionine. Polypeptides were separated by electrophoresis in the presence of lithium dodecyl sulfate on 10% to 20% polyacrylamide gradient gels (Chua, 1980); 6 M urea was added to the resolving gel.

Thylakoid Preparation and Incubation for Protein Integration

Chloroplasts were isolated from leaves of Vicia faba L. as described (Friemann and Hachtel, 1988). The chloroplast pellet was frozen in liquid nitrogen and resuspended in hypotonic Tris-buffer to break the organelles. Thylakoid membranes separated from stroma by centrifugation at 27,000 g for 10 min were washed with Tris-buffer and then suspended in 40 mM Tris-acetate, pH 8.3, containing 750 mM K-acetate, 5 mM Mg-acetate and 1 mM dithiotreitol to dissociate membrane bound RNA. Pelleted thylakoids were resuspended in 10 mM Tris-acetate, pH 8.0, and added to the in vitro (transcription-)translation mixtures at the beginning of the translation reaction. The final concentration was 20 µg chlorophyll per 30 µl assay volume. The incubation was at 37°C for 30 min. Alternatively, in vitro protein synthesis was run in the absence of thylakoid membranes. Thylakoids were added after translation was stopped by RNase A and a chase with cold methionine. Thylakoids were reisolated from the incubation mixtures by centrifugation (15,000 g for 10 min) and analysed as well as the supernatant by gel electrophoresis. Etiochloroplast membranes were prepared from plants germinated under natural light-dark conditions, then further cultivated in the dark for one week, and again illuminated for 5 hours before the isolation of plastids.

RESULTS

Synthesis and Integration into Thylakoids of D-2 Protein in an E. coli Lysate

The structural gene psbD for the tobacco D-2 protein is located on a 1414 bp HgiAI-PstI fragment (Shinozaki et al., 1986) that was inserted into the pKK233-2 vector to give the plasmid pKD. When the purified pKD plasmid DNA was added to the transcription-translation system from E. coli cells it directed the synthesis of one major product that could be identified as D-2 protein according to its apparent molecular mass of 32 kD (Neumann, 1988) and comigration with authentic D-2 protein of the thylakoid membrane. The expression of the pKD vector in the absence of thylakoid membranes yielded radioactive product patterns identical to those obtained in the presence of thylakoids. No radioactive bands were visible on the autoradiographs from control experiments in which only thylakoid membranes were incubated in the transcription-translation reaction mixture (data not shown).

For studies of the integration of proteins into thylakoids, the pKD plasmid in a first set of reactions was expressed in the E. coli lysate in the presence of thylakoid membranes. As shown in Fig. 1, a large amount of newly formed labelled D-2 protein was bound to the membranes and couldn't be removed by washing with either 10 mM Na-pyrophosphate, pH 7.3 (lane 5), 2 M NaBr (lane 3), or 0.1 M NaOH (not shown). 2 M NaBr and 0.1 M NaOH extracted preferentially the minor translation products from the thylakoids. Almost no effect was observed with 1 mM EDTA that would dissociate polypeptides bound via bivalent cations.

In another set of reactions, transcription-translation was run in the absence of thylakoids, and the membranes were added posttranslationally. Again, the bulk of the labelled D-2 protein associated to the thylakoid membranes and was resistant to washes with Na-pyrophosphate, NaOH (Fig. 1, lanes 7, 9), NaBr, or EDTA. For a comparison, the patterns of stainable proteins obtained from thylakoid membranes treated with these substances indicate that mainly the alpha and beta subunit of the chloroplast coupling factor CF1 was dissociated by NaBr whereas NaOH additionally removed most other membrane proteins (not shown).

A membrane-free chloroplast lysate, that is a stroma fraction, did not stimulate membrane integration of in vitro synthesized D-2 protein. To test whether the integrity of the thylakoid membranes affects the integration activity, we added thylakoids pretreated with either NaBr, EDTA, or NaOH to the transcription-translation system programmed by pKD. These treatments didn't significantly reduce integration activity and stability (not shown).

Between plastids at different developmental stages, the ratio of radio-labelled D-2 protein integrated into the thylakoid membranes and remaining in the supernatant of the incubation mixture differed. Membranes obtained from etiochloroplasts integrated a higher proportion of labelled D-2 protein than did thylakoid membranes from mature chloroplasts. Almost no labelled polypeptide remained in the supernatant after reisolation of the etiochloro-plast membranes following either cotranslational or posttranslational in-cubation (data not shown).

Membrane Integration of D-2 Protein Synthesized in a Reticulocyte Lysate

One possible explanation for the observed ineffectiveness of a stroma fraction is that the E. coli lysate itself contains soluble factors which support membrane integration and are able to replace putative chloroplast stromal factors in the heterologous system. To investigate this possibility the E. coli lysate was replaced by a rabbit reticulocyte lysate as a trans-

273

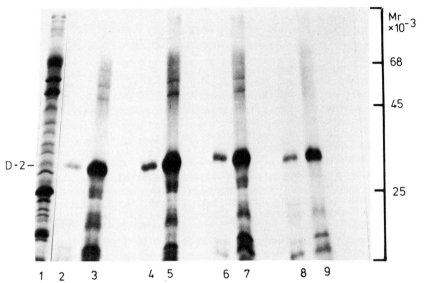

Fig. 1. In vitro expression of psbD in an E. coli lysate using the pKD recombinant plasmid as template, and integration of the translation products into isolated thylakoid membranes. 35-S-methionine labelled translation products were autoradiographed after separation on a 10%-20% polyacrylamide gel. Lane 1 (stained gel): Protein pattern of thylakoid membranes as used in lane 5. Lanes 2-5: Transcription-translation for 30 min in the presence of thylakoids. Lanes 6-9: Transcription-translation for 30 min in the absence of thylakoids that were added posttranslationally to the lysate and further incubated for 30 min. Lanes 2, 4, 6, 8: Supernatant after reisolation of thylakoid membranes by centrifugation of the lysate. Reisolated thylakoids were washed with either 10 mM Na-pyrophosphate (lanes 5, 7), 2 M NaBr (lane 3), or 0.1 M NaOH (lane 9) for 30 min at 4°C.

lation system. We programmed this system by the transcripts obtained from a recombinant dual promoter plasmid that contained the coding sequence of the psbD gene. The plasmid was linearized by cleaving within the 3'-tailing sequence of the gene and transcribed by SP6 RNA-polymerase. Transcripts were identified by electrophoresis and northern blot analysis. The observed size of the transcript is identical to that expected from the sequence data. This indicates that we have obtained fullength transcripts.

In vitro translation of the psbD transcript in the reticulocyte lysate predominantly yielded the 32 kD D-2 polypeptide (not shown). Only traces of labelled polypeptide products were observed at about 50 kD and at 30 kD. Other faint bands were due to the endogenous translation activity of the lysate.

Unfortunately, the translational activity of the reticulocyte lysate was drastically reduced in the presence of thylakoid membranes. Thylakoids, therefore, were only added after the synthesis reaction was finished. As in the E. coli lysate, gene specific translation products in the reticulocyte lysate bound to the thylakoid membranes. Binding again was resistant to a wash with 2 M NaBr or 0.1 M NaOH indicating the integration of the newly synthesized protein into the membrane.

DISCUSSION

Besides the predominant 32 kD product, low amounts of other polypeptides were detected by autoradiography in our experiments. The 50 kD product is probably a dimer of the 32 kD polypeptide (Neumann, 1988). Most of the polypeptides below 32 kD as obtained in the E. coli lysate were probably initiated from internal AUG codons (unpublisched results) as it was reported for in vitro expression of the barley psbD gene (Neumann, 1988), or arose from out-of-phase AUG codons as observed after in vitro expression of the psbE gene of barley (Krupinska, 1988).

To determine the nature of binding of in vitro expressed D-2 protein, the membranes were treated with 2 M NaBr and 0.1 M NaOH, respectively. These two substances have been widely used to extract peripheral proteins from thylakoid membranes (Kamienietzki and Nelson, 1975) and to confirm membrane integration of in vitro synthesized thylakoid membrane proteins (Chitnis et al., 1987). As was demonstrated in the control experiments, treatment of isolated Vicia thylakoids with NaBr and NaOH removed a number of proteins whereas others resisted. Therefore, binding of in vitro-translation products to the added thylakoids that was resistent to these treatments represents integration into membranes rather than association with the membrane surface.

The psbD-mRNA was predominantly found in thylakoid-associated polysomes in Vicia faba (Friemann and Hachtel, 1988). The D-2 protein was found to be synthesized by rough thylakoids in Chlamydomonas (Herrin et al., 1981). For these reasons, it is belived that in vivo this protein is cotranslationally inserted into the membrane. At variance to this, our in vitro experiments show integration of this protein also into membranes that were added after translation was finished. Thus, we succeeded to dissociate integration into the thylakoid membrane from translation. Integration which is normally cotranlationally may proceed in the absence of chain elongation. Integration events can now be studied in the absence of ongoing protein synthesis.

Many of the studies published recently have demonstrated that in order to be inserted into a membrane, proteins must have the proper conformation (Meyer, 1988). In vivo, the ribosome is probably involved in maintaining the integration competence of newly synthesized D-2 protein. It is not known whether other factors in the chloroplast stroma exist that are involved either in cotranslational stabilization of the nascent polypeptides in an integration competent conformation or in posttranslational unfolding that returns the mature protein to a competent conformation. Chitnis et al. (1987) observed the integration of in vitro synthesized precursor of the light-harvesting complex II protein into isolated thylakoids only in the presence of stroma. In our reconstitution experiments, a stroma fraction did not further stimulate the in vitro integration of D-2 protein. However, the integration mixtures contained lysates derived from E. coli or from rabbit reticulocytes. It cannot be excluded that factors in these lysates were able to replace putative chloroplast factors.

Unfolding activities might also exist on the stromal surface of the thylakoid membranes. In our experiments, integration was not reduced when thylakoids extracted by 2 M NaBr or 0.1 M NaOH were used. However, it must be emphasized that thylakoids isolated from mature chloroplasts were washed with 0.75 M K-acetate to eliminate endogenous translational activity before they were added to the translation mixtures. At variance, the endogenous activity of etiochloroplast membrane preparations was low and needed not to be removed by high salt washes. In reconstitution experiments where etiochloroplast membranes were employed we observed higher integration capacity with untreated membranes as compared to etiochloroplast membranes washed with 0.75 M K-acetate. Possibly, the high salt treatment dissociated losely bound proteins that are involved in the integration process.

ACKNOWLEDGEMENTS

This work was supported by a grant of the Deutsche Forschungsgemeinschaft. A.F. was a recipient of a Graduiertenstipendium. We are grateful
to Dr. M. Sugiura (Nagoya, Japan) for providing the clones of Nicotiana
tabacum chloroplast DNA. We thank K. Bahr, C. Buchholz and H. Geitmann for
technical assistance.

REFERENCES

Chitnis, P. R., Nechushtai, R., and Thornber, J. P., 1987, Insertion of pre-
    cursor of the light-harvesting chlorophyll a/b-protein into the
    thylakoids requires the presence of a developmentally regulated
    stromal factor, Plant Mol. Biol., 10:3.
Chua, N. H., 1980, Electrophoretic analysis of chloroplast proteins,
    Methods Enzymol., 69:434.
Ellis. R. J., 1977, Protein synthesis by isolated chloroplasts, Biochim.
    Biophys. Acta, 463:185.
Friemann, A., and Hachtel, W., 1988, Chloroplast messenger RNAs of free and
    thylakoid-bound polysomes from Vicia faba L., Planta, 175:50.
Hachtel, W., 1987, Synthesis and assembly of thylakoid membrane proteins in
    isolated pea chloroplasts: the chlorophyll a-proteins of the photo-
    system II reaction center, Plant Sci., 48:43.
Herrin, D., Michaels, A., and Hickey, E., 1981, Synthesis of a chloroplast
    membrane polypeptide on thylakoid-bound ribosomes during the cell
    cycle of Chlamydomonas reinhardii, Biochim. Biophys. Acta, 655:136.
Herrmann, R. G., Westhoff, P., Alt, J., Tittgen, J., and Nelson, N., 1985,
    Thylakoid membrane proteins and their genes, in: "Molecular Form and
    Function of the Plant Genome," L. van Vloten-Doting, G. S. P. Groot
    and T. C. Hall, eds., Plenum Press, New York.
Jagendorf, A. T., and Michaels, A., 1990, Rough thylakois: translation on
    photosynthetic membranes, Plant Sci., 71:137.
Kamienietzky, A., and Nelson, N., Preparation and properties of chloroplasts
    depleted of chloroplast coupling factor 1 by sodium bromide treat-
    ment, Plant Physiol., 55:282.
Krieg, P. A., and Melton, D. A., 1988, In vitro RNA synthesis with SP6-RNA
    polymerase, Methods Enzymol., 155:397.
Krupinska, K., 1988, Characterization and in vitro expression of the cyto-
    chrome b-559 genes in barley, Carlsberg Res. Commun., 53:233.
Meyer, D. I., 1988, Preprotein conformation: the year's major theme in
    translocation studies, TIBS, 13:471.
Neumann, E. M., 1988, In vitro transcription and translation of the psbD
    gene encoding the D-2 protein of photosystem II in barley, Carls-
    berg Res. Comm., 53:395.
Pratt, J. M., 1984, Coupled transcription-translation in procaryotic cell-
    free systems, in: "Transcription and Translation," B. D. Hames and S.
    J. Higgins, eds., IRL Press, Oxford.
Sambrook, J., Fritsch, E. F., and Maniatis, T., 1989, "Molecular Cloning -
    A Laboratory Manual," Cold Spring Harbor Laboratory Press, New York.
Shinozaki, K., Ohme, M., Tanaka, M., Wakasugi, T., Hayashida, N., Matsu-
    bayashi, T., Zaita, N., Chunwongse, J., and Sugiura, M., 1986, The
    complete nucleotide sequence of the tobacco chloroplast genome: its
    gene organization and expression, EMBO J., 5:2043.
Wickner, W. T., and Lodish, H. F., 1985, Multiple mechanisms of protein in-
    sertion into and across membranes, Science, 230:400.
Zubay, G., 1973, In vitro synthesis of protein in microbial systems, Ann.
    Rev. Genet., 7:267.

CHARACTERIZATION OF THE SEQUENTIAL LIGHT-REGULATED

ASSEMBLY OF PHOTOSYSTEM I CORE COMPLEX

O. Lotan, Y. Cohen, S. Yalovsky, D. Michaeli and R. Nechushtai

Botany Department
The Hebrew University of Jerusalem, Jerusalem 91940, Israel

INTRODUCTION

The biogenesis and assembly of the photosynthetic complexes of higher plants proceed by a multi-step process in the two intercellular compartments; the chloroplast and the cytoplasm. Hence a high degree of coordination is required between these different compartments in each step of the process. The process of biogenesis initiates with the transcription and translation of the various polypeptide subunits taking place both in the chloroplast (for the chloroplast-encoded subunits) and in the nucleus and cytoplasm (for the nuclear-encoded subunits) (Archer and Keegstra, 1991; Chitnis and Thornber, 1988). The polypeptide subunits synthesized in the cytoplasm are made as precursors having a leader (transit) sequence in their amino-terminus (Von-Heijne et al., 1989). These precursors are post-translationally imported into the chloroplasts in an energy dependent process (requiring ATP), probably via a receptor situated in the envelope membrane (Archer and Keegstra, 1991; Chitnis and Thornber, 1988). In the chloroplast the precursors enter the thylakoid membrane. They are processed into their mature form and associate with their cofactors (pigments and metal clusters) and with the chloroplast-encoded subunits to form the fully active complex. The order and mechanism in which these events occur, are not yet fully understood. Many experimental systems have been developed in order to try to dissect and follow the different stages of these complex processes and thereby determine the temporal sequence of the assembly of the complexes found in the membrane.

Photosystem I (PSI) is one of the two photochemically active chlorophyll-protein complexes embedded in the thylakoid membrane. It functions in the reducing side of the photosynthetic electron transfer chain and acts as plastocyanin-ferredoxin oxidoreductase. In higher plants and green algae the PSI complex is composed of a core complex (CCI) and a light-harvesting complex (LHCI) (Thornber, 1986). The CCI complex, in which the charge separation occurs, contains at least 12 polypeptides, about 60-100 chlorophyll a molecules, 1-5 molecules of β-carotene, 2 vitamin K_1 molecules, and 3 [4Fe-4S] clusters. Some of the CCI polypeptides are plastom-encoded, while the remaining CCI subunits are nuclear-encoded (For recent reviews on PSI see; Almog et al., 1991; Chitnis and Nelson, 1991; Scheller and Møller, 1990).

The assembly of the CCI complex seems to be different from that of other thylakoid membrane complexes. While the ATPase and the cytochrome b_6f complexes are assembled in a one-step mechanism (Nelson, 1987), and PSII seems to be assembled in two steps (Liveanu et al., 1986), the assembly of PSI seems to occur in a sequential manner (Herrmann et al., 1985; Nechushtai and Nelson, 1985). This special mode of PSI assembly raises the questions of how the sequential accumulation of the subunits and the actual assembly of the complex in the thylakoid membranes are regulated.

The present work addresses some of these questions. The results indicate that light plays a major role in regulating the steady state level of the mRNA of the CCI nuclear subunits. However, the data imply that the regulation of the sequential assembly itself probably occurs post-transcriptionally. Since CCI-subunit II (*psa*D) was shown to play an important role in the assembly of the entire complex (Herrmann *et al.*, 1985; Nechushtai and Nelson, 1985, Chitnis *et al.* 1989), a special attention was given to the characterization of its role in the biogenesis and assembly.

MATERIALS AND METHODS

Plant material

Spinach (*Spinacea oleracea* L. cv. vyropholly), pea (*Pisum sativum* L. cv. Alaska) and barley (*Hordenum vulgare* L. cv. Neomi) plants were grown at 25°C as described in (Cohen *et al.*, submitted; Lotan *et al.*, submitted; Yalovsky *et al.*, 1990).

RNA extraction and northern blot analysis

Total RNA was extracted from cotyledons by phenol extraction as described in (Leutwiler *et al.*, 1986). Northern blots were performed using the formaldehyde method as recommended by the manufacturer (Gene Screen, NEN-Reaserch products). The specific probe fragments were isolated and hybridized as described in (Lotan *et al.*, submitted).

Phytochrome activity

Spinach seedlings were illuminated with red and/or far-red light. Red light was achieved by passing the projected beam through two filters with a transition peak at 660 nm (Scholl 115 and Corning 2-58) for 40 sec and far-red light with a filter having a peak at 720 nm (Corning 7-69) for 10 min. Plants were incubated for an additional 2 hours in the dark and RNA isolations were carried out under dim green light.

Thylakoids purification and westerns blot analysis

Following 13 days of etiolation and during the greening process, spinach cotyledons were harvested and their thylakoids were purified as previously described (Lotan *et al.*, submitted).

Western blots were carried out using the specific antibodies raised against each subunit of CCI as previously described (Nechushtai and Nelson, 1985).

Quantitation of mRNA and protein levels

The relative level of a specific mRNA or a polypeptide subunit was measured by quantitating the density of each band on the autoradiogram with a soft laser densitometer (ZEMEI - soft laser densitometer, model: SLR -2D/1D) as described in (Lotan *et al.*, submitted).

In-vitro transcription and translation of clones

The plasmid p6SocPSI2-4 (carrying the gene encoding CCI-subunit II) was linearized with Hind III and transcribed with T_7 polymerase, as described in (Cohen *et al.*, submitted). The recovered mRNA was translated *in-vitro* in a wheat germ extract (Promega) using ^{35}S-methionine (Amersham) according to the manufactorer procedure.

Import and insertion of the precursor of CCI-subunit II

Import into the isolated pea plastids was performed as previously described (Kohorn *et al.*, 1986) with slight modification (for details see (Cohen *et al.*, submitted)). The insertion reaction into isolated spinach thylakoids was performed according to the procedure developed for the insertion of the pLHCP (Chitnis *et al.*, 1987) with the modification described in (Yalovsky *et al.*, 1990).

Processing of the precursor of CCI-subunit II by isolated stromal fraction

During the greening of barley seedlings, stromal fractions of isolated plastids were purified as describe in (Yalovsky *et al.*, 1990). The processing of the pre-subunit II by these fractions was analyzed by SDS-PAGE after incubation of the precursor with the total stromal extracts (containing about 10 µg protein) for 30 min at 25°C.

RESULTS AND DISSCUSION

The expression and accumulation of the mRNA of the nuclear-encoded CCI-subunits

The role of light in regulating the accumulation of the CCI subunits has been demonstrated in previous studies. It has been observed that during the greening process the polypeptide accumulation of CCI-subunits in whole leaf extracts is sequential, i.e. they appear one after the other (Herrmann *et al.*, 1985; Nechushtai and Nelson, 1985). To characterize the nature of this sequential appearance, the mRNA accumulation of the CCI nuclear-encoded subunits was investigated. Figure 1 shows the accumulation of these mRNAs; it is observed that the accumulation of all mRNA studied (subunits II-VI) shows a specific two-peak pattern. It shows that the mRNAs are present in dark grown seedling, as oppose to the polypeptides subunits (see figure 3). In the first 2 to 3 hours of illuminaton the level of the mRNA remaines constant, however after a stationary phase, a large increase in the mRNA level is observed. This increase reached its peak after 8 hours of greening. Following the first peak a slight decrease and a second major increase followed by a decrease in the level of mRNA is observed until a stabilization is reached after 18 to 20 hours of greening. This specific pattern is shared by all subunits and is observed during greening under different light intensities (Lotan *et al.*, submitted). It is therefore suggest that the specific pattern represents an endogenous characteristic of the expression of the nuclear-encoded CCI subunits. The unique pattern of mRNA accumulation may be the result of internal circadian rhythem (Kloppstech, 1985) or due to a specific stage of the chloroplast biogenesis which influence the expression of the nuclear-encoded genes (Taylor, 1989).

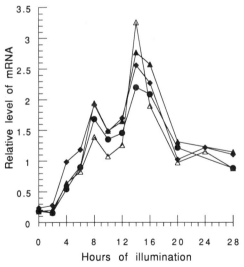

Fig. 1.   The mRNAs accumulation of the CCI nuclear-encoded subunits II-VI (*psaD-psaH*) during greening of etiolated spinach seedlings. Following12-13 days of growth in the dark, spinach seedlings were exposed to white light. Total RNA was isolated, separated on an agarose gel, transfered to nylon membrane and hybridized with probes specific for CCI-subunit III (Steppuhn *et al.*, 1988) (–•–), CCI-subunit IV (Munch *et al.*, 1988) (–□–), CCI-subunit V (Steppuhn *et al.*, 1988) (–▲–), CCI-subunit VI (Steppuhn *et al.*, 1989) (–△–). Relative levels were calculated as described in Materials and Methods.

As phytochrome is known to be the main photoreceptor involved in the light regulated expression of many photosynthetic genes (Tobin and Silverthorne, 1985) the question of its involvement in the expression of the nuclear-encoded genes of CCI was investigated. Figure 2 shows that phytochrome is involved in the light-regulated expression of the mRNA coding for CCI-subunit II. Since CCI-subunit II plays a structural role in the assembly of the entire complex (Chitnis *et al.*, 1989; Herrmann *et al.*, 1985) phytochrome is probably involved in the regulation of the assembly of the whole CCI complex.

The accumulation of the nuclear-encoded subunits in the thylakoid membranes

In order to reveal some of the steps involved in the assembly process itself, and the relationships between the later and the transcription of the corresponding genes, the accumulation of the nuclear-encoded subunits in the thylakoid membrane was followed. Figure 3 shows that their accumulation in the membrane occurs sequentially. The accumulation of each subunit has a linear feature which is substantially different from that of the corresponding mRNA. Subunits II and V were the first to be detected; subunit VI accumulation preceded that of subunits III and IV.

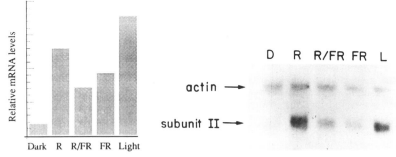

Fig. 2.   Phytochrome induction of mRNA accumulation of CCI - subunit II (*psa*D).
Following 13 days of growth in the dark, spinach seedlings were exposed to: 40 s of red light (R); 40 s of red light followed by 10 min of far-red light (R/FR) or 10 min of far-red light (FR). Dark grown (D) and light grown (L) seedlings were used for control. Total RNA was isolated and the level of CCI-subunit II mRNA in each treatment was detected as described in Materials and Methods. (Left)- Northern blot with CCI-subunit II and rehybridization with actin gene. (Right)- relative level of the mRNA.

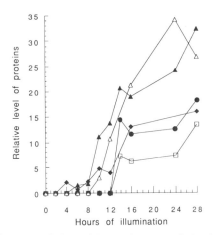

Fig. 3.   The polypeptide accumulation of the nuclear-encoded subunits II-VI of CCI in the thylakoid membrane during greening of etiolated spinach seedlings. After 13 days of etiolation, spinach seedlings were exposed to white light for increasing lengths of time. Thylakoid proteins were isolated and separated by SDS PAGE. Western blot analysis with antibodies raised against each CCI subunit {(✦) subunit II; (●) subunit III; (◊) subunit IV; (▲) subunit V and (△) subunit VI} was preformed as described in Materials and Methods.

280

These results indicate that the sequential assembly of CCI occurs in the thylakoid membrane itself. The two different patterns of the mRNAs and the corresponding polypeptides accumulation suggest the existence of post-transcriptional control mechanism, required to ensure that mRNAs are retained at their steady-state levels until translation occurs. Moreover, the existence of partially-assembled complexes in the thylakoid membranes during the greening process may imply that photosynthetic activity of PSI is carried out by these partially-assembled complexes.

The fact that subunit II was the first to accumulate in the thylakoids confirmed the previous observations obtained with whole leaves extract (Herrmann *et al.*, 1985; Nechushtai and Nelson, 1985) and indicated once again the central role of CCI subunit II in the assembly of the complex. In order to better characterize the later, an *in-vitro* molecular approach was taken. The precursor of CCI-subunit II was incubated with isolated plastids and/or thylakoids and its path was followed.

The import of CCI-subunit II into isolated chloroplasts

Following import of the *in-vitro* synthetized precursor of CCI-subunit II into intact pea plastids, a high level of the mature form of subunit II and a much lower level of the precursor form were found in their thylakoid membranes (Fig. 4 -1). When HgCl$_2$ (3mM) was present during the import reaction, the processing activity was inhibited. Within the thylakoids of these plastids only the precursor form of subunit II was detected (Fig. 4 -2). These findings of the precursor in the membranes resemble those obtained previously with other import systems. When etiochloroplasts (not fully developed) imported the precursor of the major light harvesting complex of photosystem II (pLHCP), the pLHCP was detected in the membranes (Chitnis *et al.*, 1986; Chitnis and Thornber, 1988). Similarly, the results obtained with subunit II also imply that the processing of the precursor form may occur after the insertion of the pre-subunit II into the thylakoids.

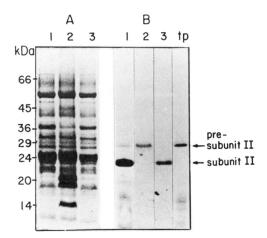

Fig. 4. Import of the precursor of CCI - subunit II into isolated pea plastids *In-vitro* produced pre-subunit II was incubated with isolated pea plastids as described in Materials and Methods. Following the import reaction the plastids were treated with thermolysin, intact plastids were recovered and their thylakoids isolated. Samples of thylakoid membranes were analyzed by SDS-PAGE. The gel was stained (Panel A), and fluorographed (Panel B). *Lane 1*: Thylakoids recovered from a control import reaction. *Lane 2*: HgCl$_2$ was added to a final concentration of 3mM immediately before the beginning of the import reaction. *Lane 3*: The reaction was carried out for 3 min after which 3mM HgCl$_2$ were added.

The insertion of CCI-subunit II into isolated thylakoid membranes

        In order to better characterize the events occuring in the thylakoids, i.e. whether insertion of pre-subunit II precedes or succeeds its processing, *in-thylakoido* system was utilized. Figure 5 illustrates the manner in which the precursor of CCI-subunit II is inserted into isolated thylakoids. It is shown that the insertion of the precursor of subunit II into the membrane does not require the presence of any stromal protein factor. This opposed the insertion of pLHCP in which the presence of a stromal protein factor is essential (Chitnis *et al.*, 1987; Fulsom and Cline, 1988). The differences between pLHCP and subunit II in their dependence on the presence of the stromal factor may arise from the fact that subunit II is a peripheral membrane protein and not an integral one as LHCP. Therefore, it is conceivable that subunit II does not require a stabilized (unfolded) conformation in order to be inserted into the membrane. By the same token, subunit II may not need to undergo a special modification prior to its insertion, and the stromal factor may be redundant.

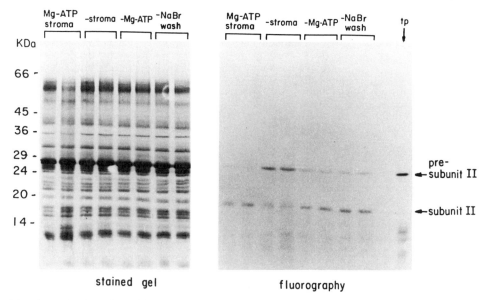

Fig. 5   The insertion of the precursor of CCI-subunit II into isolated spinach thylakoids.
The labeled precursor was incubated with isolated thylakoids (20 µg chl), stroma (60 µg protein), 10mM Mg-ATP, and 10mM non-labled methionine. Following 30 min of incubation at 25°C the membranes were pelleted and washed with 2M NaBr. Left panel - stained gel; Right panel - fluorography. Two left lanes - control; insertion was performed in the presence of stroma and Mg-ATP; (-) stroma = insertion without addition of the stromal fraction; (-) MgATP = insertion without addition of Mg-ATP; (-) NaBr wash = control insertion, but the membranes were not washed with NaBr; tp = translation products (the precursor of subunit II).

        On the other hand, the presence of the stromal fraction proved essential for the processing of the precursor form of subunit II to yeiled its mature form. Results presented in figure 5 leave no doubt that the enzyme which cleaves the precursor is located in the soluble stromal fraction. In the absence of this fraction (fig. 5, - stroma), there is complete loss of processing activity. Further investigation revealed that the precursor was assembled into the pigmented PSI complex and that insertion preceded processing (Cohen *et al.*, submitted).

The processing of CCI-subunit II during the greening process

In order to better characterize the nature of the processing event and to correlate the later with the greening process, the *in-vitro* produced precursor of CCI-subunit II was subjected to purified stroma. The stroma fraction was isolated from barley seedlings at different stages of greening. Figure 6 shows that the processing the CCI-subunit II changes during the greening process.

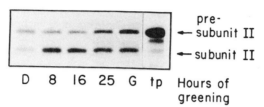

Fig. 6.   The processing of CCI-subunit II during the greening of barley seedlings. The *in-vitro* synthesized precursor of CCI-subunit II was incubated with the stromal fraction (10 μg protein) isolated from barley plastids during the greening process. The reaction which was carried out for 30 min at 25°C was terminated by the addition of dissociation buffer and the samples were analyzed by SDS-PAGE. The gel was stained, fluorographed and the amount of precursor *vs.* mature forms were determined as described in figure 5 and Materials and Methods.

CONCLUSIONS

The present study on the biogenesis of CCI reveals that light regulates the accumulation of the CCI subunits (II-VI) at both the transcriptional and the translational levels. However, the unique sequential assembly of CCI is regulated post-transcriptionally. The assembly of the complex occurs in the thylakoid membrane itself, and subunit II is the first nuclear-encoded subunit that inserts into the membrane. This insertion, which does not require the presence of any stromal protein factor, precedes the processing of the pre-subunit II which is carried out by a soluble stromal protease. The activity of this protease changes during the greening process.

ACKNOWLEDGMENTS

The research described was supported by grants from The Israeli-US Binational Science Foundation {BSF-# 87-049}, the Schonbrunn Research Fund, The Unaited states - Israel Binational Agricultural Research and Development Fund {Bard -#IS-1455-88R} and the Bat-Sheva de-Rothchild Fund.

REFERENCE

Almog, O., Shoham, G. and Nechushtai R. (1991). Photosystem I - Composition, Organization and structure. In: "Current Topics in Photosynthesis" eds: Barber, J; Vol 8.

Archer, E. K. and Keegstra, K. (1991). Current Views on Chloroplast Protein Import and Hypotheseses on the Origin of the Transport Mechanism. J. Bioenerg. Biomem. 22, 789-810.

Chitnis, P. R., Harel, E., Kohorn, B., Tobin, E. M. and Thornber, J. P. (1986). Assembly of the precursor and processed light-harvesting chlorophyll *a/b*-protein of *Lemna* into the light-harvesting complex II of barley etiochloroplasts. J. Cell Biol. 102, 982-988.

Chitnis, P. R., Nechushtai, R. and Thornber, J. P. (1987). Insertion of the precursor of the light-harvesting chlorophyll a/b-protein into thylakoids requires the presence of a develop-mentally regulated stromal factor. Plant Mol. Biol. 10, 3-11.

Chitnis, P. R. and Nelson, N. (1991). Photosystem I. In: "Cell Culture and Somatic Cell Genetics of Plants" eds; Bogorad L. et al., review in press.

Chitnis, P. R., Reilly, P. A. and Nelson, N. (1989). Insertional inactivation of the gene encoding subunit II of photosystem I of the cyanobacterium *Synechocystis* sp. PCC 6803. J. Biol. Chem. 264, 18381-18385.

Chitnis, P. R. and Thornber, J. P. (1988). The major light-harvesting complex of photosystem II: Aspects of its molecular and cell biology. Photosynth. Res. 16, 41-63.

Fulsom, D. R. and Cline, K. (1988). A soluble protein factor is required in vitro for membrane insertion proteolytic processing, assembly into LHCII and localizatoin to appressed membranes occurs in chloroplast lysate. Plant Physiol. 86, 1120-1126.

Herrmann, R. G., Westhoff, P., Alt, J., Tittgen, J. and Nelson, N. (1985). Thylakoid membrane proteins and their genes. In: "Molecular Form and Function of the Plant Genome" (L. van Vloten-Doting, G. S. P. Groot and T. L. ed.) 233-256. Plenum, Amsterdam.

Kloppstech, K. (1985). Diural and circadian rythmicity in the expression of light-induced plant nuclear messenger RNA. Planta 165, 502-506.

Kohorn, B. D., Harel, E., Chitnis, P. R., Thornber, J. P. and Tobin, E. M. (1986). Functional and mutational analysis of the light-harvesting chlorophyll a/b protein of thylakoid membranes. J. Cell Biol. 102, 972-981.

Leutwiler, S., Meyerowitz, E. M. and Tobin, E. (1986). Structure and expression of three light-harvesting chlorophyll *a/b* binding protein genes in *Arabidopsis thaliana*. Nucleic Acid Res. 14, 4051-4063.

Liveanu, V., Yocum, C. F. and N, N. (1986). Polypeptides of the oxygen-evolving photosystem II complex. J Biol Chem. 261, 5296-5300.

Munch, S., Ljungberg, U., Steppuhn, J., Schneiderbauer, A., Nechushtai, R., Beyreuther, K. and Herrmann, R. G. (1988). Nucleotide sequences of cDNAs encoding the entire precursor polypeptides for subunits II and III of the photosystem I reaction center from spinach. Curr. Genet. 14, 511-518.

Nechushtai, R. and Nelson, N. (1985). Biogenesis of photosystem I reaction center during greening. Plant. Mol. Biol. 4, 377-384.

Nelson, N. (1987). Structure and function of protein complexes in the photosynthetic membrane. In: "New Comprehensive Biochemistry, Vol. 15, Photosynthesis" (J. Amsez, ed.), 213-231. Elsevier Science Publishers, Amsterdam.

Scheller, H. V. and Møller, B. L. (1990). Photosystem I Polypeptides. Physiologia Plantarum 78, 484-494.

Steppuhn, J., Hermans, J., Nechushtai, R., Herrmann, G. S. and Herrmann, R. G. (1989). Nucleotide sequence of cDNA clones encoding the entire precursor polypeptide for subunit VI and of the plastome-encoded gene for subunit VII of photosystem I reaction center from spinach. Curr. Genet. 16, 99-108 .

Steppuhn, J., Hermans, J., Nechushtai, R., Ljungberg, U., Thummler, F., Lottspeich, F. and Herrmann, R. G. (1988). Nucleotide sequence of cDNA clones encoding the entire precursor polypeptides for subunits IV and V of photosystem I reaction center from spinach. FEBS Lett. 237, 218-224.

Taylor, W. C. (1989). Regulatory interaction between nuclear and plastid genoms. Annu. Rev. Plant Physiol. Plant Mol. Biol. 40, 211-233.

Thornber, J. P. (1986). Biochemical characterization and structure of pigment-proteins of photosynthetic organisms. In: Encyclopedia of Plant Physiol. 19, 98-142.

Tobin, E. M. and Silverthorne, J. (1985). Light regulation of gene expression in higher plants. Ann. Rev. Plant Physiol. 36, 569-593.

Von-Heijne, G., Steppuhn, J. and Herrmann, R. G. (1989). Domain Structure of mitochondrial and chloroplast targeting peptides. Eur. J. Biochem. 180, 535-545.

Yalovsky, S., Schuster, G. and Nechushtai, R. (1990). The apoprotein precursor of the major light-harvesting complex of photosystem II (LHCIIb) is inserted primarily into stromal lamellae and subsequently migrates to the grana. Plant Mol. Biol. 14, 753-764.

BIOGENESIS OF PHOTOSYSTEM I

SUBUNIT PsaE IS IMPORTANT FOR THE STABILITY OF PS I COMPLEX

Parag R. Chitnis and Nathan Nelson

Roche Institute of Molecular Biology
Roche Research Center, Nutley NJ 07110

Biogenesis of the photosynthetic membranes of chloroplasts and cyanobacteria involves intricate interplay between many complex processes such as translocation of lumenal proteins across the membranes, integration of transmembrane and peripheral proteins in the lipid bilayer, their association with pigment and metal cofactors, and assembly of individual subunits into larger complexes. Assembly of photosystem I (PS I) involves most of these complexities, thus providing a model system to study biogenesis of protein complexes in the thylakoid membranes. PS I catalyzes the photooxidation of plastocyanin and photoreduction of ferredoxin (Bengis and Nelson, 1975). The PS I preparations isolated from thylakoid membranes of chloroplasts and cyanobacteria comprise of seven or more polypeptides (Fig. 1), approximately 100 chlorophyll a molecules, several beta-carotenes, a pair of vitamin K1 molecules and three iron-sulfur clusters (Chitnis and Nelson, 1991). PS I is localized predominantly in the nonappressed regions of thylakoids in chloroplasts of higher plants.

Some Features of Assembly of PS I

The eukaryotic PS I is assembled from components synthesized inside and outside the chloroplasts. Certain subunits of PS I are encoded by nuclear genes and synthesized in the cytoplasm while others are products of plastom genes and are translated on chloroplast ribosomes. The nuclear-encoded subunits are synthesized as larger precursors with amino terminal transit peptides. These precursors are imported into chloroplasts via protein receptor complexes in the chloroplast envelope contact zones (Pain, et al., 1988). During or after their import the transit peptide is proteolytically cleaved and the subunits are assembled in the chlorophyll-protein complex by a sequence of steps which remains to be determined. The subunits synthesized in the chloroplasts do not contain typical leader sequences and they are not processed (Fish and Bogorad, 1986). The transmembrane subunits synthesized in the chloroplasts, e.g. subunits Ia and Ib, are presumably inserted into the membrane cotranslationally (Herrmann, et al., 1985).

Sequence of Assembly of PS I in Developing Plastids : The step-by-step mechanism by which PS I is assembled, appears to be different than that of other chloroplast membrane complexes (Nechushtai and Nelson, 1985). Immunological detection of individual subunits in the membranes isolated from oat, bean or spinach plastids at various stages of growth revealed the

Subunit	Molecular mass (kDa)	Gene	Location
Ia (PsaA)	83	psaA	Transmembrane
Ib (PsaB)	82	psaB	Transmembrane
II (PsaD)	15.6	psaD	Cytoplasmic side
III (PsaF)	15.7	psaF	Lumenal side
V (PsaL)	13	psaL	Cytoplasmic side
IV (PsaE)	8.1	psaE	Cytoplasmic side
VII (PsaC)	8	psaC	Cytoplasmic side

Fig. 1. Subunit Composition of Cyanobacterial PS I. Subunits of PS I isolated from *Synechocystis* sp. PCC 6803 were separated by SDS-PAGE after denaturing the preparation in the presence of 2, mercaptoethanol and SDS at room temperature for 5 h. Nomenclature for the subunits and for the genes encoding them is shown.

following order of appearance of subunits of PS I: I, II, III, IV and others. Similar gradual and differential appearance of the subunits has also been studied during light-induced accumulation of PS I subunits in thylakoids of plastids from the bundle-sheath and mesophyll cells of maize (Vainstein, et al., 1989). In maize plastids subunit VII and II are the first two polypeptides to be detected after illumination. Thus the levels of low molecular weight subunits and PS I activity increase significantly only after subunit II appears in the membrane, suggesting that subunit II may serve a central role during the assembly of the peripheral subunits of PS I.

Physiological Requirements for the Assembly of PS I : Reconstitution of individual steps involved in the assembly of a protein-complex in an *in vitro* system has been a widely used approach to analyze the biogenesis of a complex. *In vitro* import of radiolabeled amino acids or the precursors of chloroplast proteins into plastids and their subsequent incorporation into the protein-complex of interest is a well defined system for such studies (Mullet and Chua, 1983). Recently *in vitro* reconstitution systems have been developed to study integration of radiaolabeled subunits of PS I into the isolated thylakoid membranes of chloroplasts (Cohen et al., 1991) and cyanobacteria (Chitnis and Nelson, unpublished results). These experiments have identified some of the physiological and environmental factors required for the integration of subunits of PS I into thylakoid membranes and for the assembly of PS I (Table 1). The post-translational integration of PsaD into thylakoids and into PS I appears to be a spontaneous process, not requiring ATP or chaperones. These components are required for partial unfolding of some transmembrane proteins like the apoprotein of the major light-harvesting proteins of PS II (Chitnis et al. 1987, Nechushtai, personal communication).

Targeting Information : Irrespective of the site of synthesis each subunit should have the complete information not only for orienting different electron carriers within the complex but also for its destination and its precise arrangement in the complex with respect to the others. The information required for targeting of a subunit to a specific subcompartment lies mainly in the transit sequences of the nuclear-encoded subunits. The

Table 1. Effect of Some Physiological and Environmental factors on the Biogenesis of PS I

Factor	Effect	Ref.
Light	no direct effect on import or assembly	1,4
Chlorophyll	new chlorophyll not required for assembly	1
and precursors	required for release of translation arrest of PsaA-B	2
ATP	required for import	1
	not required for integration of PsaD into PS I	3, 4
Electrochemical potential	not required for integration of PsaD into thylakoids	4
hsp70, cpn60	do not enhance integration of PsaD into membranes	4

References :   (1) Mullet and Chua, 1980; (2) Mullet , 1988 (3) Cohen et al., 1991
                       (4) Chitnis and Nelson, unpublished results

informational elements required for the targeting of a subunit to PS I and for the assembly of the PS I complex are still to be identified. It has been proposed that the transit sequence of PsaD contains information for its correct targeting to PS I (Cohen et al., 1991). However, the absence of transit sequences for cytoplasmic side peripheral subunits of PS I, including PsaD, in cyanobacteria suggests that the information for targeting of these subunits to PS I is present in the mature part of these subunits (Chitnis et al., 1989a;1989b). It is more likely that PsaA and PsaB are first cotranslationally integrated into the membranes forming a core and then the overall three dimensional conformation of the other subunits rather than specific sequence information in them is involved in their proper targeting to the PS I complexes being synthesized.

Effect of Mutations on Assembly and Turnover of PS I

Mutants lacking one or more subunits of PS I have been isolated or generated in various organisms. An important inference from the study of mutants is that the low molecular weight subunits of PS I, most of which are encoded in the nucleus, play a significant role in the functional assembly of PS I. The subunits of PS I coordinating electron transfer centers (PsaA, PsaB and PsaC) are synthesized in the chloroplast. However, a number of nuclear mutations of maize (Miles et al., 1979), barley (Moller et al., 1980) and *Chlamydomonas* (Girard et al., 1980), many presumably affecting synthesis of nuclear-encoded low molecular weight subunits, result in the loss of functional PS I in the thylakoid membranes. PS I-deficient mutants of maize, isolated on the basis of high chlorophyll fluorescence (Miles, et al., 1979), generally lack or have highly reduced amounts of the P700 reaction center complex (subunits Ia and Ib) and one or more lower molecular weight subunits. A number of barley mutants lacking PS I activity have also been identified from the collection of chlorophyll-deficient mutants (Machold and Hoyer-Hansen, 1976; Moller, et al., 1980; Simpson and von Wettstein, 1980) Recessive lethal mutations at five nuclear loci have been studied in detail. The thylakoids from mutant plants show decreased P700 activity and reduced amounts of subunits Ia-Ib, as well as some lower molecular weight polypeptides associated with PS I. Other protein complexes in the thylakoids of these mutants showed normal subunit compositions and activities. Genetic and biochemical analysis of about 25 nuclear mutants of *Chlamydomonas*, belonging to 13 complementation groups demonstrated that the thylakoids of these mutants lacked the major PS I subunits (Ia and Ib) as well as six low molecular weight polypeptides (Girard, et al., 1980). The molecular weights of the latter polypeptides correspond to those of subunits of PS I and LHC I isolated biochemically.

Recently, targeted mutagenesis has been used to generate subunit-specific mutants of PS I in the cyanobacterium *Synechocystis* sp. PCC 6803 (Chitnis, et al., 1989a; Chitnis, et al., 1989b) The mutant strain ADK3, in which the gene *psaD* has been inactivated, cannot grow photosynthetically but exhibits P700 activity and normal amounts of subunit I in the thylakoids. PS I reaction center isolated from this mutant contains reduced amounts of other low molecular weight subunits. In addition, the P700 activity in the PS I from mutant is reduced to about 40 % of that of wild-type. This may indicate the presence of inactive PS I reaction centers in the membranes of the mutant. Both observations indicate that although PS I is assembled in the membrane, it may be turned over more rapidly. This may be caused or triggered by the greater accessibility of mutant reaction center to oxidation. It is also possible that the process of assembly of the lower molecular weight subunits into complete PS I reaction center is impaired in the mutant. This possibility is consistent with biogenesis studies in higher plants (Nechushtai and Nelson, 1985).

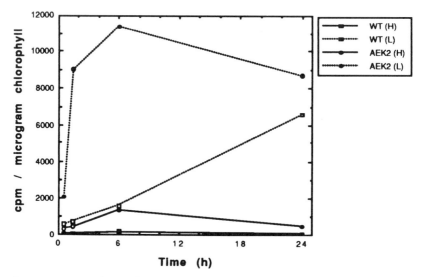

Fig. 2. *Incorporation of tritiated amino acids in the light (L) and heavy (H) fractions of PS I from wild type and AEK2 strains of Synechocystis sp. PCC 6803.* Cells of the wild type and mutants strains of *Synechocystis* sp. PCC 6803 were grown in BG 11 medium supplemented with 5mM TES-KOH and glucose (5 mM). Cultures during the exponential phase of growth were harvested by centrifuging at 5000 g for 10 min and were resuspended in buffered BG11 medium supplemented with glucose to OD730 of 5.00. [3]H labeled amino acids (Amersham) were added to the final concentration of 50 μCi/ml and the cells were incubated at room temperature in light with constant and vigorous shaking. After the labelling, cells were pelleted, resuspended in cold STNM buffer (0.4 M sucrose, 10 mM Tricine -NaOH pH 7.9, 10 mM NaCl, 1 mM $MgCl_2$) and transferred to 2-ml polypropylene microfuge tube with a screw cap. The cells were washed once again with STMN buffer and suspended in 0.5 ml buffer. Glass beads (400-625 μM size) of approximately 0.5 ml volume were added. The cells were broken by vigorous shaking on a mini bead beater for 2 min, followed by cooling on ice for 2 min. The breaking-cooling cycle was repeated for three more times. The tube was then punctured at the bottom with a hot needle and centrifuged in a 15 ml conical tube to remove cell lysate from the glass beads. The cell lysate was centrifuged at 5000 g for 5 min. to remove unbroken cells. Thylakoid membranes were pelleted by microcentrifugation at the top speed for 10 min. The thylakoids were washed twice with TM buffer (20 mM Tricine-NaOH pH 7.9, 1 mM $MgCl_2$) and resuspended in the same buffer to a final concentration of 1-2 mg/ml. The thylakoids were solubilized with 0.5 % Triton X-100 and then solubilized fraction was layered on the top of a 7 to 30 % sucrose gradient containing 20 mM Tricine and 0.2% Triton X-100 and centrifuged at 100,000 g for 16 h. PS I was obtained in two fractions; the heavier fraction probably represented a trimeric form. The chlorophyll content and radioactive counts in each fraction were determined.

Thus the loss of low molecular weight subunits in the isolated mutants of higher plants and algae and in the cyanobacterial targeted mutant ADK3 affects synthesis and/or turnover of PS I reaction center. These subunits, therefore, play important role in the stable assembly of PS I. From the preliminary evidence, the mutant AEK2 of *Synechocystis* sp. PCC 6803 lacking the gene *psaE* seemed to be an exception. The mutant strain lacks subunit IV (PsaE) but shows normal function of PS I reaction center. Its photoautotrophic growth is comparable to that of wild type strain. The PS I complex isolated from the thylakoid membranes of this mutant contains all subunits of PS I except PsaE. To determine effect of this mutation of the assembly and turnover of PS I, we studied incorporation of radioacive amino acids into PS I of wild type and mutant strains of *Synecocystis* sp. PCC 6803.

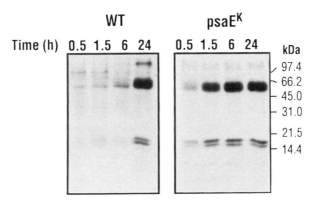

Fig. 3. Labeling of PS I subunits in wild type and AEK2 strains of Synechocystis sp. PCC 6803. The cells were labeled as described in the text, PS I fractions were isolated, denatured by incubating at room temperature in the presence of SDS and 2,mercaptoethanol. Polypeptide subunits of the light fraction of PS I were separated by PAGE and fluorographed. PS I fractions containing two microgram chlorophyll were loaded on each lane in the wild type (WT) panel while the PS I fraction in each lane of the AEK2 (psaEk) panel contains one microgram chlorophyll.

Figure 2 shows the extent of labeling of PS I after different time intervals. Although PS I is a major protein complex in the thylakoids membranes, the incorporation of labeled amino acids in PS I represented only a minor fraction (about 2-3 %) of that in the thylakoids. Since incorporation of labeled amino acids in mature thylakoids is mostly a function of turnover rather than complete *de novo* synthesis, it seems that PS I, unlike PS II, is turned over very slowly in the wild type thylakoids. In thylakoids of AEK2 strain, however, PS I was labeled ten times more rapidly than in the wild type membranes. The quantity of label incorporated in the light fractions (presumably monomeric form) of PS I was five times greater than that incorporated in the heavy fractions (presumably trimeric form) in both wild type and mutant strains. However, the labeling trends were similar in the both heavy and light fractions, indicating that there may not be a precursor-product relationship between these two fractions.

When subunits of the labeled PS I fractions were separated by SDS-PAGE and flurographed, subunits PsaA+PsaB, PsaD+PsaF and PsaL were the highly labeled polypeptides in PS I obtained from wild type as well as mutant strains (Fig. 3). Furthermore, these subunits were labeled simultaneously in both wild type and mutant PS I. In the developing plastids, these subunits appear in thylakoids in a step-by-step pattern. Therefore, incorporation of these subunits into PS I does not follow a similar pattern during the steady-state condition, indicating that there is very little synthesis of completely new PS I reaction centers and the newly synthesized subunits are integrated during turnover of individual subunits of PS I. Alternatively, but less likely, the indicating that there is very little synthesis of completely new PS I sequence of incorporation of various subunits is different during assembly of PS I under steady-state conditions in cyanobacteria. Integration of newly synthesized subunits PsaE and PsaC was not significant during 24 h period.

In summary, the data shown in Fig.2 and 3 shows that the rates of integration of many subunits into PS I is significantly more in AEK2 mutant, indicating higher rate of turnover of PS I lacking subunit PsaE. Therefore, subunit IV (PsaE) is important for the stable assembly of PS I, although its loss does not significantly affect the function of PS I. These studies provide yet another example of the versatility of cyanobacteria as a tool to study biogenesis of photosynthetic apparatus using genetic and biochemical techniques.

References

Bengis, C. , and N. Nelson, 1975, Purification and properties of photosystem I reaction center from chloroplasts, *J. Biol. Chem.*, 250:2783

Chitnis P. R., Nechushtai, R., and Thornber, J. P.,1987, Insertion of the precursor of the light-harvesting chlorophyll *a/b*-protein into the thylakoids requires the presence of a developmentally regulated stromal factor, *Plant Mol. Biol.*, 10:3

Chitnis, P. R., and N. Nelson, 1991, Photosystem I, *Cell Culture and Somatic Cell Genetics of Plants* , 7B:177

Chitnis, P. R., P. A. Reilly, M. C. Miedel , and N. Nelson, 1989, Structure and targeted mutagenesis of the gene encoding 8-kDa subunit of photosystem I of the cyanobacterium *Synechocystis* sp. PCC 6803, *J. Biol. Chem.*, 264:18374

Chitnis, P. R., P. A. Reilly , and N. Nelson, 1989, Insertional inactivation of the gene encoding subunit II of photosystem I of the cyanobacterium *Synechocystis* sp. PCC 6803, *J. Biol. Chem.*, 264:18381

Cohen, Y., Steppuhn J., Herrmann R. G., Yalovsky, S. and Nechushtai, R. ,1991, Does the insertion and assembly of the precursor of photosystem I -subunit II precede its processing ?, *EMBO J.* In press

Fish, L. E., and L. Bogorad, 1986, Identification and analysis of the maize P700 chlorophyll *a* apoproteins PSI-A1 and PSI-A2 by high pressure liquid chromatography analysis and partial sequence determination, *J. Biol. Chem.*, 261:8134

Girard, J., N.-H. Chua, P. Bennoun, G. Schmidt , and M. Delosme, 1980, Studies on mutants deficient in the photosystem I reaction centers in *Chlamydomonas reinhardtii*, *Curr. Genet.*, 2:215

Herrmann, R. G., P. Westhoff, J. Alt, J. Tittgen , and N. Nelson, 1985, Thylakoid membrane proteins and their genes,

Miles, C. D., J. P. Markwell , and J. P. Thornber, 1979, Effect of nuclear mutation in maize on photsynthetic activity and content of chlorophyll protein complexes, *Plant Physiol.*, 64:690

Moller, B., R. Smillie , and G. Hoyer-Hansen, 1980, A photosystem I mutant in barley, *Carlsberg Res. Commun.*, 45:87

Mullet, J. E., 1988, Chloroplast development and gene expression., *Ann. Rev. Plant Physiol.*, 39:475

Mullet, J. E., A. R. Grossman , and N.-H. Chua, 1982, Synthesis and assembly of the polypeptide subunits of photosystem I, *Cold Spring Harbor Symp.*, 46:979

Nechushtai, R. , and N. Nelson, 1985, Biogenesis of photosystem I reaction center during greening, *Plant. Mol. Biol.*, 4:377

Pain, D., Y. S. Kanwar , and G. Blobel, 1988, Identification of a receptor for protein import into chloroplasts and its localization to envelope contact zones, *Nature*, 331:232

ASSEMBLY OF THE PHOTOSYSTEM I MULTIPROTEIN COMPLEX AND THE OLIGOMERIC FORM OF THE MAJOR LIGHT-HARVESTING CHLOROPHYLL a/b-PROTEIN IN PEA SEEDLINGS GROWN IN FLASHED LIGHT FOLLOWED BY CONTINUOUS ILLUMINATION

James T. Jaing[1], Beth A. Welty, Daryl T. Morishige and
J. Philip Thornber

Department of Biology, University of California, Los Angeles,
California 90024-1606, USA

SUMMARY

Intermittent-light grown plants were exposed to continuous illumination, and after various periods of time in the light, thylakoids were isolated from the leaves. A two-dimensional PAGE system was used to examine changes in the protein content of the photosynthetic apparatus and of the individual multiprotein complexes of photosystems I and II in the developing plastids.

1. The light-harvesting complex of photosystem I (LHC I) was added to the P700-containing core complex (CC I) to yield a complete photosystem I unit by 24 hours in continuous light. Its addition correlated with (a) increases in the size and number of 11 - 24kDa polypeptides in the photosystem I component, and (b) a shift to longer wavelengths of the $77^{O}K$ fluorescence peak.

2. Synthesis of the major LHC II pigment-protein (LHC IIb) occurs via addition of pigments to the monomeric polypeptide prior to assembly of the pigmented monomer into the trimer.

3. Many polypeptides of unknown function, some of which are constituents of large multiprotein complexes, disappear as the protochloroplast membranes develop into chloroplast thylakoids.

INTRODUCTION

There have been many studies of the development of the photosynthetic apparatus in thylakoid membranes of eukaryotic organisms over the past 40 years

----

[1]Present address: Department of Medical Genetics, Indiana University School of Medicine, Indianapolis, Indiana, 46202-5251, USA

(summarized in Baker 1984; Hoober 1984; Leech 1984).   Most of them focused on higher plants; however, Chlamydomonas reinhardtii and Euglena gracilis have also been studied (Gershoni et al. 1982; Brandt and Winter 1987; Malnoe et al. 1988). Development of photosystem I (Mullet et al. 1981; Bredenkamp and Baker 1988), photosystem II (Jensen et al. 1986; Minami et al. 1986), the cytochrome b_6/f complex (Willey et al. 1988), the oxygen-enhancing polypeptides (Ryrie et al. 1984), and the timing of the appearance of various other photosynthetic activities (Akoyunoglou and Argyroudi-Akoyunoglou 1986; Alberte et al. 1972; Bredenkamp and Baker 1988; Grumbach 1981) have all been examined. Knowledge gained from these studies has increased our understanding not only of organellar development but also of the photosynthetic mechanism in general. At present, we know, for example, that 1) greening of proplastids in intermittent light causes them to develop into protochloroplasts, which themselves develop into chloroplasts under continuous light (cf. Hoober 1984), 2) protochloroplasts contain chlorophyll a but very little chlorophyll b; 3) the heart of PS I or PS II becomes functional before the major chlorophyll a and b-containing antenna proteins are added (Argyroudi-Akoyunoglou and Akoyunoglou 1979), 4) the appearence of oxygen evolution precedes that of P700 turnover (Alberte et al. 1972); and, 5) carbon dioxide fixation starts near the end of plastid development (Leech, 1984).

It is now well established that each photosystem in fully developed thylakoids is composed of a core component (CC) and a light-harvesting component (LHC)[2]. Each component is a multiprotein complex containing several different pigment-proteins (Thornber et al. 1991). CC I and CC II are involved in trapping light energy and converting it into chemical energy. The LHC's funnel their absorbed light energy to the core component of each photosystem. While much is known about the timing of the appearance of PS I and PS II activities, little is known about how these multiprotein complexes are assembled. It may be that 1) each component of each CC or LHC is added in a sequential order to form the mature PS I or PS II units, or 2) that the components of each CC or LHC assemble into their respective multiprotein complex first and then these complexes (CC and LHC) assemble with each other to form the mature PS I or PS II entities. We show here that the latter may be the more probable route.

Recently, we have developed an electrophoretic fractionation system (Peter and Thornber 1990; cp. Bass and Bricker, 1988;) which can detect whether an individual pigment-protein, or other proteins involved in photosynthetic energy conversion, is present in thylakoids. Furthermore, their association with one or more of the single pigment-proteins or multiprotein complexes (e.g. PS I, CC II-monomer or dimer (Peter and Thornber, 1991a)) can be determined. We thought it timely to apply this system to extend the information already available about developing thylakoid membranes. In particular, we wished to examine 1) the assembly of the PS I complex from its core and light-harvesting components, and 2) the order of assembly of the trimeric form of the major light-harvesting pigment-protein of PS II (LHC IIb) from its monomeric apoprotein and pigments.

----

[2]Abbreviations: PS I and PS II, photosystems I and II; CC I and CC II, core complexes of PS I and II; LHC I and LHC II, light-harvesting complexes of PS I and PS II; PAGE, polyacrylamide gel electrophoresis; SDS, sodium dodecyl sulfate.

MATERIALS AND METHODS

Plant growth conditions and preparation of thylakoids

Pea seeds (Pisum sativum var. Alaska) were imbibed overnight in water and planted in vermiculite. They were grown for 5 days in the dark and then for 3 days under flashed light (2 min light, 118 min dark) which was provided by two incandescent light bulbs one foot away from the seedlings (3.75 uE m^{-2} s^{-1}). The plants were then placed under continuous light provided by fluorescence tubes (76 uE m^{-2} s^{-1}) and then collected after various periods of constant illumination; i.e., 0, 6, 12, 24 and 36 h. Mature pea plants of the same age were grown in the greenhouse under normal day(15h) /night(9h) conditions. Thylakoid membranes were isolated from leaves at the different time points, essentially as described in Nechushtai et al. (1987) but without a 2M NaBr wash. The thylakoid samples were stored at -70°C before further use.

Polyacrylamide gel electrophoresis systems

To separate the pigment-protein complexes, samples of the thylakoid membranes (1.1 mg chlorophyll/ ml) were solubilized in a 5% (w/v) decyl-maltoside - 5% (w/v) octyl-glucopyranoside solution at a detergent:chlorophyll (w/w) ratio of 32:1 (0 and 6h samples) or 25:1 (12, 24, 36h and mature), for 5min at room temperature. The samples were then spun in a microcentrifuge for 30sec to remove any insoluble material. When we loaded each supernatant onto a 10% non-denaturing polyacrylamide gel on the basis of equal amounts of chlorophyll (Peter and Thornber 1990), poor resolution of the green bands occurred for the early time point samples. However, we noted that we got good resolution when we loaded less material, and under such circumstances the quantity of protein proved to be about the same in each sample. Electrophoresis was performed at 100V for approx. 1.5h at 4°C. Thereafter, lanes were excised either as an intact strip or as individual pigmented bands for a second-dimension analysis (cf. Peter and Thornber, 1990). To separate the proteins in the second dimension, each lane of a non-denaturing gel or each pigmented band was incubated at 37°C for 20min in denaturing buffer containing 4% SDS and 100mM DTT. Dissected bands were placed in a well, and whole lanes (after rotating them through 90°) were placed along the top of a 10%-16% fully-denaturing SDS-polyacrylamide gel (Laemmli, 1970) containing 4M urea. Electrophoresis was performed at 12mA for approximately 15h. Proteins were stained with Coomassie blue. Since not all the pigmented bands seen in the mature sample were apparent in the early time-point samples, a region of gel corresponding to the location of a colored band in the mature pea sample was dissected from the early time-point lane and analyzed.

Fluorescence spectrum of intact thylakoid membranes

Whole thylakoid membranes were analyzed at 77°K in an Aminco SPF 500 spectrophotometer. The samples were excited by 437nm light, and the emission spectra recorded using a 4nm bandpass. Spectra were not corrected for the response of the phototube.

RESULTS

Pigment-Protein Complexes in Developing Pea Leaves

The pigmented multiprotein complexes in pea plastids grown under intermittent light and then exposed to continuous light for increasing times were resolved by non-denaturing gel electrophoresis. Depending on the length of exposure to continuous light, between two-to-five prominent pigmented bands were observed (Fig. 1). Each of the pigmented bands was cut out and its polypeptide subunit composition confirmed by fully-denaturing SDS-PAGE (cf. Peter and Thornber, 1990). The uppermost band (Fig. 1) is a multiprotein complex of PS I having both chlorophylls a and b. Its subunits (Fig. 2) are the P700-chlorophyll a-apoprotein migrating at ~ 68kDa, six other colorless CC I proteins (Scheller and Moeller 1990) and the LHC I apoproteins migrating at 11-24kDa (cf. Deng et al. 1989). Migrating slightly faster than PS I on the non-denaturing gel is the pigmented core complex of photosystem I (CC I), which had chlorophyll a but no chlorophyll b. Its subunits are the ~ 68kDa apoprotein and some of its six closely associated colorless polypeptides; it lacks the LHC I polypeptides. There is virtually no CC I in the mature sample (Fig. 1). The major colored band below CC I is a 72kDa component, which contains three proteins of 28, 26 and 25kDa (Fig. 2), and has a chlorophyll a/b = 1.3 (Peter and Thornber 1991b). This band is an oligomeric form of LHC IIb, the major antenna component of PS II (Thornber et al. 1991). Below the 72kDa component is another substantial, colored band of ~ 30kDa (Fig. 2) which contains the same three LHC IIb apoproteins plus smaller amounts of the apoproteins of the minor LHC II components (LHC IIa(CP 29), LHC IIc(CP 27), and LHC IId(CP 24)) (Peter and Thornber 1990;1991b). The presence of the 28, 26, and 25kDa polypeptides and the essentially similar but not identical absorption spectra of the 72 and 30kDa bands (data not shown) indicate that the LHC IIb apoproteins must be associated with pigments in the 30kDa band. Note that if non-pigmented LHC IIb apoproteins are present in the thylakoid, they would migrate faster than the 30kDa band. The fastest migrating colored band is free pigment.

<u>Changes in the pattern of green bands and polypeptides present during plastid development</u>

At the early time points (0-12h), the CC I, but none of the PS I band was present (Fig. 1). Later, a PS I band was apparent later (24h - Mature). The percentage of the thylakoid's total chlorophyll and protein in the PS I band increases with longer periods of exposure to light, whereas a decrease occurs in their percentage in the CC I band (Figs. 1 and 2). In mature tissue essentially all of the photosystem I material is contained in the PS I band. When both bands are present in a sample (i.e., at 24 and 36h), those polypeptides that are present in both the CC I and PS I bands are seen as a wider zone of protein stain extending further to the right on the fully-denaturing second-dimensional gels (Fig. 2) than those that are present only in the PS I band. Thus these "elongated" polypeptides are CC I not LHC I components. The LHC I apoproteins

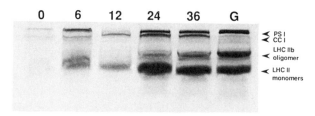

Fig. 1. Electrophoretic pattern of pigment-binding protein complexes in developing thylakoid membranes of pea. Plants were grown under intermittent light for three days and then under continuous light for various times (0, 6, 12, 24, 36 h. and Mature). Membranes were solubilized in decyl-maltoside/octyl-glucoside and subjected to electrophoresis in a 10% polyacrylamide gel at 4°C.

(20-24kDa) are first detected after 24h exposure to continuous light as components of the PS I band only (Fig. 2). Their proportion with respect to other proteins in PS I increases with length of exposure to light. We conclude that the decrease in the amount of CC I with increased time of exposure to light is due to the addition of LHC I to CC I to form the PS I band. The PS I band initially formed (i.e., in the 24 and 36h. samples) may not be of exactly the same size as that in the mature plants.

In fully mature chloroplasts, LHC IIb occurs largely, perhaps entirely, as a trimeric or larger oligomeric form of a monomeric pigment-protein (Butler and Kuhlbrandt, 1990; Thornber et al. 1991; Kuhlbrandt and Wang, 1991). Since we can see the purportedly monomeric and trimeric forms of LHC IIb on our two-dimensional PAGE system, we investigated LHC IIb's assembly. How its apoproteins, chlorophylls a and b, and xanthophyll molecules combine to form the pigmented trimer is not known: One possibility is that the pigments are added to the monomer polypeptide prior to its assembly into the trimeric form; alternatively, the monomeric apoproteins may assemble into the trimer before the pigments are added. Neither the monomeric or trimeric form of LHC IIb is observed in the photosynthetic membranes prior to exposure to continuous light (Figs. 1 and 2). As the exposure to continuous light lengthens, there is first an increase in the amount of LHC IIb pigment and apoproteins in the ~30kDa band (monomeric LHC II's). Only later, after 24h, is a colored 72kDa band (trimeric LHC IIb) detectable. The proportion of photosynthetic pigment and LHC IIb apoproteins in the 72kDa band, compared to those in the ~30kDa band, increases further by 36h and even more so in the mature sample (Figs. 1 and 2). Thus, it is likely that LHC IIb is made as a pigmented monomer prior to its incorporation into any oligomeric form. It was also noted qualititatively that the relative proportions of the three apoproteins of LHC IIb vary slightly during development. This is particularly so for the smallest apoprotein in the trimer (Type III, Morishige and Thornber 1990), but interpreting the functional significance of this observation requires further study; however, see Peter and Thornber, 1991b. The time of the initial appearance of the minor LHC II's coincides with that of the trimeric form of LHC IIb (Fig. 2).

The major polypeptides in the thylakoid preparations from leaves at early stages of plastid development are almost entirely different from those in the fully mature material (Fig. 2) (cp. Zhang and Ishida 1988). Among the more obvious changes are: a large molecular weight band, containing more than 20 polypeptides, in the most immature of the thylakoid membrane samples that largely disappears after 12h exposure to continuous light. And, the two most rapidly migrating proteins in the non-denaturing dimension which are therefore seen on the extreme right of the photos of 0 - 12h. thylakoids at about 18 and 22kDa, are also not present in mature thylakoids. Their possible function in early plastid development and their unusual rate of migration on non-denaturing electrophoresis merit further study.

Fluorescence of Developing Thylakoids

The 77°K emission spectra of developing thylakoid membranes were recorded (Fig. 3). A shift from 715nm to 735nm occurred in the longest wavelength of emission as greening progressed. This fluorescence band arises mainly from PS I components. The core complex is thought to fluoresce at ~725nm, which shifts to 735nm when LHC I is associated with the core complex (cf. Nechushtai et al. 1987). The shift we observed is therefore concomitant with the addition of LHC I to CC I after 24h or more exposure to continuous light (cf Bredenkamp and Baker 1988). Furthermore, the shift occurs gradually which reflects the changing ratio of the amounts of PS I to CC I during the later stages of plastid development.

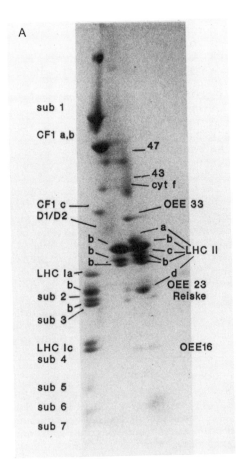

Fig. 2. Results of two-dimensional gel electrophoresis of proteins from pea thylakoid membranes grown under intermittent light followed by continuous light for various times (i.e. 0, 6, 12, 24, 36 h. and Mature). Separation in the first dimension was under non-denaturing conditions (See Fig. 1). A lane from the first dimension was turned through 90° and laid along the top of a fully denaturing 10%-16% SDS-polyacrylamide gel. The individual proteins were visualize by Coomasie blue. The figure above gives an identification of the function of the major polypeptides

B

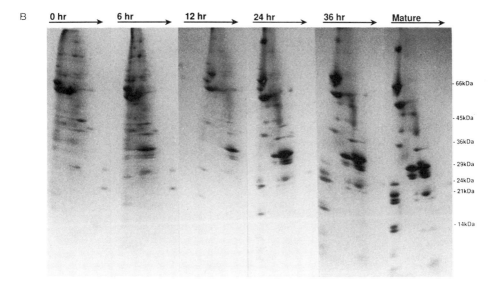

0 hr ⟶   6 hr ⟶   12 hr ⟶   24 hr ⟶   36 hr ⟶   Mature ⟶

- 66kDa
- 45kDa
- 36kDa
- 29kDa
- 24kDa
- 21kDa
- 14kDa

DISCUSSION

We have used a two-dimensional non-denaturing/ fully-denaturing PAGE system (Peter and Thornber 1990) to study the polypeptides present in thylakoid membranes during biogenesis of the photosynthetic apparatus. The advantages of this system over a one-dimensional separation are that it allows us to observe more precisely changes in the relative amount of any thylakoid polypeptide in plants grown under slightly different conditions (cf. Deng et al. 1989), to determine with which multiprotein complex, if any, a thylakoid polypeptide of interest is associated, and to examine changes in the subunit composition of multiprotein complexes during their assembly.

Thylakoids at various stages of maturity were obtained from plants raised initially under intermittent light and then exposed to continuous light for various lengths of time. This plastid developmental system was pioneered by Akoyunoglou et al. (1967) (see also Argyroudi-Akoyunoglou 1970). In flashing light, the functional hearts (core complexes) of photosystems I and II are synthesized, chlorophyll a but not chlorophyll b is made, and very little of the LHC pigment-proteins is present. LHC's are synthesized when such plants are placed in continuous light (Argyroudi-Akoyunoglou et al. 1979). The reason for our obtaining plastids from intermittent-light grown plants rather than from plants which had been grown in complete darkness followed by a few hours of light, was that the former gave a much better banding pattern on non-denaturing PAGE (Fig. 1). Hence, the pigmented multiprotein complexes in immature plastids could be seen, whereas those from etiolated tisssue gave, at best, a smear of pigment.

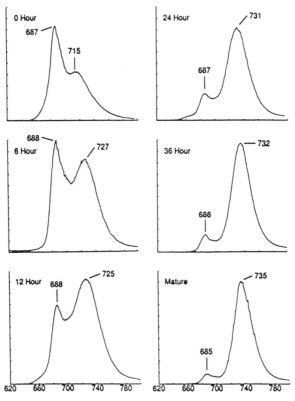

Fig. 3. Fluorescence spectra at 77°K of pea thylakoid membranes grown under intermittent light and then under continuous light for various amounts of time.

We observed the addition of the light-harvesting component to the core component of photosystem I to yield a PS I band after 24 hours further development of the flashed leaves and plastids. In view of the reported biochemical differences between leaves grown in flashed light and those in mature plants (Akoyunoglou et al. 1967; Argyroudi-Akoyunoglou et al. 1979), this result was to be expected; however, it had not been directly demonstrated before. In addition, the changes in steady-state fluorescence during plastid development (Fig. 3) are concomitant with the changing ratio of the PS I to the CC I bands (see also Argyroudi-Akoyunoglou et al. 1984).

LHC IIb occurs in situ as an oligomeric form of a several very similar pigment-proteins (Butler and Kuhlbrandt, 1990; Lyon et al. 1988). Our data on the ratio of the monomeric to the trimeric forms of LHC IIb during greening (Figs. 1 and 2) indicate that the oligomer assembles via addition of the pigments to the monomeric apoprotein which is then incorporated into a developing oligomeric structure. The minor LHC II components, LHC IIa, c and d (CP29, CP27 and CP24, respectively), may function as a connector between CC II and the trimeric LHC IIb structure (Peter and Thornber 1991b). The similar timing of their appearance and that of the oligomeric form of LHC IIb lends some support to this proposal.

Many protein spots in the two-dimensional gels disappear as maturation of the protochloroplast proceeds (Fig. 2) (see also Zhang and Ishida (1988)). For example, a large multiprotein complex of 300-500kDa containing more than 20 polypeptides is present in flashed leaf plastids but not in the mature ones. Other smaller multiprotein complexes appear to behave similarly. Their identity is not known but they appear to be thylakoid-associated complexes. We are further characterizing their apoproteins in the hope of gaining more idea about their function(s). It is possible that the largest multiprotein complex might be involved in chlororespiration (Bennoun 1982). In particular, it may be a NAD(P)H-PQ oxidoreductase (Complex I); Complex I contains over 20 polypeptides in the mitochondrion (Heron et al. 1979).

ACKNOWLEDGEMENTS

Research supported by a grant from the United States Department of Agriculture (88-37262-3557). An REU supplement to NSF grant (87-16320) is gratefully acknowledged.

BIBLIOGRAPHY

Akoyunoglou, G., and Argyroudi-Akoyunoglou, J. H. (1986). Organization of the photosynthetic units, and onset of electron transport and excitation energy distribution in greening leaves. Photosynth. Res. 10: 171-180.
Alberte, R. S., Thornber, J. P., and Naylor, A. W. (1972). Time of appearance of photosystems I and II in chloroplasts of greening jack bean leaves, J. Exp. Bot. 23: 1060-1069.
Argyroudi-Akoyunoglou, H. H., and Akoyunoglou, G., Michel-Wolwertz, M. R., and Sironval, C. (1967). Chlorophyll a as a precursor for chlorophyll b. Synthesis in barley leaves. Chimika Chronika 32: 5-8.
Argyroudi-Akoyunoglou, J. H., and Akoyunoglou, G. (1970). Photoinduced changes in the chlorophyll a to chlorophyll b ratio in young bean plants. Plant Physiol. 46: 247-249.
Argyroudi-Akoyunoglou, J. H., and Akoyunoglou, G. (1979). The chlorophyll-protein complexes of the thylakoid in greening plastids of Phaseolus vulgaris. FEBS Lett. 104: 78-84.

Argyroudi-Akoyunoglou, J. H., Castorinis, A., and Akoyunoglou, G. (1984). Biogenesis and organization of the pigment-protein complexes: relation to the low temperature fluorescence characteristics of developing thylakoids. Isr. J. Bot. 33: 65-82.

Baker, N. R. (1984). Development of chloroplast functions. In: Chloroplast Biogenesis, Vol 5 (Baker, N. R. and Barber, J. eds.), pp 207-251, Elsevier Science Publishers, Amsterdam.

Bass, W.T., and Bricker, T.M. (1988). Dodecyl-maltoside - sodium dodecyl sulfate two-dimensional polyacrylamide gel electrophoresis of chloroplast thylakoid membrane proteins. Anal. Biochem. 171: 330-338.

Bennoun, P. (1982). Evidence for a respiratory chain in the chloroplast. Proc. Natl. Acad. Sci. USA 79: 4352-4356.

Brandt, P., and Winter, J. (1987). The influence of permanent light and of intermittent light on the reconstitution of the light-harvesting system in regreening Euglena gracilis. Protoplasma 136: 56-62.

Bredenkamp, G.J., and Baker, N.R. (1988). The changing contribution of LHC I to photosystem I activity during chloroplast biogenesis in wheat. Biochim. Biophys. Acta. 934: 14-21.

Butler, P.J.G. and Kuhlbrandt, W. (1988) Determination of the aggregate size in detergent solutions of the light-harvesting chlorophyll a/b-protein complex from chloroplast membranes. Proc. Natl. Acad. Sci., U.S. 85: 3797-3801.

Deng, X., Tonkyn, J. C., Peter, G. F., Thornber, J. P., and Gruissem, W. (1989). Post-transcriptional control of plastid mRNA accumulation during adaptation of chloroplasts to different light quality environments. Plant Cell 1: 645-654.

Gershoni, J. M., Shochat, S., Malkin, S., and Ohad, I. (1982). Functional organization of the chlorophyll-containing complexes of Chlamydomonas reinhardtii. A study of their formation and interconnection with reaction centers in the greening process of the y-1 mutant. Plant Physiol. 70: 637-644.

Grumbach, K. H. (1981). Formation of photosynthetic pigments and quinones and development of photosynthetic activity in barley etioplasts during greening in intermittent and continuous white light. Physiol. Plant. 51: 53-62.

Heron, C., Smith, S., and Ragan, C. I., (1979). An analysis of the polypeptide composition of bovine heart mitochondrial NADH-Ubiquinone Oxido-Reductase by two-dimensional polyacrylamide-gel electrophoresis. Biochem J. 181: 435-43.

Hoober, J. K. (1984). In: Chloroplasts, Plenum Press, New York. Chapter VII.

Jensen, K. H., Herrin, D. L., Plumley, F. G., and Schmidt, G. W. (1986). Biogenesis of photosystem II complexes: transcriptional, translational, and posttranslational regulation. J. Cell Biol. 103:1315-1325.

Kuhlbrandt, W., and Wang, D. A. (1991) Three-dimensional structure of plant light-harvesting complex determined by electron crystallography. Nature, 350: 130-134.

Laemmli, U. K. (1970). Cleavage of structural proteins during the assembly of the head of bacteriophage T4. Nature 227: 680-685.

Leech, R. M. (1984). Chloroplast development in angiosperms: Current knowledge and future prospects. In: Chloroplast Biogenesis, (Baker, N. R. and Barber, J.,eds.), pp 1-21, Elsevier Science Publishers, Amsterdam.

Lyon, M. K., and Unwin, P. N. T. (1988). Two-dimensional structure of the light-harvesting chlorophyll a/b complex by cryoelectron microscopy. J. Cell Biol. 106: 1515-1523.

Malnoe, P., Mayfield, S. P., and Rochaix, J. D. (1988). Comparative analysis of the biogenesis of photosystem II in the wild-type and y-1 mutant of Chlamydomonas reinhardtii. J. Cell Biol. 106: 609-616.

Minami, E. I., Shinohar., K., Kuwabara, T., and Watanabe, A. (1986). In vitro synthesis and assembly of photosystem II proteins of spinach chloroplasts. Arch. Biochem. 244: 517-527.

Morishige, D. T., and Thornber, J. P. (1990), The major light - harvesting chlorophyll a/b protein (LHC IIb): the smallest subunit is a novel cab gene product. In Current Research in Photosynthesis (ed. M. Baltscheffsky), Kluwer Acad. Pub., Dordrecht, II: 261-264.

Mullet, J. E., Grossman, A. R., and Chua, N. H. (1981). Synthesis and assembly of the polypeptide subunits of photosytem I. Cold Spring Harbor Symp. Quant. Biol. 46: 979-84.

Nechushtai, R., Peterson, C. C., Peter, G. F., and Thornber, J. P. (1987). Purification and characterization of a light-harvesting chlorophyll a/b-protein of photosystem I of Lemna gibba. Eur. J. Biochem. 164: 345-350.

Peter, G.F., and Thornber, J. P. (1990), Electrophoretic procedures for fractionation of Photosystem I and II pigment - proteins of higher plants and for determination of their subunit composition. Methods in Plant Biochemistry, Volume 5: Amino acids, Proteins and Nucleic acids (ed. L.J. Rogers), Academic Press, 194-212.

Peter, G. F., and Thornber, J. P. (1991a), The photosystem II core complex is organized as a dimer in thylakoid membranes, The Plant Journal, in press.

Peter, G. F., and Thornber, J. P. (1991b) Biochemical composition and organization of higher plant photosystem II light - harvesting pigment - proteins. J.Biol.Chem., 266: in press.

Ryrie, I. J., Young, S., and Andersson, B. (1984). Development of the 33-, 23- and 16-kDa polypeptides of the photosynthetic oxygen-evolving system during greening. FEBS Lett. 177: 269-273.

Scheller, H. V., and Moeller, B. L. (1990). Photosystem I polypeptides. Physiol. Plant., 78: 484-494.

Thornber, J. P., Morishige, D. T., and Anandan, S., and Peter, G. F. (1991). Chlorophyll-Carotenoid-Proteins of Higher Plant Thylakoids. In: Chlorophylls. H. Scheer, ed. CRC Uniscience Series. pp. 549-585, CRC Press, Boca Raton.

Willey, D. L., and Gray, J. C. (1988). Synthesis and assembly of the cytochrome b-f complex in higher plants. Photosynth. Res. 17: 125-144.

Zhang, J - Z., and Ishida, M. R. (1988). The molecular assembly of chloroplasts II: a comparison of the photosynthetic membranes polypeptide composition between etioplasts and chloroplasts in Vicia faba. Annu. Rep. Res. Reactor Inst., Kyoto Univ., 21: 141-146.

PS II INHIBITOR BINDING, Q_B-MEDIATED ELECTRON FLOW AND RAPID DEGRADATION ARE SEPARABLE PROPERTIES OF THE D1 REACTION CENTRE PROTEIN

Marcel A.K. Jansen[1], Alexandra R.J. Driesenaar[2], Hadar Kless[1], Shmuel Malkin[2], Autar K. Mattoo[3], and Marvin Edelman[1]

Departments of Plant Genetics[1], and Biochemistry[2], Weizmann Institute of Science, Rehovot 76100, Israel, and Plant Molecular Biology Laboratory[3], USDA/ARS, Beltsville, MD 20705, USA

ABSTRACT

The D1 protein of the photosystem II reaction centre rapidly turns over in the light. A quantitative comparison of D1 degradation in *Spirodela oligorrhiza* in red and far red light shows that in the absence of PS II activity there is still a high rate of D1 degradation. By using several PS II herbicides, we demonstrate that also in visible light PS II electron flow to the plastoquinone pool cannot be directly correlated to D1 degradation. We conclude that PS II inhibitor binding, Q_B-mediated electron flow to the PQ-pool and light-driven degradation of the D1 reaction centre protein can be discriminated as separate properties.

INTRODUCTION

The protein components of the photosystem II (PS II) reaction centre are D1, D2, cytochrome b_{559} and the psbI gene product (1, 2). A model of the PSII reaction centre structure has been proposed (3, 4) following resolution of the crystal structure of the functionally-analogous reaction centre of photosynthetic bacteria (5) and molecular characterizations of herbicide resistant mutants (6). D2 and D1 are, respectively, the apoproteins of the primary quinone electron acceptor Q_A and the secondary quinone electron acceptor Q_B. During photosynthetic electron flow, Q_B passes through three redox forms: quinone, semiquinone anion radical and quinol (7). The D1 protein is thought to mediate electron flow to the plastoquinone pool by binding and unbinding Q_B in its various redox states. Quinone and quinol are easily exchangeable, while semiquinones are more firmly attached at the Q_B binding niche on the D1 protein (7).

The Q_B binding site on D1 is close to the binding sites for several structurally and functionally distinct herbicides (4, 8, 9). PS II herbicides block electron flow to the

Regulation of Chloroplast Biogenesis
Edited by J.H. Argyroudi-Akoyunoglou, Plenum Press, New York, 1992

plastoquinone pool by seemingly competitive displacement of the plastoquinone QB from the D1 protein (10, 11, 12, 13). The mutual displacement of chemically different inhibitors and quinones (14) together with differences in cross-resistance and functional behavior (15), has supported the concept of different though overlapping binding sites (16, 17) for various PS II inhibitors near the QB binding niche.

A characteristic feature of the PS II reaction centre in comparison to the photosynthetic bacterial centre, is the rapid, light-driven turnover of the D1-protein (18, 19). The D1 protein is unstable in the light, and it is degraded with rates exceeding those for other PS II proteins (20, 21, 22). Degradation results in the formation of a 23.5 kDa membrane-associated breakdown product (23). The site of cleavage is presumably localized in a region between helices IV and V of the D1 protein, somewhere between Arg 238 and Ile 248 (23). This region of the protein is in close proximity to domains which bind QB and PS II herbicides (3, 4, 24, 25).

The location of a proposed cleavage site, and the herbicide and quinone binding domains in one phylogenetically conserved region (26) raises questions about possible structural and functional links among them. It has been speculated that reactive species like the semiquinone anion radical or the doubly reduced quinone, both of which are formed during photosynthetic electron flow (7), bring about D1 degradation (9, 27, 28, 29, 30). This would imply interlinkage between electron flow to the plastoquinone pool and D1 degradation. Such a suggestion is supported by the findings that the PS II herbicides diuron and atrazine, which inhibit electron flow to the plastoquinone pool, also inhibit D1 protein degradation (20). The latter inhibition was thought (21, 29, 31, 32) to be due to the known displacement of quinone from D1 by these inhibitors. The possibility that PS II inhibitors cause inhibition of D1 degradation due to binding to proteins other than D1 was ruled out by experiments conducted with atrazine resistant and sensitive *Solanum nigrum*. The resistant biotype does not bind atrazine but does bind diuron. Atrazine was found to inhibit D1 degradation only in the sensitive biotype where it also inhibited photosynthetic electron flow, while diuron inhibited D1 degradation in both biotypes (33). However, other data do not support the existence of such a correlation between linear electron flow through PS II and D1 degradation. For instance, D1 protein degradation is driven by non-photosynthetic UV light (28) and by far red light (FR) (21). FR (> 700 nm) supports almost exclusively PS I (cf. 21, 34), raising questions about the degradation mechanism involved. In addition, it was shown that the PS II herbicide bromoxynil inhibited PS II electron flow to the plastoquinone pool in visible light without affecting D1 degradation (35).

In this report we study the question whether the low levels of PS II activity in FR are sufficient to mediate D1 degradation. We further examined correlations among electron flow to the plastoquinone pool, PS II inhibitor binding and D1 degradation using different PS II herbicides. It is concluded that these 3 properties can be discriminated as separate entities.

MATERIALS AND METHODS

Spirodela oligorrhiza (Kurtz) was grown photoautotrophically under cool white fluorescent lamps. Radiolabeling with [^{35}S]methionine was carried out as previously described (28). Chase conditions are described in the figure legends. Membrane proteins were fractionated by SDS-PAGE and visualized by fluorography. D1 protein degradation was quantified by densitometry (28). Oxygen evolution was measured *in vivo* by photoacoustic spectroscopy (36) as previously described (35). Photosynthetic energy storage was measured photoacoustically in a similar setup (34). Measuring light (715 nm, 15 $\mu E.m^{-2}.s^{-1}$) was modulated at a frequency of 109.5 Hz. A maximal photothermal signal was obtained by applying strong white light (1600 $\mu E.m^{-2}.s^{-1}$). Steady-state chlorophyll fluorescence was studied using a pulse fluorometer (37) as previously described (35). Photochemical quenching of fluorescence (Q-quenching), representing the fraction of Q_A oxidized, was calculated according to Schreiber (38). Chlorophyll fluorescence induction kinetics were studied using the same pulse fluorometer as above, in combination with a storage oscilloscope (35). The degree of inhibition of electron transport between Q_A and Q_B was calculated from the decrease of the complementary area above the induction curves, which was normalized to maximal fluorescence (39).

Herbicides (chemical purity 99%) were obtained from Riedel-de Haën AG, FRG (atrazine), and Chem Service, PA, USA (ioxynil). Pyridate (> 90% pure) was a kind gift of Chemie Linz AG, Austria.

RESULTS AND DISCUSSION

The occurrence of FR-driven degradation of the D1 protein in *Spirodela oligorrhiza* raised questions about the links between PS II activity and D1 degradation. PS II electron flow in *Spirodela* incubated in far red radiation was described to be less than 3% of that in red light (21). To obtain a quantitative relationship between D1 degradation and PS II activity, we measured the rate of D1 degradation under various absorbed flux densities of red or FR (Fig. 1). Increasing rates of D1 degradation were observed with increasing flux densities in both red and FR. At the highest flux densities applied, a saturation plateau seems to be reached for each light condition. Throughout the flux density range studied, differences between the rates of red and FR light-driven D1 degradation were only 1.5 to 2.5-fold. We observed that FR-driven D1 degradation saturated at equal or possibly lower flux densities than red light driven D1 degradation (Fig. 1). This excludes a scheme whereby the absorption of small amounts of FR by PS II causes D1 degradation, as this would have resulted in a shift of the saturation curve for FR-driven degradation towards higher flux densities. The data show that FR-driven D1 degradation is not connected to PS II activity.

Does FR-driven D1 degradation require PS I activity? Such activity is difficult to measure directly *in vivo* (21). Recently, a novel way of measuring PS I activity *in vivo* was described (34). We show that photosynthetic energy storage, presumably representing PS I cyclic electron flow, can be directly observed in FR in *Spirodela* (Fig. 2). Quantitatively, energy storage is in the same range as was reported for several other species (34). The PS II herbicide diuron inhibits both visible, light-driven and FR-driven D1 degradation (Fig. 3) (21, 31) while PS I cyclic-electron-flow in FR is not measurably impeded by diuron concentrations that abolish D1 degradation (Fig. 2). The presence of PS I activity in the absence of D1 degradation might indicate that these two processes are unrelated. However, it is possible that inhibition of D1 degradation by diuron is not directly linked to inhibition of electron flow but to its binding to the D1 protein itself (33). This is supported by the observation that the diuron concentration-dependence of inhibition of FR-driven D1 degradation is similar to that observed for inhibition of white-light-driven D1 degradation (Fig. 3A). Thus, we have not excluded a possible link between PS I activity and FR-driven D1 degradation.

To probe in more detail the possibility that inhibitor binding itself can result in impaired rates of D1 degradation, we studied the effects of several chemically different PS II inhibitors on both linear electron flow and D1 degradation in *Spirodela*. PS II

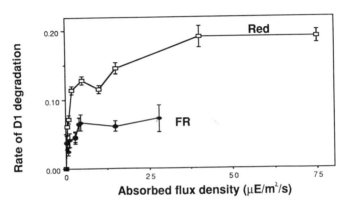

Fig. 1. The rate (h^{-1}) of D1 protein degradation in red and FR light. Red light was generated by a tungsten-halogen projector fitted with a heat absorbing glass and an interference filter (Schott, 660 nm). Broad band FR (peak 720 nm) was generated by filtering the light of a halogen floodlamp through a 180 mm thick water filter, a 2 mm thick KG3 (Schott) heat absorbing glass and a layer of Wratten 89B (Kodak) filter. Flux densities were corrected for absorbance by *Spirodela*. Values represent averaged data from several experiments. Standard errors are shown.

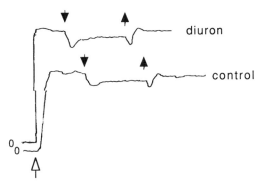

Fig. 2. Photoacoustic signal in FR radiation. Open, upward-pointing arrow, both measuring light (715 nm) and strong saturating light (white) on, resulting in a maximal thermal signal. Filled, downward-pointing and filled, upward-pointing arrows indicate respectively strong saturating light off and on. The resulting dip in signal represents photochemical energy storage. Diuron concentration, 10 μM. Shown is one representative recording.

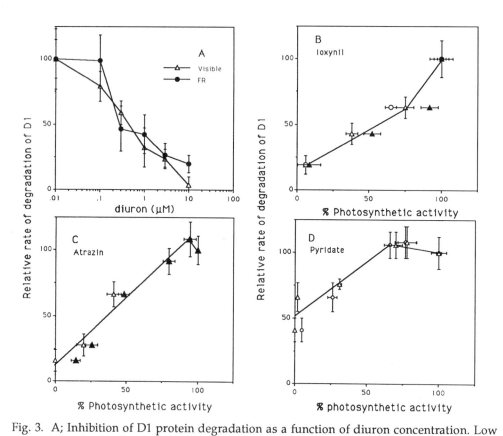

Fig. 3. A; Inhibition of D1 protein degradation as a function of diuron concentration. Low intensity (6 µE.m^{-2}.s^{-1}) light was generated by a halogen floodlamp (FR) as in fig. 1. o by fluorescent bulbs (visible). Values represent averaged data from several experiments. Standard errors are shown.
B, C, D; Inhibition of linear PS II electron flow versus inhibition of D1 degradation i the presence of various PS II herbicides. D1 degradation was measured at 6 µE.m^{-2}.s^{-} visible light. Photosynthetic activity was measured as Q-quenching (▲), oxygen evolution(△) and area above fluorescence induction curves (o). Control in the absenc of inhibitor is shown as 100% photosynthetic activity and 100% D1 degradation. Increasing herbicide concentrations were used to inhibit photosynthetic activity and/or D1 degradation. Values represent averaged data from several experiments. Standard errors are shown.

inhibitors are classified as belonging to a variety of functionally and structurally different groups. We speculated that if inhibition of D1 degradation is not simply due to inhibition of linear electron flow, but requires specific interactions of inhibitor molecules with the Q_B binding niche, we might find compounds that do not interfere with D1 degradation, but still bind to D1 and interrupt electron transport due to quinone displacement.

Previously, it was shown that in the presence of diuron, inhibition of electron flow to the PQ pool was matched by an equally strong inhibition of D1 degradation. However, in the presence of bromoxynil, inhibition of linear electron transport did not match inhibition of D1 degradation (35). We have now tested several additional PS II inhibitors. In the presence of ioxynil and atrazine (Fig. 3B, C), a linear correlation was found between inhibition of photosynthetic electron flow and inhibition of D1 degradation, while pyridate (Fig. 3D) inhibited D1 degradation with a considerably lower efficiency than it did photosynthetic electron flow. These data support the conclusion (35) that inhibition of D1 degradation by PS II inhibitors is not simply related to inhibition of electron flow to the plastoquinone pool and displacement of Q_B. Possibly, the efficiency of inhibition of D1 degradation reflects interactions of specific PS II inhibitors with specific amino acid residues involved in D1 degradation. This implies that the binding niches for atrazine, ioxynil (Fig. 3B, C), diuron and especially dinoseb (35) overlap with the site for D1 degradation, while the niches for bromoxynil (35) and pyridate (Fig. 3D) overlap less.

Another way to look at the relationship between PS II electron flow and D1 degradation is via mutation analysis. Preliminary studies revealed that in several *Synechocystis* 6803 mutants, obtained by site specific mutagenesis of the psbA gene in the region between helices IV and V, oxygen evolution was not detectable while D1 protein degradation was apparently accelerated. Degradation was diuron sensitive as in the wildtype. At first sight, this can be viewed as supporting an uncoupling of electron flow and D1 degradation. However, it is possible that the accelerated degradation in the mutants is due to the activation of some other e.g. 'housekeeping' pathway. This requires further study.

In conclusion, D1 degradation proceeds in FR light in the virtual absence of PS II activity, and in visible light while PS II linear electron flow is inhibited by some PS II herbicides. Also, the reversal is true; propylgallate and uric acid were found to retard D1 degradation without affecting electron transport (31). Therefore, D1 degradation, electron flow to the plastoquinone pool and PS II herbicide binding seem to be distinct processes located in the same general region of the protein. We speculate that there are 3 different albeit overlapping niches in the loop between helices IV and V of the D1-protein. The D1 degradation niche overlaps more strongly with the binding niches of diuron, atrazine, ioxynil and dinoseb than with those of bromoxynil and pyridate. Although we show that electron flow via Q_B is not required for D1 degradation, it cannot be excluded that the Q_B-molecule, by sitting in its pocket, has some role in regulating degradation. Based on trypsination experiments, Trebst proposed that Q_B might stabilize D1 (15). However, such a stabilization does not readily explain the strict requirement of light for degradation of the D1 protein. Yet, it could be that the different redox states which Q_B undergoes in the light, affect D1 degradation in different ways. Additionally, as D1 degradation seems to saturate at flux densities as low as 80 $\mu E.m^{-2}.s^{-1}$ (Fig. 1), non-light dependent regulatory mechanisms may be of importance for determining the rate of D1 degradation as well. Thus, the possible involvement of Q_B in regulating D1 degradation remains to be elucidated.

REFERENCES

1. Nanba, O. and Satoh, K. (1987) Proc. Natl. Acad. Sci. USA, 84, 109-112.
2. Marder, J.B., Chapman, D.J., Telfer, A., Nixon, P.J. and Barber, J. (1987) Plant Mol. Biol., 9, 325-333.
3. Trebst, A. (1986) Z. Naturforsch., 41C, 240-245.
4. Trebst, A. (1987) Z. Naturforsch., 42C, 742-750.
5. Deisenhofer, J., Epp, O., Miki, K., Huber, R. and Michel, H. (1985) Nature, 318, 618-624.
6. Rochaix, J.-D. and Erickson, J. (1988) TIBS, 13, 56-59.
7. Crofts, A.R. and Wraight, C.J. (1983) Biochim. Biophys. Acta, 726, 149-185.
8. Gressel, J.G. (1985) in Herbicide Physiology, Duke, S.O. ed., CRC Press, Boca-Raton, Fl, pp. 159-189.
9. Kyle, D.J. (1985) Photochem. Photobiol., 41, 107-116.
10. Velthuys, B.R. (1981) FEBS Lett., 126, 277-281.
11. Vermaas, W.F.J., Arntzen, C.J., Gu, L.-Q. and Yu, C.-A. (1983) Biochim. Biophys. Acta, 723, 266-275.
12. Pfister, K., Steinback, K.E., Gardner, G. and Arntzen, C.J. (1981) Proc. Natl. Acad. Sci. USA, 78, 981-985.
13. Oettmeier, W. and Soll, H.J. (1983) Biochim. Biophys. Acta, 724, 287-290.
14. Vermaas, W.F.J., Renger, G. and Arntzen, C.J. (1983) Z. Naturforsch., 39C, 368-373.
15. Trebst, A., Depka, B., Kraft, B. and Johanningmeier, U. (1988) Photosynth. Res., 18, 163-177.
16. Pfister, K. and Arntzen, C.J. (1979) Z. Naturforsch., 34C, 996-1009.
17. Van Rensen, J.J.S. (1982) Physiol. Plant., 54, 515-521.
18. Edelman, M. and Reisfeld, A. (1978) in Chloroplast Development, Akoyunoglou, G. and Argyroudi-Akoyunoglou, J.H. eds., Elsevier/North-Holland, Amsterdam, pp. 641-652.
19. Ellis, R.J. (1981) Annu. Rev. Plant Physiol. , 32, 111-137.
20. Mattoo, A.K., Hoffman-Falk, H., Marder, J.B. and Edelman, M. (1984) Proc. Natl. Acad. Sci. USA, 81, 1380-1384.
21. Gaba, V., Marder, J.B., Greenberg, B.M., Mattoo, A.K. and Edelman, M. (1987) Plant Physiol., 84, 348-352.
22. Gounaris, K., Pick, U. and Barber, J. (1987) FEBS Lett., 211, 94-98.
23. Greenberg, B.M., Gaba, V., Mattoo, A.K. and Edelman, M. (1987) EMBO, 6, 2865-2869.
24. Wolber, P.K., Eilmann, M. and Steinback, K.E. (1986) Arch. Biochem. Biophys., 248, 224-233.
25. Dostatni, R., Meyer, H. and Oettmeier, W. (1988) FEBS Lett., 239, 207-210.
26. Mattoo, A.K., Marder, J.B. and Edelman, M. (1989) Cell, 56, 241-246.
27. Kyle, D.J., Ohad, I. and Arntzen, C.J. (1984) Proc. Natl. Acad. Sci. USA, 81, 4070-4074.
28. Greenberg, B.M., Gaba, V., Canaani, O., Malkin, S., Mattoo, A.K. and Edelman, M. (1989) Proc. Natl. Acad. Sci. USA, 86, 6617-6620.
29. Greenberg, B.M., Gaba, V., Mattoo, A.K. and Edelman, M. (1989) Z. Naturforsch., 44, 450-452.
30. Ohad, I., Kyle, D.J. and Arntzen, C.J. (1984) J. Cell Biol., 99, 481-485.
31. Sopory, S.K., Greenberg, B.M., Mehta, R.A., Edelman, M. and Mattoo, A.K. (1990) Z. Naturforsch., 45C, 412-417.
32. Shochat, S., Adir, N., Gal, A., Inoue, Y., Mets, L. and Ohad, I. (1990) Z. Naturforsch., 45C, 395-401.
33. Mattoo, A.K., Marder, J.B., Gaba, V. and Edelman, M. (1986) in Regulation of Chloroplast Differentiation. Akoyunoglou, G. and Senger, H. eds., Alan R. Liss Inc., pp. 607-613.
34. Herbert, S.K., Fork, D.C. and Malkin, S. (1990) Plant Physiol., 94, 926-934.

35. Jansen, M.A.K., Malkin, S. and Edelman, M. (1990) Z. Naturforsch., 45C, 408-411.
36. Bults, G., Horwitz, B.A., Malkin, S. and Cahen, D. (1982) Biochim. Biophys. Acta, 679, 452-465.
37. Schreiber, U. (1986) Photosynth. Res., 9, 261-272.
38. Schreiber, U., Schliwa, U. and Bilger, W. (1986) Photosynth. Res., 10, 51-62.
39. Malkin, S. and Kok, B. (1966) Biochim. Biophys. Acta, 126, 413-418.

REGULATION OF PHOTOSYNTHETIC UNIT ASSEMBLY: REGULATION OF LIGHT-HARVESTING
APOPROTEIN TRANSCRIPTION, TRANSLATION AND STABILIZATION IN THYLAKOIDS

Joan Argyroudi-Akoyunoglou, Thalia Bei-Paraskevopoulou,
Kostas Triantaphyllopoulos and Rania Anastassiou

Institute of Biology, NRC "Demokritos", Athens, Greece

INTRODUCTION

The assembly of the photosynthetic units depends on close cooperation
of the genetic information of the nuclear and plastid genome. The plastome
coded reaction center apoproteins are cotranslationaly inserted into the
membrane, while the nuclear-coded light-harvesting apoproteins, after synthesis on cytoplasmic ribosomes, as higher molecular weight precursors, have to be imported into the plastid, processed to the mature size and incorporated into the thylakoid membrane. The accumulation of the LHC apoprotein
in thylakoids, however, and consequently the photosynthetic unit size, does
not depend on the amount of LHC-mRNA present, but it is rather controlled
by a complicated post-translational mechanism - which regulates LHC apoprotein stabilization via Chl-protein complex formation. Control of LHC expression, is thus exerted at diverse levels: transcription, translation, import
and protein stabilization, but also coordination with RC-plastid coded protein and cofactor biosynthesis for photosynthetic unit assembly.

TRANSCRIPTIONAL CONTROL

The transcription of the CAB gene is under the dual control of the circadian clock and phytochrome photoreception. This has become evident from
numerous studies (1-5), among which work from our laboratory with Phaseolus
vulgaris leaves (6,7). In these studies it was found that exposure of etiolated leaves to a Red-Light (RL) pulse induces LHC-mRNA accumulation (as monitored by the p-LHC-II protein produced in in vitro translation products
of the poly (A) mRNAs isolated, or by hybridization experiments with LHC-II
c DNA), which is reversed by Far-Red (FR), immediately following the RL-pulse. The LHC-mRNA accumulation depends on wavelength of light exposure, and
actually the action spectrum of transcription, as monitored by a number of
techniques (see Fig. 1, and Triantaphyllopoulos and Argyroudi-Akoyunoglou,
this Volume) coincides closely with the absorption spectrum of phytochrome
(Pr). Furthermore, the action spectrum of the LHC-II mRNA accumulation reversal by FR also coincides with the difference absorption spectrum reported
for the purified phytochrome protein (Pr-Pfr)(8,9). Phytochrome is therefore
the photoreceptor involved. The LHC-mRNA accumulation in the leaves transferred to the dark after the RL-pulse follows circadian oscillations. The
phase of the rhythm can be shifted by a second RL pulse, given to the plants
many hours after initiation of the rhythm by a first RL pulse, while this
phase shift can be prevented by FR immediately following the second RL pulse.

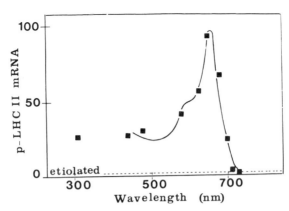

Fig. 1. Action spectrum of LHC-II mRNA transcription, as monitored by dot blot hybridization of the poly (A) mRNA samples with LHC-II nick-translated c DNA. Poly (A) mRNAs were isolated from etiolated bean leaves, pretreated by light pulses of equal dose (1.17 mmoles/m^2) at various wavelengths, and then kept in the dark for 5 additional hours. Percentages are based on the value found at 650 nm (100%)

Phytochrome, therefore, seems to act also as synchronizer of the endogenous circadian clock. Finally, a FR-pulse, given to the plants many hours after initiation of the rhythm by a first RL pulse, prevents further LHC mRNA accumulation, without preventing the rhythm. A long lasting Pfr form, seems to be therefore responsible for sustaining transcription.

COORDINATION IN TRANSCRIPTION-TRANSLATION

The question as to whether LHC-gene transcription is coordinated or not with the translation of its mRNA in vivo, was recently studied in bean. In these studies we tried to see on one hand whether the action spectrum of transcription is in any way correlated with the appearance of the protein in total leaf protein extracts, and on the other, whether the circadian rhythm in LHC-II mRNA accumulation in the dark, following a RL-pulse, may be in some way   followed by the appearance of the final gene product.

Immunoblot analyses of total-LDS-solubilized protein extracts of leaves against the LHC-II antibody, show that the appearance of the protein in leaves or thylakoids indeed depends on light exposure (9,10). An almost undetectable amount of LHC-II apoprotein is found in etiolated bean leaves; this amount is drastically increased after a RL pulse, and again reduced in leaves exposed to FR immediately following the RL-pulse. The mature protein appears in this case, suggesting that even a small amount of Chla (equal to that produced by PChlide fhototransformation) is sufficient to stabilize the protein in prothylakoids. The appearance of the protein depends greatly on wavelength of light exposure, and follows closely the action spectrum of its gene transcription (9). In addition, the LHC-II apoprotein appearance in etiolated bean leaves kept in the dark, following exposure to the RL pulse, is of rhythmical nature, and parallels closely the pattern of its mRNA accumulation (fig. 2) (10). Translation therefore of this protein seems to be closely coupled to gene transcription.

The appearance of the protein, however, correlates also with the action spectrum of PChlide photoconversion (9). It is therefore, difficult to say whether the protein appearing in immunoblots monitors translation of its mRNA, or rather stabilization of the translated protein by the Chla

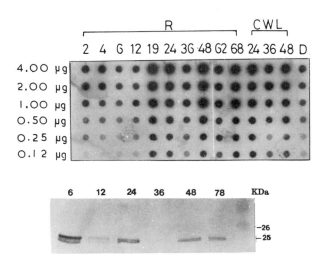

Fig. 2. A. Dot hybridization of poly(A)mRNAs isolated from etiolated,(D),
leaves kept either in the dark after exposure to RL-pulse (R) for
the hours indicated, or transferred to continuous white light (CWL)
Hybridization with LHC-II c DNA.
B. Immunoblot analysis aginst LHC-II antibody of total LDS-protein
extracts of leaves kept in the dark following exposure to RL-pulse

available. Based on the findings concerning the circadian rhythm in the protein appearance in leaves kept in the dark following a RL pulse (10, fig. 2)
and assuming that the Chla formed under these conditions remains constant
(equal to that produced by PChlide photoconversion), we may conclude that
what is monitored by the protein appearance in Western blots is indeed the
translation of the apoprotein rather than its stabilization by the Chla.
Our results therefore suggest that translation of the protein does not depend on new Chla synthesis (unless Chl synthesis occurs in the dark).

The capacity of the leaf for PChlide photoconversion to Chl(ide) in
leaves kept in the dark following a RL pulse, also follows a similar circadian rhythm (10). The question, therefore arises, whether translation of
the protein is triggered by the PChlide regeneration. We tried to give an
answer to this question by administering to the etiolated plants levulinic
acid (LA), the PChlide synthesis inhibitor, 3 hours prior to exposure to RL.
Our results showed (Fig. 3) that under conditions where PChlide resynthesis,
as monitored by Chl accumulation, is completely inhibited, the appearance
of the protein in the leaf extracts is not affected. This therefore suggests
that triggering of translation does not depend on PChlide regeneration.

To this conclusion lead also preliminary experiments which show that
in dark-grown bean leaves pretreated by a Heat- shock, the amount of LHC-II
apoprotein present in LDS-extracts of leaves kept in the dark, following the
Heat-shock is also enhanced, and oscillates in a rhythmical manner (Fig. 4).
The oscillations in this case are slightly phase-shifted as to the oscillations induced by a RL pulse; this may be due to the lag required for recovery of the plant from the heat-shock. Since in this case no Chla is expected
to be formed in the dark, it becomes clear that new Chl synthesis is not required for translation of the protein.

We can therefore conclude that 1. what we monitor by the appearance
of the protein in immunoblots reflects LHC-II apoprotein translation, rather

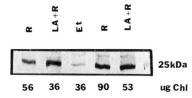

R   LA+R   Et   R   LA+R

25kDa

56   36   36   90   53   ug Chl

Fig. 3. The effect of levulinic acid (LA) administration to etiolated bean
leaves prior to exposure to a Red Light pulse on the translation
of the LHC-II apoprotein, as monitored by immunoblot analysis of
total LDS-extracts of leaves. 150 mM LA solution, pH 7.2 was used.
Leaves attached to one cotyledon were dipped in LA solution for 3
min, prior to the RL pulse exposure, and then kept in the dark for
24 hours;then leaves were used either for LDS solubilization, and
immunoblotting, or they were transferred to continuous light for
1 hour and their Chl extracted in 80% acetone. The ug Chl shown
represent values per g fr w leaf tissue. Et: etiolated.

than stabilization by the Chla available; 2. LHC-II apoprotein translation
is closely coupled to its mRNA transcription; 3. translation of the protein
is not regulated by new Chla synthesis, nor by the regeneration of PChlide;
and 4. the Chla formed by the Red-light pulse is barely sufficinet to rescue
the apoprotein from peptidase action. The later is suggested by the gradual
reduction of the protein amount, as time in the dark, following the red-
light pulse, is prolonged, in spite the enhanced LHC-mRNA, which is observed
under these conditions (fig. 2).

STABILIZATION AND ACCUMULATION OF LHC-II APOPROTEIN- COMPETITION OF REACTION
CENTER AND LHC APOPROTEINS FOR CHLOROPHYLL

The amount of LHC-II protein present in LDS-solubilized protein extracts of etiolated leaves or their thylakoids following a RL pulse, is very low as compared to a green leaf sample (see 9). For the accumulation of
the protein in thylakoids a large amount of Chl (larger than that formed
by a RL pulse) is required; in addition, Chlb may be also required for the
accumulation of the protein, as suggested by the paralle appearance of the
protein and the formation of Chlb. Earlier work has shown, however, that
it is not the absolute amount of Chl available in thylakoids, which regulates LHC accumulation, but rather its relative concentration to RC or LHC
apoproteins. In etiolated leaves, greened in intermittent light (ImL)

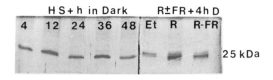

H S + h in Dark   R±FR+4h D

4   12   24   36   48 | Et   R   R-FR

25 kDa

Fig. 4. The effect of a Heat-shock pretreatment (30 min at 40°C) of dark-
grown bean plants on the expression of the LHC-II apoprotein, as
monitored by Western blot analysis of LDS-solubilized leaf proteins
(200 ug protein per slot). Leaf samples were harvested prior to the
heat-shock and at 4 to 60 hours thereafter. Western blots as in (9).

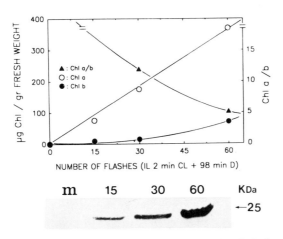

Fig. 5. The appearance of the LHC-II apoprotein in ImL-leaf extracts, as monitored by Immunoblot analysis against LHC-II antibody; the accumulation of Chl a and b in these leaves is also shown for comparison. m: marker protein; 15, 30, 60: the number of light-dark cycles given to the plants, prior to leaf protein solubilization.

of 2 min light every 98 min dark, for example, where Chl synthesis is limited to the light-phase of each cycle, the LHC apoproteins, even though translated and detected in immunoblots (fig. 5), do not accumulate in thylakoids; in contrast, the Reaction center apoproteins CPI and CPa, and core-photosynthetic units selectively accumulate (11-15). By changing, however, the rate of Chl accumulation, i.e. by shortening or prolonging the dark phase of the cycles, it was possible to show that for the same amount of Chla accumulated (after an equal number of light-dark cycle exposures), the thylakoids of plants exposed to ImL with short dark intervals accumulate LHC components; in contrast, the thylakoids of plants exposed to ImL with long dark intervals Reaction center complexes selectively accumulate (16). It should be noted here that the time period necessary for the same number of light exposures (the same number of light-dark cycles) is longer in ImL with longer dark intervals; the time provided for rection center apoprotein synthesis is thus in this case prolonged. We have thus explained these findings as showing that the affinity of the Reaction center apoproteins for Chla is higher than that of the LHC apoproteins, so that the small amount of Chl available is used for rection center apoprotein stabilization, and less Chl remains available for LHC binding and stabilization (18). This working hypothesis has been tested in plants exposed to ImL of long-dark-intervals in the presence of the plastid-protein synthesis inhibitor, chloramphenicol (17). In this case, the reaction center apoprotein synthesis was expected to be inhibited, and thus some Chla was expected to be available for stabilization of the LHC apoproteins. Indeed, under these conditions, the LHC-apoproteins were stabilized in thylakoids (17).

A similar situation was found to occur in plants transferred to the dark after a brief preexposure to continuous light (19-21). It was found that in the dark, where Chl synthesis ceases, the preexisting LHC apoprotein and its Chlb component are degraded, while their Chla is used by the Reaction center apoproteins, continuously synthesized, for the assembly of core photosynthetic units. This situation prevails under conditions where the photosynthetic activity of the plants, preexposed to continulus light, has not reached the full activity of the mature green plant, and thus new photosynthetic unit synthesis is required. This situation is also found in bean leaves, to which LA is administered after a brief preexposure to the light, and then

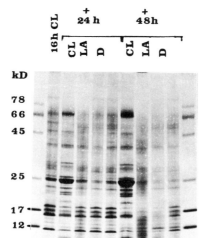

Fig. 6. Effect of LA, administered to etiolated bean leaves after exposure
to 16 h of continuous light, on the degradation of the LHC-II in
subsequent light (CL) or dark (D). LA was administered by immersing the leaves attached to one cotyledon to 150 mM solution for 3
min; the leaves were blotted on filter paper, and at the times
shown their thylakoid protein analyzed by SDS-PAGE and coomassie
stained.

the leaves are kept in the light (Fig. 6). In this case Chl synthesis is
inhibited by LA (in place of the dark); the LHC-II apoprotein formed during
the preceding exposure to continuous light, is degraded in the light in the
LA-inhibited leaves. The photosynthetic unit size in these leaves is smaller
as judged by the light-intensity requirement for saturation of the DPC-DCIP
photoreaction (Anastassiou, unpublished results).

This type of work, therefore, has suggested that the reaction center
apoproteins in bean leaves have higher affinity for Chl than the LHCs, bind
on the Chl available, and are rescued from peptidase action, while the LHCs
in the absence of Chl binding are digested. Why, and how this higher affinity is regulated, is a matter of speculation. However, it is well known that
the Chl/CPI molar ratio is much higher (about 20-25) than that for the isolated LHCP (about 5-6) (22). Furthermore, this ratio changes as greening of
the plants proceeds (25,26). The increase in the molar ratio (Chl/apoprotein)
for a reaction center apoprotein (CPI) and a light-harvesting one (LHCP[3]),
isolated by SDS-sucrose density gradient ultracentrifugation (22), from
plants during their exposure to ImL, to Continuous light, or continuous illumination following ImL, is shown in Fig. 7. As shown, and in contrast to the
constant pigment to pigment ratio found in the isolated complexes irrespective of thylakoid stage of development (25,26), the pigment to Protein molar
ratio for both complexes increases gradually as greening proceeds, to reach
a plateau at 25 Chl/Mr of CPI, or 5 Chl/Mr of LHC-II. Extrapolation to zero
light, suggests that the Chl binding by CPI, even at very early stages of
greening, is higher than that in LHC-II.

This type of post-translational control mechanism, also requires an
active protease, acting on apoproteins not rescued by Chl. A thylakoid-bound
protease, monitored by the azo-dye release from an azocoll substrate (23),
has been detected in bean leaves (24). The protease activity increases parallel to thylakoid development; it comes to a plateau after exposure of etiolated plants to 65 hours of continuous light, or 50-60 cycles of ImL (24).

INCREASE OF PROTEASE ACTIVITY IN THYLAKOIDS OF BEAN LEAVES
TREATED WITH CHLOROPHYLL SYNTHESIS INHIBITORS (LA, SA).

	SAMPLE	INHIBITOR (mM)	AZO-DYE RELEASE (OD 520)*	Chl ug/g fr w
A.	16h CL	0	0.086	340
	16h CL + 24h CL	0	0.275	945
	16h CL + LA + 24h CL	50	0.277	719
	16h CL + LA + 24h CL	150	0.488	407
B.	16h CL	0	0.163	384
	16h CL + 48h CL	0	0.562	1834
	16h CL + SA + 48h CL	55	1.042	490

* Incubation with azocoll substrate was for 2 h (A), 6r 3 h (B) at 37°C.

In plants transferred to the dark, after brief preexposure to continuous light, the protease activity, estimated per mg protein in isolated
thylakoids, is greatly increased; this occurs only in plants preexposed to
brief continuous light illumination and not in those exposed to light for
60 hours or more. This again suggests that the protease activity develops
closely to the formation of the photosynthetic unit centers. This may indicate that the peptidase activity is regulated via the electron transport
capacity of the thylakoid. The protease activity, on the other hand, seems
to be sensitive to Chl accumulation. Thus, in plants exposed to light in
the presence of Chl synthesis inhibitors (Levulinic acid, LA; or succinyl
acetone, SA) the activity estimated per mg thylakoid protein, is enhanced
(Table 1). In this case however, it should be noted that the LHC apoprotein
which represents a major proportion of the thylakoid protein (see also fig.
6) is degraded. Thus, we can not say whether the increase in protease activity is due to activation of the enzyme itself (the protease may be sensitive to photodynamic inactivation by chloroplast pigments) or rather to enrichment of the thylakoids in protease protein.

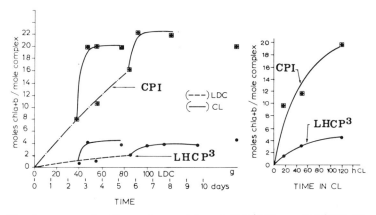

Fig. 7. The increase in the molar ratio of Chl/CPI or Chl/LHC-II complexes,
isolated by SDS-sucrose density gradient centrifugation (22) from
SDS-solubilized thylakoids, obtained from etiolated leaves exposed
to continuous light (CL) prior or after exposure to ImL cycles (LDC)

319

In any case, the close correlation found between the enhancement of the protease activity and the degradation conditions for the LHC-apoprotein suggests that the protease, monitored by the azocoll assay, may be the one involved in the regulation of the photosynthetic unit size.

ACKNOWLEDGEMENT: The partial support by a Volkswagen grant to J. H. Argyroudi-Akoyunoglou and K. Kloppstech is gratefully acknowledged.

REFERENCES

1. E. M. Tobin, Phytochrome mediated regulation of messenger RNAs for the small subunit of ribulose 1,5-bisphosphate carboxylase and the light harvesting Chl a/b protein in Lemna gibba, Plant Mol. Biol. 1:35 (1981)
2. K. Apel and K. Kloppstech, The plastid membranes of barley. Light-induced appearance of the mRNA for the apoprotein of the light-harvesting Chl a/b protein, Eur. J. Biochem., 85: 581 (1978).
3. K. Kloppstech, Diurnal and circadian rhythmicity in the expression of light-induced plant nuclear messenger RNAs, Planta, 165: 502 (1985).
4. G. Giuliano, N.E. Hoffman, K. Ko, P.A. Scolnik and A. Cashmore, A circadian oscillator controlling plant gene transcription, EMBO J., 7: 3635 (1988).
5. F. Nagy, S.A. Kay and N.-H. Chua, Gene regulation by phytochrome, Trends in Genetics 4: 37 (1988)
6. P. Tavladoraki, K. Kloppstech and J.H. Argyroudi-Akoyunoglou, Circadian rhythm in the expression of the mRNA coding for the apoprotein of the light-harvesting complex of PSII. Phytochrome control and persistent Far Red reversibility. Plant Physiol., 90: 665 (1989).
7. P. Tavladoraki and J.H. Argyroudi-Akoyunoglou, Phytochrome controlled circadian rhythm in LHC-I gene expression, FEBS Lett., 225: 305 (1989)
8. R. D. Viestra and P. Quail, Purification and initial characterization of 124-kilodalton phytochrome from Avena, Biochemistry, 22:2498 (1983)
9. K. Triantaphyllopoulos and J.H. Argyroudi-Akoyunoglou, Action spectrum of LHC-II apoprotein transcription and coordinated translation, this Volume.
10. T. Bei-Paraskevopoulou and J.H. Argyroudi-Akoyunoglou, Circadian rhythm in light-harvesting protein II mRNA transcription, translation and Protochlorophyllide regeneration, this Volume.
11. J. H. Argyroudi-Akoyunoglou and G. Akoyunoglou, The Chlorophyll-protein complexes of thylakoids in greening plastids of Phaseolus vulgaris, FEBS Lett. 104: 78 (1979)
12. G. Akoyunoglou, Development of the Photosystem II unit in plastids of bean leaves greened in periodic light, Arch. Biochem. Biophys., 183: 571 (1977).
13. G. Akoyunoglou, Assembly of functional components in chloroplast photosynthetic membranes, in: Photosynthesis, G. Akoyunoglou, ed., Vol. V, Balaban, Pa., pp. 356-366 (1981).
14. G. Akoyunoglou and J.H. Argyroudi-Akoyunoglou, Control of thylakoid growth in Phaseolus vulgaris, Plant Physiol., 61: 834 (1978).
15. G. Akoyunoglou, Assembly, Biosynthesis and properties of the pigment-protein complexes, Proc. 16th FEBS Congrss, VNU Sci Press, Utrecht, pp 35-46 (1985).
16. G. Tzinas, G. Akoyunoglou and J.H. Argyroudi-Akoyunoglou, The effect of the dark-interval in intermittent light on thylakoid development: photosynthetic unit formation and light-harvesting xomplex accumulation, Photos. Res., 14: 241 (1987).
17. G. Tzinas and J.H. Argyroudi-Akoyunoglou, Chloramphenicol-induced stabilization of light-harvesting complexes in thylakoids during development, FEBS Lett., 229: 135 (1988).

18. G. Akoyunoglou and J.H. Argyroudi-Akoyunoglou, Post-translational regulation of Chloroplast differentiation, in: Regulation of chloroplast differentiation, G Akoyunoglou and H. Senger, eds, Liss NY 571 (1986)

19. J. H. Argyroudi-Akoyunoglou, A. Akoyunoglou, K. Kalosakas and G. Akoyunoglou, Reorganization of the PSII unit in developing thylakoids of higher plants after transfer to darkness: changes in Chlb, Light-Harvesting Chlorophyll-protein content and grana stacking, Plant Physiol., 70: 1242 (1982).

20. A. Akoyunoglou and G. Akoyunoglou, Reorganization of thylakoid components during chloroplast development in higher plants after transfer to darkness. Changes in PSI unit components and in cytochromes, Plant Physiol., 79: 425 (1985).

21. J. H. Argyroudi-Akoyunoglou, Photoregulation of LHC-II accumulation in thylakoids during chloroplast development, in: Techniques and New Developments in Photosynthesis Research, Barber, J. and Malkin, R., eds., NATO ASI Series, Volume 168, Plenum, N.Y., pp. 99-110 (1989).

22. J. H. Argyroudi-Akoyunoglou and H. Thomou, Separation of thylakoid pigment-protein complexes by SDS-sucrose density gradient centrifugation, FEBS Lett., 135: 177 (1981).

23. R. Chavira Jr., T.J. Burnett and J.H. Argyroudi-Akoyunoglou, Assaying proteinases with azocoll, Anal. Biochem. 136: 446 (1984).

24. R. Anastassiou and J.H. Argyroudi-Akoyunoglou, Thylakoid-bound peptidase activity as affected by the developmental stage of the chloroplast, this Volume.

25. G. Akoyunoglou and J.H. Argyroudi-Akoyunoglou, Pigmentprotein complexes of thylakoid membranes: assembly, supramolecular organization, In: Recent Advances in Biological Membrane studies, L. Packer, ed., Plenum, N.Y., pp. 205-236 (1985).

26. P. Antonopoulou and G. Akoyunoglou, Changes in the pigment composition of the thylkaoids of Phaseolus vulgaris leaves during chloroplast development, in: "Advances in Photosynthesis Research, Vol IV", C. Sybesma, ed., Martinus Nijhoff/Dr.W. Junk Publishers, the Hague (1984).

ROLE OF THE CHLOROPLAST ENVELOPE IN THYLAKOID BIOGENESIS

J. Kenneth Hoober[§], Dawn B. Marks[§],
Jerome L. Gabriel[§] and Laurie G. Paavola[¶]

[§]Departments of Biochemistry and [¶]Anatomy and Cell
Biology, Temple University School of Medicine
Philadelphia, PA 19140

SUMMARY

Thylakoid biogenesis in Chlamydomonas reinhardtii y-1
occurred at a linear rate when degreened cells incubated at
38°C were exposed to light. Photosystem 2 (PS2) activity,
which was negligible in degreened cells, increased in parallel
with chlorophyll (Chl). Kinetic parameters suggest that these
PS2 units were larger and more efficient in light capture than
those made at 25°C. Membranes made during the initial minutes
of greening emanated from the chloroplast envelope. Kinetics
of accumulation of chlorophyll a/b-binding (Cab) proteins
implied that these proteins were degraded immediately upon
import into the plastid unless Chl was available. Molecular
modeling studies suggest that Chl is required to convert these
proteins from an extended, protease-sensitive conformation
into a compact, folded complex. A protease was partially
purified that may be involved in degradation of the
uncomplexed Cab proteins. These studies support the
hypothesis that assembly of thylakoid membranes in this
organism occurs in association with the chloroplast envelope.

INTRODUCTION

Greening of Chlamydomonas reinhardtii y-1 cells at 38°C
is a superb system to examine biogenesis of thylakoid
membranes. The kinetics of greening are sufficiently rapid
that biochemical and morphological changes can be measured
over a time span of several minutes. When degreened cells are
exposed to light, Chls a and b increase linearly, without a
lag, at a rate of about 4 μg total Chl/h and an a/b ratio of
2.0 (1,2). The Chls initially accumulate as the
geranylgeranyl esters, which are subsequently reduced to the
mature, phytylated species (2). Light energy absorbed by Chl
b is quantitatively transferred to Chl a, which implies that
these Chls are bound to Chl a/b-binding (Cab) proteins that
comprise the bulk of the light-harvesting complexes (LHCs) of
PS2 (2,3). The immediate onset of greening when cells are

exposed to light results, at least in part, from induction of
cab gene expression by the elevated temperature (4). During
an hour of preincubation in the dark at 38°C, the rate of
synthesis of Cab proteins apparently achieves levels that are
found in the light (5). However, accumulation of Cab proteins
is low unless Chl is present to stabilize the proteins.
Because Chl synthesis requires light in this organism, the
biogenic response is dependent upon light.

DEVELOPMENT OF FUNCTIONAL MEMBRANES AT 38°C

     To determine whether the initial membranes formed during
linear greening were photochemically functional, photoreduction of dichlorophenol indophenol (DCI), an assay of PS2,
by samples of broken cells was measured (3). In dark-grown
cells, which contain 1-2 μg Chl/10^7 cells, PS2 activity was
negligible (Fig. 1). Upon exposure to light, the activity
increased at a linear rate. When corrected for the residual
Chl in the degreened cells, PS2 activity increased in parallel
with Chl. Shown in Table 1 are the maximal activities at
saturating light fluence and the half-saturating light
fluences for samples from cells greened at 38°C as compared
with those greened at 25°C. Maximal DCI photoreduction on a
Chl basis by membranes made at 38°C was about half that at
25°C. Also, the half-saturation fluence was about 40% that of
membranes made at 25°C. These results suggested that PS2
units formed at 38°C contained more Chl per reaction center.
Perhaps larger light harvesting antennae contribute to the
more efficient capture of light energy (3).

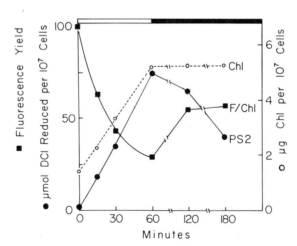

Figure 1. Development of PS2 activity during greening. Dark-
grown cells were incubated 1.5 h at 38°C in the dark and then
exposed to light. Broken cell samples were assayed for (●),
PS2 activity by DCI photoreduction; (■), fluorescence at 678
nm with excitation at 472 nm as a function of Chl (F/Chl); and
(o), total Chl. The bar at the top of the figure indicates
that cells were returned to the dark after 1 h of light.

Table 1. PS2 activity parameters as a function of light fluence and temperature of greening.

Temperature of greening	Saturation velocity	Half-saturation fluence
	nmol DCI/μg Chl·min	ergs/cm^2·s
25°C	30.3 ± 3.3	16.0 ± 0.2 x 10^4
38°C	16.6 ± 1.3	6.2 ± 2.2 x 10^4

Chls in samples from degreened cells were fluorescent, and the transfer of energy from Chl b to Chl a indicated that they were organized within LHCs (2). The level of fluorescence increased only 5-10% when the complexes were dissociated with detergent (3). Thus, nearly all the energy absorbed by the intact complexes was emitted as fluorescence, which suggested that these LHCs were not connected to reaction centers. Interestingly, during greening the total fluorescence on a cellular basis did not increase. When expressed on a Chl basis, the fluorescence declined as predicted by a simple dilution curve (marked F/Chl in Fig. 1). These results suggest that newly formed LHCs, in contrast to preexisting LHCs, were quantitatively connected with reaction centers and that absorbed light energy was dissipated through nonradiative processes. This characteristic may also be a factor in the enhanced efficiency of PS2 units formed at 38°C (Table 1).

When greening cells were returned to the dark, a loss of PS2 activity was accompanied by an increase in total fluorescence (Fig. 1). Possibly, light is required to maintain the integrity of the complexes. However, light alone is not sufficient to achieve connection of LHCs to reaction centers. When Chl synthesis in cells exposed to light was blocked with gabaculine, the F/Chl ratio did not change (3).

The greater apparent efficiency of PS2 complexes in membranes made at 38°C suggests that plants may assemble more productive photosystem units when chloroplasts develop at elevated temperatures. Consequently, if the enhanced energy conversion is translated into a higher rate of CO_2 fixation, this phenomenon may provide a strong attenuating factor for the anticipated elevation in atmospheric temperature associated with the "greenhouse effect."

SITE OF ASSEMBLY OF THYLAKOID MEMBRANES

Because Chlamydomonas cells show measurable biochemical changes associated with thylakoid development within minutes, the developing membranes containing these activities should generate detectable morphological changes within the chloroplast. The location of these initial membranes would, therefore, mark the site of assembly. We examined greening cells after 0, 5 and 15 min of light by electron microscopy. As shown in Fig. 2A, the chloroplast in dark grown cells incubated an hour at 38°C was nearly devoid of thylakoid membrane material, and the chloroplast envelope was

unremarkable in appearance. However, within 5 min of exposure
to light, an abundance of membrane material emanated from the
envelope (Fig. 2B). Small membranous vesicles in the stroma
also appeared, which possibly were derived from similar
structures that appeared to bud from the envelope. These
microscopic studies support the hypothesis (6) that thylakoid
biogenesis involves expansion and invagination of the inner
membrane of the envelope.

The detailed composition of newly formed membranes, such
as shown in Fig. 2A, is not known. Light induces a dramatic
increase in lipid synthesis to provide the matrix of the
expanding membrane at 25°C (7) and presumably does so also at
38°C. Because the membrane contained newly assembled PS2
units, these membranes also apparently contained reaction
center proteins synthesized in the chloroplast. Moreover, the
absence of fluorescence suggests that new LHCs were connected
to these reaction centers and that newly synthesized Cab
proteins were also integrated into these complexes.

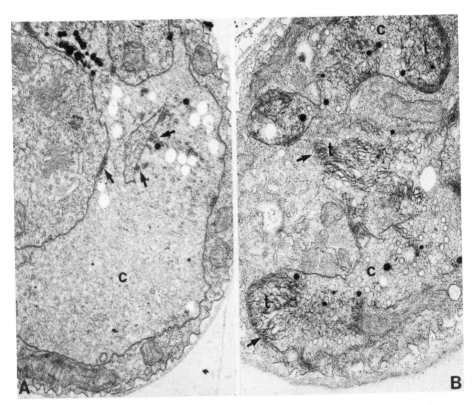

Figure 2. Structural changes that accompany chloroplast
development in cells exposed to light at 38°C. A, A degreened
cell at the end of 1.5 h incubation in the dark contains
occasional membrane-like material on the inside surface of the
envelope (arrows). B, After 5 min of exposure to light, an
abundant array of thylakoid membranes (t) extend into the
chloroplast stroma (c) from the chloroplast envelope (arrows).

A question that remains is whether the Cab proteins,
which are synthesized in the cytoplasm (8), become integrated
into the expanding membranes concomitant with transport into
the chloroplast or whether they enter the chloroplast stroma
and in a later step become integrated into the membrane (3,6)
The linear kinetics of accumulation of the Cab proteins in
membranes in this system (5) suggest that integration is
concomitant with transport. In our studies, a soluble,
stromal form of the Cab proteins could not be detected. The
kinetic studies further suggest that if association with Chl
and integration into the membrane does not occur during
import, the Cab proteins are rapidly degraded (3).

PURIFICATION OF A PROTEASE POSSIBLY INVOLVED IN Cab PROTEIN
DEGRADATION

Cells of C. reinhardtii y-1 did not accumulate Cab
proteins in the dark even under conditions in which Cab mRNA
is abundant (4,5). In contrast, cells of the pg-113 strain,
which is Chl b-deficient (9), accumulated nearly the same
amount of the Cab proteins in the dark at 38°C as in the light
(5). We reasoned that pg-113 cells may be deficient in a
protease that normally degrades the Cab proteins in the
absence of sufficient Chl. Sucrose-density gradient analysis
of membranes from y-1 cells revealed a membrane-bound protease
that was nearly absent in pg-113 cells (5). The protease is
active with leucyl p-nitroaniline, but with alanyl p-
nitroaniline has about 10% of the activity with the former
substrate. Membranes containing this protease had a density
somewhat lower than thylakoid membranes, a characteristic
expected for membranes of the chloroplast envelope (10). The
protease was purified over 100-fold from a membrane fraction
of yellow y-1 cells. Its activity was not affected by
inhibitors of serine proteases (phenylmethylsulfonyl fluoride,
3,4-dichloroisocoumarin, leupeptin or aminobenzamidine),
metalloproteases (1,10-phenanthroline, EDTA or leucine
hydroxamate) or aspartate proteases (pepstatin A). Inhibitors
of cysteine proteases, such as E-64, antipain, iodoacetate or
$HgCl_2$, did not inhibit, although p-chloromercuriphenyl
sulfonate inhibited 50% of the activity at 0.5 mM. The
protease was very sensitive to photodynamic inhibition in the
presence of light and 0.1-0.5 μM rose bengal. No inhibition
was observed under similar conditions with 5 μM methylene blue
or neutral red. Whether this protease is sensitive to
photodynamic inactivation by pigments in the developing
chloroplast remains an interesting question.

Addition of the partially purified protease to thylakoid
membranes from pg-113 cells resulted in selective degradation
of the Cab proteins (Fig. 3). Although the cellular location
of this protease has not been determined, it possibly plays a
role in regulating accumulation of the Cab proteins.

MOLECULAR MODELING OF FOLDING OF THE Cab PROTEINS

In an attempt to understand development of protease-
resistant forms of the Cab proteins, we examined by molecular
modeling the folding of the amino acid sequence deduced from
the cabII-1 gene cloned from C. reinhardtii (11). Secondary
structure was predicted for this sequence by published

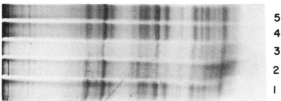

Figure 3. Digestion of Cab proteins (marked by arrows) in membranes from pg-113 cells by a membrane protease purified from y-1 cells. Lanes 1-3, incubation with protease for 0, 60 and 90 min at 25°C. Lanes 4 and 5, buffer control incubated for 0 and 90 min. Migration was from right to left. In lane 3, the major band remaining between the arrows was polypeptide 12, shown previously (5) to be protease resistant.

procedures (12,13). The resulting structure was the starting point for molecular dynamics simulations and energy minimization using the Biograf software, as described previously (14).

In the absence of Chl, the protein remained in an extended conformation except for the N- and C-termini, which folded into small globular domains (Fig. 4A). Two groups of Chl hexamers, each containing 3 Chl a/Chl b heterodimers, were constructed based on energy transfer data (15,16) and the stoichiometry of Chls within the Chl-protein complex (17,18). Introduction of the Chl hexamers initiated further folding of the protein to yield a stable tertiary structure in which the chlorin rings were on the putative luminal surface and the phytyl chains extended into the interior of the protein (Fig. 4B). The final model was consistent with published data on the orientation of the protein in the membrane, with three membrane-spanning regions (19). These findings suggest the hypothesis that folding of the protein requires sufficient Chl to provide "nucleation" structures. The model also suggests that the requirement for Chl b in stabilization of the protein is in the generation of heterodimers, the basic unit of these Chl structures. Without the Chls, the protein possibly remains in an open conformation that is susceptible to proteolysis.

The electron micrographs show that the envelope is the major membrane system in the plastid of degreened cells. The activities that convert protochlorophyllide to the chlorophyllides reside in membranes (20). Thus, because the rate of Chl synthesis is maximal when these cells are first exposed to light, the final steps in Chl synthesis apparently initially reside on envelope membranes. As membrane expansion occurs as the result of Chl and lipid synthesis, proteins synthesized in the plastid may interact with the expanding inner membrane of the envelope and generate functional reaction centers. We propose that, as the Cab proteins are transported through the chloroplast envelope from the cytoplasm, integration into the membrane, association with Chl and connection of the consequent light-harvesting complexes with reaction centers also occurs at the level of the envelope. Fission of membranes from the envelope would carry thylakoid activities deeper into the chloroplast.

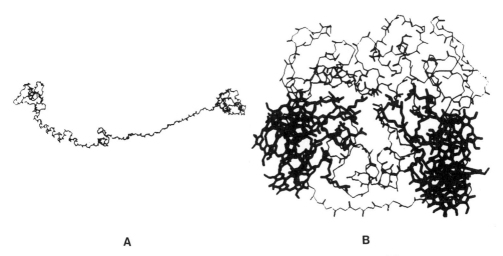

A   B

Figure 4. Models of the <u>Chlamydomonas reinhardtii</u> <u>cabII-1</u>
gene product after simulated folding <u>A</u>, in the absence of Chl
and <u>B</u>, after introduction of two hexamers of Chls <u>a</u> and <u>b</u> into
the original, unfolded structure.

ACKNOWLEDGEMENT

This research was supported by grants from the National
Science Foundation.

REFERENCES

1. J. K. Hoober and W. J. Stegeman, Kinetics and regulation
   of synthesis of the major polypeptides of thylakoid
   membranes in <u>Chlamydomonas reinhardtii y-1</u> at elevated
   temperatures, J. Cell Biol. 70:326 (1976).
2. M. A. Maloney, J. K. Hoober, and D. B. Marks, Kinetics of
   chlorophyll accumulation and formation of chlorophyll-
   protein complexes during greening of <u>Chlamydomonas
   reinhardtii y-1</u> at 38°C, Plant Physiol. 91:1100 (1989).
3. J. K. Hoober, C. O. Boyd, and L. G. Paavola, Origin of
   thylakoid membranes in <u>Chlamydomonas reinhardtii y-1</u> at
   38°C, Plant Physiol. 96 (1991), in press.
4. J. K. Hoober, D. B. Marks, B. J. Keller, and M. M.
   Margulies, Regulation of accumulation of the major
   thylakoid polypeptides in <u>Chlamydomonas reinhardtii y-1</u>
   at 25° and 38°C, J. Cell Biol. 95:552 (1982).
5. J. K. Hoober, M. A. Maloney, L. R. Asbury, and D. B.
   Marks, Accumulation of chlorophyll <u>a</u>/<u>b</u>-binding
   polypeptides in <u>Chlamydomonas reinhardtii y-1</u> in the
   light or dark at 38°C, Plant Physiol. 92:419 (1990).
6. D. R. Janero and R. Barrnett, Thylakoid biogenesis in
   <u>Chlamydomonas reinhardtii</u> 137+: cell cycle variations in
   the synthesis and assembly of polar glycerolipid, J. Cell
   Biol. 91:126 (1981).
7. J. K. Hoober, Sites of synthesis of chloroplast membrane
   polypeptides in <u>Chlamydomonas reinhardtii</u>, J. Biol. Chem.
   245:4327 (1970).

8. J. K. Hoober, The molecular basis of chloroplast development, in: The Biochemistry of Plants, Vol. 10, M. D. Hatch and N. K. Boardman, eds., Academic Press, Orlando (1987).

9. H. Michel, M. Tellenbach, and A. Boschetti, A chlorophyll b-less mutant of Chlamydomonas reinhardtii lacking in the light-harvesting chlorophyll a/b protein complex but not in its apoproteins, Biochim. Biophys. Acta 725:417 (1983).

10. L. Mendiola-Morganthaler, W. Eichenberger, and A. Boschetti, Isolation of chloroplast envelopes from Chlamydomonas. Lipid and polypeptide composition, Plant Sci. 41:97 (1985).

11. P. Imbault, C. Wittemer, U. Johanningmeier, J. D. Jacobs, and S. H. Howell, Structure of the Chlamydomonas reinhardtii cabII-1 gene encoding a chlorophyll-a/b-binding protein, Gene 73:397 (1988).

12. P. Y. Chou and G. D. Fasman, Prediction of protein conformation, Biochemistry 13:222 (1974).

13. J. Garnier, D. J. Osguthorpe, and B. Robson, Analysis of the accuracy and implications of simple methods for predicting the secondary structure of globular proteins, J. Mol. Biol. 120:97 (1978).

14. J. L. Gabriel and J. K. Hoober, Molecular modeling of phytochrome, J. Theoret. Biol. (1991), in press.

15. R. M. Van Metter, Excitation energy transfer in the light-harvesting chlorophyll a/b-protein, Biochim. Biophys. Acta 462:642 (1977).

16. W. T. Lotshaw, R. S. Alberte, and G. R. Fleming, Low-intensity subnanosecond fluorescence study of the light-harvesting chlorophyll a/b protein, Biochim. Biophys. Acta 682:75 (1982).

17. J. E. Mullet, The amino acid sequence of the polypeptide segment which regulates membrane adhesion (grana stacking) in chloroplasts, J. Biol. Chem. 258:9941 (1983).

18. P. J. G. Butler and W. Kühlbrandt, Determination of the aggregate size in detergent solution of the light-harvesting chlorophyll a/b-protein complex from chloroplast membranes, Proc. Natl. Acad. Sci. USA 85:3797 (1988).

19. W. Kühlbrandt and D. N. Wang, Three-dimensional structure of plant light-harvesting complex determined by electron crystallography, Nature (London) 350:130 (1991).

20. D. P. Bednarik and J. K. Hoober, Synthesis of chlorophyllide b from protochlorophyllide in Chlamydomonas reinhardtii y 1, Science 230:450 (1985).

THYLAKOID-BOUND PROTEASE ACTIVITY AS MONITORED BY AZO-DYE

RELEASE FROM AZOCOLL SUBSTRATE IS AFFECTED BY THYLAKOID

DEVELOPMENT AND CHLOROPHYLL SYNTHESIS INHIBITORS

Rania K. Anastassiou and Joan H. Argyroudi-
Akoyunoglou

Institute of Biology, NRC Demokritos
Athens, Greece

INTRODUCTION

Young etiolated bean leaves exposed to Continuous Light (CL),
rapidly accumulate Chla and Chlb, used by the Reaction Center (RC) and
the Light-Harvesting (LHC) apoproteins of the Photosynthetic units (PS)
for Chlorophyll-protein complex formation, leading to stabilization in
thylakoids. Plants exposed to Intermittent light, accumulate selectively Chla, which is bound on RC proteins. In this case, only RC core
compexes are stabilized in thylakoids, (1,2,3) although the LHC-II mRNA
is present and can be translated _in_ vitro (4) and in _vivo_ (5).
Furthermore, plants transferred to the dark after a brief preexposure
to CL, form small-sized PS units with concomitant degradation of LHC
complexes, which had accumulated during illumination (6,7,8).

These results have been attributed to a regulatory competition
mechanism operating between LHC and RC apoproteins for Chl, whenever
the amount of Chl accumulated is limited (ImL), or completely stopped
(D), but the RC apoproteins continue to be synthesized. According to
this mechanism, the stabilization of the LHC or RC polypeptides depends
on Chl-protein complex formation. Under limited Chl accumulation, the
RC apoproteins which have higher affinity for Chl, bind to the Chl
available, (ImL plants) or remove it from preexisting LHCs, (plants
transferred to darkness) and become stabilized, while the LHC apoproteins, in the absence of Chl binding, are unstable and digested (9).

This working hypothesis requires the existence of a proteolytic
enzyme which acts on unassembled apoproteins not rescued by Chl.
Earlier work has shown that such enzyme exists in chloroplasts (10,11).
Also it has been demonstrated that a proteolytic mechanism operates on
the small subunit of Rubisco when the synthesis of the large subunit is
inhibited and the active enzyme cannot be assembled (12).

In the present study we tried to assess the possible presence of
a thylakoid-bound protease in bean, follow its development during
greening and find the conditions of its activation.

MATERIALS AND METHODS

Six-day old etiolated <u>Phaseolus vulgaris</u> leaves were grown as in
(13). The leaves with one cotyledon removed, were either exposed to CL,
or to ImL (2 min white light alternating with 98 min dark in cycles,
Light-Dark Cycles, LDC), or to CL for 16 hours and then transferred to
the dark. Levulinic acid (LA) and 4,6-Dioxoheptanoic acid (succinylace-
tone, SA) were administered to leaves by immersing 3 times (1 min each
time every 4 hours), in the respective solutions, after their exposure
to 16 hours of CL. The leaves were dried on filter paper and kept in CL
for another 24 or 48 hours. Leaves handled in the same manner but
immersed in water were used as controls. Chlorophyll was extracted as
in (13) and determined according to (14). Intact chloroplasts were
isolated according to (15) and intact etioplasts according to (16).
Thylakoids were isolated either from osmotically broken intact plastids, or according to (7) with omission of BSA. Protein was assayed
according to (17). Protease activity was measured according to (18) by
incubatig 5 mg azocoll in 50 mM Tris-HCl pH 7.8, 1mM $CaCl_2$, with 1 mg
thylakoid protein for 3 hours at 37°C with vigorous agitation. The
reactions were stopped by transferring the tubes on ice and by removing
the remaining azocoll by filtration in a Pasteur pipette through
glasswool. Samples were then centrifuged for 8 min in an Eppendorf
centrifuge. Degradation of azocoll was measured as absorbance at 520 nm
of the supernatant. Blank values obtained in the absence of thylakoids
and in the presence of thylakoids, but in the absence of azocoll were
subtracted in each case. The rate of azocoll hydrolysis was linear as a
function of time. (Fig. 1). The assays were performed with thylakoid
protein in the concentration range where the rate of azocoll hydrolysis
was linear also as a function of thylakoid protein concentration.

RESULTS AND DISCUSSION

Thylakoids isolated either from intact or from non-intact plastids
both show proteolytic activity when assayed in a suspension with
insoluble azocoll substrate.

This proteolytic activity depends highly on the developmental
stage of the plant. The protease detected in the prothylakoids obtained
from intact etioplasts increases gradually and reaches a maximum 10
days after sowing of the plants. (Fig. 2a). The activity also increases
gradually in plants exposed to CL for up to 72 hours; thereafter it
reaches a plateau. (Fig. 2b).

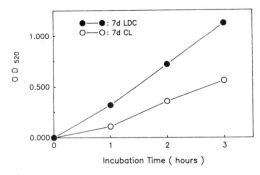

Fig. 1. Azo-dye release (OD 520) from azocoll substrate incubated with 1 mg of thylakoid protein at 37°C.

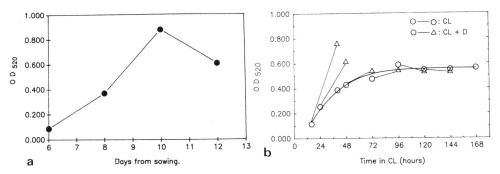

Fig. 2. Development of protease activity as monitored with azocoll
substrate in: (a) prothylakoids obtained from etiolated leaves
and (b) thylakoids obtained from six-day old etiolated
leaves exposed to Continuous Light (CL), or transferred
to the dark after preexposure to CL (CL+D).

Transfer of the plants in the dark induces increase in protease
activity (the 16h CL + 24h D sample shows 97% increase when compared
with the 16h CL + 24h CL sample) however, this occurs provided that,
the preexposure to CL is brief and only under conditions where new PS
units are formed in the dark. Transfer of the plants in the dark after
prolonged preexposure to CL is ineffective in enhancing the protease
activity. In plants exposed to ImL, the protease activity is gradually
increasing parallel to ImL exposure up to about 50-60 LDC, i.e. 3 to 4
days (Fig. 3).

In all cases, the protease activity was found to increase parallel
to thylakoid development and PS unit formation (2, 19). This may
suggest that this protease may be involved in the PS unit assembly by
degrading LHC apoproteins when new PS units are required,and Chl
synthesis is limited.

To test this possibility, we tried to imitate conditions under
which, LHCs are degraded or not assembled at all and tested whether the
protease activity is affected. Since it is known, that LHC stabilization requires Chl, we tried to see the effect of Chl synthesis inhibitors on the activity.

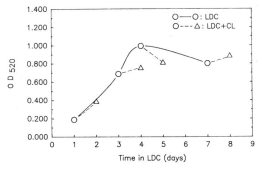

Fig. 3. Development of protease activity in thylakoids obtained from
six-day old etiolated leaves exposed to Intermittent Light
(LDC), or transferred to Continuous Light after their
exposure to LDC (LDC+CL).

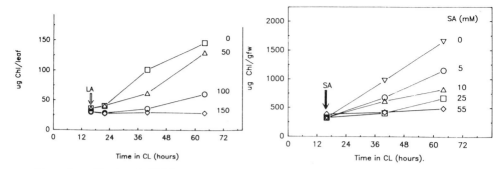

Fig. 4. Effect of different concentrations of Chlorophyll synthesis inhibitors, Levulinic acid (LA) and 4,6-Dioxoheptanoic acid (SA), on Chlorophyll synthesis.

Figure 4 shows the inhibition of Chlorophyll synthesis by LA or SA given to plants 16h after exposure to CL. As shown, the inhibition is complete at 150 mM LA or at 55 mM SA, after further exposure of the plants for 48h to light. Under these inhibitory conditions, it was shown (5) that the LHC apoprotein is degraded in CL. When these thyla- koids were assayed for proteolytic activity, we indeed found an 85% increase in the thylakoids isolated from inhibited plastids (based on the 16h + 48h CL untreated sample). (Fig. 5).

Since these proteins are degraded under our experimental condi- tions, it is difficult to conclude whether this increase is due to a real protease activation, or to the enrichment of the thylakoid in protease, since the activity we measure is based on 1 mg thylakoid protein and the major proportion of protein in thylakoids isolated from green plants, is accounted for by LHC apoproteins. We can not exclude, however the possibility that the protease is inactivated by the photo- dynamic action of pigments. A protease activity has recently been im- plicated in the degradation of LHC-II apoproteins in Chlamydomonas (20) and in barley (21) under conditions where Chl is not available for binding.

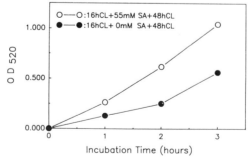

Fig. 5. Effect of SA on thylakoid protease activity. SA was admin- istered to bean leaves after 16h preexposure to CL. Thylakoids were isolated after another 48h in CL.

Table 1. Effect of different concentrations of ATP on protease activity in thylakoids isolated from plants exposed to a. 16h CL, b. 72h CL. (% remaining of control).

Sample	ATP (mM)	Activity (%)
a. 16h CL	0	100
	5	49
	10	20
	20	–
	50	0
b. 72h CL	0	100
	5	100
	10	100
	20	100
	50	–

Furthermore, we have found that the protease activity is affected by ATP added to the thylakoid suspension medium. Table 1 shows experimental results from thylakoids obtained from leaves exposed to CL for 16h or 72h. The 16h CL thylakoid protease is affected (inhibited) by increasing concentrations of ATP, while, the enzyme isolated from thylakoids exposed to 72h CL is completely unaffected.

This differential dependence on ATP of the enzyme obtained from plants of different developmental stages needs further investigation.

ACKNOWLEDGEMENT

The partial support by a Volkswagen grant to JAA and K.Kloppstech is gratefully acknowledged.

REFERENCES

1. J. H. Argyroudi-Akoyunoglou, and G. Akoyunoglou, The Chlorophyll-protein complexes of thylakoids in greening plastids of Phaseolus vulgaris, FEBS Lett. 104:78 (1979).
2. G. Akoyunoglou, Development of Photosystem II unit in plastids of bean leaves greened in periodic light, Arch. Biochem. Biophys. 183:571 (1977).
3. G. Akoyunoglou, Assembly of functional components in chloroplast photosynthetic membranes, in: "Photosynthesis" G. Akoyunoglou, ed., Balaban, Pa. (1981).
4. M. Viro and K. Kloppstech, Expression of genes of plastid membrane proteins in barley under intermittent light conditions, Planta. 154:18 (1982).
5. J. H. Argyroudi-Akoyunoglou et al., Regulation of photosynthetic unit assembly, this volume.
6. J. Bennett, Biosynthesis of the light-harvesting Chl a/b protein. Polypeptide turnover in darkness, Eur. J. Biochem. 118:61 (1981).

7. J. H. Argyroudi-Akoyunoglou, A. Akoyunoglou, K. Kalosakas, and G. Akoyunoglou, Reorganization of the PSII unit in developing thylakoids of higher plants after transfer to darkness: Changes in Chl b, LHC protein content and grana stacking, Plant Physiol. 70:1242 (1982).

8. A. Akoyunoglou, and G. Akoyunoglou, Reorganization of thylakoid components during chloroplast development in higher plants after transfer to darkness: Changes in PSI unit components and in cytochromes, Plant Physiol. 79:425 (1985).

9. G. Akoyunoglou, and J. H. Argyroudi-Akoyunoglou, Post-translational regulation of chloroplast differentiation, in: "Regulation of Chloroplast Differentiation", G. Akoyunoglou and H. Senger eds., Liss, N.Y. (1986).

10. B. Lagoutte, and J. Duranton, New enzymatic activity associated with thylakoid membranes, in: "Chloroplast development". G. Akoyunoglou and J. H. Argyroudi-Akoyunoglou, eds., Elsevier/North-Holland, N.Y. (1978).

11. L. Malek, L. Bogorad, A. Ayers, and A. Goldberg, Newly synthesized proteins are degraded by an ATP-stimulated proteolytic process in isolated pea chloroplasts, FEBS Lett. 166:253 (1984).

12. G. W. Schmidt, and M. L. Mishkind, Rapid degradation of unassembled ribulose 1,5-biphosphate carboxylase small subunits in chloroplasts, Proc. Natl. Acad. Sci. USA. 80:2632 (1983).

13. J. H. Argyroudi-Akoyunoglou, and G. Akoyunoglou, Photoinduced changes in the Chl a to Chl b ratio in young bean plants, Plant Physiol. 46:247 (1970).

14. G. McKinney, Absorption of light by chlorophyll solutions, J. Biol. Chem. 140:315 (1941).

15. J. Pfisterer, P. Lachmann, and K. Kloppstech, Transport of proteins into chloroplasts, Eur. J. Biochem. 126:143 (1982).

16. C. Schindler, and J. Soll, Protein transport in intact, purified pea etioplasts, Arch. Biochem. Biophys. 247:211 (1986).

17. O. H. Lowry, N. J. Rosenbrough, A. L. Farr, and R. J. Randal, Protein measurement with folin phenol reagent, J. Biol. Chem. 193:265 (1951).

18. R. Chavira, T. Burnett, and J. Hageman, Assaying proteinases with azocoll, Anal. Biochem. 136:446 (1984).

19. S. Tsakiris, and G. Akoyunoglou, Formation and growth of photosystem I and II units in old bean leaves, Plant Sci. Lett. 35:97 (1984).

20. J. K. Hoober, D. B. Marks, J. L. Gabriel, and L. G. Paavola, Role of the chloroplast envelope in thylakoid biogenesis, this volume.

21. T. Honda, H. Ito, Y. Tanaka, A. Tanaka, and H. Tsuji, Proteolytic digestion of apoproteins of light-harvesting chlorophyll a/b-protein complexes in barey leaves, this volume.

PROTEOLYTIC DIGESTION OF APOPROTEINS OF LIGHT-HARVESTING

CHLOROPHYLL a/b-PROTEIN COMPLEXES IN BARLEY LEAVES

T. Honda, H. Ito, Y. Tanaka, A. Tanaka and H. Tsuji

Department of Botany, Faculty of Science
Kyoto University
Kyoto 606, Japan

INTRODUCTION

Large amount of NADPH-protochlorophyllide oxidoreductase (NPR) accumulates in etioplasts (Hanser et al, 1984). When dark-grown tissues are illuminated, NPR rapidly decreases (Hanser et al., 1984) probably through degradation by some proteases in the etioplast membrane. When chlorophyll (Chl)-protein complexes (CPs) are formed in greening tissues under various conditions, for example, under light but a limited supply of Chl or during dark incubation after a brief illumination, proteases would also play an important role that they digest unassembled apoproteins. During greening under intermittent illumination where the rate of Chl synthesis is low, newly synthesized apoproteins of light-harvesting Chl a/b-protein complex of photosystem (PS) II (LHCII) fail to bind with Chl and are soon degraded (Bennett, 1981). On the other hand, when the rate of Chl synthesis is raised by feeding 5-aminolevulinic acid, a precursor of Chl synthesis, the apoproteins bind to Chl and accumulate (unpublished data). When greening tissues are placed in darkness, Chl is removed from LHCII and transferred to the apoproteins of P700-Chl a-protein complexes (CP1) and the Chl a-protein complex of PSII (CPa). Then LHCII apoproteins are decomposed in the thylakoid membrane (Tanaka et al., 1991). Apoproteins of CPa are also decomposed completely when their Chl is released in Ca-treated samples (Tanaka et al., 1991). These observations indicate that apoproteins of CPs are soon degraded when they have not yet bind with Chl or when they have been deprived of Chl by other apoproteins of CPs. Coordinate accumulation of Chl and apoproteins of different CPs would be achieved through rapid digestion of apoproteins which are not bound with Chl.

In contrast to abundant information about proteolysis of ribulose-1,5-bisphosphate carboxylase/oxigenase in vitro (Peoples and Dalling, 1978; Ragster and Chrispeels, 1981; Nettleton et al., 1985), there are very few reports on the breakdown of LHCII in vitro. Difficulty of these studies is that LHCII is an integrated membrane complex and would be digested by proteases embedded in the thylakoid membrane. In the present study, we succeeded in solubilizing proteases which digest exogenously added LHCII. Using this system, we showed that LHCII were actually stabilized by binding with Chl.

Regulation of Chloroplast Biogenesis
Edited by J.H. Argyroudi-Akoyunoglou, Plenum Press, New York, 1991

MATERIALS AND METHODS

Barley was grown in the dark for 5 days and illuminated with continuous light as described previously (Ohashi et al., 1989). Primary leaves harvested before or after illumination were homogenized in isolation buffer containing 50 mM HEPES-NaOH (pH 7.3), 0.35 M sorbitol, 5 mM dithiothreitol, 5 mM MgCl$_2$, 5 mM EDTA and 10 mM sodium pyrophosphate. The homogenates were centrifuged at 1,000xg for 10 min. Chloroplasts or etioplasts were purified by centrifugation through 35% Percoll.

Intact etioplasts or chloroplasts were washed with the isolation buffer. They were ruptured in 10 times dilution of the isolation buffer and centrifuged at 10,000xg for 10 min (washed membranes). The washed membranes were solubilized with 1% Triton X-100 at 4°C for 30 min and centrifuged at 10,000xg for 10 min (solubilized membranes). The supernatant was used for the assay of proteolytic activity. LHCII was purified from green barley leaves according to Burke et al. (1979). LHCII was treated with 80% acetone and centrifuged at 10,000xg for 10 min. The pellet was repeatedly washed with 80% acetone until all Chl was removed. Then the pellet (apoproteins) was solubilized with 1% SDS and dialyzed against 1% Triton X-100. The apoproteins and original LHCII were used as the substrates for proteolysis.

Etiolated seedlings were illuminated for 6 h. Primary leaves were cut and kept stand in vials containing water for 24 h in the dark at 28°C. After that, thylakoid membranes were isolated and their proteins were electrophoresed on a 14% polyacrylamide gel (Laemmli, 1970). After staining with Coomassie Blue, the gel were scanned. The 43 kDa protein of CPa was identified as described previously (Tanaka et al., 1991). To examine the degradation of LHCII during autolysis, washed membranes from 4-h illuminated or fully greened leaves were suspended in 50 mM Tris-MES (pH 7.0) and incubated at 35°C in the dark for 24 h. When exogenous LHCII and its apoproteins were used as substrates, they were added to the solubilized or non-solubilized membranes and incubated for 4 h at 35°C in the dark. After incubation, the samples were electrophoresed on a 20% polyacrylamide gel (Laemmli, 1970). The digestion products were identified by western blotting using antiserum to apoproteins (Tanaka et al., 1991).

RESULTS

Small amounts of apoproteins of CP1, LHCII and the 43 kDa protein of CPa were found in thylakoid membranes from etiolated barley leaves after 4 h illumination (Fig. 1). When these leaves were incubated in the dark for 24 h, CP1 apoproteins and the 43 kDa protein increased with a concomitant loss of LHCII apoproteins (Fig. 1A, B). We previously showed that LHCII in the early phase of greening was labile: it readily lost its Chl and the apoproteins were soon degraded in vivo (1991). Then we examined whether LHCII was digested in vitro incubation of thylakoid membranes. Digestion products were detected after 4 h of dark incubation of thylakoid membranes from 4-h illuminated leaves (Fig. 2). However, no such products were detected with thylakoid membranes from fully greened leaves.

It is unlikely that LHCII, an integral membrane protein, is digested by soluble proteases in the stroma. To determine whether exogenously added LHCII is digested by membrane-bound proteases, we used etioplast membranes instead of thylakoid membranes from 4-h illuminated leaves as an enzyme preparation, because there is no LHCII in etioplast membranes. When etioplast membranes were incubated with LHCII, no digestion products were detected by immunoblotting (Fig. 3). Then etioplast membranes were solubilized with Triton X-100 and incubated with LHCII. Digestion products

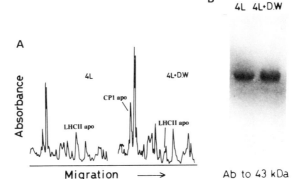

Fig. 1 Changes in thylakoid membrane proteins during dark incubation after
4 h of illumination of etiolated barley leaves. Thylakoid proteins were
electrophoresed on a 14% polyacrylamide gel. A. The gel was stained with
Coomassie Blue and scanned. B. Thylakoid proteins were blotted onto
nitrocellulose filter and probed with antiserum against the 43 kDa protein
of the CPa. 4L, 4-h light; 4L+D.W, 4-h light followed by 24-h dark
incubation with water.

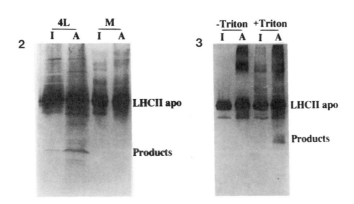

Fig. 2 (left) Autolysis of LHCII in thylakoid membranes. Thylakoid
membranes from 4-h illuminated (4L) or mature (M) leaves were incubated at
35°C for 24 h. Before (I) and after (A) incubation, samples were subjected
to electrophoresis on a 20% polyacrylamide gel and blotted onto a
nitrocellulose filter. Antiserum against LHCII apoproteins was used as an
immunological probe.

Fig. 3 (right) Digestion of exogenous LHCII by etioplast membranes or
those solubilized with Triton X-100. LHCII was incubated with etioplast
membranes (-Triton) or Triton X-100 solubilized etioplast membranes
(+Triton) at 35°C for 4 h. Before (I) or after (A) incubation, samples
were electrophoresed and subjected to western blotting as in Fig. 2.

derived from exogenous LHCII appeared after 4 h of incubation (Fig. 3).
Thus, proteases in the thylakoid membrane could not digest exogenous LHCII,
but could only after solubilized with Triton X-100. In experiments of
autolysis of thylakoid membranes from 4-h illuminated leaves (Fig. 2),
however, digestion products of LHCII were found without solubilization of
the membranes. Both endogenous LHCII and proteases were embedded in the
thylakoid membrane and there may be no physical barriers between them.

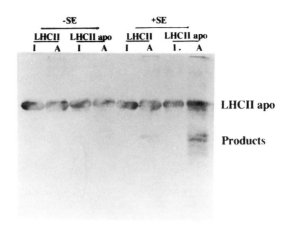

Fig. 4 Digestion of LHCII or LHCII apoproteins by solubilized etioplast membranes. LHCII or LHCII apoproteins were incubated with (+SE) or without (-SE) solubilized etioplast membranes at 35°C for 4 h. Before (I) or after (A) incubation, samples were electrophoresed and subjected to western blotting as in Fig. 2.

In order to see whether LHCII apoproteins are actually stabilized by binding with Chl, LHCII or its apoproteins without Chl were incubated with solubilized etioplast membranes and the digestion products were identified immunologically (Fig. 4). Larger number and amount of digestion products were found in incubation with the apoproteins than with LHCII itself.

DISCUSSION

When etioplast membranes from leaves of dark-grown plants and thylakoid membranes from greening leaves were incubated, NPR and LHCII apoproteins decreased, respectively, but the release of amino acid was not observed (unpublished data). It suggests that NPR and LHCII underwent endoproteolytic degradation in the membrane. However, when thylakoid membranes from 4-h illuminated leaves were incubated, proteolytic products of LHCII were detected by western blotting. Digestion products were also detected when etioplast membranes solubilized with Triton X-100 were incubated with LHCII added as the substrate. However, unsolubilized etioplast membranes did not digest exogenously added LHCII. By treatment with Triton X-100, proteases would be released from the membranes and allowed to access the substrate in the medium. Together, these results suggest that LHCII was digested by membrane-bound endoproteases in autolysis. Digestion products may be released from the membrane into the stroma and then digested to amino acids by exopeptidases. The presence of aminopeptidases in the stroma has been reported (Liu and Jagendorf, 1986; Thayer et al. 1988).

Results of in vivo experiments suggest that LHCII apoproteins without Chl are labile. When LHCII apoproteins are newly synthesized and have not yet bind with Chl (Bennett, 1981) or when they lose their Chl (Tanaka et al., 1991), they are degraded. As for the number of digestion products, when LHCII was used as the substrate, a single band of digestion product was detected. However, when the apoproteins were used, the number of digestion products increased. An electron crystallographic study shows that polypeptides of LHCII apoproteins are surrounded by Chl (Kuhlbrandt and Wang, 1991), which would protect the polypeptides from proteases.

Without Chl, more sites which are susceptible for proteases in the apoproteins would be exposed to be digested by proteases, resulting in production of various lengths of fragments.

The accumulation of Chl and apoproteins must be coregulated strictly, because free Chl would generate $^1O_2$ and the accumulation of free apoproteins may lead to the formation of malformed photosystems. Free apoproteins have to be degraded for the coordinate accumulation of Chl and apoproteins (Eichacker et al., 1990; Tanaka et al., 1991).

ACKNOWLEDGEMENTS

This research was supported by a Grant-in-Aid from the Ministry of Education, Science and Culture of Japan.

REFERENCES

Bennett, J., 1981, Biosynthesis of the light-harvesting chlorophyll a/b protein complex. Polypeptide turnover in darkness, Eur. J. Biochem., 118:61.

Burke, J. J., Ditto, C. L., and Arnzen, C. J., 1979, Involvement of the light-harvesting complex in cation regulation of excitation energy distribution in chloroplasts, Arch. Biochem. Biophys., 187:252.

Eichacker, L. A., Soll, J., Lauterbach, P., Rudiger, W., Klein, K. K., and Mullet, J. E., 1990, In vitro synthesis of chlorophyll a in the dark triggers accumulation of chlorophyll a apoproteins in barley etioplasts, J. Biol. Chem., 265:13566.

Hanser, I., Dehesh, K., and Apel, K., 1984, The proteolytic degradation in vitro of the NADPH-protochlorophyllide oxidoreductase of barley (Hordeum vulgare L.), Arch. Biochem. Biophys., 228:577.

Kühlbrandt, W., and Wang, D. N., 1991, Three-dimensional structure of plant light-harvesting complex determined by electron crystallography, Nature, 350:130.

Laemmli, O. K., 1970, Cleavage of structural proteins during the assembly of the head of bacteriophage T4, Nature, 227:689.

Liu, X., and Jagendorf, A. T., 1986, Neutral peptidases in the stroma of pea chloroplasts, Plant Physiol., 81:603.

Nettleton, A. M., Bhalla, P. L., and Dalling, M. J., 1985, Characterization of peptidase hydrolase activity associated with thylakoids of the primary leaves of wheat, J. Plant Physiol., 119:35.

Ohashi, K., Tanaka, A., and Tsuji, H., 1989, Formation of the photosynthetic electron transport system during the early phase of greening in barley leaves, Plant Physiol., 91:409.

Peoples, M. B., and Dalling, M. J., 1978, Degradation of ribulose-1,5-bisphosphate carboxylase by proteolytic enzymes from crude extracts of wheat leaves, Planta, 138:153.

Ragster, L., and Chrispeels, 1981, Autodigestion in crude extracts of soybean leaves and isolated chloroplasts as a measure of proteolytic activity, Plant Physiol., 67:104.

Tanaka, A., and Tsuji, H., 1985, Appearance of chlorophyll-protein complexes greening barley leaves, Plant Cell Physiol., 26:893.

Tanaka, A., Yamamoto, Y., and Tsuji, H., 1991, Formation of chlorophyll-protein complexes during greening. 2. Redistribution of chlorophyll among apoproteins, Plant Cell Physiol., 32:195.

Thayer, S. S., Choe, H. T., Rausser, S., and Huffaker, R. C., 1988, Characterization and subcellular localization of aminopeptidases in senescing barley leaves, Plant Physiol., 87:894.

341

RECONSTITUTION OF LHCP-PIGMENT COMPLEXES WITH MUTANT LHCP AND
CHLOROPHYLL ANALOGS

Harald Paulsen, Stephan Hobe, and Christoph Eisen

Botanisches Institut III der Universität,
Menzinger Str. 67, W-8000 München 18, Germany

INTRODUCTION

The major PSII-associated light-harvesting complex (LHCII)
is essential for the efficiency of photosynthesis in higher
plants. Its light-harvesting function is dependent upon a specific arrangement of the pigments in the complex, facilitating
rapid energy transfer between them. It is unclear how this specific arrangement is brought about although it seems likely that
a major function of the apoprotein of LHCII (LHCP) is to ensure
proper orientation of the non-covalently bound pigments. We are
interested in defining the specific interactions between pigments and LHCP that are essential for the light-harvesting function of LHCII.

Our experimental approach is to reconstitute LHCP-pigment
complexes in vitro in the presence of a detergent, following a
procedure designed by Plumley and Schmidt (1987). We isolate
LHCP, or its precursor (pLHCP), from an Escherichia coli strain
that overexpresses the protein from a pea cab gene. The reconstituted complexes are structurally very similar to LHCII monomers isolated from pea thylakoid membranes, judging from their
stability, spectroscopic properties, and from the very specific
pigment requirements for their formation (Paulsen et al., 1990).

This experimental system enables us to introduce structural
alterations, both into the protein by mutating the cab gene used
for LHCP overexpression, and into the pigments by using pigment
derivatives in the reconstitution experiments. These structural
alterations can then be correlated with changes in functionally
important parameters of the pigment-protein complexes, such as
stability and spectroscopic characteristics.

Here we demonstrate that the deletion of some segments of
LHCP does not impair its ability to form stable complexes with
apparently the complete set of pigments present in LHCII. However, in other, more critical domains of the protein, the deletion of a few amino acids abolishes detectable reconstitution
altogether. Introducing structural alterations into the chlorophyll molecules significantly changes the stability and

Regulation of Chloroplast Biogenesis
Edited by J.H. Argyroudi-Akoyunoglou, Plenum Press, New York, 1991

343

stoichiometry of the resulting reconstituted complexes in most cases.

RESULTS AND DISCUSSION

A striking structural feature of LHCP is an intramolecular homology region comprising the first (N-proximal) and third hydrophobic domains (Fig. 1C). This homology is shared by all other chlorophyll a/b-binding proteins and therefore, is possibly involved in pigment binding by these proteins. It has been speculated that this homology originates from the duplication of an ancient gene, possibly for a pigment-binding antenna protein (Green et al. 1991).

In order to test whether the homologous regions in LHCP indeed provide pigment binding sites, we constructed pLHCP deletion mutants in which either the first or the third hydrophobic domains were missing (amino acids 60-127 and 154-196,

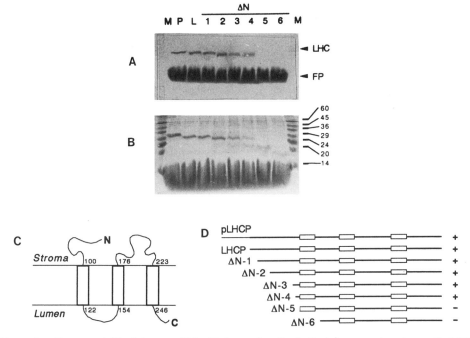

Fig. 1. Reconstitution with N-terminal deletion mutants of LHCP. Mutant LHCP carrying various N-terminal deletions (ΔN-1 to ΔN-6) was reconstituted with pigments and electrophoresed on a partially denaturing (0.1 % LDS, 0°C), "green" gel (A). The same gel was stained with Coomassie (B) LHC and FP: positions of LHCII and free pigment, respectively. M: molecular weight marker; molecular weights (in kDa) and approximate positions are indicated in the right margin. P: pLHCP. L: LHCP. C and D are schematic representations of LHCP and its deletion mutants, respectively. The boxes indicate hydrophobic domains of the protein.

respectively). Although these mutants still contain one or two of the conserved hydrophobic domains, neither of them forms stable complexes in the reconstitution (not shown). On the basis of this result we can exclude that the conserved homology domains in LHCP stably bind pigments independently from the rest of the protein molecule. We cannot exclude the formation of complexes that are too unstable to survive the isolation procedure on a partially denaturing gel.

We then addressed the question of whether modular pigment-binding segments of LHCP could be defined by introducing increasing deletions from either the N or C terminus and measuring which pigments still bind to the remaining protein. Fig. 1 shows that N-terminal deletion mutants still reconstitute until the deletion end point reaches the N-proximal hydrophobic region. On the C terminus, reconstitution is lost when about two thirds of the C-terminal hydrophilic sequence are deleted (Fig. 2).

In order to characterize the sequences whose deletion causes the mutant protein to fail in the reconstitution assay, we sequenced the deletion end points of the last mutant that still reconstitutes and the first mutant that does not reconstitute any more in both N and C-terminal deletion series.

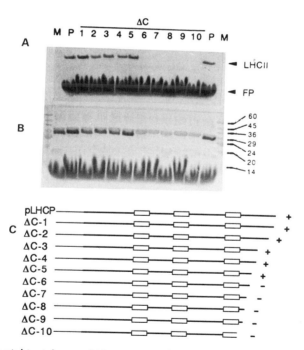

Fig. 2. Reconstitution with C-terminal deletion mutants of LHCP. Various C-terminal deletion mutants of pLHCP (ΔC-1 to ΔC-10) were reconstituted with pigments and electrophoresed on a partially denaturing, "green" gel (A). B: same gel after staining with Coomassie. C: Sketch of C-terminal deletions. Abbreviations as in Fig. 1.

As shown in Fig. 3, the "borderline" mutants differ by only 5 amino acids in length in either deletion series. Due to clone construction, some of the mutants differ from the wild-type LHCP by one or three amino acids adjacent to the deletion end point. It should be noted, however, that these exchanges are presumably conservative concerning their influence on protein structure. The critical sequences of 5 amino acids do not contain His, Gln, or Asn, amino acids that have been suggested to be involved in complexing chlorophylls in LHCII (Peter and Thornber, 1988). More detailed analysis is needed in order to find out whether it is a single amino acid within these sequences that is essential for stable complex formation.

The reconstitution experiments with N and C-terminal deletion mutants of LHCP (Figs 1 and 2) suggest that, until the deletions extend to the critical sequences, the mutants still form apparently normal pigment-protein complexes. This has been confirmed by pigment analysis of the reconstituted complexes and by CD spectroscopy (not shown). In both cases the data obtained from the shortest still reconstituting mutant were very similar to those obtained from full-length pLHCP. The deletions appear to have an all-or-nothing effect on complex formation, although on the basis of our data, we cannot exclude the loss of one or two pigments before deletions extend to the critical region. The only more gradual effect we observed was a lowered thermal stability of complexes reconstituted from ΔN-4. These complexes would disintegrate during electrophoresis unless the latter was run at 0°C. This strict dependence of complex stability on the presence of a single or limited structural component, like the critical 5 amino-acid sequences in this deletion analysis, is indicative of a highly cooperative binding of components into the complex. The notion of synergetic interactions in LHCII is

N-terminal Deletions

....DTAGLSADPETFSKNRELE|VIHSRWAMLGAL-....
    ΔN-4 ⊕   S----------------...
    ΔN-5 ⊖   M------------...

C-terminal Deletions

....QAIVTGKGPL|ENLADHLADPVNNNAWSYATNFVSRK....
....------------------------LIS ⊕ ΔC-5
....------------------------ ⊖ ΔC-6

Fig. 3. Deletion end points of borderline mutants in both the N and C-terminal deletion series. In the N-terminal deletion mutants ΔN-4 (+) and ΔN-5 (-) (see Fig. 1) the first 6 amino acids were sequenced. In the C-terminal deletion mutants ΔC-5 (+) and ΔC-6 (-) the deletion end points were determined by DNA sequencing. The corresponding sequences of the wild-type LHCP are shown. Boxes indicate hydrophobic domains.

corroborated by the earlier finding that no reconstitution takes place in the absence of, e.g., lutein of which no more than one or two molecules are bound per protein monomer (Plumley and Schmidt, 1987; Paulsen et al., 1990).

Interestingly, mutant LHCP combining the deletions of ΔN-4 and ΔC-5 still forms a stable complex with pigments although the size of the protein is only 16 kDa (not shown). In this complex, pigments constitute about half of the molecular mass. This finding makes it likely that pigment-pigment interactions contribute to the stability of LHCP-pigment complexes.

In a preliminary analysis of the influence of the chlorophyll structure on molecular interactions in LHCII we reconstituted LHCP with a number of chlorophyll analogs (Fig. 4). Of the chlorophyll derivatives tested, only GG-chlorophyll b containing geranylgeraniol instead of phytol formed complexes with normal stability and pigment stoichiometry. Several of the other derivatives could replace only some chlorophyll molecules in the complexes which then typically displayed reduced stabilities and altered stoichiometries. Some analogs would not be detectable in complexes when competing with the authentic chlorophyll during

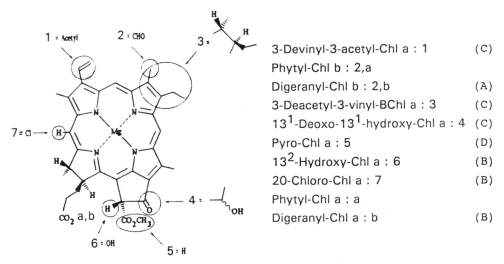

3-Devinyl-3-acetyl-Chl a : 1   (C)

Phytyl-Chl b : 2,a

Digeranyl-Chl b : 2,b   (A)

3-Deacetyl-3-vinyl-BChl a : 3   (C)

13^1-Deoxo-13^1-hydroxy-Chl a : 4   (C)

Pyro-Chl a : 5   (D)

13^2-Hydroxy-Chl a : 6   (B)

20-Chloro-Chl a : 7   (B)

Phytyl-Chl a : a

Digeranyl-Chl a : b   (B)

Fig. 4. Reconstitution with modified chlorophylls.   The structural variation of the chlorophyll molecule in the analogs is shown. The analogs can be ordered in the following categories concerning their behavior in reconstitution experiments: (A) can fully replace the analogous chlorophyll. (B) Can replace some analogous chlorophyll molecules in the complexes, usually causes reduced complex stability and altered pigment stoichiometry. (C) Competes with analogous chlorophyll in the reconstitution reaction but cannot be detected in stable complexes. (D) Is not incorporated in complexes and does not compete with analogous chlorophyll. Chl: chlorophyll; BChl: bacteriochlorophyll.

reconstitution. These results are similar to those obtained by Parkes-Loach et al. (1990) who reconstituted light-harvesting complexes of Rhodospirillum rubrum with bacteriochlorophyll analogs. In either system the structural variability of chlorophylls bound into the complexes is limited.

CONCLUSIONS

- Most of the N and part of the C-terminal hydrophilic regions of LHCP are not involved in pigment binding or stabilization of pigmented complexes. The hydrophobic domains as well as the hydrophilic sequence immediately following the third (carboxy-proximal) membrane-spanning region are essential for complex formation (or stabilization) with pigments.

- Critical regions where deletions from either end of the protein start to abolish reconstitution can be narrowed down to segments of five amino acids.

- LHCP deletion mutants either form pigmented complexes with apparently normal stability and pigment stoichiometry, or fail to reconstitute. There is no indication for a segment of LHCP forming stable complexes with a subset of the pigments normally present in LHCII. This all-or-nothing behavior indicates a highly synergetic stabilization of LHCP-pigment complexes.

- Most chlorophyll derivatives tested cannot replace chlorophyll in forming stable complexes with a correct stoichiometry of pigments.

REFERENCES

Green, B., Pichersky, E., and Kloppstech, K., 1991, Chlorophyll a/b-binding proteins: an extended family, Trends Biochem. Sci., 16:181.

Parkes-Loach, P. S., Michalski, T. J., Bass, W. J., Smith, U., and Loach, P. A, 1990, Probing the bacteriochlorophyll binding site by reconstitution of the light-harvesting complex of Rhodospirillum rubrum with bacteriochlorophyll a analogues, Biochem., 29:2951.

Paulsen, H., Rümler, U., and Rüdiger, W., 1990, Reconstitution of pigment-containing complexes from light-harvesting chlorophyll a/b-binding protein overexpressed in Escherichia coli, Planta, 181:204.

Peter, G. F. and Thornber, J. P., 1988, The antenna components of photosystem II with emphasis on the major pigment-protein, LHC IIb. in: "Photosynthetic Light-Harvesting Systems", H. Scheer and S. Schneider eds., Walther de Gruyter & Co., Berlin, New York.

Plumley, F. G. and Schmidt, G. W., 1987, Reconstitution of chlorophyll a/b light-harvesting complexes: Xanthophyll-dependent assembly and energy transfer, Proc.Natl.Acad.Sci. USA, 84:146.

BIOGENESIS OF CHLOROPLASTS AND PIGMENT-PROTEIN COMPLEXES IN

BARLEY SEEDLINGS WITH BLOCKED BIOSYNTHESIS OF CAROTENOIDS

Navassard V. Karapetyan, Marina G. Rakhimberdieva,
Yulia V. Bolychevtseva, Andrei A. Moskalenko*,
Nina Yu. Kuznetsova*, Victor I. Popov**

Biochemistry Institute, Moscow 117071; Institute
of Soil Science and Photosynthesis*, Institute
of Biophysics**, Pushchino, 142292; USSR

INTRODUCTION

The photosynthetic apparatus in higher plant chloroplasts contains three main supramolecular pigment-protein complexes: photosystem 1 complex (CP1), photosystem 2 complex (CPa) and the light-harvesting Chl a/b-protein (LHCP) (Anderson, 1986). Though the composition, structural organization and location of the complexes in membranes are elucidated, the mechanisms regulating biogenesis and assembly of these complexes from polypeptides and chlorophylls (Chl) are still unclear.

An integral part of all types of pigment-protein complexes of green plants and photosynthetic bacteria is carotenoids (Lichtenthaler et al., 1981; Cogdell and Frank, 1987). In the LHCP, carotenoids function as auxillary pigments and protectors of excited antenna Chl from photodestruction; in the reaction centers carotenoids protect the reaction center Chl (Cogdell and Frank, 1987). The absence of carotenoids has no effect on the biogenesis and activity of the reaction center of photosystem 1 (PS1) and bacteria (Siefermann-Harms, 1985). Since a pure reaction center of PS1 has not been isolated, its location within the center is not known. In PS1 carotenoids may play a structural role and are necessary for binding the core complex and the peripheral antenna (Humbeck et al., 1989; Damm et al., 1990). The structural role of carotenoids in the biogenesis of LHCP was suggested by Hladik et al. (1982); the carotenoids are necessary for the correct assembly of the LHCP apoproteins (Dahlin, 1988).

Carotenoids were found in pure photosystem 2 (PS2) reaction centers but its location and role are unclear. Using the inhibition of the carotenoid biosynthesis by herbicides we have shown that barley seedlings grown with norflurazon have no PS2 activity and suggested that carotenoids are involved in the formation of active PS2 in higher plants (Karapetyan et al., 1989). The same suggestion was reported Humbeck et al. (1989) for a _Scenedesmus_ mutant which showed no PS2 in the absence of carotenoids. This paper presents data on the pigment-protein complexes which form in carotenoidless barley leaves and discusses the biogenesis of chloroplasts in carotenoid-deficient leaves.

MATERIALS AND METHODS

Barley seeds were germinated and grown in the presence of 100 μM norflurazon at 20 C in the dark. 6-days etiolated seedlings were grown for 24 h under flash illumination: 2.5 ms flash-12 min dark, intensity of flash was about 120 J m^{-2}. Etiochloroplasts were isolated from the leaves as described in (Bolychevtseva et al., 1987), but using the centrifugation at 3500 g. The polypeptide composition was determined with PAGE according to a modified Laemmli technique (Bolychevtseva et al., 1988). The pigment-protein complexes of flash-grown seedlings were identified by PAGE with some modifications; samples were preliminary treated with a mixture of dodecyl-beta-D-maltoside and Deriphat-160 (Moskalenko, 1990). Zones of pigment-protein complexes were then cut off and gel discs were used for spectral analysis. PS1 reaction centers were detected as light-induced absorption changes at 696 nm, PS2 reaction centers were revealed as variable fluorescence at 77K. The absorption spectra of discs were recorded on spectrophotometer UV-160, while fluorescence spectra on a spectrofluorometer MPF-4. The conditions of leaf fixation and preparation of samples for electron microscopy are described in (Popov et al., 1981); measurements of partitions were performed as in (Axelsson et al., 1982).

RESULTS AND DISCUSSION

Norflurazon inhibits by 97% the biosynthesis of carotenoids in barley seedlings but photodestruction of chloroplasts took place even if the seedlings are grown at a very low light intensity: Chl bleaches at first, followed by the destruction of chloroplast lipids and 70S ribosomes (Bolychevtseva et al., 1987, 1988). The absence of PS2 in the barley seedlings grown with norflurazon at low light intensity was not a result of norflurazon-induced inhibition of PS2 polypeptides but it was caused by the destruction of the PS2 pigment-protein complex containing no carotenoids.

To prevent the destruction of the photosynthetic apparatus during the growth of norflurazon-treated seedlings, they were flash illuminated. Although the total light dose for 24 h of flash illumination was similar to that got by seedlings for 5 days at a low light intensity, no photodestruction of chloroplasts was observrd in flash-grown seedlings. The seedlings grown with norflurazon contained even more Chl a and Chl b than control seedlings, but the content of carotenoids in the treated seedlings was less than 3% of control (Table 1). The flash-grown seedlings contain no PS2 reaction centers according to the absence of variable fluorescence at 77K, but they have active PS1. Since these seedlings fail to form LHCP (Akoyun-oglou et al., 1966; Siefermann-Harms and Ninnemann, 1981) and no CPa (see later) the whole Chl is bound mainly to CP1. However, the norflurazon-treated seedlings are enriched with P700, since the Chl:P700 ratio is 60 (CP1 core complex) as com-

pared with 150 for control seedlings, which besides CP1 contain
also CPa. The same polypeptide profile of control and treated
seedlings indicates that norflurazon does not inhibit the biosynthesis of PS2 polypeptides (Karapetyan et al., 1991) As the
polypeptides and Chl are present in the membrane showing no PS2
activity, we suggested that the assembly of CPa required some
regulators or structural components.

Table 1. Characteristics of Pigments, Reaction Centers and
Pigment-Protein Complexes of Etiochloroplasts from
Barley Seedlings Grown with and without Norflurazon
at Flash Illumination (2.5 ms Flash-12 min Dark)

	Control	Norflurazon treated
Chlorophyll a	51.8	59.3
Chlorophyll b	4.1	5.3
Carotenoids	103.3	3.0
F_v/F_o at 77K	0.5	-
Chlorophyll/P700	150	60
Pigment-protein complexes	CP1, CPa	CP1
Absorption maximum, CP1	670 nm	673 nm
Fluorescence maximum, CP1	684 nm	675 nm
Absorptiom maximum, CPa	670 nm	-
Fluorescence maximum, CPa	681 nm	-
Partition length, $\mu m/\mu m^{-2}$	0,18	0.06

The content of chlorophylls and carotenoids are in $\mu g/g$ w.w.
F_v/F_o-variable fluorescence.

Carotenoids may be an essential structural component in
the assembly of CPa. The norflurazon-induced absence of carotenoids (and possible modification of lipids) may influence the
biogenesis of chloroplasts. Etiochloroplasts in flashed seedlings show margin stacking and the length of partition in
flashed carotenoidless etiochloroplasts is about three times
less than that in etiochloroplasts from control seedlings
(Table 1). It is noteworthy that the similar results were
obtained for etioplasts from seedlings grown with and without
norflurazon in the dark, although the reason of the partition
in plastids of dark-grown leaves is unclear (Axelsson et al.,
1982). Probably norflurazon-induced changes in prothylakoids
may disturb the assembly of pigment-protein complexes.

Since flashed seedlings grown with norflurazon contain PS2
polypeptides but show no PS2 activity, it was important to
reveal the CPa in these seedlings. But it is rather difficult
to determine the composition of the pigment-protein complex of
plastids isolated from these seedlings since the complexes even
in control are modified due to a change in lipid/protein ratio.
This may prevent the total solubilization of the membrane.
Besides carotenoidless complexes are very sensitive to light
and to detergents. Dodecyl-beta-D-maltoside was used to stabilize the PS2 reaction center during PAGE and to decrease the
amount of free Chl, and Deriphat-160 was added for better sepa-
aration of zones (Moskalenko, 1990).

PAGE showed three bands (I-III) for control seedlings and only two bands (I and III) for treated ones (Fig. 1). The red absorption maximum for all the complexes from control seedlings is at 670 nm (Table 1), but they have different absorption spectra in the blue region and different fluorescence spectra. Complex I contains P700 and has a fluorescence maximum at 684 nm. The protein with the molecular mass 70 kDa (PS1 apoprotein) found in band I proves the presence of CP1. No proteins are found in band III, that also shows a short-wavelength fluorescence maximum at 674 nm. This indicates on free Chl in this band. Thus CPa from control seedlings has to be present in band II which corresponds to position of pure PS2 reaction center (Fig. 1,A).

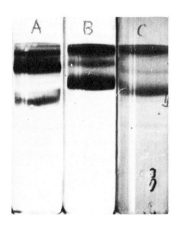

Fig. 1.   Electrophoretic patterns of PS2 reaction centers from spinach (A); and pigment-protein complexes isolated from flashed barley seedlings grown without (B) or with norflurazon (C). I-CP1, II-CPa, III-free Chl.

PAGE of the membranes from norflurazon-treated seedlings gave two bands (I and III). Band II appears only on a photo, because of very low intensity (Fig. 1,C). This is in accordance with the data on the absence of PS2 activity in flashed seedlings grown with norflurazon (Karapetyan et al., 1989). The amount of carotenoids is considerably diminished in complexes I and III. Complex I has the red absorption maximum at 673 nm and the Soret band at 434 nm which is typical of Chl a protein complex. The absorption spectra of complexes III are similar for control and treated seedlings except the carotenoid band at 476 nm found only in control (data not shown).

Although bands I (CP1) and III (free Chl) of treated seedlings have different red absorption maxima they show the same fluorescence maximum at 675 nm which may be emitted by solubilized Chl. The fluorescence maximum of the CP1 of treated seedlings is shifted by 9 nm towards the shortwave region as compared with control. Thus, the complex with the decreased content of carotenoids is easily modified with detergents. Note that PS1 in seedlings grown with norflurazon at flash illumination was already modified: shift of fluorescence band 735 to 720 nm and the Chl/P700 ratio is 60 in contrast to 150 in untreated leaves, which indicates the absence of the peripheral antenna in treated PS1 (Karapetyan et al., 1991). In the carotenoidless <u>Scenedesmus</u> mutant, the PS1 antenna is not synthesized (Römer et al.,1990).

Thus the absence of carotenoids prevents the formation of CPa, although norflurazon does not inhibit the biosynthesis of Chl and PS2 polypeptides. This means that carotenoids are necessary component for assembly and stabilization of CPa and PS2 reaction center. It is quite possible, however, that besides carotenoids other substances (i.e. lipids) and regulators are required for the formation of PS2, and the biosynthesis of these components may be blocked by norflurazon. Since norflurazon inhibits activity of phytoene desaturase it may also inhibit nonspecifically desaturase of some fatty acids and, therefore, change the lipid composition and thus affect the assembly of CPa.

REFERENCES

Akoyunoglou, G., Argyroudi-Akoyunoglou, J., Michel-Wolwertz, M.R. and Sironval, C., 1966, Effects of intermittent light and continuous light on chlorophyll formation in etiolated plants, <u>Physiol</u>. <u>Plant</u>., 19:1101

Anderson, J.M., 1986, Photoregulation of the composition, function and structure of thylakoid membranes, <u>Annu</u>. <u>Rev</u>. <u>Plant Physiol</u>., 37:93

Axelsson, L., Dahlin, C. and Ryberg, H.,1982, The function of carotenoids during chloroplast development. Y. Correlation between carotenoid content, ultrastructure and chlorophyll <u>b</u> to chlorophyll <u>a</u> ratio, <u>Physiol</u>. <u>Plant</u>., 55:111

Bolychevtseva, Yu.V., Chivkunova, O.B., Merzlyak, M.N. and Karapetyan, N.V., 1987, Effect of norflurazon on the contents of chlorophyll, fatty acid and lipid peroxidation products in barley seedlings grown under different illumination conditions, <u>Biochemistry</u> (USSR), 52:160

Bolychevtseva,Yu.V., Turishcheva, M.S., Bezsmertnaya, I.N. and Karapetyan, N.V., 1988, Polypeptide composition and structure of chloroplasts from barley seedlings grown in the presence of norflurazon in dim light, <u>Biochemistry</u> (USSR) 53:677

Cogdell, R.J. and Franck, H.A., 1987, How carotenoids function in photosynthetic bacteria, <u>Biochim</u>. <u>Biophys</u>. <u>Acta</u>, 895:63

Dahlin, C., 1988, Correlation between pigment composition and apoproteins of the light-harvesting complex II (LHC II) in wheat, <u>Physiol</u>. <u>Plant</u>., 74:342

Damm, I., Steinmetz, D. and Grimme, L.H., 1990, Multiple functions of beta-carotene in photosystem I, <u>in</u>: "Current Research in Photosynthesis" (M.Baltscheffsky ed., v.II, p.607 Kluwer Acad. Publ., Dordrecht

Hladik, J., Pančoská, P. and Sofrova, D., 1982, Influence of carotenoids on the conformation of chlorophyll protein complexes isolated from the cyanobacterium Plectonema boreanum, Biochim. Biophys. Acta, 681:263

Humbeck, K., Römer, S. and Senger, H., 1989, Evidence for an essential role of carotenoids in the assembly of an active photosystem II, Planta, 179:242

Karapetyan, N.V., Bolychevtseva, Yu.V. and Rakhimberdieva, M.G., 1989, Why do barley seedlings grown with nor-flurazon have no Photosystem 2? Abstracts of the 3rd ESP Congress, S17, Budapest

Karapetyan, N.V., Bolychevtseva, Yu.V. and Rakhimberdieva, M.G., 1991, The necessity of carotenoids for the assembly of active photosystem 2 reaction centers, in: "Light in Biology and Medicine", Duglas, R.H., Moan, J. and Rönto, G., eds., v.2, Plenum Publ. Co London, p.45

Lichtenthaler, H.K., Prenzel, U. and Kuhn, G., 1981, Carotenoid composition of chlorophyll-carotenoid-proteins from raddish seedlings, Z.Naturforsch., 37C:10

Moskalenko, A.A., 1990, Isolation of stable photosystem 2 reaction centers, Biological membranes (USSR) 7:736

Popov, V.I. Matorin, D.N., Gostimsky, S.V., Tageeva, S.V. and Allakhverdov, B.L., 1981, Ultrastructural organization of chloroplast membranes in mutants of Pisum sativum L. with impaired activity in the photosystem, Planta, 151:512

Römer, S., Humbeck, K. and Senger, H., 1990, Relationship between biosynthesis of carotenoids and increased complexity of photosystem I in mutant C-6D of Scenedesmus obliquus, Planta, 182:216

Siefermann-Harms, D., 1985, Carotenoids in photosynthesys. I. Location in photosynthetic membranes and light-harvesting function, Biochim. Biophys. Acta, 811:325

Siefermann-Harms, D. and Ninnemann, H., 1981, Chlorophyll protein complexes of flashed bean leaves, in: "Photosynthesius. III. Structure and Molecular Organisation of the Photosynthetic Apparatus", Akoyunoglou, G., ed., p.655, Balaban, Philadelphia

DEVELOPMENT OF PROPLASTIDS AND ACCUMULATION OF LHCP II APO-PROTEIN IN GOLGI AND THYLAKOIDS OF DARK-GROWN WAX-RICH CELLS OF EUGLENA GRACILIS AT LOW LIGHT INTENSITIES SEEN BY IMMUNO-ELECTRON MICROSCOPY

Tetsuaki Osafune, Shuji Sumida, Tomoko Ehara,
Jerome A. Schiff[1] and Eiji Hase[2]

Dept. Microbiol., Tokyo Med. Coll., Shinjuku, Tokyo 160;
[1]Photobiol. Group, Dept. Biol., Brandeis Univ., Waltham,
MA. 02254, U.S.A.
[2]Lab. Chem., Fac. Med., Teikyo Univ., Tokyo 192-03 Japan

INTRODUCTION

When dark-grown resting cells of Euglena are exposed to light at the low intensity threshold of chloroplast development(7 ft-c) the antennas, including the LHCP II 26.5 kD apoprotein are very low or undetectable, but on exposure of these cells to more normal intensities of light for development (80-150 ft-c) the antennas, including the LHCP II apoprotein and chlorophyll b rapidly accumulate[1]. Thin sections of similar cells at 3-7 ft-c do not show appreciable staining of either Golgi or thylakoids with rabbit immune serum against the 26.5 kD apoprotein plus protein A-gold, but when these cells are incubated at 80-150 ft-c specific staining with immunogold rapidly appears in both the Golgi and the thylakoids[2,3].

As shown previously[4-6], dark-grown, wax-rich cells of Euglena gracilis contain very rudimentary proplastids with no internal structure except for a single prothylakoid lying close to the envelope. When these cells are transferred to an inorganic medium containing an ammonium salt as nitrogen source and aerated in darkness for 5 to 6 days, an early development of proplastids occurs using wax (and paramylum) as sources of carbon and energy[5]. The early development includes formation of a rudimentary pyrenoid adjacent to the prolamellar body, which appears earlier in the peripheral region of the proplastid. No further development of proplastids was observed during prolonged dark incubation of these Euglena cells. This paper describes the localization of LHCP II apoprotein when these dark-incubated Euglena cells are exposed to light at the low intensity threshold (3-7 ft-c) for chloroplast development[6].

MATERIAL AND METHODS

Cells of <u>Euglena</u> <u>gracilis</u> Z and <u>E. gracilis</u> var. <u>bacillaris</u> maintained in darkness for many generations, were used in the present work.

Culture conditions

Cells of <u>E. gracilis</u> strain Z and var. <u>bacillaris</u> were grown in darkness in standing culture without agitation at 28 ℃ in 500 ml Erlenmeyer flasks containing 400 ml of Hutner's pH 3.5 growth medium. The flasks were inoculated with 50 ml of a dark-grown culture containing 10^6 cells/ml. After 5 days of growth, the cells were centrifuged for 15 min at 500 X g. The cells were then washed twice with inorganic medium (Cramer & Myers)[7] and were finally resuspended in inorganic medium. The suspension was bubbled continuously from this point on with 2 % (v/v) CO_2 in air at 26 ℃ in darkness for up to 6 days[4-6].

Light conditions

Six day dark-incubated cells were exposed to light at indicated intensities with continued aeration in a flat oblong vessel at 25 °C. The source of light was fluorescent tubes (Fishlux, Toshiba), whose spectral energy distribution was shown in a previous paper[8]. The light intensity was measured with a Photocell Illumino-meter API-5 (Toshiba). The foot candle unit was employed to correlate the present observations with previous studies by Stern et al.[9] and Spano et al.[1] using this unit.

Immunoelectron Microscopy

The methods were the same as those described previously [2,3].

RESULTS

When the dark-incubated cells of the Z strain are exposed to light at low intensities (3-7 ft-c) for 4 hrs, there is considerable LHCP Ⅱ antibody reaction (gold particles: see MATERIAL AND METHODS) in the Golgi and compartmentalized osmiophilic structure (COS)[10], but only a small amount of reaction in the proplastids as shown in Fig. 1. Figure 2 shows that gold particles seen in Fig. 1 are localized over both the Golgi cisternae and the dense material that lies between the cisternae when observed at high magnification. After this time (at 7 hrs of low light exposure), gold particles are localized over prolamellar body, propyrenoid and girdle thylakoid in the proplastid (Fig. 3) as well as the Golgi (not shown). Figure 4 shows that gold particles became more concentrated in both Golgi and proplastid at 12 hrs. After about 14 hrs, newly formed thylakoids began to show intricately curled forms of random orientation and gold particles also became markedly concentrated over the thylakoids, e.g., 48 hrs (Fig. 5). Even if the incubation at low intensity light is prolonged to 168 hrs, no further development is seen[11]. Fig. 6 illustrates that dark-grown wax-rich cells of <u>E. gracilis</u> var. <u>bacillaris</u> after transfer to an inorganic medium and aeration in darkness followed by exposure to low intensity light (3 ft-c) for 48 hrs also show a strong immunoreaction with LHCP Ⅱ antibody and protein A-gold.

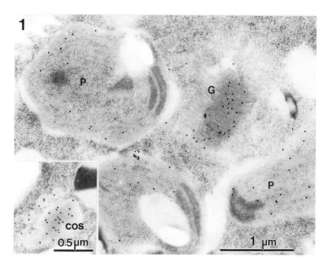

Fig. 1. Dark-grown wax-rich cells of _Euglena gracilis_ Z after transfer to an inorganic medium and aeration with 2% (v/v) CO_2 in air in darkness followed by exposure to low intensity (3-7 ft-c) light for 4 hrs. Scale=1 μm here and subsequently. Gold particles accumulated over the compartmentalized osmiophilic structure (COS)(inset) and Golgi (G) but less over proplastids.

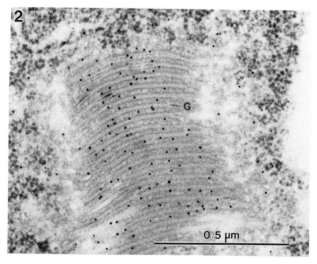

Fig. 2. Dark-grown wax-rich cells of _E. gracilis_ Z after transfer to an inorganic medium and aeration in darkness followed by exposure to low intensity light for 4 hrs. A high magnification of a Golgi apparatus. Note that gold particles are localized over both the Golgi cisternae and the dense material that lies between the cisternae.

357

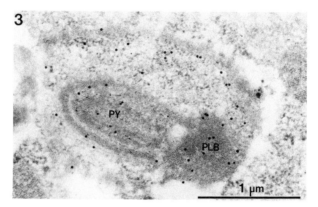

Fig. 3. Dark-grown wax-rich cells of E. gracilis Z after transfer to an inorganic medium and aeration in darkness followed by exposure to low intensity light for 7 hrs. Note that gold particles are seen over prolamellar body (PLB), propyrenoid (PY) and girdle thylakoids.

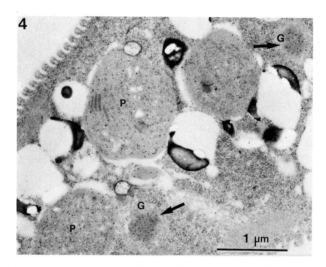

Fig. 4. Dark-grown wax-rich cells of E. gracilis Z after transfer to an inorganic medium and aeration in darkness followed by exposure to low intensity light for 12 hrs. Gold particles are concentrated over Golgi (arrows) and thylakoids.

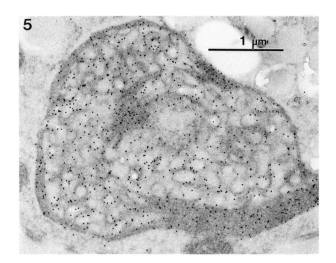

Fig. 5. Dark-grown wax-rich cells of E. gracilis Z after transfer to an inorganic medium and aeration in darkness followed by exposure to low intensity
light for 48 hrs. A proplastid with gold particles over curly thylakoids is
shown.

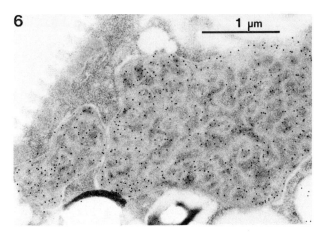

Fig. 6. Dark-grown wax-rich cells of E. gracilis var. bacillaris after transfer to an inorganic medium and aeration in darkness followed by exposure to
low intensity light for 48 hrs. Note immunogold deposition (LHCP Ⅱ immunoreaction) over curly thylakoids of proplastid.

SUMMARY

Anti LHCP II 26.5 kD apoprotein antibody and protein A-gold were used to localize the apoprotein in dark-grown wax-rich cells of E. gracilis strain Z and var. bacillaris after transfer to an inorganic medium and aeration in darkness followed by exposure to low intensity light (3-7 ft-c). Unlike ordinary dark-grown cells at low intensity, apoprotein is accumulated and the gold particles are seen in the COS, then in the Golgi (at about 3 hrs of light) and as the immunoreaction in the Golgi increases, deposition is seen over the plastid thylakoids. Between 14 hrs and 168 hrs in low light, the plastids were global or oval with curled forms of thylakoids in random orientation that were strongly labelled with gold particles. Thus wax-rich cells under aeration on inorganic medium appear to be able to accumulate LHCP II apoprotein at low light intensities unlike ordinary cells. Thus energy mobilization in aerated wax-rich cells may facilitate the accumulation of LHCP II and overcome the limitation imposed by low light intensity in the formation of photosynthetic antennas.

REFERENCES

1. Spano AJ, Ghaus H, Schiff JA: Chlorophyll-protein complexes and other thylakoid components at the low intensity threshold in Euglena chloroplast development. Plant Cell Physiol. 28:1101-1108 (1987)
2. Osafune T, Schiff JA, Hase E: Immunolocalization of LHCP II apoprotein in the Golgi of Euglena. In: Baltscheffsky M, ed. Current Research in Photosynthesis,Vol.III. Netherlands, Kluwer Academic Publishers,735-738, 1989
3. Osafune T, Schiff JA, Hase E: Immunogold localization of LHCP II apoprotein in the Golgi of Euglena. Cell Struct. Funct. 15:99-105 (1990)
4. Shihira-Ishikawa I, Osafune T, Ehara T, Ohkuro I, Hase E: An early light-independent phase of chloroplast development in dark-grown cells of Euglena gracilis Z. I. Dependence of the plastid development on previous culture conditions. Plant Cell Physiol. (Special issue) 445-454 (1977)
5. Osafune T, Ehara T, Sumida S, Hase E, Schiff JA: Light-independent processes in the formation of thylakoids and pyrenoids in proplastids of dark-grown cells of Euglena gracilis.J Electron Microsc.39:245-253 (1990)
6. Osafune T, Sumida S, Ehara T, Ueno N, Hase E, Schiff JA: Lipid (wax) and paramylum as sources of carbon and energy for the early development of proplastids in dark-grown Euglena gracilis cells transferred to an inorganic medium. J Electron Microsc. 39:372-381 (1990)
7. Cramer M, Myers J: Growth and photosynthetic characteristics of Euglena gracilis. Arch. Mikrobiol. 17:384-402 (1970)
8. Osafune T, Ehara T, Tahara M, Hase E: Effects of blue and red light on chloroplast development in dark-grown wax-rich cells of Euglena gracilis transferred to inorganic medium. J Electron Microsc. 39:254-259 (1990)
9. Stern AI, Schiff JA, Epstein HT: Studies of chloroplast development in Euglena. V. Pigment biosynthesis, photosynthetic oxygen evolution and carbon dioxide fixation during chloroplast development. Plant Physiol. 39:220-226 (1964)
10.Osafune T, Schiff JA: $W_{10}BSmL$, a mutanat of Euglena gracilis lacking plastids. Exp. Cell Res. 148:530-535 (1983)
11.Sumida S, Osafune T, Hase E: Light-induced development of thylakoids in proplastids of dark-grown wax-rich cells of Euglena gracilis transferred to an inorganic medium. J Electron Microsc. 40: 1991 in press.

LOW INTENSITY LIGHT-INDUCED DEVELOPMENT OF THYLAKOIDS IN

PROPLASTIDS OF DARK-GROWN WAX-RICH CELLS OF EUGLENA GRACILIS

Tetsuaki Osafune[1], Shuji Sumida[1] and Eiji Hase[2]

[1]Department of Microbiology, Tokyo Medical College, Shinjuku,
Tokyo 160, and [2]Laboratory of Chemistry, Faculty of Medicine,
Teikyo University, Hachiouji, Tokyo 192-03, Japan

INTRODUCTION

As shown previously[1,2], dark-grown, wax-rich cells of Euglena gracilis
contain profoundly degenerate proplastids with no internal structure except
for a single prothylakoid lying close to the envelope.   When these cells are
transferred to an inorganic medium containing ammonium salt as nitrogen source
and aerated in darkness for 5 to 6 days, an early development of proplastids
occurs using wax (and paramylum) as sources of carbon and energy [3]. The early
development includes formation of rudimentary pyrenoid (propyrenoid) at the
site adjacent to the prolamellar body, which appears earlier in the peripheral
region of the proplastid.   The peripheral single prothylakoid becomes paired
along part of its length, and a portion of the paired prothylakoid becomes extended and enfolded in the propyrenoid.   No further development of proplastids
was observed during prolonged dark incubation of these Euglena cells.   This
paper describes the observations made when these dark-incubated Euglena cells
were exposed to light at different intensities including the low intensity
threshold (3-7 ft-c) for chloroplast development[4].

MATERIAL AND METHODS

Euglena gracilis Klebs strain Z Pringsheim was obtained from the Algal
Culture Collection of the Institute of Applied Microbiology, the University
of Tokyo (IAM Z-6).
Culture conditions
The culture methods of Euglena cells were the same as those described
previously[1].
Light conditions
Six day dark-incubated cells were exposed to light at indicated intensities
with continued aeration in a flat oblong vessel at 25 °C.   The source of light
was fluorescent tubes (Fishlux, Toshiba), whose spectral energy distribution
was shown in a previous paper[5].The light intensity was measured with a Photocell Illumino-meter API-5 (Toshiba).   The use of foot candle unit was made to
correlate the present observations with previous studies by Stern et al.[4] and
Spano et al.[6] using this unit.
Measurements of chlorophyll and paramylum
Chlorophyll was determined with 80% acetone extract of cells by Mackinney's
method, and paramylum was separated and purified according to the method used
by Freyssinet et al.[7].

Regulation of Chloroplast Biogenesis
Edited by J.H. Argyroudi-Akoyunoglou, Plenum Press, New York, 1992

Measurement of photosynthetic O_2-evolution of cells
The measurement was made at 15,000 lux (saturating) with a Clark oxygen electrode.
Electron microscopy
The methods were the same as those described previously [8].

RESULTS

Formation of chlorophyll, degradation of paramylum and development of photosynthetic activity at different light intensities
Figure 1 shows time courses of the formation of chlorophyll and degradation of paramylum observed when 6 day dark-incubated Euglena cells(see Material and methods) were exposed to light at 3 ft-c and 150 ft-c.　While substantial degradation of paramylum occurred at the lower intensity of light, the chlorophyll accumulation was considerably smaller when compared with that at the higher intensity.　The low chlorophyll content of cells exposed to the low intensity light remained unaltered even when these cells were illuminated for 190 h, and the content started to increase only when these cells were exposed to higher intensity light, as seen from Fig. 1.　The photosynthetic O_2-evolution of the cells exposed to 3 ft-c light was about half that of those cells under 150 ft-c light (Fig. 2, upper).　It should be noted, however, that the photosynthetic activity in terms of per chlorophyll basis was virtually the same at the two different intensities of light (Fig. 2, lower), suggesting that the chlorophyll accumulation is a limiting process during the early development of photosynthetic activity under the conditions used.

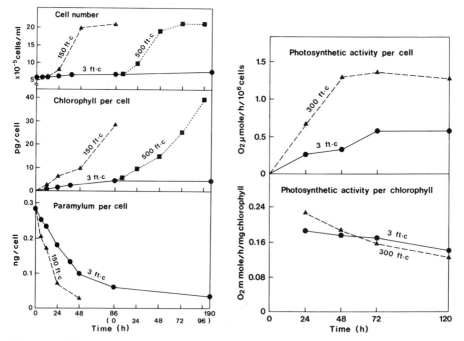

Fig.1.　Time courses of cell division, chlorophyll accummulation and paramylum degradation at different intensities of light.　Six day dark incubated cells were exposed to light at indicated intensities.
Fig.2.　Changes in photosynthetic O_2-evolution of cells exposed to light at low and higher intensities.　Photosynthetic O_2-evolution was measured at saturating light intensity (15,000 lux) with a Clark oxygen electrode.

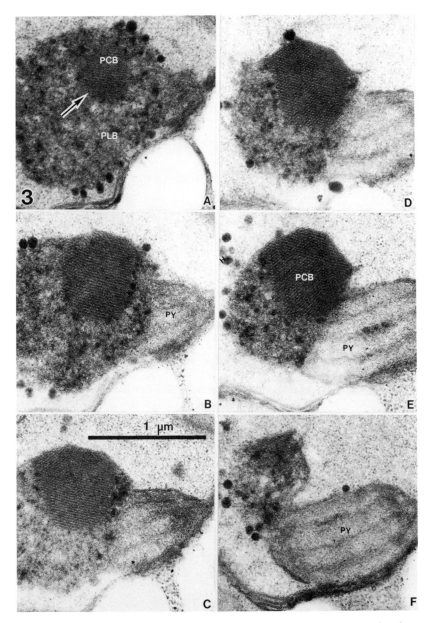

Fig. 3. Serial sections of a paracrystalline structure appearing in the prolamellar body in close association with the propyrenoid in the proplastid of a cell exposed to 3 ft-c light for 2 h.

Low intensity light-induced formation of thylakoids

The observations at 3 ft-c gave almost the same results as those at 7 ft-c. When the dark-incubated cells were exposed to light either at low (3 ft-c) or higher intensities (150 ft-c), a paracrystalline structure appeared rapidly at a site adjacent to PLB and the pyrenoid, as shown in Fig.3. In low light this structure was observed for 2-4 hours, while it disappeared within one hour in higher intensity light. The prolamellar body was detectable for a period of about 10 hours in low light,while at higher intensities of light the body disappears in a few hours. In low light, new thylakoids were observed to extend into stroma from PLB as well as from the pyrenoid as outgrowths of prothylakoids embedded in it, while at higher intensities the extension of thylakoids occurred mostly from the pyrenoid. After about 14 hours in low intensity light, newly formed thylakoids, originated either from the prolamellar body or from the pyrenoid,began to show intricately curled forms of random orientation (Fig.4). The conversion of these forms into normal straight thylakoids could be achieved only when the light intensity was shifted to fairly higher levels (500 ft-c), as seen from Figs. 5 and 6: the conversion was not observed by shifting to 150 ft-c for 96 h.

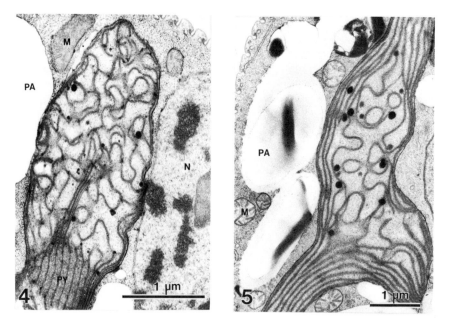

Fig.4. Curled thylakoids of random orientation in a plastid. The cell sample was in 3 ft-c light for 86 h.

Fig.5. Thylakoids in a process of recovery from curled to straight forms. The cells placed in 3 ft-c light for 86 h were transferred to 500 ft-c light and illuminated for 8 h. The chlorophyll accumulation at 500 ft-c of light is shown in Fig. 1.

SUMMARY

When dark-grown wax-rich cells of _Euglena gracilis_ are transferred to an inorganic medium containing ammonium salt as nitrogen source and the cell suspension is aerated in the dark, the wax is oxidatively metabolized, providing carbon compounds and energy for some dark processes of plastid development. Under these conditions, two or more propyrenoids are formed at the sites adjacent to the prolamellar bodies (PLBs), and a part of the paired prothylakoids lying close to the envelope is extended and enfolded in the propyrenoids. After 6 day dark incubation, these cells were exposed to low (3-7 ft-c) and higher (150-300 ft-c) intensities of light. The chlorophyll content was considerably lower in low light (3 ft-c) and the photosynthetic O_2-evolution per cell was about one half that of cells in higher intensity light (300ft-c). The photosynthetic activity per chlorophyll, however, was virtually the same between these cells placed at two different intensities of light. A paracrystalline structure appeared temporarily in or near PLB as a rapid response to light exposure at either low or higher intensities. At the low intensity, PLB was detectable for about 10 hours, while at higher intensities it disappeared in a few hours. In low light, new thylakoids were observed to extend into stroma from PLB as well as from the pyrenoid as outgrowths of prothylakoids embedded in it, while at higher intensities the extension of thylakoids occurred mostly from the pyrenoid. After about 14 hours in low light, the extending thylakoids began to show intricately curled forms of random orientation. These curled forms could be converted to normal straight thylakoids only when strong light (500 ft-c) was applied in higher intensity light, the pyrenoids with thylakoids extending from two opposite sides seemed to move, becoming one in number, toward the center of chloroplast.

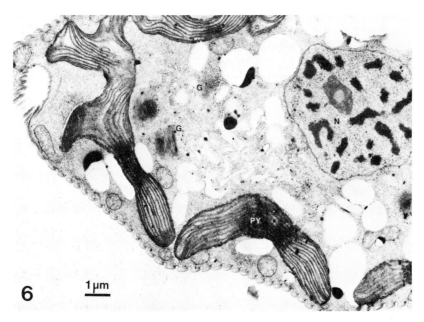

6 1 µm

Fig.6. Fully developed chloroplasts in a cell placed in 500 ft-c light for 48 h after 86 h exposure to 3 ft-c light.

REFERENCES

1) Shihira-Ishikawa I, Osafune T, Ehara T, Ohkuro I, Hase E: An early light-independent phase of chloroplast development in dark-grown cells of Euglena gracilis Z. 1. Dependence of the plastids development on previous culture conditions. Plant Cell Physiol (Special issue) 445-454 (1977)

2) Osafune T, Ehara T, Sumida S, Hase E, Schiff JA: Light-independent processes in the formation of thylakoids and pyrenoids in proplastids of dark-grown cells of Euglena gracilis. J Electron Microsc 39:245-253 (1990)

3) Osafune T, Sumida S, Ehara T, Ueno N, Hase E, Schiff JA: Lipid (wax) and paramylum as sources of carbon and energy for the early development of proplastids in dark-grown Euglena gracilis cells transferred to an inorganic medium. J Electron Microsc 39:372-381 (1990)

4) Stern AI, Schiff JA, Epstein HT: Studies of chloroplast development in Euglena. V. Pigment biosynthesis, photosynthetic oxygen evolution and carbon dioxide fixation during chloroplast development. Plant Physiol 39:220-226 (1964)

5) Osafune T, Ehara T, Tahara M, Hase E: Effects of blue and red light on chloroplast development in dark-grown wax-rich cells of Euglena gracilis transferred to inorganic medium. J Electron Microsc 39:254-259 (1990)

6) Spano AJ, Ghaus H, Schiff JA: Chlorophyll-protein complexes and other thylakoid components at the low intensity threshold in Euglena chloroplast development. Plant Cell Physiol 28:1101-1108 (1987)

7) Freyssinet G, Heizmann P, Verdier G, Trabuchet G, Nigon V: Influence des conditions nutritionnelles sur la réponse à l'éclairement chez les euglènes étiolées. Physiol Vég 10:421-442 (1972)

8) Ehara T, Sumida S, Osafune T, Hase T: Interactions betweeen the nucleus and cytoplasmic organelles during the cell cycle of Euglena gracilis in synchronized cultures: I. Associations between the nucleus and chloroplasts at an early stage in the cell cycle under photoorganotrophic conditions (Part I) Plant Cell Physiol 25:1133-1146 (1984)

ORGANIZATION AND INTERACTIONS OF
PHOTOSYNTHETIC UNITS

RECONSTITUTION OF THE PHOTOSYSTEM I COMPLEX WITH GENETICALLY

MODIFIED PsaC & PsaD POLYPEPTIDES EXPRESSED IN *ESCHERICHIA COLI*

John H. Golbeck

Department of Biochemistry
University of Nebraska
Lincoln, NE 68583 (USA)

INTRODUCTION

There are, in general, two ways to use the techniques of molecular biology to study the functions of the low molecular mass polypeptides in Photosystem I. One method is to induce modifications of the targeted polypeptide by altering the appropriate gene *in vivo*, and then to study the effect of structure on function after isolating the Photosystem I complex from the thylakoid membrane. A second method is to purify the Photosystem I core by dissociating the low molecular mass polypeptides *in vitro* and then to study the effect of structure on function by reconstituting the complex with modified polypeptides that have been expressed in *Escherichia coli*. We chose the latter strategy because we have developed techniques for the isolating the intact P700 and F_X-containing Photosystem I core [1] and for rebinding the PsaC, PsaD and PsaE proteins onto the purified Photosystem I core, thereby reestablishing electron flow from P700 to the terminal acceptors F_A/F_B [2]. This methodology might have been of limited utility had we not also developed a protocol for reinserting the F_A and F_B iron-sulfur clusters into the isolated PsaC apoprotein [3]. In theory, this technique allows a modified PsaC apoprotein to be expressed in *Escherichia coli*, isolated and purified, reconstituted with the F_A/F_B iron-sulfur clusters, and rebound to the *Synechococcus* sp. PCC 6301 Photosystem I core for functional analyses.

In practice, this isolation and reconstitution protocol has provided an excellent experimental tool to study the function of the PsaD and PsaE proteins as well as the PsaC protein in Photosystem I. The method has recently allowed us to reconstitute the Photosystem I complex with modifications of the PsaC protein which disallow the formation of either the F_A or F_B iron-sulfur cluster. This has lead to an understanding of the sequence of electron flow from F_X through F_A/F_B to soluble ferredoxin. Here, we use this approach to demonstrate:

- The PsaC, PsaD and PsaE proteins are removed by chaotrope treatment and are rebound to the Photosystem I core only after reinserting the F_A/F_B iron-sulfur clusters,
- The stable rebinding of the PsaC holoprotein to the Photosystem I core requires the presence of PsaD but not PsaE,
- The binding of a PsaC fusion protein (containing an MEHSM amino-terminal extension) to the Photosystem I core absolutely requires the presence of PsaD,
- The replacement of aspartic acid for cysteine 51 in PsaC (C51D) leads to the formation of an altered cluster in the F_A site, and a normal [4Fe-4S] cluster in the F_B site,
- The replacement of aspartic acid for cysteine 14 in PsaC (C14D) leads to the formation of an altered cluster in the F_B site, and a normal [4Fe-4S] cluster in the F_A site,
- The F_B cluster is not required for low temperature electron flow to F_A, thereby disallowing the pathway $F_X \rightarrow F_B \rightarrow F_A$, and implying that electron flow is direct from $F_X \rightarrow F_A$ or F_B.

Regulation of Chloroplast Biogenesis
Edited by J.H. Argyroudi-Akoyunoglou, Plenum Press, New York, 1992

RESULTS

Polypeptide Composition of the Photosystem I Complex, the Photosystem I Core, and the Reconstituted Photosystem I Complex from *Synechococcus* sp. PCC 6301 with Ning Li, Donald Bryant, Gerald Frank, Herbert Zuber and Patrick Warren [Ref. 4].

The strategy outlined above depends on the isolation of an intact Photosystem I core containing components P700 through F_X, and on its ability to be reconstituted with the low molecular mass proteins to restore electron flow to the F_A/F_B iron-sulfur clusters. We had shown earlier that chaotropic agents were successful in stripping the PsaC protein from the *Synechococcus* sp. PCC 6301 Photosystem I complex [1]. We also showed that the PsaC protein could be rebound to the Photosystem I core after reinserting the iron-sulfur clusters [2]. The functional defintions of these preparations, termed the 'Photosystem I complex' and 'Photosystem I core protein' were based on spectroscopic characteristics: the former contained components P700 through F_A/F_B, while the latter contained components P700 through F_X (reviewed in [5]). In order to identify the the low molecular mass polypeptides that were lost upon chaotrope treatment, we analyzed the polypeptides in the Photosystem I complex and the Photosystem I core by SDS-PAGE and N-terminal amino acid sequencing.

As shown in Fig. 1 (top), the *Synechococcus* sp. PCC 6301 Photosystem I complex contains the PsaA/PsaB heterodimer, PsaC, PsaD, PsaE, PsaF, PsaK, PsaL, and three <8-kDa polypeptides, which are probably PsaJ, the "4.8-kDa" protein (tentatively labeled PsaM), and PsaI. At the low concentrations of Triton X-100 used in this study, the PsaF protein was retained quantitatively on the Photosystem I complex. The Photosystem I complex from *Synechococcus* sp. PCC 6301 also contains several unexpected polypeptides - the P_{II} protein derived from the *glnB* gene and a 9.3-kDa protein, identified as 'X', which is similar in sequence to the putative products of two open reading frames which occur upstream from the *rbcLS* operon in *Synechococcus* sp. PCC 6301 and 7942, and which have been implicated in carboxysome assembly [6,7]. Since these proteins are missing in Photosystem I complexes isolated with longer exposure times and higher concentrations of Triton X-100, both are considered co-purifying contaminants.

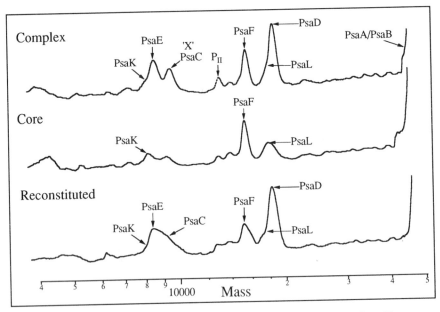

Figure 1. Low Molecular Mass Polypeptides in the Photosystem I Complex, Photosystem I Core, and Reconstituted Photosystem I Complex. The polypeptides were resolved on SDS-polyacylamide gels, scanned with a laser densitometer, and subjected to N-terminal amino acid sequencing for identification. See ref. [4] for details.

The Photosystem I core (Fig. 1, middle) contains the PsaA/PsaB heterodimer, PsaF, PsaL, PsaK, and the three <8-kDa polypeptides which are present in the Photosystem I complex. The hydrophilic PsaC, PsaD and PsaE proteins are dissociated from the hydrophobic Photosystem I core, and can be recovered after passage of the treated complex through a YM-100 ultrafiltration membrane. Upon the readdition of the low molecular mass polypeptides to the Photosystem I core in the presence of $FeCl_3$, Na_2S and β-mercaptoethanol, the reconstituted Photosystem I complex (Fig. 1, bottom) contains not only the F_A/F_B-containing PsaC protein, but also the PsaD and PsaE proteins. There is little or no rebinding of the PsaD or PsaE proteins in the absence of the PsaC holoprotein (not shown). The reconstitution of the F_A/F_B iron-sulfur clusters thus appears to be a necessary precondition for rebinding of the PsaC, PsaD and PsaE proteins to the Photosystem I core.

PsaD is Required for the Stable Binding of PsaC to the *Synechococcus sp.* PCC 6301 Photosystem I Core
with Ning Li, Jingdong Zhao, Patrick Warren, Joseph Warden and Donald Bryant [Ref. 8].

While the above study showed that the PsaC holoprotein was required for the binding of the PsaD and PsaE proteins to the Photosystem I core, it did not address the related question of whether the PsaC holoprotein can rebind in the absence of the PsaD and PsaE proteins. Because of the large quantities of the PsaC, PsaD and PsaE proteins necessary for these studies, and because of the presence of a contaminating protease which cleaves the PsaD protein [4], we chose to produce the cyanobacterial PsaC, PsaD, and PsaE proteins in a bacterial expression system. The strategy employed was to overproduce the *psaC* gene product from *Synechococcus* sp. PCC 7002 in *E. coli*, purify the PsaC apoprotein to homogeneity, and reconstitute the F_A/F_B iron-sulfur clusters *in vitro* to produce a competent PsaC holoprotein. The *psaD* and *PsaE* gene products from *Nostoc* sp. PCC 8009 and *Synechococcus* sp. PCC 7002 were similarly over-produced in *E. coli* and the PsaD and PsaE proteins were purified to homogeneity.

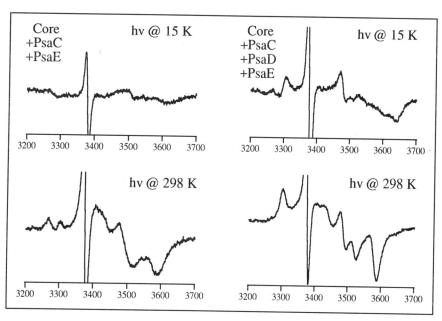

Figure 2. EPR Spectra of the Photosystem I Core Reconstituted with PsaC and PsaE (left) and with PsaC, PsaD and PsaE (right) in the Presence of $FeCl_3$, Na_2S and β-mercaptoethanol. The samples were illuminated at 15 K to transfer one electron from P700 to an acceptor (top) and at 298 K to allow photoaccumulation of both F_A and F_B (bottom). See ref. [8] for details.

When the PsaC apoprotein was incubated with the *Synechococcus* sp. PCC 6301 Photosystem I core in the presence of FeCl$_3$, Na$_2$S and ß-mercaptoethanol, F$_B$ and F$_A$ were photoreduced about equally when the sample was frozen in darkness and illuminated at 19 K (see ref. [8]). Even though the ESR resonances of F$_A$ and F$_B$ were shifted upfield, they remained as broad as in the free PsaC holoprotein. Analysis by denaturing polyacrylamide gel electrophoresis showed that the PsaC protein was dislodged by ultracentrifugation in a 0.1% Triton-containing sucrose gradient. The further addition of PsaE leads to no discernable spectral changes over the addition of PsaC alone [Fig. 2, left]. When the purified PsaD protein was added to the incubation mixture containing PsaC, F$_A$, rather than F$_B$, became the preferred electron acceptor when the sample was frozen in darkness and illuminated at 19 K (see ref. [8]). In the presence of PsaD, the ESR resonances of F$_A$ and F$_B$ moved upfield and sharpened to become nearly identical to those in a control Photosystem I complex. Analysis by denaturing polyacrylamide gel electrophoresis showed that the PsaC and PsaD proteins remained quantitively bound to the Photosystem I core after ultracentrifugation in a 0.1% Triton-containing sucrose gradient. The further addition of PsaE leads to no further changes over the addition of PsaC and PsaD alone [Fig. 2, right].

These results indicate that although PsaC is partially effective in restoring low-temperature electron flow from P700 to the F$_A$/F$_B$ clusters, the protein binds loosely to the Photosystem I core and can be removed by Triton X-100. Under these conditions, several docking orientations may occur, placing the F$_A$ or F$_B$ cluster in proximity with the preceding electron donor (presumably F$_X$). The addition of PsaD leads to a stable Photosystem I complex which is spectroscopically and biochemically indistinguishable from the native complex. The binding of PsaD must orient and lock the PsaC protein into a specific conformation, which leads to the selective photoreduction of F$_A$ rather than F$_B$. It should be noted that the addition of PsaE had little or no effect on the binding of either PsaC or PsaD to the Photosystem I core. A structure compabible with these data would place the PsaC protein in contact with the PsaA/PsaB heterodimer and overlaid with the PsaD and (possibly) PsaE proteins. This potential shielding of PsaC from the solvent would allow for the increased stability of the iron-sulfur clusters in a reconstituted Photosystem I complex. We therefore conclude that the rebinding of PsaC to the PsaA/PsaB heterodimer is potentiated by insertion of the F$_A$/F$_B$ clusters, and stabilized by the presence of PsaD.

Reconstitution of Electron Transport in Photosystem I with a PsaC1 Fusion Protein Expressed in *Escherichia Coli*
with Jingdong Zhao, Patrick Warren, Ning Li, and Donald Bryant [Ref. 9].

In a first attempt to create a genetically modified PsaC protein (PsaC1), the *psaC* gene from *Synechococcus* sp. PCC 7002 was altered *in vitro* to code for a fusion protein containing an amino-terminal extension of five amino acids (MEHSM...). The PsaC1 and PsaD proteins were then combined with the Photosystem I core in the presence of FeCl$_3$, Na$_2$S and β-mercaptoethanol, and the reconstituted Photosystem I complex was analyzed by optical and ESR spectroscopy. The flash-induced absorption change in the Photosystem I core protein decayed with a half-time of 1.2 ms, which is diagnostic of the back-reaction between P700$^+$ and F$_X^-$ in the absence of the terminal iron-sulfur centers, F$_A$/F$_B$ (see ref. [9]). When the PsaC1 protein was added to the Photosystem I core at a molar ratio of 80:1 in the presence of FeCl$_3$, Na$_2$S, and DTT and incubated for 24 hr there was a minor suppression of the 1.2 ms back-reaction, but when this sample was washed by ultrafiltration over a YM-100 membrane, nearly all of the long-lived absorption change was lost.

When the PsaD protein was added at a molar ratio of 20:20:1 of PsaC to PsaD to Photosystem I core, the 1.2 ms P700$^+$ F$_X^-$ transient was replaced in 24 hr with a long-lived transient due to the P700$^+$ [F$_A$/F$_B$]$^-$ back-reaction (see ref. [9]). Low temperature EPR studies, shown in Fig. 3 (right), indicate the restoration of the F$_A$/F$_B$ resonances at their proper peak positions and linewidths relative to the control complex (Fig. 3, left). The extreme lowfield resonance remains unidentified, and it is possible that a metal has bound to the N-terminal extension. These studies show that unlike PsaC, the presence of PsaD is an absolute requirement for the binding of the PsaC1 fusion protein to a *Synechococcus* sp. PCC 6301 core. This implies that the PsaC1 holoprotein does not bind to the PsaA/PsaB heterodimer but rather that its interaction with these proteins is mediated through PsaD.

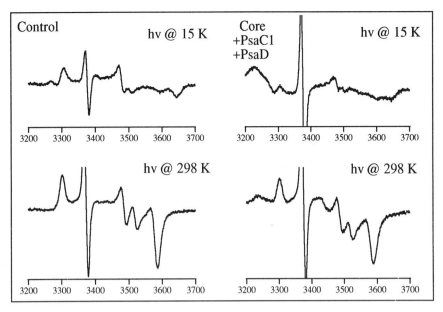

Figure 3. ESR Spectra of a Control Photosystem I Complex (left) and a Photosystem I Core Reconstituted with PsaC1 and PsaD in the Presence of FeCl₃, Na₂S and β-mercaptoethanol (right). The samples were illuminated at 15 K to transfer one electron from P700 to an acceptor (top) and at 298 K to allow photoaccumulation of both F_A and F_B (bottom).

Reconstitution of the Cyanobacterial Photosystem I Complex with Site-Directed Mutants of PsaC Expressed in *Escherichia Coli*
with Jingdong Zhao, Ning Li, Patrick Warren, Donald Bryant [Ref. 10].

There is considerable uncertainty surrounding the roles of iron-sulfur centers F_B and F_A in forward electron flow in Photosystem I. The redox potentials and low temperature behavior after illumination imply a serial flow of electrons between F_B and F_A, as does the marked inhibition of the low temperature photoreduction of F_A after selective chemical inactivation of F_B by diazonium compounds [11]. In contrast, the selective denaturation of F_B by urea-ferricyanide [12] and mercurials [13] has shown little effect on the room- and low-temperature photoreduction of F_A. There is also no reliable data on the physical identification of the F_A and F_B clusters relative to the cysteine ligands within the PsaC protein. We decided, therefore, to employ a new strategy to address these problems which involved genetically engineering an altered iron-sulfur cluster into the PsaC protein.

The presence of an aspartic acid residue in lieu of a cysteine leads to the formation of a [3Fe-4S] cluster in two naturally-occurring ferredoxins from *Pyrococcus furiosus* and *Desulfovibrio africanus* [14,15]. We reasoned that the altered properties of a [3Fe-4S] cluster in the PsaC protein might render that cluster incapable of serving as an electron transfer agent in Photosystem I forward electron flow. Additionally, this substitution would be expected to preserve closely the 3-dimensional structure of the PsaC protein since only a corner of the cubic [4Fe-4S] cluster would be missing. We selected the 'second' cysteine in each cysteine binding motif (CxxCxxCxxxCP) because the two naturally-occurring proteins occur with aspartate in this position and because the 3-dimensional structure of ferredoxins with [XFe-4S] clusters (where X = 3 or 4) such as *Pseudomonas aerogenes*, *Azotobacter vinelandii*, *Desulfovibrio gigas* and *Bacillus thermoproteolyticus* show that this cysteine is exposed to the surface of the protein in a bend region. This could allow the additional carbonyl oxygen to point into the solvent, thereby avoiding disruption of the polypeptide backbone and adding stability in terms on an additional H-bond to water.

Site-directed mutants of the PsaC protein were generated in which cysteine 14 (C14D) and cysteine 51 (C51D) were replaced by aspartic acid residues. Incubation of the C51D apoprotein with the PsaD protein and the *Synechococcus* sp. PCC 6301 Photosystem I core in the presence of $FeCl_3$, Na_2S and ß-mercaptoethanol results in a transition in the room-temperature, flash-induced absorption change from a 1.2-ms, $P700^+ F_X^-$ backreaction to a long-lived backreaction (see ref. [10]). ESR studies show that when the reconstituted sample is frozen in darkness, a nearly-isotropic resonance at $g=2.021$ is present which disappear upon reduction with $S_2O_4^{2-}$ or freezing of the sample under continuous illumination [Fig. 4, left]. This is the characteristic signature of a [3Fe-4S] cluster [16], and we infer that it is located in the aspartate-51 site. When the reconstituted sample is reduced with $S_2O_4^{2-}$ or frozen under illumination, a new set of resonances appear at $g=2.064$, 1.935 and 1.880 concomitant with the appearance of the narrow $P700^+$ radical at $g=2.002$ [Fig. 5, right]. Although the midfield resonance is ambiguous, the lowfield and highfield resonances resemble those of F_B (the $g=2.021$ cluster is non-paramagnetic when reduced; hence, the highfield resonance of F_A would have appeared at $g=1.856$ rather than 1.880). When the reconstituted sample is frozen under darkness and illuminated at 15 K, the $g=2.021$ cluster remains completely oxidized [Fig. 4, right], and only about 20% of the $g=2.064$, 1.935 and 1.880 cluster becomes photoreduced [Fig. 5, left]. This indicates that i) the [4Fe-4S] cluster resembles closely the behavior of F_B in the control and ii) the putative [3Fe-4S] cluster does not function as an electron acceptor at low temperature.

Incubation of the C14D apoprotein with the PsaD protein and the *Synechococcus* sp. PCC 6301 Photosystem I core in the presence of $FeCl_3$, Na_2S and ß-mercaptoethanol also results in a transition in the room-temperature, flash-induced absorption change from a 1.2-ms, $P700^+ F_X^-$ backreaction to a long-lived backreaction (see ref. [10]). ESR studies show that when the reconstituted sample is frozen in darkness, the dark signal disappears upon reduction with $S_2O_4^{2-}$, but it does not disappear during freezing under continuous illumination. We infer that this resonance is also due to the presence of a [3Fe-4S] cluster, but now located in the aspartate-14 site. When the reconstituted sample is reduced with $S_2O_4^{2-}$ or frozen under illumination, a new set of resonances appear at $g=2.043$, 1.940, 1.883 and 1.855 concomitant with the appearance of the narrow $P700^+$ radical at $g=2.002$ [Fig. 5, right]. Although the midfield resonance is ambiguous, the lowfield and highfield resonances resemble those of F_A (the resonance at $g=1.883$ indicates the presence of magnetic interaction with another paramagnetic species). When the reconstituted sample is frozen under darkness and illuminated at 15 K, the putative [3Fe-4S] center remains completely oxidized, and about 80% of the $g=2.044$, 1.942 and 1.856 center becomes photoreduced [Fig. 5, right]. This indicates that i) the [4Fe-4S] cluster resembles closely the behavior of F_A in the control, and ii) the putative [3Fe-4S] cluster does not function as an electron acceptor at either room or low temperature.

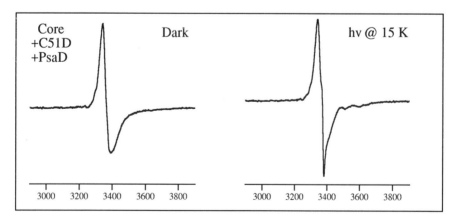

Figure 4. ESR Spectra of a Photosystem I Complex Reconstituted with C51D and PsaD in the Presence of $FeCl_3$, Na_2S and β-mercaptoethanol. The spectra were taken in darkness (left) and after illumination at 15 K (right) to transfer one electron from P700 to an acceptor.

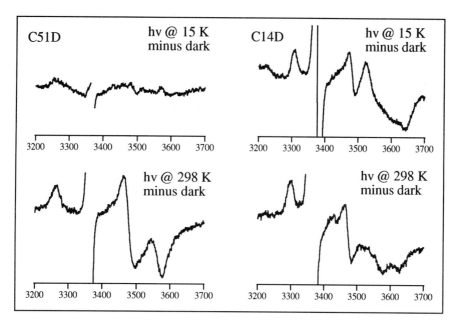

Figure 5. ESR Spectra of a Photosystem I Complex Reconstituted with C51D (left) and with C14D (right) in the Presence of PsaD and FeCl₃, Na₂S and β-mercaptoethanol (right). The samples were illuminated at 15 K to transfer one electron from P700 to an acceptor (top) and at 298 K to allow photoaccumulation of both F_A and F_B (bottom).

Based on these data, we conclude that cysteines 11, 14, 17 and 58 ligate the F_B cluster and that cysteines 21, 48, 51 and 54 ligate the F_A cluster. The mutant data indicate that F_B is not an obligatory carrier from P700 to F_A during low temperature electron flow. This eliminates the possibility of sequential electron flow from F_X -> F_B -> F_A for this region of the electron transport chain at these temperatures. It is interesting to note that the same proportion of iron-sulfur clusters become photoreduced in the C14D and C51D mutant proteins as in the control Photosystem I complex, indicating that the 80:20 ratio of F_A to F_B photoreduction is fixed at low temperature by considerations other than kinetics. Clearly, there is no 'spillover' of electrons at low temperature from P700 to the functional iron-sulfur cluser in the presence of the non-functional iron-sulfur cluster.

DISCUSSION

The experiments described here demonstrate the utility of the *in vitro* mutagenesis system and indicate that this technique may be superior to *in vivo* mutagenesis systems for studying certain aspects of Photosystem I. Attempts at interposon mutagenesis of the *psaA* and *psaB* genes of *Synechococcus* sp. PCC 7002 have failed, presumably because Photosystem I function in this cyanobacterium is required for viability. Similar problems, as well as undesired secondary mutations, could arise during attempts to probe the structural and functional properties of the low molecular mass polypeptides of the Photosystem I complex by site-directed mutagenesis. Expression of mutant proteins in *E. coli*, combined with *in vitro* reconstitution of these proteins onto Photosystem I cores, has provided a mechanism to study protein alterations that could severely perturb electron transport and thereby prevent viability in the cyanobacterium. This study demonstrates the feasibility of this approach and has defined a new role for the PsaD protein in Photosystem I. It has also permitted an identification of the F_A and F_B clusters relative to the cysteine ligands in the PsaC protein and allowed us to show that F_B is not required for low temperature electron flow from P700 to F_A. Future work will be concerned with the room temperature properties of these mutants, including the role of F_A and/or F_B in forward electron flow to ferredoxin.

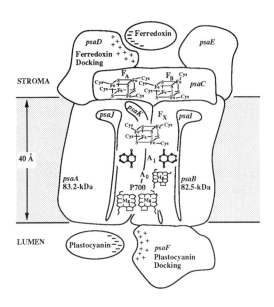

ACKNOWLEDGMENTS

This material was based upon work supported by the Cooperative State Research Service, U.S. Department of Agriculture under Agreement No. 87-CRCR-1-2382 and by the National Science Foundation, Biophysics Division under Grant # DMB-9043333.

REFERENCES

1. Parrett, K. G., Mehari, T., Warren, P. V., and Golbeck, J. H. (1989) Biochim. Biophys. Acta (1989) 973, 324-332.
2. Parrett, K. G., Mehari, T. and Golbeck, J. H (1990) Biochim. Biophys. Acta 1015, 341-352.
3. Mehari, T., Parrett, K. G., and Golbeck, J. H (1991) Biochim. Biophys. Acta 1056, 139-148.
4. Li, N., Bryant, D.A., Frank, G., Zuber, H., Warren, P.V. and Golbeck, J.H. (1991) Biochim. Biophys. Acta (in press).
5. Golbeck, J.H. and Bryant, D. (1991) "Photosystem I" in Current Topics in Bioenergetics, Vol 16 (Lee, C.P., ed) pp. 83-177.
6. Shinozaki, K. and Sugiura, M. (1985) Mol. Gen. Genet. 200, 27-32.
7. Friedberg, D., Kaplan, A., Ariel, R., Kessel, M. and Seijffers, J. (1989) J. Bacteriol. 171, 6069-6076.
8. Li, N., Zhao, J.D., Warren, P. V., Warden, J.T., J., Bryant, D. and Golbeck, J.H. (1991) Biochemistry (in press).
9. Zhao, J., Warren, P.V., Li, N., Bryant, D.A. and Golbeck, J.H. (1990) FEBS Lett. 276, 175-180.
10. Zhao, J., Li, N. Warren, P.V., Bryant, D.A. and Golbeck, J.H. (in preparation).
11. Malkin, R. (1984) Biochim. Biophys. Acta 764, 63-69.
12. Golbeck, J.H. and Warden, J.T. (1982) Biochim. Biophys. Acta 681, 77-84.
13. Sakurai, H., Inoue, K., Fujii, T. and Mathis, P. (1991) Photosyn. Res. 27, 65-71.
14. Armstrong, F.A., George, S.J., Cammack, R., Hatchikian, E.C. and Thomson, A.J. (1989) Biochem. J. 264, 265-273.
15. Conover, P.R., Kowal, A.T., Fu, W., Park, J-B., Aono, S., Adams, M.W.W. and Johnson, M.K. (1990) J. Biol. Chem. 265, 8533-8541.
16. Beinert, H. and Thomson, A. J. (1983) Arch. Biochem. Biophys. 222, 333-361.

ORGANIZATION OF THE PHOTOSYSTEM II REACTION CENTER

Kimiyuki Satoh

Department of Biology, Okayama University, Okayama 700,
Japan

INTRODUCTION

The efficient energy transformation in the primary process of
photosynthesis is ensured by the highly ordered organization of photochemical reaction centers, in physical, chemical and biological sense.
In photosystem II (PSII), the primary charge separation takes place in a
pigment-protein complex consisted of D1 and D2 proteins and some other
components, which has recently been isolated and the chemical and physical properties have been analyzed[1-4]. Present article deals with the
organization of PSII reaction center in two aspects. In the first part,
the basic chemical architecture of the isolated PSII reaction center
will be described. This includes stoichiometries of protein subunits
and of pigments and redox cofactors. In the second part, the discussion
will be focussed on some aspects of light-regulated turnover of D1
subunit of the PSII reaction center, which include the mechanism of
light-regulated synthesis of D1 protein in isolated chloroplasts and the
characterization of an enzyme involved in the C-terminal processing of
D1 precursor protein.

PART I. STRUCTURE OF ISOLATED COMPLEX

Polypeptide subunits

The PSII reaction center can now be prepared from thylakoids of
higher plants, algae and cyanobacteria with a variety of methods using
different detergents[1, 5-7]. The complex usually contains five protein
subunits; i.e., D1 and D2 proteins, α- and β-subunits of cytochrome
b-559 and psbI gene product (Fig. 1). The amino acid composition of
each subunit is now predicted based on the nucleotide sequence of chloroplast genes[8, 9] and the C- & N-terminal analysis of mature proteins in
the isolated complex[10]. Each subunit has characteristic amino acid
compositions. For example, D1 protein is enriched in Ala, Gly, Leu and
Ile, but lacks Lys. Ala is also one of the major components in D2
protein. On the other hand, psbI gene product is highly enriched in Phe
(16. 7 molar per cent), but lacks Ala. Taking advantage of these biased
distribution of amino acids in the individual subunit, we can determine
the stoichiometry in the isolated reaction center based on the amino
acid analysis. Table 1 shows an experimentally obtained ratio between

Ala and Phe in the isolated complex and compares the experimental value with those calculated for different stoichiometries. The experimental data can best be interpreted by assuming that each subunit is present in an equimolar ratio in the complex rather than by assuming more than two copies for lower molecular-weight components.

The function of each subunit of PSII reaction center has not been clearly elucidated at present. However, recent preliminary results indicated that a complex consisting of D1 and D2 proteins, but depleted in cytochrome b-559 and psbI gene product, carries all of the pigments in the PSII reaction center and furnishes site for the primary charge separation in PSII[11].

Stoichiometry of pigments and redox cofactors

Significant disagreements exist concerning to the number of pigments associated with PSII reaction center between different research groups[1, 5-7, 12]. In the first report in 1987, we evaluated, by a reversed-phase HPLC, the chlorophyll a pheophytin a β-carotene molar ratio in the complex from spinach to be around 5:2:1 (see ref. 1). The PSII reaction center complex of spinach purified by isoelectric focussing in the presence of digitonin, conducted after Triton X-100 extraction and partial purification, contains six chlorophyll a, two β-caro-

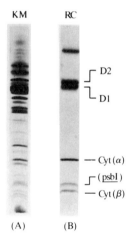

Fig. 1. Polypeptide profile of the PSII reaction center complex (RC) isolated from spinach grana thylakoids (KM).

Table 1. Experimental and calculated ratios between Phe and Ala in the isolated PSII reaction center.

Molar ratio			Phe/Ala ratio
D1-D2	Cyt b559	(psbI)	
1	1	1	1. 05
1	1	2	1. 13
1	2	1	1. 11
1	2	2	1. 18
Experimental			0. 99 ± 0. 04

tene, one cytochrome b-559 heme per two pheophytin a[12]. In this preparation, only one out of the two pheophytin a molecules in the complex is
engaged in the primary photochemistry as shown by the steady-state
kinetics of photoreduction. Thus, two pheophytin a molecules corresponds to one reaction center in this type of preparation, as in the
intact membranes. This result, together with the atomic ratio between
Mg and Fe of six to one in this complex which coincides well with the
molar ratio between chlorophyll a and cytochrome b-559 heme (deta not
shown), convinces us of the pigment stoichiometries of 6:2:2:1:1 for
chlorophyll a, pheophytin a, β-carotene, cytochrome b-559 heme and
pheophytin a active in the primary photochemistry.

On the other hand, a preparation of PSII reaction center isolated by
an alternate method which avoids Triton X-100 and uses a combination of
dodecyl-maltoside and a chaotropic reagent, LiClO$_4$, was reported to
contain ten to twelve chlorophyll a, two to three pheophytin a, along
with two cytochrome b-559 heme[6]. The authors claimed that this number
of cytochrome corresponds to one reaction center and thus the antenna
size of this reaction center is much larger than that of Nanba and Satoh
type preparations. However, the kinetic characteristics and the
chemical composition of this preparation were er-examined by Gounares et
al., and they pointed out that the pigment stoichiometries of this
preparation are essentially identical with those of the preparation
described above[13].

Another type of preparation contains four chlorophyll a and one β-
carotene per two pheophytin a[5]. These stoichiometric ratios can be
explained by partial depletion of pigments caused by detergent
treatment. We have recently prepared two types of pigment-protein
complex of PSII with different pigment stoichiometries by an intensive
treatment with higher concentration of Triton X-100 (0.5-1.0 %) in the
washing medium for DEAE-Toyopearl chromatography. One type of
preparation contains four chlorophyll a, two pheophytin a, one β-
carotene, one cytochrome b-559 heme per one photoactive pheophtin a, and
the other contains four chlorophyll a, one pheophytin a, one β-
carotene, one cytochrome b-559 heme per one photoactive pheophytin a.
The latter type of preparation is interesting in that almost all of the
pheophytin molecule in the complex can be photoreduced (Fig. 2). There

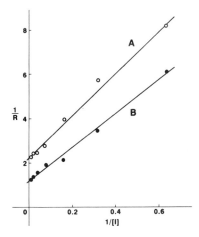

Fig. 2. Reciprocal plots of the light-saturation curves of photoaccumulation of reduced pheophytin a monitored at 545 nm for two
preparations of PSII reaction center complex. The molar ratio
between the photochemically reduced and the chemically estimated pheophytin is plotted against the actinic light intensity.

are two possibilities to explain the results: (i) the pheophytin molecule in the inactive branch which originally is not directly involved in the primary photochemistry of PSII is modified to engage in the steady-state photochemistry as a result of alteration in the protein structure, or (ii) the pheophytin molecule on the inactive branch of PSII reaction center is selectively removed during preparation. The later possibility seems to be more probable, since there is no reason to expect that two cytochrome b-559 hemes are preserved in the complex after the intensive treatment with Triton X-100 and that we could not detect any heterogeneity in the kinetic behavior of the photoreduction of pheophytin a. Thus we speculate that the reaction center complex reported here was produced by dissociation of a pheophytin a molecule on the inactive branch, as well as of the two additional chlorophyll a and one β-carotene molecules from the complex.

Structure of P-680

It is well documented that organization of PSII reaction center is similar in some respects to that of purple bacterial reaction center[14-16]. However, the similarity does not fully extend into the details. The donor side of PSII is quite unique. Redox-potential of the primary donor, P-680, is approximately 600-800 mV more positive than that of the equivalent compomemt of purple bacteria. Thus we should expect some basic differences between these two systems in the structure of the primary donor.

A preliminary Stark effect measurement suggested that P-680 is more like a monomeric chlorophyll species[17]. Another support for the monomeric nature of P-680 was provided by an EPR measurement[18, 19]. Due to the absence of the primary quinone acceptor (Q_A)[1], the triplet yield is high without preincubation under reducing conditions in the isolated PSII reaction center. The complexes were oriented on mylar strips and EPR spectra of the triplet state were taken during illumination at liquid helium temperature. Essentially the same spectral features as in PSII membranes were observed[19]. Despite the lower quality of the data[19], it was still observable that the orientation maximum for the Z peak deviates 20° - 30° from 0°. This is consistent with the triplet residing on a chlorophyll molecule that is tilted approximately 30° out of the plane of the mylar, as clearly demonstrated for oriented membranes[19].

The observations reported above exemplify that the structure of P-680 is somehow different from that of purple bacterial counterpart, in spite of the fact that the His ligands in L and M subunits are preserved in the D1 and D2 proteins. A model is favored in which P-680 is a chlorophyll, structurally analogous to one of the monomeric bacterio-chlorophyll of the purple bacterial reaction center[16].

PARTS II. SOME ASPECTS OF LIGHT-REGULATED TURNOVER OF D1 SUBUNIT OF THE PSII REACTION CENTER --DYNAMIC ASPECTS OF ORGANIZATION

One of the unusual characteristics of D1 subunit of PSII reaction center is the extraordinarily rapid rate of light-dependent synthesis[20, 21]. The phenomenon is hypothesized to represent a process sustaining the efficiency of energy conversion under environmental conditions; the repair of photodamaged PSII reaction centers[22]. The light-dependent step in the protein synthesis is now believed to be at the stage of translation, since transcripts of psbA gene are abundant in dark-adapted chloroplasts[23]. The translation is followed by a C-terminal cleavage before organizing into functional oxygen evolving complex.

Mechanism of light-regulated synthsis of D1 protein in isolated chloroplasts

The light-dependent translation of D1 protein can be observed in
isolated chloroplasts. The synthsis was inhibited by the presence of
atrazine (Fig. 3), and the inhibition curve was practically identical
with that for photosynthetic electron transport measured under similar
conditions. Under the concentration range, atrazine elicites a dramatic
effect on the stromal level of ATP in chloroplasts (Fig. 3, Table 2).

The effect of some kinds of inhibitors of photosynthetic phosphorylation which interrupt at different steps in the process was examined
both on the ATP level in chloroplasts and the light-induced synthesis of
D1 protein (Table 2). Phloridzin is known to be an energy transfer
inhibitor in the photosynthetic phosphorylation and to reduce the ATP
level in illuminated chloroplasts without affecting the electrochemical
gradient. The addition of phloridzin at 8 mM caused inhibition of the
synthesis of D1 protein in the light. At the same time, the ATP level
was reduced to the dark control level. Both CCCP, which uncouples the
energy transfer reaction from photosynthetic electron transport by
carrying H^+ across thylakoid membranes, and nigercin, an ionophore also
known to abolish ΔpH, completely inhibited the synthesis of D1 protein
in illuminated chloroplasts. The results thus suggest that the light-
inducible, stromal level of ATP is solely responsible for sustaining the
translation of D1 protein in isolated chloroplasts.

In order to test validity of the above mentioned assumption that
only stromal ATP level is responsible for regulation of the translation
of D1 protein in isolated chloroplasts, the ATP level in chloroplasts
was manipulated by adding metabolites which are expected either to
reduce or to create the internal ATP levels. In order to deplete the
stroma of ATP, glycerate, which is transported into the stroma via
glycerate translocator and then phosphorylated inside chloroplasts by
glycerate kinase at the expose of internal ATP[24], was added to
illuminated chloroplasts. Measurement of actual level of ATP indicated
that glycerate resulted in a depletion of the ATP level to approximately
one-fourth in the illuminated chloroplasts, to almost to the dark
control level, and at the same time, the synthesis of D1 protein was
completely inhibited even in the presence of saturating actinic light.
On the other hand, the combined addition of dihydroxyacetone phosphate
(DHAP) and oxaloacetate (OAA) to chloroplasts is expected to facilitate
the Calvin-Benson cycle to reverse at the step where ATP is normally

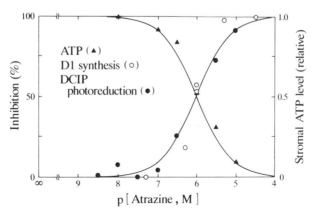

Fig. 3.   Effect of atrazine on the synthesis of D1 protein, DCIP
photoreduction and stromal ATP level in isolated spinach
chloroplasts.

Table 2. Synthesis of D1 protein and ATP level in isolated chloroplasts under different conditions

Conditions	ATP level (nmol mg Chl^{-1})	Synthesis of D1 (%)
Dark	2.36 ± 0.11	0
Light	7.85 ± 0.16	100
Light, Phloridzin (8 mM)	2.56 ± 0.18	0
Light, CCCP (10 μM)	2.96 ± 0.01	0
Light, Valinomycin (1 μM)	7.49 ± 0.07	99
Light, Nigericin (5 μM)	2.33 ± 0.04	0
Light, Glycerate (10 mM)	2.40 ± 0.11	0
Dark, DHAP (1 mM)/OAA (1 mM)	5.26 ± 0.14	25

utilized in the conversion of 3-phosphoglycerate to 1,3-bisphosphoglycerate, thereby generating ATP in chloroplasts[25]. Measurements indicated that the ATP level in chloroplasts actually be elevated in the dark approximately 2-fold by the addition of 1 mM dihydroxyacetone phosphate and 1 mM oxaloacetate (Table 2). Under these conditions, substantial amounts of D1 protein was synthesized even in the complete darkness.

Present study clearly demonstrated that, in isolated chloroplasts, the translation of D1 protein is regulated by light-inducible level of ATP in chloroplasts which presumably corresponds to the soluble fraction in stromal compartment. This fraction of ATP in chloroplasts can easily be manipulated by externally added metabolites such as glycerate or dihydroxyacetone phosphate/oxaloacetate, by inhibitors of photosynthetic electron transport and photosynthetic photophosphorylation, or by actinic light intensity. Under these varied conditions of manipulation, the rate of the synthesis of D1 protein was proportional to the stromal level of ATP irrespective of the dark-light conditions[32].

If we accept the above interpretation, a question arises whether or not the mechanism is operating in vivo. Recent study by Michaels and Herrin (1990) using synchronously growing Chlamydomonas demonstrated that ATP availability is one of the regulators in the translation of D1 protein in light-dark grown cells[26]. Physiological regulation of chloroplast translation of D1 protein by energy supply has also been proposed by Mattoo et al (1984) in Spirodela[27]. These observations exemplify that the phenomenon observed in isolated chloroplasts is operating in vivo under certain circumstances. In this case, however, the rate of translation of D1 protein is far more exceeded than that of the D2 protein. Therefore, this process must be coupled in vivo with preferential degradations of D1 protein which is taking place prior or after the complex formation of PSII reaction center[28,29], since D1 and D2 subunits exist in approximately an equimolar ratio, in functional thylakoidal membranes. However, possibilities exist that the ATP-regulated mechanism observed in isolated chloroplasts is not operating in vivo and the translation of D1 protein is regulated by some kinds of light-dependent message other than ATP.

Pulse labeling experiments using isolated chloroplasts revealed that D1 protein is arrested in the dark after illumination as an intermediate of about 23 kDa, which disappears upon light illumination (Fig. 4). This suggests that at least a part of light regulation is at the stage of the elongation of D1 protein.

Processing of D1 precursor protein

The maturation of newly synthesized precursor D1 protein involves a C-terminal cleavage at Ala-344 on the amino acid sequence deduced from

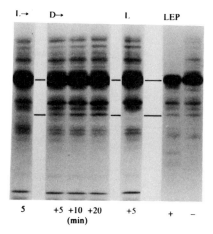

L.→   D→   L.   LEP

5   +5 +10 +20   +5   +   −
        (min)

Fig. 4. Appearance of a polypeptide band during dark incubation after
light exposure (50 Wm^{-2}). The band disappears by the subsequent illumination and is resistant to the lysylendopeptidase
(LEP) treatment.

psbA gene of spinach[34, 37]. The processing seems to be essential for the
assembly of catalytic center of water cleavage, but not for the primary
photochemistry of PSII[35]. The enzyme involved in the maturation was
solubilized from spinach thylakoids by Triton X-100 treatment and highly
purified from the extracts by a series of chromatographic procedures[36-38]. The spinach enzyme can process the D1 protein found in LF1 mutant
of a green alga S. obliquus. A synthetic oligopeptide corresponding to
the 29 amino acid of the C-terminal region of the precursor D1 protein
can also be cleaved at the predicted site indicating that the C-terminal
29 amino acids is sufficient for the recognition[37, 38]. An precursor of
24 kDa protein of Mn-stabilizing function of PSII which also requires
translocation through thylakoid membranes can not be cleaved by this
enzyme. The enzyme was suggested to be a monomeric protein of about 45
kDa by SDS-polyacrylamide gel electrophoresis conducted in the presence
of 6 M urea. The enzymatic activity was highly sensitive to the ionic
strength in the medium,, but was resistant to the prevailing protease
inhibitors such as PMSF, o-phenanthrolin, thiol reagents and heavy metal
ions.

Acknowledgements

This work was supported in part by Grant-in-Aid for Scientific
research on Priority Areas (03262210): "Molecular Mechanism of Plasticity and Signal Response in Plant Gene Expression" and for Scientific
Research (02454013) from the Ministry of Education, Science and Culture
of Japan and in part by Iwatani Naoji Foundation.

REFERENCES

1. O. Nanba and K. Satoh (1987) Proc. Natl. Acad. Sci. USA 84, 109-112.
2. K. Satoh (1989) in "Photochemical Energy Conversion (J. R. Norris
   ed.)", Elsevier Biomed. Press (Amsterdam) pp. 238-250.
3. J. Barber and A. Melis (1990) Biochim. Biophys. Acta 1020, 285-289.
4. M. R. Waiselewski, D. G. Johnson, M. Seibert and Govindjee (1989)
   Proc. Natl. Acad. Sci. USA 86, 524-528.

5. J. Barber, D. J. Chapman and A. Telfer (1978) FEBS Lett. 220, 67-73.

6. D. F. Ghanotakis, J. C. de Paula, D. M. Demetriou, N. R. Bowlby, J. Petersen, G. T. Babcock and C. F. Yocum (1989) Biochim. Biophys. Acta 974, 44-53.

7. K. Satoh and H. Nakane (1989) in "Current Research in Photosynthesis (M. Baltschefesky ed.)", Kluwer Academic Pub. (Dordrecht) Vol. 1, 271-274.

8. K. Ohyama, H. Fukuzawa, T. Kohchi, H. Shirai, T. Sano, S. Sano, K. Umesono, Y. Shiki, M. Takeuchi, Z. Chang, S. Aota, H. Inokuchi and H. Ozeki (1986) Nature 322, 572-574.

9. K. Shinozaki, M. Ohme, M. Tanaka, T. Wakasugi, N. Hayashida, T. Matsubayashi, N. Zaiat, J. Chunwongse, J. Obokata, K. Yamaguchi-Shinozaki, C. Ohto, K. Yorazawa, B. Y. Meng, M. Sugita, H. Deno, T. Kamogashira, K. Yamada, J. Kusuda, F. Takaiwa, A. Kato, N. Tohdoh, H. Shimada and M. Sugiura (1986) EMBO J. 5, 2043-2049.

10. Y. Takahashi, H. Nakane, H. Kojima and K. Satoh (1990) Plant & Cell Physiol. 31, 273-280.

11. X.-S. Tang, K. Fushimi and K. Satoh (1990) FEBS Lett. 273, 257-260.

12. M. Kobayashi, H. Maeda, T. Watanabe, H. Nakane and K. Satoh (1990) FEBS Lett. 260, 138-140.

13. K. Gonaris, D. J. Chapman, P. Booth, B. Crystall, L. B. Giorgi, D. R. Klug, G. Porter and J. Barber (1990) FEBS Lett. 265, 88-92.

14. A. Trebst (1986) Z. Naturforsh. 41C, 240-245.

15. J. Deisenhofer, O. Epp, K. Miki, R. Huber and H. Michel (1985) Nature 318, 618-624.

16. H. Michel and J. Deisenhofer (1988) Biochem. 27, 1-7

17. M. Losche, K. Satoh, G. Feher and M. Y. Okamura (1988) Biophys. J. 53, 270a.

18. A. W. Rutherford (1985) Biochim. Biophys. Acta 807, 189-201.

19. F. J. E. van Mieghem, K. Satoh and A. W. Rutherford (1991) Biochim. Biophts. Acta 1058, 379-385.

20. R. J. Ellis (1981) Annu. Rev. Plant. Physiol. 32, 111-137.

21. K. Satoh, H. Y. Nakatani, K. E. Steinback, J. Watson and C. J. Arntzen (1983) Biochim. Biophys. Acta 724, 142-150.

22. D. J. Kyle, I. Ohad and C. J. Arntzen (1984) Proc. Natl. Acad. Sci. USA 81, 4070-4074.

23. E. Minami and A. Waranabe (1984) Arch. Biochem. Biophys. 172, 51-58.

24. U. I. Flugga and G. Hinz (1986) Eur. J. Biochem. 160, 563-570.

25. Y. Inoue, Y. Kobayashi, K. Shibata and U. Heber (1978) Biochim. Biophys. Acta 504, 142-152.

26. A. Michaelis and D. L. Herrin (1991) Biochem. Biophys. Res. Commun. 170, 1082-1088.

27. A. K. Mattoo, H. Hoffman-Falk, J. B. Marder and M. Edelman (1984) Proc. Matl. Acad. Sci. USA 81, 1380-1384.

28. A. K. Mattoo and M. Edelman (1987) Proc. Natl. Acad. Sci. USA 84, 1497-1501.

29. M. Wettern and I. Ohad (1984) Israel J. Bot. 23, 253-263.

30. H. W. Heldt (1969) FEBS Lett. 5, 11-14.

31. T. Akazawa (1991) Plant Physiol. in press.

32. H. Kuroda, N. Inagaki and K. Satoh, submitted.

33. N. Inagaki, H. Kuroda and K. Satoh, submitted.

34. M. Takahashi, T. Shiraishi and K. Asada (1988) FEBS Lett. 240, 6-8.

35. B. A. Diner, D. F. Ries, B. N. Cohen and J. G. Metz (1988) J. Biol. Chem. 263, 8972-8980.

36. N Inagaki, S. Fujita and K. Satoh (1989) FEBS Lett. 246, 218-222.

37. S. Fujita, N. Inagaki and K. Satoh (1989) FEBS Lett. 255, 1-4.

38. S. Fujita and K. Satoh, in preparation.

EXTRINSIC POLYPEPTIDES AND THE INORGANIC COFACTORS OF PHOTOSYSTEM II

D.F. Ghanotakis, A. Bakou and K. Kavelaki

Department of Chemistry, University of Crete
Iraklion, Crete, GREECE

INTRODUCTION

The oxygen-evolving complex of Photosystem II (PSII) is a highly ordered structure of various polypeptides which provide the binding sites for the inorganic cofactors required for catalytic activity. Highly purified PSII preparations capable of O_2-evolution activity consist of the D_1, D_2 and cytochrome b_{559} species augmented by three extrinsic polypeptides with apparent molecular masses of 33, 23 and 17 kDa (Ghanotakis and Yocum, 1990).

Of the three extrinsic polypeptides of Photosystem II (17, 23 and 33 kDa) the 33 kDa protein is tenaciously bound to PSII, but can be removed by various treatments. Experimental evidence has led to a model which attributes to the 33 kDa species a stabilizing role. The 17 and 23 kDa proteins are amenable to extraction/reconstitution experiments, and so it has been possible to demonstrate that although they are not directly involved in the catalytic mechanism of the photosynthetic water cleavage, the 23 and 17 kDa proteins play an important role under physiological conditions. On one hand they act as "concentrators" of the inorganic cofactors calcium and chloride which are essential for oxygen evolution activity (Ghanotakis et al., 1984a; Andersson et al., 1984; Miyao and Murata, 1985), and on the other hand they play an important role in the structural arrangement around the manganese complex (Ghanotakis et al., 1984b; Tamura et al., 1986).

It is now widely accepted that manganese, chloride and calcium are necessary for the sequential four-electron oxidation of water to molecular oxygen. Of the three types of ions involved in the photosynthetic splitting of water, manganese has been extensively characterized. Various studies have placed the locus of Cl^- action within the OEC (Homman, 1988; Sandusky and Yocum, 1984). The third ion required for oxygen-evolution activity is Ca^{2+} (Ghanotakis et al., 1984a).

Regulation of Chloroplast Biogenesis
Edited by J.H. Argyroudi-Akoyunoglou, Plenum Press, New York, 1992

The possible role of Ca^{2+} in the advancement of the S-state sequence of the OEC has been investigated by a series of different experimental approaches. Although various proposals have been made regarding its role, its binding site(s) has (have) not been characterized. It was previously shown that calcium and lanthanides compete for binding sites at the oxidizing side of Photosystem II (Ghanotakis et al., 1985). That particular discovery led Ghanotakis et al. (1985) to suggest a Concanavalin A-type structure in which Ca^{2+} binds in close proximity to the Mn complex and affects its structure. Such a structural arrangement is supported by recent EPR data. It has been shown that inhibition of O_2 evolution by extraction of Ca^{2+} from PSII membranes causes a structural change of the manganese cluster as demonstrated by the formation of a new modified multiline signal, which is attributed to an S_2-oxidation state of the OEC. This structural change, generated either by Ca^{2+} extraction with citric acid at pH 3 (Ono and Inoue, 1990) or by NaCl/EDTA washing (Bousac et al., 1990), causes enhanced kinetic stability of the manganese cluster.

Using a series of extraction-reconstitution experiments we have been able to get new information regarding the structural arrangement of the extrinsic polypeptides. In addition, selective substitution of Ca^{2+} with various paramagnetic and diamagnetic lanthanide ions has allowed us to extract information about the relative location of Ca^{2+}-binding sites with respect to other redox species of the oxidizing side of PSII.

MATERIALS AND METHODS

Photosystem II membranes were prepared as described by Ghanotakis et al. (1984a). Extraction of the 17 kDa and the 23 kDa polypeptides was carried out by exposure of intact PSII membranes to 2 M NaCl. Selective reconstitution with the various extrinsic proteins was carried out by the dialysis technique described by Ghanotakis et al. (1984c). Substitution of calcium by various lanthanide ions was carried out by treatment of intact PSII membranes with 2 M NaCl in the presence of 2 mM $LnCl_3$ (Ln: lanthanide in general). This treatment results in a PSII system which has all its calcium replaced by the lanthanide and moreover has been depleted of the extrinsic 17 and 23 kDa polypeptides. Reconstitution of this system with the 17 and 23 kDa polypeptides results in a preparation denoted as [Ln]-PSII). If the rebinding of the 17 and 23 kDa species was preceded by a 2 mM EDTA treatment, a system which retains all three extrinsic proteins but contains neither calcium nor a lanthanide was prepared ([Ln/EDTA]-PSII). The oxygen-evolution activity of the various PSII preparations was measured by a YSI Clark-type oxygen electrode. The manganese and calcium content of the various preparations was determined by atomic absorption spectroscopy (Perkin-Elmer, model 4000). The number of Dy^{3+} per PSII was estimated by EPR spectroscopy.

RESULTS AND DISCUSSION

A. Effect of the manganese complex on the binding of the 17 and 23 kDa polypeptides.

Exposure of intact Photosystem II membranes to 2 M NaCl results in release of the water soluble 17 and 23 kDa polypeptides (Fig. 1, lane 2) (Akerlund et al., 1982; Ghanotakis et al., 1984c). The manganese complex of the salt-washed PSII systems is easily destroyed by exposure to an exogenous reductant such as hydroquinone (Table I); under these conditions destruction of the manganese complex is not accompanied by removal of the 33 kDa protein (Fig. 1). The PSII system which retains both the 33 kDa polypeptide and the manganese complex has completely different properties when compared to the 33 kDa-retaining but Mn-depleted PSII system. Although the 17 and 23 kDa species easily rebind to the (33 kDa/(Mn)$_4$-retaining) PSII (Fig. 1, lane 4) no rebinding of these polypeptides is observed in the Mn-depleted system which retains the 33 kDa species (Fig. 1, lane 5).

It was previously shown that manganese removal by hydroxylamine results in partial loss of the 23 and 17 kDa polypeptides (Tamura and Cheniae 1985). As shown in Fig. 2, under our conditions a concentration of hydroxylamine as high as 6 mM is required for destruction of oxygen evolution capacity. The effect of such a treatment on the polypeptide content of PSII membranes is shown in Fig. 3. Treatment of PSII with 6 mM NH$_2$OH results in almost total release of the 17 kDa species and partial release of the 23 kDa polypeptide. Subsequent washing of the NH$_2$OH-treated PSII with a medium containing 100 mM NaCl removes the remaining traces of the 17 kDa, and most of the 23 kDa species. Treatment of intact PSII membranes with a medium of the same ionic strength (100 mM NaCl) removes a fraction of the 17 kDa polypeptide, but has no significant effect on the 23 kDa species (data not shown). Apparently, the binding of the 23 kDa polypeptide after destruction of the manganese complex by NH$_2$OH is modified, as indicated by the fact that it can be removed by a medium of relatively low ionic strength.

Table I. Effect of hydroquinone on the activity and Mn content of PSII membranes

Treatment	Mn-content (No./PSII R.C.)	O$_2$-evolution activity -CaCl$_2$	+CaCl$_2$
-None	4	690	700
-1 mM HQ	4	680	690
-2 M NaCl	4	180	520
i) 2 M NaCl ii) 1 mM HQ	1	0	0

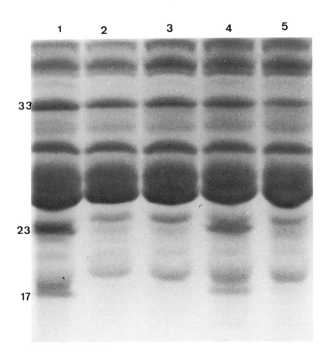

Figure 1. Gel electrophoresis patterns of various PSII preparations. Lane 1, control PSII; lane 2, 2 M NaCl-treated PSII; lane 3, as lane 2, then treated with 2 mM HQ; lane 4, as lane 2, then incubated with 17, 23 kDa; lane 5, as lane 3, then incubated with 17, 23 kDa.

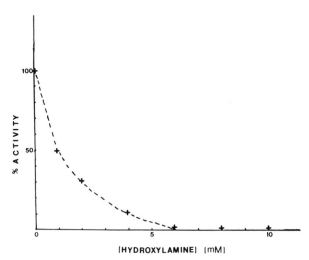

Figure 2. Effect of hydroxylamine on oxygen evolution activity of intact PSII membranes.

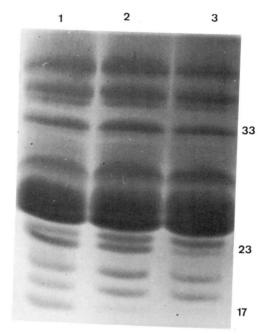

Figure 3. Gel electrophoresis patterns of PSII membranes: effect of NH₂OH on the binding of the 17 and 23 kDa polypeptides. Lane 1, control PSII membranes; lane 2, PSII membranes treated with 6 mM NH₂OH; lane 3, PSII membranes treated with 6 mM NH₂OH and then exposed, for 45 min, to 100 mM NaCl.

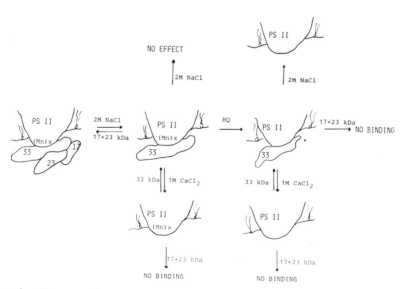

Figure 4. Proposed model for the interaction of the extrinsic polypeptides with the hydrophobic part of Photosystem II

The results presented above and a series of extraction-reconstitution experiments with PSII membranes depleted of all three extrinsic polypeptides (17, 23 and 33 kDa) (data not shown) support a model according to which destruction of the Mn complex causes structural changes at the oxidizing side of PSII, which result in release of the 17 and 23 kDa polypeptides (Fig. 4). Such a requirement was elegantly demonstrated by the photoreactivation experiments of two different groups (Becker et al., 1985; Ono et al., 1986) which demonstrated that photoligation of manganese to a (Mn-depleted/33 kDa retaining) PS II system is accompanied by tight rebinding of the 23 and 17 kDa species.

B. Calcium binding site(s) of PSII probed by lanthanides

The various Ln-treated PSII preparations were characterized by spectroscopic and polarographic techniques; the results are summarized in Table I. As shown in Table I, the [Ln]-PSII (Ln: La^{3+} or Dy^{3+}) and [Dy/EDTA]-PSII preparations retain all four atoms of the Mn complex but they are unable to oxidize water. When these preparations were incubated for at least 10 min at 25 °C with 10 mM $CaCl_2$, we observed no reactivation of the [Ln]-PSII but the [Ln/EDTA]-PSII system recovered 43% of its oxygen-evolving capacity. The fact that external calcium failed to reactivate the [Ln]-PSII preparations is in agreement with the results of Ghanotakis et al. (1985), which demonstrated that lanthanides bind to the PSII membrane much more strongly than calcium.

Although the preparations [La]-PSII and [Dy]-PSII have basically the same properties, there is a major difference in the EPR signal of the dark-stable Tyr radical (due to Tyr $_D^+$)

Table II. Characterization of various PSII preparations

Preparation[a]	polypeptide content[b]			Ca/PSII	Mn/PSII	Ln/PSII	VO_2 (+$CaCl_2$)
	17	23	33 kDa				
Control PSII	+	+	+	15	4	-	100%
[La]-PSII	+	+	+	0	4	n.d[c]	0
[Dy]-PSII	+	+	+	0	4	14-15	0
[Dy/EDTA]-PSII	+	+	+	0	4	0	43%
Tris[Dy]-PSII	-	-	-	0	0	1-3	0
Tris[Dy/EDTA]-PSII	-	-	-	0	0	0	0

[a] see text for description of various preparations
[b] see Fig. 1
[c] not determined

(Barry and Babcock, 1987). As shown in Fig. 5, in the presence of the ion La^{3+}, the EPR signal of Tyr$_D^+$ is the same as the one observed in control PSII membranes. In contrast, when calcium has been replaced by the paramagnetic ion Dy^{3+}, the EPR signal of the tyrosine radical is different; more specifically the hyperfine structure of the signal is lost, when observed under the same conditions. This observation

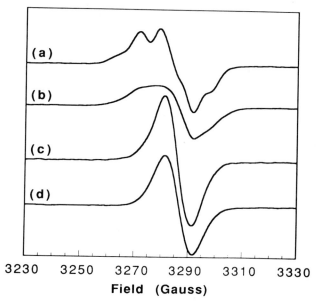

Figure 5. Comparison of the g=2 radical EPR signals in dark-adapted and illuminated [La]-PSII and [Dy]-PSII: (a) dark adapted [La]-PSII, (b) dark adapted [Dy]-PSII, (c) illuminated (200 K, 3 min) minus dark difference spectrum of [La]-PSII, and (d) illuminated (200 K, 3 min) minus dark difference spectrum of [Dy]-PSII. EPR spectrometer conditions: microwave frequency, 9.05 GHz; microwave power, 7.2 x 10^{-4} mW; field modulation frequency, 100 kHz; field modulation amplitude, 4.0 G; temperature, 15.0 K.

shows that there is a strong interaction between the paramagnetic Dy^{3+} and Tyr$_D^+$ (Innes and Brudvig, 1989)).

Since Ca^{2+} lacks both electronic transitions that give rise to accessible absorption spectra and unpaired electrons for studies through magnetic resonance techniques, its substitution with trivalent lanthanides might prove a useful approach for the characterization of the Ca^{2+}-binding sites of Photosystem II.

Acknowledgements

This research was supported by a grant from the Greek Ministry of Industry, Energy and Technology (PENED 89/7). Low temperature EPR experiments were carried out in collaboration with Dr. G. Brudvig and Ms. C. Buser at Yale University, USA (supported by a NATO travel grant (0326/88)).

REFERENCES

-Andersson, B., Critchley, C., Ryrie, I.J., Jansson, C., Larsson, C. and Anderson, J. (1984) FEBS Lett. 168, 113-17.

-Barry, B. and Babcock, G.T. (1987) Proc. Natl. Acad. Sci. USA 84, 7009-7103.

-Boussac, A., Zimmermann, J.-L. and Rutherford, A. W. (1990) Curr. Res. Photosynth. 1, 713-716.

-Ghanotakis, D.F., Babcock, G.T. and Yocum, C.F. (1984a) FEBS Lett. 167, 127-130.

-Ghanotakis, D.F., Topper, J. and Yocum, C.F. (1984b) Biochim. Biophys. Acta 767, 524-531.

-Ghanotakis, D.F., Topper, J., Babcock, G.T. and Yocum, C.F. (1984c) FEBS Lett. 170, 169-173.

-Ghanotakis, D.F., Babcock, G.T. and Yocum, C.F. (1985) Biochim. Biophys. Acta 809, 173-180.

-Ghanotakis, D.F. and Yocum, C.F. (1990) Annu. Rev. Plant Physiol. Plant Mol. Biol. 41, 255-276.

-Hirsh, D.J., Beck, W.F., Innes, J.B. and Brudvig, G.W. (1991) Biophys. J. 59, 180a.

-Homman, P.H. (1988) Biochim. Biophys. Acta 934, 1-13.

-Miyao, M. and Murata, N. (1985) FEBS Lett. 180, 303-8.

-Ono, T.-A. and Inoue, Y. (1990) Curr. Res. Photosynth. 1, 741-744.

-Sandusky, P. and Yocum, C.F. (1984) Biochim. Biophys. Acta 766, 603-611.

-Tamura, N. and Cheniae, G. (1985) Biochim. Biophys. Acta 809, 245-59.

GLYCINE BETAINE, SUCROSE, OR CHLORIDE IONS PROTECT ISOLATED PHOTOSYSTEM 2

PARTICLES FROM DENATURATION AND INACTIVATION

Katerina Kalosaka and George C. Papageorgiou

National Research Center Demokritos, Institute of Biology
Athens, Greece 153 10

INTRODUCTION

The oxygen-evolving photosystem 2 complex (PS2) of chloroplast thylakoids comprises the light harvesting Chl proteins, the core protein complex which houses the photochemical reaction center and the 4-nuclear Mn cluster, and three regulatory extrinsic proteins of molecular masses of 18-kDa, 23-kDa and 33-kDa. Furthermore, for O_2 evolution 2-3 Ca^{2+} ions and few Cl^- ions per reaction center are necessary (Hansson and Wydrzynski, 1990).

The Cl^- requirement for O_2 evolution is evoked by treatments that either displace the Cl^- bound to the PS2 complex (Itoh and Uwano, 1985) or they dissociate the 17 and 24 kD proteins, which function as Cl^- concentrators (Miyao, M. and Murata, N., 1985). Such treatments result in reduced rates of photosynthetic O_2 evolution. However, O_2 evolution can be stimulated again by adding monovalent anions, including Cl^-, to the depleted samples (Damoder et al. 1986). The precise role of Cl^- in photosynthetic O_2 evolution is still unclear (Wydrzynski et al. 1990). At least two roles are recognized: as electron transport cofactor and as protein structure stabilizer.

In this work, we used high concentrations of glycine betaine (betaine; Papageorgiou et al. 1989; Papageorgiou et al. 1991) and sucrose in order to stabilize the structure of isolated PS2 complexes and thus to distinguish between the two roles of Cl^- ions.

MATERIALS AND METHODS

Highly active PS2 particles (400-450 µmoles O_2/mg Chl.h) were prepared as in Kuwabara and Murata (1982) and were stored at -60 °C. The composition of these particles was as described by Murata et al. (1984). Partial depletion of Cl^- was performed according to Wydrzynski et al. (1990), with the following differences. After a quick thawing, the PS2 particles were diluted with buffer A containing less than 0.2 mM Cl^- (20 mM Mes-NaOH, pH 6.3) to 0.5 mg Chl/ml and were washed twice with the same buffer. Depending on the experiment, the washing and resuspension medium could include also 0.8 M betaine, or 0.8 M sucrose. Such samples were considered to be partially depleted of Cl^- ions. The corresponding Cl^--sufficient controls were similarly prepared samples that were suspended in the presence of 10 mM NaCl.

Regulation of Chloroplast Biogenesis
Edited by J.H. Argyroudi-Akoyunoglou, Plenum Press, New York, 1992

All treatments were performed in darkness and at 0 °C. However, incubations were carried out at 25 °C.

O2 evolution was assayed with a Clark O_2 concentration electrode. The samples were made in buffer A and contained 5-6 µg Chl/ml and 0.3 mM phenyl-p-benzoquinone. In some experiments, the assay mixtures contained also 10 mM NaCl. The protein composition of the samples was analysed by 15% SDS-polyacrylamide gel electrophoresis in the buffer system of Laemmli (1970). Aliquots taken out of the incubator were washed with buffer A supplemented with 10 mM NaCl and were resuspended in the same medium at 1.5 mg Chl/ml before solubilization with SDS.

RESULTS

Figure 1 illustrates the following experiment: PS2 particles were washed and resuspended in Cl⁻ free buffer A, in the absence or presence of 0.8 M betaine, or 0.8 M sucrose, and were then incubated at room temperature. The time course of O_2 evolution activity was determined during the incubation. The most rapidly deteriorating sample was the suspension in the absence of Cl⁻, sucrose or betaine. Its inactivation half-time was 20 h. On the other hand, the inactivation half-times of particles suspended in the presence of 0.8 M betaine, 0.8 M sucrose or 10 mM NaCl were typically in the range of 80 h. Below 0.8 M the organic osmolytes exerted reduced protection, while above 0.8 M there was no additional protection. Particles washed with Cl⁻-free buffer A are depleted of the essential for O_2 evolution Cl⁻. This becomes evident upon comparing initial rates obtained of particles suspended in buffer A alone (open circles) and in buffer A plus 10 mM NaCl (solid circles).

These data are consistent with previous reports that Cl⁻-depleted PS2 particles are unstable when suspended in Cl⁻-free media (Wydrzynski et al., 1990). However, we also show here that betaine or sucrose are even better stabilizers of the functionality of the PS2 particles.

Figure 2 illustrates a similar experiment as that of Fig. 1. Here, however, O_2 evolution was assayed in the presence of 10 mM NaCl. The surviving activity represents the sum of the fractions of the still active

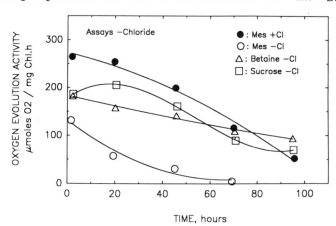

Fig. 1. Effect of betaine (0.8 M) or of sucrose (0.8 M) on the decay of O_2 evolution activity of PS2 particles during incubation in buffer A at room temperature and darkness. Assays in Cl⁻-free reaction mixtures.

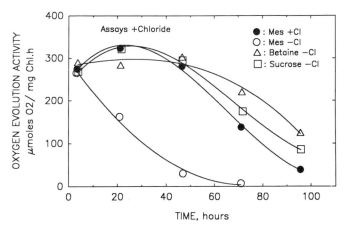

Fig. 2. Effect of betaine (0.8 M) or of sucrose (0.8 M) on the decay of O$_2$ evolution activity of PS2 particles during incubation in buffer A at room temperature and darkness. Assays in the presence of 10 mM NaCl.

particles plus those that can be reactivated by mM levels of Cl⁻ ions. The fact that activities recorded in this experiment exceeded those of the Fig. 1 experiment indicates that the fraction of Cl⁻reactivatable PS2 particles is non-negligible.

To obtain a better comparison of the function-stabilizing efficiencies of betaine, sucrose and Cl⁻, we plotted in Fig. 3 the rate differences between the plus Cl⁻ and the minus Cl- oxygen evolution assays against the duration of the incubation. The ascending branches of the resulting curves reflect the rise of the Cl⁻-reactivatable particle population while the descending branches the rise of the irreversibly inactivated particle population. In the absence of Cl⁻, betaine, or sucrose, only irreversible inactivation was observed (open circles). On the other hand, the bell-shaped time courses were recorded in their presence. This is indicative of processes

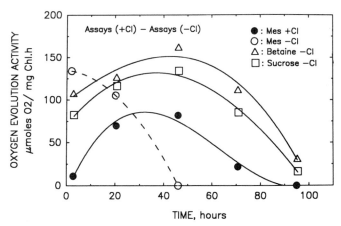

Fig. 3. The differences between O$_2$ evolution rates by PS2 particles measured in the presence and in the absence of 10 mM NaCl in the reaction mixture as a function of the duration of the incubation.

393

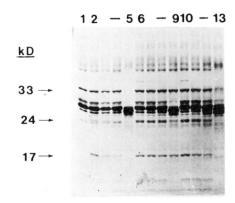

1 2 — 5 6 — 910 — 13

kD

33 →

24 →

17 →

Fig. 4. Protein composition of PS2 particles washed and incubated in the
following media for various time periods: (lane 1), 2 M NaCl-treated PS2 particles ; (lanes 2-5), Cl⁻-free (buffer A). Incubation
for 0, 8, 22, 42 h, respectively ; (lanes 6-9), buffer A supplemented with 1 M betaine. Incubation for 0, 22, 42, 72 h, respectively ; (lanes 10-13), buffer A supplemented with 10 mM NaCl. Incubation for 0, 22, 42 and 72 h, respectively.

that lead first to the accumulation of Cl⁻-reactivatable particles and subsequently to their irreversible inactivation. The figure clearly shows the
superiority of betaine as protector of PS2 particle function.

Figure 4 shows the variation of the protein composition of PS2 particles during prolonged incubation at room temperature and darkness. When
incubated in Cl⁻free medium, and in the absence of betaine or sucrose, the
PS2 particles lost first the 18-kDa and 23-kDa extrinsic proteins and then
they experienced more drastic changes, including the loss of the 33-kDa
protein. This is evident by the protein profile of the sample incubated for
0 h (lane 2), 8 h (lane 3), 22 h (lane 4) and 42 h (lane 5). In the presence of 1 M betaine, the 18-kDa and 23-kDa proteins remain attached to the
particles. This is evident for particles incubated for 0 h (lane 6), 22 h
(lane 7) and 42 h (lane 8). However, after 72 h (lane 9) further changes in
the protein profile became discernible. These results are consistent with
previous observations (Papageorgiou et al., 1991; Murata, N., Mohanty, P.
Hayashi, H. and Papageorgiou, G. C., unpublished experiments). A similar
protection of the 18-kDa and the 23-kDa extrinsic proteins was observed
when 1 M sucrose was included in the incubation medium (data not shown).
Presence of 10 mM NaCl also protected the PS2 particles against the dissociation of the extrinsic proteins (lanes 10-13). However, its protective
effect was inferior to betaine or sucrose.

DISCUSSION

Wydrzynski et al. (1990) have shown that, when suspended in Cl⁻-free
medium, the PS2 particles lose gradually their O_2 evolution activity. O_2
evolution can be restored by adding 10 mM NaCl to the suspension, but after
prolonged aging the activity loss is irreversible. On the basis of such
data, they raised the question whether Cl⁻ is indeed a cofactor of pre-
photosystem 2 electron transport ("Cl⁻-effect; Miyao and Murata, 1985) or
it is an artifact due to the aging of the particles.

In the present work, we have demonstrated the feasibility of studying the Cl⁻-induced stimulation of electron transport independently of its structure-stabilizing effect. This is possible by stabilizing PS2 particles suspended in Cl⁻-free media with high concentrations of glycine betaine, or of sucrose. Indeed, these compounds (particularly betaine) were observed to be superior to the Cl⁻ ion in maintaining the functional intergity (Fig. 3) and the structural integrity (Fig. 4) of PS2 particles in suspension. Moreover, these compounds appear to prevent the depletion of essential Cl⁻ upon washing of the PS2 particles with Cl⁻-free medium. This becomes evident from the comparison of the initial rates of the washed particles (Figs. 1 and 2).

It is interesting to note that three vastly different solutes, an inorganic anion, a bulky electroneutral disaccharide, and a bulky zwitterion exert, qualitatively, a similar protective effect on higher-order protein structure. Betaine is the superior performer, possibly because it is both electrically charged and bulky. Indeed, we have proposed earlier (Papageorgiou et al., 1989) that betaine may form layers of preferentially oriented dipoles on protein surfaces. These would shield them from chaotropic solutes both sterically and electrostatically.

REFERENCES

Damoder, R., Klimov, V. V. and Dismukes G. C. 1986, The effect of Cl⁻ depletion and X⁻ reconstitution on the oxygen evolution rate, the yield of multiline line manganese EPR signal and EPR signal II in the isolated photosystem II complex, Biochim. Biophys. Acta, 891: 378.

Hansson, O. and Wydrzynski, T. 1990, Current perceptions of photosystem II, Photosynth. Res., 23: 131.

Itoh, S. and Uwano, S. 1986, Characteristics of the Cl⁻ action site in the O2 evolution reaction in PSII particles: electrostatic interactions with ions, Plant Cell Physiol., 27: 25.

Kuwabara, T. and Murata, N. 1982, Inactivation of photosynthetic oxygen and the release of polypeptides and manganese in the photosystem II particles of spinach chloroplasts, Plant Cell Physiol., 23: 533.

Laemmli, U.K. 1970, Cleavage of stuctural proteins during the assembly of the head of bacteriophage T4, Nature, 227: 680.

Miyao, M. and Murata, N. 1985, The Cl⁻ effect on photosynthetic oxygen evolution: interaction of Cl⁻ with 18-kDa, 24-kDa and 33-kDa proteins, FEBS Lett., 180: 303.

Murata, N., Miyao, M., Omata, T., Matsunami, H. and Kuwabara, T. 1984, Stoichiometry of components in the photosynthetic oxygen evolution system of photosystem II particles prepared with Triton X-100 from spinach chloroplasts, Biochim. Biophys. Acta, 765: 363.

Papageorgiou, G. C. , Fujimura, Y. and Murata, N. 1989, On the mechanism of betaine protection of photosynthetic structures in high salt enviroment in: "Current Research in Photosynthesis," M. Battscheffsky, ed., De Kluyver, Dordrecht.

Papageorgiou, G. C., Fujimura, Y and Murata, N. 1991, Protection of the oxygen-evolving Photosystem II complex by glycinebetaine, Biochim. Biophys. Acta, 1057: 361.

Wydrzynski T., Baumgart, F., MacMillan, F. and Renger, G. 1990, Is there a direct chloride factor requirement in the oxygen-evolving reactions of photosystem II? Photosynth. Research., 25: 59.

ASSEMBLY OF THE AUXILIARY CHLOROPHYLL *ab* LIGHT-HARVESTING ANTENNA OF PHOTOSYSTEM-II IN CHLOROPLASTS

Michael A. Harrison, Jeff A. Nemson and Anastasios Melis

Department of Plant Biology, 411 GPBB
University of California, Berkeley, CA 94720, U.S.A.

INTRODUCTION

The light-harvesting complex of PSII (LHC-II) contains several polypeptides with molecular masses in the range 30-25 kDa, the products of a nuclear *cab* multi-gene family [9,16]. Structural studies have led to the proposal of molecular models in which LHC-II is assembled as a trimeric complex in the thylakoid membrane [15], with each polypeptide component of the trimer proposed to bind 7 chlorophyll *a* and 6 Chl *b* molecules [26]. In addition to the basic trimeric assembly, there exist within the antenna of PSII additional Chl *ab*-binding polypeptides which are distinct in terms of their immunological cross-reactivity [7,14] and Chl complement [1,13,26]. Solubilization studies suggest that these lower abundance complexes may occupy positions between the bulk of the LHC-II antenna and the PSII-core, and may therefore function as intermediates in excitation energy transfer from antenna to reaction centre [5,6].

Mutants lacking or deficient in Chl *b* are useful in the study of the assembly of the light-harvesting antenna. Relevant in this respect are the Chl *b*-deficient *chlorina* 103 [22] and the Chl *b*-less *chlorina* f2 mutants of barley [25], which are unable to synthesize the complement of Chl *b* required for stable assembly of the bulk light-harvesting proteins. These mutants have a greatly reduced complement of light-harvesting proteins within the thylakoid membrane [5-7,20,25,27] and a concomitantly small Chl antenna size for both photosystems [11,12].

This study is directed towards understanding the hierarchy of assembly and the molecular organization of the auxiliary antenna of PSII. Polypeptide composition, quantitation of the relative amounts of the various subunits, investigation of their Chl content, and their organization within the membrane relative to each other and to the PSII-core were addressed. Measurement of polypeptide subunits of the LHC-II in wild-type, mutant *chlorina* 103 and mutant *chlorina* f2 barley were combined with measurements of the functional Chl antenna size of PSII to provide insight into the architecture of the LHC-II complex. Examination of the residual LHC-II antenna in the *chlorina* mutants enabled identification of the polypeptide of the LHC-II which has priority of assembly and which may serve as the link in excitation energy transfer from the bulk LHC-II antenna to the PSII-core.

MATERIALS AND METHODS

Wild-type, mutant *chlorina* 103 and mutant *chlorina* f2 barley thylakoids were prepared in hypotonic buffer [11] containing 1% BSA, 1 mM benzamidine and 0.1 mM PMSF as protease inhibitors. LHC-II was isolated from PSII particles [3] by a Triton X-100-solubilization procedure [4]. The concentration of reaction centre components Q_A and P700 were determined from the light-induced absorbance changes at 320 nm and 700 nm, respectively, using a laboratory-constructed split beam difference spectrophotometer [18]. The functional Chl antenna sizes of PSII and PSI in the different thylakoid membranes were determined from the rate of light absorption by each photosystem [18,24]. In the case of PSII, the kinetics of room temperature fluorescence induction were determined for thylakoids incubated in the presence of DCMU. The rate of light absorption by PSI was determined from the kinetics of P700 photooxidation in KCN-inhibited thylakoids [11,18].

Polyacrylamide gel electrophoresis was performed at $2^{\circ}C$ on 12.5 % acrylamide gels [21] containing 4 M urea. Thylakoid membrane samples were solubilized and electrophoresed at 10 mA constant current per gel for 1 hour, followed by 20 mA constant current per gel for 36 hours. Gels were stained with Coomassie Brilliant Blue and dried onto cellulose film.

Immunoblotting was performed by standard procedures [7,23], using rabbit polyclonal antibodies to the isolated PSII reaction center polypeptides and to whole spinach LHC-II. These polyclonal antibodies were raised in this lab. Monoclonal antibodies recognizing polypeptide subunit **a** (the apoprotein of CP29) or polypeptide subunit **b** of LHC-II [19] were obtained from hybridoma lines of the American Type Culture Collection [7]. Relative development of immunoblots and protein content of gels was determined by scanning densitometry using an LKB-Pharmacia XL laser densitometer having a 4.0 units dynamic absorbance range.

Table 1. Concentrations of photosystem reaction centre components in wild type and chlorina f2 barley thylakoid membranes and isolated PSII preparations.

	Chl a/b	Chl/Q_A	Chl/Cyt b_{559}	Chl/P700	Reference
Wild type thylakoids	3.1	350	--	590	[11]
chlorina 103 thylakoids	35	139	--	228	This work
chlorina f2 thylakoids	∞	106	--	310	This work, [11,12]
WT PSII prep	2.0	260	123	3,500	This work, [11]
chlo f2 PSII prep	∞	--	26	950	This work

Ratios are expressed on a mol:mol basis. Values are the means of several measurements.

RESULTS

Isolated thylakoid membranes of wild type, chlorina 103 [22] and chlorina f2 [25] mutant of barley exhibit Chl a/b ratios of approximately 3, 35 and infinity, respectively (Table 1). The chlorina 103 mutant therefore contains some 8-9 % of the Chl b found in the wild type. This contrasts with the situation in the chlorina f2 mutant, in which the complete absence of Chl b reflects a severe restriction in the complement of accessory light-harvesting complexes for both photosystems [5-7,11,25-27]. The presence of albeit reduced levels of Chl b in chlorina 103 might therefore be anticipated to reflect a somewhat larger antenna size for one or both photosystems, with respect to the antenna size in chlorina f2. To address this question, a comparative analysis of photosystem concentration, Chl antenna size, and polypeptide content was undertaken in thylakoid membranes from the three samples.

Quantitation of the reaction centre components Q_A (PSII) and P700 (PSI) were performed from the measurement of the light-induced absorbance change at 320 nm and 700 nm, respectively. Wild type thylakoids showed a total membrane chlorophyll to photoreducible

Q_A ratio of about 350/1 [11]. The *chlorina* 103 mutant showed a Chl/Q_A ratio 139 ± 2, whereas that of *chlorina* f2 was about 106 ± 4 (Table 1).

Wild type thylakoids showed a total membrane Chl to photooxidizable P700 ratio of about 590/1 [11] whereas the corresponding values for the *chlorina* 103 and *chlorina* f2 mutants were 228 ± 12 and 310 ± 11, respectively (Table 1). These results clearly suggest a smaller auxiliary Chl antenna size for both PSII and PSI in the mutants versus the wild type. They also indicate dissimilar effects of the *chlo* 103 and *chlo* f2 mutations on the antenna organization of PSII and PSI. To address the assembly of the auxiliary light-harvesting Chl antenna in the mutants and wild type barley, we measured (a) the functional Chl antenna size of PSI and PSII in each of the three samples with biophysical and biochemical techniques; (b) the polypeptide composition of the auxiliary antenna of PSII via quantitative SDS-PAGE and immunoblot analysis.

Table 2. Functional Chl antenna size of photosystem-I and photosystem-II in wild type, *chlorina* 103 and *chlorina* f2 mutants of barley.

Chlorophyll antenna size	Wild type	*chlorina* 103	*chlorina* f2
PSI	210	137 ± 8	150 ± 15
PSII	250	58 ± 3	50 ± 2

Values are the means of four individual determinations, with standard deviations given.

The absolute chlorophyll antenna sizes of PSI and PSII in the wild type were 210 and 250 Chl ($a + b$) molecules, respectively [11] (Table 2). In the *chlorina* 103 mutant, the functional antenna size of PSI was about 137 Chl molecules and for PSII was about 58 Chl molecules (Table 2). In the *chlorina* f2 mutant the functional antenna sizes were about 150 Chl *a* molecules for PSI and 50 Chl *a* molecules for PSII [12]. Thus, the PSII antenna size of 58 Chl ($a + b$) molecules found in the *chlorina* 103 mutant represents approximately a 15% increase over the corresponding value for the Chl *b*-less *chlorina* f2 mutant. It is tempting to suggest that the difference by about 8 Chl molecules is principally due to the presence of the low levels of Chl *b* in the PSII antenna of the *chlorina* 103 mutant. This in turn might suggest the presence of specific LHC-II polypeptides in *chlorina* 103 which are not present in *chlorina* f2.

Figure 1 shows SDS-PAGE analysis of thylakoid membrane polypeptides from wild type barley and from both the *chlorina* 103 and *chlorina* f2 mutants. Also shown are the resolved constituent polypeptides of LHC-II, isolated by detergent solubilization of PSII-enriched membranes. It can be seen that five distinct polypeptide subunits are present in the isolated LHC-II complex, with apparent relative molecular masses 30, 28, 27, 24 and 20 kDa in this gel system.

Fig. 1. SDS-PAGE analysis of thylakoid membranes isolated from wild type (WT), *chlorina* 103 (clo 103) and *chlorina* f2 (clo f2) barley. Barley LHC-II, isolated from PSII-enriched membranes, is also shown (LHC-II). The **d** subunit of LHC-II is indicated by the arrow. Molecular masses of marker proteins (M) are given in kDa.

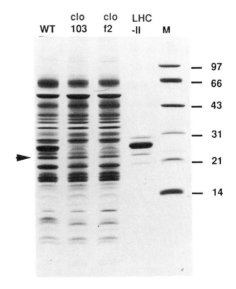

The top four subunits (of slower electrophoretic mobility) have been termed **a, b, c** and **d** [19]. Subunit **a** is the apoprotein of the Chl-protein CP29 [7,19]. Although the apparent molecular mass of the **d** subunit (indicated by the arrow) is 24 kDa in this gel system, suggesting it could be CP24 [2,8], other reports [26] indicate absence of CP24 but relatively high levels of CP26 in *chlorina* f2. The **b** and **c** subunits represent the major constituent polypeptides of the bulk LHC-II antenna [19].

Comparison of lanes 1-3 from Fig. 1 revealed differences in polypeptide composition between wild-type and mutant thylakoid membranes. Levels of subunit **a** were lower in the two mutants compared with the wild type. Levels of the **b** and **c** subunits were conspicuously lower in the thylakoid membrane of the *chlorina* 103 and *chlorina* f2 mutants, consistent with previous observations [5,20,27]. The residual **b** and **c** polypeptides in mutant thylakoids were found to be sensitive to proteolysis and were not observed unless protease inhibitors were included in membrane isolation buffers (data not shown), implying that these were not stably integrated into the membrane. In contrast, the **d** subunit occurred at relatively high levels in the two mutants, as indicated by the presence of a polypeptide migrating to 24 kDa (marked by the arrow). SDS-PAGE-analyzed proteins from WT and mutant thylakoids were scanned with an LKB-Pharmacia XL laser densitometer. To account for the slightly variable loading of the different lanes, the integrated area of the scans corresponding to LHC-II polypeptides was normalized to that of three distinct polypeptides associated with the PSII-core complex. These were the apoproteins of the CP47 and CP43 (organelle *psb*B and *psb*C gene products) and the so-called 33 kDa polypeptide of the oxygen evolving complex (nuclear *psb*O gene product). The normalized scans suggested approximately equal amounts of subunit **d** per PSII-core in wild type, *chlorina* 103 and *chlorina* f2 thylakoids.

The results showed there are no significant quantitative differences in the LHC-II polypeptide contents of thylakoid membranes from either the *chlorina* 103 or *chlorina* f2 mutants, a point reinforced by the immunoblot analyses shown in Fig. 2. These immunoblots, with gel lanes containing approximately equal PSII-core, were probed with polyclonal antibodies to whole spinach LHC-II (isolated from the PSII enriched membranes of the grana partition regions), or with monoclonal antibodies to polypeptide subunits **a** or **b**. It is clear from the immunoblots probed with antibody to whole LHC-II (Fig. 2, left panel) that the **d** subunit (indicated by the 24 kDa marker) is present in approximately equal amounts in all three thylakoid membranes examined, an observation confirmed by densitometric scanning (not shown). Immunoblotting with monoclonal antisera to subunits **a** (Fig 2, centre panel) and **b** (Fig. 2, right-hand panel) showed that these polypeptide subunits occur in the *chlorina* 103 mutant, but at greatly reduced levels with respect to those in wild type (only 30% and 10% for **a** and **b** respectively), as was also the case for *chlorina* f2 (not shown).

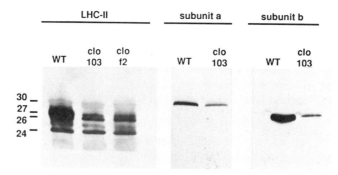

Fig. 2. Immunoblot analysis of thylakoid membranes from wild type (WT), *chlorina* 103 (clo 103) and *chlorina* f2 (clo f2). Gel lanes were probed with polyclonal antibodies raised against isolated spinach LHC-II (left panel), monoclonal antibody to the **a** subunit (CP29) (centre panel) or monoclonal antibody to the **b** subunit of LHC-II (right panel). Relative molecular masses of the four subunits are indicated.

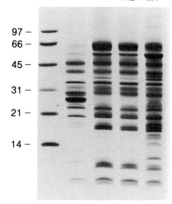

Fig. 3. SDS-PAGE analysis of PSII preparations isolated from wild type (WT) and *chlorina* f2 mutant barley by solubilization with n-octyl glucopyranoside (M1) or with Triton X-100 (M2). Unfractionated *chlorina* f2 mutant thylakoid membranes (MT) are included for comparison. Marker proteins are shown and their molecular masses in kDa indicated.

Analysis of PSII enriched fractions, isolated from wild-type and *chlorina* f2 mutant thylakoids by solubilization with non-ionic detergents, was used to further test for the relative abundance of LHC-II polypeptide subunits. Membrane fractions enriched in PSII were isolated from wild-type thylakoid membranes by solubilization with Triton X-100 ("BBY particles") [3] and from *chlo* f2 mutant membranes either by solubilization with n-octyl glucopyranoside and sucrose density gradient centrifugation or by solubilization with Triton X-100 and differential centrifugation [2]. Since photochemical quantitation of Q_A is difficult to obtain in detergent-derived membranes, we employed the use of cytochrome b_{559} as a diagnostic component for the concentration of PSII. Each PSII-core is believed to contain two copies of this cytochrome [10]. Reduced (dithionite) *minus* oxidized (ferricyanide) absorbance difference measurements indicated Chl/RCII ratios of 246/1 (Chl/Cyt b_{559} = 123) for the wild-type and 52/1 (Chl/Cyt b_{559} = 26) for the *chlorina* f2 PSII preps (Table 1), in close agreement with the respective PSII antenna sizes measured in unfractionated thylakoids through biophysical approaches.

The complement and relative abundance of LHC-II polypeptides in the PSII preparations from wild type and mutant barley were examined by SDS-PAGE and immunoblot analysis. This is illustrated in Fig. 3, which shows SDS-PAGE analyses of PSII-enriched particles from wild type (WT) and *chlorina* f2 (M1 and M2). For comparison, whole thylakoids from the *chlorina* f2 mutant are also shown in Fig. 3 (MT). Assessed by densitometric scanning, the **d** subunit was present at approximately equivalent amounts per PSII-core in all samples examined (i.e., wild type and mutant PSII preparations). All other LHC-II polypeptides occurred at significantly reduced levels both in the *chlorina* 103 and in the *chlorina* f2 mutants.

Evaluation of the stoichiometries of LHC-II polypeptide subunits based on scanning densitometry of the SDS-PAGE lanes (Fig. 3) and of the immunoblots (not shown) revealed that polypeptide subunits **a** and **d** consistently occur in a stoichiometric ratio of 1:1 in the LHC-II of wild type barley. In addition, the **b** subunit consistently occurred in an approximate stoichiometry of 5:1 with the **a**, **c** and **d** subunits in barley. Stoichiometry data derived from densitometric scans of SDS-PAGE and immunoblots are summarized in Table 3. The results suggest a stoichiometric ratio of LHC-II polypeptide subunits in the wild type barley **a:b:c:d** = 1:5:1:1 and in the *chlorina* 103 and f2 mutants **a:d** = 0.3:1. Based on the generally accepted notion of 12-14 Chl molecules bound to each LHC-II polypeptide, we suggest a stoichiometry of **a:b:c:d** = 2:10:2:2 per PSII-core in WT, yielding a model in which each of the two **a** and two **d** subunits connect the bulk of the LHC II with the PSII-core. We postulate trimeric assemblies of **b** and **c** subunits [15] which might be in the form of two homotrimers of **b** subunits (2 x **b₃**) acting as the LHC II-inner, and two heterotrimers each comprising two **b** and one **c** subunit (2 x **b₂c**) for the LHC II-peripheral [19].

Table 3. Stoichiometric ratios of LHC-II polypeptide subunits in thylakoid membranes and isolated PSII preparations of wild type, *chlorina* 103 and *chlorina* f2 mutants of barley.

LHC-II subunit	a	b	c	d
Isolated LHC-II	1.0 ± 0.1	5.4 ± 0.5	1.2 ± 0.2	1.0
Wild type thylakoids	0.9 ± 0.1	4.8 ± 0.4	0.8 ± 0.2	1.0
Wild type PSII	1.2 ± 0.1	4.5 ± 1.0	1.1 ± 0.2	1.0
chlorina f2 thylakoids	0.3	≤0.1	≤0.1	0.9
chlorina f2 PSII	0.25	≤0.1	≤0.1	1.0
chlorina 103 thylakoids	0.3	0.1	≤0.1	1.0

For wild type barley preparations, stoichiometries are expressed relative to the **d** subunit content in each sample. In the case of the *chlorina* mutants, levels of each polypeptide subunit are expressed relative to the content in the corresponding wild type material.

DISCUSSION

The minimum Chl antenna size for PSII is approximately 37 Chl *a* molecules, representing the chlorophylls bound to the PSII-core polypeptides CP43 and CP47 and to the D1/D2 reaction centre heterodimer [12,17]. This is the Chl antenna size of PSII in cyanobacteria [17] and in chloroplasts developed under intermittent illumination which limits the rate of Chl biosynthesis [12]. The thylakoid membrane of the Chl *b*-deficient *chlorina* 103 and Chl *b*-less *chlorina* f2 do not show any difference in their respective complements of LHC-II polypeptides (Figs. 1 and 2). However, PSII in the *chlo* 103 contains 58 Chl molecules whereas the antenna size for *chlo* f2 is only 50 Chl molecules. This indicates that existing polypeptide subunits bind the increment of about 8 Chl molecules in the PSII antenna of the *chlorina* 103 mutant.

Based on the polypeptide subunit stoichiometry reported above, we suggest that it is mostly the **d** subunit of LHC-II which is responsible for binding the approximately 13 Chl *a* molecules of the auxiliary antenna in *chlorina* f2 and also the approximately 21 Chls of the auxiliary antenna in *chlorina* 103. This assertion is based on our calculation that subunit **a**, occurring at a concentration of about 1/3 copy per subunit **d**, cannot possibly accommodate the 21 Chl molecules associated with the auxiliary antenna of PSII in *chlorina* 103 mutant. Although some Chl may be associated with subunit **a** (CP29) in the *chlorina* mutants [14,26,27] our results suggest that it is subunit **d** which has priority in the hierarchy of assembly, and which is the component of the auxiliary antenna of PSII most proximal to the PSII-core. Subunit **d** would therefore be responsible for the conductance of excitation energy from the bulk LHC-II antenna to the PSII-core in the wild type thylakoid membrane. We have estimated that two copies of the **d** subunit may bind a total of about 12-14 molecules of Chl *a* per PSII-core in *chlorina* f2. An additional of about 8 Chl *b* molecules in the *chlorina* 103 would have to be shared between the two **d** subunits, resulting in a total of about 10-11 Chl *a+b* molecules per **d** polypeptide and a Chl *a*/Chl *b* ratio of approximately 2. This latter figure is consistent with published values for the Chl *a*/Chl *b* ratio in the isolated auxiliary Chl-proteins of PSII (1,13,26).

In summary, the results show that polypeptide subunit **d** is present at an equal copy number per PSII-core in mutant and wild-type thylakoids. In contrast, all other polypeptides associated with the LHC-II occurred at much lower amounts in the mutants. It is suggested that polypeptide subunit **d** binds most of the Chl in the residual auxiliary antenna of PSII in the *chlorina* mutants. This protein will stably assemble in the thylakoid membrane independent of the presence of Chl *b* and in precedence to all other components of the auxiliary LHC-II. This property is quite unlike that of the bulk LHC-II where presence of Chl *b* is an absolute requirement for stable assembly [25,26]. We propose that subunit **d** serves as a link in the process excitation energy transfer from the bulk LHC-II antenna to the PSII-core.

Acknowledgements: We thank Dr. David J. Simpson for provision of the barley *chlorina* 103 mutant seeds. The work was supported by a grant from the National Science Foundation.

REFERENCES

1 Bassi R (1990) Photochem Photobiol 52 (6), 1187-1206
2 Bassi R, Høyer-Hansen G, Barbato R, Giacometti GM, Simpson DJ (1987) J Biol Chem 262, 13,333-341
3 Berthold DA, Babcock GT, Yocum CF (1981) FEBS Lett 134, 231-234
4 Burke JJ, Ditto CL, Arntzen CJ (1978) Arch Biochem Biophys 187, 252-263
5 Burke JJ, Steinback KE, Arntzen CJ (1979) Plant Physiol 63, 237-243
6 Camm EL, Green BR (1989) Biochim Biophys Acta 974, 180-184
7 Darr SC, Somerville SC, Arntzen CJ (1986) J Cell Biol 103, 733-740
8 Dunahay TG, Staehelin LA (1986) Plant Physiol 80, 429-434
9 Dunsmuir P, Smith SM, Bedbrook J (1983) J. Mol. Appl. Genet. 2, 285-300.
10 Ghanotakis DF, Yocum CF (1990) Annu. Rev. Plant Physiol. Mol. Biol. 41, 255-276.
11 Ghirardi ML, McCauley SW, Melis A (1986) Biochim Biophys Acta 851, 331-339
12 Glick RE, Melis A (1988) Biochim Biophys Acta 934, 151-155
13 Green BR (1988) Photosynth Res 15, 3-32
14 Høyer-Hansen G, Bassi R, Hønberg LS, Simpson DJ (1988) Planta 173, 12-21
15 Kühlbrandt W (1984) Nature 307, 478-480.
16 Kung S, Thornber JP and Wildman SG (1972) FEBS Lett. 24, 185-188.
17 Manodori A, Alhadeff M, Glazer AN, Melis A (1984) Arch. Microbiol. 139, 117-123.
18 Melis A, Anderson JM (1983) Biochim Biophys Acta 724, 473-484
19 Morrissey PJ, Glick RE, Melis A (1989) Plant Cell Physiol 30, 335-344
20 Ryrie IJ (1983) Eur. J. Biochem. 131, 149-155.
21 Schägger H, von Jagow G (1987) Anal Biochem 166, 368-379
22 Simpson DJ, Machold O, Høyer-Hansen G, von Wettstein D (1985) Carlsberg Res Commun 50, 223-238
23 Smith BM, Morrissey PJ, Guenther JE, Nemson JA, Harrison MA, Allen JF, Melis A (1990) Plant Physiol. 93, 1433-1440
24 Thielen APGM, van Gorkom HJ (1981) Biochim. Biophys. Acta 635, 111-123
25 Thornber JP, Highkin HR (1974) Eur J Biochem 41, 109-116
26 Thornber JP, Peter GF, Chitnis PR, Nechushtai R, Vainstein A (1988) In SE Stevens Jr, DA Bryant,eds, Light-energy Transduction in Photosynthesis: Higher Plant and Bacterial Models. The American Society of Plant Physiologists, pp 137-154.
27 White MJ, Green BR (1987) Eur J Biochem 165, 531-535

ORGANIZATION OF THE PHOTOSYSTEM II ANTENNA SYSTEM OF MAIZE
PLANTS GROWN UNDER INTERMITTENT LIGHT CONDITION

Jürgen Marquardt and Roberto Bassi

Dipartimento di Biologia
Università di Padova
Via Trieste 75, 35121 Padova (Italy)

INTRODUCTION

In angiosperms, chlorophyll synthesis is light dependent. Thus growth under intermittent light conditions results in a limitation of chlorophyll available. Chlorophyll, however, seems to be necessary for the stabilisation of chlorophyll-binding apoproteins[1,2] and there are many reports on a reduced antenna size in plants grown in intermittent light[3,4]. To have more detailed information about the antenna system of these plants we investigated the stoichiometry of chlorophyll-binding proteins of PSII and the pigment composition of the light-harvesting complexes. Since limitation of chlorophyll is characteristic also for an early stage of chloroplast development these data might serve to give an insight in PSII development even under normal growth conditions.

MATERIAL AND METHODS

Maize seedlings were grown either under intermittent light (IL) conditions (2 min light, 90 min dark) for 11 days or in continuous light (CL) with a light/dark cycle of 12/12 h. Thylakoids were prepared as described before[5]. The Chl a/b ratio was 9.0 for IL thylakoids and 3.3 for IL thylakoids. For immunoblot assay the samples were separated by electrophoresis using a high Tris buffer system[5] with 12% acrylamide. The gels were transferred onto nitrocellulose filters (Millipore) and immunoassays were carried out as described by Di Paolo et al.[6] using alkaline phosphatase coupled anti-rabbit IgG or anti-mouse IgG, respectively. All antibodies used with the exception of MLH1 were polyclonals. For isoelectrofocussing (IEF) thylakoids were solubilized in 2% dodecyl maltoside with a detergent to chlorophyll ration of 20:1. Non-denaturing IEF was performed as described by Dainese et al.[7] using Ampholine carrier amphilites in the pH range of 3.5-6. The IEF bands were eluted and polypeptide composition examined by electrophoresis on a gradient gel (12-18% acrylamide) containing 6 M Urea[7]. Chlorophyll, carotenoid and protein determinations were carried out as described previously[5].

RESULTS AND DISCUSSION

Stoichiometry of Chlorophyll-Proteins

In order to determine the relative amount of PSII chlorophyll-proteins we prepared a series of dilutions from our thylakoid samples with each dilution step containing half the amount of protein of the preceding one. The protein content of the starting material from IL thylakoids was approximately four times higher than that of the CL sample. The dilutions were loaded onto a number of gels and each gel containing CL samples was blotted together with a gel containing the IL material onto one filter. The filters were assayed with antibodies against chlorophyll-binding PSII proteins and antibody binding was visualized using alkaline phosphatase coupled anti-rabbit or anti-mouse IgG. For illustration immunoblots assayed with polyclonal antibodies against CP24 and CP43 are shown in Fig. 1. After the immunoassay we determined, up to which lane (i.e. dilution step) of the gel with CL material and of that with IL material antibody binding was detectable. In these lanes we must have an equal amount of the polypeptides in question.

A summary of the results is given in Table 1. It shows that antibody binding to CP26 and CP29 is detectable up to the same dilution step of our CL and our IL preparation, while after immunoassay with antibodies against D1, D2, CP43 and CP47 bands are visible up to one further dilution step in the IL sample. CP24 and LHCII, however, are detectable up to a further dilution step in the CL preparation. No antibody binding with IL thylakoids was detected when we tested MLH1, a monoclonal antibody raised against LHCII polypeptides, although antibody binding with CL thylakoids was visualized up to lane 4, i.e. the 3rd dilution step.

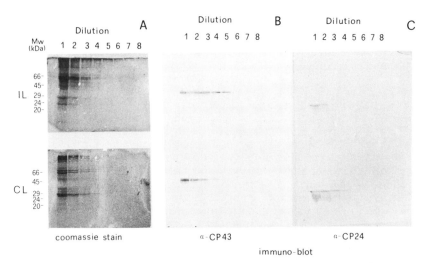

Fig. 1. Examples for the immunoblot assays carried out to determine the stoichiometry of the chlorophyll-protein complexes of PSII. (A) Coomassie stained gels with a dilution series of IL thylakoids (above) and CL thylakoids (below); (B) immunoblot of gels as shown in (A) assayed with an antibody against CP43; (C) immunoblot assayed with an antibody against CP24.

Table 1. Summary of the results of the immunoassay experiment with a dilution series of CL and IL thylakoids.

| Antibodies | Lanes with detectable antibody binding | | Difference |
	Gel with CL samples	Gel with IL samples	(IL-CL)
Polyclonals			
D1	6	7	+1
D2	4	5	+1
CP43	7	8	+1
CP47	6	7	+1
CP24	6	5	−1
CP26	5	5	+/−0
CP29	5	5	+/−0
LHCII	5	4	−1
Monoclonal			
MLH1	4	−	−4

If we take D1 as an internal standard, we find the same amounts of the core complex polypeptides CP47, CP43 and D2 relative to D1 in CL and IL thylakoids, suggesting an identical composition of the PSII core complex in CL and IL plants. In IL thylakoids, however, we find only half the amount of the minor Chl a/b proteins CP29 and CP26 and 1/4 of the amount of CP24 and LHCII per reaction centre as compared to CL thylakoids. Proceeding on the assumption that the stoichiometry of LHCII:CP29:CP26:CP24 per reaction centre in CL plants is 12.4:1.6:1.4:1.6[8] the stoichiometry in of these chlorophyll-proteins in IL plants is 3.1:0.8:0.7:0.4 per PSII core complex. Note that the relatively less diminished complexes CP29 and CP26 are thought to be closer connected to the core complex than CP24 and the major part of LHCII[9].

In addition to the different stoichiometry of the antenna complexes there must also be a difference in the composition of LHCII, since the epitopes recognized by MLH1 are absent or below detection in the LHCII of IL plants. This could be due to the expression of different gene products. This is not unprobable since a large number of different genes coding for chlorophyll a/b-proteins (cab genes) could be detected. Stayton et al.[9] report on a minimum of 17 cab genes in petunia. Another possibility is that identical gene products in CL and IL plants could have been modified post-translationally, e.g. by different processing as described by Lamppa and Ahad[10].

<u>Pigment Content</u>

In order to examine the pigment composition of the antenna complexes we separated thylakoid membranes by non-denaturing flat bed IEF. We obtained 9 green bands (CL b1-b9) in the case of CL thylakoids and 5 bands (IL b1-b5) in the case of IL thylakoids (Fig.2a). IL b1 was brownish, b2 and b3 yellowish green, b4 and b5 green. In the case of IL thylakoids a rather high amount of green material did not migrate but remained at

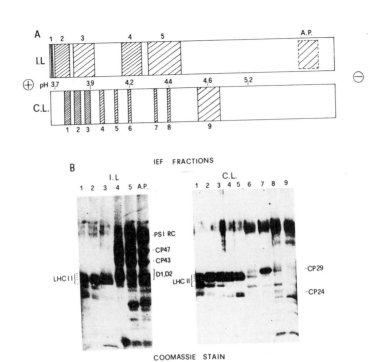

Fig. 2. Separation of thylakoid membranes by non-denaturing isoelectrofocussing. (A) Schematic representation of the separation of CL thylakoids and IL thylakoids by non-denaturing IEF; (B) electrophoretic separation of the IEF fractions under denaturing conditions and identification of some of the bands.

the application point. IL b1 and b2 migrated significantly closer to the anode than any of the CL bands, indicating a more acidic isoelectric point of the IL complexes. To identify the IEF fractions the bands were eluted and their polypeptide composition examined by electrophoresis (Fig.2b). IL b1-b3 contained LHCII polypeptides, CP24 and, as revealed by immunoassay (not shown), also CP26 which comigrates with the LHCII polypeptides in the gel system we used. IL b3 contained an additional low molecular weight polypeptide which migrated near the front. IL b4 contained PSII core complex polypeptides and CP29, while b5 contained a mixture of several PSI and PSII reaction centre polypeptides. In the material that remained at the application point we found a large number of polypeptides belonging to PSI as well as to PSII. CL b1-b5 showed a polypeptide composition similar to IL b1-b3. In CL b6 we found a mixture of some reaction centre polypeptides and minor Chl a/b complexes, whereas b7 consists mainly in CP29. CL b8 and b9 contain a number of different PSI as well as PSII reaction centre polypeptides.

Protein and pigment content of the eluted fractions were determined as mentioned above and the molar ratios were calculated. The results are given in Table 2. If we compare the IL and CL bands which consist mainly in LHCII (IL b1-b3 and CL b1-b5) we find a much lower pigment to protein ratio in the IL samples. This suggests that the IL complexes are not saturated in chlorophylls, but it cannot be excluded that in the IL samples there are also some polypeptides different from those

Table 2. Characteristics of the fractions obtained by IEF.

IEF fraction	pigment content (µMol/g protein)			Chl a/b	Chl/Car
	Chl a	Chl b	Car	(molar)	(molar)
IL b1	44.5	8.6	37.6	5.2	1.4
IL b2	80.8	27.9	61.5	2.9	1.7
IL b3	52.4	11.1	46.3	4.7	1.4
IL b4	97.7	4.2	40.8	>20	2.5
IL b5	74.5	8.0	16.2	9.3	5.1
application point	41.1	5.9	11.6	7.0	4.1
CL b1	134.3	66.6	61.1	2.0	3.3
CL b2	123.7	74.4	62.3	1.7	3.2
CL b3	166.2	105.9	80.0	1.6	3.4
CL b4	160.6	102.0	72.3	1.6	3.6
CL b5	96.0	45.5	50.6	2.1	2.8
CL b6	66.6	11.0	35.0	6.1	2.2
CL b7	73.2	10.6	31.0	6.9	2.7
CL b8	78.1	5.9	27.8	>10	3.0
CL b9	57.1	1.5	27.0	>30	2.2

of the CL samples which do not bind pigments. Contradictory to the latter explanation is that the electrophoretic profile of IL and CL bands is rather similar. However, there must be some diversity in the composition of LHCII fom CL and IL thylakoids as revealed by the immunoassay with MHL1 (see above). Further work is necessary to solve this problem.

Besides the lower pigment content in the IL fractions b1-b3 also the higher Chl a/b ratio and the lower Chl/Car ratio in these bands are obvious, i.e. the IL bands contain per Chl a less Chl b but more carotenoids than the CL bands. Thus we find a diminished Chl b content together with a reduced amount of LHCII (see above). Note that a decrease in the LHCII level is also a characteristic of Chl b-deficient mutants[11]. Plumley and Schmidt[12] showed that Chl a as well as Chl b and also xanthophylls are required for a stable assembly of LHCII in vitro.

In order to see if there is a relationship between the Chl b and the carotenoid content we calculated the ratio of Chl a / (Chl b + Car) for the LHCII containing fractions. For IL as well as for CL samples we found a rather constant value in the range of 0.97 +/- 0.08. Possibly the carotenoids serve as substitutes for a part of Chl b under IL conditions when the chlorophyll synthesis is limited, thus stabilizing the complexes.

ACKNOWLEDGMENTS

A grant of the Deutsche Forschungsgemeinschaft (DFG) to J. M. is gratefully acknowledged. P. Dainese is thanked for advices and collaboration in the preparation of figures.

REFERENCES

1. N. M. Mathis and K. O. Burkey, Light intensity regulates the accumulation of the major light-harvesting chlorophyll-protein in greening seedlings, Plant Physiol., 90:560 (1989)
2. J. E. Mullet, P. Gamble Klein and R. R. Klein, Chlorophyll regulates accumulation of the plastid-encoded chlorophyll apoproteins CP43 and D1 by increasing apoprotein stability, Proc. Natl. Acad. Sci. USA, 87:4038 (1990)
3. J. H. Argyroudi-Akoyunoglou and G. Akoyunoglou, The chlorophyll-protein complexes of the thylakoid in greening plastids of Phaseolus vulgaris, FEBS Lett., 104:78 (1979)
4. R. E. Glick and A. Melis, Minimum photosynthetic unit size in system I and system II of barley chloroplasts, Biochim. Biophys. Acta, 934:151 (1988)
5. P. Dainese and R. Bassi, Subunit stoichiometry of the chloroplast photosystem II antennasystem and aggregation state of the component chlorophyll a/b binding proteins, J. Biol. Chem., 266:8136 (1991)
6. M. L. Di Paolo, A. Dal Belin Peruffo and R. Bassi, Immunological studies on chlorophyll a/b proteins and their distribution in thylakoid membrane domains, Planta, 181:275 (1990)
7. P. Dainese, G. Hoyer-Hansen and R. Bassi, The resolution of chlorophyll a/b binding proteins by a preparative method based on flat bed isoelectric focusing, Photochem. Photobiol., 51:693 (1990)
8. R. Bassi and P. Dainese, The role of the light harvesting complex II and of the minor chlorophyll a/b proteins in the organization of the photosystem II antenna system, in: "Current Research in Photosynthesis", M. Baltscheffsky, ed., Kluwer Academic Publishers, Dordrecht (1990)
9. M. M. Stayton, M. Black, J. Bedbrook and P. Dunsmuir, A novel chlorophyll a/b binding (Cab) protein from petunia which encodes the lower molecular weight Cab precursor protein, Nucl. Acids Res., 14: 9781 (1986)
10. G. K. Lamppa and M. S. Abad, Processing of wheat light-harvesting chlorophyll a/b protein precursor by a soluble enzyme from higher plant chloroplast, J. Cell Biol., 105:2641 (1987)
11. B. A. Greene, L. A., Staehelin and A. Melis, Compensatory alterations in the photochemical apparatus of a photoregulatory, chlorophyll b-deficient mutant of maize, Plant Physiol., 87:365 (1988)
12. F. G. Plumley and G. W. Schmidt, Reconstitution of chlorophyll a/b light harvesting complexes: xanthophyll-dependent assembly and energy transfer, Procl. Natl. Acad. Sci USA, 84:146 (1987)

STRUCTURE OF THE CYANOBACTERIAL PHOTOSYSTEM II:

AN INDICATION OF DIFFERENT FUNCTIONS OF CP47 AND CP43

Josef Komenda, Jiří Masojídek and Eva Šetlíková

Department of Autotrophic Microorganisms, Institute of
Microbiology, CAS, CS-379 81 Třeboň, Czechoslovakia

INTRODUCTION

Photosystem II, a multicomponent chlorophyll-protein complex, is
very labile and easily disintegrates during isolation and other manipulation at temperatures above 0°C. PSII preparations from thermophilic
cyanobacteria are characterized by a higher stability and, therefore, they
are a suitable object for structural and functional studies. Zwitterionic
detergents (e.g. LDAO, sulfobetain 12) selectively extract PSII complexes
from these cyanobacteria (Stewart and Bendall, 1979; Schatz and Witt, 1984)
yielding active and stable PSII preparations relatively little contaminated
with other thylakoid proteins.

In this paper, we describe our results concerning the structure of
the PSII complex from thermophilic cyanobacterium Synechococcus elongatus.
We employed non-denaturing polyacrylamide gel electrophoresis (PAGE) and
diagonal re-electrophoresis of chlorophyll-proteins in combination with
Western blotting and cross-linking of polypeptide subunits. Our results
suggest the existence of the dimeric form of PSII in vivo and indicate
different functions of CP47 and CP43 in the energy transfer from phycobilisomes to PSII reaction centre.

MATERIAL AND METHODS

The thylakoid membranes of the thermophilic cyanobacterium Synechococcus elongatus were prepared according to Schatz and Witt (1984) and
resuspended in 25 mM MES, pH 6.5 containing 10 mM $MgCl_2$, 20 mM $CaCl_2$,
1 mM aminocaproic acid and 0.5 M mannitol (buffer A). Then, the membranes
were treated with sulfobetain 12 (SB 12/chlorophyl = 3.5) and centrifuged
at 360,000xg for 25 min. To purify the O_2 evolving complex (PSII-O) the
supernatant was loaded on a continuous sucrose gradient (15-45 %) and
separated at 180,000xg for 15 h. The major dark green band (Dekker et al.,
1988) with an oxygen evolving activity of 120-160 µmol O_2 mg^{-1}Chl.h^{-1}
(30°C, 0.5 mM p-benzochinon and 1 mM ferricyanide as the electron acceptors)
was used as the O_2 evolving PSII complex (PSII-O). For preparation of the
non-O_2 evolving PSII (PSII-N), the SB extract was mixed with the equal
volume of 2 M $CaCl_2$, stirred for 15 min on ice and centrifuged at 360,000xg
for 2 h. The sediment was washed again, resuspended in buffer A and spun

down. Finally, the sediment (PSII-N) was resuspended in buffer A + 10 % glycerol).

Analysis of chlorophyll-proteins was carried out under non-denaturing conditions (Nedbal et al., 1990). The upper reservoir buffer contained either 0.2 % Deriphat 160 or 0.075 % SDS. The samples for cross-linking were solubilized with octylglucoside (OG/chlorophyll = 20, w/w) and 1 % dithiobis (N-succinimidyl propionate) (DSP) in dimethyl-sulfoxide was added (DSP/chlorophyll = 15, w/w). After 15 min incubation 1/10 of the volume of 1 M Tris and 1 % SDS were added and the sample was immediately analysed. For diagonal re-electrophoresis, strips cut from non-denaturing gel were incubated for 30 min in 0.3 M Tris, pH 9.0 containing 10 % SDS, 8 M urea and 5 % dithiothreitol. Then, they were placed on the top of a 10-20 % gradient polyacrylamide gel containing 6 M urea and electrophoresed in the buffer system developed by Laemmli (1970). Western blotting was made as described by Nedbal et al. (1990).

RESULTS

Electrophoretic analysis of the PSII-O preparation (Fig. 1 - track a) showed the presence of polypeptides representing apoproteins of CP47 (ACP47, 49 kDa) and CP43 (ACP43, 43 kDa), oxygen evolving enhancer (OEE1, 35 kDa), D2 and D1 proteins (34 and 32 kDa), and products of the genes psbE (PsbE, 9 kDa), psbH (PsbH 7 kDA), psbF (PsbF, 5 kDa) and psbI (PsbI, 4 kDa) (Ikeuchi et al., 1989). In addition to these typical PSII core subunits the preparation contained components of phycobilisomes (PB), mostly the 17-18 kDa allophycocyanin (APC) binding proteins and 75 kDa "anchor" protein, and ATPase subunits (60-65 kDa). To assess a role of PB component in the PSII structure we prepared another PSII preparation by the treatment of the crude SB12 extract with 1 M $CaCl_2$. The treatment removed most of PB and ATPase subunits but also OEE1 (Fig. 1 - track b) and, therefore, this preparation designated as PSII-N had no oxygen evolving activity.

Both preparations were first solubilised with octylglucoside and SDS (OG/SDS/chlorophyll = 20/1/1) and analysed by non-denaturing electrophoresis in the presence of the zwitterionic detergent Deriphat 160. Under these conditions most in vivo interactions among PSII subunits remain preserved. Two chlorophyll-proteins (CP) CP2a and CP2b were resolved in both preparations (Fig. 1 - tracks c,d). They had almost identical polypeptide composition containing ACP47, ACP43, D2, D1 and PsbE, PsbH, PsbF and PsbI. In addition, CP2a contained also the "anchor" protein and the small APC binding subunits. The M_r of CP2a and CP2b (about 300 and 150 kDa) in the "green" gel and their polypeptide composition suggest that they represent a dimeric and monomeric form of the complete PSII core complex.

A different electrophoretic pattern was obtained when SDS (0.075 %) was used instead of Deriphat in the upper electrode buffer (Fig. 2). Under these conditions, some PSII subunits became dissociated from the core complexes. Four CPs designated as CP2b', CP2c, CP2d and CP2e were resolved in both preparations (Fig. 2 - tracks a,c). In the gel loaded with PSII-O, another green band designated as CP2a' was sometimes observed. Diagonal electrophoresis (Fig. 3, slab A) and Western blotting showed that CP2b' consisted of ACP 47, D2, D1, PsbE, PsbF and PsbI; CP2a' contained in addition 75 kDa "anchor" protein. The composition and the M_r (about 200 and 95 kDa) suggest that CP2a' and CP2b' represent a dimeric and monomeric form of CP47-D1-D2-cytb559 complex. CP2d had qualitatively the same polypeptide composition as CP2b' but differed in the electrophoretic mobility and in the ratio ACP47/D1+D2. This ratio was much lower in CP2d as confirmed by the comparison of protein band intensity after Coomassie blue staining and the intensity of D1 and D2 bands after Western blotting.

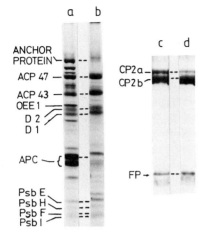

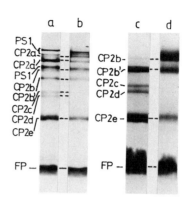

Fig. 1. Composition of PSII-O (a,c) and PSII-N (b,d) as revealed by denaturing SDS-urea PAGE (a,b) and non-denaturing Deriphat PAGE (c,d). FP=free pigments.

Fig. 2. Non-denaturing SDS PAGE of PSII-O (a,b) and PSII-N (c,d) non-cross-linked (a,c) and cross-linked with DSP (b,d) FP=free pigments, PS1=Photosystem I.

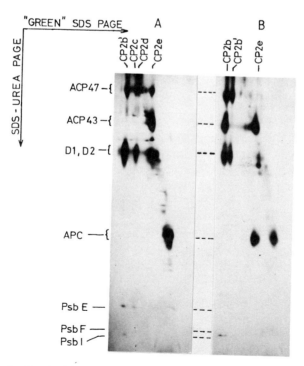

Fig. 3. Diagonal electrophoresis of PSII-N (slab A) and PSII-N cross-linked with DSP (slab B).

Therefore, we propose CP2d to be the D1-D2-cyt b559 complex contaminated by comigrating CP47. CP2c contained only ACP47 and probably represents a dimeric CP47 while CP2e consisted of ACP43 as the major and ACP47 as the minor components. We assume that during electrophoresis a portion of CP47 was released from PSII core complexes by the action of SDS and moved in the form of the dimer in CP2c or monomer in CP2e, or comigrated with the D1-D2-cyt b559 in CP2d band. This assumption was confirmed by experiments in which a cross-linking agent, DSP was used to link neighbour subunits in the PSII complex. The addition of DSP to PSII-N preparation prevented the release of CP47 from the PSII core complexes as indicated by the complete disappearance of CP2c and CP2d from the "green" SDS gel (Fig. 2 - track d) and by the absence of ACP47 in CP2e (Fig. 3, slab B). However, only a portion of CP43 was linked to the CP47-D1-D2-cyt b559 complex forming the complete PSII core complex CP2b. The remaining part of CP43 moved in CP2e band. This band also contained a 15-17 kDa APC binding polypeptides probably cross-linked to CP43. The effect of DSP on the PSII-O preparation was similar as in PSII-N. CP2c and CP2d again disappeared from the gel and new bands CP2a and CP2b representing the dimeric and monomeric form of the complete PSII core complex appeared (Fig. 2 - track b).

DISCUSSION

Bassi et al. (1989) proposed the dimeric organization of the PSII complex in vivo on the basis of data from two-dimensional PSII crystals from higher plants. Dekker et al. (1988) isolated dimeric and monomeric form of PSII from the thermophilic cyanobacterium Synechococcus sp. by sucrose density gradient centrifugation and also suggested the existence of the dimeric form in vivo. The results here are in agreement with the proposed dimeric structure of PSII in vivo but in addition we suggest that the PB components found in CP2a (but not in CP2b) play a structural role in the dimer because their removal by salt washing decreased the content of the dimeric form detected in the non-denaturing gel.

We obtained the dimeric and monomeric form of the complete PSII core complex on the non-denaturing gel using the zwitterionic detergent Deriphat 160 in the electrode buffer. It confirms the higher stability of the PSII complex from thermophilic cyanobacteria in comparison with green algae or higher plants. For example, in Chlamydomonas reinhardtii Adir et al. (1990) identified only PSII subcore complexes lacking CP43 but no complete PSII in the same gel.

During non-denaturing SDS electrophoresis, CP43 was completely released from the core complex by the action of SDS while most of CP47 remained bound in the CP47-D1-D2-cytb 559 complex. We identified the monomeric (CP2b') and probably dimeric form (CP2a') of this complex in the SDS "green" gel. The cyanobacterial complex with similar polypeptide composition has been ident- ified on the non-denaturing gel by Yamagishi and Katoh (1984). Their CP2b is probably identical with our monomeric form CP2b' but they did not observe a complex lacking CP43 but containing the "anchor" protein from PB indicates that CP43 is not engaged in the formation of the dimer and that the "anchor" protein interacts only with CP47 but not with CP43.

The "anchor" protein was shown to be involved in the regular energy transfer from PB to PSII while the 17-18 kDa allophycocyanin B (APC-B) binding polypeptide might participate in a bypass as indicated by fluor- escence kinetic data (Yamazaki et al., 1984). Our results suggest a close structural relationship between the "anchor" protein and the CP47-D1-D2- -cyt b559 complex on the one hand, and between 15-17 kDa APC binding subunits and CP43 on the other hand. Therefore, we propose that only CP47 is involved in the main energy-transfer pathway from phycobilisomes to the PSII reaction centre while CP43 plays a role in the distribution of excess energy either to other PSII or to PSI. This hypothesis is in agreement with the recent results concerning State I - State II transition in cyanobacteria (Biggins and Bruce, 1989). A variety of data indicates that membrane conformational changes occur during this process which probably change a distance or orientation between the photosystems. In this case CP43 weekly bound to the rest of the PSII core complex could be an appropriate candidate for the flexible membrane component.

We conclude that: (1) The PSII complex exists in vivo as the dimer in which the polypeptides of phycobilisomes play a structural role. (2) CP43 can be released from the dimeric PSII form without the decomposition of the dimeric structure. (3) A structural relationship between CP43 and 15-17 kDa APC binding subunits and between CP47 and "anchor" protein of PB suggests the different roles of CP47 and CP43 in the energy transfer from PB to PSII RC.

ACKNOWLEDGEMENTS

The autors thank Ms H.Pumprová and Ms E.Kubečková for technical assistence, Dr. I.Šetlík for critical reading of the manuscript and valuable discussion and Prof. Y.Inoue for kind gift of the D1 and D2 antibodies.

REFERENCES

Adir, N., Shochat, S., and Ohad, I., 1990, Light-dependent D1 Protein Synthesis and Translocation Is Regulated by Reaction Center II, J. Biol. Chem., 265:12563.

Bassi, R., Ghiretti Magaldi, A., Tognon, G., Giacometti, G.M., and Miller, K., 1989, Two dimensional crystals from the photosystem II reaction centre from higher plants, Eur.J.Cell.Biol., 50:84.

Biggins, J., and Bruce, D., 1989, Regulation of excitation energy transfer in organisms containing phycobilins, Photosynth.Res., 20:1.

Dekker, J.P., Boekema, E.J., Witt, H.T., and Roegner, M., 1988, Refined purification and further charcterization of oxygen-evolving and Tris-treated Photosystem II particles from the thermophilic cyanobacterium Synechococcus sp., Biochim.Biophys.Acta, 936:307.

Ikeuchi, M., Koike, H., and Inoue, Y., 1989, Identification of psbI and psbL gene products in cyanobacterial photosystem II reaction center preparation, FEBS Lett., 251:155.

Laemmli, U.K., 1970, Cleavage of structural proteins during the assembly of the head of bacteriophage T4, Nature, 227:680.

Nedbal, L., Masojídek, J., Komenda, J., Prášil, O., and Šetlík, I., 1990, Three types of Photosystem II photoinactivation. 2. Slow processes., Photosynth.Res., 24:89.

Schatz, G.H., and Witt, H.T., 1984, Extraction and characterization of oxygen evolving Photosystem II complexes from a thermophilic cyanobacterium Synechococcus spec., Photobiochem.Photobiophys., 7:1.

Stewart, A.C., and Bendall, D.S., 1979, Preparation of an active, oxygen--evolving photosystem II particle from a blue-green alga, FEBS Lett., 107:308.

Yamagishi, A., and Katoh, S., 1984, A photoactive photosystem II-reaction center complex lacking a chlorophyll-binding 40 kilodalton subunit from the thermophilic cyanobacterium Synechococcus sp., Biochim.Biophys.Acta, 765:118.

Yamazaki, I., Mimuro, M., Murao, T., Yamazaki, T., Yoshihara, K., and Fujita, I., 1984, Excitation energy transfer in the light harvesting antenna system of the red alga Porphyridium cruentum and the blue-green alga Anacystis nidulans: analysis of time resolved fluorescence spectra, Photochem.Photobiol., 39:233.

GLYCOLIPIDS ARE PROSTHETIC GROUPS OF POLYPEPTIDES OF THE REACTION CENTER

COMPLEX OF PHOTOSYSTEM II

R. Voβ, A. Radunz and G. H. Schmid

Universität Bielefeld, Fakultät für Biologie
D-4800 Bielefeld 1, Germany

INTRODUCTION

In a series of publications we have shown, that photosynthetic electron transport is not only inhibited by antisera to electron transport
components (1-3), but also by polypeptide antisera (4) and by monospecific antisera to lipids (5-7) and carotenoids (8). The inhibition sites
of photosynthetic electron transport of the different glycolipid antisera
(5,6) as well as those of the phospholipid antisera (7) have been found
to be for both the photosystem I and the photosystem II region on the
respective donor side (Fig. 1). The degree of inhibition was found to
vary between 20 and 60% of the maximum reaction. The inhibitory effects
of the protein and lipid antisera were dependent on the osmotic condition of the thylakoid membrane, on the pH of the reaction medium, on the
amount of antibodies bound and on the reaction temperature. These observations demonstrate that in order to inhibit photosynthetic electron
transport by antibodies two conditions have to be fullfilled:

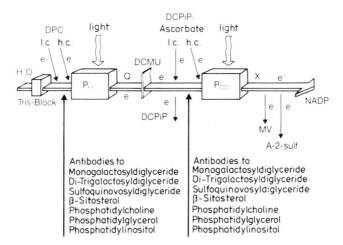

Fig. 1. Electron transport scheme with the approximate sites of inhibition of the antisera to glycolipids and phospholipids in the thylakoid
membrane.

First the thylakoid membranes has to be in an osmotically active condition and second physiological reaction conditions have to be such as to offer a high binding affinity to the antibodies.

The inhibitions observed can only be explained, if one assumes that lipids are bound as prosthetic groups onto defined polypeptides of the functional centers. Binding of antibodies onto these antigenic determinants apparently induces conformational changes in the lipid binding protein structures, causing in turn a modification of electron flow.

By means of the method of Western-Blotting, we were able to demonstrate that glycolipids and the anionic sulfolipid are bound onto the D_1- and D_2-peptides of the reaction center of photosystem II of the Nicotiana tabacum mutant Su/su. Monogalactolipid appeared to be bound onto polypeptides of the light harvesting complex of photosystem II as well as on chlorophyll binding peptides of the reaction center complex.

MATERIALS AND METHODS

Electrophoresis: Photosystem II-preparations from the yellow-green tobacco mutant N. tabacum Su/su were prepared according to Berthold et al. (9). The polypeptides of these preparations were analysed in the SDS-polyacrylamide gel electrophoresis according to Laemmli (10) using a 3.8% collection gel and a 12.5% separation gel. Samples were incubated before electrophoresis for 2 hours at room temperature with 1 M Tris buffer, pH 6.8 containing 9 M urea, 0.14 M SDS and 4% mercaptoethanol. Electrophoresis was carried out with 0.12 mA/cm^2 at 4°C in an electropho resis buffer containing 32 mM Tris, 0.25 M glycine, 4 M urea and 8.7 mM SDS.

Transfer of the polypeptides to nitrocellulose membranes: The polypeptides of the developed gels were washed in transfer buffer (10 mM Tris, pH 8.8; 2 mM EDTA; 50 mM NaCl and 0.1 mM DTE) and transferred by diffusion according to the Renart et al. procedure (11) to nitrocellulose membrane (Schleicher & Schüll, BA 85, Stärke 0.45 mm)) at room temperature during 20-30 hours or at 4°C during 50-60 hours. As the used lipid antisera contained also antibodies to methylated bovine serum albumin saturation of the free nitrocellulose areas was done with 2.5% fish gelatine solution from Sigma. Thereafter the nitrocellulose membranes were incubated for 1 hour with the glycolipid antisera. Dilution of the sera was carried out with CMF-PBS-buffer (5.5 mM Na_2HPO_4, 1.5 mM KH_2PO_4, 0.14 M NaCl and 2.6 m KCl). The dilution was usually between 1/25 to 1/300. Incubation with the peroxidase labeled anti-rabbit-IgG (from pig) was carried out with a 1/100 diluted serum for 30 minutes in the dark. The antigen-antibody complexes were made visible by developing with H_2O_2 and the substrate 4-chloro-1-naphtole.

Glycolipid antisera: Antisera were obtained by immunization of rabbits (12-15). The monospecificity of the antisera to mono-, digalacto- and sulfolipid was demonstrated by means of the passive heme agglutination test (12-15), the Elisa-test and the so-called "Dot-blot" procedure (16). Cross reactions were not observed.

Treatment of the polypeptides with lipase and periodate: Before the reaction with the sulfolipid antiserum the nitrocellulose membranes with the polypeptides were incubated for 30 minutes with lipase. For this test 50 µl lipase from Rhizopis arrhizus (5000 units/ml Boehringer), were diluted in 5 ml CMF-PBS-buffer, bringing the lipase activity to 500 units /ml. After saturation of the free nitrocellulose membrane areas with fish gelatine, the reaction with the antiserum was carried out. For the oxidation of the sugars with sodium periodate the isolated photosystem II preparations were shaken for 12 hours at room temperature in Hepes buffer (20 mM Hepes, pH 6.5; 25 mM CaCl; 5 mM $MgCl_2$ and 400 mM sucrose) containing 5% sodium periodate. Thereafter the assay was treated with SDS and the electrophoresis was carried out.

RESULTS AND DISCUSSION

The isolated photosystem II-preparation from the N. tabacum mutant Su/su consists, as shown in figure 2 by means of the analysis in the SDS polyacrylamide gel electrophoresis, of the center complex D_1 and D_2 (30 and 32 kDa), the two chlorophyll-binding polypeptides (48 and 42 kDa), the apoprotein of cytochrome b_{559} (10 and 4 kDa), the three extrinsic peptides (33, 23 and 16 kDa) and finally of two polypeptides of the light harvesting complex (28 and 26 kDa). Thus, this analysis of the N. tabacum mutant Su/su is identical to those obtained with other higher plants (17-19). Besides the listed polypeptides the majority of our analyses shows a further polypeptide with a molecular mass of 66 kDa.

The transfer of these polypeptides to a nitrocellulose membrane and subsequent incubation with the monospecific antisera to monogalactolipid, digalactolipid and sulfolipid in the Western-Blot procedure leads to the result (Figure 2) that the antiserum to monogalactolipid labels the two core peptides D_1 and D_2 but also other peptides with molecular masses 48, 42, 33, 28, 26, 23, 21 and 18 kDa. However, the antiserum to the digalactolipid and to the sulfolipid reacts only with the 66 kDa polypeptide. In order to verify the observation, whether the antibodies react with glycolipids on the polypeptides, polypeptides on the nitrocellulose membrane in the case of the sulfolipid were subjected to a lipase treatment, which revealed that the label of the 66 kDa band was eluminated. An oxidative sugar destruction with sodium periodate, however, did not prevent the glycolipid antibody labeling of the described peptide band. This experiment shows that the glycolipids were the reacting antigenic determinants and that the antibody is not solely directed toward the sugar component of the lipids but also beyond the sugar-glycerol region (13).

At first glance these analyses let the question open, whether the observed 66 kDa peptide belongs to photosystem II, and is indeed a hetero dimer of D_1 and D_2 or whether it is an impurity originating from photosystem I. If it was a heterodimer, it should be expected, that a stronger occurrence of this peptide in different analyses correlates with a proportional decrease in D_1 and D_2. A densitometric analysis of the gel with the peptide patterns of the photosystem II complex confirms this view. Therefore, the 66 kDa peptide should be a dimer of D_1 and D_2. Incubation of the nitrocellulose membrane, bearing the transferred polypeptides, with an antiserum to a 66 kDa peptide, that was shown to functionally react with the reaction center of photosystem II (20), only labeled this very peptide (Figure 3). These observations support the view that the peptide is a dimer which belongs to the photosystem II complex. An open question is why these "diffuse" migrating hydrophobic D_1- and D_2-peptides on the nitrocellulose membrane do not react neither with the glycolipid, nor with the antiserum to the 66 kDa peptide. It must be assumed that the aggregation of D_1 and D_2 leads to a conformational change, leading to a rearrangement, alteration or covering-up of certain determinants. In the aggregated form of these hydrophobic peptides antigenic determinants must become accessible, which in the condition of the D_1 and D_2 cannot react. From these observations it is concluded that the two galactolipids monogalactosyldiglyceride and digalactosyldiglyceride and the sulfolipid are bound as prosthetic groups onto the two core peptides

D_1 and D_2. The monogalactosyldiglyceride appears to be bound in addition onto the light harvesting complex and the two chlorophyll-binding polypeptides.

Our earlier results on the inhibitory effect of the antisera to glycolipids on the donor side of photosystem II find an unequivocal interpretation in the sense that the bound galactolipids react as antigenic determinants with antibodies inducing via this complex formation a conformational change of the core peptides leading in turn to an inhibition of electron flux. Apparently lipids do not only form the bimolecular double membrane determining the fluidity of the membrane, but also have as prosthetic groups a functional character. Monogalactolipids have besides their activity-stimulating effect on the coupling factor of photophosphorylation (21) a functional role in the light harvesting complex. According to studies of Siefermann-Harms et al. (22) this lipid might play a role in the energy transfer in the light harvesting complex.

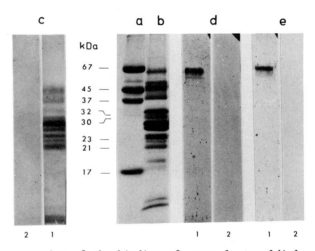

Fig. 2. Demonstration of the binding of monogalactosyldiglyceride, digalactosyldigyceride and sulfoquinovosyldiglyceride onto peptides of photosystem II of the <u>Nicotiana</u> <u>tabacum</u> mutant Su/su in the Western-blot assay.
a. Marker proteins in a 12.5% polyacrylamide gel: Bovine serum albumine (BSA) 67 kDa, catalase 60 kDa, glutamate dehydrogenase 56 kDa, ovalbumine 45 kDa, D-aminoacid oxidase 37 kDa, trypsin 24 kDa, myoglobin 17 kDa.
b. The polypeptides of photosystem II in the polyacrylamide gel.
c. Nitrocellulose membrane with polypeptides of photosystem II after reaction with: 1. the antiserum to monogalactolipid; 2. the control serum (serum dilution 1:100).
d. Nitrocellulose membrane with polypeptides of photosystem II after reaction with: 1. the antiserum to digalactosyldiglyceride; 2. the control serum (serum dilution 1:25).
e. Nitrocellulose membrane with polypeptides of photosystem II after reaction with: 1. the antiserum to sulfoquinovosyldiglyceride; 2. the control serum (serum dilution 1:25).

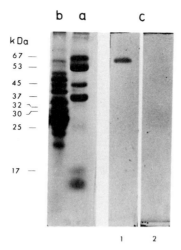

Fig. 3. Reaction of a 66 kDa polypeptide antiserum with the polypeptides of photosystem II on a nitrocellulose membrane according to the Western-blot assay.

a. Marker proteins in a 7 to 12% polyacrylamide gradient gel: Bovine serum albumine 67 kDa, glutamate dehydrogenase 56 kDa, ovalbumine 45 kDa, D-Aminoacid oxidase 37 kDa, hexokinase 25 kDa, myoglobin 17 kDa.

b. Polypeptides of the photosystem II-preparation in the polyacrylamide gel.

c. Nitrocellulose membrane with the polypeptides of photosystem II after the reaction with 1. the antiserum to the 66 kDa peptide (serum dilution 1:1000); 2. the control serum (serum dilution: 1:100).

REFERENCES

1. G. H. Schmid and A. Radunz, Reactions of a Monospecific Antiserum to Ferredoxin-NADP-Reductase with chloroplast Preparations, Z. Natur-forsch. **29c**, 384 (1974).

2. G. H. Schmid, A. Radunz and W. Menke, The Effect of an Antiserum to Plastocyanin on Various Chloroplast Preparations, Z. Naturforsch. **30c**, 201 (1975).

3. G. H.Schmid, A. Radunz and W. Menke, Localization and Function of Cytochrome f in the Thylakoid Membrane, Z. Naturforsch. **32c**, 271 (1977).

4. G. H. Schmid, W. Menke, A. Radunz and F. Koenig, Polypeptides of the Thylakoid Membrane and their Functional Characterization, Z. Natur-forsch. **33c**, 723 (1978).

5. A. Radunz, K. P. Bader and G. H. Schmid, Serological Investigations of the Function of Galactolipids in the Thylakoid Membrane, Z. Pflanzenphysiol. **114**, 227 (1984).

6. A. Radunz, K. P. Bader and G. H. Schmid, Influence of Antisera to Sulfoquinovosyl Diglyceride and to β-Sitosterol on the Photosyn-thetic Elektron Transport in Chloroplasts from Higher Plants. In: Struct., Funkt. & Metabol. Plant Lipid 9 (A. Siegenthaler and W. Eichenberger, eds.) Elsevier Science Publishers B.V., 479 (1984).

7. A. Radunz, Serological Investigations on the Function of Phospholipids in the Thylakoid Membrane. In: Proceedings of the VIth International Congress on Photosynthesis, August 1983, Brussels, Belgium (C. Sybesma, ed.) Martinus Nijhoff/Dr. W. Junk Publishers, The Hague, The Netherlands, Vol. III.2, 151 (1984).

8. A. Radunz and G. H. Schmid, On the Localization and Function of the Xanthophylls in the Thylakoid Membrane, Ber. Deutsch. Bot. Ges. **92**, 437 (1979).

9. D. A. Berthold, G. T. Babcock and C. F. Yocum, A highly resolved, oxygen-evolving photosystem II-preparation from spinach thylakoid membranes, FEBS Lett. **134**, 231 (1981).

10. K. K. Laemmli, Cleavage of Structural Proteins during the Assembly of the head of Bacteriophage T_4, Nature **227**, 680 (1970).

11. J. Renart, J. Reiser and G. H. Stork, Transfer of Protein from Gels to Diazobenzyloxymethyl-paper and with antisera: A method for studying antibody specifity and antigen structure, Proc. Natl. Acad. Sci. USA, **76**, 3116 (1979).

12. A. Radunz and R. Berzborn, Antibodies against Sulfoquinovosyl-diacyl glycerol and their Reactions with Chloroplasts, Z. Naturforsch. **25b**, 412 (1970).

13. A. Radunz, Lokalisierung des Monogalaktosyldiglycerids in Thylakoid-membranen mit serologischen Methoden, Z. Naturforsch. **27b**, 822 (1972)

14. A. Radunz, Localization of the Tri- and Digalactosyldiglyceride in the Thylakoid Membrane with serological methods, Z. Naturforsch. **31c**, 589 (1976).

15. A. Radunz, Binding of Antibodies onto the Thylakoid Membrane. IV. Phosphatids and Xanthophylls in the Outer Surface of the Thylakoid Membrane, Z. Naturforsch. **33c**, 941 (1978).

16. R. Voß, Immunologischer Nachweis von Glykolipiden an Peptiden des Photosystem II Höherer Pflanzen mit Hilfe des Western-Blot-Verfahrens, Diplomarbeit Universität Bielefeld (1991).

17. S. Specht, E. K. Pistorius and G. H. Schmid, Comparison of Photo system II Complexes isolated from tobacco and two chlorophyll decient tobacco Photosystem II mutants. Photosynth. Res. **13**, 47 (1987)

18. B. Andersson, Proteins Participating in Photosynthetic water oxidation. In: Encyclopedia of Plant Physiology, New Series, **Vol. 19**, Photosynthesis III (L.A. Staehelin and C.H. Arntzen, eds.) p. 447, Springer Verlag, Berlin (1986).

19. C. J. Arntzen and H. B. Pakrasi, Photosystem II Reaction Center: Polypeptide Subunits and Funcional Cofactors. In: Encyclopedia of Plant Physiology, New Series, **Vol. 19**, Photosynthesis III (L.A. Staehelin and C.J. Arntzen, eds.) p. 457, Springer Verlag, Berlin (1986).

20. F. Koenig, W. Menke, A. Radunz and G. H. Schmid, Localization and Functional Characterization of Three Thylakoid Membrane Polypeptides of the Molecular Weight 66 000, Z. Naturforsch. **32c**, 817 (1977).

21. U. Pick, K. Gounaris, A. Admon and J. Barber, Activation of the CF_0-CF_1, ATP Synthase from <u>Spinacia</u> Chloroplasts by chloroplast lipids, Biochim. Biophys. Acta **765**, 12 (1984).

22. D. Siefermann-Harms, J. W. Ross, K. H.Kaneshiro and H. Y. Yamamoto, Reconstitution by monogalactosyldiacylglycerol of energy transfer from light-harvesting chlorophyll a/b-protein complex to the photosystems in Triton X-100-solubilized thylakoids, FEBS Lett. **149**, 191 (1982).

THE Fo AND THE O-J-I-P FLUORESCENCE RISE IN HIGHER PLANTS AND ALGAE

Reto J. Strasser and Govindjee

Bioenergetics Laboratory, University of Geneva
1254 Jussy-Lullier, Switzerland, and
Biophysics Division, University of Illinois
Urbana, IL 61801, USA

Abstract.   The variable chlorophyll (Chl) a fluorescence yield is related to the photochemical activity of photosystem II (PS II) of the oxygen evolving organisms. The kinetics of the fluorescence rise from the minimal yield Fo to the maximal yield Fm is a monitor of the accumulation of net reduced Qa with time in both active (Qb-containing) and inactive (non-Qb) PS II centers. The measurements of true Fo and that of the complete fluorescence transient from true Fo to Fm are useful in obtaining a kinetic picture of PS II activity. Using a shutter-less system (Plant Efficiency Analyzer, Hansatech, UK) that is capable of providing the first measured point at about 20 microseconds and that allows data accumulation over several orders of magnitude of time, we have measured the complete fluorescence transient in low and moderate (up to 700 W m^{-2}) light intensities in several photosynthetic systems (higher plant leaves and chloroplasts; and the cell suspensions of green alga Chlamydomonas reinhardtii and several of its herbicide-resistant mutants, altered in single amino acids in its D1 protein). In all cases, the fluorescence transient follows a regular pattern of O-J-I-P--T, where two intermediate inflections J (at about 2 ms) and I (at about 20 ms) appear between Fo and Fm levels. Furthermore, the ratio of Fm to Fo is about 5 in all cases and the lowered published ratio in several cases is suggested to be due to the J level being mistaken for Fo. We also present data on the effects of varying the dark times between preillumination and measurements of the transient, on the intensity dependence, and on the effect of the addition of diuron. The relationship of the O-J rise to the fast fluorescence rise observed by other investigators will also be discussed.

Chlorophyll (Chl) a fluorescence transients provide information on the photochemical efficiency of photosystem II (PS II) of the oxygen evolving organisms[1,2,3,4]. Fo (or the O level) level is the "instantaneous" low fluorescence when photochemical efficiency is maximal; here, the concentration of the electron acceptor Qa is maximal. Measurement of true Fo level is not a trivial problem in instruments that use camera shutters. Most mechanical shutters have a full opening time of one or more ms. On the other hand, light emitting diode has the advantage that no shutter is needed and the time resolution can be as short as 500 ns.

With a new commercial instrument (Plant Efficiency Analyzer,

Regulation of Chloroplast Biogenesis
Edited by J.H. Argyroudi-Akoyunoglou, Plenum Press, New York, 1991

Hansatech, UK) Chl a fluorescence transient was recorded in a time span ranging from 20 microseconds to 20 minutes. The data was then plotted on a logarithmic scale covering 6 orders of magnitude. All photosynthetic samples tested showed a typical transient from the "O" to the "P" level with two intermediate inflections that we call J and I, respectively. Thus the regular transient is OJIP, not simply OIP.

Under moderate light intensities (200-700 W m^{-2}), the first rise from O to J is highly sigmoidal and levels off at about 2 ms, the time at which most camera shutters open. However, the I phase levels at about 20 ms, and the peak P at about 200 ms (the times being dependent upon light intensities). Fig. 1A shows our data for pea leaves on a log scale, and fig. 1B,C shows data on two linear time scales. A preillumination of 1 s followed by 5 s darkness before the onset of the continuous exciting light shows an increased Fo and a much faster fluorescence rise to the J level followed by a distinct decline and then an increase to the P level. After several minutes in light, the fluorescence reaches the terminal level T. Interrupting the light for 2 s after T reveals the absence of variable fluorescence in this state.

In the green alga Chlamydomonas reinhardtii, that had been measured by instruments with camera shuttters to have hight Fo (indicated as F at 4 ms) yields (see e.g.[5]), our experiments show that both the wild type and the herbicide-resistant D1 mutant DCMU-4 (S264A) had high variable fluorescence; the ratio of Fm to true Fo was 5 as in leaves and chloroplasts (fig. 2). In S264A mutant, there is a rapid Fo to J rise. It appears that instruments using camera shutters often mistake J level to be true Fo level unless "O" level is marked by calculation from measurements at low light intensities.

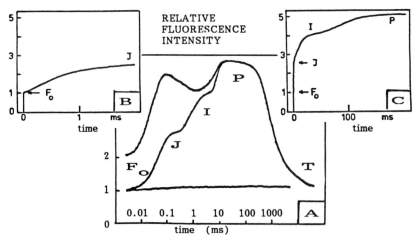

Fig. 1. OJIP--T Chlorophyll a fluorescence transient of an attached pea leaf, excited with red (650nm) LED's giving an intensity of 650 W m^{-2}, are shown on a logarithmic (A) or on a linear (B,C) time scale. Fig. 1A middle curve and Fig. 1B and 1C were plotted from the same data points of a dark adapted sample. Fig 1A upper curve shows a dark adapted sample which was preilluminated for 1s followed by 5s dark before the onset of continuous light. Fig. 1A lower curve shows the fluorescence trace when the light in the steady state T was turned off for 2s.

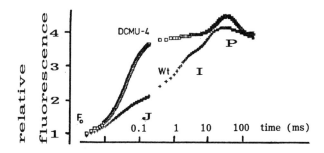

Fig. 2. OJIP Chlorophyll a fluorescence transient of Chlamydomonas reinhardtii (wild-type wt) and the herbicide-resistant D1 mutant DCMU-4 (S264A).

To gain an insight into the dynamics of the O-J-I-P transient in pea leaves, three sets of experiments were performed: (1) a preillumination (the same as measuring light) of 1 s was given to dark-adapted pea leaves followed by a variable dark time before the transients were measured; (2) after the pea leaves were adapted to continuous light (several minutes) and the fluorescence had reached the terminal level T, a variable dark time was given before a succeeding set of transients were measured; and (3) fluorescence transients were measured in dark-adapted leaves upon excitation with different light intensities. Our results showed that the first O to J phase is very sensitive to preillumination and to the dark time between preillumination and measurements, whereas the slower I to P phase was not. On the other hand, in the light-adapted state, the fast OJ rise is regenerated in one or two minutes dark, whereas the P level required about 20 minutes. Furthermore, the J level requires much higher light intensities than the P level to saturate. As expected, the addition of diuron, that blocks electron flow beyond Q_a, provoked a fast fluorescence rise from the O to the J level. Studies on the heterogeneity of photosynthetic units have been reported earlier by one of us[12,13] on fluorescence transient curves in the presence of diuron. They can be deconvoluted in two exponential and one hyperbolic functions. According to the grouping (cooperativity) concept[14], these fluorescence kinetics were attributed to big (with LHCII's), small (without LHCII's) and grouped (with cooperativity) PSU's.

The O to J rise measured here at moderate light intensities is obviously related to the fast fluorescence rise measured at high light intensities by the use of specialized methods to get a shutter opening time in µs range[6,7,8] or using, like we do, a shutterless system and fast date acquisition[9]. By the use of modulated fluorescence methods the appearence of a new peak in the fluorescence transient has been reported[10,11] and called I_1. However, no true kinetics of the OI_1 phase has been measured and this I_1 appeared as a peak only at extremely high light intensities (up to about 15.000 W m^{-2}), whereas in our experiments a short preillumination of moderate light (e.g. 1s and 300-600 W m^{-2}) was enough to create the OJ phase as a peak as high as the peak fluorescence P. Further research is needed to correlate the OI_1 phase of high light conditions to the regular OJIP-transient seen in all analyzed plants.

In conclusion, the method used in the current study, that allows the measurement of the complete transient from the O level to the T level over several orders of time scale from microseconds to minutes, provides the measurement of the true Fo level and the kinetics of all the transient changes (OJIP) in moderate light intensities. We can see the entire OJIDP--T-transient at one look when plotted on a logarithmic time scale.

Govindjee thanks the University of Geneva for support as a visiting Professor during the summer of 1991. We thank Professor J.D. Rochaix for the Chlamydomonas cells.

References

1. C. Papageorgiou, Chlorophyll fluorescence: an intrinsic probe of photosynthesis, in: "Bioenergetics of Photosynthesis", Govindjee, ed., Academic Press, New York, 320 (1975).
2. J. Lavorel and A.-L. Etienne, In vivo chlorophyll fluorescence, Topics in Photosynthesis 2:203 (1977).
3. Govindjee and K. Satoh, Fluorescence properties of chlorophyll b and chlorophyll c containing algae, in: "Light Emission by Plants and Bacteria", Govindjee, J. Amesz and D.C. Fork, ed., Academic Press, New York, 497 (1986).
4. G.H. Krause, and E. Weis, Chlorophyll fluorescence and photosynthesis: the basics, Ann. Rev. Plant Physiol. Plant Mol. Biol. 42:313 (1991).
5. Govindjee, B. Schwarz, J.-D. Rochaix, and R.J. Strasser, The herbicide-resistant D1 mutant L275F of Chlamydomonas reinhardtii fails to show the bicarbonate-reversible formate effect on chlorophyll a fluorescence transients, Photosynth. Res. 27:199 (1991).
6. P. Morin, Etude des cinétiques de fluorescence de la chlorophylle in vivo dans les premiers instants qui suivent le début de l'illumination, J. Chim. Phys. 671:674 (1964).
7. A. Joliot and P. Joliot, Etude cinétique de la réaction photochimique libérant l'oxygène au cours de la photosynthèse, C.R. Acad. Sc. Paris 258:4622 (1964).
8. R. Delosme, Etude de l'induction de fluorescence des algues vertes et des chloroplastes au début d'une illumination intense, Biochim. Biophys. Acta 143:108 (1967).
9. B. Genty, J. Harbinson, J.-M. Briantais and N.R. Baker, The relationship between non-photochemical quenching of chlorophyll fluorescence and the rate of photosystem 2 photochemistry in leaves, Photosynth. Res. 25:249 (1990).
10. C. Neubauer, and U. Schreiber, The polyphasic rise of chlorophyll fluorescence upon onset of strong continuous illumination: I. Saturation characteristics and partial control by the photosystem II acceptor side, Z. Naturforsch. 42c:1246 (1987).
11. U. Schreiber and C. Neubauer, The polyphasic rise of chlorophyll fluorescence upon onset of strong continuous illumination: II. Partial control by the photosystem II donor side and possible ways of interpretation, Z. Naturforsch. 42c:1255 (1987).
12. R.J. Strasser and H. Greppin, Primary reactions of photochemistry in higher plants, in: "Photosynthesis III. Structure and Molecular Organisation of the Photosynthetic Apparatus", G. Akoyunoglou, ed., Balaban Internat. Science Serv., Philadelphia, 717 (1981).
13. R.J. Strasser, The grouping model of plant photosynthesis: heterogeneity of photosynthetic units in thylakoids, in: "Photosynthesis III. Structure and Molecular Organisation of the Photosynthetic Apparatus", G. Akoyunoglou, ed., Balaban Internat. Science Serv., Philadelphia, 727 (1981).
14. R.J. Strasser, The grouping model of plant photosynthesis, in: Chloroplast Development, G. Akoyunoglou et al., ed., Biomedical Press, Elsevier/North Holland, 513 (1978).

CHARACTERIZATION OF THE OXYGEN EVOLVING COMPLEX OF

PROCHLOROTHRIX HOLLANDICA

Tsafrir S. Mor, Anton F. Post* and Itzhak Ohad

Department of Biological Chemistry and
*Department of Molecular and Microbial Ecology
The Institute of Life Sciences
The Hebrew University of Jerusalem, Jerusalem, 91 904. ISRAEL

INTRODUCTION

Prochlorophytes are oxygenic photosynthetic prokaryotes, distinguished from cyanobacteria by the properties of their photosystem (PS) II antenna and their pigment composition (Lewin, 1976). Prochlorophytes do not have phycobilisomes but contain chlorophyll b which is bound together with chlorophyll a to an intrinsic membrane protein resembeling the chlorophyll a/b binding proteins of higher plants and green algae (Bullerjahn et al., 1987). These features gave rise to the hypothesis that prochlorophytes are related to the progenitor of the chloroplasts of higher plants and green algae (Barber, 1986). However, in the prochlorophyte Prochlorothrix hollandica the chlorophyll a/b binding protein was shown to associate with PS-I (Bullerjahn et al., 1987) and it is probably not related to the light harvesting complex II, since they are not immunological cross-reactive (Bullerjahn et al., 1987; Bullerjahn et al., 1990).

Besides the differences in the antenna structure mentioned above, PS-II of chloroplasts diverges from that of cyanobacteria in the organization of the oxygen evolving complex (OEC). In chloroplasts three soluble proteins located in the thylakoid lumen are associated with PS-II. These are the 33 kDa manganese stabilizing protein (MSP) and two smaller proteins of 23 and 16 kDa. In cyanobacteria only the MSP was found (Burnap and Sherman, 1991).

In this work we have analysed the OEC of Prochlorothrix in order to further establish the relation between Prochlorothrix and the two archetypes of oxygenic photosynthetesis examplified in this research by the cyano-bacterium Synechocystis 6803 and the green alga Chlamydomonas reinhardtii.

MATERIALS AND METHODS

Prochlorothrix hollandica was grown in liquid medium (Burger-Wiersma et al., 1989) under limiting light conditions (20 μE·m^{-2}· S^{-1}).
Chlamydomonas reinhardtii y-1 strain and Synechocystis 6803 strain were grown as described by Schuster et al., 1988 and Burnap and Sherman, 1991 respectively. Prochlorothrix and Synechocystis thylakoid membranes were

obtained as folllows. cells were suspended in 50 mM Tricine pH 7.8
containing 10 mM NaCl and 5 mM MgCl₂ and then shaken with glass beads in
the Brown homogenizer. The homogenate was loaded on a sucrose step gradient
(60%, 45% and 30% w/v) in the same buffer and centrifuged at 4°C for 30 min
at 27,000 rpm in a Beckman TST SW28 rotor. Membranes were collected at the
interface of the 45-60% layer. Membranes of Chlamydomonas were obtained as
described (Schuster et al., 1988).

Triton X-114 phase partitioning was done according to Bricker and
Sherman (1982). Thylakoids were suspended to a protein concentration of 2
mg/ml in buffer containing 150 mM NaCl, 1mM EDTA, 10 mM Tris (pH=7.6) and 1%
(W/W) Triton X-114 (Sigma). The suspension was incubated in ice for 30 min
then at 37°C for 5 min and centrifuged for 3 min at top speed in an
Eppendorf minifuge. Phases were then separated and the procedure was
repeated twice using the same buffer for washing the aqueous phase and
Triton-less buffer for washing the detergent phase. Proteins were
precipitated by 5% (w/v) final concentration of trichloroacetic acid. The
pellet was then suspended in lithium dodecylsulphate sample buffer. Tris
washing of thylakoid membranes was done as previously described (Schuster et
al., 1988).

For tryptic digestion the membranes were suspended to a protein
concentration of 1.2 mg/ml in 50 mM tricine (pH=7.8), 10 mM NaCl and 5 mM
MgCl₂. Proteolysis was carried out at 10°C using Sigma purchased trypsin,
and stopped by adding double molar amounts of soy bean trypsin inhibitor
also purchased from Sigma.

PS-II electron transport activity was assayed by spectrophotometric
measurement of ferricyanide reduction. Electrophoresis and electrotransfer
of proteins to nitrocellulose filters were done by standard methods.
Antibodies were visualised with the ECL Western blotting detection system
(Amersham) and subsequent exposure to X-ray film. Antibodies against the MSP
and 23 kDa OEC protein and CF1-β subunit of the proton ATPase were the
generous gift of R. Herrman and R. Nechushtai.

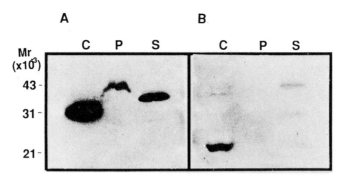

Fig. 1. Prochlorothrix MSP cross-reacts with an antibody raised against MSP
of spinach thyalkoids. Thylakoid proteins of Synechocystis 6803 (S),
P. hollandica (P) and C. reinhardtii (C) were separated by electrophoresis, transffered to nitrocellulose filters and immunodecorated
by antibodies raised against spinach MSP (panel A) or the 23 kDa
polypeptide (panel B).

RESULTS AND DISCUSSION

An antibody raised against the MSP of spinach reacted with a
polypeptide in the thylakoid membranes of Prochlorothrix. This polypeptide
has a molecular mass of 33 kDa similar to that of the MSP of Synechocystis,
but higher than the MSP of Chlamydomonas (Fig. 1a).
However, the antibody against the 23 kDa protein of the OEC reacted only
with a polypeptide from the membranes of the green alga (Fig. 1b). The weak
bands produced by this antibody in the lanes representing the thylakoid
proteins of Synechocystis and Prochlorothrix must be considered the results
of non-specific reaction, since they were absent from blots overlayed with
another antibody raised against the same polypeptide (data not shown). We
therefore conclude that a protein similar to the cyanobacterial MSP is
present in Prochlorothrix. These results also suggest that Prochlorothrix
OEC lacks the smaller polypeptides and thus is similar to that of
cyanobacteria.

To check whether the anti-23 kDa antibody failed to react because a
homologous protein is absent in Prochlorothrix or because of antigenic
variance, the thylakoids of Prochlorothrix and Chlamydomonas were washed
with 0.8 M Tris (pH=8.0), a procedure which extracts these extrinsic
membrane proteins from the thylakoids (Schuster et al., 1988). This
treatment efficiently removed the three OEC polypeptides from the membranes
of Chlamydomonas (Fig. 2, "Pellet") which were recovered from the
supernatant fluid (Fig. 2, "Sup"). Surprisingly, polypeptides of molecular
masses corresponding to those of the OEC proteins could not be washed off
the membranes of Prochlorothrix, although other extrinsic polypeptides could
be removed. The presence of MSP in the supernatant could not be traced even
with immunoblots (data not shown). Thus, the MSP of Prochlorothrix is more
strongly bound to the membrane than the corresponding protein of

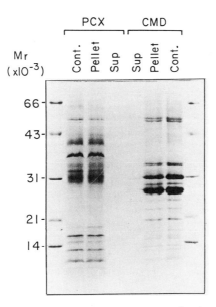

Fig. 2. MSP of P. hollandica adheres more strongly to the thylakoid
membranes than the MSP of C. reinhardtii. Thylakoids of P.hollandica
(PCX) and of C. reinhardtii (CMD) were washed with 0.8 M Tris.
Proteins in the supernatant fluid were precipitated by 80% acetone.
Equivalent portions were electrophoresed on a 10-17.5% acrylamide
gel and stained with coomassie briliant blue. Control, untreated
membranes; pellet, washed membranes; sup, supernatant fluid.

Chlamydomonas. This finding was further strengthened by the fact that other protocols (such as washing with 2 M NaCl) for removing peripheral membrane proteins in general and the OEC proteins in particular were ineffective in the case of Prochlorothrix (data not shown).

The stronger association of the MSP of Prochlorothrix with the thylakoids can be a consequence of its more hydrophobic nature. This hypothesis was tested by subjecting the membranes to Triton X-114 phase partitioning. In Chlamydomonas and Prochlorothrix hydrophobic proteins (e.g. D1, not shown) were found in the detregent phase and hydrophyllic proteins were found in the aqueouos phase (e.g. the ß subunit of CF1, Fig 3). Accordingly, MSP of Chlamydomonas (Fig. 3) and of Synechocystis (data not shown) was found in the Triton phase. This however was not the case for Prochlorothrix. The MSP of this organism was preferentially separated

	Chlamydomonas		Prochlorothrix	
	Water	DPC	Water	DPC
Control membranes	13.7	13.1	12.6	14.4
Trypsinized memb,	8.9	16.1	20.5	28.1

Fig. 3. The MSP of P. hollandica partitions with hydrophobic proteins. Thylakoid membranes of P. hollandica and of C. reinhardtii were fractionated with Triton X-114 to hydrophillic proteins found in the aqueous phase (AQ), and hydrophobic proteins found in the Triton X-114 phase (TR); C, unfractionated membranes. Samples were electorphoresed, electrotransfered to nitrocellulose filters and reacted with the respective antibodies as indcated.

Table 1. Effect of trypsin digestion on PS II electron transport activity. Membranes were digested with 200 µg trypsin/ml. Ferricyanide reduction is expressed as µmol Ferricyanide/mg chl/hr. Water, water as electron donor; DPC, Diphenyl carbazide as electron donor.

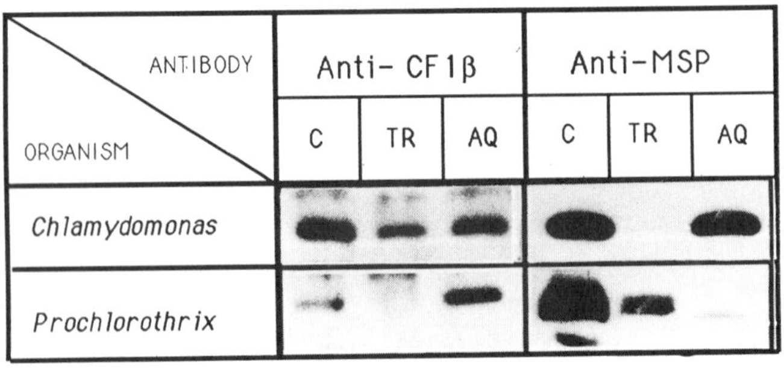

ANTIBODY ⁄ ORGANISM	Anti- CF1β			Anti-MSP		
	C	TR	AQ	C	TR	AQ
Chlamydomonas						
Prochlorothrix						

together with the hydrophobic proteins (Fig. 3). These results clearly demonstrate that the MSP of Prochlorothrix is more hydrophobic as compared to the same protein from Chlamydomonas and Synechocystis.

The data presented thus far indicate that the MSP of Prochlorothrix is at least partially embedded in the membrane. This is in striking contrast to the situation in other organsims examined. Since the MSP is assumed to be located on the inner surface of the thyalkoid membrane in close proximity to the Mn cluster it was of interest to show that the MSP of Prochlorothrix is exposed to proteolytic activity. To this end thylakoids of Prochlorothrix and Chlamydomonas were subjected to controlled trypsin digestion. The proteaase should effect only exposed or partially exposed membrane proteins. In Both cases, the 32.5 kDa D1 and the 33 kDa MSP polypeptides were lost as a result of the trypsinization since both could no longer be recognized by specific antibodies in immunoblots (data not shown). These results indicate that the MSP of Prochlorothrix is indeed protruding from the membrane so as to account for its effective digestion by trypsin.

The different structural and organizational characteristics of the MSP of Prochlorothrix demonstrated above may also have functional manifestations. This is clearly demonstrated by results of the experiment shown in Table 1 in which PS-II electron transport activity was assayed in trypsinized thylakoids. As expected from the loss of the MSP, proteolytic digestion had a marked inhibitory effect when water was used as an electron donor in thylakoids of Chlamydomonas. When the water splitting reaction was circumvented by using the artificial electron donor diphenylcarbazide (DPC), PS-II activity was hardly effected by the trypsin treatment. Loss of DCMU sensitivity of PS-II activity demonstrates that the detrimental effect of trypsin is not restricted to the donor side of PS-II as previously reported and as suggested by the trypsin effects on the D1 protein (Regitz and Ohad, 1976). In contrast to theses results, trypsinization of Prochlorothrix membranes stimulated the PS-II activity when either water or DPC were used as electron donors. Loss of DCMU sensitivity was also observed in the case of Prochlorothrix. We attribute the persistance of electron donation to PS-II despite the protease induced damage to its protein constituents to a more protected manganese binding niche which is less accessible to trypsin. The actual rise in electron flow activity can be explained by an increase in the accesibility of the Q_A site, the electron donor, to the electron acceptor, Ferricyanide, in the trypsin treated membranes.

Since we could not detect trypsin fragments of the MSP, we cannot exclude the possibility that a trypsin resistant fragment of this protein is responsible to the seeming insenstivity of the water oxidation process to proteolytic digestion of Prochlorothrix membranes. Such a fragment was indeed found in trypsinized PS-II particles of spinach when proteolysis was carried out at pH equal to or lower than 7.25 (Volker et al., 1985). Such particles also maintained water splitting activity. Prochlorothrix membranes trypsin treated a wide pH range (5.5-9.0) maintained activity throughout (data not shown) supportive of the notion that Prochlorothrix has a more resistant donor side. Another explanation is suggeted by the recent finding that MSP is not strictly needed for electron donation to PS-II by the manganese cluster. The cluster is bound to the PSII reaction center by the core polypeptides including D1. Thus it is possible that the part of the D1 polypeptde that is responsible for the manganses binding is more shielded against trypsin in Prochlorothrix than in Chlamydomonas.

Either way the PS-II donor side of Prochlorothrix is clearly different in its properties as compared to Synechocystis and Chlamydomonas. The functional significance of these differences is yet to be investigated.

Interestingally Prochlorothrix was shown to be more sensitive to photoinhibition than Chlamydomonas (Mor et al., 1991). Since the molecular mechanism of this phenomenon is considered to include deterioration of donor side activity as a result of light inflicted damage to the acceptor side (Prasil et al., 1991), it is not unlikely that the unique features of the PS-II donor side of Prochlorothrix play a role in its increased sensitivity to light.

AKNOWLEDGEMENTS

This work was supported by a research grant awarded by the binational Israel-America foundation (BSF) to I.O. in cooperation with L.Sherman.

REFERENCES

Barber, J., 1986, New organisms for elucidating the origin of higher plant chloroplast, TIBS, 11:234.

Bricker, T.M., and Sherman, L.A., 1982, Triton X-114 phase-fractionation of maize thylakoid membranes in the investigation of thylakoid protein topology, FEBS LETT., 149:197-202.

Bullerjahn, G.S., Matthijs, H.C.P., Mur, L.R., and Sherman, L.A., 1987, Chlorophyll-protein complexes of the thylkoid membranes form Prochlolrothrix hollandica, a prokaryote containing chlorophyll a and b, Eur. J. Biochem., 168:295-300.

Bullerjahn, G.S., Jansen, T.C., Sherman D.M., and Sherman, L.A., 1990, Immunological characterization of the Prochlorothrix hollandica and Prochloron sp. chlorophyll a/b antenna proteins, FEMS Microbiol. Lett., 67:99-106.

Burger-Wiersma, T., Stal, L.J., and Mur, L.R., 1989, Prochlorothrix hollandica gen. nov., sp. nov., a fillamentous oxygenic photoautotrophic procaryote containning chlorophyll a and b: Assignment to Procholrotrichaceae fam. nov. and order Prochlorales Florenzano, Balloni, and Materassi 1986, with emendation of the ordinal description, Int. J. Syst. Bacteriol., 39:250-257.

Burnap, R.L., and Sherman, L.A., 1991, Deletion mutagenesis in Synechocystis sp. PCG6803 indicates that the Mn-stabilizing protein of photosystem II is not essential for O₂ evolution, Biochemistry, 30:440-446.

Lewin, R.A., 1976, Prochlorophyta as a proposed new devision of algae, Nature, 261:697-698.

Mor, T.S., Post, A.F., and Ohad, I., 1991, Procholorothrix hollandica is more sensitive to photoinhibition than Chlamydomonas reinhardtii, This volume.

Post, A.F., Gal A., Ohad I., Milbauer, K.M., and Bullerjahn, G.S., 1991, Characterization of light activated reversible phosphorylation of a chlorophyll a/b antenna apoprotein in the photosynthetic prokaryote Prochlorothrix hollandica, submitted.

Prasil, O., Adir, N., and Ohad, I., 1991, Dynamics of photosystem II: Mechanisms of photoinhibition and recovery processes, in Topics in Photosynthesis, vol. 11, Barber, J., ed., in press.

Regitz, G. and Ohad I., 1976, Trypsin sensitive photosynthetic activities in chloroplast membranes from Chlamydomonas reinhardtii y-1, J. Biol. Chem., 251:247-252.

Schuster, G., Timberg, R. and Ohad, I., 1988, Turnover of thylakoid photosystem II proteins during photoinhibition of Chlamydomonas reinhardtii, Eur. J. Biochem., 177:403-410.

Volker, M., Ono, T., Inoue, Y. and Renger, G., 1985, Effect of trypsin on PS II particles. Correlation between Hill-activity, Mn abundance and peptide pattern, Biochim. Biophys. Acta., 806:25-34.

PROCHLOROTHRIX HOLLANDICA IS MORE SENSITIVE TO PHOTOINHIBITION

THAN CHLAMYDOMONAS REINHARDTII

Tsafrir S. Mor, Anton F. Post* and I. Ohad

Department of Biological Chemistry and
*Department of Molecular and Microbial Ecology
The Institute of Life Sciences
The Hebrew University of Jerusalem, Jerusalem 91 904, Israel

INTRODUCTION

Prochlorothrix hollandica belongs to Prochlorophyta (Lewin, 1976), a group of oxygenic photosynthetic prokaryotes assumed to be related to the progenote of the chloroplasts of higher plants and green algae (Barber, 1986). Unlike cyanobacteria, P. hollandica lacks phycobilisomes and possesses a chlorophyll a/b antenna complex which is however different from the light harvesting complex of photosystem (PS) II from chloroplasts of higher plants (Bullerjahn et al., 1987; Bullerjahn et al., 1990). In P. hollandica this membrane intrinsic pigment-protein complex appears to be associated with PS I (Bullerjahn et al., 1987; Bullerjahn et al., 1990). To date, no peripheral antenna was found to be bound to PS II of P. hollandica. Another important feature characterizing the P. hollandica thylakoids is the membrane morphology. In the chloroplast the photosynthetic membranes are structurally and functionally differentiated into appressed (grana) and non-appressed regions (stroma membranes) and their various membrane complexes are segregated in these two different membrane domains. PS II is mostly localized in the grana whereas PS I is restricted to the stroma membranes. In cynaobacteria the thylakoids do not display such spatial differentiation and segragation of the two photosystems is not observed. P. hollandica contains very limitted grana formations (Burger-Wiersma et al., 1986) and hence uneven distribution of PS I and PS II is very unlikely.

The electron donor side of PS II was shown to have some unique structural and functional properties (Mor et al., 1991). P. hollandica contains a more hydrophobic manganese stabilizing protein and the water oxidation activity of PS II is more resistant to proteolytic digestion.

One of the important features of the oxygen evolving photochemical reaction center of PS II is its inactivation by light. The process, termed photoinhibition, is a result of a series of complex molecular events influencing both the reducing side and the oxidizing side of PS II which leads to the deterioration of the reaction center polypeptides notably D1 and to a lesser extent D2 (reviewed by Prasil et al., 1991).

Since P. hollandica possesses a PS II antenna and an oxygen evolving system different from that of cyanobactria or the higher plant chloroplast it was of interest to examine the light sensitivity and efficiency of

Regulation of Chloroplast Biogenesis
Edited by J.H. Argyroudi-Akoyunoglou, Plenum Press, New York, 1992

asbsorbed light utilisation relative to the other types of photosynthetic organisms. In this work we have compared the sensitivity of P. hollandica to photoinhibitory light treatment to that of Chlamydomonas reinhardtii.

MATERIALS AND METHODS

Growth of P. hollandica and C.reinhardtii cells was as described (Mor et al., 1991). Cells were harvested by centrifugation and suspended in fresh growth medium to a cell density equivalent to 30 µg chlorophyll/ml. The cell suspensions were kept for 30-45 min in growth conditions prior to the beginning of high light treatment. Cells were then moved to glass tubes (3.5 cm diameter) and illuminated at various light intensities and duration as indicated. Light was supplied by a tungsten-halogen lamp (600 W) filtered through a heat filter using a Cabin professional projector. Suspensions were kept at 25°C and were stirred constantly. Where indicated, chloramphenicol (CAP, d-threo form, Sigma) was added to a final concentration of 200 µg/ml (Chlamydomonas) and 10 µg/ml (Prochlorothrix). For the recovery experiments, photoinhibied cells were returned to growth light intensity conditions. Oxygen evolution was measured with a Clarck type oxygen electrode.

RESULTS AND DISCUSSION

Measurements of the light saturation of oxygen evolution demonstrated that P. hollandica activity is saturated at 40 µE/m²s significantly lower than C. reinhardtii which required 250-300 µE/m²s to reach saturation at equal chlorophyll concentrations. In contrast to the high efficiency of light utilisation, the Vmax of oxygen evolution in P.hollandica was about half of that of C.reinhardtii cells (data not shown).

The results of experiments in which the sensitivities of P. hollandica and C. reinhardtii cells to strong illumination were compared are shown in Fig. 1. Cells were incubated at the indicated light intesities and their

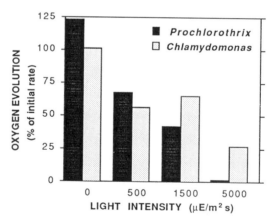

Figure 1. Photoinhibition of P. hollandica and of C. reinhardtii as a function of light intensity. Cell suspensions of P. hollandica and C. reinhardtii were exposed to the indicated light intensity for 1hr. Photosynthetic activity was assayed by measuring net oxygen evolution and is expressed as percent of the initial rate. The 100% values were 36 and 86 µmol $O_2 \cdot$mg chl$^{-1} \cdot$hr^{-1} for P.hollandica and C. reinhardtii respectively.

photosynthetic activities were measurd before (control) and after the photoinhibitory treatment. The degree of photoinhibition of P. hollandica was similar to that of C. reinhardtii when the intensity of light was lower than about 10 fold saturation intensity. However, when the cells were incubated at higher light intensity P. hollandica was photoinhibited to a greater extent than C. reinhardtii. The damage was probably restricted to the photosynthetic apparatus as respiration was not impaired under those conditions (data not shown).

Strong illumination of P. hollandica promotes the turnover of D1 as is the general case in all other photosynthetic organisms so far examined (Prasil et al., 1991). In accordance to this, prevention of prokaryotic protein synthesis by addition of CAP enhances the degree of photoinhibition. The increased sensitivity of P. hollandica to strong light is observed also when cells were photoinhibited in the presence of CAP (data not shown). Since the CAP concentrations used in these experiments completely inhibited D1 protein synthesis in both organisms the faster inactivation of P. hollandica oxygen evolution activity indicates a higher light sensitivity of PS II in this organism as compared to that of C. reinhardtii.

Furthermore when the capacity of the two organisms to recover from the photoinhibted state was compared (Fig. 2), a difference was again observed. While C. reinhardtii could still recover from severe photoinhibition, the ability of P. hollandica to recover was more limited. In experiments similar to the one depicted in Fig. 2, P. hollandica was shown to recover slower than C. reinhardtii (not shown). Moreover, P. hollandica could not recover photosynthetic activity if photoinhibition was carried for more than one hour at 5000 $\mu E \cdot m^{-2} sR-1R$ (Fig. 2) even after 20 hours of incubation at low light intensity (data not shown). In both cases recovery was dependent on de-novo protein synthesis and could not occur when CAP was added to the cell suspension during the recovery period (not shown). These results indicate that in addition to the increased light sensitivity of PS II, the replacement of damged PS II subunits and/or prosthetic groups is less efficient in P. hollandica as compared with C. reinhardtii.

As stated in the Introduction, PS II of P. hollandica lacks a distinct peripheral antenna. Yet P. hollandica is very efficient in utilization of light at low intesity. This could imply either that these cells posses a larger intrinsic PSII core antenna (i.e. more 43/47 kDa polypeptides per reaction center) or more likely a better transfer of excitation energy from antenna chlorophylls to the primary donor of PS II, P680. The net result is that the quantum yield of charge separation is probably higher than in the case of the green alga. This may contribute to the observed increase in the quantum yield of the reaction center II photoinactivation in this prochlorophyte.

Limitations in electron donation by the water oxidizing system results in a fast irreversible inactivation of photosystem II (Prasil et al., 1991). This was shown by inhibiting water oxidation of thylakoids in-vitro (e.g. Jegerschold et al., 1990) or in-vivo in mutants with impaired oxygen evolution capacity (Gong and Ohad, 1991). The features of the donor side of PS II of P. hollandica might be the limiting factor of the electorn transport chain in high light. Under such conditions the accumulation of the highly reactive radicals $P680^+$ and Z^+ (the primary and secondary donors of PS II) is enhanced. These radicals are suspected to play a major role in the degradation of the PS II polypeptide D1 (Prasil et al., 1991). This hypothesis is currently under investigation.

Chloroplasts and cyanobacteria have evolved several defense mechanism against photoinhibition including regulation of light energy distribution between PS I and PS II. Among these is the reduction of the antenna cross section by dissociation of antenna proteins from reaction center of PS II (Andersson and Styring, 1991). Such mechanisms cannot obviously operate in P. hollandica since this organism apperantly lack peripheral PS II antenna altogether.

The major chlorophyll a/b binding protein of P. hollandica is a 35 kDa protein which is found to be phosphorylated by a redox controlled kinase activity (Post et al., 1991). The kinase is activated in the light, but unlike the light harvesting II kinase of higher plant chloroplasts it is inactivated only very slowly in the dark. Phosphorylation of PS II antenna and its consequent dissociation from PS II and its migration to non-appressed regions of the thylakoids, are thought to be the underlying mechanism of the regulation of the light absorbed energy distribution between PS I and PS II in higher plants and green algae. This type of regulation, termed state-transition, may be inefficient in P. hollandica if it relies on phosphorylation of antenna components. However in cynobacteria a different molecular mechanism that does not involve protein phosphorylation leads to state-transitions. In this case the proximity of PS II to PS I increases the chance for energy spill-over from one photosystem to the

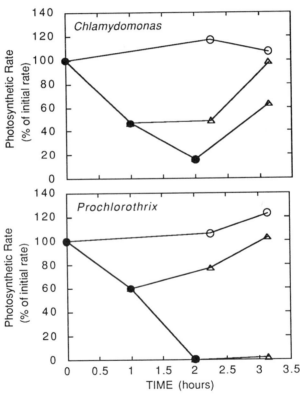

Figure 2. Recovery of photosynthetic activity after photoinhibition of P. hollandica and C. reinhardtii. Cell suspensions were photoinhibited with 5000 $\mu E \cdot m^{-2} \cdot s^{-1}$ for 60 or 120 min. At these times the suspensions were moved to the control illumination conditions: 30 $\mu E \cdot m^{-2} \cdot s^{-1}$. Photosynthetic activity was assayed by measuring oxygen evolution and is expressed as percent of the initial rates. 100% values were 32 and 47 μmol O_2/mg chl·hr for P. hollandica and C. reinhardtii respectively. Circles, control; filled circles, photoinhibition; triangles, recovery.

436

other. Whether state transitions occur in P. hollandica by a cynobacterial type of mechanism or by the chloroplast one is at present unknown (Burger-Wiersma and Post, 1989). The state transition may aleviate overexcitation of PSII only to limited extent. Considering the low light intensity required for saturation of electron flow it appears that in P. hollandica the protection ascribed by this mechanism is even more limited than in the other types of oxygenic PS II.

The data presented here clearly demonstrate that P. hollandica is more sensitive to photoinhibiton than the green alga chosen as a reference organism, C. reinhardtii. A correlation was made betweem the results and some unique features of the prochlorophyte's PS II. These correlations, as yet speculative, should serve as a working hypothesis in future studies.

ACKNOWLEDGMENT

This work was supported by a research grant awarded by the binational Israel-America foundation (BSF) to I.O. in cooperation with L.Sherman.

REFERENCES

Andersson B., and Styring S. 1991. Photosystem II - molecular organization, function and acclimation, in "Current topics in bioenergetics", vol. 16. (Lee, C.P., ed.). pp. 1-81. Academic Press, New York.
Barber, J., 1986, New organisms for elucidating the origin of higher plant chloroplast, TIBS, 11:234.
Bullerjahn, G.S., Matthijs, H.C.P., Mur, L.R., and Sherman, L.A., 1987, Chlorophyll-protein complexes of the thylkoid membranes form Prochlolrothrix hollandica, a prokaryote containing chlorophyll a and b, Eur. J. Biochem., 168:295-300.
Bullerjahn, G.S., Jansen, T.C., Sherman D.M., and Sherman, L.A., 1990, Immunological characterization of the Prochlorothrix hollandica and Prochloron sp. chlorophyll a/b antenna proteins, FEMS Microbiol. Lett., 67:99-106.
Burger-Wiersma, T., and Post, A.F., 1989, Functional analysis of the photosynthetic apparatus of Prochlorothrix hollandica (Prochlorales), a chlorophyll b containing procaryote, Plant Physiol., 91:770-774.
Burger-Wiersma, T., Veenhuis, M., Korthals, H.J., Van de Wiel, C.C.M., and Mur, L.R., 1986, A new prokaryote containing chlorophylls a and b, Nature, 320:262-264.
Gong, H., and Ohad, I., 1991, Occupancy of photosystem II-Q_B site by plastoquinone is essential for the degradation of D1 protein during photoinhibition in-vivo, J. Biol. Chem., in press.
Jegerschold, C., Virgin, I., and Styring S. 1990. Light dependent degradation of the D1 protein in photosystem II is accelerated after inhibition of the water splitting reaction, Biochemistry. 29:6179-6186.
Lewin, R.A., 1976, Prochlorophyta as a proposed new devision of algae, Nature, 261:697-698.
Mor, T.S., Post, A.F., and Ohad, I., 1991, Characterization of the manganes stabilizing protein of of Procholorothrix hollandica, This volume.
Post, A.F., Gal A., Ohad I., Milbauer, K.M., and Bullerjahn, G.S., 1991, Characterization of light activated reversible phosphorylation of a chlorophyll a/b antenna apoprotein in the photosynthetic prokaryote Prochlorothrix hollandica, submitted.
Prasil, O., Adir, N., and Ohad, I., 1991, Dynamics of photosystem II: Mechanisms of photoinhibition and recovery processes, in Topics in Photosynthesis, vol. 11, Barber, J., ed., in press.

COMPARISON OF THE O_2-EVOLVING COMPLEXES OF <u>OSCILLATORIA</u> <u>CHALYBEA</u> WITH THAT
OF DIFFERENT GREENING STAGES OF HIGHER PLANT CHLOROPLASTS:
AN AMPEROMETRIC AND MASS SPECTROMETRIC STUDY

Klaus P. Bader, Fabrice Franck* and Georg H. Schmid

Lehrstuhl Zellphysiologie, Fakultät für Biologie,
D-4800 Bielefeld 1, Germany
*Service de Photobiologie, Université de Liège, Sart-Tilman,
B-4000 Liège, Belgium

INTRODUCTION

Photosynthetic formation of reduction equivalents is achieved by the
reduction of $NADP^+$ with water as the electron donor. In this reaction water
is split and oxygen is the oxidation product. Molecular oxygen is formed
by the abstraction of 2 electrons from each of the two water molecules.

$$2\ H_2O \longrightarrow 4\ H^+ + 4\ e^- + O_2$$

The reaction requires the creation of 4 oxidation equivalents which are
produced by the successive absorption of four light quanta accumulating
4 positive charges according to the Kok model.
At present, no evidence for the occurrence of intermediary products in the
successive oxidation reactions has been presented. Such intermediary
products are easily imaginable as "oxygen precursor" or partially oxidized
water molecules, as a proton release pattern of 1:0:1:2 seems to go along
with the four light reactions.
However, evidence was presented that up to the oxidation state S_3 molecular
water is still exchangeable at the OEC, before it is oxidized to molecular
oxygen by one more flash. These experiments were carried out with
<u>Oscillatoria</u> <u>chalybea</u>, a filamentous cyanobacterium which exhibits in
comparison to higher plant chloroplasts and green algae a number of peculiarities with respect to its flash induced O_2-evolution pattern (Fig. 1).
<u>Oscillatoria</u> always shows an oxygen signal under the first flash of a sequ-
ence and although the pattern exhibits strong damping of the oscillation,

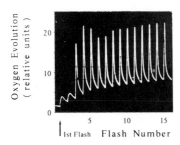

Oxygen Evolution (relative units)

20

10

0

5   10   15

1st Flash   Flash Number

Fig.1.
Polarographic recording of the oxygen
evolution in the filamentous cyanobacterium <u>Oscillatoria</u> <u>chalybea</u> upon
illumination with short saturating
light flashes. Even after excessive
dark adaption substantial oxygen
yields under the very first flash are
observed which could be shown to be
due to a metastable S_3-state in the
organism.

high steady state yields of oxygen are obtained. Moreover, slow deactivation kinetics of the highly oxidized redox states S_2 and S_3 are observed
which can be further slowed down by addition of chemicals like ABDAC
(Bader, 1989).

This made Oscillatoria a well-suited organism for the above mentioned
experiment, as slow deactivation kinetics mean more time for experiments
on S_2 and S_3. We populated the S_3-state by giving two preflashes and added
$H_2^{18}O$ onto the thus formed S_3-conditions. At various intervals after this
addition, a third analyzing flash yielded high amounts of $^{18}O_2$, an oxygen
isotope whose background was absolutely zero before the addition of $H_2^{18}O$.
We were able to conclude from our amperometric and mass spectrometric
experiments that in Oscillatoria chalybea a particularly high portion of
S_3 is metastable (resulting in the first flash signal of an oxygen
evolution pattern) and that even in this condition molecules from the
surrounding bulk water can be exchanged (Bader et al., 1987).

RESULTS AND DISCUSSION

    Observation of a phenomenon similar to the one described for
Oscillatoria came from investigations dealing with etiolated plant
material. During development of the photosynthetic apparatus very early
stages of greening were examined. With such assays we could demonstrate
that greening etioplasts of oat yield a flash pattern exhibiting
peculiarities comparable to those observed with Oscillatoria (Fig.2).

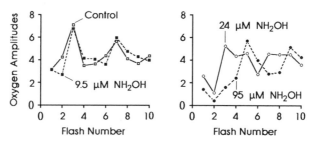

Fig.2. Oxygen evolution pattern in etio-chloroplasts from oat in the ab-
       sence and in the presence of hydroxylamine. Dark adaption time:
       2 min. Longer dark adaptation periods do not reduce the signal under
       the first flash to zero.

An interesting aspect in this context is the observation that the photosystem II performance of the phylogenetically old system of the filamentous
cyanobacteria Oscillatoria chalybea coincides with that in a defined stage
of chloroplast development during the assembly of the photosynthetic
apparatus in greening higher plants. In agreement with these considerations
we observe that completely differentiated i.e. greened systems lost these
properties and exhibited the normal flash pattern known from the
literature with no oxygen yield under the first flash and S_3 being an
extremely short-living condition (Franck and Schmid, 1989).

In the course of our experiments dealing with the S_3-state in Oscillatoria
we made an intriguing observation. Under the oxygen partial pressure of
normal air (i.e. 21% O_2) oxygen interacts with the S-state system of PS
II in Oscillatoria. When we scrutinized the related phenomena we found out
that quite obviously an oxygen uptake had occurred leading to the formation
of hydrogen peroxide in the dark which in turn was immediately decomposed
upon flash (or continuous light) illumination (Fig.3).

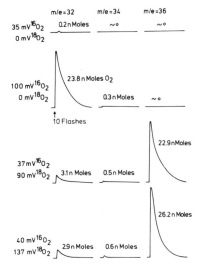

	m/e =32	m/e =34	m/e =36

35 mV $^{16}O_2$
0 mV $^{18}O_2$ — 0.2 n Moles — ~o — ~o

100 mV $^{16}O_2$ — 23.8 n Moles O_2
0 mV $^{18}O_2$ — 0.3 n Moles — ~o

↑
10 Flashes

22.9 n Moles

37 mV $^{16}O_2$
90 mV $^{18}O_2$ — 3.1 n Moles — 0.5 n Moles

26.2 n Moles

40 mV $^{16}O_2$
137 mV $^{18}O_2$ — 2.9 n Moles — 0.6 n Moles

Fig.3.
Photosynthetic oxygen evolution as the consequence of 10 flashes in thylakoids from Oscillatoria chalybea. After flushing with N_2 the gas phase over the assay, artificial gas atmosphere conditions as indicated by the different oxygen isotope background values were installed. The signals of masses 32, 34 and 36 were simultaneously recorded and the areas integrated to give the indicated absolute values.

From the observed isotope distribution and the absence of the mixed label ($^{16}O^{18}O$ = mass 34) we were able to conclude that hydrogen peroxide is decomposed according to

$$H_2O_2 \dashrightarrow 2\ H^+ + 2\ e^- + O_2.$$

No catalatic type of reaction has occurred in this case, as the mixed label which would prove the intermediate formation of $H_2^{18}O$ is missing. These experiments and considerations entail the conclusion that in Oscillatoria photosynthetic oxygen evolution in vivo consists of two distinct portions, one coming from normal water splitting and the other one -which might under certain conditions be the predominant portion- from H_2O_2-decomposition.

The described reaction seems to be the result of an interaction between photosystem II and oxygen, an interpretation which is substantiated by our experiments. Some years ago, Renger introduced the so-called ADRY-reagents into flash-induced oxygen evolution measurements. In this context the chemical Ant-2-p has been shown to specifically interact with the S_2- and S_3-states (Renger, 1972; Renger et al., 1973; Hanssum et al., 1985).

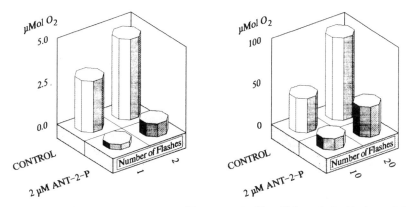

Fig.4. Effect of 2-(3-Chloro-4-trifluoromethyl)anilino-3,5 dinitrothiophene (Ant-2-p) on the oxygen evolution in Oscillatoria chalybea. In this experiment care was taken to install conditions, where light induced oxygen evolution comes preponderantly from the H_2O_2-decomposition.

Our experiments show that Ant-2-p exerts a substantial influence also on the oxygen evolution coming from the hydrogen peroxide decomposition. Hence, hydrogen peroxide formation involves the S_2- and/or· the S_3-state. Furthermore, we conclude from the absence of any oxygen uptake signal in the presence of the agent that the site of the formation and of decomposition of hydrogen peroxide are identical and should be located in the region of S_2 and S_3. Further analyses of the S-state deactivation kinetics under the influence of the reagent argue in favor of an effect on the S_2-redox state (results not shown).

The described phenomenon of hydrogen peroxide formation and decomposition which we think to be due to an interaction of the S-state system with O_2 does not take place in higher plants. Isotopic distribution of the observed flash induced oxygen evolution is completely independent on the presence or absence of $^{18}O_2$ in the gas phase and shows nearly theoretical isotope distribution values (which can easily be calculated from the $H_2^{16}O/H_2^{18}O$ ratio in the liquid phase of the assay under the assumption that only one reaction i.e. water oxidation has occurred (Table I.).

$^{16}O_2$-gas atmosphere:

Table I. Comparison of the calculated and the measured composition of oxygen evolution in a system containing 0,26 (0,265)% $H_2^{18}O$ in a $^{16}O_2$- and $^{18}O_2$ containing gas atmosphere, respectively.

Isotope	Alpha	Isotopic distribution of oxygen evolved			
	(measured fraction)	expected from the composition of H_2O	measured		
			2 flashes	3 flashes	10 flashes
$^{16}O^{16}O$		0.527	0.524	0.549	0.555
$^{16}O^{18}O$	0.26+0.008	0.398	0.410	0.396	0.384
$^{18}O^{18}O$		0.075	0.066	0.055	0.062

$^{18}O_2$-gas atmosphere:

$^{16}O^{16}O$		0.527	0.514	0.559	0.541
$^{16}O^{18}O$	0.265+0.008	0.398	0.379	0.362	0.391
$^{18}O^{18}O$		0.075	0.106	0.078	0.068

The experimental result that the observed effect is somehow related to S_2 can easily be substantiated by some trivial reflections on the reactivity of the S-states in question: Any reactivity of S_3 with oxygen must be minimal, otherwise it would not be metastable. S_1, on the other hand, is the dark stable state in virtually all investigated systems as well as in Oscillatoria. This fits into work which compares the reactivity of S-states in other systems towards various reagents of the type of NH_2OH, NH_2NH_2 and others (Table II).
Investigations by Franck and Schmid with etioplasts of oat i.e. in the developing photosynthetic apparatus have shown that the reactivity of S_2 towards reducing agents like NH_2OH exceeds that of the other S-states by far. Similar results came from Messinger and Renger (1990) for spinach chloroplasts thus confirming the observation of Franck and Schmid (1989). It therefore seems quite obvious that the relative reactivity of all S-states is the same in Oscillatoria as in higher plant chloroplasts.

Table II. Calculated S-state distributions in differently pretreated etiochloroplast preparations from oat under the assumption of 10%
misses and 0% double hits.

Pretreatment	S_0	S_1	S_2	S_3	Σ
2 hv	0.10	0.10	0.28	0.61	1.09
2 hv+2 min	0.17	0.42	0.24	0.17	1.00
2 hv+24 uM hydroxylamine +2 min	0.25	0.42	0.06	0.16	0.89
2 hv+95 uM hydroxylamine +2 min	0.17	0.12	0.02	0.10	0.41

We think that the observed differences concerning the reactivity of the
S-state system with oxygen should be of structural and evolutionary origin.
We assume that the principal absence of the two extrinsic peptides of MW
16 and 23 kD in photosystem II of cyanobacteria, which is one of the major
differences to higher plants, might be the reason. Still today, the
function of these two peptides is not yet clear. The 23 kD peptide seems
to regulate binding of the Cl-ions necessary for the water splitting
reaction. (The peptide might also regulate the affinity of its own
Cl-binding site or play a role in a process that in its absence is taken
over by high Cl-ion concentrations).

Very careful descriptions are found in the literature concerning the
conditions under which H_2O_2 is electron donor to photosystem II in
preparations from higher plants (Sandusky and Yocum, 1988; Schröder and
Akerlund, 1986). In fact, a detailed analysis of several properties and
peculiarities of Oscillatoria chalybea with those of higher plants leads
to the conclusion that under in vivo conditions many properties of Oscillatoria correspond to the ones observed in higher plants under Cl⁻-deficien-
cy. That Oscillatoria is indeed living under permanent Cl⁻-stress (i.e.
deficiency) is demonstrated by experiments in which the addition of
different chloride concentrations to more or less intact photosynthetically
active Oscillatoria protoplasts increases the oxygen evolution signals
substantially (Fig. 5).

If this interpretation goes in the good direction, it should be possible
to observe comparable interactions between the S-state system and oxygen
with inside-out particles from which the two extrinsic peptides have been
removed by NaCl-washing and which are suspended in a chloride-free medium.
And indeed our mass spectrometric assays revealed in such experiments an

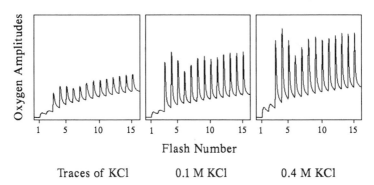

Traces of KCl   0.1 M KCl   0.4 M KCl

Fig.5. Effect of increasing concentrations of chloride on the oxygen evolution pattern in thylakoids from Oscillatoria chalybea.

oxygen uptake phenomenon which is DCMU sensitive demonstrating its relation to photosystem II (results not shown).

One of the pivotal conclusions of our experiments is that the OEC in Oscillatoria chalybea is in a conformational condition in which the S-state system interacts with oxygen. The condition might be brought about by the absence of the two extrinsic peptides and is apparently influenced by chloride ions. In the sense of this scheme H_2O_2 formation might take place via the interaction of O_2 with a sufficiently protonated redox component 'L' (Renger, 1987). In context with the influence of chloride also a picture in the sense of the Homann scheme might apply which says that chloride ions seem to have a multifunctional role with the major or principal task being that of a stabilizer of a well-defined conformational condition (Homan, 1987). In cyanobacteria (without the two extrinsic peptides) the conformation of photosystem II is obviously different. One of the main problems arising from this is an interaction with oxygen.

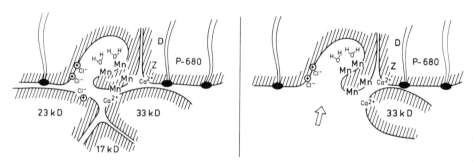

Fig.6. Schematic representation of a possible organization of the water oxidizing site of photosystem II in chloropast thylakoids (from Homann, 1987) and a corresponding model for cyanobacteria with special emphasize to a thus facilitated accessibility to the inner region.

It should be noted that at the time when organisms like Oscillatoria chalybea "invented" oxygenic photosynthesis reactions like this played obviously no role as the oxygen content of the atmosphere was certainly low at that time. The described property finds its parallel in the oxygen sensitivity of Ribulose 1.5-bisphosphate carboxylase and that of nitrogenase.

REFERENCES·

Bader, K.P. 1989, Biochim. Biophys. Acta, 975, 399-402
Bader, K.P. and Schmid, G.H. 1988, Biochim. Biophys. Acta, 936, 179-186
Bader, K.P. Thibault, P. and Schmid, G.H. 1987, Biochim. Biophys. Acta, 893, 564-571
Franck, F. and Schmid, G.H. 1989, Biochim. Biophys. Acta, 977, 215-218
Hanssum, B., Dohnt, G. and Renger, G. 1985, Biochim. Biophys. Acta, 806, 210-220
Homann, P.H. 1987, Journal of Bioenergetics and Biomembranes, Vol.19, No.2, 105-123
Messinger, J. and Renger, G. 1990, FEBS Lett. 277, 141-146
Renger, G. 1972, Biochim. Biophys. Acta, 256, 428-439
Renger, G., Bouges-Bocquet, B. and Delosme, R. 1973, Biochim. Biophys. Acta 292, 796-807
Renger, G. 1987, Angew. Chemie 99, 660-678
Sandusky, P.O. and Yocum, C.F. 1988, Biochim. Biophys. Acta 936, 149-156
Schröder, W.P. and Åkerlund, H.E. 1986, Biochim. Biophys. Acta 848, 359-363

THE EFFECT OF TEMPERATURE AND RADICAL PROTECTION ON THE

PHOTOINHIBITION OF SPINACH THYLAKOIDS

Aloysius Wild, Michael Richter, and Birgit Böthin

Institute of General Botany
Johannes Gutenberg-University
W-6500 Mainz, Germany

KEYWORDS/ABSTRACT: photoinhibition/ spinach thylakoids/ D1-protein/ Q_B-site/ radical protection/ vitamin E

Photoinhibition of spinach thylakoids was studied by examination the inactivation of different parts of the electron transport chain, the decline of the variable fluorescence and the loss of atrazine binding sites. The results obtained after photoinhibition at 20°C and 0°C revealed an inactivation at the Q_B-site of the D1-protein as the first event in the course of photoinhibition. The natural antioxidants glutathione and ascorbate as well as the enzymes SOD and catalase diminished photoinhibition to similar extents. Further protection was achieved through combination of both radical defense systems. In addition to the radical scavenging properties glutathione and ascorbate have reducing abilities towards oxidized vitamin E preventing thus damage within the thylakoid membrane.

INTRODUCTION

Photoinhibition occurs when light energy absorption through pigments largely exceeds the energy consumption of photosynthetic reactions. The loss of electron transport capacity, the decline of the variable part of the chlorophyll fluorescence and the degradation of D1-protein are well documented consequences of photoinhibition, indicating that the photosystem II complex is predominantly affected. The mechanism of photoinhibition, however, is still obscure. At the moment there is no general agreement concerning the nature and the site of primary events. The reaction center II (Cleland 1988) and the PS II acceptor side (Ohad et al. 1990) are currently most discussed in this context.

Our results point to the Q_B-binding niche of the D1-protein as the site of primary attack. By lowering the temperature to 0°C the inactivation of Q_B-function can be observed separately from D1-degradation and the concurrent loss of reaction centre II function. The strong protection by antioxidants signifies a major contribution of activated oxygen species to the mechanism of in vitro photoinhibition. The interaction of the soluble antioxidants glutathione and ascorbic acid with the membrane bound alpha-tocopherol is discussed.

MATERIALS AND METHODS

All experiments were carried out with isolated spinach thylakoids. The isolation procedure and the photoinhibitory treatment have been previously described (Richter et al. 1990 a). The D1-protein content of thylakoids was determined through the binding of ^{14}C-atrazine. The reaction centre II was characterized by the capacity for Q_B-independent reduction of silicomolybdate (SiMo) and the measurement of variable fluorescence (F_v/F_m).

Light saturated electron transport activity of different parts of the electron transport chain was measured polarographically in an oxygen electrode as described by Richter et al. (1990 a). The Q_A-Q_B-electron transfer was further examined by means of fluorescence relaxation kinetics. The experimental procedure of fluorescence induction and relaxation studies has been described (Richter et al. 1990 a).

The possible involvement of oxygen radicals in the mechanism of photoinhibition was investigated through radical scavenging experiments, in which thylakoids were exposed to high-intensity light in the presence of saturating amounts of either glutathione (10 mM) and ascorbate (5 mM) or SOD (100 U/ml) and catalase (2000 U/ml) or a combination of both the antioxidants and the enzymes. Further details are given in Richter et al. (1990 b). The alpha-tocopherol content of thylakoids was determined by high performance liquid chromatography.

RESULTS AND DISCUSSION

Photoinhibition at 20°C caused a significant loss of atrazine binding sites. This change of D1-protein properties was strongly inhibited at 0°C compared to 20°C (Tab. 1), probably due to the inhibition of a membrane bound protease activity through lowering the temperature. The decline of the functional properties of reaction centre II, i.e. PSII-photochemistry ($H_2O \longrightarrow$ SiMo) and variable fluorescence (F_v/F_m), seemed to be controlled by D1-degradation as revealed by the parallel decline at 20°C and 0°C, respectively. This is consistent with the D1-protein being a constituent of reaction centre II. The Q_B-dependent electron flow ($H_2O \longrightarrow$ p-BQ, FeCy, MV), however, declined independent from D1-degradation (Tab. 1). This inactivation could neither be localized at the water splitting enzyme nor at photosystem I and the cytochrome b_6f-complex as tested by specific electron donors and acceptors (Tab. 1). Fluorescence relaxation studies confirmed that photoinhibition at 0°C led to an impairment of the Q_A^--reoxidation as can be concluded from the slower decline of fluorescence following saturating single turnover flashes (Fig. 1). This implies a partial

Table 1. D1-protein degradation, loss of variable fluorescence and inactivation of different parts of the
photosynthetic electron transport chain during
photoinhibition of thylakoids at 20°C and 0°C.
Data are given in percent of dark controls.

Photoinhibition at	20°C	0°C	
atrazine binding sites	57	84	
F_v/F_m	51	87	RC II
H_2O --> SiMo (incl. DCMU)	45	83	Q_A-site
H_2O --> p-BQ	27	35	Q_B-site
H_2O --> FeCy	34	54	
H_2O --> MV	39	56	whole chain
DPC --> MV	39	56	without OEC
DQH_2 (DCMU) --> MV	82	89	Cyt b_6f-complex
DAD/Asc (DCMU) --> MV	85	102	PS I

inhibition of the electron transfer through the Q_B-binding
site.
Exposure of thylakoids to high-intensity light at 20°C
in the presence of the antioxidants glutathione and ascorbate
or the enzymes SOD and catalase resulted in comparable though
not total protection of atrazine binding sites and photochemistry (H_2O --> FeCy). Photoinhibition was nearly completely suppressed upon combination of the antioxidant system
with the enzymatic radical protective system though both had
been applied at saturating amounts (Tab. 2). The loss of

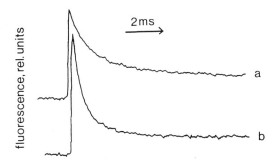

Figure 1. Fluorescence relaxation kinetics following saturating single turnover flashes of thylakoids photoinhibited at
0°C (a) and measured at 20°C compared to a dark control (b).

Table 2. The decline of atrazine binding sites and FeCy
mediated electron transport during photoinhibition
under different conditions of radical protection.
All values are given in percent of a dark control.
The thylakoids were photoinhibited at 0°C only in
the case of electron transport measurements. n.d. =
not determined.

Radical protective system	activity H$_2$O --> FeCy		atrazine binding sites
	20°C	0°C	20°C
(1) GSH/Asc	57 ± 12	80 ± 4	77 ± 1
(2) SOD/catalase	57 ± 4	n.d.	83 ± 3
(1) + (2)	79 ± 5	n.d.	97 ± 1
without additions	31 ± 9	54 ± 6	58 ± 4

FeCy mediated Q_B-dependent photochemical capacity can be
diminished by glutathione and ascorbate wether D1-degradation
does occur (20°C) or not (0°C) (Tab. 2). This first implies
that independent of temperature the inactivation of electron
transport during photoinhibition results from oxygen radical
attack. We further conclude that the radical induced loss of
activity at the Q_B-site is the primary photoinhibitory event
in relation to D1-degradation. No experiments including SOD
and catalase have been conducted at 0°C as the activity of
both enzymes should be too low at this temperature.
     The fact that the antioxidants and the protective
enzymes complement one another in radical detoxification
possibly indicates some repair properties of the antioxidants
towards oxidized amino acids (Hoey and Butler, 1984) or
results from the recycling of oxidized alpha-tocopherol by
the glutathione-ascorbate-system (Scarpa et al., 1984). The
latter is supported by the results presented in Fig. 2 and
Fig. 3. HPLC estimations of the alpha-tocopherol content of
spinach thylakoids revealed a fast decrease of the membrane
bound antioxidant during photoinhibition. Because of its
higher tocopherol content we took spinach grown under strong
light for this experiment as compared to the plants chosen
for the observation of inactivation of electron transport and
D1-degradation. Therefore, the photoinhibitory treatment had
to be prolonged to obtain a comparable degree of photoinhibition. Freshly prepared spinach thylakoids contained 1 vit E
molecule per 100 chlorophyll molecules in accordance with
data presented by Asada and Takahashi (1987). The decrease of
alpha-tocopherol was retarded by the action of SOD and
catalase. Significantly higher amounts of reduced tocopherol
were obtained when the photoinhibition was conducted in the
presence of glutathione and ascorbate (Fig. 2). Taking into
account the equal protection against photoinactivation of
electron transport by SOD and catalase as compared to glutathione and ascorbate (Tab. 2), we assume that the enzymes are
more effective in direct radical detoxification while the
antioxidants act more indirectly through reducing oxidized

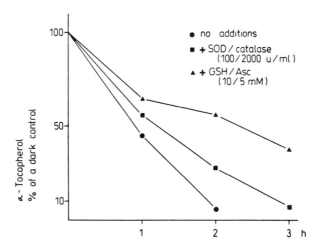

Fig. 2   The decline of reduced alpha-tocopherol of thylakoids
during photoinhibition in the presence of different radical
protective systems (data in percent of dark controls).

alpha-tocopherol. The interaction between the soluble antioxidants and the thylakoid bound vit E is possibly restricted
by the tight membrane appression in the grana partition of
thylakoids so that part of the alpha-tocopherol is hardly
accessible to the soluble antioxidants. This could be one
reason for the observed decline of alpha-tocopherol despite
the presence of glutathione and ascorbate. Vit E might also
disappear as a result of irreversible radical modification.
        Regarding the degree of photoinactivation and the extent
of D1-degradation a 30 min continuous photoinhibition led to

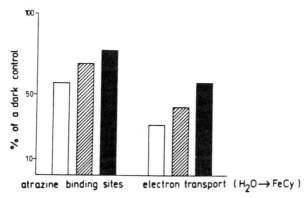

Fig. 3   The influence of an insertion of a 5 min dark
incubation with glutathione and ascorbate into the 30 min
photoinhibitory treatment of thylakoids on the decline of
electron transport capacity and atrazine binding sites (data
in percent of dark controls).

(☐)   without addition of antioxidants
(▨)   GSH and Asc present only during the dark incubation
(■)   GSH and Asc present throughout the whole treatment

the same results as when an additional 5 min dark period was inserted in between two 15 min photoinhibitory treatments. The presence of glutathione and ascorbate during the dark period, however, brought about a significant protection against loss of electron transport capacity and D1-breakdown. As the thylakoids had been washed two times after the dark treatment, this protection could not result from direct scavenging of light generated activated oxygen species by the antioxidants, but must be due to some repair ability of glutathione and ascorbate in the dark. The latter is probably mediated via reduction of oxidized alpha-tocopherol.

Summarizing our results, we propose photoinhibition to be a two step process consisting of a radical induced inactivation of the Q_B-function followed by the degradation of D1-protein turning the reaction centre II non functional. The radical modification of the protein has been made distinguishable from the degradation by lowering the temperature during photoinhibition. The alpha-tocopherol of the thylakoids is a central component of the radical protective system of chloroplasts. It is kept in the reduced state by the interaction with ascorbic acid and glutathione, both present at high amounts in chloroplasts (Asada and Takahashi, 1987).

We suppose D1-degradation to be itself part of the defense mechanism against strong photooxidative damage to the entire photosynthetic apparatus. The protein could perhaps function in thylakoids in the same way a fuse does in an electric curcuit. In case of energy overload a widespread damage being difficult to repair is prevented through quick destruction of the fuse component, thus switching off the system function.

REFERENCES

Asada, K. and Takahashi, M., 1987, Production and scavenging of active oxygen in photosynthesis, in: Topics in Photosynthesis Vol. 9 Photoinhibition, D. J. Kyle, C. B. Osmond and C. J. Arntzen, eds., Elsevier, Amsterdam.
Cleland, R. E., 1988, Molecular events of photoinhibitory inactivation in the reaction centre of photosystem II, Austr. J. Plant Physiol., 15:135.
Hoey, B. M. and Butler, J., 1984, The repair of oxidized amino acids by antioxidants, Biochim. Biophys. Acta, 723:383.
Ohad, I., Adir, N., Koike, H., Kyle, D. J., and Inoue, Y., 1990, Mechanism of photoinhibition in vivo, J. Biol. Chem., 265:1972.
Richter, M., Rühle, W., and Wild, A., 1990, Studies on the mechanism of photosystem II photoinhibition I. A two step degradation of D1-protein, Photosynth. Res., 24:229.
Richter, M., Rühle, W., and Wild, A., 1990, Studies on the mechanism of photosystem II photoinhibition II. The involvement of toxic oxygen species, Photosynth. Res., 24:237
Scarpa, M., Rigo, A., Maiorino, M., Ursini, F., and Gregolin, C., 1984, Formation of alpha-tocopherol radical and recycling of alpha-tocopherol by ascorbate during peroxidation of phosphatidylcholine liposomes, Biochim. Biophys. Acta, 801:215.

THE EFFECT OF HEAT TREATMENT ON THE CAPACITY OF THE ELECTRON TRANSPORT:

INTERACTION OF LIGHT AND STRUCTURAL ORGANIZATION

Karen Van Loven and Roland Valcke

Department SBM
Limburgs Universitair Centrum, Universitaire Campus
B-3590 Diepenbeek, Belgium

INTRODUCTION

Subjecting plants to a heat treatment causes a decrease in the photo-synthetic electron transport activity. Heat stress intensifies the influence of photoinhibitory conditions. Schuster, et al. (1989) noted that photoin-hibition in *Chlamydomonas* cells occurs after heat treatment at much lower light intensities than at normal growth temperature. They noticed that under these former conditions thylakoid proteins (among which the D1 protein) were aggregated. This process was connected with protein degradation in-duced by free radicals (Sopory, et al., 1990).
On the other hand, Havaux and Strasser (1990) showed that light offers protection to the photosynthetic apparatus. Even with very low light in-tensities the quenching of the PS2 fluorescence from attached pea leaves was limited and fully reversible, and quantum yield and maximal O_2 evolution rate were unaffected, this in contradiction to dark heated leaves.

This paper draws attention to the influence of different levels of structural organization (thylakoid membranes, leaf tissue and intact plants) on the prevention from heat injury. The comparison of the effect of a low light intensity (50 µmoles quanta PAR m^{-2}sec^{-1}) with darkness, during heat treatment, was made.

MATERIALS AND METHODS

Plant material: Tobacco plants (*Nicotiana tabacum* cv. Petit Havana SR1) were grown in vegetable soil under conditions of 12h day (25°C)/ 12h night (20°C), 150 µmoles quanta PAR m^{-2}sec^{-1} (Cool White 25 SA (Osram) and Phil-linea (Phillips) lamps), relative humidity 60%. At the age of 7-8 weeks the middle leaves were used to measure.

ABBREVIATIONS: BSA: bovine serum albumin; DAD: diaminodurene (2,3,5,6-tetramethyl-p-phenylene diamine); DCPIP: 2,6-dichlorophenol-indo-phenol; DPC: 1,5-diphenylcarbohydrazide; EDTA: ethylene diamine tetra-acetic acid; HEPES: N-2-hydroxyethylpiperazine-N'-2-ethanesulphonic acid; o.e.c.: oxygen evolving complex; PAR: photosynthetic active radiation; PS2: photosystem 2.

Regulation of Chloroplast Biogenesis
Edited by J.H. Argyroudi-Akoyunoglou, Plenum Press, New York, 1992

Thylakoid membranes were extracted rapidly: the leaf tissue was cutted with sharp razor-blades and blended in a minimal volume of chilled extraction medium (sorbitol 0.35 M; $MnCl_2$ 1 mM; $MgCl_2$ 1 mM; EDTA 2 mM; K_2HPO_4 0.5 mM; BSA 0.4% (w/v); HEPES 0.05 M - pH 8). The suspension was filtered through 1 layer of miracloth and 1 layer of nylon (80 mesh). The chlorophyll content was measured according to Bruinsma (1963).

Measurement of the PS2-mediated electron transport: The capacity of the photosynthetic electron transport over PS2 was measured as soon as possible after extracting the thylakoid membranes. It was measured with a Clark-type oxygen electrode (Hansatech, Ltd.). The membranes were diluted 20 times in the measuring medium, containing K_2HPO_4 2 mM; NaCl 2 mM; $MgCl_2$ 5 mM; NaN_3 0.25 mM; Tricine 10 mM - pH 7.8; and $K_3Fe (CN)_6$ 0.5 mM and DAD 1.2 mM as electron acceptors. The measurement was performed at 25°C and the cuvet was illuminated with a slide projector lamp (1500 µmoles quanta PAR m^{-2}sec^{-1}).

The activity of the reaction center of PS2 was measured as the photoreduction of DCPIP. The oxygen evolving complex was inhibited with NH_2OH (1 µmol / mg chlorophyll). DPC (0.67 mM) was used as artificial electron donor.

RESULTS

Incubation of isolated thylakoid membranes at elevated temperatures leads to an enhanced inhibition of the capacity of the PS2-mediated electron transport. This was measured first with the oxygen electrode (fig. 1). Storage of the membranes in darkness for 1 hour at 4°C reduced the photosynthetic capacity only slightly, in relation to the ageing of the extract. Incubation of the extracted membranes in darkness at 25°C caused an immediate decrease of the PS2-capacity for about 30%, and this difference was remained during the incubation period compared with the cold stored (4°C) membranes. Temperature rise up to 40°C resulted in a more drastic loss (-80%) of the PS2-capacity, caused by a thermal denaturation of the photosynthetic membranes.

Under low light conditions (50 µmoles quanta PAR m^{-2}sec^{-1}) at 40°C the inhibition of the PS2-mediated electron transport was complete (fig. 1). However, at 25°C the low light intensity even stabilized the PS2-capacity. Moreover, it exceeded slightly the stable storage conditions at 4°C in darkness.

The measurement of the photoreduction of DCPIP with the use of DPC as electron donor, skips the oxygen evolving complex and focuses consequently on the capacity of the reaction center of PS2 (fig. 2). No inhibition of the electron transport capacity was observed when extracted thylakoid membranes were incubated at 25°C, neither in darkness, nor at 50 µmoles quanta PAR m^{-2}sec^{-1}. At 40°C the capacity was decreased with 40%, but again no difference between dark and light incubation could be observed.

Detached tobacco leaves were immersed in water at 25°C and 40°C. This allowed us a fast temperature control in the leaf tissue. Incubation of the leaves at 25°C in darkness caused initially a slight inhibition of the PS2-capacity, which was partly restored after one hour (fig. 3). When the leaves were submerged in darkness at 40°C, compared to 25°C, the inhibition of the PS2-capacity was aggravated severely. Incubation of the leaves at low light intensities (50 µmoles quanta PAR m^{-2}sec^{-1}) did not affect the course of the PS2-capacity at 25°C. However, at 40°C these low light intensities prevented PS2 from inhibition.

452

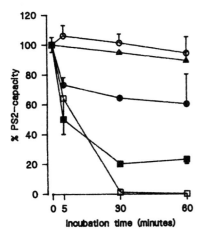

Fig. 1. PS2-capacity of treated thylakoid suspensions, measured with the O_2-electrode ($H_2O \longrightarrow DAD-K_3Fe(CN)_6$).

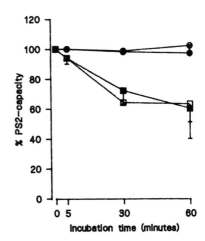

Fig. 2. PS2-capacity of treated thylakoid suspensions, measured as photoreduction of DCPIP ($DPC \longrightarrow DCPIP$).

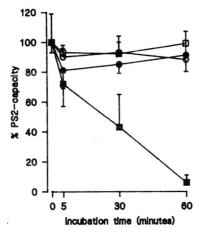

Fig. 3. PS2-capacity of treated, submerged leaves, measured with the O_2-electrode.

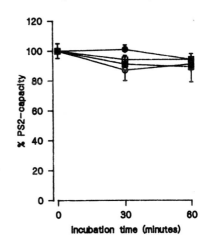

Fig. 4. PS2-capacity of treated intact plants, measured with the O_2-electrode.

Legend: ▲ = 4°C-dark incubation; ● = 25°C-dark incubation; ■ = 40°C-dark incubation; O = 25°C-light incubation; □ = 40°C-light incubation.

Intact plants were also incubated up to 1 hour at an air temperature of 25°C or 40°C (fig. 4). In darkness at 25°C, the PS2-capacity was hardly affected. At 40°C the capacity was slightly reduced. Incubation under low light intensities (50 µmoles quanta PAR m^{-2}sec^{-1}) caused little inhibition (ca. 10%) at 25°C and had even less effect at 40°C (ca. 6% inhibition).

DISCUSSION

Heat treatment of a suspension of isolated thylakoid membranes changes the capacity of the photosynthetic electron transport. The PS2-mediated electron transport decreases in darkness with increasing incubation temperature (fig. 1). However, the relationship between the PS2-inhibition and temperature is not linear, but appeares to be biphasic (data not shown). The combined effect of protein degrading reactions (e.g. Critchley and Chopra, 1988) and membrane alterations (Yordanov, et al., 1987; Süss and Yordanov, 1986) is responsible for that.

The measurements of the photoreduction of DCPIP show that the reaction center of PS2 itself is sensitive (40% inhibition) to heat treatment (40°C) (fig. 2). O_2-electrode measurements (fig. 1) demonstrate that the oxygen evolving complex is even though more sensitive to heat stress (Katoh and San Pietro, 1967) (compare fig. 1 and fig. 2). Thompson, et al. (1989) suggested a better accessibility of the Mn-site of the o.e.c. to the reactive OH-, upon heat treatment. The oxygen evolving capacity is lost due to a subsequent reduction of the Mn by OH-.

Photoreduction of DCPIP shows that the low light intensity (50 µmoles quanta PAR m^{-2}sec^{-1}) does not affect the PS2 rection center. Consequently differences between dark and light incubation detected with the O_2-electrode, will be caused by alterations in the o.e.c. . At 25°C the light is evidently beneficial in stabilizing the o.e.c. (Weis, 1982). On the contrary at 40°C the o.e.c. is totally inhibited. This last reaction is still not understood.

Heat shock treatment by submergeing the leaf tissue affects the photosynthetic activity in a different way. The injurious effect on the PS2-mediated electron transport is prevented by a low light intensity. Such improvement could not be obtained at 25°C.

In the intact plants the photosynthetic capacity of PS2 is unaffected by the heat treatment. This is in contradiction to the thermal injury in the dark on the photosynthetic apparatus in detached, submerged leaves and to the effect on the PS2 fluorescence characteristics in intact plants as measured by Havaux and Strasser (1990). In both cases the temperature at the leaf surface could be altered rapidly. During our heat treatment of the intact plants, which started with a 15 minute temperature rise (from 25°C to 40°C), the air temperature was kept at 40°C. The temperature at the leaf surface was though not controlled and the stress effect might have been tempered.

From this it becomes evident that with increasing structural organization the protection against heat stress is improved. Low light intensity stabilizes the photosynthetic apparatus. The mechanisms involved seem to be different from those causing photoinhibition. They need further examination for a better comprehension of this process.

ACKNOWLEDGEMENTS

We thank Katleen Raeymaekers and Greet Clerx for excellent technical assistance.
This work was financially supported by the Belgian Fonds voor Kollektief Fundamenteel Onderzoek (grant nr. 290009.87).

REFERENCES

Bruinsma, J., 1963, The quantitative analysis of chlorophylls a and b in plant extracts, Photochem. Photobiol. 2: 241-250.
Critchley, C. and Chopra, R.K., 1988, Protection of photosynthetic O_2 evolution against heat inactivation: the role of chloride, pH and coupling status, Photosynth. Res. 15: 143-152.
Havaux, M. and Strasser, R.J., 1990, Protection of photosystem II by light in heat-stressed Pea leaves, Z. Naturforsch. 45c: 1133-1141.
Katoh, S. and San Pietro, A., 1967, Ascorbate-supported NADP photoreduction by heated Euglena chloroplasts, Arch. Biochem. Biophys. 122: 144-150.
Schuster, G., Shochat, S., Adir, N. and Ohad, I.,1989, Inactivation of photosystem II and turn-over of the D1-protein by light and heat stresses, in: "Techniques and new developments in photosynthesis research", Barber, J. and Malkin, R., eds., Plenum Publishing Corporation. Pp. 499-510.
Sopory, S.K., Greenberg, B.M., Mehta, R.A., Edelman, M. and Mattoo, A.K., 1990, Free radical scavengers inhibit light-dependent degradation of the 32 kDa photosystem II reaction center protein, Z. Naturforsch. 45c: 412-417.
Süss, K.-H. and Yordanov, I.T., 1986, Biosynthetic cause of in vivo acquired thermotolerance of photosynthetic light reactions and metabolic responses of chloroplasts to heat stress, Plant Physiol. 81: 192-199.
Thompson, L.K., Blaylock, R., Sturtevant, J.M. and Brudvig, G.W., 1989, Molecular basis of the heat denaturation of photosystem II, Biochem. 28: 6686-6695.
Weis, E., 1982, Influence of light on the heat sensitivity of the photosynthetic apparatus in isolated spinach chloroplasts, Plant Physiol. 70: 1530-1534.
Yordanov, I., Goltsev, V., Stoyanova, T. and Venediktov, P., 1987, High-temperature damage and acclimation of the photosynthetic apparatus, I. Temperature sensitivity of some photosynthetic parameters of chloroplasts isolated from acclimated and non-acclimated bean leaves. Planta 170: 471-477.

DOES A HEAT-SHOCK TREATMENT AFFECT PHOTOSYNTHESIS IN NICOTIANA TABACUM?

Roland Valcke and Karen Van Loven

Dept. S.B.M., Limburgs Universitair Centrum
B-3590 Diepenbeek, Belgium

INTRODUCTION

An exposure of cells and organisms to temperatures well above the normal growing temperature induces rapid and dramatic changes in both transcription and translation. The heat-shock response is characterized by the expression of a set of nuclear and organelle genes, coding for the synthesis of heat-shock proteins (Lindquist 1986).

In plants, photosynthetic activity is strongly reduced by heat-shock treatment, due to specific damage to photosystem II (Nash et al. 1985). This damage can be enhanced when heat treatment is accompanied by light stress (Schuster et al. 1989). Protection mechanisms, such as synthesis of chloroplast heat-shock proteins (HSPs) have been reported (Schuster et al. 1988). An overview of the complexity and abundance of families of low molecular weight heat-shock proteins in plants and the regulation of their transcription has been published recently (Gurley and Key, 1991).

In the last decade, several papers has been published dealing with the structural and functional changes in the photosynthetic electron transport in response to heat stress. Heating bean chloroplasts to 40°C causes extensively unstacking of grana without changing membrane particle distribution (Gounaris et al., 1983). Subfractionation of thylakoids above 35°C induces laterally migration of the PSII core complex and of the light-harvesting complex (Sundby and Andersson, 1985). Exposing maize leaves to temperatures above 36°C, inhibits fluorescence changes indicating a block in a state intermediate between state 1 and state 2 (Havaux, 1988).

In an accompanying paper (Van Loven and Valcke, this proceedings), we show that the inhibitory effect of heat and light on the electron transport capacity of PSII is assessed by the structural organization in which the PSII complex is located.

In the present work, we report that the effect of a heat treatment on a whole plant depends on the growth conditions of that plant and on its physiological status. Using a leaf disc electrode, photosynthetic parameters were probed on tobacco leaves, subjected to heat stress under two different light conditions.

Regulation of Chloroplast Biogenesis
Edited by J.H. Argyroudi-Akoyunoglou, Plenum Press, New York, 1992

457

MATERIALS AND METHODS

        Nicotiana tabacum L. cv. petit Havana SR1 was grown in pot soil
under two different conditions:
        - in growth chambers, under a light regime of 12h light, 12h dark.
Light intensity (I(1)) was 150 μmoles quanta m-^{2}sec-^{1}, temperature was 25°C
in light and 20°C in dark; relative humidity was 60%.
        - in a greenhouse, under direct sunlight supplemented with incandes-
cent lamps. Light intensity (I(2)) was 450 μmoles quanta m-^{2}sec-^{1}. Temperature and relative humidity depends slightly on the weather conditions.

        Heat-shock treatment of the first series of plants was performed
in a home build growth chamber in which the temperature can be raised and
fall from 25 to 40°C and back in about 15 minutes. The light intensity in
the growth chamber was 50 μmoles quanta PAR m-^{2}sec-^{1}. Plants were treated
for one hour at 40°C. The second series of plants were heat shocked in a
special build compartiment in the greenhouse. The light intensity was kept
at 450 μmoles quanta PAR m-^{2}sec-^{1}.

        Oxygen evolution, quantum yield, photosynthetic rate, nominal
light utilisation capacity, light compensation point and fluorescence were
performed with a leaf disc electrode essentially as described by Walker
(1989). Chlorophyll concentrations were determined in 80% aceton (Lichtenthaler and Wellburn, 1983).

RESULTS

        Figure 1 shows the maximal photosynthetic rate (MPR) (expressed
in μmoles O$_2$ m-^{2}sec-^{1}mg-^{1}chl) as a function of time. All results are expressed as % of the first measured rate, one hour after the start of the photoperiod. The heat-shock treatment (temperature 40°C) starts at this time and

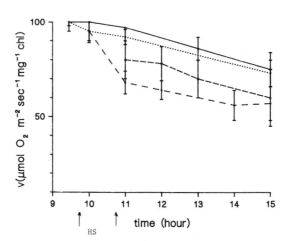

Fig.1   maximal photosynthetic rate
         ——— I(1), control; ------- I(2), HS
         ..... I(2), control; - - - - I(2), HS.

proceeds for 1 hour (taken into account the time needed to warm-up and cool-down the experimental chamber). During the day, the MPR of the control plants, irrespective of their growth environment (growth chamber or green-house), decreases. This probably reflects the gradual accumulation of photo-synthetic intermediates and the regulation of the electron transport and carbon fixation. Heat shock treatment of the whole plant results in a lower MPR at every point measured. The inhibition of the MPR is slightly enhanced by higher light intensities (compare open and closed symbols). In both cases, the MPR of heat-shock plants, compared to the control, remains lower during the subsequent illumination period. There is no indication of a repair me-chanism.

The quantum yield is strongly influenced by the growing condition (table 1.) Under high light intensity, the quantum yield of the control plants is 0.106; under low light only 0.079. Heat-shock treatment under high light condition results in a significant decrease in quantum yield. Under low light it even increases, although not statistically significant. This results indicate that heat treatment affects the maximal efficiency (theoretically 0.111). The observation that under low light a lower quantum yield is obtained compared to high light, remains an open question.

Tables 2 to 5 give the values of the dark respiration (DR), the light compensation point (LCP), the photon flux density (PFD, to give 50% of the rate obtained at 800 μmoles quanta PAR m^{-2}sec^{-1}) and the nominal light utilisation capacity (NLUC). In contrast to the general accepted concept of "sun" and "shade" leaves, the dark respiration and the light compensation point of the leaves of tobacco plants grown under low light intensities is higher compared to the leaves of plants grown under high light. Remarkable is also that there is no difference in the 50% PFD-value, nor in the values of the nominal light utilisation capacity. Heat-shock treatment induces a reduction in the DR, LCP and the 50%-PFD but not in the NLUC, irrespective of the light intensity used.

Figure 2 illustrates the evolution of the total chlorophyll (mg chl in a 10 cm^2 leaf disc) during the measuring period and the effect of heat treatment under low light conditions (the same pattern is obtained under high light). Heat shock clearly retards the chlorophyll synthesis, normally observed during the illumination period.

Table 1. Quantum yield

	control	HS
I(1)	0.079 $\pm$ 0.003	0.083 $\pm$ 0.003
I(2)	0.106 $\pm$ 0.004	0.083 $\pm$ 0.005

I(1): 150 μmoles quanta PAR m^{-2}sec^{-1}
I(2): 450 μmoles quanta PAR m^{-2}sec^{-1}
HS: heat-shock.

Table 2. Dark respiration

	control	HS
I(1)	2.5 + 0.2	2.0 + 0.2
I(2)	2.1 + 0.2	1.4 + 0.3

Symbols same as in table 1.

Table 3. Light compensation point

	control	HS
I(1)	33 + 3	25 + 2
I(2)	20 + 2	15 + 3

Symbols same as in table 1.

Table 4. Photon flux density

	control	HS
I(1)	110 + 10	86 + 5
I(2)	110 + 20	70 + 10

Symbols same as in table 1.

Table 5. Nominal light utilisation capacity

	control	HS
I(1)	87 + 2	90 + 1
I(2)	88 + 4	85 + 2

Symbols same as in table 1.

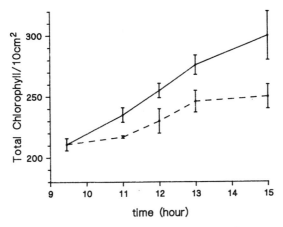

Fig.2 total chlorophyll/10cm^2
— control; - - - - HS

The chl a/b ratio, nor the total carotenoid content is affected by heat treatment (results not shown). This results could indicate that the heat-shock inhibits the light driven enzym system of the chlorophyll synthesis.

DISCUSSION

The effect of elevated temperature on the photosynthetic system depends on the plant material used, the physiological condition of the growing plant and the structural organization in which the target system is localized. Heat treatment of Chlamydomonas cells in dark does not affect the photosynthetic electron transport (Schuster et al., 1989). Inhibition is obtained with light intensities about ten times lower than required for saturating photosynthesis. In contrast, Havaux and Strasser (1991) demonstrated on basis of their fluorescence measurements, that heat stress in the dark causes irreversible damage to PSII and that a light trigger at high temperature has a protecting effect.

The question how plants regulate the excitation energy reaching the reaction centre and how they develop a system to avoid photoinhibitory damage has only been attended recently (Bilger and Björkman, 1991). Various mechanisms have been proposed, one of them thought to be mediated by the carotenoid zeaxanthin as demonstrated extensively in a series of studies by Demmig-Adams and co-workers (for a review, see Demmig-Adams and Adams III, 1991).

In anaother paper (Van Loven and Valcke, this proceedings), we demonstrate that the effect of heat treatment depends on the structural organization in which PSII is located. The results of the current study indicate that heat treatment of a whole plant affects photosynthesis in a complex manner, depending on the growth conditions of the plant and on the way heat shock is applied. Fluorescence measurements indicate a shift in the oscillations and in the position and amplitude of the "M-peak" (results not shown). This points towards a possible mechanism on the level of the phosphate and metabolite balances. Further experiments, including quenching analysis, are needed to improve the current results.

ACKNOWLEDGEMENTS

The authors wish to thank Katleen Raeymaekers for technical assistance. The work is supported by grant n° 290009.87 of the "Fonds Kollektief
Fundamenteel Onderzoek (FKFO)"-Belgium.

REFERENCES

Bilger, W., and Björkman, O., 1991, Role of the xanthophyll cycle in photo-
    protection elucidated by measurements of light-induced absorbance chan-
    ges, fluorescence and photosynthesis in leaves of Hedera canariensis.
    Photosyn.Res., 25(3), 173-186.
Burke, J.J., 1990, High temperature stress and adaptations in crops, in:
    "Stress Responses in Plants: Adaptation and Acclimation Mechanisms,"
    R.G. Alscher and J.R. Cumming, Eds., Plant Biology, vol 12, 295-309,
    Wiley-Liss, N.Y.
Demmig-Adams, B., and Adams III, W.W., 1991, The carotenoid zeaxanthin and
    'high-energy-state quenching' of chlorophyll fluorescence, Photosyn.
    Res., 25(3), 187-198.
Gurley, W.B., and Key, J.L., 1991, Transcriptional regulation of the heat-
    shock response: a plant perspective, Biochemistry, 30, 1-12.
Havaux, M., 1988, Effects of temperature on the transition between state 1
    and state 2 in intact maize leaves, Plant Physiol.Biochem., 26, 245-251.
Havaux, M., and Strasser, R.J., 1991, Protection of photosystem II by light
    in heat-stressed pea leaves, Z.Naturforsch., 45c, 1133-1141.
Kobza, J., and Edwards, G.E., 1987, Influence of leaf temperature on photo-
    synthetic carbon metabolism in wheat, Plant Physiol., 83, 69-74.
Lichtenthaler, H.K., and Wellburn, A.R., 1983, Determination of total caro-
    tenoids and chlorophylls a and b of leaf extracts in different solvents,
    Biochem.Soc.Trans., 11(5), 591-592.
Lindquist, S., 1986, The heat-shock response, Ann.Rev.Plant Physiol., 55,
    1151-1191.
Nash, D., Miyao, M., and Murata, N., 1985, Heat inactivation of oxygen
    evolution in photosystem II particles and its acceleration by chloride
    depletion and exogeneous manganese, Biochim.Biophys.Acta., 807, 127-133.
Schuster, G., Shochat, S., Adir, N., and Ohad, I., 1989, Inactivation of
    Photosystem II and turnover of the D1-protein by light and heat stresses,
    in: "Techniques and New Developments in Photosynthesis Research",
    Barber, J., and Malkin, R., Eds., Plenum Press, N.Y., pp 499-510.
Schuster, G., Even, D., Kloppstech, K., and Ohad, I., 1988, Evidence for
    protection by heat-shock proteins against photoinhibition during heat-
    shock, EMBO J., 7, 1-6.
Walker, D.A., 1989, Automated Measurement of leaf photosynthetic O$_2$ evolution
    as a function of photon flux density, Phil.Trans.R.Soc.Lond., B323,
    313-326.

IMMUNOGOLD-LOCALIZATION OF TUBULIN-LIKE PROTEIN IN THE INTRA-SKELETAL

STRUCTURE OF PEA CHLOROPLASTS

Nikola Valkov[1] and Alexander G. Ivanov[2*]

[1]Institute of Molecular Biology, [2]Central Laboratory of
Biophysics, Bulgarian Academy of Sciences, 1113 Sofia,
Bulgaria

INTRODUCTION

Initial observations of individual tubular-like elements (microtubules)
have been reported in algal chloroplasts (Hoffman, 1967; Picket-Heaps,
1968). Tubular elements arranged in parallel hexagonal packing were also
described in chloroplasts of some CAM (Crassulacean acid metabolism) plants
(Brandao and Salema, 1974; Santos and Salema, 1981). It was demonstrated
that the appearance of microtubules in the stroma of chloroplasts could
be induced by subjecting CAM-facultative plants to salt stress (Salema and
Brandao, 1978). Extensive arrays of plastid microtubules which appear to
connect the thylakoid apparatus to the plastid envelope have been observed
also in other higher plant chloroplasts (Sprey, 1968; Jhamb and Zalik,
1975; Vaughn and Wilson; 1981).

In the present study, the immunogold anti-β-tubulin labeling of thin
sections was used for detailed localization of tubulin-like protein(s) in
the photosynthetic apparatus of mature pea chloroplasts. A method for
visualization of chloroplast fibrilar structures in their "native" state
which involves extraction of previously fixed material was also employed.

MATERIALS AND METHODS

Extraction of fixed material: Thin slices of mature pea (Pisum sativum L., var. Ran) leaves were fixed with 1% glutaraldehyde in 20 mM TEA
buffer (pH 7.5). Afterfixing for 10 min at room temperature the slices
were thoroughly washed. Glutaraldehyde fixed material was extracted over-
night at room temperature with 8 M urea, 1 % SDS, 40 mM β-mercaptoethanol
and 20 mM EDTA. After postfixation specimens were embedded in Lowicril K4M.

Post-embedding immunoelectron microscopy: Ultrathin sections of both
extracted and non-extracted material were collected on 400-mesh nickel
grids. Grids were floated onto drops of 1:100 diluted monoclonal antibody
specific to the β-subunits of brain tubulin (Amersham) in 1 % BSA, PBS
for 4 h at room temperature. After three washes on 0.1 % Triton X-100,

*To whom all correspondence should be sent at: Central Laboratory of
Biophysics, Bulgarian Academy of Sciences, Acad.G. Bonchev Str. 21,
1113 Sofia, Bulgaria.

1 % BSA, PBS, grids were incubated for 1 h with Protein A-gold (10 nm particle size, Sigma) diluted 1:20 with 1 % BSA, PBS. After three final rinses with 0.1 % Triton X-100, 1 % BSA; PBS; PBS and water, grids were lightly contrasted. Electron microscopy was performed on JEM 100B at 80 kV.

RESULTS AND DISCUSSION

There is a number of electron microscopic data indicating that tubular-like elements ("microtubules"), arranged in parallel hexagonal packing (Fig. 1B) exist in the chloroplast interior in a variety of plant species (Picket-Heaps, 1968; Santos and Salema, 1981; Jhamb and Zalik, 1975; Vaughn and Wilson, 1981). Moreover, highly purified chloroplast tubular fractions were shown to consist of protein which exhibit similar electriphoretic mobility to α- and β-tubulin, although the proteolitic cleavage pattern of tubulin subunits and tubular protein fractions were different (Knoth, et al., 1984). Although these data, the arrangement of chloroplast "microtubules", their localization and function of tubulin-like protein(s) remain not fully elucidated at present.

The extraction procedures with nonionic detergents, high molarity salt solutions, etc. have been widely used for studying of the "nuclear matrix" or/and fibrous networks which remain after extraction of cell nuclei (Van der Velden and Wanka, 1987; Nigg, 1988; Beven, et al., 1991). Similar approach was used in the present study (see in the Material and Methods) for visualization of the internal arrangement of chloroplast tubular elements.

It is demonstrated that after extraction of previously fixed material, the thylakoid membranes are absent, but the overall shape of the organelle is preserved (Fig. 1C). The inner space of the organelle is composed of fibrilar structure. The fibrous (skeletal-like) network appears continous and seems quite similar, if not identical, to those observed in cell nucleus after the same fixation/extraction procedure (Fig. 2B). These data are consistent with the earlier transmission (Vaughn and Wilson, 1981) and scanning (Barnes and Blackmore, 1984) electron microscopic observations that "microtubules" are connected to the plastid envelope.

Bearing in mind that proteins of the intermediate filament/lamin-related family have been recognized in the internal fibrous structure of plant nuclear matrix (Beven, et al., 1991), it is possible to speculate, that the skeletal-like network observed in extracted chloroplast have similar nature.

Immunogold anti-β-tubulin labeling observations of thin sections of non-extracted material have indicated that the main portion of gold label is concentrated to the cpDNA-containing nucleoids (Fig. 3A), which appear scattered throughout the chloroplast interior (Kuroiwa, et al., 1981). A few gold particles are also found over the thylakoid membranes. Furthermore, Fig. 3B and C. show that gold particles, indicating the presence of β-tubulin are present mainly over the isolated morphologically intact chloroplast nucleoids (Nemoto, et al., 1990). These data are in close relation to the earlier suggestion of Kuroiwa, et al. (1981) that cpDNA could be compactly organized in plastid nucleoids by some DNA-binding proteins.

Since the functional role of cpDNA-binding proteins in the regulation of cpDNA replication and gene expression during the the chloroplast differentiation has been emphesized recently (Nemoto, et al., 1990), the possible involvment of tubulin-related protein(s) in these processes could not be excluded. The expression of β-tubulin genes in the initial

464

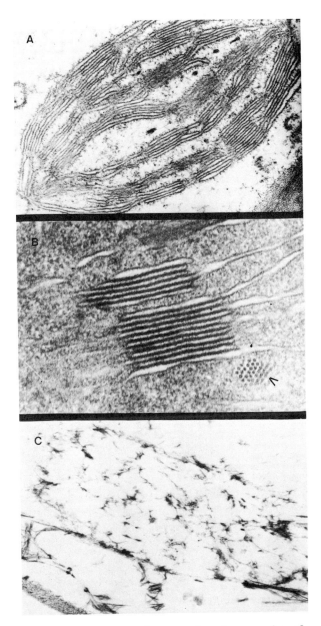

Fig. 1. A.- Electron microscopic micrographs of
control chloroplast (x 34000); B. - Tubulin-
like structures (arrow) in the intrathylakoid space of control chloroplast
(x 60000); C. - Electron micrograph of
fixed and extracted chloroplast (x 20000).

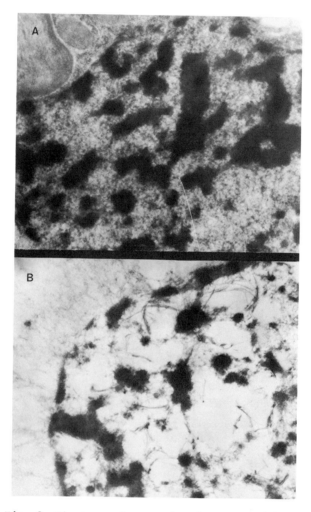

Fig. 2. Electron micrographs of control (A) and
        extracted after fixation (B) nucleus –
        (x 30000).

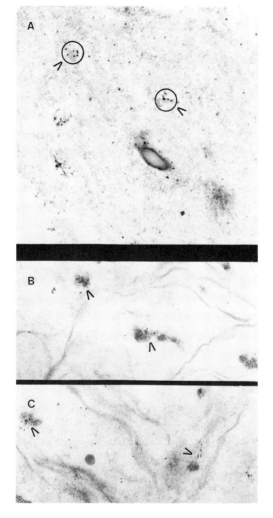

Fig. 3. Immunoelectron microscopic localization
of β-tubulin with the help of monoclonal
antibody and Protein A-gold (10 nm).
A. - The label is confined presumably
to intrathylakoid space containing
chloroplast nucleoids (x 45000); B. and C. -
Immunogold localization of β-tubulin in
isolated chloroplast nucleoids (x 34000).

phase of blue-light-induced chloroplast differentiation reported by
Kaldenhoff and Richter (1990) support this assumption.

The non-random distribution of gold labels clearly indicates that
the chloroplast fibrilar structure could not be composed only of tubulin-
related protein(s) and support the assumption mentioned above that some
protein(s) from intermediate filaments type could be also involved.

ACKNOWLEDGEMENTS

We thank Dr. Alexander M. Christov for excellent assistance. This work was supported by the Bulgarian Science Committee under the Research project No. 519.

REFERENCES

Barnes, S.H. and Blackmore, S., 1984, Scanning electron microscopy of chloroplast ultrastructure, Micron Microsc. Acta 15: 187-194.

Beven, A., Guan, Y., Peart, J., Cooper, C. and Shaw, P., 1991, Monoclonal antibodies to plant nuclear ±atrix reveal intermediate filament-related components within the nucleus, J. Cell Sci. 98: 293-302.

Brandao, I. and Salema, R., 1974, Microtubules in chloroplast of a higher plant (Sedum sp.), J. Submicrosc. Cytol. 6: 381-390.

Hoffman, L.R., 1967, Observations on the fine structure of Oedogonium.III. Microtubular elements in the chloroplasts of Oe. cardiacum, J. Physiol. 3: 212-221.

Jhamb, S. and Zalik, S., 1975, Plastid development in a vivescens barley mutant and chloroplast microtubules, Can. J. Bot. 53: 2014-2025.

Kaldenhoff, R. and Richter, G., 1990, Light induction of genes preceding chloroplast differentiation in cultured plant cells, Planta 181: 220-228.

Knoth, R., Klein, P. and Hausmann, P., 1984, Morphological and chemical studies on the crystalloid-forming "succulent protein" from normal and ribosome-deficient Aeonium domesticum plastids, Planta 161: 105-112.

Kuroiwa, T., Suzuki, T., Ogawa, K. and Kawano, S., 1981, The chloroplast nucleus: distribution, number, size and shape, and a model for the multiplication of the chloroplast genome during chloroplast development, Plant Cell Physiol. 22: 381-396.

Nemoto, Y., Kawano, S., Koudah, K., Nagata, T. and Kuroiwa, T., 1990, Studies of plastid-nuclei (nucleoids) in Nicotiana tabacum L. III. Isolation of chloroplast-nuclei from mesophyll protoplasts and identification of chloroplast DNA-binding proteins, Plant Cell Physiol. 31: 767-776.

Nigg, A.E., 1988, Nuclear function and organization: the potential of immuno-chemical approaches, Int. Rev. Cytol. 110: 27-92.

Picket-Heaps, J.D., 1968, Microtubule-like structure in the growing plastids or chloroplasts of two algae, Planta 81: 193-200.

Salema, R. and Brandao, I., 1978, Development of microtubules in chloro-plasts of twa halophytes forced to follow crassulacean acid metabolism, J. Ultrastruc. Res. 62: 132-136.

Santos, I. and Salema, R., 1981, Chloroplast microtubules in some CAM-plants, Bol. Soc. Brot. Ser. 2, 53: 1115-1122.

Sprey, B., 1968, Zur feinestructur des plastidenstroma von Hordeum vulgare, Protoplasma 66: 469-479.

Van der Velden, H.M.W. and Wanka, F., 1987, The nuclear matrix - its role in the spatial organization and replication of eucaryotic DNA, Molec. Biol. Rep. 12: 69-78.

Vaughn, K.C. and Wison, K.G., 1981, Improved visualization of plastid fine structure: plastid microtubules, Protoplasma 108: 21-27.

A PLATELET ACTIVATING FACTOR-LIKE COMPOUND ISOLATED FROM HIGHER PLANTS

Catherine Vakirtzi-Lemonias, Vassilis Tsaoussis and
Joan H. Argyroudi-Akoyunoglou

Institute of Biology, NCSR "Demokritos", Aghia Paraskevi
Attiki, Greece

INTRODUCTION

Platelet Activating Factor, PAF, (1-0-alkyl-2-acetyl-sn-glycero-3-phosphocholine), an ether phospholipid with hormone-like properties in higher organisms, is biosynthesized by a variety of animal cell types and elicits a series of different responses in mammals. It is the most potent platelet aggregating factor known todate, its signal transduction is mediated via the phosphoinositide cycle and recently it has been reported to cause a biphasic pHi change in platelets (1-3). In addition it has been found to stimulate ATP-dependent H^+ transport in plant microsomes (4).

Previous studies from our laboratories have shown that PAF, added exogenously to chloroplast or subchloroplast fractions, derived from stroma or grana lamellae, induces a dose-dependent increase of the 77 K fluorescence emission ratio F685/F730. The effect was specific for PAF and more pronounced in the grana derived fraction. PAF induced also an increase in the Chl a fluorescence yield and enhanced the association of chlorophyll in the supramolecular pigment-protein complexes of the thylakoids, especially those of PSII. The effects were attributed to alterations in the organization of the photosynthetic unit (5,6).

In view of these findings we investigated the possibility that plant tissues, especially thylakoids and their subfractions, have as a natural constituent PAF or a PAF-like compound. Once characterized the potential role of the compound in the regulation of chloroplast development and in signal transduction will be studied.

It may be added that ether lipids of a more complex structure have been isolated from photosynthesizing green algae and have been shown to constitute a significant proportion of the algal thylakoid membrane (7). Structurally these lipids are diacylglycero-4'-0-(N,N,N-timethyl) homoserines, (8 and cited references).

METHODOLOGY

Chloroplasts, (thylakoids), and sub-chloroplast fractions were isolated from 3 week-old pea leaves. The isolation of the 10K, (grana) and the 40K, (stroma lamellae) fractions was accomplished by differential centrifugation of a French press disrupted chloroplast suspension (9). Extraction of total lipids from leaves, the thylakoids and the thylakoid subpellets was achieved by homogenization in $CHCl_3:MeOH:H_2O$ mixtures, originally establishing a 1:2:0:8, one phase system and then partitioning

Regulation of Chloroplast Biogenesis
Edited by J.H. Argyroudi-Akoyunoglou, Plenum Press, New York, 1992

the lipids in the CHCl$_3$ phase by changing the ratio to 1:1:0:9. Lipid classes with emphasis on the separation of PAF from PC was accomplished by a two-stage, one dimentional tlc on precoated preparative silica gel G plates. The solvent systems were PE:EE:AcOH 80:20:1 and CHCl$_3$:MeOH: AcOH:H$_2$O 85:15:10:3.5 v/v for the 1st and 2nd stages of development. The plates were thoroughly dried between developments. Visualization of PAF and PC standards was done with the Dittmer-Lester reagent (10). The PAF and PC zones of the silica were scrapped off the plates, extracted x3 with CHCl$_3$:MeOH:H$_2$O 1:2:0.8, the combined extracts were made 1:1:0.9 and "PAF" and PC were partitioned in the CHCl$_3$ phase. After evaporation of the solvents the lipids were taken up in a bovine serum albumin, BSA, solution, 2.5mg/ml and used in the various assays.

The following tests were used for the identification-quantitation of the PAF-like compounds. a) TLC migration in the solvent systems described above. b) Ability to aggregate washed rabbit platelets (11). The reaction was initiated by addition to the platelet suspension (3x10^5 cells/ml), aliquots of the BSA-"PAF" and PC extracts. For quantitation studies a calibration curve was constructed by plotting % aggregation elicited by known concentrations of animal PAF vs concentration. c) Inhibition of the aggregatory ability of "PAF" inhibitors WEB 2086 and BN 52021 (2). d) Inactivation of "PAF" by PAF Acetyl-hydrolase (EC 3.1.1.48), present in plasma (1). We used mouse plasma which has a very active PAF AH (12), and assayed the residual activity by the rabbit platelet aggregation assay. e) Inhibition by "PAF" of the enzymic hydrolysis of [^{3}H-acetyl] PAF. In a final volume of 100ul, [^{3}H-acetyl]PAF 30x10^{3}cpm, was incubated, (2min, 37°C) with mouse plasma (4 µg protein), in the presence of animal PAF or leaf, thylakoid, 10K or 240K extracted "PAF". Hydrolyzed ^{3}H-acetate was measured as previously described (12).

RESULTS

In order to determine whether a plant "PAF" was a natural constituent of higher plant thylakoids we developed a simple method for tlc isolation of "PAF", involving a one-dimentional, two-step development procedure. The first step removed the pigments to the front of the plate and the second separated PAF from PC (Rfs .20 and .30 respectively). The silicic acid, scraped from the lanes containing the PAF-like compound(s), was extracted and used for the characterization and quantitation studies.

Washed rabbit platelets are very sensitive to PAF, responding with aggregation to concentrations as low as 10^{-12}M. We used therefore this assay to detect the presence of PAF-like acting compounds in the extracts obtained from leaves, thylakoids and the 10K and 240K fractions. Figure 1 shows that "PAF" isolated from these samples elicits a strong aggregation reaction in washed rabbit platelets. PC isolated from leaves or the 240K fraction, elicits also a small and reversible aggregation of the platelets and gives a measure of the relative quantities of "PAF" and PC in the preparations (it is known that the potencies of aggregation of platelets by the two compounds, PAF and PC, differ by a factor of about 10^5). Such a comparison in quantities however assumes that "PAF" is similar in structure and as active as the authentic PAF of animal origin.

WEB 2086 and BN 52021 are two compounds which inhibit in a specific manner the aggregation of platelets elicited by PAF exerting their effects at the PAF receptor level (2). We tested therefore the ability of these compounds to inhibit the aggregatory ability of "PAF" and Table 1 summarizes the results. WEB 2086 is a more potent inhibitor compared to BN 52021 for PAF (2) and the same difference was found for "PAF".

For further functional characterization of "PAF" we took advantage of the fact that blood plasma has a specific PAF AH which inactivates PAF. We incubated therefore "PAF" isolated from leaves with mouse plasma, which has an active enzyme (12) for 5 or 30 min and then tested

470

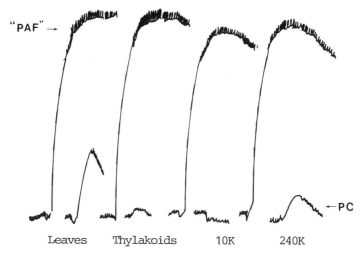

"PAF" →

←PC

Leaves   Thylakoids   10K   240K

Fig. 1. Aggregation of washed rabbit platelets by "PAF" and PC isolated
from pea leaves, thylakoids and their 10K and 240K subfractions.

the mixture for residual aggregating activity. The results, Fig. 3, show
significant inhibition after 5 min and loss of activity after 30 min
incubation. Analogous results were obtained for "PAF" isolated from
thylakoids and thylakoid subfractions. In another experimental approach
we tested the ability of "PAF" to inhibit the hydrolysis of [^{3}H-acetyl]
PAF by the mouse plasma enzyme. Again, as shown in Table 2, a reduction
in the recovered radioactivity comparable to that caused by 1µH and 10µH
PAF, was observed after a 2min incubation with various concentrations of
"PAF". It must be pointed out here that in both hydrolytic assays we
observed that the mouse enzyme was markedly less active towards "PAF"
when compared to its known high activity towards PAF (12). This suggests
that "PAF" must have the structure requirements for function but the two
molecules are not identical.

Table 1. Inhibition of the aggregatory activity of "PAF" by BN 52021 and
WEB 2086. The inhibitors or solvent were preincubated with the
platelets for 2min and aggregation was initiated by the addition of "PAF" aliquots. "PAF" concentration in arbitrary units
was calculated from the heights of aggregation curves.

Source of "PAF"	"PAF" concentration Arbitrary units	Inhibition %
Leaves	118	
Leaves+20µM BN 52021	66	56
Thylakoids	75	
Thylakoids+30µM BN 52021	48	64
Thylakoids+20µM WEB 2086	0	100
10K	40	
10K+20µM BN 52021	0	100
240K	59	
240K+20µM BN 52021	0	100

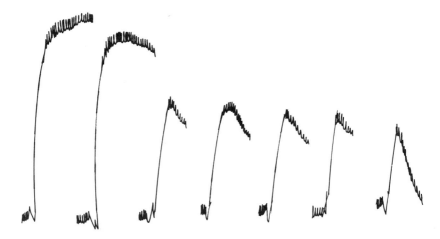

Fig. 2. Effect of PAF acetylhydrolase on the activity of "PAF" assayed
by platelet aggregation. "PAF" was tested for aggregatory
activity before and after incubation with mouse plasma. I-III,
"PAF" corresponding to 100, 50 and 12.5mg fresh leaf weight. IV
and V, "PAF" I and II after incubation for 5 and 30min with the
enzyme.

For the quantitation of the "PAF" concentration in thylakoids and
the 10K and 240K fractions we used the washed rabbit platelet aggregation
assay (13). A calibration curve with known quantities of animal PAF was
constructed and the concentration of "PAF" from the different preparation
was calculated assuming a 1:1 aggregation activity of PAF to "PAF". Fig.
3 shows the results of an experiment used to quantify the "PAF" content
of the 240K thylakoid subfraction and Table 3 gives the distribution of
"PAF" in thylakoids, grana and stroma lamellae. It appears that the
latter is the most rich in "PAF" on a chlorophyll basis.

DISCUSSION

Our findings show that the photosynthetic machinery of plants
contain a lipid component which by several criteria may be identified as
a PAF-like bioactive molecule. Since the two compounds, PAF and "PAF",
share similar functional properties they must also have similar

Table 2. Enzymic hydrolysis of [^{3}H-acetyl]PAF in the presence of plant
"PAF" or animal PAF. For experimental details see Methodology.
Chl content of the aliquots assayed for "PAF" is shown.

Tissue	Chl, (µg)	<-cpm	Chl, (µg)	<-cpm
Leaves	102	2450	204	860
Thylakoids	85	1445	170	540
10K	147	2540	294	1700
240K	13	6330	26	4785
PAF, animal, 1µM		2650		
PAFn animal, 10µM		2150		

472

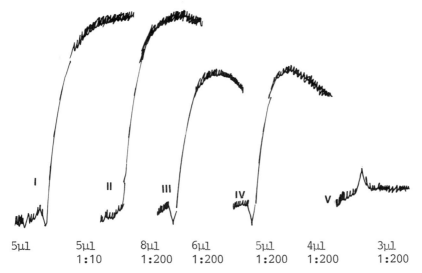

| 5μl | 5μl | 8μl | 6μl | 5μl | 4μl | 3μl |
| | 1:10 | 1:200 | 1:200 | 1:200 | 1:200 | 1:200 |

Fig. 3. Quantitation of the "PAF" isolated from the 240K pellet, (800μg Chl). Various aliquots of "PAF" from a 50 μl total volume were assayed.

functional groups. We have not yet conducted a physicochemical charac- terization of the "PAF" molecule. There are indications however suggest- ing that the structure of the plant "PAF" may differ from the animal PAF, the strongest one coming from the low activity of PAF AH towards "PAF". A rough estimate by a RIA approach (14), of the quantity of "PAF" present in the incubation mixture during the [^{3}H-acetyl]PAF hydrolysis assay, gives a figure about 10^3 higher as compared to the one obtained by the aggregation assay. It must be emphasized again that both reference curves were constructed using animal PAF and a 1:1 activity for PAF and "PAF " was adopted for calculations. Small differences in the structure of the molecule may affect in a different extent platelet aggregation -as exemplified by the difference in aggregating activity of PAF C16 and C18 (1) - and its effectiveness as a substrate for the PAF acetylhydrolase.

The structural-functional relationship for the PAF molecule, as assayed by platelet aggregation are: a) an alkyl moiety at the sn-1 position. As mentioned, alkyl chains of C16 are more effective than the C18 analogs while analogs with an acyl substituent are inactive (e.g. PC 10^{-7}M). b) a short chain O-acyl group, acetate or propionate, is needed for activity at the sn-2 position. Activity decreases with increasing chain length and the C6 compounds are inactive. c) a phosphocholine group is necessary at the sn-3 position. Thus the activity progressively

Table 3. Concentration of the PAF-like compound in leaves, thylakoids and the 10K and 240K fractions, as estimated by the rabbit platelet aggregation assay. Chl content of the aliquot assayed for "PAF" is shown.

Tissue	Chl content (mg)	"PAF" pmoles/μmol Chl
Leaves	30.52	4.7
Thylakoids	14.50	3.6
10K	14.70	1.6
240K	0.96	86.8

decreases when the methyl groups are removed to produce the inactive PE. On the other hand the PAF PAF AH is a specific esterase for the PAF molecule and the relative activity of hydrolysis of PAF acetate is much higher as compared to that of PAF propionate (C2>>C3=C4>C6). The substrate recognition by the enzyme is not markedly dependent on the type of the bond, i.e. ether or ester, at the sn-1 position of PAF, and although there is preference for the choline group, sequential removal of its methyl groups results only in a modest decrease of the enzyme specificity (15). Thus, from the requirements for activity described above, it appears that the plant "PAF" in all probability has an ether bond with a long enouph carbon chain at the sn-1 position and a rather short chain at the sn-2 position and a rather short chain at the sn-2 position.. The structure of the moiety at the sn-3 position may differ considerably from animal PAF. Thus the physicochemical characterization of the molecule is indispensable.

It may be concluded that pea leaves, thylakoids and their 10K and 240K subfractions have as constituent a PAF-like compound(s) with the following characteristics. a) It aggregates washed rabbit platelets in a dose-response manner. b) Its aggregatory ability is inhibited by the specific PAF inhibitors WEB 2086 and BN 52021 in a dose-dependent relationship. c) It loses its aggregating activity after incubation with mouse plasma which contains a specific PAF acetylhydrolase. d) It inhibits in a concentration dependent the hydrolysis of [^{3}H-acetyl]PAF by mouse plasma. The plant PAF-like compound appears to be preferably concentrated in the 240K subfraction of thylakoids in relation to their chlorophyll content. These features are characteristic of the PAF molecule which is biosynthesized by stimulation of animal cells. Thus the PAF-like compound which was isolated from the plant tissues must resemble PAF and must at least have structural moieties analogous to those of the animal PAF.

REFERENCES

1. T. Lee and F. Snyder, in "Phospholipids and Cellular Regulation",
    Vol. II, J.F. Kuo, ed., CRC Press, Boca Raton, Florida (1986).
2. P. Braquet, L. Touqui, T.Y., Shen and B.B. Vargaftig, Pharmacol.
    Reviews 39:97 (1987).
3. M.L. Borin, V.G. Pinelis, O.A. Azizova, Y.V. Kudinov, C.M. Markov,
    E.J. Cragoe and B.I. Khodorov, J. Lipid Mediators 1:257 (1989).
4. G.F.E. Scherer, B. Stoffel and G. Martiny-Baron, Biol. Chem.
    Hoppe-Syeler 369:7 (1988).
5. J.H. Argyroudi-Akoyounoglou and C .Vakirtzi-Lemonias, Arch. Biochem.
    Biophys. 253:38 (1987)
6. J.H. Argyroudi-Akoyounoglou and C. Vakirtzi-Lemonias, Arch. Biochem.
    Biophys. 275:271 (1989).
    7. D.R. Janero and R. Barrnett, J. Lipid Res. 23:307(1982).
8. N. Weber, D. Bengenthal, C.K. Kokate and H.K. Mangolg, J. Lipid
    Mediators 1:37 (1989).
9. P.V. Jane, D.J. Goodchild and R.B. Paru, Biochim. Biophys. Acta
    216:162 (1970).
10. J.D. Dittmer and R.L. Lester, J. Lipid Res. 5:126 (1964).
11. M.B. Zucker, Methods in Enzymol. 169:117 (1989).
12. C. Vakirtzi-Lemonias, Biochem. Soc. Trans. 12:689 (1984).
13. M. Bossant, E. Ninio, D. Delautier and J. Benveniste, Methods in
    Enzymol. 187:125(1990).
14. D.R. Bangham in "Standarization in RIA Design and Quality Control".
    J.I. Thorell, ed. Pergammon Press, Paris, Oxford (1982).
15. D.M. Stafforini, S.M. Prescott and T.M. McIntyre, J. Biol. Chem.
    262:4223 (1987).

ADAPTATION MECHANISMS IN PLANTS AND ALGAE

A REGULATORY FEEDBACK MECHANISM FOR LIGHT ACCLIMATION OF THE PHOTOSYNTHETIC

APPARATUS: ARE PHOTOSYSTEMS II AND I SELF-REGULATORY LIGHT SENSORS?

Jan M. Anderson and W.S. Chow

CSIRO, Division of Plant Industry
GPO Box 1600, Canberra ACT 2601
Australia

The coordinated interaction between light-harvesting, energy conversion, electron transport, proton translocation and carbon assimilation is exquistitely matched in photosynthesis. Nowhere is this more clearly evident than in the sun/shade acclimation of the composition, function and structure of the photosynthetic apparatus[1-3]. Plants have sophisticated, multiple molecular mechanisms for sensing and responding to enormous fluctuations in light environment. This light acclimation has evolved because (1) plants are immobile and unable to escape from potentially damaging levels of high light, and (2) light is the driving force for photosynthesis and a trigger for photomorphogenesis. Many regulatory processes, both short-term and long-term, contribute to photosynthetic control whereby the production and consumption of ATP and NADPH closely match carbon assimilation[4]. On the one hand, there are many dynamic short-term changes, such as state transitions, which correct a temporary imbalance in the distribution of excitation energy between the photosystems, and ensure maximum quantum effeciency under limiting light, while at high irradiance photosynthetic control serves to dissipate excess excitation energy when the potential rate of ATP and NADPH formation exceeds the rate at which they are required for carbon assimilation. On the other hand, there are long-term changes in the relative amounts of the thylakoid complexes and mobile electron transport carriers, associated with acclimation to sun or shade[1-3].

DYNAMICS OF SHADE/SUN ACCLIMATION

The shade environment of understory plants is made up of very low continuous irradiance which is deficient in red and blue light but enriched in far-red light, together with direct sunlight via intermitent sunflecks or larger canopy gaps. The photosynthetic apparatus is faced with two major sets of compromises, those required to bring about the most effective utilisation of light quantity and quality, and those required to withstand both short-term and long-term fluctuations in light quality and quantity. Since these responses of the photosynthetic apparatus to ever-changing light conditions are crucial determinants of plant growth and survival, they are amongst the most dynamic of all plant functions.

Compared to sun plants, shade plants have larger, thinner leaves with a greater area to mass ratio. Their larger chloroplasts have greatly

Regulation of Chloroplast Biogenesis
Edited by J.H. Argyroudi-Akoyunoglou, Plenum Press, New York, 1992

increased amounts of thylakoid membranes and smaller stromal volumes. In shade, where light is the limiting factor, much of the photosynthetic resources must be invested in the synthesis and maintenence of the light-harvesting assemblies of both photosystems[3]. Shade plants have more chlorophyll per chloroplast and more chlorophyll b. Relative to PSI, shade plants have fewer, but larger, PS II units to enhance light capture at low irradiance. The PSII/PSI reaction centre ratios are also lower (1.2-1.4) than those of sun plants (1.8-2.2). This increase in light-harvesting assemblies in shade plants, which occurs at the expense of electron transport and photophosphorylation components and carbon-fixation enzymes, results in much lower maximal photosynthetic capacities which saturate at low irradiance. Conversely, with sun or high-light plants, the rate of electron transport steps is limiting, rather than light absorption by the two photosystems. Thus, acclimation to sun or high irradiance leads to greater amounts of cytochrome b/f complex, plastoquinone, plastocyanin and ferredoxin as well as ATP synthase (on a chlorophyll basis) which support faster rates of NADPH and ATP synthesis, and in turn, higher maximal photosynthetic capacity. The photosystem stoichiometry is adjusted also so that sun plants have higher PSII/PSI stoichiometries, with thylakoids having more, though smaller, PSII units, relative to PSI. Nevertheless, under nonstressed environmental conditions, shade and sun plants have the same high quantum efficiency of photosynthesis, irrespective of their growth irradiance[5].

ADJUSTMENTS OF PHOTOSYSTEM STOICHIOMETRY IMPROVE THE QUANTUM EFFICIENCY OF PHOTOSYNTHESIS

How do shade plants achieve the desirable characteristics of being as effecient as sun plants under limiting light? Recently we have provided direct evidence[6] for the notion that adjustments of the photosystem stoichiometry (the PSII/PSI reaction centre ratio) in response to variable light quality serve to correct uneven light absorption between the photosystems[7,8]. Comparison of peas, grown in yellow light to preferentially excite PSII (PSII-light) and red light enriched in far-red to mainly excite PSI (PSI-light), showed that plants grown in PSI-light had lower Chl a/Chl b ratios and less chlorophyll per unit leaf area than PSII-light plants[6]. PSI-light-grown peas had a greater PSII concentration, as determined by DCMU-binding sites, compared with PSII-light thylakoids (Fig. 1B). Conversely, the concentration of photo-oxidisable P700 was greater in PSII-light thylakoids than in PSI-light thylakoids (Fig.1A). This adjustment of the photosystem stoichiometry is very substantial. The PSII/PSI ratio of PSI-light-grown thylakoids increased to 2.5, while that of PSII-light-grown thylakoids was only 1.1, compared with 1.8 for control peas[6].

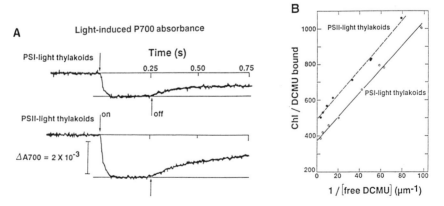

Figure 1. P700 and PSII content in PSI-light and PSII-light pea thylakoids.

When peas grown in PSI-light or PSII-light were probed with saturating white light, PSII-light plants had a greater maximum photosynthetic capacity per unit leaf area than PSI-light plants[6]. However, more interesting differences were observed under limiting light. The initial slopes of the light response curves of PSI and PSII-light-grown leaves (Fig. 2), plotted as a function of absorbed light intensity, define the quantum yields of photosynthesis. When probed with PSI irradiance (i.e. the light those plants were grown under), the PSI-light-grown peas have a 19% better quantum yield of photosynthesis than PSII-light-grown peas. Conversely, under PSII irradiance, the PSII-light-grown plants have a 21% better quantum yield of photosynthesis compared to the PSI-light-grown peas.

These results provide the first direct evidence for a cause and effect relationship between altered PSII/PSI reaction centre ratios and the high quantum yield of photosynthesis of PSI-light-grown plants or PSII-light-grown-plants, when measured under the same light quality as in the growth environment. A remarkable feature of diverse higher plants growing in many varying light environments is their high, and constant quantum yield of 0.106 mol of O_2 evolved per mol of photons absorbed[5]: these values are close to the theoretical maximum of 0.125. The above results directly prove that the dynamic adjustments of the PSII/PSI reaction centre ratios in chloroplasts are a compensatory strategy designed to retain a high quantum efficiency of photosynthesis near the theoretical maximum. This adjustment and optimisation of the photosystem stoichiometry in thylakoid membranes is important, since most choloroplasts are actually functioning for most of the time under light-limiting conditions, due to the pronounced attenuation of light both within leaves and within the canopy of a single tree or forest.

THE INFLUENCE OF BRIEF OR PROLONGED FAR-RED IRRADIANCE ON THE PHOTOSYNTHETIC APPARATUS

To probe the possible involvement of phytochrome or other photosensors in the response by the photosynthetic apparatus to far-red irradiance, we compared the effects of brief and prolonged supplementary far-red irradiance on the composition, function and structure of pea chloroplasts. Peas were grown in controlled environments (12 h fluorescent light, 12 h dark) without, or with (1) 15 min supplementary far-red illumination at the end of each photoperiod (brief FR), or (2) high levels of supplementary far-red light (R/FR ratio of 0.04) during the entire photoperiod (long-term FR)[9]. Brief FR simulated the decrease in the red/far-red ratio near the end of the day which is sensed by phytochrome, while long-term FR simulates extreme shade conditions under the forest canopy[10].

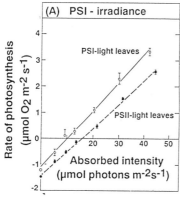 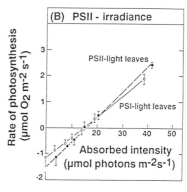

Figure 2. Initial linear portion of the light response curve of photosynthesis in pea leaves probed with (A) PSI- or (B) PSII-irradiance.

Table 1. Photosynthetic acclimation to supplementary brief far-red light. Peas were grown in white fluorescent light (12 h light/12 h dark) without, or with 15 min supplementary far-red light at end of photoperiod (**brief FR**).

	PSII	cyt f	PSI	ATP synthase	Rubisco activity	P_{max}
	(mmol/mol Chl)			(mmol/mol Chl/s)		
+FR	1.89	0.78	1.53	353	62	41.1
-FR	2.09	0.87	1.62	369	67	39.2
+FR/-FR	0.90	0.89	0.94	0.96	0.92	1.05

Brief FR: Brief FR treatment led to marked morphological effects on peas compared to control plants, with an increase in internode lengths, but a decrease in leaflet area, chloroplast size, and chlorophyll content. These classical effects on morphology are attributible to regulation by phytochrome[10]. Significantly, however, compared to control irradiance, **brief FR** had little, or a slightly negative, effect on the relative amounts of electron transport complexes, ATP synthase and Rubisco activities when based on either chlorophyll or PSI content[9] (Table 1). Relative to P700 (for which the accumulation of the P700 apoprotein is not regulated by phytochrome[11]), the amounts of chloroplast components were unaffected by the same brief illumination which produced marked morphological effects. We propose[9] that phytochrome exerts a neglible effect on the relative amounts of the major thylakoid complexes, or indeed on maximal leaf photosynthetic capacity per unit chlorophyll, in the mature chloroplast. This proposal is supported by previous results[12,13].

Long-term FR: In order to simulate canopy shade, we supplemented fluorescence light with high levels of far-red (so that the red : far-red ratio was 0.04, comparable to extreme shade values[10]). Prolonged supplementary far-red irradiance also led to the morphological phytochrome-regulated effects listed above for **brief FR**. However, in marked contrast to **brief FR** (Table 1), there was a marked increase in photosynthetic capacity per unit chlorophyll (Table 2). The increased photosynthetic capacity under **long-term FR** was associated with comparable enrichment of Rubisco and ATP synthase activities, as well as cytochrome f and to a lesser extent PSII, but a decrease in PSI on a chlorophyll basis (Table 2).

Can the *increases* in these photosynthetic components in **long-term FR** (Table 2) be attributed to the action of phytochrome? As for **brief FR**, we propose also that the main signals which control the relative amounts of major chloroplast components are not controlled by phytochrome, although phytochrome regulation of the mRNA levels of some individual thylakoid proteins cannot be excluded. Further, Jenkins and Smith[13] found, using continuous fluorescent light, with or without supplementary far-red light, no significant effect of varying levels of far-red light on rbcS RNA levels in pea. Although brief far-red illumination, via phytochrome, had a pronounced effect of decreasing the rbcS RNA levels in etiolated tissues[14-17] and dark-treated light-grown tissues[17], none of these studies provides any evidence of how phytochrome in the Pr form (with far-red illumination) could possibly lead to an increase in photosynthetic components.

Apart from phytochrome, which is not exerting any positive effect on the composition of the photosynthetic apparatus in **brief FR** (Table 1), photosystem I itself is the only other obvious far-red sensing pigment

Table 2 Photosynthetic acclimation of peas to prolonged supplementary far-red light. Peas were grown (12 h light/12 h dark) in white fluorescent light without, or with supplementary far-red light in the 12 hr photoperiod (**long-term FR**).

	PSII	cyt f	PSI	ATP synthase	Rubisco activity	P_{max}
	(mmol/mol Chl)			(mmol/mol Chl/s)		
+FR	2.38	1.34	1.64	357	89	50.4
-FR	2.21	1.11	1.78	256	70	38.5
+FR/-FR	1.08	1.21	0.92	1.39	1.29	1.31

system present in chloroplasts. We hypothesize that the observed modulations of chloroplast composition and photosynthetic capacity in **long-term FR** (Table 2) were induced by the far-red light acting through PSI, and increasing photosynthesis by extra cyclic electron transport around PSI. Additionally, or alternatively, the far-red light drives chloroplasts into transition State 1, whereby spillover of excitation energy from PSII to PSI is minimised; the resultant increased energy partitioned into PSII, together with that into PSI, speeds up noncyclic electron transport compared to the control. In either case, the supply of ATP would be increased, leading to an increase in protein synthesis, protein translocation across membranes, and assembly of functional complexes *in a post-transcriptional manner, without the involvement of phytochrome*. We postulate that prolonged far-red enriched illumination induced an increase in chloroplast components (Table 2) by a photosynthetic feedback mechanism.

Further evidence for this hypothesis is provided by an experiment where pea leaves were floated on an ATP solution for 22 h at 25° C under low white irradiance[9]. We observed an increase in the maximum Rubisco activity (a measure of a key functional enzyme) as compared with comparable leaf discs floated on water under the same irradiance. Here a direct increase of the ATP supply helped to increase the abundance of photosynthetic components, consistent with our proposed photosynthetic feedback mechanism. More generally, the pools of the photosynthetic energy storage metabolites, ATP and NADPH, could play a prominent role in regulating the abundance of appropriate photosynthetic comoponents by exerting a critical influence on the photosynthetic feedback mechanism[9].

PHOTOSYSTEMS II AND I ACT AS SELF-REGULATORY SENSORS OF LIGHT QUALITY AND LIGHT QUANTITY?

As the foregoing evidence shows, brief or prolonged supplementary far-red irradiance gave rise to very different effects on the composition and function of the photosynthetic apparatus (Tables 1 & 2), but not on plant morphology, suggesting the involvement of different photosensors, either photosystem I itself or phytochrome, respectively. Can this notion be extended further to include a feedback regulatory mechanism being involved in the adjustments of light-harvesting components, thylakoid complexes and stromal enzymes in response to changes in the prevailing environment?

Although features of a generalised regulatory mechanism are largely unknown, a number of points are clear. Firstly, this regulatory mechanism is highly conserved throughout the plant kingdom: dynamic adjustments of thylakoid complexes are a common feature of light acclimation from cyanophytes[18,19] through to higher plants[3,20,21]. Secondly, the mechanism is always "switched on". For example, if mature plants germinated and grown in low irradiance are transferred to high irradiance, they will eventually have the same composition, function and structure as plants germinated and grown in high irradiance[22,23]. Further, when *Alocasia*, a shade plant, is slowly acclimated to high irradiance, it will eventually exhibit the characteristic features of a sun plant[24]. Thirdly, such dynamic changes in response to varying genvironmental conditions are not restricted to responses to light quantity and light quality. Consider mutations in chlorophyll b synthesis: with partial or complete deficiency in Chl b, the amounts of the thylakoid complexes of the mutants are very different to wild type species[4,21]. Sublethal dosages of PSII herbicides will also modulate the chlorophyll composition leading to enhanced chlorophyll synthesis, particularly Chl b, resulting in apparent "shade" acclimation[20,21]. Fourthly, there is no evidence for the existence of genes for shade tolerance. The chloroplast genome of the shade plant, *Marchantia*[25] is very similar to those found in sun species, rice[26] and tobacco[27]. The light responses of photosynthesis that are controlled by the chloroplast genome seem to occur during protein translation and post-translation[28]. Further, it appears that sun or shade responses occur not by specific nuclear genes, but rather through varying degrees of transcription and translation[28]. Clearly, there is enormous plasticity in the photosynthetic apparatus, but the differences exhibited are a matter of degree rather than absolute.

In the feedback regulatory mechanism for dynamic modulation of the photosynthetic apparatus, we suggest that the above environmental signals, namely light quality, light quantity, Chl b deficiency and sublethal PSII herbicide dosages, are all sensed by the photosystems themselves. We propose that the light-harvesting chlorophylls, together with the reaction centres of each photosystem, act as self-regulating sensors. The photosystems would be auto-regulated in a somewhat analagous manner to the self-regulatory photoreceptor, phytochrome, which is down-regulated during development[29].

Our proposal[9], that environmental signals for this regulatory mechanism are sensed by the photosystems themselves acting as the primary sensor differs from proposals by Melis et al.[18,21] and Fujita and Murakami[19,30,31] who identify the primary signal as adjustments to electron transport at the level of cytochrome b/f complex. In our scheme (Fig. 3), the need for adjustment of the thylakoid complexes sensed by the photosystems, is then identified at the common level of electron transport between the photosystems, perhaps the redox state of cyt f as suggested by Murakami and Fujita[31]. In turn, the energystorage pools of ATP and NADPH will be modulated by the relative extent of noncyclic electron tranport (which produces both ATP and NADPH) and cyclic electron flow (leading to enhanced ATP formation). These energy storage pools will regulate gene expression, as postulated previously by Melis et al.[18,21] and Chow et al.[9]. They suggested that the environmental signal transmission of gene expression occurs via the pools of ATP and/or NADPH in the chloroplast stroma, with

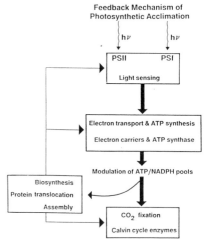

Figure 3.

480

enhanced transcription/translation and biosynthesis/assembly activity being directed towards those components of the electron transport chain that are rate-limiting, together with degradation/disassembly lowering the concentration of other components (e.g. PSI in PSI-light-grown plants)

We conclude, from the above considerations, that phytochrome has a negligible effect in modulating the stoichiometries of the major thylakoid membrane and stromal components of fully-greened mature leaves. Rather, our results strongly suggest that the relative amounts of chloroplast components are mainly regulated by post-transcriptional events, which are driven primarily by the two photosystems acting as light sensors.

REFERENCES

1   O. Björkman, Responses to different quantum flux densities, in: Encyclopedia of Plant Physiology, New Series, Vol. 12A O.L. Lange et al. eds, pp.57-107, Springer-Verlag, Berlin (1981).
2   J.M. Anderson and C.B. Osmond, Shade-sun responses: compromises between acclimation and photoinhibition, in: "Photoinhibition." D.J. Kyle et al. eds, pp.1-38, Elsevier, Amsterdam (1987).
3   J.M. Anderson, W.S. Chow and D.J. Goodchild, Thylakoid membrane organisation in sun/shade acclimation, Aust. J. Plant Physiol. 15: 11-26 (1988).
4   C. Foyer, R. Furbank, J. Harbinson and P. Horton, The mechanisms contributing to photosynthetic control of electron transport by carbon assimilation in leaves, Photosynth. Res. 25: 83-100 (1990).
5   O. Björkman and B. Demmig, Photon yield of oxygen evolution and chlorophyll fluorescence characteristics at 77K among vascular plants of diverse origin, Planta 170: 489-504 (1987).
6   W.S. Chow, A. Melis and J.M. Anderson, Adjustments of photosystem stoichiometry in chloroplasts improve the quantum efficiency of photosynthesis, Proc. Natl. Acad. Sci. USA 87: 7502-7506 (1990).
7   R.E. Glick, S.W. McCauley and A. Melis, Effect of light quality on chloroplast membrane organization and function in pea, Planta 164: 487-494 (1985).
8   X.W. Deng, J.C. Tonkyn, G.F. Peter, J.P. Thornber and W. Gruissem, Post-transcriptional control of plastid mRNA accumulation during adaptation of chloroplasts to different light quality environments, Plant Cell 1: 645-654 (1989).
9   W.S. Chow, D.J. Goodchild, C. Miller and J.M. Anderson, The influence of high levels of brief or prolonged supplementary far-red illumination during growth on the photosynthetic characteristics, composition, and morphology of *Pisum sativum* chloroplasts, Plant Cell Environ. 13: 135-145 (1990).
10   H. Smith, Light quality, photoperception and plant strategy, Ann. Rev. Plant Physiol. 33: 481-518 (1982).
11   W. Laing, K. Kreuz and K. Apel, Light-dependent, but phytochrome-independent, translational control of the accumulation of the P700 chlorophyll-a protein of photosystem I in barley (*Hordeum vulgare* L.), Planta 176: 269-276 (1988).
12   W.S. Chow, W. Haehnel and J.M. Anderson, The composition and function of thylakoid membranes from pea plants grown under white or green light with or without far-red light, Physiol. Plant. 70: 196-202 (1987).
13   G.I. Jenkins and H. Smith, Red: far-red ratio does not modulate the abundance of transcripts for two major chloroplast polypeptides in light-grown *Pisum sativum* terminal shoots, Photochem. Photobiol. 42: 679-684 (1985).
14   Y. Sasaki, T. Sakihama, T. Kamikubo and K. Shinozaki, Phytochrome-mediated regulation of two mRNAs, encoded by nuclei and chloroplasts,

of ribulose-1,5-bisphosphate carboxylase/oxygenase, Eur. J. Biochem. 133: 617-620 (1983).

15 L.S. Kaufman, W.F. Thompson and W.R. Briggs, Different red light requirements for phytochrome-induced accumulation of cab RNA and rbcS RNA, Science 226: 1447-1449 (1984).

16 J. Silverthorne and E.M. Tobin, Demonstration of transcriptional regulation of specific genes by phytochrome action, Proc. Nat. Acad. Sci. USA 81: 1112-1116 (1984).

17 R. Fluhr and N.-H. Chua, Developmental regulation of two genes encoding ribulose-bisphosphate carboxylase small subunit in pea and transgenic petunia plants: phytochrome response and blue-light induction, Proc. Natl. Acad. Sci. USA 83: 2358-2362 (1986).

18 A. Melis, Adaptation of photosystem stoichiometry in oxygen-evolving thylakoid membranes, in: "Current Research in Photosynthesis", M. Baltscheffsky ed., IV: 291-298, Kluwer Academic Press, The Netherlands (1990).

19 Y. Fujita and A. Murakami, Regulation of electron transport composition in cyanobacterial photosynthetic system: stoichiometry among photosystem I and II complexes and their light-harvesting antennae and cytochrome b/f complex, Plant Cell Physiol. 28: 1547-1553 (1987).

20 J.M. Anderson, Photoregulation of the composition, function and structure of thylakoid membranes, Ann. Rev. Plant Physiol. 37: 93-136 (1986).

21 A. Melis, A. Manodori, R.E. Glick, M. Ghirardi, S.W. McCauley and P.J. Neale, The mechanism of photosynthetic membrane adaptation to environmental stress conditions: a hypothesis on the role of electron transport capacity and of ATP/NADPH pool in the regulation of thylakoid membrane organization and function Physiol. Vég. 23: 757-765 (1985).

22 W.S. Chow and J.M. Anderson, Photosynthetic responses of *Pisum sativum* to an increase in irradiance during growth. I. Photosynthetic activities, Aust. J. Plant Physiol. 14: 1-8 (1987).

23 W.S. Chow and J.M. Anderson, Photosynthetic responses of *Pisum sativum* to an increase in irradiance during growth. II. Thylakoid membrane components, Aust. J. Plant Physiol. 14: 9-19 (1987).

24 W.S. Chow, L. Qian, D.J. Goodchild and J.M. Anderson, Photosynthetic acclimation of *Alocasia macrorrhiza* (L.) G. Don to growth irradiance: structure, function and composition of chloroplasts, Aust. J. Plant Physiol. 15: 107-122 (1988).

25 K. Ohyama et al., Chloroplast gene organization deduced from complete sequence of liverwort *Marchantia polymorpha* chloroplast DNA, Nature 322: 572-574 (1986).

26. J. Hiratsuka, et al., The complete sequence of the rice (*Oryza sativa*) chloroplast genome, Mol. Gen. Genet. 217: 185-194 (1989).

27. K. Shinozaki, et al., The complete nucleotide sequence of the tobacco chloroplast genome: its gene organization and expression, EMBO J 5: 2043-2049 (1986).

28. W. Gruissem, Chloroplast gene expression: how plants turn their plastids on, Cell 56: 161-170 (1989).

29. J.L. Lissemore and P.H. Quail, Rapid transcriptional regulation by phytochrome of the genes for phytochrome and chlorophyll a/b-binding protein in *Avena sativa*, Mol. Cell. Biol. 8: 4840-4850 (1988).

30 A. Murakami and Y.Fujita, Regulation of photosystem stoichiometry in the photosynthetic system of a cyanophyte *Synechocystis* PCC 6714 in response to light-intensity, Plant. Cell Physiol. 32: 223-230 (1991).

31 A. Murakami and Y. Fujita, Steady state of photosynthetic electron transport in cells of the cyanophyte *Synechocystis* PCC 6714 having different stoichiometry between PSI and PSII: analysis of flash-induced oxidation-reduction of cytochrome f and P700 under steady state of photosynthesis, Plant Cell Physiol. 32: 213-222 (1991).

DYNAMIC LIGHT ACCLIMATION OF THE PHOTOSYNTHETIC APPARATUS OF HIGHER PLANTS

Jan M. Anderson, W.S.Chow, Heather Adamson[*] and A Melis[**]

CSIRO, Division of Plant Industry
GPO Box 1600, Canberra ACT 2601
Australia

INTRODUCTION

Higher plants have sophisticated, multiple molecular mechanisms that enable them to sense information about their light environment, and subsequently to evoke not only the appropriate developmental and physiological responses to that environment, but also to cope with momentary, daily and seasonal fluctuations in light quantity and light quality[1-4]. These changes in the light environment have a profound effect on the composition, structure and function of the photosynthetic apparatus. Indeed, responses of the photosynthetic apparatus to changing light quality and light quantity are amongst the most dynamic of all plant functions.

An important feature of light acclimation involves the dynamic long-term regulation of the relative composition of the chlorophyll a-proteins and chlorophyll a/b-proteins of both photosystem (PS) II and PSI. Changes in the pigment-protein composition represent a dual strategy in response to changes in light quantity and light quality[3,4]. There is a dynamic regulation of the amounts of both PSII and PSI reaction centres, i.e. photosystem stoichiometry. Further, the number of chlorophyll molecules associated with each photosystem (the light-harvesting antenna size) is also regulated. These long-term acclimative responses that require the synthesis and assembly of new membrane components, coupled with the disassembly and degradation of other components, are regulated by nuclear and chloroplast gene expression. In addition there are many short-term acclimative strategies (e.g. state transitions), which help to ensure a balanced distribution of excitation energy to both photosystems, and prevent damage due to sudden light stresses[5]. These short-term strategies which involve only a rearrangement of existing components within the photochemical apparatus, and not its basic composition, although not discussed here, are also important for coping with ever-fluctuating light conditions.

---

[*] School of Biological Sciences, Macquarie University, North Ryde, NSW 2109, Australia
[**]Department of Plant Biology, University of California, Berkeley, CA 94720, USA

Regulation of Chloroplast Biogenesis
Edited by J.H. Argyroudi-Akoyunoglou, Plenum Press, New York, 1992

Table 1. Effect of light quantity on the photosynthetic apparatus of pea and spinach grown under constant light quality.

Light quantity	Spinach		Pea	
	High	Low	High	Low
Chl a/Chl b	3.3	2.7	3.3	2.6
PSII/(1000 Chl)	3.0	2.0	3.0	1.8
PSI/(1000 Chl)	1.6	1.6	1.5	1.5
Photosystem stoichiometry (PSII/PSI)	1.8	1.2	2.0	1.2

LIGHT QUANTITY ACCLIMATION

Many studies[1-4], with plants grown in cabinets under varying light quantity of constant quality, demonstrate that as the growth irradiance decreases, there is a marked decrease in the Chl a/ Chl b ratios, reflecting a change in the chlorophyll-protein composition with more Chl a/b-proteins and less Chl a-proteins (Table 1). As seen in Table 1, it is remarkable, however, that the amount of P700 on a chlorophyll basis remains constant. Rather, it is the relative amount of PSII reaction centre that is varied, with a decrease in functional PSII reaction centres as the growth irradiance is decreased. Clearly, the photosystem stoichiometry is regulated by light quantity, with high-light plants having higher PSII/PSI reaction centre ratios compared to low-light plants[4] (Table 1). Relative to PSI, high-light plants have more, but smaller, PSII units to optimise electron transport and proton translocation at high irradiance. On the other hand, low-light plants have fewer, but larger, PSII units to maximise photon absorption and utilisation at low irradiance.

LIGHT QUALITY ACCLIMATION

The dynamic long-term regulation of the photosynthetic apparatus by light quality is also well established[6-8]. Plants grown in light preferentially absorbed by PSI (eg red light enriched in the far-red component; termed here PSI-light) have lower Chl a/Chl b ratios than those grown in light preferentially absorbed by PSII (yellow light; termed here PSII-light)[9,10] (Table 2). Again, the lower Chl a/Chl b ratios of plants grown in PSI-light reflect less chlorophyll associated with PSI Chl-proteins and more with PSII Chl-proteins. Chloroplasts from plants grown

Table 2. Effect of light quality on the photosynthetic apparatus of spinach and peas. Plants were germinated and grown in PSII-light (yellow) and PSI-light (red) of constant quantity[+].

Light quality	Spinach		Pea	
	PSII	PSI	PSII	PSI
Chl a/Chl b	3.72	2.99	2.24	1.97
P680/1000 Chl	2.84	2.88	1.97	2.67
P700/1000 Chl	1.84	1.34	1.73	1.05
PSII/PSI	1.54	2.15	1.10	2.50

[+]Red plus far-red light provided by incandescent light filtered by red plexiglass; yellow light provided by fluorescent light filtered by yellow plexiglass.

in PSI-light have much higher PSII/PSI ratios than those from plants grown in PSII-light.

Due to the lateral heterogeneity in the distribution of thylakoid protein complexes, with most of PSII being located in the grana domains and PSI being located in the nonappressed domains[11], there should be marked changes in thylakoid structure under the two light quality growth conditions. Comparison of the electron micrographs of chloroplasts of spinach grown under PSI-light and PSII-light, reveals two significant differences (Fig.1). Chloroplasts from plants grown in PSI-light, have both an increased ratio of appressed to nonappressed membranes, and also more membrane layers per granal stack than those from plants grown in PSII-light, confirming earlier results[7,12]. Under PSI irradiance, peas grown in PSI-light had a 19% better quantum yield of photosynthesis than PSII-light-grown plants; conversely under PSII irradiance, PSII-light-grown plants had a 21% better quantum yield than PSI-light-grown leaves[9]. These results prove directly that the photosystem stoichiometry is adjusted and optimised as a compensatory strategy to correct imbalance in the light absorption between the photosystems, thereby allowing plants to maintain a high quantum efficiency of photosynthesis in response to different light quality.

CANOPY SHADE ACCLIMATION

Plants living in deep shade on the forest floor receive diffuse shade light, whose photosynthetically active radiation (PAR) may be less than 1% of the PAR incident above the canopy[13]. The spectral quality of this weak shade light has been altered due to absorption of light by the chlorophylls and carotenoids in the leaf canopy. The shade light is depleted in red and blue wavelengths, and enriched in far-red light (> 700 nm), which is preferentially absorbed by PSI. However, shade light is periodically punctuated by sunflecks, lasting from seconds to tens of minutes, and longer periods of direct sunlight through larger canopy gaps, which although present for only a small fraction of day light, may nevertheless contribute up to 40-70% of the PAR received by understorey plants[3,14].

Since there were some discrepancies in the existing data on the content of PSII on a chlorophyll basis in shade plants, we recently re-examined this point in a wide range of Australian shade-adapted species[15]. As expected, all plants had the characteristic enrichment of chlorophyll b, leading to low Chl a/Chl b ratios (Table 3)[15]. On a leaf area basis, the chlorophyll contents were high, ranging from 440 to 780 μmol Chl m^{-2}. All of the species examined exhibited the usual ultrastructural characteristics of shade plants. Their large chloroplasts had wider granal stacks each containing more thylakoid membranes, with shorter interconnecting stroma thylakoids than found in sun plants.

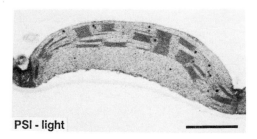
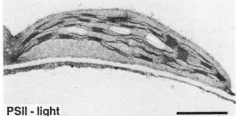

PSI - light   PSII - light

Figure 1. Comparison of ultrastructural changes in thylakoid membrane structure of chloroplasts from thin-sections of leaves of spinach plants grown under PSI-light or PSII-light.

The P700 content of the shade species was rather low, typically 1.1 to 1.54 mmol P700 per mol chlorophyll (Table 3), so that the Chl/P700 ratios varied from about 600 to 900. These results confirm earlier findings that the amounts of P700 on a Chl basis are lower in shade plants than in sun plants. The concentration of functional PSII reaction centres *in vivo* was determined in leaves by the O_2 yield of leaf discs exposed to 1% CO_2 and repetitive flashes at 4 Hz, in the presence of background far-red light to ensure no limitation in P700 turnover[16]. We have shown that this direct and convenient assay gives a valid measure of the number of functional PSII reaction centres in leaf discs, and that in non-stressed leaves, this number agrees with that of DCMU-binding sites in isolated thylakoids[16]. The PSII content on a chlorophyll basis was also rather low in the shade species (Table 3) with Chl/PSII ratios of 444 to 595. The diverse shade-adapted species studied here do not have fixed PSII/PSI reaction centre ratios ascribable to shade-adapted species in general, but rather show distinct differences in their photosystem stoichiometry (Table 3). However, the photosystem stoichiometry of these shade species is rather low, and generally lower than those of sun species (1.8-2.2).

Table 3. Chl a/Chl b ratios and the photosystem reaction centre stoichiometry of some Australian shade-adapted species.

Species	Chl a Chl b	PSII (mmol (mol Chl)$^{-1}$)	P700	PSII PSI
Aristolelia australasica	1.90	2.15	1.27	1.69
Blechnum wattsii	1.70	1.78	1.29	1.38
Cordyline stricta	1.84	1.68	1.11	1.51
Dicksonia antarctica	1.79	1.79	1.51	1.19
Doodia aspera	1.65	2.25	1.34	1.68
Hedycarya angustifolia	2.08	2.14	1.54	1.39
Helmholtzia glaberrima	1.74	1.83	1.14	1.61
Pollia crispata	2.02	2.14	1.14	1.88
Tasmannia purpurascens	1.97	1.88	1.46	1.29

To explore the reason for the variations in the photosystem stoichiometry amongst diverse shade species it is necessary to consider the light environment beneath the canopy. Compared with the exposed canopy, the light received by plants growing on the forest floor is a blend of weak diffuse light enriched in far-red irradiance (absorbed mainly by PSI), and sunflecks or gap sunlight. Thus, it would be expected that both light quantity and light quality would regulate the chlorophyll-protein composition and the photosystem stoichiometry in shade species. In order to balance the input of excitation energy to both photosystems in shade plants, the weak low irradiance will tend to *decrease* the PSII content on a chlorophyll basis (Table 1). However, the concomitant increase in far-red irradiance mainly absorbed by PSI, will tend to *increase* the PSII content on a Chl basis (Table 2). We conclude, therefore, that it is the opposing interplay of both light quantity and light quality that modulates the PSII/PSI reaction centre stoichiometry in species growing in shaded terrestrial habitats. Since the photosystem stoichiometry of shade species is usually lower than that of sun or high-light species it seems that the marked decrease in the light quantity may play a greater role than the altered light quality of shade light in the adjustment of PSII reaction centre content and photosystem stoichiometry of shade plants.

Table 4. Effect of light quantity on the photosynthetic apparatus of *Tradescantia albiflora* grown under varying irradiance. Plants grown on an open bench in the glasshouse represent 100% relative growth irradiance; the irradiance was reduced by varying layers of shade cloth.

Relative growth irradiance level (%)	$\frac{Chl\ a}{Chl\ b}$	PSII	P700 $(mmol\ (mol\ Chl)^{-1})$	$\frac{PSII}{PSI}$
100	2.12	2.24	1.74	1.29
35	2.22	2.27	1.75	1.30
12.5	2.31	2.30	1.67	1.38
4.0	2.29	2.21	1.69	1.31
1.4	2.24	2.18	1.63	1.34

LIGHT ACCLIMATION OF TRADESCANTIA ALBIFLORA: AN EXCEPTION TO THE RULE

Light quantity: *Tradescantia albiflora* is a facultative shade plant which is native to the rainforests of South America. However, it can acclimate to drier, sunnier habitats and flourishes as a weed around Sydney, and on the central coast of New South Wales, Australia. The distribution of chlorophyll between the chlorophyll-proteins of *Tradescantia* grown in varying irradiance is most unusual. Contrary to the well-established pattern of modulation of the Chl a/Chl b ratios with light quantity, with low-light plants having more chlorophyll per chloroplast and lower Chl a/Chl b ratios (2.2-2.4) compared to those of high-light plants (>2.8)[1-4], thylakoids isolated from *Tradescantia* grown under varying light intensities had constant Chl a/Chl b ratios[17] (Table 4). The distribution of chlorophyll amongst the Chl a/b-proteins and Chl a-proteins of both photosystems, as determined by non-denaturing gel electrophoresis, was also constant, consistent with the lack of change in the Chl a/Chl b ratios with growth irradiance. This lack of light quantity acclimation of the composition of the chlorophyll-protein complexes, means that the sizes of the light-harvesting antennae of PSII and PSI are also constant. Shade *Tradescantia* chloroplasts had three times more chlorophyll per chloroplast, and twice the length of thylakoid membranes compared to plants grown in sunlight[18]. As expected from the lateral heterogeneity of distribution of thylakoid complexes in appressed versus nonappressed membrane domains[11], the invariant Chl a/Chl b ratios of *Tradescantia* in response to growth irradiance meant that all chloroplasts had the same amount of appressed membranes relative to nonappressed membranes[18].

Even though the light-harvesting antenna size of each photosystem is fixed at varying growth irradiance, there may have been variations in the stoichiometries of the PSII and PSI reaction centres, since the PSII/PSI ratio in other species is modulated both by light quantity and light quality (Tables 1 and 2). However, P700, determined as the light-induced absorbance change at 703 nm, was remarkably constant with ~1.7 mols of P700 per 1000 Chl, i.e. Chl/P700 ratio of 588 under all growth irradiances[17] (Table 4). Furthermore, the amount of functional PSII reaction centres of *Tradescantia* leaves was also constant with ~2.2 P680 per 1000 Chl, i.e. Chl/P680 ratio of 444 (Table 4). Thus, regardless of the growth irradiance, *Tradescantia* thylakoids had similar PSII/PSI reaction centre ratios of ~1.3, a typical shade plant value.

Thus, *Tradescantia* is atypical with both inflexible photosystem stoichiometry and "shade type" light-harvesting PSII units at all growth irradiances. However, the amount of cytochrome f increases with increasing growth irradiance, as do ATP synthase activity and Rubisco activity,

Table 5. Effect of light quality on the photosystem stoichiometry of isolated thylakoids from *Tradescantia*, grown in red plus far-red (PSI-light) or yellow (PSII-light) light of constant quantity.

Light quality	$\frac{\text{Chl } a}{\text{Chl } b}$	PSII	P700	$\frac{\text{PSII}}{\text{PSI}}$
		(mmol (mol Chl)$^{-1}$)		
Red + far-red, (PSI-light)	1.88	2.88	1.15	2.50
Yellow, (PSII-light)	2.23	2.68	1.57	1.71

showing acclimation of electron transport, photophosphorylation and photosynthetic capacity[17].

Light quality: *Tradescantia albiflora* was also grown under two light quality environments chosen to favour excitation of one photosystem over the other. Yellow light (**PSII-light**) was used to excite preferentially PSII, while red light (**PSI-light**) preferentially excited PSI[9]. Under these different light quality conditions of equal light quantity, there was a change in Chl a/Chl b ratios (Table 5) with thylakoids isolated from plants grown in PSI-light having a very low ratio 1.9, compared to 2.2 for plants grown in PSII-light. Growth in PSI-light led to a marked reduction (27%) in P700 content compared to growth in PSII-light, while there was a slight increase in functional PSII content (Table 5). Thus, in contrast to light quantity, light quality has modulated the photosystem reaction centre stoichiometry for *Tradescantia*. These results confirm earlier findings of marked enrichment of P700 content under PSII-light growth conditions, and an enhancement of PSII content under PSI-light[6,7,9,10]. These acclimative changes help to optimize energy distribution between the two photosystems, and minimise the marked imbalance of electron flow generated by the prevailing absorption of light by a particular photosystem[9].

A MOLECULAR REGULATORY FEEDBACK MECHANISM FOR LIGHT ACCLIMATION OF THE PHOTOCHEMICAL APPARATUS

At present, the regulatory mechanisms responsible for the symphonic orchestration of the multiple signals and responses leading to the regulation of the relative concentrations of PSII and PSI discussed here (Tables 1-5), as well as cytochrome b/f complex, ATP synthase and carbon assimilation enzymes, is unknown. Clearly, long-term changes in the photosystem stoichiometry imply that a feedback regulatory mechanism is functioning that adjusts and optimises the light reactions under varying environmental light climates. In nature, attenuation in light quantity is invariably accompanied by changes in light quality whether in aquatic[8] or terrestrial[13] environments (eg across a leaf and within the canopy of a single tree or forest). Thus, it is vital that this regulatory mechanism is highly conserved throughout the plant kingdom: indeed adjustment of the photosystem stoichiometry is a characteristic feature of cyanophytes[8] through to vascular plants[3,6].

Recently we proposed a regulatory feedback mechanism of light acclimation[19], whereby the environmental light signals responsible for sun/shade acclimation are sensed by the photosystems themselves, rather than by phytochrome. We suggest that the chlorophyll pigments of the light-harvesting antennae and the reaction centres of the photosystems, and not any photomorphogenic receptors, are the primary light sensors. The need for the adjustment of the photosystem stoichiometry could then be identified by the redox state of steady electron transport, which in turn will modulate the ATP and NADPH pools of energy storage metabolites. The response then occurs through gene expression with enhanced energy-dependent transcription/translation, polypeptide translocation across membranes, and assembly of complexes being directed towards those components of the electron transport chain that become predominantly rate-limiting. Concomitantly, there may be disassembly of complexes and degradation of some components.

Our hypothesis of the photosystems themselves being the primary light sensors[19] differs from those of Melis et al.[20,21] and Fujita and Murakami[22] who postulate that the first step in signal identification is an imbalance in the rate of electron transport. While we agree with Melis[20,21] that the common thread for the regulatory mechanism is through the rates of electron and proton flow, which in turn determine the pools of the main photosynthetic metabolites, ATP and NADPH, we suggest[19] that it is the photosystems themselves that act as light sensors of both light quality and light quantity. Following the primary sensing of the light signal, this signal is then reflected as an imbalance in the rate of electron transport between the photosystems, the redox state of the cytochrome b/f complex, and then the pool of ATP and NADPH. In this way, the adjustment and optimisation of the relative concentrations of PSI and PSII would be self-regulatory, provided that events downstream of the primary sensing step are not impaired. Since *Tradescantia* responds to variations in light quality (Table 5) but not quantity (Table 4) during growth, an impairment to the complete, and as yet unknown, signal transmission and transduction responses is likely in *Tradescantia* during growth in light of varied irradiance, but constant quality.

REFERENCES

1. O. Björkman, Responses to different quantum flux densities, in : "Physiological Plant Ecology 1: Responses to the physical environment", E.O. Lange et al., eds. Springer-Verlag, Berlin, 12A: 57-107 (1981).
2. J.M. Anderson, Photoregulation of the composition, function and structure of thylakoid membranes, Ann. Rev. Plant Physiol. 37: 93-136 (1986).
3. J.M. Anderson and C.B. Osmond, Shade-sun responses: compromises between acclimation and photoinhibition, *in*: "Photoinhibition," D.J. Kyle et. al. eds. pp.1-38, Elsevier, Amsterdam (1987).
4. J.M. Anderson, W.S. Chow and D.J. Goodchild, Thylakoid membrane organisation in sun/shade acclimation, Aust. J. Plant Physiol. 15: 11-26 (1987).
5. J.M. Anderson and B. Andersson, The dynamic photosynthetic membrane and regulation of solar energy conversion. Trends Biochem. Sci. 9: 351-355 (1988).
6. A. Melis, Light regulation of photosynthetic membrane structure, organization and function. J. Cell. Biochem. 24: 271-285 (1984).
7. R.E. Glick, S.W. McCauley and A. Melis, Effect of light quality on chloroplast membrane organisation and function in pea. Planta 164: 487-494 (1985).
8. Y. Fujita, K. Ohki and A. Murakami, Chromatic regulation of photosystem composition in the photosynthetic system of red and blue-green algae. Plant Cell Physiol. 26: 1541-1548 (1985).
9. W.S. Chow, A. Melis and J.M. Anderson, Adjustments of photosystem stoichiometry in chloroplasts improve the quantum efficiency of photosynthesis, Proc. Natl. Acad. Sci. USA 87: 7502-7506 (1990).
10. W.S. Chow, C. Miller and J.M. Anderson, Surface charges, the heterogeneous lateral distribution of the two photosystems and thylakoid stacking. Biochim. Biophys. Acta 1057: 69-77 (1991).

11. B. Andersson and J.M. Anderson, Lateral heterogeneity in the distribution of chlorophyll-protein complexes of the thylakoid membranes of spinach chloroplasts, Biochim. Biophys. Acta 593: 427-440 (1980).

12. X.W. Deng, J.C. Tonkyn, G.F. Peter, J.P. Thornber and W. Gruissem, Post-transcriptional control of plastid mRNA accumulation during adaptation of chloroplasts to different light quality environments, The Plant Cell 7: 645-654 (1989).

13. O. Björkman and M.M. Ludlow, Characterization of the light climate on the floor of a Queensland rainforest, Carnegie Inst. Wash. Yearbk. 71: 85-94 (1972).

14. R.L. Chazdon, Sunflecks and their importance to forest understorey plants, Adv. Ecol. Res. 18: 1-63 (1988).

15. W.S. Chow, J.M. Anderson and A. Melis, The photosystem stoichiometry in thylakoids of some Australian shade-adapted plant species. Aust. J. Plant Physiol. 17: 665-674 (1990).

16. W.S. Chow, A.B. Hope and J.M. Anderson, Further studies on quantifying photosystem II *in vivo* by flash-induced oxygen yield from leaf discs, Aust. J. Plant Physiol. 18: in press (1991).

17. W.S. Chow, H.Y. Adamson and J.M. Anderson, Photosynthetic acclimation of *Tradescantia albiflora* to growth irradiance: lack of adjustment of light-harvesting components and its consequences, Physiol. Plant. 81: 175-182 (1991).

18. H.Y. Adamson, W.S. Chow, J. M. Anderson, M. Vesk and M.W. Sutherland, Photosynthetic acclimation of *Tradescantia albiflora* to growth irradiance: morphological, ultrastructural and growth responses, Physiol. Plant. 82: in press (1991).

19. W.S. Chow, D.J. Goodchild, C. Miller and J.M. Anderson, The influence of high levels of brief or prolonged supplementary far-red illumination during growth on the photosynthetic characteristics, composition, and morphology of *Pisum sativum* chloroplasts, Plant Cell Environ. 13: 135-145 (1990).

20. A. Melis, A. Manodori, R.E. Glick, M.L. Ghirardi, S.W. McCauley and P.J.Neale, The mechanism of photosynthetic membrane adaptation to environmental stress conditions: a hypothesis on the role of electron-transport capacity and of ATP/NADPH pool in the regulation of thylakoid membrane organization and function, Physiol. Vég. 23: 757-765 (1985).

21. A. Melis, Adaptation of photosystem stoichiometry in oxygen - evolving thylakoid membranes, in "Current Research in Photosynthesis," M. Baltscheffsky ed. Vol.IV, 291-298, Kluwer Academic Publishers, The Netherlands (1990).

22. A. Murakami and Y. Fujita, Steady state of photosynthetic electron transport in cells of the cyanophyte *Synechocystis* PCC 6714 having different stoichiometry between PSI and PSII: analysis of flash-induced oxidation-reduction of cytochrome f and P700 under steady state of photosynthesis, Plant Cell Physiol. 32: 213-222 (1991).

MODIFICATION OF CHLOROPLAST DEVELOPMENT BY IRRADIANCE

Anastasios Melis

Department of Plant Biology, 411 GPBB
University of California, Berkeley, CA 94720, U.S.A.

INTRODUCTION

Vascular plants and green algae respond to changes in the light environment in which they grown. Acclimation of plants to different irradiance regimes affects chloroplast development as well as the composition, structure and function of the photochemical apparatus [1,28]. The response may involve the *cab* gene expression, affecting both the size and composition of the chlorophyll (Chl) antenna of photosystem I (PSI) and photosystem II (PSII), the PSII/PSI stoichiometry in the thylakoid membrane, and the operation of the PSII repair cycle which is responsible for the degradation and replacement of damaged D1/32 kDa (organelle *psb*A gene product) proteins in chloroplasts [25].

A manifestation of plant responses to irradiance variation was first provided in measurements of the chlorophyll content and of the Chl *a*/Chl *b* ratio among "high-light" and "low-light" grown plants [4,5,20,21,27,30,41]. That light intensity changes during plant growth induce changes in the size and composition of the light harvesting antenna of the photosystems is also evidenced in measurements of the functional light-harvesting Chl antenna size of sun and shade plants [22], of "high-light" and "low-light" algae [20,30,33,37], and of high-light and low-light vascular plant chloroplasts [1,25]. In general, low irradiance promotes a large Chl antenna size for both PSII and PSI (large photosynthetic unit size). High irradiance elicits a smaller Chl antenna size. This adjustment in the Chl antenna size of the photosystems comes about because of changes in the size and composition of the auxiliary Chl *a-b* light-harvesting complex (LHC II and LHC I) of PSII and PSI, respectively [18,19,29]. The response appears to be well conserved in all photosynthetic organisms examined. The regulation of this phenomenon at the molecular and membrane levels is currently unknown.

The photosynthetic performance of a plant can be severely diminished following exposure to light intensities in excess of that required to saturate photosynthesis [32]. The threshold intensity for the onset of this *photoinhibition* can be very low, as in shade-adapted species and in plants where other environmental stresses (e.g., chilling, drought, lack of CO_2, heat stress) have a synergistic effect with irradiance stress [35]. Photoinhibition adversely affects the function of PSII and is manifested as lower rates of electron transport and oxygen evolution, resulting in a lower quantum yield and a lower light-saturated rate of photosynthesis [11,16,31,32,35]. Most investigators now believe that primary damage to PSII by irradiance involves a functional component at the donor side of the complex, probably the photochemical reaction center molecule P680 or the secondary electron donor Tyr(Z) [3,6-9,38].

It was proposed [31] that such damage to PSII occurs even under physiological plant growth conditions, and that it is the underlying reason for the frequent turnover of the D1/32

Regulation of Chloroplast Biogenesis
Edited by J.H. Argyroudi-Akoyunoglou, Plenum Press, New York, 1992

kDa protein of PSII [23,24]. Under optimal growth conditions, the rate of photodamage does not exceed the rate of repair, therefore, no adverse effect on photosynthesis is manifested. Under transient or steady-state irradiance stress, however, the rate of photodamage might exceed the rate of repair in the chloroplast, resulting in accumulation of damaged PSII centers in the thylakoid membrane and causing a decline both in the quantum yield and light saturated rate of photosynthesis [11,31,36]. A chloroplast recovery from "damage" necessitates the replacement of both structural and functional components in the D1/32 kDa reaction center protein, thereby explaining the need for a frequent *in vivo* turnover of this protein [12].

This work presents a biochemical and biophysical analysis of chloroplast development under moderate and high irradiance conditions. Chloroplast development is assessed in terms of size and composition of the auxiliary Chl antenna in PSI and PSII, PSII/PSI ratios in the thylakoid membrane, and function of the PSII repair cycle in the chloroplast. The results revealed degradation products from the LHC II proteins and accumulation of damaged PSII centers in the thylakoid membrane of plants under high irradiance.

MATERIALS AND METHODS

Plant material

Dunaliella salina cultures were grown in an artificial hypersaline medium containing 2.0 M NaCl [34]. Carbon was supplied as $NaHCO_3$ at an initial concentration of 20 mM. Cultures were grown at $30^\circ C$ under incandescent illumination at 200 μmol photon.m^{-2}·s^{-1} (control cells) or at 2,000 μmol photon m^{-2}·s^{-1} (high light: HL-grown cells). Cells were harvested at the end of the logarithmic phase of growth and stored at $-20^\circ C$.

Thylakoid membrane isolation

Working in dim light, cells were suspended in hypotonic buffer containing 50 mM Tricine (pH 7.8), 10 mM NaCl and 5 mM $MgCl_2$ supplemented with 0.5% (w/v) each sodium ascorbate, bovine serum albumin, and polyvinyl pyrrolidone. Cells were broken by passing once through a Yeda press at 13.7 MPa. Following Yeda press treatment, the slurry was centrifuged at 3,000xg for 3 min to remove unbroken cells and large cell fragments. The supernatant was then centrifuged at 50,000xg for 30 min. The thylakoid membrane pellet was resuspended in hypotonic buffer. Chlorophyll concentrations were determined in 80% acetone according to [2]. Thylakoid membrane preparations were kept in the dark at $0^\circ C$ until analysis.

Photosystem Concentrations

PSI (P700) and PSII (Q_A and pheophytin) quantitation measurements were made with a laboratory-constructed split-beam difference spectrophotometer [26]. The concentration of P700 was determined from the amplitude of the light-induced absorbance change at 700 nm (ΔA_{700}) using a differential extinction coefficient of 64 mM^{-1}.cm^{-1} [14]. The reaction mixture contained approximately 100 μM Chl, 0.02% SDS, 250 μM methylviologen and 2.5 mM Na-ascorbate. The concentration of Q_A was determined from the amplitude of the light-induced absorbance change at 320 nm (ΔA_{320}) applying a differential extinction coefficient of 13 mM^{-1}.cm^{-1} [40]. The reaction mixture contained approximately 100 μM Chl, 20 μM DCMU, and 2.0 mM $K_3Fe(CN)_6$. The concentration of the primary electron acceptor pheophytin of PSII was determined from the light-induced absorbance change at 685 nm (ΔA_{685}) [9,15]. A differential extinction coefficient of 65 mM^{-1}.cm^{-1} was applied [9]. The reaction mixture contained approximately 10 μM Chl suspended in 20 mM Tris-HCl (pH 7.8) containing 35 mM NaCl, 2 mM $MgCl_2$, 2 μM methylviologen, 2 μM indigodisulfonate and sufficient sodium dithionite to lower the redox potential to -490 mV.

Functional Chlorophyll Antenna Size

Photosystem II (fluorescence induction) and PSI (P700 photooxidation) kinetic measurements were performed using the above described difference spectrophotometer.

Broad-band actinic excitation of 40 $\mu mol.m^{-2}.s^{-1}$ in the green region of the spectrum was provided by a combination of Corning CS 4-96 and CS 3-68 filters. The rate of light absorption by PSII was determined under light-limiting conditions from the kinetics of the area growth over the fluorescence curve of DCMU-treated membranes. The reaction mixture contained 50 μM Chl and 20 μM DCMU. The rate of light absorption by PSI was determined from the kinetics of the absorbance change at 700 nm. The reaction mixture contained 100 μM Chl, 20 μM DCMU and 200 μM $K_3Fe(CN)_6$. Functional antenna size estimates for PSI, PSI$_\alpha$, PSII β and PSII$_\gamma$ were made from the solution of a system of equations as described elsewhere [26,39].

Table 1. Response of Chloroplasts in *Dunaliella salina* to High Irradiance

	CONTROL $200\,\mu mol.m^{-2}.s^{-1}$	HL-grown $2,000\,\mu mol.m^{-2}.s^{-1}$
Rate of cell growth	$\sim 3\ d^{-1}$	$\sim 2\ d^{-1}$
Chl *a*/Chl *b*	4/1	20/1
Chl/PSI	620/1	1090/1
Chl/PSII (photochemical)	445/1	355/1
PSII/PSI (photochemically competent PSII)	1.4/1	$\sim 3/1$
PSII/PSI (total PSII)	1.4/1	$\sim 12/1$
PSII/PSI (photochemically inert PSII)	0/1	$\sim 9/1$

The rate of cell growth is given as cell divisions per day. Pigment ratios and photosystem quantitations are given (mol:mol basis) for control (200 μmol photons.m$^{-2}.s^{-1}$) and HL (2,000 $\mu mol.m^{-2}.s^{-1}$) grown *D. salina*. Photochemically competent PSII was measured from the light-induced reduction of Q_A and pheophytin. Total PSII content (photochemically competent and damaged centers) was estimated from quantitative SDS-PAGE and immunoblot analysis).

Thylakoid Membrane Polypeptide Analysis

Thylakoid membrane proteins were resolved by SDS-PAGE using the discontinuous buffer system of Laemmli [17] with a 12.5% resolving gel and a 4.5% stacking gel. The samples were solubilized in an equal volume of 200 mM Tris-HCl (pH 6.8) buffer containing 20% glycerol, 7% SDS, 2 M urea and 10% β-mercaptoethanol, and incubated at 50°C for 3 min to disperse the proteins. Electrophoresis on 0.15x14x16 cm slab gels was performed at a constant current of 10 mA for 15 h. Gels were stained with 0.1% Coomassie brilliant blue R for protein visualization.

Immunoblot Analysis

Identification of reaction center polypeptides was accomplished with immunoblot analysis using rabbit polyclonal antibodies (raised in the lab) against proteins of the spinach LHC II and the D1/D2 32/34 kDa heterodimer of the PSII reaction center. Electrophoretic transfer of the SDS-PAGE resolved polypeptides to nitrocellulose, and the subsequent incubations with the above antibodies and with alkaline-phosphate conjugated antibodies were performed as described previously [36]. Cross-reaction was quantitated upon scanning the nitrocellulose strips with an LKB-Pharmacia XL laser densitometer.

RESULTS

The level of irradiance during cell growth brought about significant changes in the development of chloroplasts. Control *Dunaliella salina* cultures had a physiologically green color and, in the logarithmic phase of growth, exhibited approximately 3 cell divisions per day. Cells in HL-grown cultures had a distinct yellowish color and the rate of cell growth in the logarithmic phase was only about 2 cell divisions per day.

Table 1 presents a summary of the photochemical apparatus component quantitation in control and HL-grown *D. salina*. The results show that cells respond to high irradiance in several different ways. Chloroplast development under HL conditions induced changes in the Chl a/Chl b ratio, from about 4/1 in the control to about 20/1 in the HL-grown cells [33,37]. The overall Chl/P700 ratio increased from 620/1 in the control, to 1090/1 at HL intensities. This response of the apparent photosynthetic unit size to HL conditions was unexpected and did not conform with the "model" response of plants to increased light-intensity. The underlying cause of the greater Chl/P700 in the HL-grown cells was investigated in more detail (see below).

The Chl/PSII was measured independently as Chl/Q_A and Chl/pheophytin. It was slightly lower in HL than control thylakoids (355 vs 445, Table 1). The resulting PSII/PSI stoichiometry (based solely on functional measurements) was 1.4/1 for control cells and about 3/1 for HL-grown *D. salina*. To gain further insight into the underlying cause of the changes observed, the functional Chl antenna size of each photosystem was determined (see Materials and Methods).

In control cells, the size of the chlorophyll antenna of PSI was 210 Chl $(a+b)$ molecules. The size of the Chl antenna of PSII$_\alpha$ and PSIIβ was 500 and 130 Chl $(a+b)$ molecules, respectively. In HL-grown cells, the functional Chl antenna size of PSI was estimated to be 105 Chl molecules (Table 2). The dominant form of PSII under HL (PSII$_\gamma$) contained about 60 Chl molecules. The small number of PSIIβ centers which were present in the thylakoid membrane of HL-grown cells (about 5% of the total PSII) contained 130 Chl $(a+b)$ molecules.

Table 2. Relative Concentration and Absolute Chlorophyll Antenna Size of the Photosystems in *Dunaliella salina* Chloroplasts

	CONTROL $200\,\mu mol.m^{-2}.s^{-1}$	HL-grown $2,000\,\mu mol.m^{-2}.s^{-1}$
PSI		
Relative Concentration	1.0	1.0
Chlorophyll Antenna	210	105
PSII$_\alpha$		
mol:mol Ratio with PSI	0.9/1	0/1
Proportion of Total PSII	65%	0%
Chlorophyll Antenna	500	--
PSIIβ		
mol:mol Ratio with PSI	0.5/1	0.15/1
Proportion of Total PSII	35%	5%
Chlorophyll Antenna	130	130
PSII$_\gamma$		
mol:mol Ratio with PSI	0%	2.85/1
Proportion of Total PSII	0%	95%
Chlorophyll Antenna	--	60

The relative concentration of different forms of PSII, (PSII$_\alpha$, PSIIβ and PSII$_\gamma$) in the thylakoid membrane of control and HL-grown cells is given either relative to PSI ([PSI] = 1.0), or as percent of the photochemically competent PSII reaction centers. The absolute number of functional Chl $(a+b)$ antenna molecules associated with PSI, PSII$_\alpha$, PSIIβ and PSII$_\gamma$ is also given. Estimates were based on measurements of light utilization by each photosystem (please see Materials and Methods).

Interestingly, PSII$_\alpha$ centers were totally absent from the thylakoid membrane of HL-grown cells. Thus, in spite of the significantly greater Chl/PSI ratio in HL-grown cells (Table 1), the *functional* Chl antenna size of both PSI and PSII was considerably smaller than that of control cells. This paradox could be explained by the hypothesis that a sizable fraction of Chl in the thylakoid membrane of HL-grown cells was photochemically inert. This phenomenon was investigated in more detail.

Quantitation measurements of P700, Q_A and pheophytin provide information on the concentration of photochemically competent photosystems in the thylakoid membrane (Table 1). However, these light-induced spectrophotometric measurements cannot sense photochemically inert units. The presence of photochemically inert PSII complexes with their associated chlorophyll in the thylakoid membrane would explain the unexpectedly high Chl/PSI and Chl/PSII ratios measured in HL-grown cells (Table 1).

To address the question of the total concentration of PSII centers in the thylakoid membrane, quantitative SDS-PAGE and immunoblot analysis of thylakoid membrane polypeptides was undertaken. Levels of the D1/D2 32/34 kDa reaction center polypeptides and of the Chl *a-b* LHC II apoproteins were compared in HL-grown and control cells. Figure 1 compares the thylakoid membrane polypeptide profile in control (lane 1) and HL-grown (lane 2) *D. salina*. Fig. 2 shows the relative cross-reaction of proteins associated with the Chl *a-b* light-harvesting complex of PSII (LHC II) in control (lane 1) and HL-grown cells (lane 2). The results of Fig. 2 support the earlier observation [13,37] that, under physiological conditions, *D. salina* thylakoids contain large quantities of at least 5 different polypeptides associated with the LHC II. These polypeptides constitute the auxiliary light-harvesting antenna of PSII$_\alpha$ and PSII$_\beta$ in control cells (Table II). High-light grown cells (Fig. 2, lane 2) show considerably fewer proteins associated with the LHC II than those of the control cells (lane 1). The smaller quantity of LHC II polypeptides in HL-grown cells probably originates from the apoproteins associated with the LHC II of PSII$_\beta$ in these cells (Table 2). Alternatively, they may represent apoproteins that bind to the thylakoid membrane but fail to assemble due to lack of chlorophyll. Of interest in this respect is the distinct cross reaction between the polyclonal antibodies and a polypeptide of about 18.5 kDa in lane 2. This novel protein may be a partial proteolysis product of unstable LHC II apoproteins appearing exclusively under HL-growth conditions.

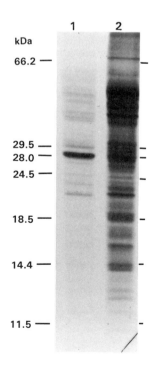

Fig. 1. Coomassie stained profile of polypeptides from *D. salina* thylakoid membranes. Lanes were loaded with 2.6 nmol Chl from control (lane 1) and HL-grown samples (lane 2).

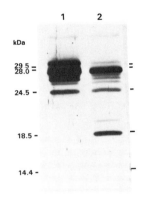

Fig. 2. Immunoblot analysis of *D. salina* thylakoids. Both lanes were probed with antibodies against the Chl *a-b* light-harvesting complex of PSII (LHC II). The significantly lower cross reaction of this antibody with proteins from the HL-grown cells suggested a greatly reduced PSII auxiliary antenna size relative to that of control cells. Note the cross reaction of the antibody with a polypeptide of $M_r = 18.5$ kDa in the HL-grown samples.

Figure 3 shows the immunoblot analysis of the D1/D2 32/34 kDa reaction center polypeptides of PSII. Lane 1 (control) was loaded with 8.4 nmol Chl $(a+b)$ whereas lane 2 (HL) was loaded with 4.2 nmol Chl. On the basis of the results in Table 1, lane 1 (control) contained about 19 pmol of photochemically competent PSII-RC. Lane 2 (HL) contained about 12 pmol of photochemically competent PSII-RC. The proteins on nitrocellulose in Fig. 3 were probed with the combined antibodies against the D1/D2 32/34 kDa PSII reaction center polypeptides. Quantitation of the cross reaction in the two lanes with an LKB-Pharmacia XL laser densitometer (not shown) revealed an estimated 2.5 times greater cross reaction between antibodies and HL-grown sample, suggesting the presence of 2.5 times more PSII-RC protein in lane 2 (HL) than in lane 1 (control). Note that about a third of the total PSII-RC protein in the HL sample migrated at about 160 kDa, indicating the presence of PSII-RC aggregates. These results suggested that HL-grown *D. salina* thylakoids contain an excess of PSII complexes which are photochemically inert and, as such, they can not be detected by the Q_A or pheophytin photoreduction measurements (Table 1). Correcting for the dissimilar loading of PSII-RC in the two lanes of Fig. 3, a ratio of 4:1 total/photochemically competent PSII was estimated in the thylakoid membrane of HL-grown *D. salina*. Thus, about 75% of all PSII centers could not generate a stable charge separation in HL-grown cells. This estimate is consistent with the Chl and PS quantitations (Table 1) which suggested that only a fraction of the total Chl in HL-grown samples was coupled to a photochemically competent reaction center. This unusual phenomenon is probably the result of steady-state photoinhibition of PSII occurring under the HL-conditions (see below).

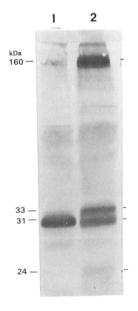

Fig. 3. Immunoblot analysis of *D. salina* thylakoids, probed with antibodies against the D1/D2 32/34 kDa reaction center polypeptides of PSII. Lane 1 (thylakoids from control cells) was loaded with 8.4 nmol Chl (19 pmol of photochemically active Q_A) and lane 2 (HL-grown thylakoids) was loaded with 4.2 nmol Chl (12 pmol Q_A). The much greater cross reaction with proteins in the HL-grown cells suggested the presence of photochemically inert PSII units in these samples.

496

DISCUSSION

In summary, the results show distinct modifications in the development of chloroplasts under high irradiance:

(a) The functional Chl antenna size of PSI was reduced from 210 Chl ($a+b$) molecules in the control to only about 105 Chl a molecules under HL, suggesting the selective lack of LHC I assembly in the HL-samples. This interpretation is consistent with the finding that 95 Chl a molecules constitute the smallest Chl antenna size associated with the PSI-core in the thylakoid membrane of green plants [10].

(b) The light-harvesting antenna size of PSII showed a similar response to high irradiance. In control cells, the dominant form of PSII was PSII$_\alpha$ (Chl antenna size >250 Chl $a+b$ molecules). The smaller antenna size PSII$_\beta$ contained only about 130 $\pm$ 10 Chl ($a+b$) molecules. In HL-grown cells, PSII$_\alpha$ was absent and only a few PSII$_\beta$ centers were present. The dominant form of PSII in these samples contained about 60 Chl molecules suggesting the presence of the PSII-core complex along with a small complement of the auxiliary LHC II. This PSII antenna size is similar to that found in the Chl b-less chlorina f2 mutant of barley [10] and in the chlorophyll b-deficient chlorina 103 [see contribution by Harrison and Melis] which lacked the bulk of the LHC II complex from the thylakoid membrane. Thus, it appears that a PSII unit with 50-60 Chl ($a+b$) molecules can assemble stably in the thylakoid membrane of vascular plants and green algae. We propose to denote this photosystem-II unit as PSII$_\gamma$. A detailed analysis of the polypeptide content and hierarchy of assembly in the auxiliary antenna of PSII$_\gamma$ is given by Harrison and Melis (please see relevant contribution in this volume).

(c) The molecular mechanism for the regulation of assembly in the LHC II is not understood. The results of Fig. 2 suggest that the regulation probably occurs at the level of Chl biosynthesis, evidenced by the fact that LHC II apoproteins are synthesized under high irradiance and probably degraded due to lack of chlorophyll (Fig. 2, 18.5 kDa polypeptide). The down-regulation of Chl biosynthesis at HL intensities is meaningful given that the rate of photosynthesis is limited by the dark reactions of carbon fixation and/or by carbon dioxide availability to the cell. Under such conditions, the biosynthesis/assembly of a large Chl antenna size for PSI and PSII would be both unnecessary and metabolically expensive. A unregulated biosynthesis/assembly of LHC may even be counterproductive for the plant since a large antenna size under HL would result in increased non-photochemical dissipation of excitation at the reaction centers, leading to further photoinhibition and/or other photodamage [7,35]. Therefore, a light-intensity dependent attenuation of biosynthesis/assembly of the LHC allows the plant to conserve significant amounts of metabolic energy for other cellular functions, and to minimize adverse effects due to the non-photochemical dissipation of excitation.

(d) The substantially different PSII/PSI stoichiometry ratio and the large fraction of photochemically inert PSII complexes in HL-grown cells are probably interrelated phenomena. Under HL-conditions, the rate of photosynthesis is limited by carbon fixation and/or carbon dioxide availability to the cells. The high-irradiance condition apparently causes a steady-state photoinhibition of the PSII reaction centers and results in the accumulation of damaged PSII units in the thylakoid membrane. These units must contain a non-functional D1/32 kDa protein with a damaged photochemical reaction center and are unable to perform a stable charge separation reaction [9,23,36]. The reason they accumulate in the thylakoid membrane is attributed both to the accelerated rate of damage under HL and to the apparent inability of the PSII repair mechanism [12] to keep up with the greatly enhanced cellular demand for reaction center repair.

REFERENCES

1   Anderson JM (1986) Annu. Rev. Plant Physiol. 37, 93-136
2   Arnon DI (1949) Plant Physiol. 24, 1-15
3   Arntz B, Trebst A (1986) FEBS Lett. 194, 43-49

4   Bjorkman O, Boardman NK, Anderson JM, Thorne SW, Goodchild DJ, Pyliotis NA (1972) Carnegie Inst. Washington Yearbook, 71, 115-135

5   Boardman NK (1977) Annu. Rev. Plant Physiol. 28, 355-377

6   Callahan FE, Becker DW, Cheniae GM (1986) Plant Physiol. 82, 261-269

7   Cleland RE, Melis A (1987) Plant, Cell and Environment 10, 747-752

8   Cleland RE, Neale PJ, Melis, A (1986) Photosynth. Res. 9, 79-88

9   Demeter S, Neale PJ, Melis A (1987) FEBS Lett. 214, 370-374.

10 Glick RE, Melis A (1988) Biochim. Biophys. Acta 934, 151-155

11 Greer DM, Berry JA, Bjorkman O (1986) Planta 168, 253-260

12 Guenther JE, Melis A (1990) Photosynth. Res. 23, 105-109

13 Guenther JE, Nemson JA , Melis A (1988) Biochim. Biophys. Acta 934, 108-117

14 Hiyama T, Ke B (1972) Biochim. Biophys. Acta 267, 160-171

15 Klimov V, Klevanik A, Shuvalov V, Krasnovsky A (1977) FEBS Lett. 82, 183-186

16 Kyle DJ, Ohad I, Arntzen CJ (1984) Proc. Natl. Acad. Sci. USA 81, 4070-4074

17 Laemmli U (1970) Nature 227, 680-685

18 Larsson UK, Anderson JM , Andersson B (1987) Biochim. Biophys. Acta 894, 69-75

19 Leong TA, Anderson JM (1984) Photosynth. Res. 5, 105-115

20 Ley AC, Mauzerall DC (1982) Biochim. Biophys. Acta 680, 95-106

21 Lichtenthaler HK, Kuhn G, Prenzel U, Buschmann C, Meier A (1982) Z. Naturforsch. 73, 464-474

22 Malkin S, Fork DC (1981) Plant Physiol. 67, 580-583

23 Mattoo AK, Edelman M (1987) Proc. Natl. Acad. Sci. USA 84, 1497-1501

24 Mattoo AK, Pick U, Hoffman FH, Edelman M (1981) Proc. Natl. Acad. Sci. USA 78, 1572-1576

25 Melis A (1991) Biochim. Biophys. Acta in press

26 Melis A, Anderson JM (1983) Biochim. Biophys. Acta 724, 473-484

27 Melis A, Harvey GW (1981) Biochim. Biophys. Acta 637, 138-145

28 Melis A, Manodori A, Glick RE, Ghirardi ML, McCauley SW, Neale PJ (1985) Physiol. Veg. 23, 757-765

29 Morrissey PJ, Glick RE, Melis A (1989) Plant Cell Physiol. 30, 335-344

30 Neale PJ, Melis A (1986) J. Phycol. 22, 531-538

31 Ohad I, Kyle DJ, Arntzen CJ (1984) J. Cell Biol. 99, 481-485

32 Osmond, CB (1981) Biochim. Biophys. Acta 639, 77-78

33 Pick U, Gounaris K, Barber J (1987) Plant Physiol. 85, 194-198

34 Pick U, Karni L, Avron M (1986) Plant Physiol. 81, 92-96

35 Powles SB (1984) Annu. Rev. Plant Physiol. 35, 15-44

36 Smith BM, Morrissey PJ, Guenther JE, Nemson JA, Harrison MA, Allen JF, Melis A (1990) Plant Physiol. 93, 1433-1440

37 Sukenik A, Bennett J, Falkowski P (1988) Biochim. Biophys. Acta 932, 206-215

38 Theg SM, Filar LJ, Dilley RA (1986) Biochim. Biophys. Acta 849, 104-111

39 Thielen APGM, Van Gorkom HJ (1981) Biochim. Biophys. Acta 635, 111-120

40 Van Gorkom HJ (1974) Biochim. Biophys. Acta 347, 439-442

41 Wilhelm C, Wild A (1984) J. Plant Physiol. 115, 125-135

REGULATION OF THE PHOTOSYNTHETIC ADAPTATION IN *SCENEDᴇSMUS*

OBLIQUUS DEPENDING ON BLUE AND RED LIGHT

D. Hermsmeier, E. Mala, R. Schulz, J. Thielmann,
P. Galland and H. Senger

FB Biologie/Botanik, Philipps-Universität Marburg
Lahnberge
D-3550 Marburg (Germany)

Introduction

The unicellular green alga *Scenedesmus obliquus* adapts to different irradiances and wavelengths of light by altering the molecular organization of the photosynthetic apparatus. Regulation occurs on the level of pigment-accumulation, assembly of pigment-protein complexes and gene-expression. Action spectroscopy revealed the involvement of at least two photoreceptor-systems regulating adaptation antagonistically with markedly different thresholds (Thielmann *et al.*, 1991; Thielmann and Galland, 1991). One of them was a BL-(flavin-type-) receptor, which mediates weak-light adaptation, resulting in increased Chl, LHC II (Humbeck *et al.*, 1988; Senger and Bauer, 1987) and accumulation of *cab*-mRNA (Hermsmeier *et al.*, 1991); the other one was a VL/RL-receptor with action peaks at 404 and 650 nm, inducing high-irradiance responses indicated by decreased amounts of Chl, LHC II, LHCP and *cab*-mRNA (Hermsmeier *et al.*, 1991).

Results

Autotrophic stock cultures of *Scenedesmus obliquus* wild-type strain D3 (Gaffron 1939) and the Chl *b*-less mutant WT-LHC₁ (Bishop and Öquist 1980) were grown homocontinuously under WL (20 W m^{-2}) in liquid culture medium (Bishop and Senger 1971). They were then adapted 20 - 24 h to monochromatic light obtained with interference filters (SCHOTT, Mainz, FRG). For adaptation under high-irradiance conditions (0.1 to 100 μmol m^{-2} s^{-1}) cultures were maintained autotrophically. To provide energy supply at low fluence rates (10^{-6} to 10^{-1} μmol m^{-2} s^{-1}), ranging below the photosynthetic compensation point, cells were transferred into media containing 0.5 % (w/v) glucose. All data were normalized to values obtained from dark-grown cultures, ensuring that only light-depending effects were evaluated.

Abbreviations: BL, blue light; *cab*-mRNA, transcripts coding for the chlorophyll *a/b*-binding proteins; *Cab*-proteins, chlorophyll *a/b*-binding proteins; CC II, core complex of photosystem II; Chl, chlorophyll; LHC II, light-harvesting complex of photosystem II; LHCP, light-harvesting chlorophyll *a/b*-binding protein; PCV, packed cell volume; RL, red light; VL, violet light; WL, white light.

Regulation of Chloroplast Biogenesis
Edited by J.H. Argyroudi-Akoyunoglou, Plenum Press, New York, 1992

To characterize the photoreceptors regulating adaptation, fluence rate-response curves for accumulation of Chl were measured, providing thresholds and saturations of this reaction depending on irradiance and wavelength. Chl / PCV was determined in an irradiance-range from 10^{-6} to 10^2 μmol m^{-2} s^{-1} for a number of wavelengths between 380 and 750 nm using threshold-boxes. The thresholds of these reactions were 0.5 μmol m^{-2} s^{-1} for BL and RL under autotrophic conditions. Heterotrophically grown cultures exhibited thresholds of $4 \cdot 10^{-3}$ (VL), $2 \cdot 10^{-2}$ (BL) and $6 \cdot 10^{-4}$ (RL) μmol m^{-2} s^{-1}. Proceeding from these data, action spectra for 2 μmol m^{-2} s^{-1} (autotrophic conditions; fig. 1, closed circles) and $4 \cdot 10^{-3}$ μmol m^{-2} s^{-1} (heterotrophic cultures; fig. 1, open circles) were calculated. As clearly shown by the action spectra, Chl-accumulation was regulated antagonistically by a BL- and a VL/RL-photoreceptor. While BL raised accumulation of Chl by 50 %, VL and RL decreased the amount of Chl by 35 % and 48 % referred to the dark-control.

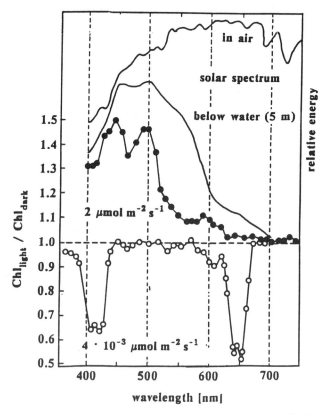

Fig. 1. Action spectra for Chl-accumulation under high (2 μmol m^{-2} s^{-1}, closed circles) and low ($4 \cdot 10^{-3}$ μmol m^{-2} s^{-1}, open circles) fluence rates. Data were normalized to the dark-control (set to 1.0). For comparison spectra of the relative energy distribution of solar light in air and below water (5 m depth) are figured.

While it has been pointed out that high-irradiance (9.4 μmol m^{-2} s^{-1}) BL induced weak-light adaptation in *Scenedesmus*, e.g. characterized by an increase in Chl b, xanthophyll / carotene, LHC II and LHC II / CC II, RL was shown to cause the reverse (Bishop *et al.*, 1989; Humbeck *et al.*, 1988; Senger and Bauer, 1987). In respect to these results it was of interest, whether strong-light adaptation was controlled by the novel VL/RL-receptor found under very low irradiance conditions. Therefore the pigment-composition in wild-type strain D3 (WTD3) and the Chl b-less mutant WT-LHC$_1$ was analysed by HPLC according to a method of Humbeck *et al.* (1989). VL- and RL-

$(4 \cdot 10^{-2} \, \mu\text{mol m}^{-2}\text{s}^{-1})$ adapted wild-type and mutant cells indicated large decreases (by 18 to 95 % of the dark-control) in Chl and carotenoids, at which Chl a/b was increased and xanthophyll / carotene diminished in WTD3. In WT-LHC$_1$, however, xanthophyll / carotene was elevated to 156 %, probably due to a loroxanthin-cycle, strongly expressed in this mutant. Overall changes in pigment-composition represented high-irradiance adaptation resting on VL- and RL-illumination. Furthermore investigation of pigment-protein complexes by lithium dodecylsulfate polyacrylamide gel electrophoresis (LDS-PAGE) according to Humbeck *et al.* (1989) demonstrated, that the loss in Chl, especially Chl b, and xanthophylls was paralleled with a decrease in LHC II in wild-type cells. The ratio LHC II / CC II was diminished to 62 % of the dark-control. In the mutant lacking Chl b and LHC II the core complex of photosystem I (CC I) was enriched in Chl relatively. Again the redistribution of Chl from the peripheral light-harvesting antenna (LHC II) to the core complexes of photosystem I and II was a typical feature of strong-light adaptation, in this case evoked by RL of only $4 \cdot 10^{-2} \, \mu\text{mol m}^{-2}\text{s}^{-1}$. That the BL-photoreceptor mediates weak-light and the VL/RL-receptor strong-light adaptation, might be due to the natural environment of the algae: the ratio BL / RL is raised with increasing depth of water (shade conditions) and vice versa (fig. 1).

To see whether the decline in detectable LHC II was paralleled by a decrease in the corresponding apoprotein (LHCP), a Western blot of total protein extracts was prepared according to a method of Towbin *et al.* (1979). Results are shown in fig. 2.

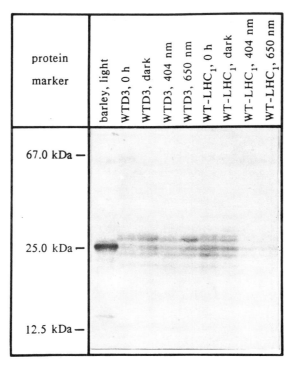

Fig. 2. Western blot of *Cab*-proteins obtained from the wild-type strain WTD3 and the mutant WT-LHC$_1$. Total protein of cells grown autotrophically under WL (20 W m^{-2}), 20 h heterotrophically in darkness and under monochromatic light (404, 650 nm; $4 \cdot 10^{-2} \, \mu\text{mol m}^{-2}\text{s}^{-1}$) respectively, was blotted on nitro cellulose membrane. Cab-proteins were immuno-stained with rabbit anti-*Cab* antibodies raised against barley-LHCP, kindly provided by K. Apel (Zürich). For reference protein from light-grown barley was blotted. The figure was taken from Hermsmeier *et al.* (1991).

Polyclonal antibodies raised against the mature 25 kDa-LHCP of light-grown barley were prepared by K. Apel (1979) and kindly placed at our disposal. After immunostaining with alkaline phosphatase three LHCP-bands with 24, 26 and 30 kDa could be detected. In comparison LHCP of the unicellular green alga *Dunaliella tertiolata* have Mr of 24.5, 28.5, 30 and 31 kDa (Mortain-Bertrand *et al.* 1990) and those of the three major apoproteins of *Chlamydomonas* were 25 to 33.5 kDa, depending on the applied PAGE (Bassi and Wollman, 1991). Compared to the dark-control, in wild-type cells only a slight decrease in LHCP could be detected after VL- and RL-adaptation. Mutant cells, lacking the green LHC II-band in LDS-gels, nevertheless contained *Cab*-proteins. VL and RL, however, inhibited accumulation of LHCP strongly, probably due to a higher turnover-rate in respect to the wild-type (fig. 2).

Photocontrol of *cab*-gene expression was investigated for VL, BL and RL, according to the absorption peaks of the photoreceptors described above. First a Northern blot (fig. 3) of total RNA, extracted from heterotrophically grown wild-type and mutant cells, was prepared according to Sambrook *et al.* (1989). Cultures were adapted 20 h to darkness, VL or RL ($4 \cdot 10^{-2}$ μmol m^{-2} s^{-1}). Transcripts were hybridized to a *cab*-gene probe from *Chlamydomonas*, cloned by Imbault *et al.* (1988) and kindly provided by U. Johanningmeier (Bochum). While the dark-control contained *cab*-mRNA to a great extent, accumulation of transcripts was strongly inhibited in VL- and RL-irradiated cultures (fig. 3). Again this effect was most pronounced in the mutant WT-LHC$_1$.

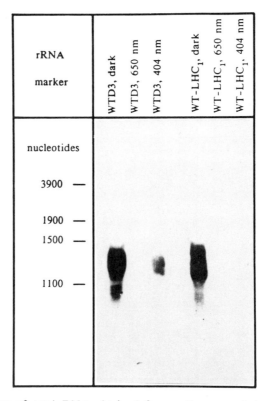

Fig. 3. Northern blot of total RNA obtained from cells grown heterotrophically during 20 h in darkness and at $4 \cdot 10^{-2}$ μmol m^{-2} s^{-1} of VL (404 nm) or RL (650 nm). *Cab*-mRNA was hybridized with a ^{32}P-labelled *cab*-gene probe from *Chlamydomonas* kindly provided by U. Johanningmeier (Bochum). The figure was taken from Hermsmeier *et al.* (1991).

502

Second the action of WL, BL and RL on *cab*-mRNA accumulation in heterotrophic and autotrophic cultures of WTD3 was investigated by Dot-blot analysis (fig. 4). In the case of heterotrophic nutrition WL-, BL- and RL-adapted cells accumulated less *cab*-mRNA compared to cells grown 20 h in darkness. BL of only $4 \cdot 10^{-2}\ \mu mol\ m^{-2}\ s^{-1}$ $(10^{-2}\ W\ m^{-2})$ had approximatly the same effect as high irradiance WL $(20\ W\ m^{-2})$, while *cab*-transcripts of RL-irradiated cultures $(4 \cdot 10^{-2}\ \mu mol\ m^{-2}\ s^{-1}; 7.4 \cdot 10^{-3}\ W\ m^{-2})$ were reduced to 45 %. The high level of *cab*-mRNA in the heterotrophic dark-culture may be due to the fact, that these cells resemble weak-WL- $(5\ W\ m^{-2})$ rather than strong-WL- $(20\ W\ m^{-2})$ adapted cells (for details see also Thielmann *et al.* (1991) and Thielmann and Galland (1991)). Under autotrophic conditions the dark-control accumulated less *cab*-mRNA than WL-cultures. BL-irradiated cells, however, indicated an increase in transcripts, while RL inhibited accumulation to an extent of 40 %.

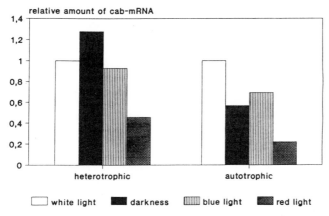

Fig. 4. Accumulation of *cab*-mRNA in heterotrophically or autotrophically grown WTD3 determined by Dot blot hybridization. On the ordinate relative amounts of *cab*-mRNA are indicated. Cells were pregrown under WL $(20\ W\ m^{-2})$ and subsequently exposed 20 h to BL (461 nm) or RL (650 nm) of $4 \cdot 10^{-2}\ \mu mol\ m^{-2}\ s^{-1}$ or maintained in darkness. Poly(A)$^{+}$RNA was extracted and hybridized with the same *cab*-gene probe as in fig. 3. The activity of each dot was quantified by liquid scintillation counting.

Regarding the nature of the VL/RL-photoreceptor, preliminary experiments provide evidence that phytochrome genes are present in the *Scenedesmus* genome. Further investigations will have to prove, if these genes are expressed and, in the case that they are, whether the VL/RL-effect depends on the action of their gene products.

Acknowledgements

This project was supported by the Deutsche Forschungsgemeinschaft (SFB 305, B1). We wish to thank Prof. Klaus Apel, who kindly provided the polyclonal antibodies against barley LHCP and Dr. Udo Johanningmeier for making the *Chlamydomonas cab*-gene probe available. We also thank Mrs. Kerstin Bölte for skillful technical assistance.

References

Apel, K., 1979, Phytochrome induced appearance of mRNA activity for the apoprotein of the light harvesting chlorophyll *a/b* protein of barley (*Hordeum vulgare*), Eur. J. Biochem, 97:183.

Bassi, R. and Wollman, F.-A., 1991, The chlorophyll-a/b proteins of photosystem II in *Chlamydomonas reinhardtii*, Planta, 183:423.

Bishop,. N. I. and Senger, H., 1971, Preparation and photosynthetic properties of synchronous cultures of *Scenedesmus*, in: "Methods in Enzymology, Vol. XXIII: Photosynthesis", San Pietro, A., ed., Academic Press, London.

Bishop, N. I. and Öquist, G., 1980, Correlation of the photosystem I and II reaction center chlorophyll-protein complexes CPa_I and CPa_{II} with photosystem activity and low temperature fluorescence emission properties in mutants of *Scenedesmus*, Physiol. Plant., 49:477.

Bishop, N. I., Humbeck, K., Römer, S. and Senger, H., 1989, The mode of adaptation of the photosynthetic apparatus of a pigment mutant of *Scenedesmus* without light harvesting complex to different light intensities, J. Plant Physiol., 135:144.

Gaffron, H., 1939, Über auffallende Unterschiede in der Physiologie nahe verwandter Algenstämme, nebst Bemerkungen über "Lichtatmung", Biol. Zentralblatt, 59:302.

Hermsmeier, D., Mala, E., Schulz, R., Thielmann, J., Galland, P. and Senger, H., 1991, Antagonistic blue- and red-light regulation of *cab*-gene expression during photosynthetic adaptation in *Scenedesmus obliquus*, J. Photochem. Photobiol., submitted.

Humbeck, K., Hoffmann, B. and Senger, H., 1988, Influence of energy flux and quality of light on the molecular organization of the photosynthetic apparatus in *Scenedesmus*, Planta, 173:205.

Humbeck, K., Römer, S. and Senger, H., 1989, Evidence for an essential role of carotenoids in the assembly of an active photosystem II, Planta, 179:242.

Imbault, P., Wittemer, C., Johanningmeier, U., Jacobs, J. D. and Howell, S. H., 1988, Structure of the *Chlamydomonas reinhardtii cab* II-1 gene encoding a chlorophyll-*a/b*-binding protein, Gene, 73:397.

Mortain-Bertrand, A., Bennett, J. and Falkowski, P. G., 1990, Photoregulation of the light-harvesting chlorophyll protein complex associated with photosystem II in *Dunaliella tertiolata*. Evidence that apoprotein abundance but not stability requires chlorophyll synthesis, Plant Physiol., 94:304.

Sambrook, J., Fritsch, E. F., and Maniatis, T., 1989, "Molecular cloning. A laboratory manual", Cold Spring Harbor Laboratory Press, Cold Spring Harbor, New York.

Senger, H. and Bauer, B., 1987, The influence of light quality on adaptation and function of the photosynthetic apparatus, Photochem. Photobiol., 45:939.

Thielmann, J., Galland, P. and Senger, H., 1991, Action spectra for photosynthetic adaptation in *Scenedesmus obliquus*. I. Chlorophyll biosynthesis under autotrophic growth, Planta, 183:334.

Thielmann, J. and Galland, P., 1991, Action spectra for photosynthetic adaptation in *Scenedesmus obliquus*. II. Chlorophyll biosynthesis and cell growth under heterotrophic conditions, Planta, 183:340.

Towbin, H., Staehelin, T. and Gordon, J., 1979, Electrophoretic transfer of proteins from polyacrylamide gels to nitrocellulose sheets: Procedure and some applications, Proc. Natl. Acad. Sci. USA, 76:4350.

LIGHT QUALITY AND IRRADIANCE LEVEL INTERACTION IN CONTROL OF CHLOROPLAST

DEVELOPMENT.

K. Eskins

Bioactive Constituents Research
USDA, ARS, Agricultural Research Service, National Center
  for Agricultural Utilization Research
1815 North University Street
Peoria, Illinois 61604 USA

The processes of greening and plant development are now reasonably
understood from a descriptive point of view. That is, some whole plant
and chloroplast developmental responses and their regulatory signals and
cues are documented. It is known that photoreceptors initiate and
continually regulate the developmental process and that the interaction
of different photoreceptors with one another depends on the quality and
irradiance level of light and on the stage of tissue development.

Development involves both whole plant and chloroplast responses to
environmental cues and available energy. However, these interdependent
systems may respond in quite different ways to the same light environment. That is, certain whole plant responses, such as flowering in
Arabidopsis may be more quickly triggered by blue light than by red
light. The mode of this blue light action is via blue light establishment of a phytochrome photoequilibrium favorable to flowering, but
significant variation in the time of flowering in red light can be
caused by small amounts of blue in the red light which do not significantly alter the phytochrome photoequilibrium (Eskins, manuscript in
preparation). At the same time, however, blue light is turning on the
expression of the chloroplast gene (psbA) which codes for the photosystem II reaction center protein. In this case, blue light operates
through the blue-light receptor. Once triggered by blue light or by
red/far-red light, the flowering response cannot be reversed by a switch
to red light; but red light will decrease the expression of the psbA
gene even in mature tissue. Thus, some gene expression is fixed by a
duration of environmental signalling.

Since UV, blue, red, and far-red are the major wavelengths absorbed
by plants, it is reasonable that these wavelengths correspond to sensor
systems which have evolved in response to each portion of the spectra as
well as to the ratios of red to far-red and the ratio of blue to
red/far-red. A distinct receptor responds to high energy UV-B light,
phytochrome responds to red, far-red and blue/UV-A light, and cryptochrome responds to UV-A/blue light. The photoreceptor, protochlorophyllide responds to both UV/blue and red light but not to far-red
light. It is turned on by light and turned off by dark. Studies of
phototropism (Loser and Schafer, 1986) have also suggested the presence
of additional blue photoreceptors and a blue/red receptor which is not

Regulation of Chloroplast Biogenesis
Edited by J.H. Argyroudi-Akoyunoglou, Plenum Press, New York, 1992

phytochrome. Therefore, it is difficult to delineate the action of separate receptors during growth in red, blue or white light (Elmlinger and Mohr, 1991).

We will consider some studies of chloroplast development in red, blue and white light which help clarify the interaction of photoreceptors. We will especially consider those different responses of the chloroplast to red and blue light which do not have a clear connection to phytochrome. We will consider these as possible cases of cryptochrome regulation, of developmental activation of phytochrome and of photoregulation by a blue/red receptor whose response to blue or red light is dependent on the irradiance level and stage of development.

Initially, greening of etiolated tissue involves the photoconversion of phytochrome (P_r to P_{fr}) and protochlorophyllide (PChlide to Chlide). This early switching is an on/off signal and results in conversion of the photoreceptor to an alternate form which is used up or degraded. In organized tissue from higher plants, photoconversion of only these two photoreceptors is necessary for greening to occur, and red light alone is sufficient. In tissue cultures and root cultures, however, greening will not occur unless a sustained blue or white light is given. The ability of red light to continue the blue-light initiated development depends on the period of pretreatment with blue light (degree of plastid development) and on the source of the callus tissue. For instance mRNA for the small subunit of Rubisco (SSU) is induced in tobacco cultures by 24 h of blue light (Richter and Wessel, 1985). Five additional days of blue light cause a fourfold increase in the message. Five days of red light, however, cause a small decrease in the message. If blue light is given for 8 days, an additional 7 days of blue light increase message by 50%, but an additional 8 days of red light also increase message by 10 to 15%. A change from negative to positive response to red light is activated only by long-term treatment with blue light. If soybean callus derived from root, hypocotyl and cotyledon are greened in white light, chl b and LHCP2 levels are highest in cotyledon tissue and lowest in root tissue (Eskins and Smith, Private Communication). If cultures are transferred to blue light, chl b increases to a higher but equal level in all callus. If transferred to equal irradiance red light, cotyledon callus decreases in chl b, but root callus increases chl b to levels produced in blue light. Activation of the red-light receptor by blue light is also tissue dependent. Our most recent work deals with the very early stages of maize development and seeks to see if signals delivered to the embryo during radicle emergence can influence the formation of chloroplast. Germinating seeds were treated with a short presignal (6 to 12 h) of blue or red light, left in the dark for 12 h, then developed for an additional 2 to 4 days under blue or red light/dark cycles (14 h light/10 h dark). Seeds which were given a red or blue presignal accumulated the same amounts of chl a and chl b when developed under red light but under blue light, blue presignaled seedlings were more advanced than red presignaled seedlings. At these earlier stages of development (3-4 days after radical emergence), continuous blue light was more effective for chloroplast development than red light.

Additional experiments measured the effects of continuous (3-4 days) red or blue light/dark cycles given during radical and mesocotyl emergence. Seedlings were then planted and irradiated by an additional 7 days of red/dark or blue/dark cycles. These seedlings were compared to planted seedlings which experienced dark during germination and emerged from the soil into red/dark or blue/dark cycles. These experiments show two results. First, under longer term developmental conditions (11 days compared to 4 days), red light is more effective than blue light for chloroplast development. Second, a short red treatment during germination and a longer treatment during mesocotyl emergence is equivalent to a dark treatment, but a short blue treatment during germination is stimulatory and a longer blue light treatment during mesocotyl emergence is repressive compared to dark (Eskins and Smith, manuscript submitted). Recent results with far-red light also show this stimulation by a far-red presignal and repression by continuous far-red (Oelmuller and Schuster, 1987). How does phytochrome interaction explain these results?

In whole plant systems, growth under long-term red or blue light gives plants which are characteristic of the light quality. That is, both red and blue light support development, but they are not equally effective. In a typical example, chloroplast from red grown plants accumulate more chl a, chl b, light-harvesting pigment proteins, and grana stacks than chloroplast from blue grown plants (Buschmann et al., 1978) Blue grown plants have chloroplast whose thylakoids contain more reactions centers, have higher Hill activity and require higher fluence rates to saturate photosynthesis. In these plants blue light gives similar types of chloroplast to those formed under high-irradiance white light, while red light gives similar chloroplast to those formed under low-irradiance white light.

Unicellular algae also respond to red and blue light by constructing distinct types of chloroplast (Humbeck et al., 1988). In these systems, however, the response is almost a direct mirror image of the response in higher plants. In algae, blue light produces higher levels of chl b, LHCP2 and grana and is similar to low-intensity white light and red light is similar to high-intensity white light. A similar mirror image response to blue and red light is found in plants adapted to high fluence and plants adapted to low fluence (Leong et al., 1985). In addition, the response to blue or red within a single species depends on the irradiance level and the degree of development (Milivojevic and Eskins, 1991). The differences in response to blue and red light found in unicellular algae and in higher plants may be due to the fact that blue light receptor is the primary photoreceptor in algae and phytochrome is the primary receptor in higher plants. Thus, in undifferentiated or undeveloped tissue and under very low light irradiance, higher plant chloroplasts exhibit the primitive response of their evolutionary progenitors, the blue-green algae. In mature or developed tissue or under high light irradiance, the phytochrome response is dominant.

Work with intermittent blue and red light provided some additional clues to the interaction of quality and irradiance level (Akoyunoglou et al., 1980). Blue light/dark cycles (LDC) repressed chl a formation compared to red or white LDC's. The formation of chl b was repressed in all qualities of intermittent light, but blue LDC's enhanced chl b compared to red or white LDC's. There was little difference in chloroplast thylakoid pigment-proteins between red and blue LDC grown plants, and both red and blue grown were depleted in light-harvesting proteins and enhanced in reaction center proteins. Interesting, white LDC's which contained both red and blue light acted the same as red LDC's. When the intermittent light grown chloroplast were transferred to continuous light (CL) of the same quality, more pigments and pigment-proteins were formed and blue light repressed the formation of both chl a and chl b. Also, however, blue CL repressed the formation of light-harvesting proteins, and here, white light acted like blue light. Thus LDC's represent a low fluence response or an early development stage, and continuous light a high fluence response or a more mature developmental stage.

Our work with maize chloroplast development adds some additional information. First, the differences between blue and red may not always be due to establishment of different ratios of P_{fr}/P_{tot} by blue and red light. This work (Eskins et al., 1986, Eskins and McCarthy, 1987) and the earlier work of Akoyunoglou (Akoyunoglou et al., 1980) showed that chloroplast development was determined by the irradiance level of red light and was independent of the ratio P_{fr}/P_{tot}. This may, however be connected in some manner to the red light very low fluence response described by Kaufman et al. (1985) which is not reversible by far-red light. Secondly, in maize, blue and red light may give similar or different responses dependent on the irradiance. A comparison of the relative effectiveness of 1-30 μmol m^{-2} s^{-1} of red and blue light on the equilibrium levels of LHCP2 protein and mRNA in maize leaves (Eskins et al., 1989) showed a maximum response of red to blue at 8 μmol m^{-2}s^{-1}. These results should be compared to those of Warpeha and Kaufman, (1990) which showed a similar bell-shaped curve for blue light stimulation of LHCP2 mRNA in pea. It is especially interesting that the bell shape response curve is dependent on growth of the plants in low red light and that a short pretreatment of dark grown plants with red light to establish a phytochrome photoequilibrium is ineffective. Thirdly, at irradiance levels where blue and red give different responses, mixtures of blue and red which have quite different P_{fr}/P_{tot} ratios all act like blue light (Eskins and Beremand, 1990). That is, a small fluence of blue light in the presence of a larger fluence of red light acts like pure blue light. Fourthly, in pine and soybeans, very low levels of blue light are stimulatory to chl b and LHCP2 and red is repressive. At higher irradiance levels which also correspond to more advanced stages of development, blue light represses and red enhances chl b and LHCP2 (Milivojevic and Eskins, 1991).

The connection between light quality, irradiance level, stage of development and chloroplast, and nuclear genomes is indeed complicated. Present evidence suggests that a blue light receptor, phytochrome, and perhaps an additional receptor which measures blue and red light as a function of irradiance operate during chloroplast development.

REFERENCES

Akoyunoglou, G., and Anni, H., Kalosakas, K., 1980, The effect of light quality and the mode of illumination on chloroplast development in etiolated bean leaves, in: "The Blue Light Syndrome," pp. 473-484, H. Senger, ed., Springer-Veerlag, Heidelberg.

Buschmann, C., Meier, D., Kleudgen, H. K., and Lichtenthaler, H. K., 1978, Regulation of chloroplast development by red and blue light, Photochem. Photobiol., 27:195-198.

Elmlinger, M. W., and Mohr, H., 1991, Coaction of blue/ultraviolet-A light and light absorbed by phytochrome in controlling the appearance of ferredoxin-dependent glutamate synthase in the Scots pine (Pinus sylvestris L.) seedlings, Planta, 111183:374-380.

Eskins, K., McCarthy, S. A., Dybas, L., and Duysen, M., 1986, Corn chloroplast development in low fluence red light and in low fluence red light plus far-red light, Physiol. Plant., 67:242-246.

Eskins, K., and McCarthy, S. A., 1987, Blue, red and blue plus red light control of chloroplast pigments and pigment-proteins in corn mesophyll cell; irradiance level-quality interaction, Physiol. Plant., 71:100-104.

Eskins, K., Westhoff, P., and Beremand, P., 1989, Light quality and irradiance level interaction in the control of expression of light-harvesting complex of photosystem II. Pigments, pigment-proteins and mRNA accumulation, Plant Physiol., 91:163-169.

Eskins, K., and Beremand, P., 1990, Light quality and irradiance level control of light-harvesting complex of photosystem 2 in maize mesophyll cells. Evidence for a low-fluence rate threshold in blue-light reduction of mRNA and proteins, Physiol. Plant., 78:435-440.

Humbeck, K., Hoffmann, B., and Senger, H., 1988, Influence of energy flux and quality of light on the molecular organization of the photosynthetic apparatus in Scenedesmus, Planta, 173:205-212.

Kaufman, L. S., Briggs, W. R., and Thompson, W. F., 1985, Phytochrome control of specific RNA levels in developing pea buds: the presence of both very low and low fluence response, Plant Physiol., 75:388-393.

Leong, T-Y., Goodchild, D.J, and Anderson, J. M., 1985, Effects of light quality on the composition, function and structure of photosynthetic thylakoid membranes of Asplenium australasicum (Sm.) Hook, Plant Physiol., 78:561-567.

Loser, G., and Schafer, E., 1986, Are there several photoreceptors involved in phototropism of Phycomyces blakesleeanus? Kinetic studies of dichromatic irradiation, Photochem. Photobiol., 43(2):195-204.

Milivojevic, D., and Eskins, K., 1991, Effects of irradiance-quality (blue, red) interaction on the synthesis of pigments and pigment-proteins in maize and black pine mesophyll chloroplast, Physiol. Plant., 80:624-628.

Oelmuller, R., and Schuster, C., 1987, Inhibition and promotion of light of the accumulation of translatable mRNA of the light-harvesting chlorophyll a/b-binding proteins of photosystem II, Planta, 172:60-70.

Richter, G., and Wessel, K., 1985, Red light inhibits blue light-induced chloroplast development in cultured plant cells at the mRNA level, Plant Mol. Biol., 5:175-182.

Warpeha, K. M. F., and Kaufman, L. S., 1990, Two distinct blue-light responses regulate the levels of transcripts of specific nuclear-coded genes in pea, Planta, 182:553-558.

REORGANIZATION OF THYLAKOID MEMBRANE LATERAL

HETEROGENEITY FOLLOWING STATE I - STATE II TRANSITION

Roberto Bassi and Paola Dainese

Dipartimento di Biologia, Universita di Padova,
via Trieste 75-35100 Padova, Italy

INTRODUCTION

The thylakoid membrane of higher plant chloroplast is
organized into two main compartments: the grana and the stroma
exposed membranes. This ultrastructural evidence corresponds
to a compartmentation of the light harvesting and the electron
transport components which are each localized in the granal or
in the stroma exposed membranes. Thus Photosystem I (PSI) has
been located in stromal membranes together with its light
harvesting complex (LHCI) and the ATPase complex, while PSII
and its major antenna LHCII has been shown to be present in
the partition of grana stacks (1). The last major thylakoid
complex, the cytochrome b6/f (cyt b6/f) has been found to be
present in both grana and stroma membranes (2), so that it has
been proposed that it is mainly placed in the frets, membranes
connecting stroma lamellae to grana (3).

While this general pattern is generally accepted, the
location of particular chloroplast membrane components can be
changed in response to specific regulation mechanisms. Thus,
PSII Reaction Centre (PSII RC) can be transferred to stromal
membranes following heat treatment of the thylakoids (4), and
there is general agreement that at least a portion of the
LHCII complex can migrate from grana to stroma membranes
during the phosphorylation mediated state I - state II
transitions.

When leaves are put into darkness or exposed to light
which preferentially excites PSI, they undergo a transition to
state I in which the energy absorbed by LHCII is preferentially transferred to PSII. This state is characterized by
high fluorescence yield. Under conditions in which light is
preferentially absorbed by PSII, a transition to a low
fluorescence state II is induced over a time scale of several
minutes allowing more light to be absorbed by PSI (5). These
changes are induced by the phosphorylation of LHCII (6)
mediated by a thylakoid bound Mg^{++}-dependent kinase (7,8)
whose activity is regulated by the redox state of the
plastoquinone pool which in turn reflects the excitation
energy distribution between PSII and PSI. The addition of

negatively charged phosphate groups to the LHCII complex
causes it to dissociate from PSII and favours its migration
out of the appressed regions of the grana into the stroma
lamellae where PSI centres are located (9).

In this study we have examined the changes in the
composition of grana and stroma membranes isolated following
state I to state II transitions. Previously we have demonstrated that only particular subpopulations of LHCII migrate
from grana to stroma membranes (10,11); this LHCII, when in
the grana, belongs to a supramolecular antenna complex which
is connected to PSII RC through the minor chlorophyll-protein
CP29 (12). Now we show that:

i)   once in the stroma membranes, phospho LHCII transfers its
      phosphate groups to a non-LHCII protein which, as a
      consequence of phosphorylation, reversibly binds to
      PSI-LHCI complex.

ii)   the cyt b6/f complex migrates from grana to stroma
      membranes following state I-state II transition.

These results are discussed in terms of reorganization of
the photosynthetic apparatus from linear to cyclic electron
flow.

MATERIALS AND METHODS

Maize plants were grown as previously described (10).
Phosphorylation in vivo was as in (10), in vitro as in (10).
Isolation of PSII membranes by detergent treatment and of
stroma membranes by Yeda press was as in (10). Stroma and
grana membranes were also isolated by digitonin (13). Sucrose
gradient centrifugation, immunoblot, preparative SDS-PAGE,
electroelution and SDS-PAGE were as in (12). Non-Denaturing
IEF was as in (14).

RESULTS

Changes in the composition of stroma membranes

We have previously shown that is possible to isolate
stroma membranes by a non-detergent method from thylakoids
adapted to state I and state II (10,11). In fig. 1A we show
the polypeptide composition of such a preparation. The chl
a/b ratio was 6.2 in the stroma from state II chloroplasts and
8.4 in the sample from state I chloroplasts, while both
preparations were depleted in PSII RC components. Immunoblotting of these two preparations clearly showed that both LHCII
and cyt b6/f complex are strongly increased in state II stroma
membranes while the concentration of the thylakoid components
such as PSII RC, PSI RC, LHCI, CP29, CP26 remained unchanged.
It should also be noted that LHCII transferred to stroma in
state II completely lacks its 26 kDa component which is only
present in grana membranes. We have isolated this polypeptide
by preparative SDS-PAGE and electroelution. N-terminal
sequence showed that the first 12 aminoacids, including two
threonines which can be phosphorylated, lacks in this
polypeptide (Fig. 2).

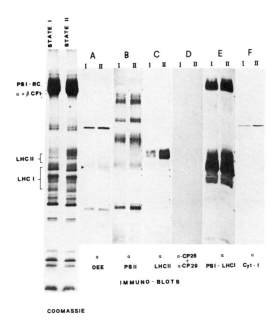

Fig. 1.   Immunoblot of isolated stroma membranes from state I
          and state II thylakoids with antibodies against the
          main thylakoid complexes.

LHCII cons.   NH$_2$-MRKTATKAKPVSSGSPWYGPDRVKYLGPFSGES.....
              - -   :::::::::::::::
LHCII 26kDa   NH$_2$-GNDLWYGPDRVKYLGPFSAQT.....

Fig. 2.   Comparison of the N-terminal sequence of the 26 kDa
          LHCII polypeptide with a consensus sequence for type
          I LHCII proteins.

The fate of phospho-LHCII in stroma membranes.

     When thylakoids were phosphorylated in the presence of
^{32}P-ATP, and then let to dephosphorylate in the dark, one
particular SDS-PAGE band (28.5 kDa) increased its labeling
with respect to the other bands in the 28.5-30 kDa range where
LHCII polypeptides migrated. Fractionation of the thylakoids
into their grana and stroma membrane components showed that
LHCII in state II stroma was about 50 times more labeled than
LHCII remaining in the grana on a protein basis. The 28.5 kDa
phosphoprotein was only present in stroma membranes although
the great majority (>90%) of LHCII was located in the grana
fraction (Fig. 4).

     The addition of NaF, although strongly decreasing the
over all dephosphorylation, did not change this pattern and
after 5' of dark the 28.5 kDa polypeptide was the major
phosphoprotein in stroma membranes (not shown), thus showing
that this polypeptide received phosphate groups from the
phospho LHCII once migrated in the stroma membranes. In fact
it was not possible to phosphorylate the 28.5 kDa polypeptide
in stroma membranes isolated from state I and state II
membranes (Fig. 4).

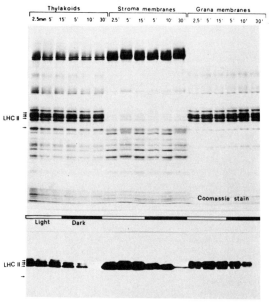

LHC II

Coomassie stain

Light   Dark

LHC II

AUTORADIOGRAPHY

Fig. 3.   Phosphorylation pattern of thylakoid membranes during illumination and subsequent incubation in the dark. At different times the membranes were fractionated in their grana and stroma domains by the digitonin method (13) and submitted to SDS-PAGE (top) and subsequent autoradiography (bottom).

Isolation and characterization of the 28.5 kDa polypeptide

Fractionation of stroma membranes by sucrose gradient ultracentrifugation showed that the 28.5 kDa phosphoprotein migrated with the PSI-LHCI complex and that it was bound to this complex only in the case of state II membranes (Fig. 5). The polypeptide was then purified from the PSI-LHCI complex by preparative SDS-PAGE and analyzed by aminoacid composition, identification of the phospho-aminoacid and immunoblot. It was shown that, although phosphorylated in a threonine as LHCII, its aminoacid composition is significantly different from that of LHCII and it is not recognized by a polyclonal antibody directed to LHCII.

The purified 28.5 kDa polypeptide was also used to raise antibodies in rabbit. Immunoblot showed that this phosphoprotein is only present in stroma membranes both in state I and state II conditions.

DISCUSSION

Heterogeneity of LHCII phosphorylation

LHCII has long been thought to be an homogeneous protein containing one or two polypeptides. Only recently a higher complexity in its composition and function has been recognized

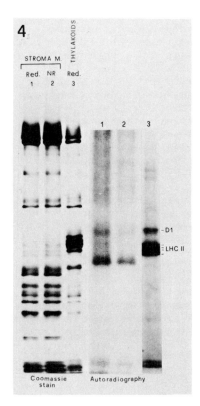

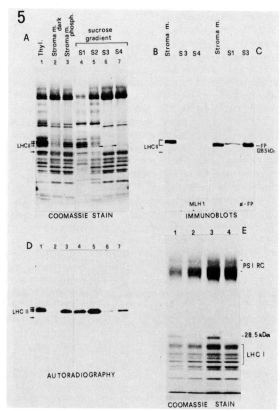

Fig.4. In vitro phosphorylation of stroma membranes isolated from state I or state II thylakoids by Yeda press treatment. The samples were first incubated in the dark to allow dephosphorylation of LHCII and then NaF and ^{32}P-ATP were added with and without 10 mM dithionite and light. Neither LHCII nor the 28 kDa phosphoprotein were labeled in these conditions.

Fig.5. Isolation of the 28.5 kDa phosphoprotein. A) S1-S4 green fractions from ultracentrifugation of solubilized state II stroma membranes. Arrows show the 28kDa phosphoprotein bound to PSI-LHCI. B) Immunoblot of gradient fractions with anti LHCII and phosphoprotein antibodies. C) autoradiography of the gel in panel (A). D) PSI-LHCI complexes purified from state II (1 and 3) and state I (2 and 4) stroma membranes. State transition was obtained in vivo (1 and 2) or in vitro (3 and 4).

at both protein and DNA level (for a review see ref. 15). With respect to its role in phosphorylation mediated state transition, heterogeneity in the degree of phosphorylation of different LHCII polypeptides has been described (16,17). In previous studies we have shown that the polypeptide composition of the mobile LHCII is different with respect to that of the non-mobile fraction (10) while only two LHCII isoforms out of six isolated from maize thylakoids are found in state II stroma membranes (11). We have previously shown that the mobile LHCII has a particular location within the PSII antenna system in state I, i.e. it belongs to a supramolecular complex composed of CP29, CP24 and LHCII which bind 30% of the PSII chlorophyll (12,18). The selectivity of LHCII phosphorylation and migration raises the question of which determinants are responsible for it. Certainly the finding that the only LHCII polypeptide which is not phosphorylated and totally confined to grana membranes lacks the N-terminal sequence which is phosphorylated in other LHCII polypeptides suggest that primary sequence determinants are important. However, the great majority of LHCII stays in the grana and is little or not phosphorylated (Fig. 2, ref. 11). Nevertheless tightly bound and mobile LHCII both contain polypeptides in the 28 to 30 kDa range which are recognized by the LHCII antibody directed to the trypsin cleavable, phosphorylatable peptide (11, 19). The recent finding that the LHCII kinase is bound to the cyt b6/f complex (20) in grana membranes (8) suggest that the degree of phosphorylation of the different LHCII subpopulations may depend on the relative organization of the cyt b6/f and PSII complexes in grana partitions. The requirement of a particular conformation for LHCII is further suggested from that isolated state II stroma membranes are unable to phosphorylate LHCII although the kinase is present together with cytb6/f as shown by the presence of a 60 kDa phosphoprotein attributed to the autophosphorylated kinase (8,20).

Driving force for LHCII and Cytochrome b6/f migration

It has been proposed that the electrostatic repulsion between the phosphate groups in the PSII RC and tightly bound LHCII on one hand and those in the mobile LHCII on the other side is the driving force for LHCII migration (6,9). Our results are consistent with this proposal since the mobile LHCII in stroma membranes carry 50 times more ^{32}P, on a protein basis, with respect to the tightly bound LHCII. The case of the cyt b6/f complex is somehow different since there is no report of its phosphorylation when in the thylakoid membrane (however some phosphorylation has been described in the isolated cyt b6/f-kinase complex by Gal et al. (20). It has been suggested that the cyt b6/f is transferred as a complex with phospho-LHCII (21) and/or that an increased affinity for PSI RC, induced by changes in the redox state, may explain its higher concentration in state II stroma membranes. We propose that a complex exists, containing the cyt b6/f complex, the kinase and the mobile LHCII, which is transferred from grana to stroma after phosphorylation. Thus the driving force for cyt b6/f migration would be provided by the highly phosphorylated LHCII. This complex is proposed to be connected to the PSI-LHCI complex in state II stroma membrane through the 28.5 kDa phosphoprotein which we have shown to receive phosphate groups from the mobile LHCII thus becoming tightly bound to PSI-LHCI complex.

Physiological role of state transitions

LHCII phosphorylation mediated state transitions has been mainly understood as a short term chromatic adaptation the aim of which is to optimize linear electron flow through the two photosystems in light conditions which would otherwise decrease the quantum yield of photosynthesis (6). In environmental conditions such as those of shade plants or inner surface of leaves, which receive far red enriched light, a mechanism able to increase the antenna size of the unfavored photosystem may be important. An alternative hypothesis has been put forward by several authors (22,23) who failed to detect an increase in PSI antenna size in state II conditions and therefore favoured the idea that protection of PSII from excess light was the primary aim of state transitions. A third hypothesis has been developed by the group of Horton (24). They suggested that cellular variations in ATP consumption contributed to state transitions with state I corresponding to oxygenic linear electron flow responsible for CO_2 fixation whereas state II would be promoted in conditions of low ATP concentrations and elicit cyclic electron flow. Thus state transitions would be controlled by the interplay of the redox poise and ATP concentrations (25). While the PSII photoprotection hypothesis has been challenged by the report of an increased PSI antenna size in state II (10,26), the finding that not only LHCII but also Cyt b6/f complex is transferred from grana to stroma membranes following state I-state II transitions (Fig. 1, ref. 21), strongly support the hypothesis that state transitions correspond to switching between cyclic and non-cyclic electron flow. Two sets of data are consistent with this view:

i) It has been recently shown that the rapid diffusion of plastoquinones which transfer electrons between PSII and cyt b6/f complex is limited to small domains containing less than 8 PSII centres (27,28). Therefore, linear electron flow should be sustained by plastocyanine diffusing in the lumenal space from its binding site on cyt b6/f complexes in the stacked regions to PSI in the stroma membranes (29). In state II conditions, when up to 80% of the cyt b6/f complex is located in unstacked membranes (21), cyclic electron flow should be strongly favored.

ii) It has been shown that the ATP requirement of the photosynthetic cell is effective in the control of state transitions (30).

Supercomplexes associating cyt bc type with cytochrome oxidase and with photochemical reaction centres have been shown in photosynthetic bacteria (35) favoring the efficiency of the electron transfer while experimental conditions close to state II produced an increase in the rate of cyt b6/f oxidation by PSI (R. Delosme, cited in ref. 21). We have obtained preliminary results that the Cyt b6/f complex is connected to PSI-LHCI in state II while it is not in state I. Therefore our results support the view that state I-state II transitions is a mechanism aimed to favor cyclic electron flow and ATP production.

ACKNOWLEDGMENTS

We like to thank Dr. F.A. Wollman (I.B.P.C., Paris) and O. Vallon (I. Monod, Paris) for helpful discussions and for the kind gift of the antibody against cytb6/f complex. Dr. G. Frank (ETH, Zurich) and Dr. D. Dalzoppo (CRIBI, Padova) are thanked for N-terminal sequencing.

REFERENCES

1. Vallon, O., Wollman, F.A. and Olive, J.,1986, Lateral distribution of the main protein complexes of the photosynthetic apparatus in Chlamydomonas reinhardtii and spinach. An immunocytochemical study using intact thylakoid membranes and a PSII enriched membrane preparation. Photobiochem. Photobiophys. 12, 203-220.
2. Olive, J., Vallon, O., Wollman, F.A., Recouvreur, M. and P. Bennoun, 1986, Localization of the Cytochrome b6/f in the thylakoid membranes. Biochim. Biophys. Acta 851, 239-248.
3. Melis, A., Svensson, P. and Albertsson, P.A., 1986, The domain organization of the chloroplast thylakoid membrane. Localization of the photosystem I and of the cytochrome b6/f complex. Biochim. Biophys. Acta 850, 402-412.
4. Sundby, C. and Andersson, B., 1985, Temperature induced reversible migration along the thylakoid membrane of photosystem II regulates its association with LHCII. FEBS Lett. 191, 24-28.
5. Bonaventura, C. and Myers, J., 1969, Fluorescence and oxygen evolution from Chlorella pyrenoidosa. Biochim. Biophys. Acta 189, 366-383.
6. Allen, J.F., Bennett, J., Steinback, K.E. and Arntzen, C.J., 1981, Chloroplast protein phosphorylation couples plastoquinone redox changes to distribution of excitation energy between the two photosystems. Nature 291, 25-29.
7. Bennett, J., 1979, Chloroplast phosphoproteins. Eur. J. Biochem. 99, 133-137.
8. Coughlan, S.J., and Hindt, G., 1986, Purification and characterization of a membrane bound protein Kinase from Spinach Thylakoids. J. Biol. Chem. 261, 11378-11385.
9. Anderson, B., Akerlund, H.E., Jergil, B. and Larsson, C., 1982, Differential phosphorylation of the light harvesting chlorophyll-protein complex in appressed and non-appressed regions of the thylakoid membrane. FEBS Lett. 149, 181-185.
10. Bassi, R., Giacometti, G.M. and Simpson, D.J., 1988, Changes in the organization of stroma membranes induced by State I-State 2 transtions. Biochim. Biophys. Acta 935, 152-165.
11. Bassi, R., Rigoni, F., Barbato, R. and Giacometti, G.M., 1988, Light Harvesting Chlorophyll a/b-proteins in phosphorylated membranes. Biochim. Biophys. Acta 936, 29-38.
12. Dainese, P., and Bassi, R., 1991, Stoichiometry of the chloroplast Photosystem II antenna system and aggregation state of the component Chl a/b proteins. J. Biol. Chem. 266, 8136-8142.
13. Leto, J.L., E. Bell, McIntosh, L., 1985, Nuclear mutation leads to an accelerated turnover of chloroplast encoded 48kDa and 34.5kDa polypeptides in thylakoids lacking photosystem II. EMBO J. 4, 1645-1653.
14. Dainese, P., Hoyer-Hansen, G. and Bassi, R., 1990, The resolution of Chlorophyll a/b binding proteins by a preparative method based on flat bed isoelectric focusing. Photochemistry and Photobiology 51, 693-703.

15. Bassi, R., Rigoni, F. and Giacometti, G.M., 1990, Chlorophyll binding proteins with antenna functions in higher plants and green algae. Photochemistry and Photobiology 51, 1187-1206.
16. Larsson, U.K. and Andersson, B., 1985, Different degree of phosphorylation and lateral mobility of two polypeptides belonging to the light harvesting complex of the photosystem II. Biochim. Biophys. Acta 809, 396-402.
17. Islam, K., 1987, The rate and extent of phosphorylation of the two light harvesting a/b binding protein complex (LHCII) polypeptides in isolated spinach thylakoids.
18. Bassi, R. and Dainese, P., 1991, A supramolecular light harvesting complex from the chloroplast photosystem II membranes (submitted).
19. Darr, S.C., Somerville, S.C. and Arntzen, C.J., 1986, Monoclonal antibodies to the light harvesting chlorophyll a/b protein complex of photosystem II. J. Cell Biol. 103, 733-740.
20. Gal, A., Hauska, G., Herrmann, R. and Ohad, I., 1990, Interaction between Light-Harvesting Chlorophyll a/b protein (LHCII) kinase and cytochrome b6/f complex. J. Biol. Chem. 265, 19742-19749.
21. Vallon, O., Bulte, L., Dainese, P., Olive, J., Bassi, R. and Wollman, F.A., 1991, Lateral redistribution of cytochrome b6/f complexes along thylakoid membranes upon state transitions. Proc. Natl. Acad. Sci. Usa in the press.
22. Hawort, P. and Melis, A., 1983, Phosphorylation of thylakoid membrane proteins does not increase the absorption cross-section of PSI. FEBS Lett. 231, 95-98.
23. Larsson, U.K., Ogren, E., Oquist, G. and Andersson, B., 1986, Electron transport and fluorescence studies on the functional interaction between phospho-LHC and PSI in isolated stroma lamellae vesicles. Photobiochem. Photobiophys. 13, 29-40.
24. Horton, P., 1985, Interaction between electron transfer and carbon assimilation. In: "Photosynthetic mechanisms and the environment", Barber, J. and Baker, N. Eds. pp. 135-187.
25. Fernyhough, P., Foyer, C.H. and Horton, P., 1983, The influence of metabolic state on the level of phosphorylation of the light-harvesting chlorophyll-protein complex in chloroplasts isolated from maize mesophyll. Biochim. Biophys. Acta 725, 155-161.
26. Forti, G. and Vianelli, A., 1988, Influence of thylakoid protein phosphorylation on Photosystem I photochemistry. FEBS Lett. 231, 95-98.
27. Joliot, P., Lavergne, J. and Beal, D., 1990, Organization of the plastoquinone pool in chloroplasts: evidence for clusters of different sizes. In: "Current research in photosynthesis", M. Baltsfcheffsky, ed. II, pp. 879-882.
28. Lavergne, J. and Joliot, P., 1991, Restricted diffusion in photosynthetic membranes. TIBS 16, 129-134.
29. Haenel, W., Ratajczac, R. and Robenek, H., 1989, Lateral distribution of plastocyanin in chloroplast thylakoids. J. Cell Biol. 108, 1397-1405.

30. Bulte, L., Gans, P., Rebeille, F. and Wollman, F.A., 1990, ATPcontrol on state transitions in vivo in Chlamydomonas reinhardtii. <u>Biochem. Biophys. Acta</u> 1020, 72-80.

31. Joliot, P., Vermeglio, A. and Joliot, A., 1989, Evidence for supercomplexes between reaction centers, cytochrome C2 and Cytochrome bc1 complex in Rhodobacter sphaeroides whole chells. <u>Biochem. Biophys. Acta</u> 975, 336-345.

LHCII KINASE ACTIVATION DURING GREENING CORRELATES WITH THE

ASSEMBLY OF CYTOCHROME b_6/f COMPLEX AND GRANA FORMATION

Yoram Soroka, Alma Gal and Itzhak Ohad

Department of Biological Chemistry,
The Hebrew University, Jerusalem 91904, Israel

Introduction

LHCII kinase activity was reported to be regulated by its association with the cytochrome b_6/f complex in vivo and in vitro in mature plants and green algae [1-5]. The goal of this work was to find out whether this relationship exists also during the greening process and thylakoid membrane formation. The assembly of the cytochrome b_6/f complex is assumed to be a gradual process consisting of synthesis of its subunits both in the cytoplasm (Rieske Fe-S) and in the chloroplast (cyt f, b_6, subunit IV) and their proper insertion in the thylakoid membrane. The natural substrates, LHCII polypeptides, are also synthesized, processed and inserted into the membrane in a gradual process concomitant with the greening and formation of the grana stacks. However there was no available information regarding the appearance of LHCII kinase during the greening. Previous data indicated that LHCII phosphorylation during the greening of Chlamydomonas reinhardtii y-1 mutant occurs 5-8 hours after the onset of the greening [6]. This is also the time at which grana stacks containing more than 2 paired thylakoids can be observed. Thus it was of interest to find out whether the rise in LHCII kinase activity was due to the gradual appearance of the enzyme or to the functional assembly of the cytochrome b_6/f complex accompanied by the structural organization of the grana stacks. The results presented in this work, based on greening systems in Chlamydomonas y-1 mutant as well as of dark grown spinach, support the concept that cytochrome b_6/f mediates the LHCII kinase activation and indicate that the assembly of the cytochrome complex is related to the activation of the pre-existing enzyme to phosphorylate the LHCII polypeptides in a redox dependent manner.

Materials and Methods

Growth of Chlamydomonas reinhardtii y-1 cells as well as their degreening and greening protocol were carried out as previously described[7]. At the desired time points cells were harvested by centrifugation (3000xg, 2 min.) and stored on ice or frozen directly in liquid N_2 until used for thylakoid membrane preparation. Spinach seedlings were grown in the dark for 14 days and then exposed to cool light illumination. After 12, 24, 36, 48 and 72 hours samples were taken and thylakoid membranes were prepared .

Phosphorylation of LHCII polypeptides was carried out in the light ($300\mu E^2m^2s^{-1}$) or in the dark with or without addition of 1 mM duroquinol. The assay system contained membrane sample (0.2 mg/ml Chl normalized per gr wet weight for spinach or equal amount of membranes normalized per cell number for Chlamydomonas), 0.2 mM $^{32}P-\gamma-ATP$ (300-500 cpm/pmol),50 mM Tris-Cl pH 8.0, 10 mM $MgCl_2$ and 5 mM NaF. After 15 min at 25 c° the assay was stopped by rapid centrifugation and the membrane pellets were solubilized by sample buffer and analyzed by SDS-PAGE followed by autoradiography. Phosphorylation of histone S-III (0.25 mg/ml) or P1 (0.3 mg/ml, a dodecapeptide analog of the N terminus of LHCII polypeptide, the generous gift of Dr. J.Bennett) was carried out as previously described[5,8]. Western blot analysis was done using $^{125}I_2$ protein A as second detector[9].

Results and Discussion

The sequential appearance of the cytochrome b_6/f complex components during the greening process of Chlamydomonas reinhardtii cells is shown in figure 1.

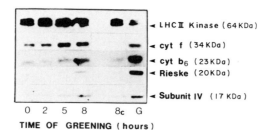

TIME OF GREENING (hours)

Figure 1: **Appearance of cytocrome b_6/f complex components and of LHCII kinase (64 kDa) during greening in Chlamydomonas thylakoids.** Thylakoid membranes were prepared from Chlamydomonas cells at different stages of chloroplast developement. SDS-PAGE was performed on samples normalized to an equal cell number. Western immunoblotting was done with the respective antibodies raised against spinach polypeptides. (8c - 8 hours of greening in presence of Chloramphenicol, 200 µg/ml).

From these data, it can be seen that cytochromes f and b_6 are clearly apparent from the onset of greening while subunit IV and the Rieske Fe-S protein are gradually accumulating. Both cytochromes exist already in their mature form as can be judged by heme staining (data not shown). Addition of CAP during the greening inhibited the increase in the amount of all the cytochrome complex components including the nuclear encoded Rieske Fe-S protein (fig.1,CAP). This may indicate the involvment of another chloroplast encoded protein which may affect specifically the integration of the Rieske protein. Absence of this subunit was demonstarted to affect the assembly of the entire complex[10].

While all the cytochrome b6/f components are gradually synthesized
and incorporated into the membrane at different rates it can be seen that
LHCII kinase (identified as a 64 kDa polypeptide [11] is present throughout
the greening process. The level of the polypeptide was not affected by
addition of CAP during the greening thus indicating that it is a nuclear
encoded protein. Similar results were obtained also in greening spinach
thylakoids (data not shown), although in the latter case the appearance of
cytochrome b6 is much slower.

It was previously demonstrated in greening _Chlamydomonas_ y-1 cells
that LHCII polypeptides were phosphorylated 5-8 hours after the onset of
greening[6]. In the present work we followed the appearance of the redox
dependent phosphorylation in isolated membranes and its relation to the
accumulation of the cytochrome b6/f complex. Results of such an experiment
are shown in Fig.2.

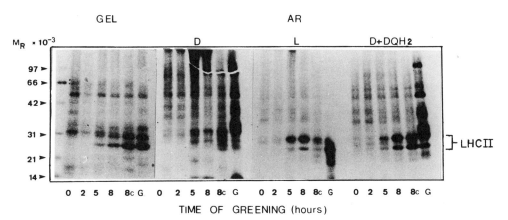

Figure 2. __In-vitro__ phosphorylation pattern of LHCII polypeptides in greening
thylakoids of __Chlamydomonas.__
During the greening process cell samples were withdrawn at the
specified time points and thylakoid membranes were prepared.The
membranes (normalised to the cell number) were incubated with
$^{32}P-\gamma-ATP$ as described in materials and methods.The
phosphorylated membranes were separated by SDS-PAGE (G) followed
by exposure to X-ray film (Autoradiogrem-AR).8c, 8 hours of
greening with CAP.

The LHCII polypeptides of 25kDa and 29kDa can be detected by
coomasie staining 5 hours after of onset of greening and they become the
major thylakoid polypeptides after 8 hours when the chlorophyll level/cell
is about 40-60% of the control cells grown in continuous light[6].In-vitro
phosphorylation of LHCII ploypeptides could be detected already 5 hours
after the onset of the greening process (Fig.2). The activity could be
demonstrated in light incubated thylakoids or in thylakoids incubated in the
dark with addition of duroquinol. The phosphorylation of the LHCII
polypetides in the dark was significantly lower than that in the light or
presence of reduced duroquinone. These results indicate that the
phosphorylation of the LHCII polypeptides occurs as soon as the substarte is
available and it is redox controled from its onset.

As mentioned above (Fig. 1) the presence of a polypeptide reacting
with antikinase antibodies raised against the spinach polypeptide could be
detected in membranes obtained from the dark grown cells. Similar results
were obtained with thylakoids obtained from spinach leaves. However the
greening process in the latter case is considerably slower. The LHCII
polypeptides could be clearly detected after 24 hours of greening and their
phosphorylation could be demonstrated in vitro only after 48 hours of
greening process which was completed after 72 hours. As in the
Chlamydomonas case, the LHCII phosphorylation in spinach thylakoids was
redox controled from the onset of its activity (data not shown).

The above results could be explained either by assuming that the
kinase present in the membranes of the dark grown cells is inactive or that
it is active but the substarte, the LHCII polypeptides is not available
before a certain time of the greening process. To distinguish between these
possibilities, thylakoids obtained at different hours of the greening
process of spinach seedlings were incubated with Histone III-S or the P1
polypeptide which was shown before[a] to serve as a specific substrate for
the redox-controled kinase. The results of these experiments demonstrated
that the total kinase activity using histone III-S as a substrate was
constant from the onset of the greening process (Fig.3) and was not redox
controled (data not shown).

The P-1 dodecapeptide was also phosphorylated by greening spinach
thylakoids. The light stimulated P1 phosphorylation was significant in
isolated thylakoids only at the end of the greening process of spinach
seedlings . While in dark incubated thylakoids there was no significant
change in P1 phosphorylation during the greening, addition of duroquinol
stimulated phosphorylation activity throughout greening process (Fig.4).
This duroquinol stimulated activity could be due to the presence of a quinol

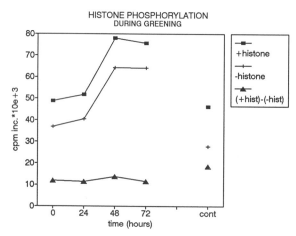

Figure 3. Histone phosphorylation by spinach thylakoids during greening.
Spinach thylakoids (0.2 mg/ml Chl normalised to wet weight)
prepared at various times from the onset of the greening were
incubated with histone S-III (0.25 mg/ml) as an external
substrate as described. For evaluation of the endogenous
phosphorylation activity of the samples thylakoids were incubated
under the same assay conditions without any addition. The net
phosphorylation activity is obtained by substraction of (+hist)-
(-hist) as shown above.

binding site of the kinase[5]. However the light stimulated (redox control)
of LHCII phosphoryaltion seems to correlate with the assembly of the
cyt.b6/f complex at the later stages of the greening process also in
spinach plants.

The interpretation of the data presented or referred to in this work
is based primarilly on the correlation between the progressive accumulation
of the substrate LHCII and that of the cytochrome complex components during
the greening process while the kinase content detected by immunobloting and
histone phosphorylation is practically constant on a membrane basis through
out the greening. The amount of kinase in mature thylakoids is very low
relative to that of the cytochrome b6/f complex[11] and only a small
proportion of the cytochrome complex is associated with the kinase[5]. Thus
it is inherent to the complexity of the system that the appearance of the
redox controled LHCII activation can not be detected as a sharp transition
from the inactive to the active state. The fact that the rise in the redox
controled phosphorylatuion of the LHCII polypeptides does not correlate with
the kinase content or histone phosphorylation activity nor with the amount
of LHCII or that of the cytochrome b_6/f componets but with the complete
activation of the electron flow and appearance of the grana system is
indicative of the requirement for the develepment of structural organisation
and interaction of these complexes for the expression of the redox controled
LHCII phosphorylation and state transition activity.

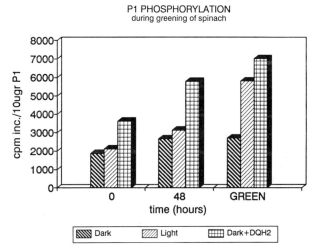

P1 PHOSPHORYLATION
during greening of spinach

Figure 4. Phosphorylation of P1 dodecapeptide by greening spinach
thylakoids.
Membranes (Chl 0.2 mg/ml, normalized to wet weight) were
incubated with P1 (0.3 mg/ml) under standard phosphorylation
conditions, in the light ($300\mu E^2 m^2 s^{-1}$) or in the dark with
or without addition of 1 mM duroquinol. After 15 min at 25 °C
the membranes were removed by centrifugation and the P-1
polypeptide was precipitated from the supernatant by addition
of 20%w/v TCA and its radioactivity counted.

Acknowledgements

         This work was supported by Grant SFB-184 from the   Deutsche Forsc-
hungsgmeineschaft to I.O. and R.Herrmann from the Botanisches
Institute,Munchen Universirty,Munchen,F.R.G.

References

1. Gal,A., Shahak,Y.,Schuster,G. and Ohad I.,Specific loss of LHCII
   phosphorylation in the Lemna mutant 1073 lacking the cytochrome b_6/f
   complex, FEBS Lett. 221,205-210.(1987).
2. Wollman,F.-A. and Lemaire,C.,Studies on kinase-controled state
   transitions in Photosystem II and b_6/f mutants from Chlamydomonas
   reinhardtii which lack quinone binding proteins, Biochim.Biophys.Acta.
   933,85-94,(1988).
3. Bennett,J.,Shaw,E.K.,and Michel,H., Cytochrome b_6/f complex is
   required for phosphorylation of Light Harvesting Chlorophyll a/b complex
   II in chloroplast photosynthetic membranes, Eur.J.Bichem. 171,91-
   100,(1988).
4. Coughlan, S.J.,Chloroplast thylakoid protein phosphorylation is
   influenced by mutartions in the cytochrome b_6/f complex,
   Biochim.Biophys.Acta. 933,413-422,(1988).
5. Gal,A., Hauska,G.,Herrmann,R. and Ohad,I.,Interaction between
   Light-Harvesting Chlorophyll-a/b protein (LHCII) kinase and cytochrome
   b_6/f complex. J.Biol.Chem. 265,19742-19749,(1990).
6. Owens,G.C. and Ohad,I.,Changes in polypeptide phosphorylation
   during membrane biogenesis in Chlamydomonas reinhardtii y-1,
   Biochim.Biophys.Acta. 722 ,234-241,(1983).
7. Ohad,I.,Siekevitz,P, and Palade,G.E.,,Biogenesis of Chloroplast
   membranes, J.Cell.Biol. 35,521-552,(1967).
8. Bennett,J.,Shaw,E.K. and Bakr, S. ,Phosphorylation of thylakoid
   proteins and synthetic peptide analogs, FEBS Lett. 210,22-26,(1987).
9. Gershoni,J.M. and Palade,,G.E., Protein blotting: principles and
   applications, Anal.Biochem. 131,1-5,(1983).
10. Bruce,D.B. and Malkin,R. , Biosynthesis of the chloroplast
    cytochrome b_6/f complex- studies in a photosynthetic mutant of Lemna,
    Plant Cell, 3, 203-212,(1991).
11. Coughlan, S.J. and Hind, G.,Purification and characrerization of a
    membrane-bound protein kinase from spinach thylakoids,J.Biol.Chem.261,
    11378-11385,(1986).

PHOSPORYLATION ON D1 AND D2 PROTEINS OF PHOTOSYSTEM IIα
INDUCES HETEROGENEITY OF Q–B SITE ACTIVITY

M.T. Giardi[*], F. Rigoni, G.M. Giacometti, and R. Barbato

[*]IREV–CNR; via Salaria Km 29.3– 00016 Monterotondo Scalo – Roma
Department of Biology, via Trieste 75, 35121 Padova, Italy

INTRODUCTION

Electron transfer on the acceptor side of photosystem two involves two quinones Q–A and Q–B. The quinone at the primary electron acceptor side (Q–A) does not exchange rapidly with the plastoquinone (PQ) pool, whereas the secondary acceptor is bound to Q–B when is in semiquinone form and exchanges readily with PQ pool when it is in the fully oxidized or reduced form (1).

PS II herbicides such as phenyluree, triazines, and phenolic compounds replace bound PQ from its binding site, thus preventing oxidation of Q–A. Exogenous quinones, often employed as photosystem II electron acceptors, have been shown to compete with herbicides for binding. Moreover quinones act as inhibitors of PS II electron transfer by binding to the same Q–B pocket and displacing the native PQ. Due to the different chemical structure of all these inhibitors it is generally accepted that they have different but overlapping binding domain probably on reaction centre II proteins (D1 and D2).

It is known that in thylakoids membranes there is at least one protein kinase which catalyses the ATP–dependent phosphorylation of several PS II associated polypeptides, and that kinase can be activated when PQ pool is photochemically reduced. It has been shown that light–harvesting chl a/b protein complex is the major site of phosphate incorporation and that phosphorylation regulates the excitation energy distribution between the two photosystems through a lateral migration of a "mobile" pool of phosphorylated LHC. It has also been proposed that the phosphorylation of PS II core polypeptides including 43, 34, 32 and 9 KD polypeptides (ascribed respectively to CP 43, D1, D2 and the phosphoprotein coded by the psbH plastidial gene which is not tightly bound to the core) helps the migration of phosphorylated mobile LHC II by electrostatic repulsion. However several other effects have been reported in the literature which might suggest that core protein phosphorylation modifies photosynthetic electron flow by directly altering PS II function. Phosphorylation of proteins seems also to have a function as a protective mechanism against photoinhibition of reaction centre II. The effect of protein phosphorylation on the activity of Q–B pocket is not clear. In fact contradictory results are reported in literature concerning this problem and, according to Hodges *et al.*, the observed increase in inhibition of Hill reaction by herbicides after phosphorylation process is not due to an increased affinity for herbicides but it is a secondary effect due to contemporary whole reduction of PQ and the consequent lack of its competition with herbicides on Q–B site (1,2).

Regulation of Chloroplast Biogenesis
Edited by J.H. Argyroudi-Akoyunoglou, Plenum Press, New York, 1992

We have recently isolated heterogeneous PS II core populations that showed different trend in the binding of herbicide dinoseb and subsequently a correlation was found between the extent of phosphorylation and the isoelectric point of these populations (3,4). Therefore the present work is directed in an effort to clarify the effect of phosphorylation on herbicide–plastoquinone binding domain trying to distinguish the effects due to the reduction of PQ pool. Our approach is to study the binding of herbicides and synthetic quinones to thylakoids, PS II preparations and the above mentioned PSII cores populations isolated from membranes adapted to conditions of low and high phosphorylation.

The results indicate that the activity of most PS II–directed herbicides as well as synthetic quinones appears to be significantly depressed in phosphorylated membranes. Our findings also contribute to interpreting data regarding heterogeneity of PSII and suggest a function for core proteins phosphorylation.

MATERIAL AND METHODS

Phosphorylation of thylakoids membranes and isolation of PSII membranes were performed as previously described (4). PSII membranes solubilized with 1% n–dodecyl–β–D–maltoside were applied to the cathode region of a flat bed granulated gel and isoelectrofocusing was run as previously reported (3). Chlorophyll, cytb$_{559}$ contents and electron transport rates from diphenylcarbazide (DPC) to 2,6–dichlorophenolindophenol (DCPIP) were measured as reported in 5. Herbicide binding was assayed respectively as reported in 6 for membranes and in 7 for isolated PSII cores.

RESULTS AND DISCUSSION

An approach to investigate if phosphorylation of protein might affect the herbicide binding site was to extract from the Q–B site the secondary plastoquinone acceptor molecule whose competition masks the searched effect with consequent measurements of apparent affinity constants. This was obtained by the use of PSII particles in which PQ is present in the stoichiometric ratio with the reaction centre II of 1–2 : 1 (one ascribed to PQ tightly bound to Q–A and the other to PQ weakly bound to Q–B). It has also been reported that by using an heptane–isobutanol extraction on aqueous suspension of membranes the plastoquinone at Q–B site can be selectively removed without disturbing significantly the functioning of Q–A site (8).

Table I shows the dissociation constants for the binding of some herbicides (rappresentative of the four classes triazines, phenyluree, hydroxybenzonitriles and phenolic compounds) in PS II particles isolated from thylakoids illuminated in the presence of ATP and NaF or in the absence of ATP and then dark adapted.

It is manifest that phosphorylation of the membranes increases systematically the binding constant for most photosynthetic herbicides indicating a reduction of affinity and that lack in affinity is more evident when plastoquinone competition is absent. The main differences are obtained with phenolic herbicides in accordance with the observation that the introduction of negatively charged groups into the proteinaceous field should affect mainly the binding of nucleophilic compounds as is the case of phenolic herbicides. However, the number of chlorophyll molecules per binding site does not change significantly after phosphorylation while extracted membranes also show an increased number of chlorophyll per sites.

The observed difference in affinity binding of phosphorylated and non–phosphorylated membranes cannot be attributed to different oxidation state of electron transport components, following illumination or dark adaptation conditions, since it was also observed in the presence of the strong reductant dithionite (data not shown).

Table I

Dissociation constants for the binding of herbicides to PS II membranes isolated from phoshorylated and non–phosphorylated thylakoids as reported in 3. –PQ, plastoquinone was extracted from particles; the error for the reported values is 10%.

Herbicides	K (nM)		Chl/site	Kb pho/nonpho
	pho	nonpho		
t–butryne	60	45	260	1.3
–PQ	27	16	260	1.7
chl–bromurone	37	20	150	1.8
–PQ	19	10	200	1.9
ioxynil	33	18	250	1.8
–PQ	19	9	270	2.1
i–dinoseb	33	16	200	2.0
–PQ	27	12	230	2.2

We have recently shown that it is possible to separate from membranes differently phosphorylated core populations by isoelectrofocusing. It has also been shown that isolated PS II cores are able to bind photosynthetic herbicides with a specific reaction even if with reduced affinity (3,7). Thus four PS II core complexes differently phosphorylated (we refer to as complexes a–d) were isolated from radiolabelled ^{32}P [ATP] thylakoids as previously reported (3) and further characterized for the relative distribution of radioactivity into the single polypeptides. We confirmed that when the level of phosphorylation on D1 and D2 proteins is low, the phosphoprotein 9KD is more tightly bound to the core (complex a). The complex with more acidic isoelectric point (pI) shows the highest content of radioactivity mainly localized on D1 polypeptide (Table II).

Table II

^{32}P–ATP radioactivity distribution into core polipeptides as measured by the area of the peaks (arbitrary units) of an autoradiography densitogram.
* represents the incorporated radioactivity ratio between D1 and D2 proteins
# represents relative radioactivity incorporation into D1 for each core complex.
The polypeptides are assigned as follows: 55 KD, D1\D2 heterodimer. 43 KD, CP43; 34 Kd, D2; 32 KD, D1; 9 KD, psbH plastidial gene product.

complex	pI	55KD	43KD	34KD	32KD	9KD	D1\D2*	ratio#
a	5.5	0.1	0.1	0.6	0.3	2.0	0.50	1.0
b	4.9	0.1	0.6	1.5	1.2	0.3	0.80	4.0
c	4.8	0	0.5	2.5	2.3	0.1	0.92	7.7
d	4.7	0	0.5	4.5	8.2	1.1	1.43	27.0

Table III
Dissociation constants for herbicide binding to differently phosphorylated cores (a–d). The error for the reported values is 20%.

	K (nM)			
herbicides	a	b	c	d
chl–bro.	70	250	180	400
sites\cytb559	1.0	1.1	1.3	0.8
i–dinoseb	94	700	400	1000
sites\cytb559	1.9	0.8	0.9	0.8
ioxynil	40	180	130	400
sites\cytb559	1.1	0.6	0.7	0.5

Table IV
Electron transport rate measured from DPC (200 μM) to DCPIP (100μM). Bigger concentrations of reagents overcome the herbicide inhibition; the activity is represented as μmol reduced DCPIP (mg chl)$^{-1}$ h^{-1}. Herbicide conc. 25 μM. The values are the average of 10 experiments.

	El.Trans.	% Inhibition		
		herbicides		
complex	none	atrazine	diuron	dinoseb
a	230	30	100	15
b	120	20	25	0
c	140	10	40	0
d	20			

The core complexes were also tested for their ability to bind photosynthetic herbicides and for the inhibition of their electron transport from DPC to DCPIP in the presence of herbicides (Table III–IV).

We found that an increased level of radioactivity on both D–1 and D–2 proteins corresponds to a decreased affinity for herbicides, among these being most affected are phenolic herbicides. In fact the differences for a phenylurea are less than an order of magnitude while for dinoseb, phosphorylation is able to increase the affinity constant of a factor more than 10. The number of chlorophyll molecules per binding site changes significantly for dinoseb in complex a that contains the 9 KD phosphoprotein. The findings suggest that 9 KD protein could be directly involved in the binding of this herbicide.

The next question is whether added quinones are able to displace herbicides in isolated PS II preparations and whether phosphorylation of the core proteins affects their competition with herbicides. In fact the affinity of quinones is expected to be lower in phosphorylated membranes because of their close structural analogy with phenolic herbicides. We tested a quinone and also an indophenolcompound used as electron acceptor that were previously shown to be efficient replacers of PS II herbicides and of the native PQ bound to the Q–B binding environment.

Both these compounds are less able to displace radiolabelled herbicides from phosphorylated than from non–phosphorylated membranes indicating that the activity of Q–B pocket in quinone binding is reduced by phosphorylation. However the number of chlorophyll molecules per herbicide site during competition does not change (data not shown). Thus in PS II membranes the interaction between herbicides and quinones is apparently competitive as demonstrated for thylakoid membranes. DCPIP, used as acceptor in electron transport measurements, shows, like the quinone, a reduction in its ability to bind to phosphorylated membranes, thus this seems to explain the reduced ability for this acceptor to oxidize the highest phosphorylated core d (Table III –V).

Table V

% displacement of radiolabelled chlorbromurone binding to PS II membranes due to 0.5 mM 2–hydroxy–1,4–naphthoquinone (NAQ) or 0.5 mM dichlorophenolindophenol (DCPIP). Membranes conc. 50 μg/ml of chl. Herbicide conc. 10^{-8} M. The values are the average of 5 experiments.

| | % displacement | | | |
| | NAQ | | DCPIP | |
Herbicides	pho	nonpho	pho	nonpho
chl–bromurone	7	21	5	15
ioxynil	30	50	25	35

It is known that phosphorylation of certain subpopulation of phospho–LHC II causes their movement from the granal regions of the thylakoid membrane to the stroma lamellae, where it transfers excitation energy to photosystem I. This is believed to be the mechanism of the so–called State 1–State 2 transitions. Although other PS II proteins can be phosphorylated by the action of a protein kinase(s), the function of this phosphorylation process is unknown. Other than the well–documented redistribution of excitation energy several other effects have been reported which could be a result of the phosphorylation of the core polypeptides, i.e. these effects include an inhibition of light–saturated PSII electron transfer, a destabilization of the anionic semiquinone Q–B⁻ and an accumulation of Q–A⁻ (9). We recently showed the existence in grana particles of PS II cores populations with different trend in phosphorylation of their polypeptides. The cores are present in an equilibrium partially regulated by light and dark conditions. It was possible to convert the cores into one other but this conversion was never complete and, in dark condition, at least two differently phosphorylated populations were always present.

Comparative studies with PS II preparations revealed that when the extent of phosphorylation on D–1 and D–2 proteins increases, the binding activity of Q–B is affected. The magnitude order of this phenomenon measured in PS II membranes is consistent with the interconversion (about 35 %) observed for core populations. We observed in accordance with other reports that protein phosphorylation results in a lower rate of PS II electron transfer from DCP to DCPIP at saturing light intensities and such an inhibition seems to be correlated to the reduction of DCPIP affinity and the consequent inability for this acceptor to oxidize the PS II when D–1 and D–2 are highly phosphorylated as in core complex d.

Our experimental results are in accordance with the observation by Hodges et al. which, on the basis of a destabilization of the semiquinone Q–B⁻ form after phosphorylation of thylakoids, suggested that herbicides should be less able to bind to phosphorylated membranes compared to non–phosphorylated particles (9).

Therefore the reduction of affinity for quinones in phosphorylated membranes suggests that when increased light energy causes the reduction of PQ pool and activates the kinase, bound PQ leaves its site both as a consequence of its modification (reduction) and as a consequence of Q–B pocket modification (phosphorylation). In this way phosphorylation of PS II core proteins seems to be a mechanism of electron transport regulation.

The observed heterogeneity comes from PS II α since our preparation does not retain PS II β that is located on stroma lamellae. Recently it has been shown that also PS IIα is heterogeneous being present at least three phases in fluorescence induction curves of DCMU–treated grana particles and some populations with different antenna size (10,11). The existence of PS II cores with different behaviour versus herbicides and quinones strongly supports our previous suggestion that phosphorylation on PS II core polypeptides explains in part the heterogeneity observed on PS II when these inhibitors are used. Heterogeneity is a difficult and controversial field and it is difficult to judge the relationship that our findings may have to the PS II heterogeneity represented by the reports on alternative acceptors and on the existence of centres not able to transfer electrons to plastoquinone (Q–B type and non–B–type centres) (12–13). On the other hand our findings contribute an important and new way in interpreting data regarding heterogeneity of PS II since for the first time a biochemical approach for this problem is being undertaken.

Acknowledgement

We would like to thank prof. Govindjie for useful suggestion.

REFERENCES

1. O. Hansson and T. Wydrzynski, Current perceptions of photosystem II , Photosynth.Res. , 23:131, (1990).
2. M.Hodges, N.K.Packam and J.Barber, Modification of photosystem II activity by protein phosphorylation, FEBS Lett., 181:83, (1985).
3. M.T.Giardi, J.Barber, M.C.Giardina and R.Bassi, Studies on the herbicide binding site in isolated photosystem II core complexes from a flat–bed isolectrofocusing method, Z.Naturforsch., 45c:366, (1990).
4. M.T.Giardi, F.Rigoni, R.Barbato and G.M.Giacometti, Relationships between heterogeneity of the PSII core complex from grana particles and phosphorylation, Biochem. Biophys. Res. Commun., 176:1298, (1991).
5. M.F.Hipkins and N.R.Baker, Photosynthesis energy transduction: a practical approach , IRL Press, Oxford, Whashington (1986)
6. W.Tischer and H.Strotmann, Relationship between inhibitor binding by chloroplasts and inhibition of photosynthetic electron transport, Biochim Biophys. Acta, 460:113, (1977).
7. M.T.Giardi, J.B.Marder and J.Barber, Herbicide binding to the isolated photosystem II reaction centre, Biochim. Biophys. Acta, 934:64, (1988).
8. T.Wydrzynski and Y.Inoue, Modified photosystem II acceptor side properties upon replacement of the quinone at the QB site with 2,5–dimethyl–p–benzoquinone and phenyl–p–benzoquinone, Biochim.Biophys.Acta, 893:33, (1988).
9. M.Hodges, A.Boussac and J.M.Briantais, Thylakoid membrane protein phosphoryaltion modifies the equilibrium between photosystem II quinone electron acceptors, Biochim.Biophys.Acta, 894:138, (1987).
10. M.Hodges and J.Barber, Analysis of chlorophyll fluorescence induction kinetics exhibited by DCMU–inhibited thylakoids and the origin of a and β centres, 848:239, (1986).
11. P.A.Albertsson, S.G.Yu and U.K.Larsson, Heterogeneity in photosystem IIα. Evidence from fluorescence and gel electrophoresis experiments, Biochim.Biophys.Acta, 1016:137, (1990).
12. Govindjee, Photosystem II heterogeneity, Photosynth. Res., 25:151, (1990).
13. M.T.Black, T.H.Brearley and P.Horton, Heterogeneity in chloroplast photosystem II, Photosynth. Res, 8:193, (1986).

REDOX-REGULATED PROTEIN PHOSPHORYLATION AND PHOTOSYSTEM II
FUNCTION

Tedd D. Elich, Marvin Edelman, and
Autar K. Mattoo

Plant Molecular Biology Laboratory, BARC-West
Building 006, USDA/ARS, Beltsville, MD,
20705 (T.D.E., A.K.M.), and Department of
Plant Genetics, The Weizmann Institute of
Science, Rehovot, Israel 76-100 (M.E.)

Oxygenic photosynthetic organisms derive electrons from
H_2O and evolve molecular oxygen as a by-product of
photosynthesis. Such organisms include higher plants, algae,
and cyanobacteria. Oxygenic photosynthesis is mediated by two
membrane-bound, pigment-protein complexes: Photosystem I (PSI)
and Photosystem II (PSII). PSII carries out the reactions
resulting in the oxidation of water and the reduction of
plastoquinone while PSI utilizes reduced plastoquinone to
drive the reduction of NADP. The reaction center of a
photosystem is defined as the minimal unit necessary for
primary charge separation and stabilization. The solution of
the x-ray crystal structure of the anoxygenic bacterial
photosynthetic reaction center (Deisenhofer *et al.*, 1985;
Allen *et al.*, 1987) led to the prediction (Trebst, 1986) and
subsequent verification (Nanba and Satoh, 1987; Marder *et al.*,
1987) that the PSII reaction center consists of a heterodimer
of homologous proteins, D1 and D2, and cytochrome b_{559}. The
D1/D2 heterodimer is thought to contain all the electron
carriers and cofactors necessary for electron transport
through the reaction center: P680; the Y_Z and Y_D electron donor
tyrosines on D1 and D2, respectively; pheophytin; the

Regulation of Chloroplast Biogenesis
Edited by J.H. Argyroudi-Akoyunoglou, Plenum Press, New York, 1992

plastoquinone electron acceptors Q_A and Q_B; and non-heme iron (for reviews see Rutherford, 1989; and Mattoo *et al.*, 1989).

Both D1 and D2 are known to be reversibly phosphorylated *in vitro* on their blocked N-terminal threonine residues by an endogenous redox-regulated thylakoid protein kinase (Michel *et al.*, 1988). In addition to these two reaction center proteins, four other PSII-associated proteins have also been demonstrated to be reversibly phosphorylated on or near their N-terminus under similar conditions: CP47, the psbH gene product and type I and II LHCII polypeptides (see Bennett, 1991, for review). While not proven, it is generally thought that two kinases are involved in catalyzing these modifications - one acting on the LHCII polypeptides and the other on the PSII core proteins. If two kinases do exist, they are similar to the extent that both activities are membrane-bound, immunologically related, and under redox control (Bennett, 1991).

Phosphorylation of the LHCII polypeptides has received the most attention since this modification is known to mediate changes in energy distribution that compensate for imbalances between the two photosystems. Under conditions that favor PSII excitation, reduction of the plastoquinone pool leads to activation of an thylakoid kinase. Phosphorylation of so called "mobile LHCII" results in the dissociation of this antennae sub-population from PSII centers located in the stacked regions, thus effectively decreasing the optical cross section of PSII. Furthermore, it is thought that the dissociated, mobile LHCII then migrates to the unstacked regions where it may act to transfer energy to PSI.

The role of phosphorylation of the PSII core polypeptides is not as well established. Although lacking experimental evidence, it has been speculated that phosphorylation of the PSII core may facilitate dissociation of the mobile LHCII due to charge repulsion (Allen and Holmes, 1986), or that the N-terminal phosphoryl groups may be important for buffering the partition gap between stacked membranes (Bennett, 1991). In addition, a number of effects on PSII function have been reported to occur upon phosphorylation of thylakoid proteins *in vitro* including: i) an increase (Shochat *et al.*, 1982; Vermaas *et al.*, 1984), decrease (Habash and Baker, 1990) or no change (Hodges *et al.*, 1985) in the affinity for PSII herbicides; ii) an increase (Jursinic and Kyle, 1983) or decrease (Hodges *et al.*, 1987) in the stability of Q_B^-; iii) a decrease in light-saturated photosynthetic electron transport (Horton and Lee, 1984; Hodges *et al.*, 1985; Packham, 1987; Habash and Baker, 1990; Giardi *et al.*, 1991); iv) protection against photoinhibition (Horton and Lee, 1985); and v) a decrease in the connectivity of PSII (Kyle *et al.*, 1982). It should be noted, however, that in none of these cases was the extent of thylakoid protein phosphorylation known nor were the observed effects correlated with phosphorylation of any specific polypeptide(s) with the exception of the recent study by Giardi *et al.*(1991) which showed a correlation between rates of electron transport and the relative extent of PSII core phosphorylation. Furthermore, all of these studies were performed on isolated thylakoids or osmotically shocked chloroplasts and thus their physiological relevance remains questionable. In this regard, to our knowledge, no studies to date have rigorously identified and characterized an *in vivo* phosphorylation of a PSII-associated protein. Therefore, it is not even clear that the same proteins are phosphorylated on the same sites *in vivo* as *in vitro*. Interestingly, in several studies where *in vivo* and *in vitro* phosphorylated thylakoids could be compared, qualitative and quantitative differences in the ^{32}P-radiolabeled protein patterns are apparent (Schuster *et al.*, 1986; Bhalla and Bennett, 1987; Bennett *et al.*, 1988)

Recently, we resolved an *in vivo*-generated electrophoretic variant of D1, designated 32*, in *Spirodela oligorrhiza* as well as four other higher plants (Callahan *et al.*, 1990). This form is generated specifically in granal-localized reaction centers and its appearance was correlated with the onset of D1 degradation (Callahan *et al.*, 1990). Formation of 32* was light-dependent and inhibited by DCMU (Callahan *et al.*, 1990) - a herbicide known to inhibit the light-dependent degradation of D1. Given these results, we hypothesized that the modification giving rise to 32* may be a signal targeting the protein for degradation.

We have now identified 32* as the *in vivo* phosphorylated form of D1 (manuscript submitted). The phosphorylation occurs on a threonine residue(s) located near the N-terminus and is identical to the phosphorylation of D1 catalyzed *in vitro* by an endogenous redox-regulated kinase (manuscript submitted). These results, together with those from our previous study (Callahan *et al.*, 1990), suggest the new and novel possibility that D1 phosphorylation may somehow be involved in mediating the well characterized light-dependent degradation of this important reaction center protein. Present studies are aimed at testing this hypothesis using a variety of approaches. First, the ability to specifically resolve the two forms of D1 is being exploited to examine *in vivo* the metabolism of the phosphorylated form under various conditions to test if this modification correlates with D1 degradation. Second, we are attempting to identify kinase inhibitors that can effectively block D1 phosphorylation *in vivo* so that we can test their effect on D1 degradation. Finally, we are planning experiments to genetically delete or change the D1 phosphorylation site and characterize the phenotype of the resulting unphosphorylatable D1 mutants.

ACKNOWLEDGMENTS

The results mentioned in this paper were supported in part by a BARD grant to M. E. and A. K. M. and a USDA/CRGO grant to A. K. M.

REFERENCES

Allen, J.P., Feher, G., Yeates, T.O., Komiya, H., and Rees, D.C. (1987) Proc. Natl. Acad. Sci. USA 84, 5730-5734.

Bennett, J., Shaw, E.K., and Michel, H. (1988) 171:95-100.

Bennett, J. (1991) Annu. Rev. Plant Physiol. Plant Mol. Biol. 42:281-311

Bhalla, P., and Bennett, J. (1987) Arch. Biochem. Biophys. 252:97-104.

Callahan, F.E., Ghirardi, M.L., Sopory, S.K., Mehta, A.M., Edelman, M., and Mattoo, A.K. (1990) J. Biol. Chem. 265, 15357-15360.

Deisenhofer, J., Epp, O., Miki, K., Huber, R., and Michel, H. (1985) Nature 318, 618-624.

Giardi, M.T., Rigoni, F., Barbato, R., and Giacometti, G.M. (1991) Biochem. Biophys. Res. Comm.176:1298-1305.

Habash, D.Z., and Baker, N.R. (1990) J. Exper. Bot. 44:761-767.

Hodges, M., Boussac, A., and Briantais, J.-M. (1987) Biochim. Biophys. Acta 894:138-145.

Hodges, M., Packham, N.K., and Barber, J. (1985) FEBS Lett. 181:83-87.

Horton, P., and Lee, P. (1984) Biochim. Biophys. Acta 767:563-567.

Horton, P., and Lee, P. (1985) Planta 165:37-42.

Jursinic, P.A., and Kyle, D.J. (1983) Biochim. Biophys. Acta 723:37-44.

Kyle, D.J., Haworth, P., and Arntzen, C.J. (1982) Biochim. Biophys. Acta 680:336-342.

Marder, J.B., Chapman, D.J., Telfer, A., Nixon, P.J., and Barber, J. (1987) Plant Molec. Biol. 9, 325-333.

Mattoo, A.K., Marder, J.B., and Edelman, M. (1989) Cell 56, 241-246.

Michel, H., Hunt, D.F., Shabanowitz, J., and Bennett, J. (1988) J. Biol. Chem. 263, 1123-1130.

Nanba, O., and Satoh, K. (1987) Proc. Natl. Acad. Sci. USA 84, 109-112.

Packham, N.K. (1987) Biochim. Biophys. Acta 893:259-266.

Rutherford, A.W. (1989) Trends Biochem. Sci. 14, 227-232.

Schuster, G., Dewit, M., Staehelin, A., and Ohad, I. (1986) 103:71-80.

Trebst, A. (1986) Z. Naturforsch. 41c, 240-245.

Vermaas, W.F.J., Steinback, K.E., and Arntzen, C.J. (1984) Arch. Biochem. Biophys. 231:226-232.

THYLAKOID PROTEIN PHOSPHORYLATION LEADS TO ORGANIZATION OF THE OLIGOMERIC

FORMS OF PIGMENT-PROTEIN COMPLEXES IN PEA GRANA AND STROMA LAMELLAE

John Georgakopoulos and Joan Argyroudi-Akoyunoglou

Institute of Biology, NRC "Demokritos", Athens, Greece

INTRODUCTION

A number of thylakoid polypeptides are known to be reversibly phosphorylated in vitro, by one or more thylakoid-bound redox-regulated protein kinases (D_1,D_2,CP_{43}, the 8kd-psbH gene product, LHC-II) (1). The phosphorylation of LHC-II has been considered to be involved in state transitions. The phosphorylation-induced increase in the low temperature (77 K) fluorescence emission ratio F730/F685 in chloroplasts, has been attributed to lateral movement of a "mobile" LHC-II fraction from grana (where it serves PSII) to stroma lamellae (where it serves PSI) (2,3). Indeed, the light-subchloroplast fraction of stroma lamellae (which is rich in Chla,but lacks LHC-II and Chlb) contains grana components (Chlb and LHC-II) when obtained from phosphorylated samples (4). On the other hand, the low-salt-induced increase in F730/F685 ratio in chloroplasts, has been shown to be correlated with an enhanced Chl binding in PSI pigment-protein complexes and enhanced organization of the LHC-I into the pigment-protein complexes of PSI (5) suggesting that the reason for these fluorescence changes may not be due to lateral movement of LHC-II, but rather to increased PSI unit absorption cross section, brought about during low-salt-induced grana unstacking. To test whether the increase in the F730/F685 ratio upon phosphorylation of chloroplasts may be due to a similar process, we studied the effect of phosphorylation on the state of pigment-protein organization, by mild SDS-PAGE analysis, in unfractionated thylakoids as well as in grana and stroma lamellae produced by French press or digitonin disruption of chloroplasts.

MATERIALS AND METHODS

Chloroplasts, obtained from pea plants(Pisum sativum), grown for 5 days in the dark and then exposed to continuous light for up to 10 days, were isolated as pellet at 1,000xg for 10 min after homogenization of 4 g,fr,w, leaves at a time with 40 ml 0.4 Sorbitol-50mM Tricine-10mM NaCl, pH 7.8 . The pellet was washed with 10mM Tricine-10mM NaCl-5mM $MgCl_2$, pH 7.8, and repelleted at 10,000xg for 10 min. Thylakoids resuspended in 0.1M Sucrose-30mM Tricine- 10mM NaCl-5mM $MgCl_2$, pH 7.8 (resuspension buffer) to a Chl concentration of 2mg Chl/ml were used for phosphorylation.

Phosphorylation: To chloroplasts, diluted to 400 µg Chl/ml with resuspension buffer (containing 10mM NaF), ATP was added (1mM final concentration); the suspension was illuminated at or above 5,000 lux for 20 min at 20°C, with white light. Controls without ATP served as non-phosphorylated samples.

Thylakoid fractionation: Grana and stroma lamellae were obtained by differential centrifugation of digitonin or French press disrupted plastids as before (6,7). Chloroplasts at 400 μg Chl/ml in resuspension buffer were disrupted with digitonin (0.5% final concentration at 0°C for 30 min) or by passage through an Aminco French press at 6,000 psi. Centrifugation of homogenates followed at 1,000xg for 10 min (1K); 10,000xg for 30 min (10K, grana); 50,000xg for 30 min (50K) and 240,000xg for 60 min (240K, stroma lamellae).

Low temperature fluorescence spectra were obtained at 77 K as before (8) in samples diluted to 30 μg Chl/ml.

The Chl-protein complexes were separated by mild SDS-PAGE (9). Thylakoids, stroma lamellae or grana fractions, resuspended in 0.05 Tricine-NaOH, pH 7.3, at 1,800 μg Chl/ml were solubilized with cold SDS (SDS/Chl=10) and immediately electrophoresed (3 mA/tube gel) at 4°C. The electrophoretic profiles of pigment-protein complexes were obtained in a Joyce-Loebl Chromoscan, and the distribution of Chl among complexes was estimated by weight on the basis of the area under each peak.

RESULTS AND DISCUSSION

In accordance with earlier results, phosphorylation of pea chloroplasts under our conditions leads to phosphorylation of a thylakoid polypeptide in the molecular range of LHC-II as verified by autoradiography with $(\gamma-^{32}P)ATP$ (data not shown). Similarly, the state transitions, known to occur upon phosphorylation of chloroplasts,are fully reproduced (see Table 3): chloroplasts prior to phosphorylation have a F730/F685 low temperature fluorescence emission ratio of 1.1 (State I); this ratio is increased to about 2 upon phosphorylation (State II).

Furthermore, phosphorylation leads to grana destacking/unstacking, as monitored by the increase in the % Chl found in the light-subchloroplast fractions (10K sup),produced after digitonin disruption of chloroplasts (Table 1).

Table 1. Grana unstacking upon phosphorylation as monitored by the % Chl distribution among subchloroplast fractions obtained after French-press disruption of pea plastids

	Chl distribution (%)*				
	1K-p	10K-p	50K-p	240K-p	10K-sup
NON PHOS	8	48	38	6	44
PHOS	8	39	44	8.6	54

* % of the chloroplast Chl (starting amount 430 μg Chl/ml)

Also significant differences in composition of the 10K and 240K subchloroplast fractions are observed upon phosphorylation of chloroplasts. As shown in Table 2, the heavy-grana fraction, upon phosphorylation, becomes enriched in PSI complexes, while the light-stroma lamellae-containing fraction becomes enriched in Chlb (lower Chla/b ratio) and in PSII complexes (mainly LHCP[1] and LHCP[3]). This enrichment of the 240K fraction in LHCPs,

Table 2. Chl distribution among complexes in 10K and 240K
subchloroplast fractions produced by digitonin
disruption of phosphorylated or non-phosphorylated
chloroplasts. The resolution profiles are shown
in Figures 1 and 2.

	10K (GRANA)-pellet		240K (STROMA LAMELLAE)-pellet	
	NON	PHOS	NON	PHOS
Chl a/b	2.2	2.4	6.2	3.9
Chl(%)*	63.0	52.3	6.5	9.2
CPIa+CPI(%)	6.0	12.0	37.0	30.1
CPIa/CPI	0.4	1.2	0.2	0.2
LHCP1+2+3(%)	74.1	72.8	34.5	54.1
LHCP1+2/3	0.6	1.1	1.0	1.6
PSI(%)**	21.4	32.5	50.5	48.7
PSII(%)	62.0	55.9	23.9	41.9

* % of chloroplast amount
** PSI=CPIa + CPI + LHCP^2,   PSII=CPa + LHCP^1 + LHCP^3

has been earlier reported and considered to show the lateral movement of
the "mobile"-LHC-II from grana to stroma lamellae where it serves the
PSI unit.

Apart from the differences described above, we have noticed, in addition
that the organization of the oligomeric forms of PSI and PSII complexes is
enhanced in all subchloroplast fractions, and French press or digitonin
homogenates (disrupted but unfractionated thylakoids), obtained from phosphorylated chloroplasts (Figures 1, 2, 3). In grana fractions, as well as in
homogenates, mainly the oligomeric form of the PSI complex, CPIa, is enhanced.

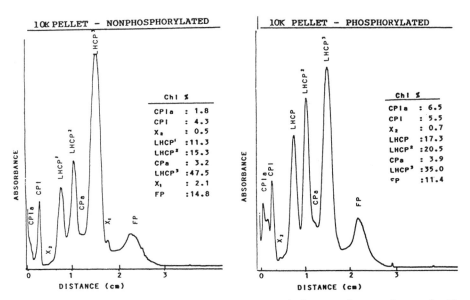

Fig. 1. Electrophoretic profiles of chlorophyl-protein complexes in the
10K (grana) fragment from phosphorylated and non-phosphorylated
samples produced by digitonin disruption.

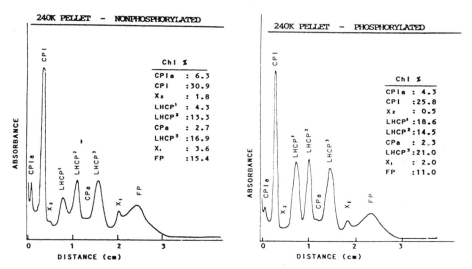

Fig. 2. Electrophoretic profiles of chlorophyl-protein complexes in the 240K (stroma lamellae) fragment from phosphorylated and non phosphorylated samples produced by digitonin disruption.

In stroma lamellar fraction the oligomeric form of LHC-II (LHCP[1]) is enhanced. However, in grana fractions and homogenates, the organization of the CPIa form correlates with the increase in the F730/F685 ratio; in contrast, phosphorylated stroma lamellar fractions show a reduction of the fluorescence ratio (see Figure 4). This indicates that the changes responsible for the increase in F730/F685 ratio (State I to State II transition) observed in chloroplasts, upon phosphorylation, probably originate in grana. As already described above, the increase in LHCP complexes observed in the stroma lamellar fraction, upon phosphorylation, has been considered to show the lateral movement of phosphorylated "mobile"-LHC-II from grana to stroma lamellae, where it serves the PSI unit. If this was the case, however, one

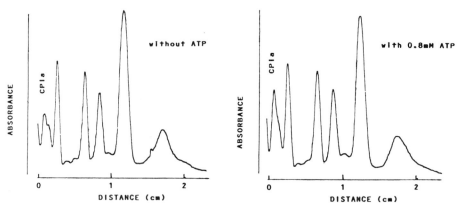

Fig. 3. Resolution of pigment-protein complexes obtained from French-press homogenates of phosphorylated (0.8mM ATP) and non phosphorylated control chloroplasts. The homogenates were pelleted at 240,000xg for 2h, and the pellets were resuspended in 30mM Tricine pH 7.3 at 1800 µg Chl/ml. SDS was added (SDS/Chl=10, w/w) and the samples electrophoresed immediately.

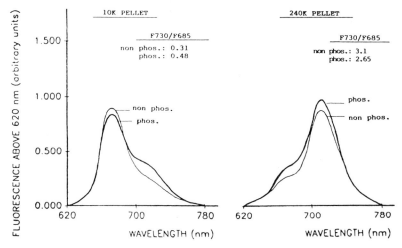

Fig. 4. Fluorescence emission spectra of 10K (grana)-pellet and 240K
(stroma lamellae)-pellet obtained from phosphorylated and non-
phosphorylated chloroplasts after their disruption with digitonin.
The pellets were suspended in 50mM Tricine, pH 7.3 to 30 µg Chl/ml,
and their low-temperature fluorescence spectra recorded.

would expect an increase in the F730/F685 ratio in this fraction, upon
phosphorylation. Our findings therefore suggest that the LHCPs appearing
in this fraction (50K-sup, or 240K pellet) for some reason do not serve
the PSI unit.

    To the same conclusion lead the fluorescence emission spectra of the
supernatants successively obtained by differential centrifugation of the
French-press homogenized chloroplasts. As shown in Table 3, the homogenate,
1K-sup, and 10K-sup, all show the increase in the F730/F685 ratio, upon
phosphorylation. In contrast, the 50K-sup (240K-stroma lamellar fraction)
does not show the expected changes in the F730/F685 ratio, upon phosphorylation; in this case, phosphorylated stroma lamellae show instead a reduction of the ratio.

Table 3.   F730/F685 ratio of chloroplasts, French-press homogenates and
           of the 1K, 10K and 50K supernatant fraction prior (non phos)
           or after phosphorylation (phos).

	non phos		phos	
	F730/F685	%	F730/F685	%
Chloroplasts	1.1	52	2.1	100
Homogenate	0.73	52	1.4	100
1K sup	0.73	53	1.38	100
10K sup	1.1	58	1.9	100
50K sup	4.1	141	2.9	100

Based on the findings on one hand that the "mobile" LHC-II does not seem to serve the PSI, and on the other that the supercomplex of PSI, CPIa, is organized upon phosphorylation, we may conclude that the state transitions are brought about by a change in the absorption cross section of PSI, which is independent of the presence of LHC-II. Alternatively the state transitions may be brought about by the decrease in the antenna size of PSII located in grana. The finding that phosphorylation leads to enhanced organization of the CPIa further suggests that phosphorylation facilitates tighter binding of PSI components (CPI, LHC-I, pigments). We do not know whether in such a binding another polypeptide may be involved. The latter however, seems also possible in view of recent findings (Bassi, this volume). Furthermore, the finding that the LHC-II components, "transferred" to stroma lamellae do not serve the PSI unit, may instead show that this LHC-II is localized on membranes not in close proximity with PSI-RC. This is also a possibility, in view of earlier findings showing that French press or digitonin disruption of "low-salt" chloroplasts leads to contamination of the 50K-sup with thylakoid membrane fragments originating in grana (10).

In conclusion, our results show that:
1) The F730/F685 ratio changes upon phosphorylation reflect mainly the changes occuring in grana fraction.
2) The LHCP appearing in the 240K fraction upon phosphorylation does not contribute to the F730/F685 ratio increase; this suggests that it does not serve the PSI unit.
3) The CPIa organization is stabilized by phosphorylation.
4) The CPIa organization is correlated with the F730/F685 enhancement, induced by phosphorylation.

REFERENCES

1. J. Bennett, Annu. Rev. Plant Physiol. Plant Mol. BIol. 42:281 (1991).
2. J. F. Allen, J. Bennett, K. E. Steinback, and C. J. Arntzen, Nature 291:21 (1981).
3. W. S. Chow, A. Telfer, D. J. Chapman, and J. Barber, Biochim. Biophys. Acta 638:60 (1981).
4. D. J. Kyle, L. A. Staehelin, and C. J. Arntzen, Arch. Biochem. Biophys. 222:527 (1983).
5. J. H. Argyroudi-Akoyunoglou, Arch. Biochem. Biophys. 286:524 (1991).
6. J. M. Anderson, N. K. Boardman, Biochim. Biophys. Acta 112:403 (1966).
7. P. V. Sane, D. J. Goodchild, and R. B. Park, Biochim. Biophys. Acta 216:162 (1970).
8. A. Castorinis, J. H. Argyroudi-Akoyunoglou, and G. Akoyunoglou, in: "Protochlorophyllide Reduction and Greening," C. Sironval, and M. Brouers, ed., Dr W. Junk, The Hague (1984).
9. J. M. Anderson, J. C. Waldron, S. W. Thorne, FEBS Lett. 92:227 (1978).
10. J. H. Argyroudi-Akoyunoglou, and G. Akoyunoglou, in: "Regulation of Chloroplast Differentiation," G. Akoyunoglou, and H. Senger, ed., Alan R. Liss, Inc., (1986).

DEVELOPMENT OF SHADE-TYPE APPEARANCE - LIGHT INTENSITY ADAPTATION -

AND REGULATION OF THE D1 PROTEIN IN <u>SYNECHOCOCCUS</u>

Friederike Koenig

Botanisches Institut
Johann Wolfgang Goethe-Universität
Frankfurt am Main, Germany

WILDTYPE

Architecture and Function of the Photosynthetic Apparatus

Shade-type appearance is induced in <u>Synechococcus</u> (<u>Anacystis</u>) besides by growth in low light also in high light intensity in the presence of sublethal concentrations of DCMU-type inhibitors. This inhibitor class was found to be the only one to give the effect (Koenig 1987a). The antenna synthesized in the presence of DCMU-type herbicides is functional: Room temperature excitation spectra of 684nm fluorescence emission clearly demonstrate the participation of the extraordinarily high concentration of phycocyanin in artificially shade-adapted cells in excitation energy transfer to chlorophyll (Koenig 1990a). With respect to pigmentation, excitation spectra of 684nm fluorescence and light saturation curves of O_2-evolution the naturally induced shade phenotype and shade cells from high light intensity are alike (Koenig 1987a, 1990a). Surprisingly however, while the natural shade phenotype decreases with light intensity, the observed shift of the phycocyanin to chlorophyll ratio in the presence of inhibitors increases with light intensity and temperature (Koenig 1987a). While blue, green and red light have the same effect at low (18 $\mu E \cdot m^{-2} \cdot s^{-1}$) quantum flux density, blue light is clearly more effective than red or even green light at medium (44 $\mu E \cdot m^{-2} \cdot s^{-1}$) and higher light intensity. With increasing quantum flux density the special role of blue light becomes more and more evident.

The development of shade-type appearance under conditions of inhibitor presence in strong light is not simply the consequence of an imbalance of electron transport, since an addition of thiosulphate to cultures growing in high light in the presence of DCMU-type inhibitors can only partially prevent or revert the change from sun to artificial - herbicide-induced - shade phenotype (Koenig 1990a). Growth of <u>Anacystis</u> in the presence of thiosulphate but absence of inhibitors had been shown to cause a decrease of the phycocyanin to chlorophyll ratio, even in low light (Koenig 1990b).

Due to the fact that, from all inhibitors tried, only DCMU-type herbicides were observed to induce shade character in high light intensity, their binding site, the reaction center polypeptide D1 of photoystem II, responsible for the reversible binding of the secondary plastoquinone electron acceptor Q_B (Michel and Deisenhofer 1988, Trebst 1987, 1988) gained attention for possibly being involved in giving a signal for the change in architecture and function of the photosynthetic apparatus - also in natural light intensity adaptation. Several more candidates with signal function

Regulation of Chloroplast Biogenesis
Edited by J.H. Argyroudi-Akoyunoglou, Plenum Press, New York, 1992

in photosynthetic electron transport are currently discussed (Chow et al. 1990, Fujita et al. 1987, Heber et al. 1988, Melis et al.1989, Mullineaux and Allen 1990, Murakami and Fujita 1991).

Not only low light intensity and inhibitors in strong light are effective, but also temperature has an influence on the architecture of the photosynthetic apparatus: When cells precultured at 32°C are transferred to low temperature (ice bath) in high light intensity ($110 \, \mu E \cdot m^{-2} \cdot s^{-1}$) for 8 hours and thereafter returned to the original conditions, they change the ratio of phycocyanin to chlorophyll after a long period of delay to higher rates, before they slowly approach the original ratio of pigments (Fig. 1.). Such a change in pigmentation together with a "memory effect" after chilling in high light intensity is not seen in cultures containing atrazine in sublethal concentrations.

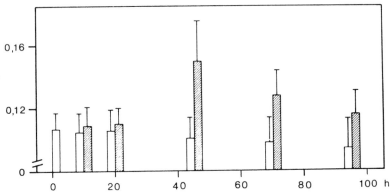

Fig. 1.   Influence of chilling on the pigmentation (molar ratio of phycocyanin to chlorophyll) of Synechococcus PCC 6301 precultured at 32°C and $110 \, \mu E \cdot m^{-2} \cdot s^{-1}$. White bars: Control cells - 32°C throughout the experiment; striped bars: cells kept at 4°C for 8 hours in the beginning of the experiment and afterwards transferred to 32°C. Light intensity was $110 \, \mu E \cdot m^{-2} \cdot s^{-1}$ for all cultures during the experiment. Mean errors of seven independent series are indicated (two parallels per series).

Dynamics of the Photosynthetic Apparatus

For higher plants and green algae it had been found that the D1 protein, one of the photosystem II reaction center polypeptides reversibly binding the secondary plastoquinone electron acceptor Q_B, was characterized by a light-dependent turnover which could be halted by DCMU-type inhibitors (Mattoo et al. 1981, 1984, Wettern and Ohad 1984), and that growth in the presence of sublethal concentrations of DCMU-type inhibitors gave rise to an adaptive reorganization of the photosynthetic apparatus (Mattoo and Edelman 1985). The observed turnover was faster in blue and UV than red or even green light (Greenberg et al. 1989). The similarity of the cyanobacterial system was demonstrated later by Koenig (1987b) and Golubinoff et al. (1988) who could show that also the cyanobacterial D1 protein has a light dependent turnover. In contrast to the chloroplast system, however, which has only one gene (psbA) for the Q_B binding protein (D1), cyanobacteria have a gene family which - in the case of Synechococcus (Anacystis) PCC 6301 and 7942 - consists of three psbA genes - psbAI, psbAII and psbAIII - which code for two different proteins - form I derived from psbAI and form II derived from psbAII and psbAIII (Schaefer and Golden 1989b). Different transcript levels are observed for the different psbA genes in dependence on light intensity during growth of the cells (Bustos et al. 1990, Schaefer and Golden 1989a). Total mRNA for the D1 protein decreases with decreasing light intensity (Fliegmann and Koenig unpublished, Lönneborg et al. 1988, Mohamed and Janssen 1989). In Synechococcus, both proteins - form I and form II - are observed in different light intensities, their ratio varying, again in dependence on light intensity (Schaefer and Golden 1989b).

Until recently, no detailed information on synthesis and break-down of the cyanobacterial D1 protein was available. The corresponding results were obtained lately by labelling the proteins with radioactive sulphur like in higher plants and green algae. Both ^{35}S methionine and ^{35}S sulphate were used for comparison. In case of ^{35}S sulphate the cells were transferred to sulphate-free media two hours prior to the addition of radioactivity. The pulse was ended by addition of cold sulphate up to 10mM. In case of ^{35}S methionine cells were labelled in fresh sulphur-containing media without prior deprivation of the sulphur source. Pulse was ended by addition of an excess (1mM) of cold methionine. Cultures normalized to equal cell density were compared under different conditions of light intensity and inhibitor presence. No grossly qualitative differences in the labelling pattern were found for ^{35}S methionine or ^{35}S sulphate, respectively.

Synthesis of the D1 protein: Growth of Anacystis PCC 6301 in high light in the presence of sublethal concentrations of DCMU-type inhibitors leads to an increased synthesis of phycocyanin paralleled by a reduced rate of ^{35}S methionine incorporation into the D1 protein compared to the high light controls, as is characteristic for naturally induced shade phenotype. On the contrary, sun phenotype is characterized by a low rate of antenna synthesis, but a high rate of ^{35}S methionine incorporation into the D1 protein (Koenig 1990a). Typically, together with low rates of D1 synthesis, high synthesis rates of two as yet unidentified low molecular weight components are observed. The respective polypeptides are highly labelled within minutes, long time before elements of the antenna show incorporation of ^{35}S methionine (Koenig 1990a).

Degradation of the D1 protein: Degradation of the D1 protein was observed to be equally slow in artificially shade-adapted cells and in cells grown in deep shade (half life time 129 min), high light cells showing fast (60 min) and cells grown at intermediate light intensity intermediate (100 min) rate of break-down (Table 1). Presence of thiosulfate had no influence on D1 turnover in artificially shade-adapted cells.

Table 1.   Comparison of Anacystis (PCC 6301) cultures grown under different conditions of light intensity and inhibitor (atrazine) presence.

light intensity $(\mu E \cdot m^{-2} \cdot s^{-1})$	additions during preculture	chase	half life time of D1 (min)
110	–	–	60
15	–	–	111
5	–	–	129
110	–	10^{-6}M atr.	100
110	10^{-6}M atr.	–	129

temperature 32°C
no additional CO_2

The dynamics of the D1 protein in relation to the "memory effect" (Fig. 1.) remains to be studied.

HERBICIDE-TOLERANT MUTANTS

The following series of mutants of Anacystis R2 (Synechococcus spec. PCC 7942) with amino acid exchanges exclusively in positions 255 and/or 264 of the D1 protein - with different tolerances to DCMU-type inhibitors - was studied with respect to architecture and dynamics of the photosynthetic apparatus in comparison to the wildtype (Table 2.):

```
wildtype - Phe 255, Ser 264 (Hirschberg et al. 1987a,b)
Di1      - Phe 255, Ala 264 (Hirschberg et al. 1987a,b)
G264     - Phe 255, Gly 264 (Ohad and Hirschberg 1990)
Tyr5     - Tyr 255, Ser 264 (Horovitz et al. 1989)
D5       - Tyr 255, Ala 264 (Hirschberg et al. 1987a,b)
Di22     - Leu 255, Ala 264 (Hirschberg et al. 1987a,b)
```

All of these mutants were described to have gradually reduced electron transport (Gleiter et al. 1989, Ohad et al. 1990) and moreover, all of these mutants were shown to be characterized by a gradually increasing ratio of phycocyanin to chlorophyll, beginning with the wildtype and mutant Tyr5, which has nearly wildtype appearance, via mutants Di22, Di1 and G264 to mutant D5 (Koenig 1990c, Table 2.). Parallel to a gradual decrease in electron transport (Gleiter et al. 1989) there is a gradual increase in the ratio of phycocyanin to chlorophyll (Koenig 1990c).

Dynamics of the Photosynthetic Apparatus

Sythesis and degradation of the D1 protein. From the data obtained for shade adaptation in the wildtype one would expect similar characteristics concerning the dynamics of the D1 protein for the mutants. However, while the naturally and the artificially induced shade character in Anacystis wildtype is correlated to slow synthesis and break-down of the D1 protein, the constitutive shade appearance (Koenig 1987a, 1990c) of mutants of Anacystis R2 PCC 7942, tolerant to different DCMU-type herbicides (Hirschberg et al. 1987a,b, Ohad and Hirschberg 1990) is characterized by accelerated synthesis (Koenig 1990c) and degradation (Ohad et al. 1990) of the D1 protein. Break-down of the D1 protein was reported earlier to be faster in all mutants compared to the wildtype (Ohad et al. 1990), no distinctions were made between the different mutants, however. A careful reivestigation yielded the following result (Table 2.): While mutant D5 - Tyr 255, Ala 264 - (Hirschberg et al. 1987a,b) with the highest phycocyanin to chlorophyll ratio (Koenig 1990c) was found to have a very instable D1 protein (half life time 41 min) compared to the wildtype - Phe 255, Ser 264 - (74 min), in mutant Di1 - Phe 255, Ala 264 - (Hirschberg et al. 1987a,b) with an intermediate phycocyanin to chlorophyll ratio between wildtype and mutant D5 (Koenig 1990c) the half life time of the D1 protein was determined to be 56 min.

Table 2.   Comparison of Anacystis R2 (PCC 7942) wildtype and herbicide-tolerant mutants with respect to rate of D1 synthesis, rate of D1 breakdown and PC/chl ratio, when grown in air without additional CO_2 at 32°C in strong white light (110 $\mu E \cdot m^{-2} \cdot s^{-1}$ PAR). Cultures carefully normalized to equal cell density were compared.

	D1 half life time (min)	D1 synthesis in 18 min (arb. units)	PC/chl (molar ratio)
wildtype	74	16.1	0.508
Tyr5	66	11.8	0.504
Di22	67	17.5	0.528
Di1	56	20.7	0.540
G264	41	28.6	0.570
D5	49	38.7	0.615

CONCLUSION

At present, there is no unifying concept between shade-type appearance, electron transport capacity and the dynamics of the D1 protein in Anacystis (Synechococcus) wildtype on one hand and in herbicide-tolerant mutants on the other hand. The D1 protein may play different roles under different conditions, as already suggested by Heber et al. (1988). In the case of the mutants, the instable D1 protein (Koenig 1990c, Ohad et al. 1990) may be the reason for the reduced electron transport (Gleiter et al. 1989, Ohad et al. 1990) which in turn may cause the adaptive reorganization of the photosynthetic apparatus. There is increasing evidence, however, that part of the dynamics of this protein is independent on electron transport (Aro et al. 1990, Jansen et al. 1989, Koenig 1990a, Mattoo and Edelman 1985) and may by itself give a signal for adaptive alterations in architecture and function of the photosynthetic apparatus. For the case of the mutants it is suggested that - on the background of an instable D1 protein due to mutational amino acid exchanges - low light intensity and presence of sublethal concentrations of DCMU-inhibitors will equally decrease the rate of synthesis as well as the rate of degradation, just like in wildtype cells.

ACKNOWLEDGEMENTS

Herbicide-tolerant mutants were kindly provided by Dr. J. Hirschberg, The Hebrew University of Jerusalem. This work was supported by the Deutsche Forschungsgemeinschaft.

REFERENCES

Aro, E.-M., Hundal, T., Carlberg, I., and Andersson, B., 1990, In vitro studies on light-induced inhibition of photosystem II and D_1-protein degradation at low temperatures, Biochim. Biophys. Acta, 1019:269.

Bustos, S. A., Schaefer, M. R., and Golden, S. S., 1990, Different and rapid responses of four cyanobacterial psbA transcripts to changes in light intensity, J. Bacteriol., 172:1998.

Chow, W. S., Melis, A., and Anderson, J. M., 1990, Adjustments of photosystem stoichiometry in chloroplasts improve the quantum efficiency of photosynthesis, Proc. Natl. Acad. Sci. USA, 87:7502.

Fujita, Y., Murakami, A., and Ohki, K., 1987, Regulation of photosystem composition in the cyanobacterial photosynthetic system: the regulation occurs in response to the redox state of the electron pool located between the two photosystems, Plant Cell Physiol., 28:283.

Gleiter, H. M., Ohad, N., Hirschberg, J., Fromme, R., Renger, G., Koike, H., and Inoue, Y., 1989, An application of thermoluminescence to herbicide studies, Z. Naturforsch., 45c:353.

Goloubinoff, P, Brusslan, J., Golden, S., Haselkorn, R., and Edelman, M., 1988, Characterization of the photosystem II 32 kDa protein in Synechococcus PCC7942, Plant Mol. Biol., 11: 441.

Greenberg, B. M., Gaba, V., Canaani, O., Malkin, S., Mattoo, A. K., and Edelman, M., 1989, Separate photosensitizers mediate degradation of the 32-kDa photosystem II reaction center protein in the visible and UV spectral regions, Proc. Natl. Acad. Sci. USA, 86:6617.

Heber, U., Neimanis, S., and Dietz, K.-J., 1988, Fractional control of photosynthesis by the Q_B protein, the cytochrome f/b_6 complex and other components of the photosynthetic apparatus, Planta, 173:267.

Hirschberg, J., Ohad, N., Pecker, I., and Rahat, A., 1987a, Isolation and characterization of herbicide resistant mutants in the cyanobacterium Synechococcus R2, Z. Naturforsch., 42c:758.

Hirschberg, J., Ben Yehuda, A., Pecker, I., and Ohad, N., 1987b, Mutations resistant to photosystem II herbicides, in: "Plant Molecular Biology," D. von Wettstein and N.-H. Chua, eds., Plenum Publ. Corp., New York.

Horovitz, A., Ohad, N., and Hirschberg, J., 1989, Predicted effects on herbicide binding of amino acid substitutions in the D1 protein of photosystem II, FEBS Letters, 243:161.

Jansen, M. A. K., Malkin, S., and Edelman, M., 1989, Differential sensitivity of 32 kDa-D1 protein degradation and photosynthetic electron flow to photosystem II herbicides, Z. Naturforsch., 45c:408.

Koenig, F., 1987a, A role of the Q_B binding protein in the mechanism of cyanobacterial adaptation to light intensity? Z. Naturforsch., 42c:727.

Koenig, F., 1987b, General occurrence of a pool of the M_r 32000 Q_B binding polypeptide - D1 protein - in photosynthetic membranes? Biol. Chem. Hoppe Seyler, 368: 1260.

Koenig, F., 1990a, Shade adaptation in cyanobacteria, Photosynth. Res., 26:29.

Koenig, F., 1990b, Growth of Anacystis in the presence of thiosulphate and its consequences for the architecture of the photosynthetic apparatus, Botanica Acta, 103:54.

Koenig, F., 1990c, Anacystis mutants with different tolerances to DCMU-type herbicides show differences in architecture and dynamics of the photosynthetic apparatus, depending on site and mode of the amino acid exchanges in the D1 protein, Z. Naturforsch., 45c:446.

Lönneborg, A., Kalla, S. R., Samuelsson, P., Gustafsson, P., and Öquist, G., 1988, Light-regulated expression of the psbA transcript in the cyanobacterium Anacystis nidulans, FEBS Letters, 240:110.

Mattoo, A. K., Pick, U., Hoffman-Falk, H., and Edelman, M., 1981, The rapidly metabolized 32.000-dalton polypeptide of the chloroplast is the "proteinaceous shield" regulating photosystem II electron transport and mediating diuron herbicide sensitivity, Proc. Natl. Acad. Sci. USA, 78:1572.

Mattoo, A. K., Hoffman-Falk, H., Marder, J. B., and Edelman, M., 1984, Regulation of protein metabolism: Coupling of photosynthetic electron transport to in vivo degradation of the rapidly metabolized 32-kilodalton protein of the chloroplast membranes, Proc. Natl. Acad. Sci. USA, 81:1380.

Mattoo, A. K. and Edelman, M., 1985, Photoregulation and metabolism of a thylakoid herbicide-receptor protein, in: "Frontiers of membrane research in agriculture," E. Berlin and P.C. Jackson, eds., Rowman and Allanheld, Totowa.

Melis, A., Mullineaux, C. W., and Allen, J.F., 1988, Acclimation of the photosynthetic apparatus to photosystem I or photosystem II light: Evidence from quantum yield measurements and fluorescence spectroscopy of cyanobacterial cells, Z. Naturforsch., 44c:109.

Michel, H. and Deisenhofer, J., 1988, Relevance of the photosynthetic reaction center from purple bacteria to the structure of photosystem II, Biochem., 27:1.

Mohamed, A. and Jansson, C., 1989, Influence of light on accumulation of photosynthesis-specific transcripts in the cyanobacterium Synechocystis 6803, Plant Mol. Biol., 13:693.

Mullineaux, C. W. and Allen, J. F., 1990, State 1-state 2 transitions in the cyanobacterium Synechococcus 6301 are controlled by the redox state of electron carriers between photosystems I and II, Photosynth. Res., 23:297.

Murakami, A. and Fujita, Y., 1991, Regulation of photosystem stoichiometry in the photosynthetic system of the cyanophyte Synechocystis PCC 6714 in resonse to light intensity, Plant Cell Physiol., 32:223.

Ohad, N., Amir-Shapira, D., Koike, H., Inoue, Y., Ohad, I. and Hirschberg, J., 1990, Amino acid substitutions in the D1 protein of photosystem II affect Q_B^- stabilization and accelerate turnover of D1, Z. Naturforsch., 45c:402.

Ohad, N. and Hirschberg, J., 1990, A similar structure of the herbicide binding site in photosystem II of plants and cyanobacteria is demonstrated by site specific mutagenesis of the psbA gene, Photosynth. Res., 23:73.

Schaefer, M. R. and Golden, S. S., 1989a, Differential expression of members of a cyanobacterial psbA gene family in response to light. J. Bacteriol., 171:3973.

Schaefer, M. R. and Golden, S. S., 1989b, Light availability influences the ratio of two forms of D1 in cyanobacterial thylakoids, J. Biol. Chem., 264:7412.

Trebst, A., 1987, The three-dimensional structure on the herbicide-binding niche of the reaction center polypeptides of photosystem II, Z. Naturforsch., 42c:742.

Trebst, A., Depka, B., Kraft, B., and Johanningmeier, U., 1988, The Q_B site modulates the conformation of the photosystem II reaction center polypeptides, Photosynth. Res., 18:163.

Wettern, M. and Ohad, I., 1984, Light-induced turnover of thylakoid polypeptides in Chlamydomonas reinhardi, Israel J. Bot., 33:253.

WHAT DO WE KNOW ABOUT THE BIOGENESIS AND THE DEVELOPMENT OF SECONDARY

THYLAKOIDS (GRANA FORMATION) IN HIGHER PLANT SYSTEMS ?

J.A. De Greef

Laboratory of Plant Physiology and Biochemistry, Department
of Biology, University of Antwerpen, U.I.A. - R.U.C.A.,
Universiteitsplein 1, B-2610 Antwerpen-Wilrijk (Belgium)

KEYWORDS / ABSTRACT : etiolated bean seedlings / greening / broad-band
far red illumination / electron microscopy / chloroplast development /
phylogenetic hierarchy / ontogenetic stages / grana formation / stacking
/ primary and secondary thylakoids / thylakoid membrane biogenesis /
chloroplast architecture / 3-dimensional model.

When etiolated bean seedlings were greened with prolonged, broad band far
red light, the development of the chloroplast was much prolonged, resulting in a greater temporal resolution of sequential processes. With electron microscopy we could show that only primary thylakoids were built up
under far red light. Grana formation occurred when the seedlings were
exposed to white light. From these data a new model for the initiation of
secondary thylakoids from primary thylakoids is proposed.

INTRODUCTION

On the occasion of the International Meeting on Chloroplast Development, organized by George AKOYUNOGLOU and Joan ARGYROUDI-AKOYUNOGLOU in
1978 at the Greek island of Spetsai[1], a new model of 3-dimensional
chloroplast infrastructure was presented and published by Mustardy and
Brangeon[2] attempting to establish the formation and spatial relationship
of grana and the interconnecting fretwork during plastid differentiation
by serial section analysis of *Lolium multiflorum* L. leaves at 4 developmental stages. This model of organisation in higher plant chloroplasts is
the most recent concept of membrane stacking in grana with regard to the
progressive stages of elucidation suggested by Menke, Weier, Wehrmeyer,
Heslop-Harrison, Paolillo, Thomson and others as reviewed by Coombs and
Greenwood[3].

The structure of the internal lamellae system of mature chloroplasts
is extremely complex. According to the conventional view of thin sections
two distinct features of the membranous structure can be recognised : the
first, closely packed regions correspond to the grana seen under the
light microscope, and the second comprises the less dense but extensive

Regulation of Chloroplast Biogenesis
Edited by J.H. Argyroudi-Akoyunoglou, Plenum Press, New York, 1992

interconnecting stroma lamellae (=fretwork complex). All 3-dimensional models of chloroplast infrastructure are based on extrapolations from sectioned views of electron micrographs and have emphasized the continuity of the grana-fretwork complex.

On the basis of early E.M.-information showing aspects of chloroplast organization it was suggested that the grana consisted of a series of separate sac-like discs between 200 and 400 nm in diameter, stacked to form discrete cylindrical granal structures linked one with another by more extensive intergranal lamellae. Subsequent studies showed that the intergranal (= stroma) lamellae were perforated so forming a fretwork system connected to thylakoids at several levels within the same granum. Further studies indicated spiral formation of grana thylakoids and a spiral fretwork arranged around and interconnecting individual grana. In the Mustardy-Brangeon's model grana initiation takes place at the perforated sites of the primary thylakoids via overlapping lamellar growth originating from growing points at the margins of the holes. Due to this thylakoid overgrowth, the parent lamellae are tilted vis à vis the appressed discs and constitute the first helix around the grana. Secondary grana build-up occurs by infolding of grana discs or by the fret overtopping the grana stacks. The associated helix multiplication of frets proceeds by splitting of lamellae or helices and their eventual fusion with frets of neighbouring grana. This secondary differentiation process of the thylakoid system could be the basis of the flexibility and dynamic nature of the chloroplast architecture, allowing rapid changes in the organization and density of the membrane network in response to different environmental factors.

Coinciding with the latest vogue of plant molecular biology in the eighties, most recent studies addressing the biogenesis of thylakoid membranes deal with the import and processing of nuclear encoded polypeptides synthesized in the cytosol as precursors across the chloroplast envelope before being incorporated into the thylakoids (see Staehelin[4] for early review and Keegstra et al.[5], Cline et al.[6,7], and Lamppa et al.[8], these Proceedings) and with the identification and characterization of components of the protein import machinery of chloroplasts (Schnell et al.[9], these Proceedings).

In contrast to the role of the chloroplast envelope in thylakoid membrane biogenesis above mentioned, it should be noted that during chloroplast development, and during the greening of etioplasts in particular, conspicious and sometimes, quite long infoldings are seen in plastids. These observations suggested to early workers that thylakoids may arise from "budding" of the inner envelope membrane (von Wettstein[10]). Since no conclusive experimental evidence in support of this hypothesis has been reported, the original idea of von Wettstein was abandoned with progression of time.

At present there is renewed interest in the role of the chloroplast envelope membranes in thylakoid formation. When cells of dark-grown Chlamydomonas reinhardtii y-1, which are deficient in thylakoid membranes, were exposed to light for only several minutes, newly formed thylakoid membranes were observed emanating from the chloroplast envelope. At the same time chl-protein complexes increased at linear rates and most of the chl was recovered in rapidly sedimenting membrane fractions from broken cells. The inability to detect a soluble, processed form of the proteins implied that these remained associated with membranes. These results of Hoober et al.[11] (these Proceedings) suggest that thylakoid membranes develop in association with the chloroplast envelope membranes at sites of import of cytoplasmically made proteins. In their studies on pigments of the plastid envelope membranes Joyard et al.[12] (these Proceedings) were able to demonstrate that this membrane system could be involved in the synthesis of chlorophyll and in carotenogenesis.

From all these data it is clear that there is quite a lot of confusion about thylakoid membrane biogenesis for the present.

In this contribution we present another alternative/additional approach to understand the process of grana stacking in relation to the primary thylakoids.

METHODOLOGY

Several strategies can be used to elucidate the fine structure of photosynthetic membranes and their disposition either within the cytoplasm or within differentiated regions of chloroplasts :
(1) Comparative studies of phylogeneticly related organisms as shown in Fig.1, illustrating the general organization of photosynthetic membranes. In prokaryotes the photosynthetic apparatus is confined either to the cytoplasmic membrane, or to cell-membrane derived vesicles, or to flattened vesicles or thylakoids which can be differentiated into stacked and unstacked regions, traversing the cytoplasm. In eukaryotes the photosynthetic apparatus is compartmentalized in plastids. Rhodophyta exhibit the simplest organization of eukaryotic plastids. The structure, organization and chemistry of their thylakoids carries conviction of a primitive homology with the lamellae of Cyanobacteria ; they are single, evenly spaced, mostly parallel to the main axis of the organelle and lack appressed membrane regions, but they seem all to be interconnected within a given plastid. The arrangement of thylakoids varies systematically from simpler systems in lower plant classes to a more complex arrangement, basically of the same type, that has emerged as a more advanced form in photosynthetic organisms of higher ranking phylogenetic positions. Indeed, in the Phaeophyta thylakoids tend to be organized into bands of three with extensive regions of membrane adhesion (= giant grana). The further specialization of girdle bands of thylakoids is found in many species. All green algae and higher plants possess stacked membrane regions, but the arrangement of these stacked regions within chloroplasts is quite variable, suggesting that thylakoid membrane stacking is of great functional importance. In the Chlorophyta adjacent thylakoids (between 2 and 6) form stacked membrane regions (= pseudo-grana).
Invaginations of the inner membrane of the envelope are a common feature, particularly in developmental stages. They usually become less numerous towards maturity. Although the term "stack" is in common usage for all forms of thylakoid apposition a distinction has to be drawn between the type of less firm association between adjacent thylakoids that characterize the "bands" of the Phaeophyta (= giant grana) and the Chlorophyta (= pseudo-grana), on the one hand, and the type of firm bonding displayed by higher plants with aligned stacks of secondary thylakoids forming the partitions of true grana that are interconnected through typical, single, primary (stroma) thylakoids.
(2) Ontogenetic studies by serial section analysis of samples taken at determined intervals along the length of a grass leaf from the base (youngest part) up to the leaf tip (adult stage)[2].
(3) Prolongation of the period of chloroplast development by greening the leaves under different light régimes such as continuous low intensity white light[13] (WL), a series of brief flashes of red light[14] (R), preexposure to periodic WL illumination[15], prolonged irradiation with far red light[16,17] (FR). One advantage of extending the period of development is the greater temporal resolution of sequential processes.

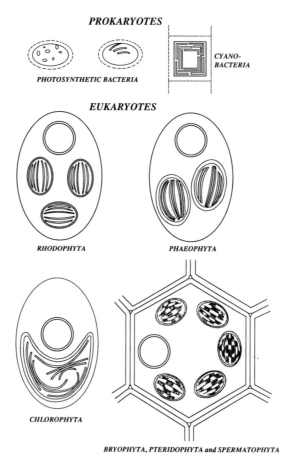

Fig. 1.   Schematic representation showing general trends in the
morphological diversity of photosynthetic membranes as they
are present in phylogeneticly related organisms.

RESULTS

When we exposed 8-day old dark-grown been seedlings to prolonged FR illumination (each day had 12 h FR and 12 h dark), etioplasts developed into photosynthetically active chloroplasts[17]. All development processes measured (leaf expansion, chloroplast differentiation, chlorophyll accumulation, photosynthetic activity and photophosphorylation) were much slower in FR light than in white (or red) light.

Electron micrographs of the photosynthetic apparatus greened with FR light revealed primarily the absence of fusion between well differentiated primary thylakoids and an obvious lack of true grana formation[17]. After 12 h FR the electron micrographs show essentially the same picture as that of etiolated tissue with large crystalline prolamellar bodies, but they contained much more lamellae. These lamellae or prothylakoids were discontinuous and appeared to consist of numerous small segments as if the process by which small vesicles coalesce to form large thylakoids, was not completed. Short regions of fusion between adjacent lamellae occurred at all stages including the dark grown leaf. Up to 60 h FR prolamellar bodies were visible, but they became gradually smaller. Irradiation with 30 min WL at any of these stages caused their complete disappearance. At the time of onset of photosynthetic activity (oxygen evolution, after 24 h FR, large uniform primary thylakoids began to appear. By 42 h FR the number of primary thylakoids had increased. They were single, evenly spaced and lined up parallel to each other and to the main axis of the organelle resembling very well the lamellae organization of the Rhodophyta plastid. By 60 h FR several bands of unfused parallel primary thylakoids were present, similar in appearance to the plastids of the Phaeophyta. These parallel stacks of primary thylakoids could be induced to form the grana-fretwork complex of mature chloroplasts by placing the FR treated leaves in white light.

A detailed E.M. study of the effect of continuous WL after a pretreatment of the 8-day old dark-grown bean seedlings with 60 h FR upon chloroplast development revealed the following features :
1) During the first hours of WL illumination long stretches of fusion could be seen between adjacent, primary thylakoids, mostly in parallel alignment and evenly spaced throughout the stroma ; the 60 h FR bands of unfused parallel primary thylakoids were not present anylonger.
2) From 12 h WL on distinct grana formation could be observed and we noticed a gradual increase in the height of the stacks (number of secondary thylakoids) up to 72 h WL when lens shaped, mature chloroplasts were present.
3) During all the time of stacking small vesicular invaginations of the inner membrane of the plastid envelope together with small vesicles in the peripheral stroma were evident.
4) Between 3 h and 6 h exposure to WL, at the onset of the stacking process, a peculiar phenomenon occurred at the membrane level of the primary thylakoids ; most electron micrographs taken in this time lapse showed many blister-like protrusions of the thylakoid membrane facing the stroma : several of these protrusions had the shape of a truncated cone (Fig. 2, a) ; in a few cases we noticed the presence of numerous membrane bounded, empty discs throughout the stroma of some plastids (Fig. 2, b) ; the diameter of these discs varied from 0.2-0.5 µm, conformable to the diameter of the cylinderical grana structure ; therefore, it seems reasonable to us that the E.M. view of the cut shown in Fig. 2, b is a tangential section to the membrane plane of the cross-section presented in Fig. 2, a ; in Fig. 2, a and b it can also be seen that some of the blisters and

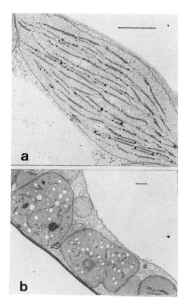

Fig. 2. The onset of grana formation of etiolated bean leaves being
exposed to 3 h WL after 60 h FR pretreatment :
a)   extensive regions of blister-like protrusions of
      thylakoids in cross-section
b)   membrane bounded discs in developing plastids
(each bar represents 1 μm)

discs show membrane infoldings. Since these blister-like protrusions are
not frequently seen in E.M. sections, they must be very short-lived and
thus unstable. They become probably stabilized when they fold into grana
structures by addition of substances from plastidic and cytoplasmic ori-
gins.

Our recent data can be interpreted either as an alternative model or
as an additional mode of display to the existing models in order to visualize the continuity of primary and secondary thylakoids (the grana-
fretwork complex in classical terms).

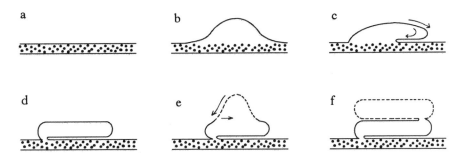

Fig. 3.   Two dimensional model of the onset of grana (secondary
          thylakoid) formation from a primary thylakoid :
          a)   primary thylakoid
          b)   blister-like protrusion with the shape of a
               truncated cone
          c)   invagination of the membrane of the cone at its
               base and flattening of the cone-shaped blister
          d)   formation of the first secondary thylakoid inter-
               connected with the primary thylakoid by a kind of
               an isthmus
          e)   blistering of the stroma-faced membrane of the
               secondary thylakoid and invagination of the cone's
               base at the opposite side of the first infolding
          f)   a new secondary thylakoid is laid down on the first
               one and interconnected by another isthmus.

In Fig. 3 the subsequent steps of secondary thylakoid formation from
a primary thylakoid (Fig. 3, a) is illustrated. A blister is formed (Fig.
3, b) and the protruded membrane section of the blister folds by invagination (Fig. 3, c) to form the first secondary thylakoid interconnected
with the parent primary thylakoid by a kind of umbilical cord or isthmus
(Fig. 3, d) ; since this connection is only a small part of the granal
disc, it cannot often be seen on cross-section. The stroma faced membrane
of this secondary thylakoid can blister again (Fig. 3, e) and the membrane of the protrusion will invaginate in the opposite direction of the
first formed grana (secondary) thylakoid until it attains the size of
the underlying disc (Fig. 3, f). This process can be repeated till an
overtopping primary thylakoid is reached and fused into the growing stack
of secondary thylakoids.

CONCLUSION

Studying the ultrastructural organization of the chloroplast in higher plants in order to make a 3-dimensional model of its architecture, based on extrapolations from sectioned views, it is useful for the interpretation of the electron micrographs to keep in mind that there is a phylogenetic relationship between the different forms of photosynthetic apparatus that has evolved in time. In its origin the basic framework of the inner membrane of the envelope, on the one hand, and that of the network of thylakoids, on the other, is primarily prokaryotic. In the symbiotic phase of chloroplast development phylogeneticly characteristic, eukaryotic, nuclear-encoded components have been inserted in the envelope and in the thylakoid membranes, regulated by both plastidic factors and gene expression. Therefore, it is quite conceivable that during the morphogenesis of chloroplasts in higher plants a number of prokaryotic features can be met, e.q. invagination and protrusion of membranes, vesicle formation, coalescence of different membrane structures into a larger assembly. It is of interest that the time course for each phase of grana formation in our model is very short. Indeed, it can be seen in Fig. 2, b the membrane bounded discs all appear at a given time, indicating a sharp onset of activity with a high degree of synchrony within and among the chloroplasts.

These short time events make it difficult to analyse the sequence of develomental processes occurring in grana formation. Everyway, it seems to us that the far red greened leaves should provide an excellent experimental system to explore the light requirements for the stacking process.

I wish to dedicate this article in remembrance of George AKOYUNO-GLOU, an outstanding scientist and a thoughtful, warm-hearted friend, who has spent so much of his life to the knowledge of chloroplast development.

REFERENCES

1. G. Akoyunoglou and J. H. Argyroudi-Akoyunoglou, "Chloroplast Development", Elsevier/North-Holland Biomedical Press, Amsterdam, New York, Oxford, (1978).
2. L. A. Mustardy and J. Brangeon, 3-Dimensional chloroplast infrastructure : developmental aspects., in: "Chloroplast Development", G. Akoyunoglou & J. H. Argyroudi-Akoyunoglou, eds., Elsevier/North-Holland Biomedical Press, Amsterdam, New York, Oxford (1978).
3. J. Coombs and A. D. Greenwood, Compartmentation of the photosynthetic apparatus., in: "The Intact Chloroplast", J. Barber, ed., Elsevier/North-Holland Biomedical Press, Amsterdam, New York, Oxford (1976).
4. L. A. Staehelin, Chloroplast structure and supramolecular organization of photosynthetic membranes., in: "Encyclopedia of Plant Physiology, New Series, Volume 19, Photosynthesis III", L. A. Staehelin & C. J. Arntzen, eds., Springer-Verlag, Berlin, Heidelberg, New York, Tokyo, (1986).
5. K. Keegstra, B. Bruce, Y. Kaneko, H. Li, J. Marshall, J. Ostrom and S. Perry, Transport of proteins into and across the envelope membranes of chloroplasts., in: "Regulation of Chloroplast Biogenesis", J. H. Argyroudi-Akoyunoglou, ed., Plenum Publishing Corporation, New York (1991).

6. K. Cline, W. F. Ettinger and S. M. Theg, Studies on the transloca-
    tion of proteins into and across the thylakoid membrane., in:
    "Regulation of Chloroplast Biogenesis", J. H. Argyroudi-
    Akoyunoglou, ed., Plenum Publishing Corporation, New York
    (1991).
7. C. Dahlin and K. Cline, The plastid protein import apparatus in
    wheat seedlings is developmentally regulated., in: "Regulation
    of Chloroplast Biogenesis", J. H. Argyroudi-Akoyunoglou, ed.,
    Plenum Publishing Corporation, New York (1991).
8. G. Lamppa, M. Abad, S. Clark, J. Oblong and L.-M. Yang, Import and
    processing of the light-harvesting chlorophyll a:b binding pro-
    tein., in: "Regulation of Chloroplast Biogenesis", J. H.
    Argyroudi-Akoyunoglou, ed., Plenum Publishing Corporation, New
    York (1991).
9. D. J. Schnell, D. Pain and G. Blobel, Identification and characteri-
    zation of components of the protein import machinery of
    chloroplasts., in: "Regulation of Chloroplast Biogenesis", J.
    H. Argyroudi-Akoyunoglou, ed., Plenum Publishing Corporation,
    New York (1991).
10. D. von Wettstein, 1959, in: "Light, physical and biological action",
    H. Seliger & W. D. McElroy, eds., Academic Press, New York
    (1965).
11. J. K. Hoober, D. B. Marks, J. L. Gabriel and L. G. Paavola,
    Thylakoid membrane biogenesis : the role of the chloroplast
    envelope., in: "Regulation of Chloroplast Biogenesis", J. H.
    Argyroudi-Akoyunoglou, ed., Plenum Publishing Corporation,
    New York (1991).
12. J. Joyard, M. A. Block and R. Douce, Pigments of the plastid enve-
    lope membranes., in: "Regulation of Chloroplast Biogenesis", J.
    H. Argyroudi-Akoyunoglou, ed., Plenum Publishing Corporation,
    New York (1991).
13. Y. Eilam and S. Klein, The effect of light intensity and sucrose
    feeding on the fine structure and chlorophyll content in etio-
    lated leaves. J. Cell Biol. 14:169 (1962).
14. C. Sironval, J. M. Michel, R. Bronchard and E. Englert-Dujardin, On
    the primary thylakoids of chloroplasts grown under a flash
    regime., in: "Progress in Photosynthesis Research", H. Metzner,
    ed., International Union of Biological Sciences, Tübingen
    (1969).
15. G. Akoyunoglou, J. H. Argyroudi-Akoyunoglou, J. H. Christias, S.
    Tsakiris and M. Tsimili-Michael, Thylakoid growth and differen-
    tiation in continuous light as controlled by the duration of
    preexposure to periodic light. in: "Chloroplast Development",
    G. Akoyunoglou & J. H. Argyroudi-Akoyunoglou, eds., Elsevier/
    North-Holland Biomedical Press, Amsterdam, New York, Oxford
    (1978).
16. M. Häcker, Der Abbau von Spreicherprotein und die Bildung von
    Plastiden in den Kotyledonen des Senfkeimlings (Sinapis alba
    L.) unter dem Einfluss des Phytochroms. Planta, 76:309 (1967).
17. J. De Greef, W. L. Butler and T. F. Roth, Greening of etiolated bean
    leaves in far red light. Plant Physiol., 47:457 (1971).

GENETIC MANIPULATION AS A MEANS TO MONITOR
ASSEMBLY AND FUNCTION

SITE DIRECTED MUTAGENESIS OF A CHLOROPLAST ENCODED PROTEIN

Elisabeth Przibilla and Raina Yamamoto

Lehrstuhl für Biochemie der Pflanzen
Ruhr-Universität
D-4630 Bochum, Germany

ABSTRACT

The *psbA* gene of wild type *Chlamydomonas reinhardtii* was mutated using the tungsten particle bombardment method. To select transformants two different approaches were used. The thylakoids of the mutant strains were tested for sensitivity to phenolic herbicides.

INTRODUCTION

The reaction center of photosystem II of oxygenic organisms consists of the D1, D2, and cytochrome b559 polypeptides. Many different types of herbicides inhibit photosynthesis by replacing the second electron acceptor Q_B from its binding site at the D1 protein. Amino acid exchanges within this binding site result in changes of the affinity of herbicides to the D1 protein and can be used to draw conclusions about its three-dimensional structure.

Since bacteria and cyanobacteria can easily be transformed it was possible to engineer their genes *in vitro* and study *in vivo* the function of specific point mutations. Recently it became possible to use the same approach also in eukaryotic organisms. Here we show the generation and characterisation of two *Chlamydomonas reinhardtii* mutants having asparagine 266 of the D1 protein replaced by threonine, either single, or in combination with a change of amino acid 264.

MATERIALS AND METHODS

Chlamydomonas reinhardtii wild type strain 11/32 b was grown in liquid HSHA medium (Sueoka, 1960) in light (7000 Lux), harvested at mid-log phase, and its chloroplasts transformed according to the procedure described by Boynton et al., 1988. Chloroplast DNA copy number was not reduced prior to transformation. One day after bombardment cells were spread onto high salt medium containing either 100 µg/ml spectinomycin or 10^{-5}M metribuzin, depending on the plasmids used for transformation.

Plasmid pcp58 containing *C. reinhardtii* chloroplast DNA EcoR1 fragment R24 was obtained from Dr. Rochaix. It was used to subclone the HindIII-KpnI

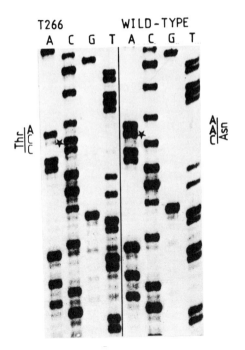

Figure 1

Sequence comparison of mutant T266 and wild type *psbA* gene of *C.reinhardtii*. Total RNA was isolated from cells and used as template for AMV reverse transcriptase after annealing of a 23 bases oligonucleotide complementary to nt. 874 to 896 of the *psbA* gene. Chains of cDNA were terminated due to the incorporation of dideoxy nucleotides, and separated on an 8% polyacrylamide gel.

1100 bp fragment including exon 5 of the *psbA* gene. The wild type sequence was changed in vitro by site-directed mutagenesis using the method of Nakamaye and Eckstein (1986).

The clone having only the change of amino acid 266 was cotransformed with plasmid P228. P228 harbours the 16S rRNA gene from a spectinomycin resistant *C.reinhardtii* strain (Harris et al., 1989) and was obtained from Dr. Randolph-Anderson. Total RNA was prepared as described (Johanningmeier et al., 1987), cDNA was produced and sequenced according to Fichot and Girard, 1990.

Thylakoids were isolated from mutant and wild type cells and electron transport measured from water to DCPIP. The pI_{50} value, i.e. the concentration of herbicide inhibiting the rate of DCPIP reduction by 50 %, was determined (Pucheu et al.,1984).

RESULTS AND DISCUSSION

There are two kinds of herbicides inhibiting photosystem II. Information has accumulated about the "classical" type that allowed to construct a computer-based model of their binding niche (Tietjen et al., 1991). On the other hand it is controversial, how the "phenol-type" herbicides are arranged within this niche because of a lack of mutants being resistant against these. There are two exceptions. In *C.reinhardtii* a mutant of Valin 219 has been described which shows resistance against short alkyl unsubstituted, but not against ring substituted bromonitrophenols (Wildner et al., 1990). In addition, a cyanobacterium mutant of amino acid 266 has been found, that is less sensitive to ioxynil (Aljani et al., 1989). It would be useful to have this same mutation within in eukaryotic organism for comparison.

The introduction of DNA into chloroplasts has been shown by complementation of a deletion mutant of the *atpB* gene and selection for photosynthetic cells (Boynton et al., 1988). In the case of the *psbA* gene such a deletion mutant is available, but causes several problems (Przibilla et al. 1991). It would be possible to use wild type cells as recipients instead, if the phenotype of the mutated strain would be known and could be selected for.

In our case this was not possible. So a second point mutation of amino acid 264 was introduced on the same plasmid which allowed selection of transformants on metribuzin. In another approach an unrelated marker conferring resistance to spectinomycin was used for cotransformation.

We isolated RNA from resistant colonies, converted it to cDNA and sequenced it. All colonies resistant against metribuzin contained the change of amino acid 266 in addition to that of 264.

Out of ten spectinomycin resistant colonies two had the mutation at amino acid 266. Figure 1 shows a comparison of mutant and wild type sequence. There are some background signals visible in several places of the autoradiogram. Probably they are the result of premature termination of newly synthesized DNA chains by the reverse transcriptase, but we can not fully exclude the possibility that there is a small percentage of wild type DNA copies in the mutant strains. On the other hand Boynton et al. (1990) found that the transformed genoms tend to segregate to produce homoplasmic cells as long as there is no selection pressure which forces the cells to keep both kinds of copies. Also the double mutant A264T266 described above is stable since more than 18 months even without any selection pressure.

The cotransformation procedure described above does not work equally well in all cases. In an experiment designed to change amino acid 237 of the *psbA* gene, out of 35 spectinomycin resistant colonies not a single one had the mutation wanted. During preparation of this work the same cotransformation approach was published by Dr. Kindle (1991). She also used a wild type strain as recipient and introduced a plasmid conferring spectinomycin resistance together with an *atpB* deletion construct. Out of fifty resistant colonies thirteen contained the deletion. In this case the recipient cells were pretreated with FdUrd to reduce plasmid DNA copy number. Such a treatment was shown to increase the frequency of transformation dramatically (Boynton et al., 1990). We did not use FdUrd because it not only reduces chloroplast DNA content but also is mutagenic in its own right (Harris, 1989), and we wanted to keep the probability low for undetected additional point mutations.

Table 1

R/S values (ratio of the pI_{50} of the mutant to that of the wild type) of thylakoids isolated from mutant and wild type cells. Electron transport was measured from water to DCPIP.

Inhibitor	T266	A264T266	A264
Hydroxyquinoline	0.8	1.6	0.5
Ioxynil	2.5	0.8	0.5
Ketonitrile	1.3	0.4	0.6
Bromonitrothymol	1.3	0.2	0.4
Benzylbromonitrophenol	1.3	0.4	n.d.
Phenyliodonitrophenol	1.3	0.5	n.d.

In table I the resistance of mutant thylakoids to six different phenolic herbicides is compared to that of wild type thylakoids. The mutants are consistently somewhat less sensitive to ioxynil, this is true also for the double mutant-the mutation of 264 alone makes cells more sensitive to phenolic herbicides. The change of asparagine to threonine in position 266 reverts this effect. Although clearly an interaction of amino acid 266 with phenolic herbicides is there, the small effect is not sufficient to assign it a major contribution to the binding strength.

ACKNOWLEDGEMENT

The authors want to thank U. Hilp for excellent technical assistance in determination of the pI_{50}-values. This work was supported by the BMFT projekt no. 0139308A.

REFERENCES

Ajlani, G., Meyer, I., Vernotte, C., and Astier, C., 1989, Mutation in phenol-type herbicide resistance maps within the *psbA* gene in *Synechocystis* 6714. FEBS Lett.246:207.

Boynton, J.E., Gillham, N.W., Harris, E.H., Hosler, J.P., Johnson, A.M., Jones, A.R., Randolph-Anderson, B.L., Robertson, D., Klein, T.M., Shark, K.B., and Sanford, J.C., 1988, Chloroplast transformation in *Chlamydomonas* with high velocity microprojectiles. Science 240:1534.

Boynton, J.E., Gillham, N.W., Harris, E.H., Newman, S.M., Randolph-Anderson, B.L., Johnson, A.M., Jones, A.R., 1990, Manipulating the chloroplast genome of *Chlamydomonas* - molecular genetics and transformation, in: Current Research in Photosynthesis, Vol.III, M.Baltscheffsky, ed., Kluver Academic Publishers, Netherlands.

Fichot, O., and Girard, M., 1990, An improved method for sequencing of RNA templates. Nucl.Acid Res.18:6162.

Harris, E.H., 1989, The *Chlamydomonas* sourcebook, Academic Press, San Diego.

Harris, E.H., Burkhart, B.D., Gillham, N.W., and Boynton, J.E., 1989, Antibiotic Resistance Mutations in the Chloroplast 16S and 23S rRNA Genes of *Chlamydomonas reinhardtii*: Correlation of Genetic and Physical Maps of the Chloroplast Genome, Genetics 123:281.

Johanningmeier, U., Bodner, U., and Wildner, G.F., 1987. A new mutation in the gene coding for the herbicide-binding protein in *Chlamydomonas*. FEBS Lett. 211:221.

Kindle, K.L., Richards, K.L., and Stern, D.B., 1991, Engineering the chloroplast genome: Techniques and capabilities for chloroplast transformation in *Chlamydomonas reinhardtii*, Proc. Natl. Acad. Sci. USA 88: 1721.

Nakamaye, K.L., and Eckstein, F., 1986, Inhibition of restriction endonuclease NciK cleavage by phosphothiorate groups and its application to oligonucleotide-directed mutagenesis. Nucl. Acids Res. 14:9679.

Przibilla, E., Heiss, S., Johanningmeier, U., and Trebst, A., 1991, Site-specific mutagenesis of the D1 Subunit of photosystem II in wild-type *Chlamydomonas*, The Plant Cell 3:169.

Pucheu, N., Oettmeier, W., Heisterkamp, U., Masson, K., and Wildner, G.F., 1984, Metribuzin-resistant mutants of *Chlamydomonas reinhardtii*, Z. Naturforsch. 39c:437.

Sueoka, N., 1960, Mitotic replication of deoxyribonucleic acid in Chlamy domonas reinhardtii, Proc.Natl.Acad.Sci.USA 46:83.

Tietjen, K.G., Kluth, J.F., Andree, R., Haug, M., Lindig, M., Müller, K.H., Wroblowsky, H.J., and Trebst, A., 1991, The herbicide binding niche of photosystem II - a model, Pestic.Sci. 31:65.

Wildner, G.F., Heiterkamp, U., and Trebst, A., 1990, Herbicide cross resistance and mutations of the *psbA* gene in *Chlamydomonas reinhardtii*, Z. Naturforsch. 45c:1142.

SITE-DIRECTED MUTAGENESIS TO PROBE THE ROLE OF THE D2 PROTEIN IN

PHOTOSYSTEM II

Beth Eggers and Wim Vermaas

Department of Botany and Center for the Study of Early
Events in Photosynthesis, Arizona State University, Tempe,
AZ 85287-1601, USA

INTRODUCTION

Photosystem II (PS II) is responsible for the light-driven redox
reactions transfering electrons from water to the plastoquinone pool
resulting in the evolution of oxygen (Vermaas and Ikeuchi, 1991;
Andersson and Styring, 1991). PS II consists of four major and several
smaller ($\leq 10,000$ M_r) integral membrane proteins and also several
peripheral polypeptides. The integral proteins include CP43, CP47, D1,
D2, PSII-I, and cytochrome b559. CP43 and CP47 are chlorophyll-binding
proteins and act as light-harvesting antenna. D1 and D2 are the two
reaction center proteins that together create the binding environment for
the reaction center components, for the quinone-type electron acceptors,
Q_A and Q_B, and presumably for Mn involved in water splitting. The
functions of cytochrome b559 and of PS II-I, the product of the *psbI*
gene, are still unknown. The most important peripheral protein is the 33
kDa manganese-stabilizing protein (MSP), also known as PS II-O or OEE-1
with OEE standing for oxygen-evolving enhancer. This 33 kDa protein is
located on the lumenal side of the membrane and is thought to be involved
in stabilizing the manganese that is necessary for the water-splitting
process. However, recent *in vivo* experiments have shown that in the
cyanobacterium *Synechocystis* sp. PCC 6803 the MSP is not absolutely
essential for the stable assembly of reaction centers and for oxygen
evolution (Burnap *et al.*, 1991; Philbrick *et al.*, 1991; S. Mayes and J.
Barber, personal communication). However, in the alga *Chlamydomonas
reinhardtii* (Mayfield *et al.*, 1987), the 33kDa protein appears to be
required for oxygenic photosynthesis and PS II stability.

In the past few years, cyanobacteria have become increasingly
attractive for use in the study of oxygenic photosynthesis. This is due
to the amenability of certain cyanobacterial strains to genetic
manipulation and the similarity of the photosynthetic complexes in
cyanobacteria and higher plants. Selected strains of cyanobacteria are
naturally transformable: these organisms take up foreign DNA and
incorporate it into their genome by homologous recombination (Grigorieva
and Shestakov, 1982; Williams, 1988; Golden *et al.*, 1987). Some species
of cyanobacteria, such as *Synechocystis* sp. PCC 6803 and *Synechococcus*
sp. PCC 7002, are not only spontaneously transformable, but also are
facultative photoheterotrophs and can be propagated on medium containing
a suitable carbon source when PS II is inactivated. This makes such
organisms suitable experimental systems for studying mutations that
impair the function of PS II.

Regulation of Chloroplast Biogenesis
Edited by J.H. Argyroudi-Akoyunoglou, Plenum Press, New York, 1992

X-ray diffraction analysis of crystallized bacterial reaction centers (Deisenhofer *et al.*, 1985; Feher *et al.*, 1989; Michel *et al.*, 1986) has led to detailed information regarding the structure of the reaction center from such bacteria. Because the D1 and D2 proteins are functionally similar to the L and M subunits of the photosynthetic purple bacterial reaction center and share structural similarity in functionally important domains (Hearst, 1986; Michel and Deisenhofer, 1988; Rutherford, 1986; 1987), information from the bacterial reaction center has yielded insight into which residues and domains of the PS II reaction center may be of functional importance. However, even though drawing analogies between reaction centers from purple bacteria and PS II can lead to valuable hypotheses regarding PS II structure and function, such hypotheses will need to be tested experimentally. To gain further information regarding PS II function, structure, and assembly, we have utilized site-directed mutagenesis in the cyanobacterium *Synechocystis* sp. PCC 6803, focusing on the D2 protein, one of the reaction center components in PS II.

THE D2 PROTEIN

On the basis of homologies with the reaction center from photosynthetic purple bacteria, the D2 protein is thought to be involved in the binding of P680 and the non-heme iron, and to create the majority of the binding pocket for Q_A and a pheophytin (Michel and Deisenhofer, 1988). From sequence comparison between D2 and the M subunit from bacterial reaction centers, D2 residues that may serve as ligands or that are close to cofactors can be identified. These residues can then be targeted by site-directed mutagenesis to determine their role in PS II.

In *Synechocystis* 6803 and other cyanobacteria, two genes, *psb*DI and *psb*DII, code for very similar D2 proteins (Williams & Chisholm, 1987). The *psb*DI gene is predominantly expressed (Yu and Vermaas, 1990). The *psb*DI gene is located in an operon with the *psb*C gene, which codes for CP43. The 3' end of *psb*DI (excluding the stop codon) overlaps with 5' end of *psb*C by fourteen bases. The techniques for *in vitro* oligonucleotide-directed mutagenesis are now relatively standard (Kunkel *et al.*, 1987; Vandeyar *et al.*, 1988; Zoller and Smith, 1983). However, to monitor the effects of the expression of an altered D2 gene in *Synechocystis*, first both wild-type genes had to be deleted from this organism. This was accomplished by replacing the *psb*DI/C operon and the *psb*DII gene with two different antibiotic-resistance cartridges. Upon deletion of *psb*DI/C and *psb*DII, PS II reaction centers no longer accumulate in the thylakoid membrane, but the phycobilin to chlorophyll ratio is not significantly changed (Fig. 1). Because most chlorophyll in cyanobacteria is associated with PS I, this implies that the phycobilisome/PS II ratio is flexible and indicates that expression of phycobilisome genes is not regulated by the presence of PS II.

Site-specific mutations in D2 were made in the *psb*DI gene in a plasmid which also contains regions up- and downstream from the *psb*DI/C operon and includes a third antibiotic-resistance cartridge. The *Synechocystis* 6803 double-deletion mutant (lacking *psb*DI and *psb*DII) is then transformed with the plasmid containing the altered *psb*DI gene. Thus, the mutated *psb*DI gene is the only gene of the *psb*D family that is present in the resulting cyanobacterial mutant. A detailed overview of procedures used in the generation of the D2 mutants in *Synechocystis* 6803 has been provided elsewhere (Vermaas *et al.*, 1990a).

THE Q_A-BINDING REGION

In the D2 protein, a histidine residue (His-214 in *Synechocystis* 6803) is in a position homologous to the His residue (M217) in the M subunit of *Rhodopseudomonas viridis* that is associated with the primary quinone (Q_A)

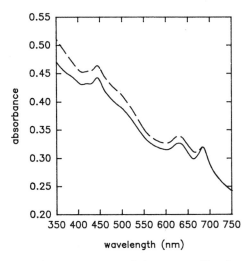

Fig. 1. Absorption spectrum of intact cells from wild-type and the *psb*DI/C/DII-deletion mutant of *Synechocystis* 6803. (solid line: wild-type; dashed line: mutant)

and with the non-heme iron. When His-214 of D2 was replaced by asparagine, the resulting mutant did not contain structurally or functionally intact PS II complexes (Vermaas *et al.*, 1987). To further study the Q_A-binding environment, residues near His-214 were targeted. In *Rps. viridis* there is a glycine residue (Gly-M218) following the histidine residue that forms a ligand to Q_A. Gly-215 of D2, the residue next to His-214, has been changed to tryptophan. The resulting mutant (G215W) was shown to be highly susceptible to photoinactivation of PS II without dramatic impairment of PS II structure and function (Vermaas *et al.*, 1990b). Several other amino acid substitutions have been made at this residue. Upon replacement of Gly-215 by an aspartic acid or a valine residue, an obligate photoheterotrophic phenotype results. In contrast to G215W, G215D and G215V are unable to evolve oxygen, even for short periods. The D2 protein is detectable in thylakoid membrane preparations from these mutants (Fig. 2), but intact PS II reaction centers can not be detected by diuron binding assays (Table I). However, fluorescence induction measurements on thylakoids from G215V and G215D show a rapid induction, reminiscent of induction in the presence of PS II herbicides. This suggests that these mutants are blocked in electron transfer from Q_A to Q_B, and that diuron binding has been lost in these mutants. Even though the precise role of Gly-215 is still obscure, it appears to be related to the structure of the Q_A binding site and to the properties of Q_A and/or the non-heme iron. For example, changes at this site could affect Q_A binding or the redox properties of Q_A resulting in the loss of electron transport.

In *Rps. viridis*, the Ala M216 residue next to His-M217 (ligand to Q_A), along with Val M263 and Thr M220, help to form the Q_A-binding site (Michel *et al.*, 1986). The Thr-217 residue, which is homologous to Thr-M220, was changed to serine. The resulting mutant had an apparently wild-type phenotype (Table I). Thus, replacement of Thr-217 by another residue with a hydroxy group does not lead to dramatic effects on PS II function but this mutant will need to be investigated further.

An important part of the Q_A-binding site in purple bacteria is formed by a tryptophan residue (Trp-M250 in *Rps. viridis*). This residue

Table I. Characteristics of selected D2 mutants and wild-type of
Synechocystis sp. PCC 6803.

Strain	Photoautotrophic growth	Presence of D2	Diuron Binding
Wild-type	+	+++	++
G215D	−	++	−
G215V	−	++	−
T217S	+	nd	nd
F252C	+	nd	nd
F252Y	+	nd	nd
W253F	+	++	+
H268Q	−	++	−
F270C	+	nd	nd
F270Y	+	nd	nd

nd = not determined

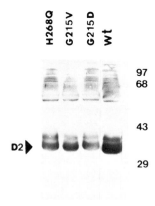

Fig. 2. Immunoblot of G215D, G215V,
H268Q, and wild-type of *Synechocystis*
6803. Blots were probed with
antisera raised against spinach D2.

is thought to be important for rapid electron transfer between pheophytin
and Q_A (Plato *et al.*, 1989). In D2 a homologous Trp residue (Trp-253)
appears to have a homologous function in PS II. Upon mutation of this
residue to Leu, a mutant without functionally or structurally intact PS
II centers was obtained (Vermaas *et al.*, 1990b). This is in contrast to
the situation in purple bacteria (Coleman and Youvan, 1990), where
mutation of Trp to Leu leads to loss of Q_A but not to a loss of reaction
centers. Thus, binding of Q_A appears much more important for reaction
center stability and assembly in PS II than in purple bacteria. When
Trp-253 of D2 was replaced by Phe, the resulting mutant was
photoautotrophic and had about 40% of the diuron binding ability as
compared to wild-type (Table I). In purple bacteria, a Trp-to-Phe
mutation leads to mutants with rather normal phenotypes (Coleman and
Youvan, 1990). Replacement of Trp-253 with another aromatic residue thus
may change the conformation of the binding pocket for Q_A enough to cause
a destabilization and a decrease in the number of PS II centers, but does
not prevent binding of Q_A or electron transfer to Q_A. The W253F mutant
has not been characterized yet in terms of redox properties of Q_A and
electron transport kinetics between Pheo and Q_A. Substitutions at Phe-
252, next to Trp-253, to cysteine or tyrosine result in photoautotrophic

mutants (Table I). Changes at Phe-252 thus seem to have little effect on the stability of the PS II complex and on the binding of Q_A.

In addition to Q_A, the non-heme iron also appears to be critical for PS II function. The non-heme iron (Fe^{2+}) is located between the two plastoquinones, Q_A and Q_B; in purple bacteria the Fe^{2+} is liganded by four histidine residues (two from L, two from M) and a glutamic acid residue in the M subunit. To probe the role of a D2 His residue analogous to one of the Fe ligands in purple bacteria, the His residue (H268 of D2) was changed to a glutamine. Immunoblots (Fig. 2) show a large amount of D2 protein present in thylakoids of the mutant (H268Q). However, H268Q is an obligate photoheterotroph that is unable to bind diuron (Table I) but does have variable fluorescence with induction kinetics similar to those seen in wild-type systems in the presence of diuron (not shown). Thus, electron transfer presumably cannot proceed beyond Q_A in this mutant, indicative of a role of the non-heme iron in electron transfer in PS II. This is in contrast to the situation in purple bacteria, where the non-heme iron can be extracted without dramatic effects on electron transfer to Q_B (Debus et al., 1986).

In Rps. viridis, a Trp residue, located two amino acids down from His-M266 (which forms one of the ligands to the non-heme iron) is known to be in reasonably close contact with Q_A and Fe^{2+}. In D2, a Phe in a homologous position (Phe-270) has been changed to a cysteine and a tyrosine resulting in photoautotrophic mutants (Table I).

Thus, comparison of the mutational effects on the acceptor side of PS II with those in purple bacteria demonstrates that, in spite of obvious homologies between the two systems, the structure and function of PS II appears much more susceptible towards mutations in putative cofactor ligands than is the case in purple bacteria.

THE PS II DONOR SIDE

D2 appears to be involved in events on the PS II donor side. It has been shown by site-directed mutagenesis and EPR analysis of mutants that a tyrosine residue of D2 (Tyr-160) is D, an electron donor in PS II (Debus et al., 1988a; Vermaas et al., 1988). A similar mutation of the homologous tyrosine residue in the D1 protein leads to an inhibition of electron transport in PS II (Debus et al., 1988b; Metz et al., 1989), and this D1 residue, Tyr-161, is considered to be Z, the physiological electron donor to P680. D2 also appears to be involved in creating the binding environment for the water-splitting system. A residue, Glu-69, on the D2 protein appears to affect the binding of Mn which is necessary for water-splitting (Vermaas et al., 1990c). Additional evidence for involvement of the D1/D2 reaction center complex in water splitting comes from analysis of a mutant, LF-1, of the green alga Scenedesmus obliquus. This mutant is defective in C-terminal processing of the D1 protein (Diner et al., 1988) and fails to functionally assemble the Mn cluster necessary for water-splitting. The mutant shows a reduction in the number of high-affinity sites for Mn binding (Seibert et al., 1989). This suggests that the C-terminal end of D1 bears Mn-binding sites.

Based on the two-fold symmetry in the PS II structure, we are interested to investigate whether the C-terminus of D2 also is involved in binding of Mn. The D2 protein has a long hydrophilic C-terminal stretch and the protein is not C-terminally processed as is D1. We used site-directed mutagenesis to introduce premature stop codons into psbDI resulting in the early termination of translation and thus a truncated D2 protein. The loss of 9 (ET-9), 14 (ET-14), 16 (ET-16), or 57 (ET-57) residues from the C-terminal end of D2 results in obligate photoheterotrophic mutants (Table I). In the mutants ET-16 and ET-57 no D2 protein was detected in the thylakoids, suggesting a destabilization of the PS II complex (Fig. 3). However,

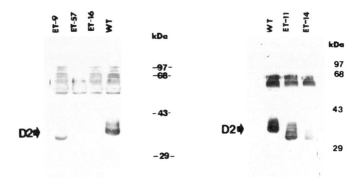

Fig. 3. Immunoblots of the early termination mutants (ET-57, ET-16, ET-9, ET-11, and ET-14) and wild-type of *Synechocystis* 6803. Blots were probed with antisera raised against spinach D2.

Table II. Characteristics of wild type and mutants with early translation termination (ET) of D2 of *Synechocystis* sp. PCC 6803. The number in the strain designation indicates the number of residues that are missing on the C-terminal end of D2 in the respective mutant.

Strain	Photoautotrophic growth	Presence of D2	Diuron Binding
Wild-type	+	+++	++
ET-9	−	++	−
ET-11	+	++	+
ET-14	−	+	+
ET-16	−	−	−
ET-57	−	−	−

the ET-9 mutant contained a reasonable amount of truncated D2 protein in its thylakoid membranes (Fig. 3) while CP43 is absent (not shown). The reason for this is that the codon change in *psb*DI causing the loss of 9 amino acids also affected the *psb*C Shine-Dalgarno sequence (not shown). The phenotype of ET-9 thus is due to the loss of CP43 rather than to the truncation of D2. The ET-11 mutant (with 11 amino acids removed from the C-terminus of D2) is photoautotrophic (Table II). Note, however, that ET-14 is an obligate photoheterotroph. Thus, the area 11-14 residues from the C-terminus of D2 appears to harbor structurally or functionally important domains. In ET-14 some D2 protein is detectable in thylakoid membrane preparations, while in ET-11 more D2 remains, but less than in wild-type (Fig. 3). Judging from diuron binding results, upon the shortening of D2 by 11 or 14 amino acids the PS II complex is able to assemble, but the number of PS II centers present is less than in wild-type (Table II). Thus, our results indicate that the region including at least part of the 16 C-terminal residues of D2 is of importance. Further analysis of this region is in progress and will help to elucidate the role of the C-terminus in the PS II assembly, stability and function.

ACKNOWLEDGEMENTS

This research is funded by a grant from the National Science Foundation (DCB-9019248). This is publication 85 from the Arizona State University Center for the Study of Early Events in Photosynthesis. The Center is funded by U.S. Department of Energy grant #DE-FG02-88ER13969 as part of the USDA/DOE/NSF Plant Science Center program.

REFERENCES

Andersson, B. and Styring, S., 1991, Photosystem II: molecular organization, function, and acclimation, in: "Current Topics in Bioenergetics," Vol. 16, C.P. Lee ed., pp. 1-81, Academic Press, San Diego.

Burnap, R.L. and Sherman, L.A., 1991, Deletion mutagenesis in *Synechocystis* sp. PCC 6803 indicates that the Mn-stablizing protein of photosystem II is not essential for O_2 evolution, Biochem., 30: 440-446.

Coleman, W.J. and Youvan, D.C., 1990, Spectroscopic analysis of genetically modified photosynthetic reaction centers, Annu. Rev. Biophys. Biophys. Chem., 19: 333-367.

Debus, R.J., Feher, G. and Okamura, M.Y., 1986, Iron-depleted reaction centers from *Rhodopseudomonas sphaeroides* R-26.1: characterization and reconstitution with Fe^{2+}, Mn^{2+}, Co^{2+}, Ni^{2+}, Cu^{2+}, and Zn^{2+}, Biochem., 25: 2276-2287.

Debus, R.J., Barry, B.A., Babcock, G.T. and McIntosh, L., 1988a, Site-directed mutagenesis identifies a tyrosine radical involved in the photosynthetic oxygen-evolving system, Proc. Natl. Acad. Sci. USA, 85: 427-430.

Debus, R.J., Barry, B.A., Babcock, G.T. and McIntosh, L., 1988b, Directed mutagenesis indicates that the donor to P^+680 in photosystem II is tyrosine-161 of the D1 polypeptide, Biochem., 27: 9071-9074.

Deisenhofer, J., Epp, O., Miki, K., Huber, R., and Michel, H., 1985, Structure of the protein subunits in the photosynthetic reaction centre of *Rhodopseudomonas viridis* at 3 Å resolution, Nature, 318: 618-624.

Diner, B.A., Ries, D.F., Cohen, B.N., and Metz, J.G., 1988, COOH-terminal processing of polypeptide D1 of the photosystem II reaction center of *Scenedesmus obliquus* is necessary for the assembly of the oxygen-evolving complex, J. Biol. Chem., 263: 8972-8980.

Feher, G., Allen, J.P., Okamura, M.Y., and Rees, D.C., 1989, Structure and function of bacterial photosynthetic reaction centres, Nature, 339: 111-116.

Golden, S.S., Brusslan, J., and Haselkorn, R., 1987, Genetic engineering of the cyanobacterial chromosome, Meth. Enzymol., 153: 215-231.

Grigorieva, G. and Shestakov, S., 1982, Transformation in the cyanobacterium *Synechocystis* sp. 6803, FEMS Microbiol. Lett., 13: 367-370.

Hearst, J.E., 1986, Primary structure and function of the reaction center polypeptides of *Rhodopseudomonas capsulata* - the structural and functional analogies with the photosystem II polypeptides of plants, in: "Photosynthesis III, Encyclopedia of Plant Physiology, New Series," Vol. 19, L.A. Staehelin and C.J. Arntzen, eds, pp. 382-389, Springer-Verlag, Berlin, ISBN 3-540-16140-6.

Kunkel, T.A., Roberts, J.D., and Zakour, R.A., 1987, Rapid and efficient site-specific mutagenesis without phenotypic selection, Meth. Enzymol., 145: 367-382.

Mayfield, S.P.. Bennoun, P., and Rochaix, J.-D., 1987, Expression of nuclear encoded OEE1 protein is required for oxygen evolution and stablity of photosystem II particles in *Chlamydomas reinhardtii*, EMBO J., 6: 313-318.

Metz, J.G., Nixon, P.J., Rögner, M., Brudvig, G. and Diner, B.A., 1989, Directed alteration of the D1 polypeptide of photosystem II: evidence that tyrosine-161 is the redox component, Z, connecting the oxygen-evolving complex to the primary donor, P680, Biochem., 28: 6960-6969.

Michel, H., Epp, O., and Deisenhofer, J., 1986, Pigment-protein interactions in the photosynthetic reaction centre from *Rhodopseudomonas viridis*, EMBO J., 5: 2445-2451.

Michel, H. and Deisenhofer, J., 1987, The structural organization of photosynthetic reaction centers, in: "Progress in Photosynthesis Research," Vol. I, J. Biggins, ed., pp. 353-362, Martinus Nijhoff, Dordrecht, ISBN 90-247-3450-9.

Michel, H. and Deisenhofer, J., 1988, Relevance of the photosynthetic reaction center from purple bacteria to the structure of photosystem II, Biochem., 27: 1-7.

Philbrick, J.B., Diner, B.A., and Zilinskas, B.A., 1991, Construction and characterization of cyanobacterial mutants lacking the manganese-stabilizing polypeptide of photosystem II, _J. Biol. Chem._, 266: 13370-13376.

Plato, M., Michel-Beyerle, M.E., Bixon, M., and Jortner, J., 1989, On the role of tryptophan as a superexchange mediator for quinone reduction in photosynthetic reaction centers, _FEBS Lett._ 249: 70-74.

Rutherford, A.W., 1986, How close is the analogy between the reaction centre of photosystem II and that of purple bacteria? _Biochem. Soc. Trans._, 14: 15-17.

-, 1987, How close is the analogy between the reaction centre of PS II and that of purple bacteria? 2. The electron acceptor side, _in_: "Progress in Photosynthesis Research," Vol. I, J. Biggins, ed., pp. 277-283, Martinus Nijhoff, Dordrecht, ISBN 90-247-3450-9.

Seibert, M., Tamura, N., and Inoue, Y., 1989, Lack of photoactivation capacity in _Scenedesmus obliquus_ LF-1 results from loss of half the high-affinity manganese-binding site. Relationship to the unprocessed D1 protein, _Biochim. Biophys. Acta_, 974: 185-191.

Trebst, A., 1986, The topology of the plastoquinone and herbicide-binding polypeptides of photosystem II in the thylakoid membrane, _Z. Naturforsch._, 41c: 240-245.

-, 1987, The three-dimensional structure of the herbicide-binding niche on the reaction center polypeptides of photosystem II, _Z. Naturforsch._, 42c: 742-750.

Vandeyar, M.A., Weiner, M.P., Hutton, C.J., and Batt, C.A., 1988, A simple and rapid method for the selection of oligonucleotide-directed mutants, _Gene_, 65: 129-133.

Vermaas, W.F.J., Williams, J.G.K., and Arntzen, C.J., 1987, Site-directed mutations of two histidine residues in the D2 protein inactivate and destablize photosystem II, in the cyanobacterium _Synechocystis_ 6803, _Z. Naturforsch._, 42c: 762-768.

Vermaas, W.F.J., Rutherford, A.W., and Hansson, Ö., 1988, Site-directed mutagenesis in photosystem II of the cyanobacterium _Synechocystis_ sp. PCC 6803: the donor D is a tyrosine residue in the D2 protein, _Proc. Natl. Acad. Sci. USA_, 85: 8477-8481.

Vermaas, W.F.J., Charité, J., and Eggers, B., 1990a, System for site-directed mutagenesis in the _psb_DI/C operon in _Synechocystis_ sp. PCC 6803, _in_: "Current Research in Photosynthesis" M. Baltscheffsky, ed., Vol. I, pp. 231-238. Kluwer, Dordrecht.

Vermaas, W.F.J., Charité, J., and Shen, G., 1990b, Q_A binding to D2 contributes to the functional and structural integrity of photosystem II, _Z. Naturforsch._, 45c: 359-365.

Vermaas, W.F.J., Charité, J., and Shen, G., 1990c, Glu-69 of the D2 protein in photosystem II is a potential ligand to Mn involved in photosynthetic oxygen evolution, _Biochem._, 29: 5325-5332.

Vermaas, W.F.J. and Ikeuchi, M., 1991, Photosystem II, _in_: "Cell Culture and Somatic Cell Genetics of Plants, Vol. 7B: The Photosynthetic Apparatus: Molecular Biology and Operation," I.K. Vasil and L. Bogorad, eds, pp. 25-111. Academic Press, San Diego.

Williams, J.G.K., 1988, Construction of specific mutations in photosystem II photosynthetic reaction center by genetic engineering methods in the cyanobacterium _Synechocystis_ 6803, _Meth. Enzymol._, 167: 766-778.

Williams, J.G.K. and Chisholm, D.A., 1987, Nucleotide sequences of both _psb_D genes from the cyanobacterium _Synechocystis_ 6803, in: "Progress in Photosynthesis Research," Vol. IV, J. Biggins, ed., pp. 809-812, Martinus Nijhoff, Dordrecht, ISBN 90-247-3453-3.

Yu, J. and Vermaas, W.F.J., 1990, Transcript levels and synthesis of photosystem II components in cyanobacterial mutants with inactivated photosystem II genes, _Plant Cell_, 2: 315-322.

Zoller, M.J. and Smith, M., 1983, Oligonucleotide-directed mutagenesis of DNA fragments cloned into M13 vectors, _Meth. Enzymol._, 100: 468-500.

PHOTOSYSTEM I IN THE NITROGEN-FIXING CYANOBACTERIUM, *ANABAENA VARIABILIS* ATCC 29413 : SUBUNIT COMPOSITION AND DIRECTED MUTAGENESIS OF COFACTOR BINDING PROTEINS

Himadri B. Pakrasi, Karin J. Nyhus, R. Mannar Mannan and Hans C.P.J. Matthijs[*]

Department of Biology, Box 1137, Washington University, St. Louis, MO 63130-4899, USA. [*]Permanent Address - Laboratorium voor Microbiologie, Universiteit van Amsterdam, Nieuwe Achtergracht 127, 1018 WS Amsterdam, The Netherlands

Photosystem I (PSI) and photosystem II (PSII) are two pigment-protein complexes in the thylakoid membranes of chloroplasts of green plants, eukaryotic algae, and cyanobacteria. Each of these two complexes has more than ten polypeptide subunits and a large number of cofactors, *e.g.*, chlorophyll, quinone, heme, iron-sulfur centers, etc. PSII mediates electron transfer from water to plastoquinone whereas PSI catalyzes transfer of electrons from plastocyanin or cytochrome c553 to ferredoxin. Reduced ferredoxin in its turn donates electrons to NADP$^+$, a reaction catalyzed by ferredoxin-NADP oxidoreductase (FNR).

In recent years cyanobacterial genetic systems have been extensively used to study the structure and function of various proteins of PSII (Pakrasi and Vermaas, 1991). In these studies a widely used organism has been the unicellular cyanobacterium, *Synechocystis* sp. PCC 6803. In the presence of glucose in the growth medium, *Synechocystis* 6803 cells can grow even in the absence of any functional PSII complex. Additionally, this strain is naturally transformable by exogenous DNA. Thus, gene replacement via homologous double reciprocal replacement events could be used to generate targeted insertion, deletion and site-directed modifications of a number of genes encoding PSII proteins. These studies have been very helpful in elucidating the function of various proteins as well as individual amino acid residues and protein domains in PSII.

Compared to PSII, little progress has been made until recently in developing similar genetic systems to study the PSI complex in cyanobacteria. One of the reasons is that *Synechocystis* 6803 and another unicellular strain, *Synechococcus* sp. PCC 7002, two transformable strains do not grow well in complete darkness (chemoheterotrophy). Attempts to create targeted inactivation mutations in genes encoding the reaction center of PSI in *Synechococcus* 7002 grown under photoheterotrophic conditions have failed (Zhao *et al.*, 1990). On the other hand, Chitnis *et al.* have been successful in inactivating genes encoding two peripheral proteins of PSI in *Synechocystis* 6803 (Chitnis, *et al.*, 1989a,b). However, these two proteins are not essential for the catalytic activity of PSI as evidenced by the ability of these mutants to grow in the absence of glucose in the growth media. Taken together, these data indicate that growth in darkness may be a necessary condition for the isolation of PSI-deficient mutants. Recently a 'light-activated' heterotrophic growth condition has been described for *Synechocystis* 6803 (Anderson and McIntosh, 1991) that may prove to be an important step toward genetic analysis of PSI in this organism.

In our laboratory, we are employing a filamentous cyanobacterium, *Anabaena variabilis* ATCC 29413, for genetic analysis of proteins of PSI for two reasons - (1) *Anabaena* 29413 grows well in complete darkness in the presence of fructose in the growth media, and (2) recently, Wolk and coworkers (Maldener, *et al.*, 1991) have been successful in creating a gene replacement mutation in a protease gene in this organism. In addition, *Anabaena* 29413 filaments are capable of fixing nitrogen in their specialized heterocysts. PSI is thought to play an important role in this energy intensive process, since PSII is nonfunctional in heterocysts cells. Thus, genetic analysis of PSI in this cyanobacterial strain will allow us to examine the dependence of nitrogen fixation on photosynthetic electron transfer processes. In this article, we describe our recent progress in creating targeted mutations in this organism that disrupt electron transport through the PSI

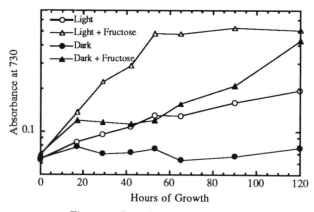

Figure 1. Growth of *Anabaena* 29413

complex. In addition, we report the isolation of a highly purified PSI complex and identification of its major protein components.

As described earlier, *Anabaena* 29413 cells are capable of growth under chemoheterotrophic conditions. Figure 1 shows the growth profile of wild type *Anabaena* 29413 cells in light (60 $\mu E/m^2/s$) and darkness, in the presence or absence of fructose in the growth media. In the absence of fructose, these cells cannot grow in complete darkness. However, they grow reasonably well in darkness in the presence of fructose.

Cyanobacteria are rich sources of the PSI complex since the ratio of PSI and PSII complexes in the thylakoid membranes of these cells ranges between 4:1 to 8:1. We have developed a simple procedure to isolate a highly purified PSI complex from *Anabaena* 29413 based on Barry, et al., 1988. Briefly, membranes isolated from cells were solubilized in β-dodecyl maltoside and the solubilized material was fractionated on a fast flow Q-Sepharose (Sigma) column on an FPLC system (Pharmacia). The green fraction eluted from this column was dialyzed and passed over the same column equilibrated with a different buffer system . The purified PSI fraction from this column was devoid of any contamination by PSII, cyt b_6f and ATP synthase complexes. All of the major protein bands of this isolated PSI preparation were subjected to N-terminal sequencing. We found eleven major protein components in this complex. Table I shows the names of the genes encoding these proteins, their nominal molecular masses and their functions, if known. Figure 2 shows a schematic diagram depicting the organization of these proteins in the PSI complex. The reaction center of PSI is localized on a heterodimer of two large proteins, PsaA and PsaB. PsaC is another protein that binds two iron-sulfur centers F_A and F_B. Protein sequence analysis revealed the presence of the PsaI protein, a small hydrophobic protein that has

Table 1. Proteins of PSI in *Anabaena* 29413

Gene	Protein (kDa)	Function
psaA	83	Reaction center
psaB	83	Reaction center
psaC	9.0	Binds iron-sulfur centers F_A, F_B
psaD	16	Binds ferredoxin and PsaC protein
psaE	8.0	Unknown, binds PsaD protein
psaF	15	Binds plastocyanin and cyt c_{553}
psaI	4.0	Unknown
psaJ	4.6	Unknown
psaK	5.7	Unknown
psaL	15.5	Unknown
psaN	4.8	Unknown

been postulated to play an important role in the PSI reaction center. This is the first, and to date only, evidence for the presence of the PsaI protein in any cyanobacterial PSI complex. In addition, PsaN, another small protein was found that is homologous to a 4.8 kDa protein in *Synechococcus vulcanus*. We also identified the PsaL protein that is distinctly different from the PsaG protein found in the PSI complex from spinach and barley. Cyanobacterial PSI complex lacks the PsaG and the PsaH proteins whereas PsaN has so far been found only in cyanobacteria. Seven of the eleven proteins of PSI are predicted to be integral membrane proteins whereas PsaC, PsaD, PsaE and PsaF are peripherally associated to the membrane.

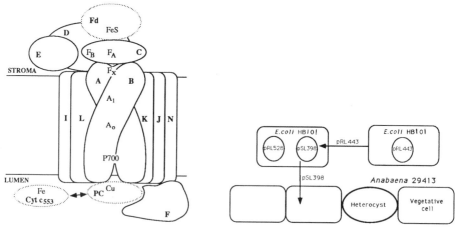

Figure 2. PSI Complex of*Anabaena* 29413   Figure 3. Triparental Mating of *Anabaena* 29413

Gene Replacement in *Anabaena* 29413

Filamentous cyanobacteria are not easily transformable. Wolk and his associates have developed a conjugation mediated DNA transfer system in many filamentous cyanobacteria, including *Anabaena* 29413 (Wolk, *et al.*, 1984). Figure 3 shows a cartoon diagram depicting the transfer of a plasmid containing a gene of interest (cargo plasmid) from donor *Escherichia coli* cells to *Anabaena* 29413 cells. In this system, pRL443 (an RP4 derivative), a conjugative plasmid in *E. coli* HB101 cells, mobilizes the cargo plasmid from HB101 cells to *Anabaena* 29413 cells. Another important component of this system is pRL528, a plasmid coresident in the same HB101 cells as the cargo plasmid. pRL528 encodes two important functions - (1) it contains the *mob* gene that encodes a nicking protein for the *bom* site on the cargo plasmid, and (2) two methylases that methylate Ava I, Ava II and Avr I restriction sites in the cargo plasmid. This latter function is essential so that the incoming cargo plasmid is not digested by the restriction enzymes in the *Anabaena* cells. Two cargo plasmids, pRL271 and pRL277 have been developed that have convenient restriction sites for the cloning of any gene of interest. In addition they contain a *sacB* gene encoding a levansucrase enzyme. In the presence of this gene product, bacterial cells are killed in sucrose-containing media. This positive selection system has been helpful in selecting double reciprocal recombination mediated gene replacement mutants in *Anabaena* 7120, a strain that has been extensively used for genetic analysis of the nitrogen fixing apparatus (Cai and Wolk, 1990). pRL277 additionally contains a Sm[r]/Sp[r] gene and pRL 271 contains a Cm[r] gene. In our experiments we have used replica-plating techniques to find exconjugants that are sensitive to Sp and Sm or Cm, implying that they have undergone double reciprocal recombination events. Both of these plasmids are unable to replicate in *Anabaena* 29413 cells.

To date we have succesfully used this sytem to create insertion mutations in three genes, *psaA* (T372-1), *psaB* (T373-2) and *psaC* (T398-1). For the mutagenesis of the *psaC* gene, a Nm[r] gene cartridge was inserted in the middle of the coding region. This interrupted gene was then cloned in pRL277 and conjugated into *Anabaena* 29413 wild type cells. Nm[r] colonies resulting from such experiments were restreaked several times on fructose containing plates and incubated in complete darkness. A number of these colonies were then checked for their sensitivity to spectinomycin. About 22% of the colonies were Sp[s] but Km[r], indicating that they have undergone a double reciprocal recombination event leading to replacement of the wild type copy of the *psaC* gene by the interrupted copy. Similar experiments were done to obtain the *psaA* and *psaB* interruption mutants.

All three of these mutants are highly light sensitive. They are unable to grow under normal growth light conditions (60 $\mu E/m^2/s$) and grow poorly under 5 $\mu E/m^2/s$ of white light. This extreme light sensitivity may be the principal

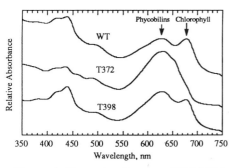

Figure 4. Absorption Spectra of WT, T372, and T398 Cells

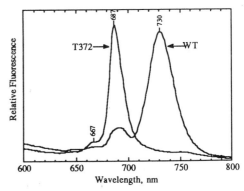

Figure 5. 77K Fluorescence Spectra of WT and T372 Cells

reason for the inability to create PSI-deficient mutants of *Synechococcus* 7002 grown under photoheterotrophic conditions, as mentioned earlier. These mutants also appeared significantly more blue than the wild type cells. Figure 4 shows the room temperature absorption spectra of intact filaments of wild type, T372 and T398 mutants. The absorption peak around 620 nm arises from phycobilin pigments whereas that around 678 nm originates from chlorophylls. The T398 mutant cells had an approximately two-fold increase in the phycobilin to Chl ratio. The most interesting spectrum is that of T372, the *psaB* insertion mutant. In these cells, the ratio of phycobilin to Chl is much higher than that in wild type cells. In cyanobacterial cells, most of the Chl molecules are associated with the PsaA and PsaB proteins as antenna molecules of PSI whereas the PSII complexes are served by the phycobilisomes as the light-harvesting complexes. In the absence of the PsaB protein in the T372 mutant, the PsaA-PsaB heterodimer can not form so that most of the Chl molecules are missing. This is supported by the 77K fluorescence spectra from intact T372 cells (Figure 5). The Chl emission peaks between 685 and 695 arise from PSII whereas the peak at 730 nm originates from PSI. As shown in this figure, T372 lacks the peak at 730 nm indicating that the Chl antenna of PSI is missing from this mutant. Interestingly, the T372 cells have functional PSII complexes and are capable of mediating oxygen evolution in the presence of artificial electron acceptors.

The *psaC* insertion mutant, T398, lacks the PsaC protein. However, these cells had fully assembled PSI reaction center. Detailed analysis of the properties of this mutant has been documented elsewhere. Figure 6 shows the difference spectra of light-induced absorption
changes in thylakoid membranes from wild type (A) and T398 (B) mutant cells. The troughs at 705 nm in WT cells and at 703 nm in T398 cells show the extent of

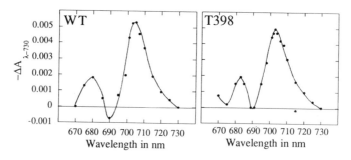

Figure 6. Light-Induced Absorption Changes

light-induced bleaching of P700, the reaction center of PSI. These data clearly demonstrate that normal amount of PSI reaction center is assembled in this mutant. Moreover, immunoblot analysis indicates that the PsaA/PsaB proteins are present in this mutant. These data are in sharp contrast to the findings of Rochaix and coworkers about a *psaC* insertion mutant of the green alga *Chlamydomonas reinhardtii* (Takahashi, *et al.*, 1991). This algal mutant lacks the entire PSI complex. The differences between these two *psaC* insertion mutants are interesting and warrants future studies.

Ferredoxin-NADP Oxidoreductase

FNR is a single subunit protein found both in soluble and membrane bound forms. The membrane bound form has been postulated to have an additional role as a dehydrogenase in the oxidation of NADPH in a chlororespiration type pathway. To investigate the functional role of FNR, we have isolated a λ clone from a λEMBL3 library of *Anabaena* 29413 DNA. Figure 7 shows a restriction map of a 2.8 kbp Sal I - Kpn I fragment from this clone that encompasses the *petH* gene. We have also inserted a Nm^r gene at a Sty I site in the coding region of the *petH* gene to create an insertional inactivation mutant of FNR. We are currently transferring the interrupted gene back into *Anabaena* cells.

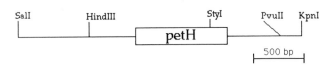

Figure 7. Restriction Map of *petH* Gene of *Anabaena* 29413

Conclusion

An efficient genetic system has been developed for targeted mutagenesis of various PSI proteins in the filamentous cyanobacterium, *Anabaena* 29413. Detailed knowledge of the protein components of the PSI complex in this organism as well as availability of shuttle vectors have made it an attractive system for site specific mutagenesis studies of the reaction center and PsaC proteins of PSI.

Acknowledgements

We thank Dr. C.P. Wolk and his lab members for providing useful plasmids and bacterial strains; Drs. M. Ikeuchi and Y. Inoue for determining the N-terminal sequence of the PSI proteins, Dr. J. Whitmarsh and P. Nyman for spectroscopic analysis and Dr. T. Thiel for helpful discussions. Ths research was supported by grants from National Institutes of Health and Markey Foundation to H.B.P. HCPM was supported for his travel by the Netherlands Organization for Scientific Research (NWO).

References

S. L. Anderson, and L. McIntosh, J. Bact. 173: 2761 (1991).

B. A. Barry, C. J. Bender, L. McIntosh, S. Ferguson-Miller, and G.T. Babcock, Israel J. Chem. 28: 129 (1988).

Y. Cai, and C. P. Wolk, J. Bact. 172: 3138 (1990).

P. R. Chitnis, P. A. Reilly, M. C. Miedel, and N. Nelson J. Biol. Chem. 264: 18374 (1989a).

P. R. Chitnis, P. A. Reilly, and N. Nelson, J. Biol. Chem. 264: 18381 (1989b).

I. Maldener, W. Lockau, Y. Cai, and C.P. Wolk Mol. Gen. Genet. 225: 113 (1991).

H. B. Pakrasi, and W. F. J. Vermass, in "Topics in Photosynthesis" J. Barber, ed., vol. 10, Elsevier Publ., Amsterdam, in press.

Y.Takahashi, M. Goldschmidt-Clermont, S.-Y. Soen, L. G. Franzen, and J.-D. Rochaix, EMBO J. 10: 2033 (1991).

C. P. Wolk, A. Vonshak, P. Kehoe, and J. Elhai, Proc. Natl. Acad. Sci. USA 81: 1561 (1984).

J. Zhao, P. V. Warren, N. Li, D. A. Bryant, and J. H. Golbeck, FEBS Lett. 276: 175 (1990).

TOWARDS PROTEIN ENGINEERING OF *CHLAMYDOMONAS REINHARDTII*

CHLOROPLAST ATP SYNTHASE

Stefan Leu, Jacob Schlesinger, Ronith Motzery, Noun Shavit
and Allan Michaels

Department of Life Sciences
Ben Gurion University of the Negev
Beer Sheva, Israel

INTRODUCTION

Membrane bound ATP synthases from chloroplasts, mitochondria and bacteria are the major catalyst in the transduction of energy derived from electron transport, whereby the proton flow across a membrane energized by electron transport is the driving force for synthesis of ATP. The chloroplast ATP synthase ($CF_o \cdot CF_1$) is one of four major protein complexes of the thylakoid membrane (1, 2). CF_o contains four subunits (I-IV) and CF_1 is composed of five different polypeptide chains designated $\alpha - \varepsilon$ in order of decreasing M_r (1). The genes encoding subunits α, β and ε (atpA, atpB and atpE) of CF_1 are located in the chloroplast DNA, while the genes encoding subunits γ and δ (atpC and atpD) are located in the nuclear DNA. All genes encoding CF_1 subunits have been identified and their DNA sequences determined in at least one species. The molecular weight of CF_1 is about 380,000 and the subunit stoichiometry is $\alpha_3\beta_3\gamma\delta\varepsilon$. An $\alpha_3\beta_3\gamma$ complex has been characterized and is a suitable model to study structure and function relationship of CF_1 (3). The $\alpha_3\beta_3$ complex is the catalytic core of F-type ATPases (4, 5). The other subunits appear to be responsible for the binding of CF_1 to CF_o and the coupling of ATP synthesis to proton translocation. In CF_1, the γ subunit regulates the enzyme activity according to the degree of energization of the membrane (6, 7). Active CF_1 complexes of spinach and maize have been reconstituted from isolated subunits (8).

Amino acids and protein domains involved in catalysis and substrate /ligand binding in F_1 type ATPases have been identified by chemical modification and mutant analysis (9). Several amino acids identified in mutants or by chemical modification have been altered in cloned genes by site directed mutagenesis, and the effects of the mutations were studied in complexes reconstituted from subunits expressed in *E. coli* (10 - 12). These studies revealed the importance of amino acid residues of the α and β subunit in the catalytic process. However, not all of the modified amino acids proved to be essential to catalysis (12).

In *Chlamydomonas reinhardtii*, the DNA sequences of the atpB (13), atpE (14) ane atpC (15) genes have been determined. We have in our laboratory clones of the atpA, atpB, and atpC genes from *Chlamydomonas reinhardtii* and have successfully overexpressed subunit β in *E. coli* (16). We intend to establish a reconstitution system for *Chlamydomonas*

Regulation of Chloroplast Biogenesis
Edited by J.H. Argyroudi-Akoyunoglou, Plenum Press, New York, 1992

CF$_1$ from subunits expressed in *E. coli*. Here we describe the molecular characterization of the *Chlamydomonas* atpA gene, preliminary functional studies with the overexpressed β subunit, and the overexpression and purification of γ subunit almost identical to the mature CF$_1$ subunit.

MATERIALS AND METHODS

Preparation of plasmid DNA, subcloning and other manipulations with DNA were performed using standard techniques, enzymes were used according to the manufacturers instructions. DNA sequencing was performed using double strand template DNA cloned into the plasmid pBS SK (-) (Stratagene). RNA was produced from genes cloned into pT7 using T7 RNA polymerase according to the manufacturers instructions, using CsCl purified plasmid linearized at the end of the gene with restriction endonuclease (17, 18). *In vitro* translations in the reticulocyte lysate, purification of *Chlamydomonas* thylakoid membranes, determination of incorporated radioactivity, analysis of proteins by SDS polyacrylamide gel electrophoresis (PAGE) and immunostaining of western blots were done as previously described (17, 18).

Overexpression of subunit β in *E. coli* BL21 was done as previously described (16). [^{3}H] 2',3'-O-(4-benzoyl)benzoyl-adenosine-5'-triphosphate (BzATP) was synthesized and used for photoaffinity labeling as described (19, 20). A DNA fragment encoding the mature *Chlamydomonas* chloroplast γ subunit was excised from an atpC cDNA clone (a generous gift from Dr. S. Merchant)(15) and subcloned into pUHE 23-1 (21). The resulting plasmid was transformed into *E. coli* JM109 and overexpressed by induction of log phase cultures of the transformants in LB with 0.4 mM IPTG in the presence of ampicillin. *E. coli* cells were harvested by centrifugation at 10000 x g and lysed by ultrasonication. Lysates were fractionated and pellets were washed by centrifugation (15000 x g) and resuspension in different buffers by ultrasonication.

RESULTS AND DISCUSSION

DNA Sequence and in vitro Transcription-Translation of the atpA Gene

The chloroplast atpA gene from *Chlamydomonas reinhardtii* was subcloned into pT7-2. A 480 bp AvaII-EcoRI fragment, shown to contain the start of the atpA gene (22), was extracted from EcoRI fragment 15 of chloroplast DNA, joined with a 2.7 kb EcoRI-XbaI fragment extracted from EcoRI fragment 7 of chloroplast DNA, and inserted into the polylinker region of pT7-2. The correct orientation of the insert was checked by restriction analysis. The DNA sequence of the atpA gene was determined. Gene fragments obtained by digesting the atpA pT7-2 plasmid DNA with the restriction endonucleases HindIII, PstI and SspI were subcloned into the polylinker site of the vector pBS SK-, and the sequence was determined using denatured doublestranded DNA as template. The DNA sequence encoding the atpA gene revealed an open reading frame encoding 508 amino acids, displaying high homology to the amino acid sequence of tobacco chloroplast α subunit (75% identical amino acids) and considerable homology to the *E. coli* α subunit (49% identical amino acids).

The atpA gene cloned into pT7-2 was transcribed *in vitro* using T7 RNA polymerase. The resulting RNA was translated in the reticulocyte lysate, in the absence or presence of *Chlamydomonas* thylakoid membranes corresponding to 0.4 mg/ml of chlorophyll. A main product of about 54 kDa, corresponding to the molecular weight of the atpA gene product, was obtained (Fig. 1). Translation of the atpA mRNA, as well as that of the

Table 1

Rates of radioactive methionine incorporations by *in vitro* translation of different *in vitro* produced mRNA's in the presence of thylakoid membranes, relative to the corresponding incorporations in the absence of membranes. TCA insoluble radioactivity (minus control not containing mRNA) incorporated in aliquots of translation mixtures with each mRNA containing thylakoid membranes were divided by the TCA insoluble radioactivity (minus control not containing mRNA) incorporated in equal aliquots of the corresponding translation mixtures not containing membranes (= n, average of at least five independent determinations).

RNA	n	S. dev
rbcL	0.80	0.12
psaB	2.04	0.55
atpA	1.1	0.3
atpB	1.3	0.1

atpB mRNA (18), was not significantly affected by the presence of thylakoid membranes, in contrast to the translations of the rbcL and the psaB mRNA's, which were inhibited or stimulated, respectively, by thylakoid membranes (Table 1). About 50% of the *in vitro* translated α subunit was recovered in the membrane fraction, when translation mixtures containing thylakoid membranes were fractionated (after translation) by centrifugation at 15000 x g and the resulting membranes were washed twice (Fig. 1, lane 8). This result suggests, that the *in vitro* translated α subunit can associate with CF_1 as was observed with *in vitro* translated β subunit (18) or with α and β subunits translated in a homologous run off translation system (23). Two possible explanations for these observations can be offered: the *de novo* synthesized subunits assemble with an existing pool of free, membrane associated subunits as previously suggested (23), or they exchange with subunits in preexisting CF_1.

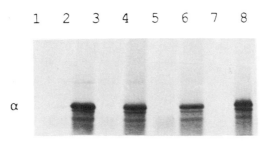

Fig. 1. Autoradiograph of the translation products of atpA mRNA *in vitro* translation analyzed by PAGE: 1: translation mixture (-RNA); 2: translation mixture containing atpA mRNA obtained by *in vitro* transcription; 3: translation mixture (- RNA) containing thylakoid membranes (400 μg/ml of chlorophyll); 4: translation mixture (+RNA) containing thylakoid membranes; 5, 6: supernatants of translation mixtures shown in lanes 3 and 4, respectively, after centifugation at 15000 x g for 10 minutes; 7, 8: Pellets of translation mixtures shown in lanes 3 and 4, respectively. Pellets obtained by centrifugation of the reaction mixtures were washed twice in reaction buffer.

Overexpression in *E. coli* of Functional CF₁ β Subunit

We have subcloned the chloroplast atpB gene of *Chlamydomonas* (13) into pT7-2. RNA obtained by *in vitro* transcription of the resulting atpBpT7 plasmid was translated in the reticulocyte lysate, and the resulting product was shown to assemble with CF₁ (18). The same plasmid was used to transform *E. coli* BL21 cells, which contain an inducible genomic T7 RNA polymerase gene (24). Upon induction in the log phase with IPTG, the transformed BL21 cells produced more than 10% of the total cell protein of *Chlamydomonas* β subunit in form of an urea soluble aggregate, if overgrowth of the culture by noninducible cells was prevented by the addition of D-cycloserine (18). The overproduced β subunit was identified by polyacrylamide gel electrophoresis (Fig. 2 A, lanes 3, 4) and immunostaining of western blots using specific antibodies (23) raised against the α and β subunits of *Chlamydomonas* CF₁ (Fig. 2 B, lanes 3, 4). The pellet obtained after centrifugation of lysed bacteria was washed in 5 x TE (50 mM Tris/HCl pH 8, 5 mM NaEDTA) containing 2 % Triton-X-100, in 5 x TE containing 2 M LiCl and in 5 x TE, yielding 50 - 75% enriched β subunit, and dissolved in 4M urea (Fig. 2, lanes 5). The urea soluble fraction was dialyzed against 2 M urea in 5 x TE and then against three changes of dialysis buffer containing 20 mM Tris/HCl pH 8.0, 0.2 mM EDTA, 0.1 mM ATP, 0.1 mM DTT, 100 mM NaCl, and 10% glycerol. Most of the preparation remained soluble after dialysis (Fig. 2, lanes 6) and could be stored for extended periods at -20 °C.

The enriched preparation of β subunit was shown to specifically bind the photoreactive ATP analog BzATP. Urea dissolved β subunit was passed through Sephadex G25 equilibrated with 10 mM Tricine/NaOH pH 8.0. 250 - 500 µg/ml of protein was incubated for 30 min at room temperature in the presence of 5 mM MgCl₂ and 50 - 100 µM of [³H] BzATP. The mixture was then exposed to UV irradiation after which protein was precipitated and the pellet was washed with 5% perchloric acid (PCA). The pellet was dissolved in 5% SDS and analyzed by PAGE and fluorography. Radioactivity was determined by liquid scintillation counting. As shown in the fluorograph (Fig. 3), the mature 52 kD β sub-

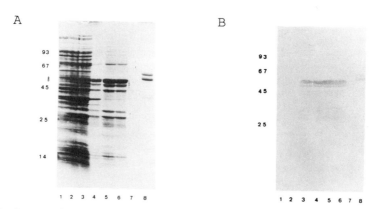

Fig. 2. Expression and purification of *Chlamydomonas reinhardtii* CF₁ β subunit from the cloned atpB gene in *E. coli* BL21.
A: analyzed by PAGE and stained with coomassie blue;
B: analyzed by immunostaining of a western blot from a gel identical to that shown in A, using β subunit specific first antibody. 1: not transformed BL21 cells; 2: BL21 cells transformed with atpBpT7, not induced; 3: BL21 cells transformed with atpBpT7, induced with IPTG; 4: Pellet of induced cells (lane 3) lysed by ultrasonication; 5: Washed pellet, dissolved in 4 M urea; 6: Solubilized pellet (lane 5), dialyzed as described in the text, supernatant after centrifugation; 7: sediment after dialysis; 8: lettuce CF₁

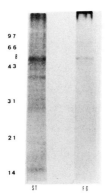

97
66
β
43

31

21

14

ST   FG

Fig. 3. Binding of BzATP to β
subunit overexpressed in *E.
coli*: 250 μg/ml of enriched
β subunit were photolabeled
in the presence of 100 mM
BzATP: ST: coomassie stained.
FG: Fluorograph.

Table 2

Competition of photoaffinity labeling by BzATP of overexpressed
β subunit with ATP: Radioactivity bound to PCA precipitated
protein (cpm/μg) was determined
by liquid scintillation counting.
Adsorbed radioactivity in non
irradiated samples (69 cpm/μg)
was subtracted.

Treatment	cpm/μg	% of control
control(-ATP)	218	100
+ 5 mM ATP	39	17
+ 20 mM ATP	0	0

unit specifically bound the radioactive ATP analog, while impurities
present in the preparation were not labeled. The β preparation bound 0.3
mol/mol of BzATP. Excess of ATP effectively displaced the analog,
inhibiting binding to the subunit (Table 2). This result indicates that
BzATP specifically reacts with the ATP binding site found in isolated
CF_1 β subunits (25, 26), and that the overexpressed β subunit can refold
to a native conformation after urea treatment. This technique will allow
us to study the ATP binding domain of the β subunit by site directed
mutagenesis.

Overexpression of Subunit γ from its Cloned Gene

A cDNA clone encoding γ subunit of *Chlamydomonas* CF_1 has been
identified and sequenced, and the N-terminus of the processed, mature
CF_1 γ subunit has been identified by amino acid sequencing (24). A 1170
bp restriction fragment obtained from the original cDNA clone encodes
the mature γ subunit, starting with the amino acid sequence Gly Leu Lys.
This fragment was ligated into the plasmid pUHE-23, yielding a fused
gene starting with three amino acids encoded by pUHE in frame with the
open reading frame in the cloned 1170 bp atpC gene fragment (Met Arg Ile
- Gly Leu Lys). JM 109 cells transformed with the resulting plasmid were
grown in the presence of ampicillin to the mid log phase, induced by
addition of 0.4 mM IPTG and grown for 6 - 12 hours. Cells were harvested
and lysed by ultrasonication and centrifuged (15000 x g, 15 min), and
cell fractions were analyzed by PAGE. An overexpressed protein of 38 kD
(10 - 20% of total cell protein) was predominantly found in the
insoluble cell fraction and was identified as γ subunit by immunostaining of western blots. This polypeptide was enriched to up to 90 %
purity by washing of the pellet with 2 % Triton-x-100, 2 M LiCl and 5 x
TE as described for the overexpressed β subunit. The γ subunit was
dissolved in 4 M urea and remained soluble after urea was removed by
dialysis.

Acknowledgements

This research was supported by a grant from the Bertie I. Black
Foundation, Great Britain (to N. S.), and by grant Nr. 823A-028429 from
the Swiss National Science Foundation (to S. L.). We thank Dr. S.
Merchant for supplying the atpC cDNA clone, Prof. H. Bujard for the pUHE
plasmid and Dr. J. M. Galmiche for a gift of *Chlamydomonas* subunit γ
antiserum.

REFERENCES

1. Marder, J. B. and Barber J. (1989) Plant, Cell and Environment 12, 595 - 614.
2. Shavit, N. (1980) Ann. Rev. Biochem. 49, 111 - 138.
3. Mitra, B. and Hammes, G. G. (1988) Biochemistry 27, 245 - 250.
4. Miwa, K. and Yoshida, M. (1989) Proc. Natl. Acad. Sci. USA, 86, 6484 - 6487.
5. Avital, S. and Gromet-Elhanan, Z. (1991) J. Biol. Chem. 266, 7067 - 7072.
6. Moroney, J. V., Andreo, C. S., Vallejos, R. H. and McCarty, R. E. (1980) J. Biol. Chem. 255, 6670 - 6674.
7. Ketcham, S. R., Davenport, J. W., Warncke, K. and McCarty R. E. (1984) J. Biol. Chem. 259 7286 - 7293.
8. Kasamo, K., Kagita, F. and Arai, Y. (1989) Plant Cell Physiol. 30, 729 - 738.
9. Futai, M., Noumi, T. and Maeda, M. (1989) Ann. Rev. Biochem. 58, 111 - 136.
10. Yohda, M., Ohta, S., Hisabori, T. and Kagawa, Y. (1988) Biochim. Biophys. Acta 933, 156 - 164.
11. Miwa, K., Ohtsubo, M., Denda, K., Hisabori, T., Date, T. and Yoshida M. (1989) J. Biochem. 106, 679 - 683.
12. Odaka, M., Kobayashi, H., Muneyeki, E. and Yoshida, M. (1990) Biochim. Biophys. Res. Comm. 168, 372 - 378.
13. Woessner, J. P., Gillham, N. W. and Boynton, J. E. (1986) Gene 44, 17 - 28.
14. Woessner, J. P., Gillham, N. W. and Boynton, J. E. (1987) Plant Mol. Biol. 8, 151 - 158.
15. Yu, L. M. and Selman, B. R. (1988) J. Biol. Chem. 263, 19342 - 19345.
16. Blumenstein, S., Leu, S., Abu-Much, E., Bar-Zvi, D., Shavit, N. and Michaels, A. (1990) in: Current research in Photosynthesis III (Baltscheffsky, M., ed.) Kluwer Academic Publishers, Dordrecht, The Netherlands, 193 - 196.
17. Leu, S. and Michaels, A. (1990) in: Current research in Photosynth. III (Baltscheffsky, M., ed.), Kluwer Academic Publishers, Dordrecht, The Netherlands, 569 - 572.
18. Leu, S., Weinberg, D. and Michaels, A. (1990) FEBS Lett. 269, 41 - 44.
19. Bar-Zvi, D. and Shavit, N. (1984) Biochim. Biophys. Acta 765, 340 - 346.
20. Bar-Zvi, D., Tiefert, M. A. and Shavit, N. (1983) FEBS Lett. 160, 233 - 238.
21. Stüber, D., Ibrahimi, I., Cutler, D., Dobberstein, B. and Bujard, H. (1984) EMBO J. 3, 3143 - 3148.
22. Hallick, R. B. (1984) FEBS Lett. 177, 274 - 276.
23. Herrin, D. and Michaels, A. (1985) Arch. Biochem. Biophys. 237, 224 - 236.
24. Studier, F. W. and Moffatt, B. A. (1986) J. Mol. Biol. 189, 113 - 136.
25. Nadaciva, S. and Harris, D. A. (1990) in: Current research in Photosynth. III (Baltscheffsky, M. ed.), Kluwer Academic Publishers, Dordrecht, The Netherlands, 41 - 44.
26. Avital, S. and Gromet Elhanan, Z. (1990) in: Current research in Photosynth. III (Baltscheffsky M. ed.). Kluwer Academic Publishers, Dordrecht, The Netherlands, 45 - 48.

ACCELERATED RATE OF TURNOVER OF THE D1 SUBUNIT OF PHOTOSYSTEM II IS

CORRELATED WITH INHIBITION OF ELECTRON TRANSFER FROM Q_A TO Q_B IN

CYANOBACTERIAL MUTANTS

Nir Ohad, Yorinao Inoue[1] and Joseph Hirschberg

Department of Genetics, The Hebrew University, Jerusalem 91904, Israel and [1]The RIKEN Institute, Wako-shi, Saitama, 35101 Japan

INTRODUCTION

Photosystem II (PSII) is a pigment-protein complex that couples the light-induced reduction of plastoquinone with oxidation of water. It is now established that a complex of only five integral thylakoid membrane polypeptides together with chlorophyll a, pheophytin and β-carotene molecules, and a non-heme ferrous atom, form the photosynthetic reaction center (RC) of PSII (reviewed in 1). It is believed that two polypeptides, D1 and D2, comprise the core of the RC in a manner that is analogous to L and M subunits of the RC in purple bacteria. The D1 subunit of PSII is a 32 kDa integral membrane polypeptide known also to bind the plastoquinone Q_B and herbicides that inhibit electron transfer in PSII (2). It is encoded by the chloroplast gene psbA and synthesized within the chloroplast as a 34 kDa precursor polypeptide, which is assembled into PSII complexes following processing.

D1 turns-over rapidly in the light at a rate that is proportional to the light intensity (3). Under extremely high intensities of light this damage can lead to photoinhibition (4). The reason for the rapid light-dependent turnover of D1 and the detailed mechanism which regulates this process are not understood. It has been proposed that a light-induced damage to the D1 protein is the primary signal for its degradation (5). A light-induced change in the stability of the secondary electron acceptor, Q_B^-, was demonstrated in Chlamydomonas cells in vivo (6). This change is followed by an irreversible covalent modification of the D1 protein (7). A modified form of the D1 protein, termed 32*, which has a slower mobility on SDS-PAGE, was identified in Spirodela following illumination (8). Since a broad spectrum of light induce rapid degradation of D1 (9) it is possible that different light receptors convey the light effect on D1 turnover in each wavelength. Previously we have reported that amino acid substitutions in the D1 polypeptide of Synechococcus PCC7942 caused Q_B^- destabilization and accelerated the turnover of D1 (10). A positive correlation was found between the degree of Q_B^- destabilization in various mutants and the rate of D1 degradation. In this report we examine the correlation between the rate of electron flow in PSII and the rate of photo-induced degradation of D1 in three mutants of Synechocystis PCC6803 which were generated by site-directed mutagenesis of psbA.

Generating mutations in the D1 polypeptide

The unicellular cyanobacterium Synechocystis PCC6803 was used throughout this research. We have employed a glucose-tolerant strain, 4D1, which contains only copy II of the psbA gene family following the deletion of copies I and III (11). This strain, which can grow either photosynthetically or photoheterotrophically on glucose-containing media, was used as a normal control in photosynthesis measurements as well as a host for the genetically engineered psbA genes.

Three mutations in the psbAII gene were created by site-specific mutagenesis in vitro (12). These point mutations result in substitutions of residue Leu 271 in the D1 polypeptide by Met (strain L271M), Ala (L271A) or Ser (L271S). Leu 271 is located in the predicted Q_B-binding site in D1 (Fig. 1).

Following transformation of 4D1 cells, the mutations were introduced into the single functional psbA gene (12). The presence of a single psbAII gene in each mutant was verified by Southern hybridization analysis of total DNA extracted from mutant cells using the psbAII sequence as a molecular probe (data not shown). Nucleotide sequence analysis of the psbAII gene that was amplified from total DNA isolated from mutant cyanobacterial cells by the polymerse chain reaction (PCR), was carried out in order to confirm that it contains only a single mutation (Fig. 2).

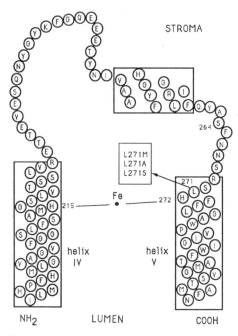

Figure 1. Schematic illustration of the Q_B binding domain in D1 (after Trebst, 22). The site-specific modifications that were introduced to D1 are indicated.

Photosynthesis in the D1 mutants

The mutations were introduced into the cyanobacterial cells that were grown under photoheterotrophic conditions namely, in a BG11 medium containing 5mM Na-thiosulfate, 5mM glucose and $20\mu M$ atrazine, as previously described (11). In order to check the photosynthetic ability of the mutants, cells were transferred to a BG11 medium devoid of glucose and atrazine. Mutants L271M and L271A grew photosynthetically though growth rate of the latter was slow. Mutant L271S was non-photosynthetic.

The photosynthetic efficiency of the mutants was estimated by measuring O_2 evolution and comparing it to the autotrophic host strain 4D1 (Table I). Oxygen evolution by whole cells was measured using a Clarck type electrode. The rate of

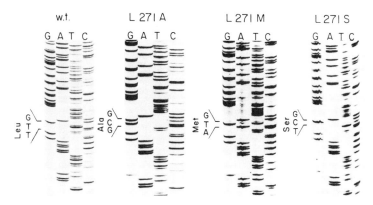

Figure 2. Mutations in psbAII of Synechocystis PCC6803. Autoradiogram of sequencing gels demonstrating the mutations in psbAII of various genetically-engineered mutants.

O_2 evolution correlates to the autotrophic growth rates in suspension cultures. In order to test the effects of the mutations on electron transfer in PSII, fluorescence induction curves were analyzed in the host strain 4D1 and in the mutants. Fluorescence was measured using a fluorimeter connected to a PC computer operating the shutter and recording the digitized signal. Cell suspension samples (1ml, $4\mu g$ chlorophyll/ml) were excited by blue light (Corning filter 6-92) at $260 \mu E.m^{-2}.sec^{-1}$. The measuring photodiode was protected by a red filter (Schott filter 6-76). The maximal fluorescence (F_m) value was recorded in the presence of the herbicides $10\mu M$ atrazine or 10μ diuron, which inhibit electron transfer in all the strains examined. F_v, variable fluorescence $= F_m-F_o)/F_o$; F_o, the initial level of fluorescence upon illumination; F_s, steady-state fluorescence. All mutants showed variable fluorescence to different degrees, indicating the existence of electron transport in PSII from H_2O to at least Q_A.

Table I. Photosynthesis in <u>Synechocystis</u> PCC6803 mutants

Strain	Photosynthetic growth	O_2 evolution[1] [% of wt]	F_v	Efficiency factor of electron transport $[(F_m-F_s)/(F_m-F_o)]$
4D1K(wt)	+++	100	0.96	0.89
L271M	++	70	0.53	0.58
L271A	+	40	0.55	0.32
L271S	-	0	0.29	0

[1] 100% equals 250μM O.$_2$/mg chlorophyll/hr.

The results summarized in Table I indicate that PSII complexes which carry out primary photochemical reactions, were assembled in all the mutants, including the non-photosynthetic one L271S.

While the value of the variable fluorescence is proportional to the number of reaction centers that accept electrons in Q_A (but blocked in Q_B), the electron flow parameter, EF, $[(F_m-F_s)/F_o]$ evaluates the proportion of reaction centers in which electron transfer proceeds beyond Q_A. This latter value correlates with O_2 evolution and CO_2 fixation (13). The reduction of the F_v and EF values in the mutants is caused by an increase in the F_o rather than decrease of F_m. Therefore, we conclude that the mutations in residue Leu 271 of D1 reduce the electron flow in PSII beyond Q_A and do not reduce the proportion of active reaction centers per chlorophyll. The ratio EF/F_v gives the fraction of reaction centers in which electron transfer procceeds beyond Q_B, among the entire population of centers that are capable of transferring electrons at least to Q_A. This ratio, termed "efficiency factor of electon transport", confirms that the lack of net flow of electrons in PSII in mutant L271S because of a block in Q_B.

Electron transport in PSII in the D1 mutants

Effects of the amino acid substitutions in position 271 of D1 on Q_B was analyzed by thermoluminescence (TL) measurements. TL is known to originate from recombination of positive and negative charges stored on the donor and acceptor sides of PSII after illumination. The B-band of TL is assigned to the recombination of S_2,S_3-Q_B^- (14-16). The peak temperature of the B band glow curve depends on the activation free energy of the recombination reaction, and thus provides a measure of the redox potential difference between the involved redox couples (17,18).

A sample of cells suspension (1 mg chlorophyll/ml) was diluted 5 fold in 50 mM Hepes buffer, pH 7.5, containing 10mM MgCl, 1M sucrose and 10mM $CaCl_2$. The sample was illuminated for 45 s (orange filter) at room temperature and then dark adapted for 10 min on ice. 80-100 μl of the sample were absorbed onto a 2 cm paper filter and placed on the specimen holder of the TL apparatus. The sample was cooled to a temperature of 20^0C below the temperature of emission of the B glow, and excited by one flash of light as described (10,19). The sample was immediately cooled in liquid N_2. TL was measured with a previously described setup using a heating rate of 1 ^{0}C/s. The B band glow curve in wt cells, obtained after a single flash excitation, had a maximum emission at about 35^0C.

Table II. Electron transfer from Q_A to Q_B, stability of Q_B^- and turnover of the D1 polypeptide in Synechocystis PCC6803 mutants

Strain	Fluorescence decay phase amplitude[1] [%]			Thermoluminescence-peak temperature of B band [^{0}C]	D1 turnover at 1000 μE.m^{-2}.s^{-1}-t½ [min]
	fast	medium	slow		
4D1K (wt)	67	22	11	35	60
L271M	46	16	38	25	50
L271A	35	29	36	20	30
L271S	0	20	80	15	20

[1] Measured at 34 ^{0}C in intact cells after 2 minutes of dark adaptation following one flash of light. The half life of the fast phase in the wt cells was 400 μs, the medium phase was 5 ms and the slow phase was 0.5 s.

The peak of the B band in all three mutants was found to be shifted to a lower temperature in all the mutants as compared to the wt (Table II). The degree of downshift was different in each mutant indicating a gradient of interference with Q_B^- stability: wt>L271M>L271A>L271S. These results substantiate the fluorescence induction data which demonstrate a similar gradual effects in these mutants.

In order to determine the rate of electron transfer from Q_A to Q_B in the mutants the kinetics of fluorescence relaxation was measured. The decay of chlorophyll fluorescence was measured using the PAM fluorimeter as described by Schreiber (20). Whole cells were dark adapted at 34 ^{0}C for 2 minutes before measurement (for detailed description see Prášil et al. in this volume). Following a single turnover flash of white light, a measuring light (650 nm) of 100 KHz was applied during the first 10 ms to resolve the fast fluorescence decay, and then changed to 1.6 KHz light. The fluorescence curves were analyzed by fitting the data to a multi-exponential model (program kindly given to us by Dr. Z. Kolber). The kinetics of the fluorescence decay were found to be composed of three exponential decay phases. In Synechocystis PCC6803 the half life of the fast phase was determined to be 400 μs, the medium phase is 5 ms and a slow phase - 0.5 s. The first fast phase is considered to represent the forward electron transfer reaction from Q_A to Q_B, and the slow phase is correlated to the back reaction (see Prášil et al., in this volume).

The results shown in Table II indicate that the contribution of the fast phase to the overall fluorescence decay is reduced in all three mutants in Leu 271, in spite of the fact that the t½ of the fast phase was not changed in the mutants. This finding indicates that the forward electron transfer from Q_A to Q_B is drastically impaired in the mutants. Concomitantly, the contribution of the slow phase to the overall fluorescence decay is increased. The increase in the back-reaction in the various mutants correlates nicely with the destabilization of QB^- as revealed by the TL measurement (Table II). The role of the amino acid residue in position 271 in the D1 polypeptide is probably analogous to that of Ile 229 in the L subunit of the reaction center in the purple bacterium Rps. viridis (discussed in 12). It is therefore implicated that a hydrophobic side chain is necessary for quinone binding and also for stabilization of the Q_B^- semiquinone.

Turnover of the D1 protein

The light-dependent turnover of the D1 protein was determined as described before (10). Cell suspensions in BG11 medium at a final concentration of 15 μg chlorophyll/ml were preincubated at 34^0C for 30 minutes under dim light after which light intensity was increased to $1000\mu E.m^{-2}sec^{-1}$ and ^{35}S-Sulfate (4 μCi/ml) was added. After pulse labeling for 30 minutes cells were washed and grown in fresh medium under the same conditions. Samples were taken at various times and thylakoid membrane polypeptides were resolved by SDS-PAGE. Equal amounts of membranes were loaded on the gel as determined by chlorophyll measurements (5 μg chlorophyll/sample). Radioactivity was recorded by autoradiography (Fig. 3). Identification of the D1 protein was confirmed by D1-specific antibodies (data not shown). Following scanning the autoradiogram in Fig. 3, the T½ of D1 under $1000\mu E.m^{-2}.s^{-1}$ was calculated to be 60 minutes in the normal D1 (strain 4D1), 50 minutes in mutant L271M, 30 minutes in L271A and 20 minutes in L271S.

The increased rate of turnover of D1 in the mutants correlates with the degree of reduction in electron transfer from Q_A to Q_B and the amplitude of destabilization

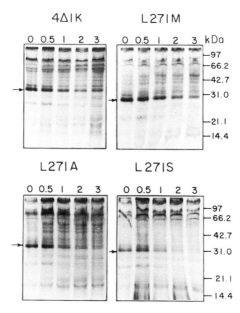

Figure 3. Turnover of D1 in the w.t. strain (4D1k) and in mutants L271M, L271A and L271S of Synechocystis PCC6803. Autoradiogram of thylakoid membrane proteins separated by SDS-PAGE following pulse labelling of cells with ^{35}S-sulphate for 30 minutes. Samples were taken at 0, 0.5, 1, 2 and 3 hours after washing the radioactivity at the end of the pulse labeling.

of Q_B^-. This result substantiates our previous observation in D1 mutants of <u>Synechococcus</u> PCC7942 D1 mutants (10) in which the rate of D1 degradation was proportional to the destabilization of the semiquinone ion Q_B^-. We hypothesized that the triggering event for D1 turnover is provoked by a damaging free radical which is created randomly at a low frequency under light, whose formation probability is increased when the equilibrium $Q_A^-Q_B \rightleftharpoons Q_AQ_B^-$ is changed by mutations that impair electron transfer from Q_A to Q_B or destabilize Q_B^-.

Surprisingly, the rate of turnover of D1 in the non-photosynthetic mutant L271S was the highest. This finding indicates that the triggering of D1 degradation does not require electron flow through PSII to the plastoquinone pool. It was recently demonstrated (21) that in mutants in which photosynthetic electron transport is blocked beyond PSII, such as mutants lacking plastocyanin, cytochrome b_6/f or PSI activity, D1 does not turnover in the light. In these mutants the Q_B site in D1 is not occupied by an oxidized plastoquinone (PQ) since the plastoquinone population is fully reduced. Our data from the D1 mutants in <u>Synechocystis</u> PCC6803 support the hypothesis that the presence of a bound plastoquinone molecule in the Q_B site in D1, is essential for the induction event leading to D1 degradation and D1 turnover.

CONCLUSIONS

Reduced efficiency of electron transfer from Q_A to Q_B correlates with increased photoinduced degradation of the D1 polypeptide. The induction of this degradation does not require net electron flow through PSII beyond Q_B. Our results support the hypothesis that damaging free radicals that are produced when electron transfer in PSII is inhibited while plastoquinone is bound to the Q_B site in D1, trigger an event that leads to D1 degradation.

The Leu 271 in the D1 polypeptide is in close interaction with the plastoquinone Q_B. Substitutions of this amino acid reduce electron transfer from Q_A to Q_B and destabilize the semiquinone union QB⁻.

ACKNOWLEDGEMENTS

We wish to thank Dr. L. McIntosh for providing the strain 4D1, Mr. O. Prášil and Dr. I. Setlik for the fluorescence decay measurements, and Dr. I. Ohad for fruitful discussions. This research was supported by a grant from NCRD, Israel and GBF, Germany, DISNAT 153.

REFERENCES

1. Hansson, Ö. & Wydrzynski, T. (1990) <u>Photosyn. Res.</u> 23,131-162.
2. Kyle, D.J. (1985) <u>Photochem. Photobiol.</u> 41, 107-116.
3. Matoo, A.K., Hoffman-Falk, H., Marder, J.B. & Edelman, M. (1984) <u>Proc. Natl. Acad. Sci. USA,</u> 81, 1380-1384.
4. Ohad, I., Kyle, D.J., & Hirschberg, J. (1985) <u>EMBO J.</u> 4, 1655-1654.

5. Kyle, D.J., Ohad, I. & Arntzen, C.J. (1984) Proc. Natl. Acad. Sci. USA 81, 4070-4074.

6. Ohad, I., Koike, H., Shochat, S. & Inoue, I. (1988) Biochim.Biophys. Acta. 933, 288-298 .

7. Adir, N., Shochat, S. & Ohad, I. (1990) J. Biol. Chem. 265, 12563-12568.

8. Ghirardi, M.L., Callahan, F.E., Sopory, S.K., Elich, T.D., Edelman, M. & Mattoo, A.K. In M. Baltscheffsky (ed.), Current Research in Photosynthesis, Vol. II, Kluwer Academic Publishers, 1990, pp. 733-738.

9. Greenberg, B.M., Gaba, V., Canaani, O., Malkin, S., Mattoo, A.K. & Edelman, M. (1989) Proc. Natl.Acad. Sci. USA 86, 6617-6620.

10. Ohad, N., Amir-Shapira, D., Koike, H., Inoue, Y., Ohad, I., & Hirschberg, J. (1990) Z. Naturforsch. 54c, 402-408.

11. Debus, R.J., Barry, B.A., Sithole, I., Babcock, G.T. & McIntosh, L. (1988) J. Biochem. 27, 9071-9074.

12. Ohad, N. & Hirschberg, J. (submitted for publication)

13. Krause, G.H. & Weiss, E. (1984) Photosyn. Res. 5, 139-157.

14. Rutherford, W., Crofts, A.R. & Inoue, Y. (1982) Biochim. Biophys. Acta 682, 45 7-465.

15. Demeter, S. & Vass, I. (1984) Biochim. Biophys. Acta 764,24-32.

16. Rutherford, A.W. Renger, G. Koike H. & Inoue, Y. (1984) Biochim. Biophys. Acta 767, 548-556.

17. Vass, I., Horvath, G., Herczeg, T. & Demeter, S. (1981) Biochim. Biophys. Acta 634, 140-152.

18. DeVault, D., Govindjee & Arnold, W. (1983) Proc. Natl. Acad. Sci. USA 80, 983-987.

19. Gleiter, H.M., Ohad, N., Hirschberg, J., Fromme, R., Renger, G., Koike, H. & Inoue, Y. (1990) Z. Naturforsch. 45c, 353-358.

20. Schreiber, U. (1986) Photosyn. Res. 9,261-272.

21. Gong, H. & Ohad, I. (1991) J. Biol. Chem. (in press).

22. Trebst, A. (1987) Z. Naturforsch. 42C, 742-750.

MUTATIONS IN THE Q_B-BINDING NICHE IN THE D1 SUBUNIT OF PHOTOSYSTEM II

IMPAIR ELECTRON TRANSPORT FROM Q_A TO Q_B

Ondřej Prášil[1], Nir Ohad and Joseph Hirschberg

Department of Genetics, The Hebrew University of Jerusalem, Jerusalem 91904, Israel and [1]Institute of Microbiology ČSAV, 37981 Třebon, Czechoslovakia

INTRODUCTION

PSII is a pigment-protein complex in the thylakoid membrane that couples the light induced reduction of plastoquinone with oxidation of water. It consists of at least 9 intrinsic polypeptides, 3 extrinsic ones and several prosthetic groups and cofactors. (for review see ref. 1). Following absorption of a photon and charge separation, an electron is transferred from the primary donor P680, to pheophytin, which transfers it to the first stable electron acceptor in PSII, a tightly bound plastoquinone called Q_A. The electron is then transferred to a second stable acceptor, a weakly bound plastoquinone called Q_B, which is a two electron acceptor. The doubly reduced Q_B^{2-} is protonated forming plastoquinol (QH_2) that leaves PSII and is exchanged by an oxidized plastoquinone. The D1 subunit of PSII is a 32 kDa integral membrane polypeptide, which bind the plastoquinone Q_B and herbicides that inhibit electron transfer in PSII (2). Together with the D2 polypeptide, which binds Q_A, it constitutes the photosynthetic reaction center (RC) of PSII. The structure of the Q_B binding site in the D1 protein was suggested to be analogous to the L subunit of the RC in purple bacteria (3,4). Q_A, Q_B and the non-heme iron atom are magnetically coupled and are believed to function together in transferring electrons from Q_A to Q_B (reviewed in 5).

This work is part of a broader attempt to characterize new mutants of the Q_B binding niche of the D1 protein in <u>Synechocystis</u> PCC6803 (6) and to understand structure-function relationship of Photosystem II. Eight mutants with different substitutions in the Q_B binding site of the D1 protein resulted in gradual changes in growth rates, ranging from fully autotrophs (photosynthetic growth) to photoheterotrophs (non-photosynthetic growth). The influence of mutations in the Q_B binding region of D1 on growth rate and correlation with fluorescence induction, oxygen evolution and thermoluminescence signals has been analyzed (6). It was found that photosynthetic electron transport in PSII was inhibited to different extent in the various mutants. This inhibition was correlated with destabilization of the semiquinone ion Q_B^- that was detected as a shift in the thermoluminescence B-band peak temperature.

Here we present a comparative analysis of electron transfer events in the acceptor side of PSII in the D1 mutants, focusing on the following questions: i) Do

the point mutations in the Q_B binding niche influence the kinetics of electron transfer and/or the redox potential difference between Q_A/Q_A^- and Q_B/Q_B^- ?
ii) Is there any electron transport from Q_A to Q_B in the non-photosynthetic mutants?
iii) What is the influence of different substitutions of the same amino acid residue in D1 on electron transfer? To answer these questions we have analyzed the decay of fluorescence yield after single turnover flash in whole cyanobacterial cells and its temperature dependence.

MATERIALS AND METHODS

Cyanobacterial strains and growth conditions

Eight different single point mutations in the psbAII gene of Synechocystis PCC6803 were created (6). These mutations were introduced into the strain 4D1, which contains only copy II of the psbA gene family (7). Strain 4D1 was used as a control for the mutants in the various measurements. Detailed description of mutants and growth conditions are given in (6). Names of the mutants consist of a single letter code of the original amino acid residue, followed by the position in the D1 polypeptide and the code for the replacing amino acid.

The strains and mutants were grown in BG11 medium as previously described (7) at light intensity of 20 μmol.m^{-2}.sec^{-1}. Heterotrophic mutants (Y254S, F255W, L271S) were grown on glucose-supported BG11 medium.

Fluorescence measurements

Kinetics of chlorophyll fluorescence was measured with computer controlled fluorometer (PAM, Walz, FRG) with standard accessories as described by Schreiber (8). The single turnover flash was given by Xe flash lamp (XST-103, Walz, FRG). In order to obtain sufficient time resolution, 100 kHz frequency for modulated measuring light was used during first 10 msec after the flash. The cells (2 ml, 10 μg chl/ml) were resuspended in their growth medium and placed in a temperature-controlled measuring chamber for 2 min dark adaptation. The fluorescence decay data were analyzed by multiexponential fitting procedure (program kindly provided by Dr. Z. Kolber). The decay kinetics of fluorescence yield (ΔF) was found to fit pattern in the form (9):

$$\Delta F(t) = \alpha_1 . \exp(-t/\tau_1) + \alpha_2 . \exp(-t/\tau_2) + \alpha_3 . \exp(-t/\tau_3)$$

Where α stands for amplitudes and τ for half-times of exponential components of ΔF decay. The amplitudes were normalized so that $\alpha_1 + \alpha_2 + \alpha_3 = 100\%$, and are used as such further on.

RESULTS AND DISCUSSION

The sequence of events after excitation by a single turnover flash is a complex one and it occurs at different time scales. During the first few μsec after flash excitation, fluorescence signal rises from its pre-flash level to its maximal flash-induced level, where it remains at maximal value for approx. 100 μsec (10). In our experimental setup the detector was automatically blocked during first 120 μsec after the flash. We therefore monitored events from 120 μsec up to 400 msec. At this time interval we observed multi-phasic decline of fluorescence yield to its pre-flash value (10). The kinetics of this decline reflects reoxidation of Q_A^- (11). The amplitude of fluorescence is not linearly related to the amount of Q_A^- (12) because of exciton migration between connected PSII units. However, as noted by Nixon et al. (13), there is little indication of energy transfer between PSII units in Synechocystis PCC6803 and so parameters obtained from kinetic analysis can be used for comparative studies. Oxidation of Q_A^- can proceed either by forward electron transfer to Q_B or as a backward recombination with positive charges on the donor side of PSII (S states of Mn cluster) (19). Each of this processes occurs on different time scales with different kinetics.

In the 4D1 (wt) strain and the mutants which grow photosynthetically, the decay curve could be resolved into three monoexponential components. Higher number of components did not yield a better fit. The non-photosynthetic mutants and DCMU treated samples of the wt showed two components. Comparison of parameters obtained from analysis of fluorescence yield decay of mutants at 34 °C is given in Table I and Fig. 1.

Table I. Kinetics of fluorescence decay and thermoluminescence B-band peak position in D1 mutants of Synechocystis PCC6803

	τ_1 [μsec]	α_1 [%]	τ_2 [msec]	α_2 [%]	τ_3 [sec]	α_3 [%]	ΔF_m	B band [°C]
4D1	380	67	4.9	22	0.49	11	0.70	35
Y254F	500	62	5.7	23	0.48	15	0.52	35
E242Q	450	62	5.5	26	0.69	12	0.46	35
L271V	340	46	7.9	26	0.48	28	0.51	30
L271M	400	46	6.5	16	0.42	38	0.35	25
L271A	500	35	15	29	0.26	36	0.42	20
L271S	-	-	3.9	20	0.47	80	0.20	15
F255W	680	20	4.8	20	0.45	60	0.23	15
Y254S	-	-	4.4	30	0.34	70	0.11	15

The mutants are arranged in the table according to their ability to grow autotrophically; α stands for relative amplitudes and τ for half-times of fluorescence yield decay components, ΔF_m represents maximal value of fluorescence yield induced by a single flash (F_m - F_0/F_0). Thermoluminescence data are taken from (6). Values of kinetic parameters of fluorescence decay represent averages of several measurements performed with different cell cultures. Slight variation in parameters, reflecting different growth conditions, was observed.

The $Q_A^- \rightleftharpoons Q_B$ electron transport

The fast component of fluorescence yield decay reflects the forward electron transport from Q_A^- to Q_B. Its half-time (τ_1) is conserved among various photosynthetic organisms (15). The half-time of Q_A^- oxidation depends on the reduction state of Q_B, being different for $Q_A^- Q_B \rightleftharpoons Q_A Q_B^-$ (100-200 μsec), $Q_A^- Q_B^- \rightleftharpoons Q_A Q_B^{2-}$ (300-500 μsec) (14) and much longer for Q_A^- (PQ) $\rightleftharpoons Q_A Q_B^-$ (second order reaction, 11). In order to detect differences between Q_A^- reoxidation by Q_B or Q_B^-, samples should be dark relaxed to oxidize most of plastoquinone pool and the measuring light should be weak and non-actinic. Since these conditions were not strictly met in our experiments we assume that the fluorescence decay component we observed in 300-600 μsec range represents the average rate of Q_A^- reoxidation with Q_B.

As can be seen in Table I, the τ_1 in the autotrophically growing mutants is in the range of 350-500 μsec. This shows that the rate of electron transport is not drastically changed by the mutations. However the relative amplitude of this component is gradually decreasing, from 67% in 4D1 to 35% in L271A. The relative amplitude (α_1) of this phase decreased concomitantly to the increase of the amplitude of the slow component ($Q_A^-S_2$ recombination) (Fig.1), suggesting changes in the difference between midpoint redox potentials of Q_A/Q_A^- and Q_B/Q_B^-. The obligatory photoheterotrophic mutants are either completely lacking the fast phase (L271S, Y254S) or exhibiting a decay which is much slower (F255W).

The msec (middle) component

The second kinetic component decayed in 4D1 with τ_2 of 5 msec and $\alpha_2 \approx 20\%$. The origin of this component of Q_A^- reoxidation is not completely clear (10). It might originate either from centers with vacant Q_B binding site, reflecting the association of PQ from the pool with the Q_B site and its consecutive reduction (11), or it could reflect heterogeneity in the PSII acceptor side, originating from non-B centers (16). Component with similar τ was also observed in formate treated, HCO_3^- depleted samples, suggesting its dependence on the state of the Q_A-Fe-Q_B complex (17). In most of the mutants α_2 did not change, remaining in the range of 20-30%. This results are in agreement with earlier observations by Etienne (18). On the other hand we observed a gradual increase of τ_2 in photosynthetic mutants, up to τ = 15 msec in L271A.

The behavior of mutant L271A (see also temperature dependence bellow) and, to a lesser extent, L271V, point to the fact that 271 site has an important role in stabilizing the Q_A-Fe-Q_B complex. Substitutions of Leu 271 by amino acids with a shorter side chain (Val and Ala) should have a strong influence on association constant of PQ to the Q_B-binding site, but at the same time they could influence ligation of the non-heme ferrous atom to the neighboring His 272.

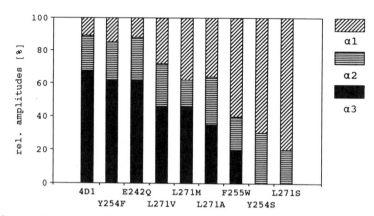

Fig. 1. Comparison of relative amplitudes (α) of mono-exponential components of fluorescence yield decay in 4D1 and D1 mutants. Measurements of fluorescence yield decay were performed at 34°C. Values represent averages of several measurements. Respective half-times τ are given in Table 1.

In the non-photosynthetic mutants (L271S, Y254S, F255W) α_2 was very similar to the photosynthetic strains ($\approx$ 20%), but τ_2 was slightly shorter ($\approx$ 3 - 4 msec). Since this value is the same as for DCMU-treated photosynthetic mutants (even for L271A, with $\tau_2 \approx$ 15 msec without DCMU and $\tau_2 \approx$ 4 msec in the presence of the inhibitor), we think that this phenomenon reflects other pathway for Q_A^- reoxidation, not involving Q_B.

$S_2Q_A^-$ recombination

The third kinetic component of the fluorescence decay reflects the recombination reaction of Q_A^- with the S_2 state of the oxygen evolving cluster on the donor side ($Q_A^- S_2 \rightleftharpoons Q_A S_1$) (19) and is observed also in the presence of DCMU. Its half-time (τ_3) is in the range of 0.4 - 0.5 sec. As expected, this value was not changed in the D1 mutants. However, pronounced changes were observed in α_3 which increased from 11% in the 4D1 up to 80% in L271S. This increase can be correlated to the gradual shift in the thermoluminescence B band peak position and decrease in oxygen evolution (6).

Temperature dependence of fluorescence decay

The decay of fluorescence yield was measured for different mutants in temperatures ranging from 0°C - 50°C. The cells were kept for 5 min at a given temperature in the temperature- controlled cuvette before fluorescence was measured. It was observed that the half-times and relative amplitudes are temperature dependent.

In all strains the τ_1 and τ_2 decreased with rising temperature from 0°C to 34°C. The values of τ_1 at 0°C in 4D1, Y254F, E242Q, and L271V were in the range of 800 - 1000 μsec. Different values were observed for L271A and L271M mutants, with τ_1 at 0°C of 2000 and 1400 μsec, respectively. The half-time of the second component, τ_2, was in the range of 8-12 msec at 0°C in all strains except L271A (20 msec). The relative amplitude α_1 exhibited only slight increase (20%) during temperature rise from 0°C to 40°C; α_2 changed in all mutants only slightly (10 -15 %) in temperature interval 0°C - 50°C; α_3 of the recombination process remained almost constant, while the half-time τ_3 showed an unexpected increase at increasing temperatures. The L271A mutant showed different behavior from the rest, with pronounced decrease in α_1, from 65% at 0°C to 15% at 50°C.

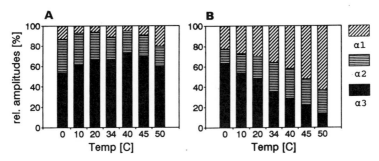

Fig. 2. Temperature dependence of relative amplitudes α for 4D1 (A) and mutant L271A (B). Cells were kept in dark at a given temperature for 5 min before measurements.

CONCLUSIONS

The decay of fluorescence yield during 400 msec after a single turnover flash in Synechocystis PCC6803 cells followed multiexponential kinetics, with three components.

The half-time (τ_1) of forward electron transport was not affected in the photosynthetic mutants, suggesting that the kinetics of this process is not influenced by the mutations in the D1. However, the relative amplitudes (α_1) showed gradual decrease, in correlation with increase in amplitude of the slowest recombination component (α_3). This is in agreement with observed shift in thermoluminescence B band peak temperature, suggesting decreased difference between the redox potentials of Q_A/Q_A^- and Q_B/Q_B^-. The fast component of the fluorescence decay was missing in mutants Y254S and L271S, which are unable to grow autotrophically, indicating a block in electron transfer from Q_A to Q_B. In the non-autotrophic mutant F255W this component was very small suggesting that some electron transfer from Q_A to Q_B is taking place in this strain.

The kinetic parameters of fluorescence yield decay showed temperature dependence. The half-times τ_1 and τ_2 decreased and τ_3 increased with rising temperature from 0°C to 50°C in all strains studied. The relative amplitudes α did not show significant temperature dependence, except for L271A. In this mutant the relative amplitude α_1 decreased with rising temperature, parallel to the increase in α_3.

Acknowledgments

We are grateful to Dr. Z. Kolber and Dr. P. Falkowski for providing programs for analysis of fluorescence decay data, Karel Knopp for software assistance and Dr. I. Setlik for fruitful discussions.

REFERENCES

1. Hansson, O. & Wydrzynski, T. (1990) *Photosyn. Res.* 23,131-162.
2. Kyle, D.J. (198 5) *Photochem. Photobiol.* 41, 107-116.
3. Michel, H. & Deisenhofer, J., (1988) *Biochemistry* 27, 1-7.
4. Trebst, A. (1987) *Z. Naturforsch.* 42C, 742-750.
5. Diner, B.A., Petrouleas, V., & Wendolski, J.J. (1991) *Physiol. Plant.* 81, 423-436.
6. Ohad, N. & Hirschberg, J. (submitted for publication).
7. Debus, R.J., Barry, B.A., Sithole, I., Babcock, G.T. & McIntosh, L. (1988) *J. Biochem.* 27, 9071-9074.
8. Schreiber, U. (1986) *Photosyn. Res.* 9,261-272.
9. Kolber, Z., Zehr, J., & Falkowski, P. (1988) *Plant Physiol.* 88, 923-929.
10. Falkowski, P.G., Wymen, K., Ley, A.C., & Mauzerall, D.C. (1986) *Biochim. Biophys. Acta* 849, 183-192.
11. Crofts, A.R. & Wraight, C.A. (1983) *Biochim. Biophys. Acta* 726, 149-185.
12. Joliot, A. & Joliot, P. (1964) *C.R. Acad. Sci. Paris* 258, 4622-4625.
13. Nixon, P.J., Rogner, M., & Diner, B.A. (1991) *Plant Cell* 3, 383-395.
14 Robinson, H.H. & Crofts A.R. , (1983) *FEBS Lett.* 153, 221-226.
15. Jansen, M.A.K. & Pfister, K. (1990) *Z. Naturforsch.* 45c, 441-445.
16. Chylla, R.A., Garab, G. & Whitmarsh, J. (1987), *Biochim. Biophys. Acta* 894, 562-571.
17. Eaton-Rye, J.J. & Govindjee (1984) *Photobiochem. Photobiophys.* 8, 279-288.
18. Etienne, A.L., Ducruet, J.M., Ajlani, G., & Vernotte, C. (1990) *Biochim. Biophys. Acta* 1015, 435-440.
19. Renger, G., Hanssum, B., Gleiter, H., Koike, H., & Inoue, Y. (1988) *Biochim. Biophys. Acta* 936, 435-446.

ISOLATION AND GENERAL CHARACTERIZATION OF CHLOROPHYLL B

DEFICIENT MUTANTS AND THEIR DEVELOPMENTAL PHENOTYPES FROM THE

GREEN ALGA, SCENEDESMUS OBLIQUUS

Norman I. Bishop and Sharon Maggard

Department of Botany and Plant Pathology
Oregon State University, Corvallis, OR, U.S.A.

INTRODUCTION

A well recognized characteristic of higher plants, green algae, Euglenaceae and Prochlorophyta is the presence of chrorophyll b (chl b) in the light harvesting systems of photosystem I (LHC-I) and photosystem II (LHC-II)[1]. The original evidence demonstrating the localization of chl b in a specific membrane protein arose from detergent-gel electrophoresis studies on chloroplast membrane proteins derived from a chl b deficient mutant of barley[2,3]. This seminal observation, plus the equally important finding that loss of PS-I in the PS-8 mutant of Schenedesmus resulted in loss of a distinct chlorophyll-protein complex[3], provided the first clear experimental evidence for separate photosynthetic reaction centers and light harvesting complexes. Concentrated research efforts by numerous laboratories have extended this fundamental theme such that it is now recognized that each photosystem has elaborated its immediate antennae system plus an additional light harvesting network to assure energy gathering potential adequate for a variable light intensity environment[1].

LHC-IIb is the most abundant of four LHC-IIs; monomer preparations generally show 5-7 chl a, 4-6 chl b and 2-3 distinct xanthophylls per polypeptide. Reconstitution studies of LHC-IIb from denatured apoproteins have shown that both chlorophylls are essential as are certain carotenoids including lutein and neoxanthin[4,5]. Since all of the chl b of the chloroplast is associated with LHC-I or LHC-II and the biosynthesis of chl b and the LHC-apoproteins appear to be coregulated, it has been speculated that genetic modification of specific sites in the polypeptide chain might account for the inability of certain mutants of higher plants and algae to accumulate chl b. However, the finding that the LHC-II apoproteins derived from Chlamydomonas strains deficient in chl b can be reconstituted[4] with a total pigment extract does not support this supposition. The near absence of neoxanthin in all chl b deficient mutants examined to date[6,7] and the requirement of neoxanthin in LHC-II reconstitution studies suggested to Plumley and Schmidt[4] that mutation of the pathway for neoxanthin biosynthesis might be the underlying cause for chl b deficiency. To address this problem, as well as the broader one of chl b biosynthesis, a variety of regular and developmental, chl b-deficient mutants have been derived from normal and diverse photosynthetic-deficient strains of Schenedesmus obliquus. Some general characteristics of these strains are described in the following pages.

PROCEDURES

Mutant Induction and Isolation: Mutants deficient in either photosynthetic capacity and/or the light-harvesting pigment complex (LHPC) were obtained for this study by treatment of the appropriate control strain of Scenedesmus with ethyl methane sulfonate[8]. Selection of strains deficient in chl b was made by visual examination of matured algal clones (14 days old) which had been grown heterotrophically on agar plates at 30°C. Strains deficient in chl b were detected by their lighter green, often bluish-green, color. Such colonies were transferred to individual culture tubes and subsequently evaluated by difference spectra analysis. Strains lacking the characteristic shoulder of chl b at approximately 650 nm were saved for further evaluation. Greening (developmental) phenotypes were obtained by a secondary mutation of each of the retained chl b-deficient strains in which yellow colonies visually comparable to mutant C-2A'9 of Scenedesmus were selected. Final selection and retention of such phenotypes was based on the greening kinetics of individual strains. Determination of photosynthetic and photoreductive capacity, separate photosystem activity by isolated chloroplast membrane preparations, LDS-gel electrophoresis, etc. were performed as previously described[8].

Total Pigment Analysis: The amounts and types of the individual chlorophylls and carotenoids contained in 1 ml packed cell volume were determined for each phenotype studied by rupturing[10] and extraction in cold (4 °C) absolute ethanol, removal of cell debris by centrifugation and partitioning into diethyl ether[10,12]. An aliquot of this extract was analyzed for total chlorophyll and the individual chlorophylls and carotenoids were identified by thin layer- and high performance liquid-chromatography. Pigments eluted from the HPLC column were collected, pooled over several runs, concentrated and redissolved in various solvents for spectral characterization. Ethanol was used for the xanthophylls, neoxanthin, loroxanthin, antheraxanthin, violaxanthin and lutein; acetone was used for the chlorophylls and hexane for the carotenes.

RESULTS AND DISCUSSION

A general description of the various types of chlorophyll b deficient mutants of Scenedesmus collected for this study is presented in Table I. The general phenotype represented by WT-LHC-I (originally designated as KO-913) behaves similarly to the wild-type (WT) Scenedesmus in that it retains the capacity for autotrophic, heterotrophic and mixotrophic growth, has near normal photosynthetic capacity[14], responds to increased light intensity during autotrophic growth by increasing the antennae size associated with the reaction

Table I. Summary of the General Features of Chlorophyll b and Light-Harvesting Complex Deficient Phenotypes of Scenedesmus

WT-LHC-I

Grows adequately autotrophically, heterotrophically or mixotrophically: adapts to light intensity gradient during autotrophic growth by increasing chl a of the reaction centers[14]. No reversion to the WT phenotype noted during extended homocontinuous growth. LDS-PAGE of chloroplast fragments shows the absence of the LHCP but the polypeptide of the LHC apoproteins are present. Difference spectroscopy shows absence of chls a and b of the LHCs. Four additional strains (WT-LHC-II, -III, -IV, & V) with this general phenotype have been independently isolated. Some of these show less reduced levels of chl a than noted in LHC-I.

WT-LHC-I-Y-3

A developmental strain of WT-LHC-I which grows yellow heterotrophically and synthesizes only chl a to the level of the original phenotype when illuminated. Greening kinetics, both time and intensity dependency, are comparable to those of the well-studied, greening mutant, C-2A', of Scenedesmus[9]. Similar greening strains have been developed from the other LHC-deficient strains listed above and from C-2A'(C-2A'-LHC-I).

PS-28-LHC-I

A unique phenotype similar to WT-LHC-I except the original phenotype, mutant PS-28, synthesizes the digeranyl chlorophyll derivatives. Strain grows heterotrophically and retains photosynthetic activity unlike autotrophic cultures. Two developmental strains have been isolated and show excellent greening kinetics.

Non-photosynthetic LHC-deficient strains

Six additionally LHC-deficient strains and their respective secondary developmental phenotypes were derived from existing photosynthetic deficient mutants of Scenedesmus lacking one or more of the major components of the photosystems. These include PS-11-LHC-I (PS-II pleiotropic strains lacking CP-47, CP-43, D1 & D2 polypeptides); LF-1-LHC-1 (Lacks water photolysis because of retention of the unprocessed form of the D1 polypeptide); LF-23-LHC-I (Deficient in the D1 polypeptide of the PS-II reaction center); CP-I-13-LHC-1 (Lacks PS-I activity and associated membrane polypeptides); B6F-50-LHC-I (Deficient in cytochrome f); and CF-15-hf-17-LHC-I (Lacks CF1 and components of PS-I. Retains excellent PS-II activity). Cells of 48 hr old heterotrophic cultures of the respective developmental strains possess a limited greening capacity because of the photosynthetic deficiency.

center cores and retains normal chloroplast membrane polypeptides including the apoproteins of the LHCs. Spectral analysis of whole cells of this mutant demonstrate the loss of both chlorophylls associated with the LHCPs. A difference spectrum determined between cells of the mutant and the WT defines very clearly the apparent total loss of both chl a and b of the LHC-II. Other chlorophyll b-deficient strains, such as WT-LHC-IV, show primarily a deficiency of only chl b when grown heterotrophically (data not shown). Although no LHCP is detected by standard LDS-PAGE electrophoresis of chloroplasts of such strains, the difference spectra analyses suggest that chl a remains bound to the apoproteins. Preliminary results with a less stringent detergent gel electrophoresis systems[15] suggest this to be true in some cases. Currently efforts are being made to optimize the electrophoretic separation of the chlorophyll-proteins of Scenedesmus to evaluate this situation further.

Secondary mutant phenotypes demonstrating a light dependency for greening, analogous to that seen for mutant C-2A' of Scenedesmus[9], were sought primarily to evaluate the kinetics of development of the photosynthesis in the absence of chl b and of certain portions of the photosynthetic apparatus. Although limited application of these strains has been made to date preliminary studies on the time and light-intensity dependency of chl a biosynthesis in LHC-I Y-1 and LHC-IV-Y-4 confirm the differences noted for chl a amounts in the static parent strains.

The major underlying similarities in the different strains of Scenedesmus listed in Table I are (1) the general deficiency of chl b and (2) the presence of the apoproteins of LHC-II in apparently normal amounts. Unlike chl b mutants in higher plants which show a strongly decreased level of these polypeptides, comparative algal mutants do not (see reference 7, for example). Their presence is apparently unmodified form in the Scenedesmus mutants suggests, but does not prove, that the inability to develop chl b is not caused by alteration of these polypeptides. As was noted earlier all mutants lacking chl b, either of higher plants[1] or algae[4,6,7] have shown a corresponding deficiency in the xanthophyll, neoxanthin. Similarly, reconstitution of LHC-II requires, in addition to chl a and b, three xanthophylls, including neoxanthin, for maximal effects[4,5]. The total pigment complementation of some of the Scenedesmus strains listed in Table I were examined by TLC and HPLC. Although most of the initial purification was performed with TLC, only data obtained with the HPLC procedure will be presented here. When freshly prepared extracts of young algal cultures were examined, the HPLC elution profile routinely showed 8 major peaks[12] whose identity was determined from published values and from standards prepared in this laboratory. The identification (by peak number, general name and chemical name and general light absorption characteristics) are summarized in Table II. With the exception of antheraxanthin, which eluted between violaxanthin and lutein, no attempt was made to identify the numerous minor peaks observed. With the column and elution program employed for the HPLC analysis, lutein and zeaxanthin only resolved when lutein was decreased in concentration; consequently, values expressed for lutein in Tables III and IV represent a summed value for the two xanthophylls. Other peaks corresponding to chls a' and

b′ were noted in samples that were either old or had been
exposed to higher temperatures. The digeranyl derivatives of
chls a and b were noted only in extracts of PS-28 and its
secondary mutations. Because of their long retention times,
quantification of alpha- and beta-carotene was determined
directly from TLC preparations. Quantitation of the content of
each of the identified components in cells of WT and WT-LHC-I
grown autotrophically, heterotrophically and mixotrophically
was performed by graphic computation of the area under each
peak and comparison against comparable values from authentic
standards. These values, expressed as moles per 100 moles chl
a, are presented in Table III. The most consistent and obvious
differences noted between the WT and mutant samples is an
approximate 70% decrease in neoxanthin concentration and a
corresponding increase in the level of violaxanthin. No other
consistent variations were noted for loroxanthin, antheraxanthin or lutein. The noted accumulation of violaxanthin,
seemingly at the expense of neoxanthin, supports a precur-
sor/product relationship as proposed in most schemes for
xanthophyll biosynthesis. Similar, but less dramatic,
responses were also noted for dark grown cultures (heterotrophic); although a near-normal pigment profile was obtained
significantly less lutein was synthesized. This is perhaps
associated with the observed increased levels of alpha-

Table II. Characterization and Identification of Pigments
Detected in Extracts of Scenedesmus by reversed-
phase HPLC

Peak #	Common Name	Chemical Name	Absorption Maxima (nm)
1	Neoxanthin	5,6-Epoxy-6′7′-didehydro 5,6,5′,6′tetrahydro beta, beta-carotene-3,3′5′-triol	465.5; 436.5; 413
2	Loroxanthin	19-Hydroxy lutein (Tri-hydroxy alpha-Carotene)	472.5; 445
3	Violaxanthin	5,6,5′,6′ diepoxy 3,3′ dihydroxy beta-Carotene	470; 439; 416
4	Lutein	3,3′dihydroxy alpha-Carotene	473; 444.5; 419
5	Chlorophyll b		645; 445
	Chlorophyll b digeranyl		645; 445
6	Chlorophyll a		661; 429
	Chlorophyll a digeranyl		661; 429
7	alpha-Carotene	beta, epsilon-Carotene	472; 444; 420
8	beta-Carotene	beta, beta-Carotene	477; 450

Table III. Pigment Composition of Wild-Type <u>Scenedesmus</u> and WT-LHC-I Cells Grown Autotrophically, Heterotrophically and Mixotrophically

Compound	Autotrophic WT(6)*	Autotrophic LHC=I(6)	Heterotrophic WT(3)	Heterotrophic LHC-I(3)	Mixotrophic WT(4)	Mixotrophic LHC-I(4)
Neoxanthin	5.6 .3	1.7 .1	5.0 .2	2.2 .2	6.0 .4	1.8 .2
Loroxanthin	1.5 .1	2.5 .1	3.0 .2	1.5 .1	3.7 .2	3.4 .2
Violaxanthin	0.5 .1	5.7 .3	1.0 .1	0.8 .1	1.7 .1	5.7 .2
Lutein	10.7 .5	9.7 .4	3.2 .2	3.2 .2	9.8 .4	8.6 .4
Chl <u>b</u>	61.5 3.2	0	64.9 3.0	0	48.7 4.6	0
Chl <u>a</u>	100	100	100	100	100	100

*The numbers in parenthesis represent the total number of analyses run on each culture type. Values given are moles/100 moles Chl <u>a</u>.

Table IV. Comparison of the Pigment Composition of Heterotrophically Grown Cells of WT and Mutant Phenotypes Deficient in Photosynthetic Activity and the Capacity for Chlorophyll <u>b</u> Synthesis

Compound	WT	PS-28 LHC	LF-1 LHC	LF-15hf-17 LHC	CP-I-13 LHC	LF-23 LHC	PS-11 LHC
Neoxanthin	5.0	0.8	2.1	2.5	2.4	1.9	2.1
Loroxanthin	3.0	3.9	5.2	6.2	5.5	4.8	5.7
Violaxanthin	1.0	1.2	5.5	4.1	4.3	5.2	4.8
Lutein	3.2	1.4	6.1	5.5	5.9	7.0	5.8
Chlorophyll <u>a</u>	100	100*	100	100	100	100	100

*This calculation is based on the concentration of the digeranyl-digeraniol derivative of chl <u>a</u> which is produced in mutant PS-28LHC. Values given are moles/100 moles Chl <u>a</u>.

carote in dark grown cultures of <u>Scenedesmus</u> (unpublished data). Comparative analysis performed on the photosynthetic- and LHC-deficient strains (Table IV) showed a similar relationship for violaxanthin and neoxanthin.

As noted earlier, the chl <u>b</u>-less strains of <u>Scenedesmus</u> retain near normal levels of the apoproteins of the LHCs like other similar mutants of <u>Chlamydomonas</u>[4,6,7]. Standard LDS-PAGE membrane polypeptide profiles of WT-LHC-I and heat-denatured WT chloroplasts have not revealed any major loses or differ- ences in the electrophoretic properties of these polypep- tides[12]. Although modelling of the light-harvesting complex based on the deduced amino acid sequence of LHC-II has focused on potential sites for chl <u>a</u> and chl <u>b</u> binding sites[1], direct experimental evidence as to their existence or identity remains to be established. Research currently in progress on reconstitution of LHC-II from apoproteins of some of the strains described here may determine if the mutation resulting in chl <u>b</u> deficiency can be ascribed to direct amino acid sequence modification of the apoproteins.

BIBLIOGRAPHY

1. J.P. Thornber, G.F. Peter, P.R. Chitnis, R. Nechushtai and A. Vainstein, The light-harvesting complex of photosystem II of higher plants. <u>in</u>: "Light-Energy Transduction in Photosynthesis; Higher Plant and Bacterial Models." S.E. Stevens, Jr. and D.A. Bryant, Eds., pp. 137-154. Amer.Soc. Plant Physiol. (1988).
2. J.P. Thornber and H.R. Highkin, Composition of the photosynthetic apparatus of normal barley leaves and a mutant lacking chlorophyll <u>b</u>. <u>Eur. J. Biochem.</u> 41:109 (1974).
3. J.P. Thornber, R.P.F. Gregory, C.A. Smith and J. Bailey, Studies on the nature of the chloroplast lamella. I. Preparation and some properties of two chlorophyll- protein complexes. <u>Biochem.</u> 6:391 (1967).
4. F.G. Plumley, and G.W. Schmidt, Reconstitution of chlorophyll a/b light-harvesting complexes: Xanthoph- yll-dependent assembly and energy transfer. <u>Proc. Natl. Acad. Sci. USA.</u> 84:146 (1987).
5. H. Paulsen, U. Rumler and W. Rudiger, Reconstitution of pigment containing complexes from light-harvesting chlorophyll a/b-binding protein over-expressed in <u>Escherichia coli</u>. <u>Planta</u> 181:204 (1990).
6. W. Eichenberger, A. Boschetti, H.P. Michel, Lipid and pigment composition of a chlorophyll <u>b</u>-deficient mutant of <u>Chlamydomonas reinhardtii</u>. <u>Physiol. Plant</u>. 66:589 (1986).
7. H. Michel, M. Tellenbach and A. Boschetti, A chlorophyll <u>b</u>-less mutant of <u>Chlamydomonas reinhardtii</u> lacking in the light-harvesting chlorophyll a/b-protein complex but not its apoproteins. <u>Biochim. Biophys. Acta</u> 725:417 (1983).
8. N.I. Bishop, Isolation of mutants of <u>Scenedesmus obliquus</u> defective in photosynthesis, <u>in</u>: "Methods in Chloroplast Molecular Biology," M. Edelman et al., Eds., Elsevier Biomedical Press, N.Y. pp 51-64 (1982).

9. H. Senger and N.I. Bishop, The development of structure and function in chloroplasts of greening mutants of <u>Scenedesmus</u>. I. Formation of chlorophyll. <u>Plant and Cell Physiol</u>. 13:633 (1972).

10. R.J. Berzborn and N.I. Bishop, Isolation and properties of chloroplast particles of <u>Scenedesmus obliquus</u> with high photochemical activity. <u>Biochim. Biophys. Acta</u> 292:700 (1973).

11. T. Braumann and H.L. Grimme, Reversed-phase high-performance liquid chromatography of chlorophylls and carotenoids. <u>Biochim. Biophys. Acta</u> 637:8 (1981).

12. S. Maggard, A study on the carotenoids and membrane polypeptides of the light-harvesting complex of mutants of <u>Scenedesmus obliquus</u> deficient in chlorophyll <u>b</u>. Master's Thesis, Oregon State University, (1990).

13. N.I. Bishop and G. Oquist, Correlation of the photosystem I and II reaction center chlorophyll-protein complexes, CP-aI and CP-aII, with photosystem activity and low temperature fluorescence emission properties in mutants of <u>Scenedesmus</u>. <u>Physiol. Plant</u>. 49:477 (1980).

14. N.I. Bishop, K. Humbeck, S. Romer and H. Senger, The mode of adaptation of the photosynthetic apparatus of a pigment mutant of <u>Scenedesmus</u> to different light intensities. <u>J. Plant Physiol</u>. 135:144 (1989).

15. G.F. Peter and J.P. Thornber, Electrophoretic procedures for fractionation of photosystems I and II pigment-proteins of higher plants and for determination of their subunit composition, <u>in</u>: "Methods in Plant Biochemistry," Vol. 5, Academic Press, pp. 195-210 (1991).

CONTRIBUTORS

611

DATE DUE

NOV 0 7 1993	
JAN 2 6 1994	
APR 2 6 2006	
APR 2 8 2006	